Nanomaterials for Water Treatment and Remediation

Emerging Materials and Technologies
Series Editor
Boris I. Kharissov

Nanomaterials in Bionanotechnology: Fundamentals and Applications
Ravindra Pratap Singh and Kshitij RB Singh

Biomaterials and Materials for Medicine: Innovations in Research, Devices, and Applications
Jingan Li

Advanced Materials and Technologies for Wastewater Treatment
Sreedevi Upadhyayula and Amita Chaudhary

Green Tribology: Emerging Technologies and Applications
T.V.V.L.N. Rao, Salmiah Binti Kasolang, Xie Guoxin, Jitendra Kumar Katiyar, and Ahmad Majdi Abdul Rani

Biotribology: Emerging Technologies and Applications
T.V.V.L.N. Rao, Salmiah Binti Kasolang, Xie Guoxin, Jitendra Kumar Katiyar, and Ahmad Majdi Abdul Rani

Bioengineering and Biomaterials in Ventricular Assist Devices
Eduardo Guy Perpétuo Bock

Semiconducting Black Phosphorus: From 2D Nanomaterial to Emerging 3D Architecture
Han Zhang, Nasir Mahmood Abbasi, Bing Wang

Biomass for Bioenergy and Biomaterials
Nidhi Adlakha, Rakesh Bhatnagar, and Syed Shams Yazdani

Energy Storage and Conversion Devices: Supercapacitors, Batteries, and Hydroelectric Cell
Anurag Gaur, A.L. Sharma, and Anil Arya

Nanomaterials for Water Treatment and Remediation
Srabanti Ghosh, Aziz Habibi-Yangjeh, Swati Sharma, and Ashok Kumar Nadda

2D Materials for Surface Plasmon Resonance-Based Sensors
Sanjeev Kumar Raghuwanshi, Santosh Kumar, and Yadvendra Singh

Functional Nanomaterials for Regenerative Tissue Medicines
Mariappan Rajan

Uncertainty Quantification of Stochastic Defects in Materials
Liu Chu

For more information about this series, please visit:
https://www.routledge.com/Emerging-Materials-and-Technologies/book-series/CRCEMT

Nanomaterials for Water Treatment and Remediation

Edited by

Srabanti Ghosh

Aziz Habibi-Yangjeh

Swati Sharma

Ashok Kumar Nadda

CRC Press
Taylor & Francis Group
Boca Raton New York London

CRC Press is an imprint of the
Taylor & Francis Group, an **informa** business

First edition published 2022
by CRC Press

6000 Broken Sound Parkway NW, Suite 300, Boca Raton, FL 33487-2742

and by CRC Press
4 Park Square, Milton Park, Abingdon, Oxon, OX14 4RN

Library of Congress Cataloging-in-Publication Data
Names: Ghosh, Srabanti, editor.
Title: Nanomaterials for water treatment and remediation / edited by
Srabanti Ghosh, Aziz Habibi-Yangjeh, Swati Sharma, and Ashok Kumar Nadda.
Description: First edition. | Boca Raton, FL : CRC Press, 2022. | Series: Emerging materials and technologies | Includes bibliographical references and index. | Summary: "This book explores recent developments in the use of advanced nanomaterials (ANMs) for water treatment and remediation. In-depth reaction mechanisms in water treatment technologies including adsorption, catalysis, and membrane filtration for water purification using ANMs are discussed in detail. It includes investigation of fabrication processes of nanostructured materials and fundamental aspects of surface at the nanoscale. The book also covers removal of water-borne pathogens and microbes through a photochemical approach. This text is aimed at researchers and industry professionals in chemical, materials, and environmental engineering, as well as related fields"-- Provided by publisher.
Identifiers: LCCN 2021035506 (print) | LCCN 2021035507 (ebook) | ISBN 9780367633073 (hbk) | ISBN 9780367633097 (pbk) | ISBN 9781003118749 (ebk)
Subjects: LCSH: Water--Purification--Materials. | Nanostructured materials. | Catalysts. | Nanofiltration.
Classification: LCC TD433 .N29 2022 (print) | LCC TD433 (ebook) | DDC 628.1/62--dc23
LC record available at https://lccn.loc.gov/2021035506
LC ebook record available at https://lccn.loc.gov/2021035507

ISBN: 978-0-367-63307-3 (hbk)
ISBN: 978-0-367-63309-7 (pbk)
ISBN: 978-1-003-11874-9 (ebk)

DOI: 10.1201/9781003118749

Typeset in Times
by SPi Technologies India Pvt Ltd (Straive)

Contents

Preface ... vii
Editors ... xi
Contributors ... xv

Chapter 1 Advancement of Nanomaterials for Water Treatment ... 1
Zahra Shariatinia

Chapter 2 Enhancement in Degradation of Antibiotics Using Photocatalytic Semiconductors under Visible Light Irradiation: A Review ... 69
Rupali Setia, Harshita Chawla, and Seema Garg

Chapter 3 Graphitic Carbon Nitride-Based Nanomaterials as Photocatalysts for Organic Pollutant Degradation ... 93
Anise Akhundi and Aziz Habibi-Yangjeh

Chapter 4 2D Materials for Wastewater Treatments ... 121
Nishanth Thomas, Amit Goswami, Kris O'Dowd, Gerard McGranaghan, and Suresh C. Pillai

Chapter 5 TiO_2-Based Nanomaterial for Pollutant Removal ... 163
M. Dhaneesha, Shahanas Beegam, and P. Periyat

Chapter 6 Nanomaterials for the Removal of Heavy Metals from Water ... 181
Subhadeep Biswas, Mohammad Danish, and Anjali Pal

Chapter 7 Photoactive Polymer for Wastewater Treatment ... 217
Ridha Djellabi, Claudia Letizia Bianchi, Muhammad Rizwan Haider, Jafar Ali, Ermelinda Falletta, Marcela Frias Ordonez, Anna Bruni, Marta Sartirana, and Ramadan Geioushy

Chapter 8 Plasmonic Nanomaterials for Remediation of Water and Wastewater ... 245
Ewa Kowalska, Kenta Yoshiiri, Maya Endo-Kimura, Tharishinny Raja-Mogan, Oliwia Paszkiewicz, Zhishun Wei, Kunlei Wang, Marcin Janczarek, and Agata Markowska-Szczupak

Chapter 9 Magnetic Nanomaterials for Wastewater Remediation....................279
Ramalingam Suhasini and Viruthachalam Thiagarajan

Chapter 10 Nanofiber Membranes for Wastewater Treatments..........................309
N.S. Jamaluddin, N.H. Alias, N.H. Othman, M.S.M Shayuti, F. Marpani, and M.H.A. Aziz

Chapter 11 Carbon Nanomaterials for Removal of Pharmaceuticals from Wastewater 333
Aydin Hassani and Alireza Khataee

Chapter 12 Metal–Organic Frameworks and Their Derived Materials in Water Purification 371
Chizoba I. Ezugwu, Srabanti Ghosh, Marta E.G. Mosquera, and Roberto Rosal

Chapter 13 Photocatalytic Mechanism in Low-Dimensional Chalcogenide Nanomaterials: An Exciton Dynamics Insight................................409
Yawei Yang and Wenxiu Que

Chapter 14 Emerging Semiconductor Photocatalysts for Antibiotic Removal from Water/Wastewater..447
Mohamad Fakhrul Ridhwan Samsudin and Suriati Sufian

Index..465

Preface

Wastewater management is one of the major challenges across the world. Various chemical and biological methods of treatment have been developed to treat contaminated water. New materials with intriguing physical and chemical properties provide opportunities to address these challenges. Nanomaterials have a large surface area to volume ratio, high surface reactivity (e.g., catalytic activity), and increased multifunctionality widely used to remove contaminants from surface water and groundwater contaminated with organic/inorganic pollutants via advanced oxidation and chemical reduction, sorption, complexation, (co)precipitation, or membrane filtration. Consequently, nanomaterials have been modified or hybridized with one or more nanomaterials to achieve multifunctionality by tuning size, morphology, dimensionality, surface properties, etc. The nanocomposites of various inorganic nanoparticles with organic material have also been proven effective in removing the various impurities from contaminated water. The major pollutants, chemical pesticides, pharmaceutical wastes, fertilizers, heavy metal ions, dyes, paints and refinery waste, sugar mills textiles fibers contaminated the major volume of drinking water on the Earth's surface. In this volume, we compile up-to-date information on the various nanomaterials-based techniques used to treat polluted water.

This book primarily focuses on the recent developments of the rational design, synthesis of advanced nanomaterials and their applications for water treatment and remediation through various techniques such as adsorption, advanced oxidation process, chemical oxidation and reduction reactions, membrane-based separation, and synergistic multifunctional all-in-one nanodevices. It includes (i) rational design of nanoporous materials with tunable pore structure; (ii) fabrication of nanomaterials by surface chemistry engineering; (iii) role of nanomaterials assisted oxidation and reduction processes; (iv) design of nanomaterial supported membrane-based separation; (v) multifunctional nanomaterials for water treatment. This volume also comprehensively discusses the types of nanomaterials and their interfacial modification effect in removal, absorption and degradation features, and the effect of various factors that could influence the mechanism, capacity, and stability of nanomaterials.

To this end, the book consists of 14 chapters that reflect the progression from introductory to focused topics. **Chapter 1** introduces the fundamentals of various water treatment techniques using nanomaterials. Starting with basic water treatment methods, the principles of adsorption, filtration, coagulation, flocculation, photocatalytic degradation, ozonation, chemical oxidation reactions, and reverse osmosis using membranes are reviewed. **Chapter 2** presents the synthesis of heterostructures, modification of semiconductor interface using co-catalysts, photodegradation of antibiotics and other organic matter using visible-light-active semiconductors. In **Chapter 3,** Akhundi and Habibi-Yangjeh summarized the latest progress in constructing g-C_3N_4-based nanocomposites and their employment for the degradation of organic pollutants. They provide an overview of advancements, existing challenges, and opportunities for future development. **Chapter 4** highlights the applications of 2D materials for advanced membrane fabrication and photocatalytic treatment of pollutants and microorganisms. Recent advances in the fabrication of thin-film composite-based membranes using 2D materials such as graphene, graphene oxide, MXenes, transition metal dichalcogenides, photo-catalytically active 2D materials (g-C_3N_4, iron oxide, manganese oxide, metal oxyhalide, boron nitride) for water purification and separation of textile dyes, metal cations, heavy metals and various pharmaceutical products are discussed in detail. The next chapter (**Chapter 5**) presents several advancements made in aspect of TiO_2 based nanomaterials for applications in water treatment and pollutant remediation. Several methods including modifying the electronic band structure of TiO_2 using different techniques such as self-doping, metal/non-metal doping, co-doping, coupling with a narrow band gap semiconductor, different surface modifications, sensitization by metal complexes dyes or organic and capping with other semiconductor or metal nanoparticles are discussed for wastewater treatment. **Chapter 6** describes the extensive research on nanoparticles for heavy metal remediation. They extensively discuss the removal of heavy metals like arsenic (As), cadmium (Cd), chromium (Cr), etc. and the mechanism behind the pollutant clean-up along with the use of various advanced characterization techniques. The reusability, metal recovery from waste effluents, prospects, and current challenges are also discussed. **Chapter 7** introduces photoactive polymers for organic pollutants oxidation and heavy metals reduction, the inactivation of bacterial species, the synergistic effects obtained through the combination of inorganic semiconductors and the photoactive polymer-based membranes. A brief overview of the latest scientific advances in the development of photoactive polymer and composites to overcome the present shortcomings for fabricating visible-light-driven photocatalysis for water treatment along with their operating principles is also presented. In the following **Chapter 8**, Kowalska's group focus on plasmonic nanomaterials for wastewater treatment and water purification, including degradation of organic compounds and microorganisms, and removal of inorganic pollutants. The photocatalysts with advanced structures, such as faceted oxides and photonic crystals, and mechanisms clarifications have been discussed in detail. They include several examples of plasmonic photocatalysts, including noble metal/semiconductor and noble metal/semiconductor/semiconductor composites, which may provide an efficient approach for removing dyes from water. Suhasini and Thiagarajan demonstrate the potential of magnetic nanoparticles for wastewater remediation water through recycling and reuse as an efficient, practical, and low-cost water treatment method. **Chapter 9** reviews the fundamental theory and concepts of magnetic nanoparticle separations and integrating various nanocomposites, polymers, and layered double hydroxides for water remediation. **Chapter 10** emphasizes the various nanofiber membrane development especially using the electrospinning technique for the removal of various pollutants. They describe several factors and parameters that affected the performance of the nanofiber membranes under experimental conditions. They provide an overview of recent applications of nanofiber membranes in removing different types of hazardous pollutants in wastewater that contains heavy metal, oil, dye, and pharmaceutical pollutant and challenges and future strategies for sustainable development and applications of the nanofiber membranes in the field of wastewater treatment. In **Chapter 11**, Hassani and Khataeeb reported various forms of carbon nanomaterials (such as carbon nanotubes, graphene, reduced graphene oxide-based materials, fullerene, activated carbon, biochar, carbon quantum dots and graphitic carbon nitride etc.) for the elimination of pharmaceuticals and personal care products by advanced oxidation processes (AOPs), including photocatalysis, sonocatalysis, Fenton and Fenton-like process, electrochemical process, sulfate radical-based AOPs, and hybrid AOPs by the aid of carbon nanomaterials. **Chapter 12** highlights the recent advances in the use of metal–organic frameworks (MOFs) and their derivatives for the efficient removal of contaminants in wastewater. In particular, photocatalysis, sulfate radical-based advanced oxidation process, heavy metals and organic pollutants removal, and MOFs-mixed matrix membranes strategies are detailed in the discussion. They present an overview of the synthesis, components, and secondary building units of MOFs together with MOF-derived materials and the interaction mechanisms between these frameworks and the contaminants. In **Chapter 13**, Yang and Que's group reviews the progress of Low-dimensional metal chalcogenide nanomaterials for photocatalytic water treatment and remediation. They demonstrate that properties of these materials can be adjusted by the degree of quantum confinement in 0D quantum dots, 1D nanorods and 2D nanoplatelets, by material composition, and by forming heterostructures. This chapter summarized the fundamental photocatalytic mechanism insight into the exciton and carrier properties and their dependences on the dimensionality and chemical composition in metal chalcogenide nanomaterials. In **Chapter 14**, Samsudin and Sufian emphasized a panorama of the recent advances related to the fabrication and construction of emerging semiconductor photocatalysts for the treatment of antibiotic residues. They describe the fundamental principles of the advanced oxidation process, semiconductor photocatalyst, the degradation pathways of antibiotics, and invigorating perspectives on the future directions. A comprehensive understanding of the engineering of advanced nanomaterials, fundamentals, challenges, and applications offers a useful guide for a broad readership in various fields of catalysis, material sciences, environment, and energy to find the latest information. This book, titled *Nanomaterials for Water Treatment and Remediation*, comprises 14 high-quality chapters written by 50 experts from 13 countries. I would like to thank all the contributors. We owe special thanks to individual authors for their excellent chapters. We would also like to express my sincere gratitude to the publisher CRC and to the team of Dr. Allison Shatkin (Senior Publisher), Prof. Boris I. Kharissov, Gabrielle Vernachio and others involved in the successful production of this book.

Editors

Srabanti Ghosh, Ph.D., is a Senior Scientist in the Energy Materials and Devices Division, CSIR-Central Glass and Ceramic Research Institute, Kolkata, India. She earned a Ph.D. in 2010 at UGC-DAE CSR, Kolkata Centre and Jadavpur University, India. She has been awarded Postdoctoral Fellowship, Marie Curie Cofund, RBUCE-UP by the European Commission and PRES UniverSud Paris. Her research interests encompass synthesis and applications of semiconductors and conducting polymer-based hybrid nanomaterials for solar light harvesting, photocatalysis, electrocatalysis and fuel cell. She has been selected as Got Energy Talent (GET)-COFUND Marie Curie Fellow at Department of Organic and Inorganic Chemistry, Universidad de Alcalá (UAH), Spain. She received the MRSI Young Scientists Award in Young Scientists Colloquium organized by Material Research Society of India (MRSI), and Young Investigator Award in Gordon Radiation Chemistry (GRC), USA. She is co-author of 80 scientific publications and two patents, 19 book chapters, and edited three books with Wiley, Elsevier. Her work has been presented at more than 72 national and international conferences. H-index and Citation in Google scholar are 31 and 3056. She routinely acts as a reviewer of SCI Journals from different editorials (RSC, ELSEVIER, ACS, Wiley, Springer Nature, MDPI, etc.).

Aziz Habibi-Yangjeh, Ph.D., earned a Ph.D. in physical chemistry at the Sharif University of Technology, Tehran, Iran, in 2001. He is a full professor of physical chemistry at the University of Mohaghegh Ardabili, Ardabil, Iran. His research interests include the preparation of different heterogeneous visible-light-driven photocatalysts based on zinc oxide (ZnO), titanium dioxide (TiO_2), and graphitic carbon nitride (g-C_3N_4) and their applications in different fields, especially wastewater decontamination and photofixation of nitrogen. He has published more than 220 JCR papers, including seven reviews and nine highly cited papers. His h-index is 50. Moreover, he is on the editorial board of three international journals. He contributed three book chapters in the field of photocatalysis. He has supervised more than 35 MSc and Ph.D. students in physical chemistry.

Ashok Kumar Nadda, Ph.D., is an Assistant Professor in the Department of Biotechnology and Bioinformatics, Jaypee University of Information Technology, Waknaghat, Solan, Himachal Pradesh, India. He has extensive research and teaching experience of more than 8 years in the field of microbial biotechnology, with research expertise focusing on various issues about nano-biocatalysis, microbial enzymes, biomass, bioenergy and climate change. Dr. Ashok is teaching, Enzymology and Enzyme technology, Microbiology, Environmental Biotechnology, Bioresources and Industrial products to undergraduate, master's, and Ph.D. students. He also trains the students for enzyme purification expression, gene cloning, and immobilization onto nanomaterials experiments in his lab. He holds international work experience in South Korea, India, Malaysia, and the People's Republic of China. He worked as a postdoctoral fellow in the State Key Laboratory of Agricultural Microbiology, Huazhong Agricultural University, Wuhan, China. He also worked as a Brain Pool researcher/Assistant Professor at Konkuk University, Seoul, South Korea. Dr. Ashok has a keen interest in microbial enzymes, biocatalysis, CO_2 conversion, biomass degradation, biofuel synthesis, and bioremediation. His work has been published in various internationally reputed journals, namely *Chemical Engineering Journal, Bioresource Technology, Scientific Reports, Energy, International Journal of Biological Macromolecules, Science of Total Environment*, and *Journal of Cleaner Production*. Dr. Ashok has published more than 100 scientific contributions in the form of research, review, books, book chapters and others at several platforms in various journals of international repute. The research output includes 71 research articles, 27 book chapters and 11 books. He is the main series editor of *Microbial Biotechnology for Environment, Energy and Health*, publishing the books under Taylor and Francis, CRC Press USA. He is also a member of the editorial board and reviewer committee of the various journals of international repute. He has presented his research findings in more than 40 national/international conferences. He has attended more than 50 conferences/ workshops/colloquia/seminars *etc.* in India and abroad. Dr. Ashok is also an active reviewer for many high-impact journals published by Elsevier, Springer Nature, ASC, RSC, and Nature Publishers. His research works have gained broad interest through his highly cited research publications, book chapters, conference presentations, and invited lectures.

Swati Sharma, Ph.D., is an Assistant Professor at the University Institute of Biotechnology, Chandigarh University, Mohali, Punjab, India. She is working extensively on waste biomass, biopolymers, and their applications in various fields. Dr. Sharma earned a Ph.D. at the University Malaysia Pahang, Malaysia. She worked as a visiting researcher in the college of life and environmental sciences at Konkuk University, Seoul, South Korea. Dr. Sharma earned an MSc from Dr. Yashwant Singh Parmar University of Horticulture and Forestry, NauniSolan H.P.

India. She has also worked as a program co-coordinator at the Himalayan action research center Dehradun and Senior research fellow at India agricultural research institute in 2013–2014. Dr. Sharma has published her research papers in reputed international journals. Presently, Dr. Sharma's research is in the field of bioplastics, hydrogels, keratin nanofibers and nanoparticles, biodegradable polymers and polymers with antioxidant and anticancer activities and sponges. Dr. Swati has published 22 research papers in various internationally reputed journals, five books, and a couple of book chapters.

Contributors

Anise Akhundi
Department of Chemistry
University of Mohaghegh, Ardabili
Ardabil, Iran

Jafar Ali
Department of Biochemistry and
Molecular Biology
University of Sialkot
Sialkot, Pakistan

N.H. Alias
Faculty of Chemical Engineering
Universiti Teknologi MARA
Selangor, Malaysia

M.H.A. Aziz
Advanced Membrane Technology
Research Center (AMTEC)
School of Chemical Engineering
Faculty of Engineering
Universiti Teknologi Malaysia
Johor, Malaysia

Shahanas Beegam
Department of Chemistry
University of Calicut
Thenjipalam, India

Claudia Letizia Bianchi
Università degli Studi di Milano
Dip. Chimica and INSTM-UdR Milano
Milan, Italy

Subhadeep Biswas
Department of Civil Engineering
Indian Institute of Technology
Kharagpur, India

Anna Bruni
Università degli Studi di Milano
Dip. Chimica and INSTM-UdR Milano
Milan, Italy

Harshita Chawla
Department of Chemistry
Amity Institute of Applied Sciences
Amity University
Noida, India

Mohammad Danish
Department of Civil Engineering
Indian Institute of Technology
Kharagpur, India

M. Dhaneesha
Department of Environmental Studies
Kannur University
Kannur, India

Ridha Djellabi
Università degli Studi di Milano
Dip. Chimica and INSTM-UdR Milano
Milan, Italy

Maya Endo-Kimura
Institute for Catalysis (ICAT)
Hokkaido University
Sapporo, Japan

Chizoba I. Ezugwu
Department of Analytical Chemistry
Physical Chemistry and Chemical
Engineering
University of Alcalá
Alcalá de Henares
Madrid, Spain

Ermelinda Falletta
Università degli Studi di Milano
Dip. Chimica and INSTM-UdR Milano
Milan, Italy

Seema Garg
Department of Chemistry
Amity Institute of Applied Sciences
Amity University
Noida, India

Ramadan Geioushy
Nanomaterials and Nanotechnology
Department of Advanced Materials
Division Central Metallurgical R&D Institute (CMRDI)
Helwan, Egypt

Srabanti Ghosh
Energy Materials and Devices Division
CSIR – Central Glass and Ceramic Research Institute
Kolkata, India

Amit Goswami
Centre for Precision Engineering
Materials and Manufacturing Research (PEM)
Institute of Technology Sligo
Sligo, Ireland

Muhammad Rizwan Haider
School of Civil and Environmental Engineering
Harbin Institute of Technology Shenzhen
Shenzhen, China

Aydin Hassani
Department of Materials Science and Nanotechnology Engineering
Faculty of Engineering
Near East University
Mersin, Turkey

N.S. Jamaluddin
Faculty of Chemical Engineering
Universiti Teknologi MARA
Selangor, Malaysia

Marcin Janczarek
Institute of Chemical Technology and Engineering
Faculty of Chemical Technology
Poznan University of Technology
Poznan, Poland

Alireza Khataee
Research Laboratory of Advanced Water and Wastewater Treatment Processes
Department of Applied Chemistry
Faculty of Chemistry
University of Tabriz
Tabriz, Iran
and
Department of Environmental Engineering
Gebze Technical University
Gebze, Turkey

Ewa Kowalska
Institute for Catalysis (ICAT)
Graduate School of Environmental Science
Hokkaido University
Sapporo, Japan

Agata Markowska-Szczupak
Department of Chemical and Process Engineering
West Pomeranian University of Technology in Szczecin
Szczecin, Poland

F. Marpani
Faculty of Chemical Engineering
Universiti Teknologi MARA
Selangor, Malaysia

Gerard McGranaghan
Centre for Precision Engineering
Materials and Manufacturing Research (PEM)
Department of Mechanical and Manufacturing Engineering
Institute of Technology Sligo
Sligo, Ireland

Marta E.G. Mosquera
Department of Organic and Inorganic Chemistry
Instituto de Investigación en Química "Andrés M. del Río" (IQAR)
Universidad de Alcalá
Madrid, Spain

Kris O'Dowd
Nanotechnology and Bio-Engineering Research Group
Department of Environmental Science
Institute of Technology Sligo
Sligo, Ireland

N.H. Othman
Faculty of Chemical Engineering
Universiti Teknologi MARA
Selangor, Malaysia

Marcela Frias Ordonez
Università degli Studi di Milano
Dip. Chimica and INSTM-UdR Milano
Milan, Italy

Anjali Pal
Department of Civil Engineering
Indian Institute of Technology
Kharagpur, India

Oliwia Paszkiewicz
Department of Chemical and Process Engineering
West Pomeranian University of Technology
Szczecin, Poland

P. Periyat
Department of Environmental Studies
Kannur University
Kannur, India

Suresh C. Pillai
Nanotechnology and Bio-Engineering Research Group
Department of Environmental Science
Institute of Technology Sligo
Sligo, Ireland

Wenxiu Que
Electronic Materials Research Laboratory
Key Laboratory of the Ministry of Education
International Center for Dielectric Research
Shaanxi Engineering Research Center of Advanced Energy Materials and Devices
School of Electronic Science and Engineering
Xi'an Jiaotong University
Xi'an, China

Tharishinny Raja-Mogan
Institute for Catalysis (ICAT)
Graduate School of Environmental Science
Hokkaido University
Sapporo, Japan

Roberto Rosal
Department of Analytical Chemistry
Physical Chemistry and Chemical Engineering
University of Alcalá
Madrid, Spain

Mohamad Fakhrul Ridhwan Samsudin
Chemical Engineering Department
Universiti Teknologi PETRONAS
Perak, Malaysia

Marta Sartirana
Università degli Studi di Milano
Dip. Chimica and INSTM-UdR Milano
Milan, Italy

Rupali Setia
Department of Chemistry
Amity Institute of Applied Sciences
Amity University
Noida, India

Zahra Shariatinia
Department of Chemistry
Amirkabir University of Technology (Tehran Polytechnic)
Tehran, Iran

M.S.M Shayuti
Faculty of Chemical Engineering
Universiti Teknologi MARA
Selangor, Malaysia

Suriati Sufian
Chemical Engineering Department
Centre of Innovative Nanostructures and Nanodevices (COINN)
Universiti Teknologi PETRONAS
Perak, Malaysia

Ramalingam Suhasini
Photonics and Biophotonics Lab
School of Chemistry
Bharathidasan University
Tiruchirappalli, India

Viruthachalam Thiagarajan
Photonics and Biophotonics Lab
School of Chemistry
Bharathidasan University
Tiruchirappalli, India

Nishanth Thomas
Nanotechnology and Bio-Engineering Research Group
Department of Environmental Science
Institute of Technology Sligo
Sligo, Ireland

Kunlei Wang
Institute for Catalysis (ICAT)
Hokkaido University
Sapporo, Japan
and
Northwest Research Institute Co. Ltd. of CREC
Lanzhou, China

Zhishun Wei
Hubei Provincial Key Laboratory of Green Materials for Light Industry
Hubei University of Technology
Wuhan, China

Yawei Yang
Electronic Materials Research Laboratory
Key Laboratory of the Ministry of Education
International Center for Dielectric Research
Shaanxi Engineering Research Center of Advanced Energy Materials and Devices
School of Electronic Science and Engineering
Xi'an Jiaotong University
Xi'an, China

Aziz Habibi-Yangjeh
Department of Chemistry
University of Mohaghegh, Ardabili
Ardabil, Iran

Kenta Yoshiiri
Institute for Catalysis (ICAT)
Graduate School of Environmental Science
Hokkaido University
Sapporo, Japan

1 Advancement of Nanomaterials for Water Treatment

Zahra Shariatinia

Amirkabir University of Technology (Tehran Polytechnic)
Tehran, Iran

CONTENTS

1.1 Introduction 1
1.2 Various Water Treatment Techniques Using Nanomaterials 4
1.3 Applications of Carbon-Based Nanomaterials in Water Treatment 10
1.4 Applications of Boron Nitride, Metal–Organic Frameworks (MOFs), and Zeolites in Water Treatment 22
1.5 Applications of Mesoporous Silica Nanomaterials in Water Treatment 31
1.6 Applications of Metals and Metal Oxides/Sulfides/Selenides Nanoparticles/Quantum Dots in Water Treatment 37
1.7 Applications of Clays and Layered Double Hydroxide (LDH) Nanoparticles in Water Treatment 41
1.8 Applications of Polymer Nanocomposites in Water Treatment 45
1.9 Conclusion 51
Acknowledgments 52
Conflicts of Interest 52
References 52

1.1 INTRODUCTION

The world's population has increased rapidly, tripling in the 20th century and is projected to grow by about 40–50% during the upcoming fifty years (Kang and Cao 2012). Therefore, population growth, urbanization, and industrialization lead to higher demands for supplying fresh water. On the other hand, the existing freshwater sources are increasingly contaminated and become unavailable for humans and industries. Moreover, issues related to producing pure water will grow within the next decades as water scarcity occurs worldwide, even in water-rich regions (Chen, Liu, et al. 2020b, Wang, Mi, et al. 2020a, Xiong et al. 2020b).

The World Water Council estimated that until 2030, almost 3.9 billion persons would live in "water-scarce" locations (Pendergast and Hoek 2011, Zhou et al. 2020). Besides the global water scarcity, low water quality is another problem in several

DOI: 10.1201/9781003118749-1

world zones. Also, the World Health Organization has predicted that around 1.1 billion people cannot receive drinking water, and 2.6 billion cannot receive suitable healthy water (Pendergast and Hoek 2011). Thus, it has been expected that 2.2 million people will die due to correlated diarrheal diseases each year because of aquatic infections (Pendergast and Hoek 2011, Zhang, Jiang, et al. 2020c). Consequently, current fresh water sources must be protected and it is required to develop more different water sources to provide the increasing demands for hygienic water. For this goal to be achieved, advanced water treatment materials and technologies must be developed (Guo et al. 2020, Liu, Khan, et al. 2020a, Nasrollahzadeh et al. 2021).

Rapid population growth and increased demands for printing, dyeing, chemical and pharmaceutical industries have enhanced the environmental pollution and resulted in the huge deposition of domestic and industrial contaminants into water, soil, and air, which can cause serious problems and adverse effects to the health of human beings, animals, and plants (Bodzek, Konieczny, and Kwiecińska-Mydlak 2020). Annually, many organic, inorganic and microbial pollutants are released into the ecosystem (Liu et al. 2019, Tahir et al. 2020, Zhang, Igalavithana, et al. 2020b). Therefore, it is essential to remove and/or destroy such contaminants by means of various effective techniques such as advanced oxidation processes, adsorption and separation using membranes (Alijani et al. 2014, 2017a, Lata and Vikas 2019, Jourshabani et al. 2018, Morcali and Baysal 2019, Jourshabani, Lee, and Shariatinia 2020).

So far, many water treatment strategies have been employed, including physical, chemical, and biological approaches using adsorption, electrocoagulation, electrochemical, H_2O_2/ultraviolet (UV)/oxygen systems, and membrane separation techniques, etc. (Alijani and Shariatinia 2017, Alijani, Shariatinia, and Aroujalian 2017b, Jourshabani, Shariatinia, and Badiei 2017a, 2017d, Afzal et al. 2019, Tahir, Tufail, et al. 2019b). The adsorption is an effective, simple, and economical process (Alijani, Shariatinia, and Aroujalian Mashhadi 2015). However, in the adsorption method, pollutants removed from water are remained in the environment (Ganta et al. 2020). The traditional wastewater treatment procedures destroying the pollutants like ozonation and chlorination are disinfection efficient methods but they produce harmful by-products (Kozari, Paloglou, and Voutsa 2020). Therefore, the synthesis of suitable materials with adsorption and photocatalytic properties under UV/visible/near-infrared light illumination is crucial as they can simultaneously adsorb and degrade contaminants (Behzadifard, Shariatinia, and Jourshabani 2018, Soltanabadi, Jourshabani, and Shariatinia 2018, Sun et al. 2020). Since the first photochemical sterilization in 1985 (Matsunaga et al. 1985), photocatalytic water purification has become one of the most favorable and sustainable techniques for wastewater treatment. In this process, reactive oxygen species, ROS, are produced using photocatalysts under UV/visible/near-infrared light irradiation, which can effectively degrade organic as well as inorganic pollutants (Jourshabani, Shariatinia, and Badiei 2017b, 2018a, Xiong, Ma, and Liu 2020a, Zyoud et al. 2019, Bargozideh, Tasviri, and Kianifar 2020, Elami and Seyyedi 2020).

It is well known that the chlorination methods used for water treatment have serious problems as the usage of chlorine substances forms harmful toxic by-product compounds (Mansor and Tay 2020). Therefore, other disinfection procedures like UV radiation and ozonation are widely examined in which the photo-oxidation process is occurred using the singlet oxygen radicals (Savino and Angeli 1985). In 1985, it was reported that >80% photocatalytic deactivation of *Escherichia coli* (10^3 CFU/mL) happened within 30 min using eosin, rose Bengal, and methylene blue covalently bound onto several supports (Savino and Angeli 1985, Bekbölet 1997). The photocatalytic water purification of TiO_2 photocatalyst was evaluated in the degrading organic contaminants and bacterial pathogens by means of solar light irradiating devices. It was indicated that the microbial cells were photoelectrochemically sterilized using TiO_2/Pt semiconductors within 60–120 min upon irradiation of metal halide lamp (Matsunaga et al. 1985). In 1988, photocatalytic antibacterial activity was investigated against *Streptococcus mutans* bacterium (Saito et al. 1992). Also, a continuous disinfection system was developed in 1988 using TiO_2 powder immobilized onto the membranes (Matsunaga et al. 1988). The deactivation of *poliovirus 1* and *coliforms* was examined in secondary wastewater, including TiO_2 suspension. Furthermore, the influences of photocatalyst dosage, gas composition and intensity of irradiated light were explored (Wei et al. 1994, Watts et al. 1995). Fast death of *E. coli* cells was achieved within dechlorinated water solutions lacking major radical trapping agents (Bekbölet and Araz 1996). A set up was made and used for photocatalytic *E. coli* sterilization based on diffuse light-emitting optical fibers (Matsunaga and Okochi 1995). The optimal conditions included >0.4 µg/mL TiO_2, 60 m Einstein/sm^2 intensity of light and 2 h irradiation of light. Such photocatalytic activity was improved when iron was added to the reaction vessel [10]. It was described that the viral decomposition was caused by oxidation attacking the hydroxyl radicals, which was assisted by the Fenton mechanism.

Alternatively, nanotechnology offers the synthesis of numerous nanomaterials for more efficient wastewater purification (Tahir, Farman, et al. 2019a, Bahamon and Vega 2019, Basso et al. 2020, Mirzaeifard et al. 2020). At present, several kinds of nanomaterials, including metal nanoparticles, semiconductors, carbon nanomaterials etc. are utilized as nanoadsorbents, nanocatalysts, and nanofillers in nanocomposite membranes for water treatment due to valuable and tunable characteristics like low toxicity, biocompatibility, biodegradability, large specific surface areas, big pore sizes, suitable band gaps, and electrical conductivities (Nagajyothi et al. 2019, Kumar et al. 2020, Saffari, Shariatinia, and Jourshabani 2020, Novakovic et al. 2020, Çağlar Yılmaz et al. 2020).

Advanced oxidation processes (AOPs) have been found a highly effective water treatment method (Li, Jiang, et al. 2020b). In these processes, extremely reactive species, such as hydroxyl ($^\bullet OH$), H_2O_2, O_3 and $^\bullet O_2^-$, are *in-situ* generated and are used in the mineralization of organic pollutants, aquatic pathogens, plus the sterilization of by-product compounds (Álvarez et al. 2020). Among diverse AOP techniques, heterogeneous photocatalysis – which uses semiconductor catalysts like ZnO, TiO_2, Fe_2O_3, V_2O_5, ZrO_2, Nb_2O_5, WO_3, CdS, ZnS, and GaP, etc. – is a very efficient method that degrades numerous organic contaminants and converts them to biodegradable materials and finally mineralizes them into harmless CO_2 and water (Han, Duan, et al. 2020a). ZnO and TiO_2 are among the most effective semiconductor photocatalysts broadly utilized in the photocatalytic water purification processes. ZnO and TiO_2 reveal the highest photocatalytic activities when the photons with energies of $300<\lambda<390$ nm are irradiated. Also, they exhibit high stability during the frequent cycles of photocatalysis, but GaP and CdS semiconductors are decomposed to yield toxic materials (Fernandes et al. 2020). Although ZnO and TiO_2 are thermally and chemically stable, resistant to chemical damages, and illustrate robust mechanical features, they are only photocatalytically active in the UV spectral region. Hence, it is necessary to develop new semiconductor photocatalysts using advanced nanomaterials to achieve very efficient photocatalysts that are active under the irradiation of visible light or near-infrared lights (Feng et al. 2020, Som, Roy, and Saha 2020).

So far, many attempts have been made to improve the performances of different water treatment processes through the development of advanced nanomaterials, which can remove/destroy the aqueous contaminants in lower time periods and also under irradiation of ultraviolet (UV), visible (Vis) and/or near-infrared (NIR) lights. Also, some advances have been achieved through using nanomaterials in a combination of various water treatment methods such as photocatalysis and oxidation using strong oxidizing agents such as ozone, H_2O_2, and $K_2S_2O_8$. The purpose of all advanced methods is to find the most efficient and advanced highly efficient nanomaterials/nanocomposites to purify water with the most cost-effective techniques.

This study presents a comprehensive review on diverse advanced nanomaterials used in various water disinfection processes, including carbon nitrides, boron nitride, carbon nanotubes (CNTs), fullerenes, graphene nanosheets, carbon quantum dots, activated carbon, metal–organic frameworks (MOFs), zeolites, mesoporous silica, Mobil composition of matter (MCM), and Santa Barbara amorphous (SBA) derivatives, metals and metal oxides/sulfides/selenides nanoparticles/quantum dots, clay nanomaterials, LDHs, and polymer nanocomposites.

1.2 VARIOUS WATER TREATMENT TECHNIQUES USING NANOMATERIALS

Several water treatment procedures are used for the water treatment using different effective nanomaterials (Scheme 1.1). There are three classes of water treatment

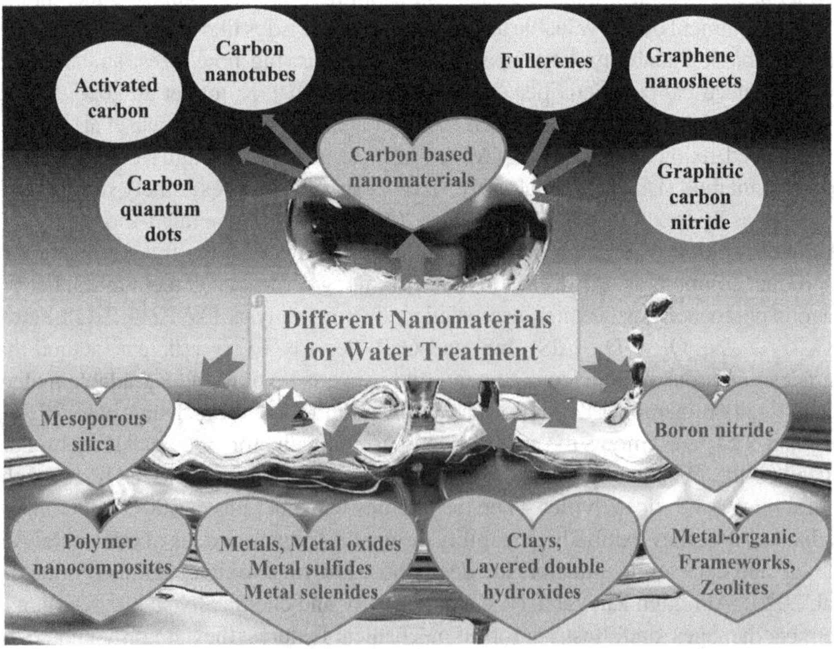

SCHEME 1.1 Different nanomaterials used in water treatment processes.

TABLE 1.1
The Strength and Weakness of Each Available Treatment Techniques for Organic Pollutants

	Strength	Weakness
Biological treatment	1. Cost-effective and environmentally friendly 2. Excellent reduction of odor and color 3. High throughput	1. It has not enough capability to remove high concentration levels of organic waste 2. Biological process is difficult to control 3. It does not diminish chemical oxygen demand values 4. Low biodegradability of some types of organic matter such as dye
Chemical precipitation	1. Low energy consumption 2. Simple process 3. A wide range of chemicals are available commercially 4. Efficient removal in total organic carbon and absorbable organic halogen especially in pulp and paper industry	1. High chemical consumption (i.e., lime, oxidants, H_2S, etc.) 2. pH value needs to be monitored 3. High sludge generation 4. Adjunction of non-reusable chemicals (coagulants, flocculants, aid chemicals) were required
Membrane filtration	1. Effective and rapid 2. Eco-friendly, non-toxic, non-corrosive, and safe to plants and animals 3. High removals of constituents	1. Relatively high cost of maintenance and operation 2. Limited feeding rate 3. Permeate flux and productivity are decreased because of membrane fouling
Traditional photocatalysis	1. Environmentally friendly 2. Low process cost, high energy saving, outstanding susceptibility 3. Adjustment of catalyst loading is easy	1. Materials are difficult to recovery and regeneration 2. Degradation effectiveness reduces due to the excess of pollutant 3. Potential photocatalyst losses under long-term operation

methods, including (a) physical/physicochemical, (b) chemical and (c) biological approaches. The details corresponding to such processes are presented next. These methods are utilized to remove/decompose inorganic, organic, biological, metal, and solid particles from both industrial and domestic wastewaters. Table 1.1 affords the strengths and weaknesses of different water treatment procedures applied to remove/decompose organic contaminants (Chen, Cheng, et al. 2020a).

(a) Physical/physicochemical methods

The physical/physicochemical methods include adsorption, air flotation, flocculation and sedimentation, extraction, and filtration/separation (using membranes).

Adsorption. The adsorption principles and the capability of diverse nanomaterials for the removal of gaseous and/or aqueous pollutants from air and/or water have been recognized by various researchers (Shen et al. 2020). The adsorption is widely employed as a cost-effective and efficient water treatment technology to produce purified water. Numerous solid nanomaterials are utilized as adsorbents for water treatment, such as flocculants and materials like iron hydroxide, aluminum oxide and synthetic/mineral nanomaterials. There are several adsorption sites on the surfaces of the nanomaterials as small pores onto them the pollutant materials are adsorbed. Activated carbon, clay minerals, and metal oxides are among several materials commonly applied as adsorbents.

Dissolved air flotation is another water treatment method extensively applied for the treatment of industrial wastewaters and yields pure water from wastewater by removing materials dispersed within the water, like solids and oil (Lee et al. 2020). In this process, the air is dissolved into wastewater upon applied pressure to remove the contaminant materials, and after that, the air is released at atmospheric pressure into flotation tank basins. The released air can form small bubbles which adhere to the materials dispersed in water and lead to their floating onto the water surface, where they are removed via skimming. Dissolved air flotation can be employed to remove pollutants from effluents of oil factories, paper mills, natural gas processing industries, chemical and petrochemical industries, domestic water treatment, and other industrial wastewaters. Induced gas flotation is a wastewater treatment process that is very similar to the dissolved air flotation method.

Coagulation, flocculation, and sedimentation are important techniques utilized to treat wastewaters (Chen, Nakazawa, et al. 2020c). The coagulation process destabilizes colloidal particles by chemical reactions occurred among colloids and coagulants. Flocculation carries the destabilized particles and results in collisions to the flocculated/aggregated particles called flocs. Principally, coagulants are added into water to aggregate all wastes existing within water (like fish eggs, earth particles and dust) as flocs. Such flocs are heavier than water; thus, they are settled down in the sedimentation tanks and remove about 90% of dispersed contaminant materials. Coagulation and flocculation are applied as an initial or intermediate step in wastewater treatments used between other treatment procedures, such as sedimentation and filtration.

Coagulation. This is a chemical procedure that includes a neutralizing charge onto the particle surface by adding a coagulant into the water, which destabilizes the colloidal suspension.

Flocculation. In this method, colloidal suspension particles form flocs spontaneously or by adding a clarifying compound. Flocculation is not only applied in water sanitization, but is also used in sewage treatment, brewing, and cheese production.

Sedimentation. In this process, particles dispersed within fluids (e.g., pollen and dust or solutions of peptides and proteins) dispose from the fluid upon applying electromagnetic forces, centrifugation, and gravitational forces.

Membrane filtration/separation. The membrane processes applied for wastewater treatment include two membrane techniques (Huang et al. 2020). One class includes nanofiltration (NF) and reverse osmosis (RO) membranes. Such membranes contain dense, non-porous separation layers cast on porous supports that are utilized to remove substances dissolved in water. Another group is membrane filtration that used membranes containing microporous separation layers as barriers to the smallest particles presenting in feed solutions while allowing dissolved substances to cross their micro-pores. The membrane filtration is usually employed as a treatment procedure; nevertheless, it can be applied as a pretreatment to the RO step. The RO and membrane filtration were initially applied in the industrial water treatment to produce ultrapure water. Also, the membrane filtration process is important for treating drinking water, removing its turbidity, and providing a disinfection barrier (predominantly against parasitic microbes like giardia and cryptosporidium).

(b) Chemical methods

The chemical methods are composed of thermal oxidation (combustion), ion exchange, chemical precipitation, electrochemical precipitation, and chemical oxidation (oxidation using systems containing H_2O_2, oxygen, ozone (ozonation), ultraviolet (UV) irradiation, Fe^{2+}/H_2O_2 systems (Fenton and Fenton-like mechanisms) and/or photocatalysis reactions). Among these methods, the AOPs are of great significance to the scientific community and the industries because such techniques reveal high performances, cost-effective, and can be used in several cycles.

Thermal treatment regulates wastewater evaporation and, at the same time, collects the contaminant materials (Lian et al. 2020a). Thermal treatment is a process that controls the wastewater evaporation, accumulates all particles, non-hazardous residuals, and finally treats hazardous air pollutants, HAPs, and volatile organic compounds, VOCs, within one instrument. Indeed, certain airstreams only comprised of contaminants of nearly large Henry's Law coefficient values are economically stripped out using air strippers. Nonetheless, wastewater streams comprise numerous contaminant substances that do not have nearly large vapor pressures or little water solubility values. Thermal oxidation is a facile method in which the stream temperature is elevated to break the chemical bonds holding the VOC molecules together. The HAPs/VOCs existing within the exhaust stream will be converted into thermal energy, water, and CO_2 (carbon dioxide) upon applying high temperatures in the combustion procedure.

Chemical precipitation (also called reagent coagulation) is a technique that can precipitate contaminations from wastewater through changing electro-oxidation potential and pH or by co-precipitating by precipitation substances/coagulants like aluminum/ferrous sulfate (Son et al. 2020). Reagent oxidation is a specific reagent coagulation process because oxidizing agents like potassium bichromate or permanganate are added into water to decompose organic pollutions and to vary the valences of multivalent cations after they are precipitated.

Electrocoagulation is utilized to precipitate impurities by means of ions transported within water by a dissolving anode upon applying an electric current (Combatt et al. 2020). Common anode materials used in electrocoagulation include aluminum andiron. Electro-oxidation process involves decomposing organic contaminations by hypochlorites under applying electrical currents between electrodes within solutions comprising chloride anions.

AOPs are a collection of non-catalytic and catalytic procedures which happen due to the great oxidation capability of the $^{\bullet}OH$ radicals (Monteoliva-García et al. 2020). Various AOPs produce $^{\bullet}OH$ radicals by different methods, but all of them principally occur via the *in-situ* creation of the $^{\bullet}OH$ radicals, which rapidly react with numerous organic substances. Also, the $^{\bullet}OH$ radicals must be produced in adequate quantities to effectively decompose the organic materials. The AOPs are categorized into heterogeneous and homogeneous processes that can be applied in presence of external energies such as electrical, ultrasonic, radiant, and light energies or without them. The $^{\bullet}OH$ radicals are non-toxic and strong oxidizing agents that are simply produced.

Furthermore, they are non-corrosive to water treatment equipment and have a very short lifetime. Accordingly, AOPs are environmentally biocompatible techniques confirming they can be developed as economical methods. The efficiencies of different AOPs are related to the effectiveness of the $^{\bullet}OH$ radicals produced in the process (Jourshabani, Lee, and Shariatinia 2020). The $^{\bullet}OH$ radicals attack the organometallic and organic pollutants existing within the wastewater in a non-selective way and perfectly cause their whole mineralization into inorganic ions, water, and CO_2. The $^{\bullet}OH$ radicals are extremely reactive species that are not accumulated in water. Consequently, they powerfully react with organic contaminants resistant to other oxidizing reagents with rate constants of about 10^6–10^{10} $M^{-1}.s^{-1}$(Yang, Li, et al. 2020b).

In most AOPs, strong oxidizing reagents like hydrogen peroxide and ozone are used with transition metal cations as catalysts/photocatalysts and/or ultrasound/UV irradiation. Upon irradiation of photons/lights, positive holes and electrons are produced in the valence band (h^{+}_{vb}) and conduction (e^{-}_{cb}) of photocatalyst (see Equation 1.1) (Chen, Cheng, et al. 2020a). The holes can generate hydroxyl radicals (Equation 1.3) that afterward oxidize the organic substances (Equation 1.6) or directly react with organic pollutant compounds (Equation 1.5). The effect of oxygen adsorbed onto the photocatalyst surface is significant as it reacts with the photo-created electrons to create superoxide radicals (Equation 1.4). The electrons can react with organic contaminants to yield reduction compounds (Equation 1.7).

$$\text{Photocatalyst} + h\nu \rightarrow e^{-}_{cb} + h^{+}_{vb} \tag{1.1}$$

$$h^{+}_{vb} + OH^{-}_{surface} \rightarrow {}^{\bullet}OH \tag{1.2}$$

$$h^{+}_{vb} + H_2O_{adsorbed} \rightarrow {}^{\bullet}OH + H^{+} \tag{1.3}$$

$$e^{-}_{cb} + O_{2adsorbed} \rightarrow {}^{\bullet}O_2^{-} \tag{1.4}$$

$$h_{vb}^{+} + \text{Organic compounds} \rightarrow \text{Oxidation products} \tag{1.5}$$

$${}^{\bullet}OH + \text{Organic compounds} \rightarrow \text{Degradation products} \tag{1.6}$$

$$e^{-}_{cb} + \text{Organic compounds} \rightarrow \text{Reduction products} \tag{1.7}$$

Different AOPs can be classified according to the method used for generating $^{\bullet}OH$ radicals as the strong oxidizing agent (Jourshabani, Shariatinia, and Badiei 2017a). The most important AOPs are photolytic, photocatalytic, ultrasound, electrochemical, ozonation, and AOPs using hydrogen peroxide. The numerous accessible procedures evidence that the AOPs can be done through versatile methods. The most valuable features of the AOPs are (1) their high capacities to mineralize organic contaminants to water and CO_2 plus oxidation of inorganic materials and ions like nitrates and chlorides. (2) The highly reactive $^{\bullet}OH$ radicals produced in the AOPs decompose numerous organic substances in a non-selective manner, particularly without producing highly toxic by-products.

Ozonation is commonly used for the disinfection/treatment of water. This process is employed to destroy organic contaminations using pure water enriched with O_3 (ozone). This technique creates $^{\bullet}OH$ (hydroxyl) radicals in water byO_3decomposition under light radiation and results in some complex reactions (Abdurahman and Abdullah 2020). In this method, $^{\bullet}OH$ radicals are generated as one of the most reactive species, and their amounts determine the process efficiency. Several kinetic models have been introduced to describe the efficacy of $^{\bullet}OH$ radicals produced in the AOPs by means of scavengers of $^{\bullet}OH$ radicals.

Catalytic ozonation is also performed by the two heterogeneous and homogeneous catalytic ozonation methods. The ozone decomposition in the homogeneous process is catalyzed using transition metal cations in water, and in the heterogeneous method, it is catalyzed using solid catalysts. Various metal cations and metal oxides such as Mn^{2+}, Fe^{3+}, Fe^{2+}, Fe_2O_3, Al_2O_3, Al_2O_3, MnO_2, TiO_2, and Ru/CeO_2 can be utilized to accelerate pollutants degradation. The properties of transition metals not only affect the reaction rates but also they control the amount of O_3 usage and the reaction selectivity.

(c) Biological methods

The biological processes can be divided into chlorination, aerobic, anaerobic, and activated mud techniques.

Chlorination is the method in which chlorine and/or chlorine compounds like sodium hypochlorite are added into water (Mazhar et al. 2020). This process is applied in order to kill all microbes, viruses, and bacteria existing within water. Particularly, chlorination can inhibit the spread of waterborne illnesses such as typhoid, dysentery, and cholera.

Aerobic treatment of wastewaters is a biological procedure that utilizes oxygen to decompose organic pollutants and other contaminants such as phosphorus and nitrogen (Liang et al. 2020b). Oxygen is constantly entered into the sewage/wastewaters using a mechanical aeration apparatus like an air compressor/blower. Next, aerobic bacteria feed the organic materials existing in the wastewater and convert them to biomass and CO_2 that are easily removed. Aerobic treatment is commonly employed to refine industrial wastewaters pretreated via anaerobic techniques. Such a method guarantees that the pollutants in the wastewater are completely destroyed, and they can safely be discharged according to strict ecological guidelines. Aerobic treatment procedures are appropriate for beverage, food, municipal, and chemical industries. Some diverse methods are used for the aerobic treatment of sewage and wastewaters, including (i) conventional activated sludge in which organic materials are broken down using aerobic bacteria in an aeration tank to form biological flocs/sludge that will subsequently be separated from water within the sedimentation tank. (ii) Moving bed biofilm reactor, MBBR, in which the biofilm is grown onto plastic carriers circulated and suspended within aeration tanks. They are preserved within the tanks through retaining sieves. (iii) Membrane bioreactor, MBR, is an advanced technique that combines membrane filtration with the activated sludge method.

The aerobic wastewater treatment illustrates several advantages as it is an effective, facile, and stable procedure which yields high-quality fluid. The produced sludge does not have any odor and can be used as exceptional fertilizer in agriculture. Besides, contaminants and nutrients are completely removed if the aerobic treatment is applied together with the anaerobic treatment. Hence, the wastewater can safely be utilized without breaking strict environmental rules.

Anaerobic treatment is an established and energetically effective technique utilized to treat industrial wastewaters (Cheng et al. 2020). This process applies anaerobic microorganisms (biomass) to transform organic contaminants and chemical oxygen demand (COD) to biogas in a milieu that lacks oxygen gas. Anaerobic microbes that are specific toward the oxygen-free environments are chosen because they can destroy organic materials presenting within industrial fluids so that they convert organic contaminants to biogas (CH_4 + CO_4) plus a low quantity of bio-solid compounds. Then, this high-energy biogas can be employed as the boiler feed to generate green heat and electricity simultaneously. Anaerobic treatment shows some benefits compared with aerobic treatment, including little energy usage, the small surface area of a reactor, lower consumption of chemicals, and lower costs of handling sludge. In order to provide treated water of a higher quality that can be discharged to a water course, aerobic treatment should be performed after the anaerobic treatment process.

Activated mud. In this technique, activated muds (such as red mud) are used as adsorbents in wastewater treatment (Ioannidi et al. 2020). Red mud is a by-product achieved in the Bayer technology, which is considered an industrial waste entered into the environment in huge amounts. In order to produce water of high quality, red mud can be utilized as an efficient adsorbent to remove pollutants from wastewaters using an inexpensive waste by an environmentally friendly process. Red mud mostly contains oxides of titanium (Ti), aluminum (Al) and iron (Fe). Therefore, combined with its textural properties, it has a great surface reactivity that is very critical to have an efficient adsorbent. Upon treatment of red mud, it is activated and reveals high performance as an adsorbent in the removal of organic contaminant substances (like formaldehyde and phenol), inorganic pollutants (like phosphates) and heavy metal cations (Lin, Kim, et al. 2020a). Before treatment of the red mud, it usually undergoes a pretreatment stage, including washing, drying, crushing, grinding, filtering, and neutralization. Several approaches can be utilized in the red mud treatment to achieve an effective adsorbent. There are two commonly used activation methods involving heat treatment and acid treatment or a combination of both. Acid treatment is a chemical treatment that includes the red mud treatment using an acid like HCl. Heat treatment is a thermal process such as calcination. The acid and heat treatments enhance the adsorption capacity of the mud obtained as the bauxite residue. Certainly, material composition and treatment factors significantly affect the adsorbent performance.

1.3 APPLICATIONS OF CARBON-BASED NANOMATERIALS IN WATER TREATMENT

Various kinds of carbon-based nanomaterials are commonly used in water treatment processes. Activated carbon, carbon nanotubes (CNTs), fullerenes, graphene nanosheets, carbon quantum dots (CQDs) and graphitic carbon nitride (g-C_3N_4) are advanced nanomaterials employed as adsorbents/extraction materials, destroying different organic and inorganic contaminants, supports of photocatalysts and membrane fillers in wastewater purification procedures. These materials indicate high specific surface areas, great sorption capabilities, and high stability at temperatures below ~400°C (Kaveh, Shariatinia, and Arefazar 2016, Zhao and Chung 2018, Vatanparast and Shariatinia 2019a, 2019b, 2018a, 2018b, Xu et al. 2019, Kommu and Singh 2020, Rahimi-Aghdam et al. 2020, Remanan et al. 2020, Reza et al. 2020). These materials are usually combined/modified with semiconductors and metals to attain nanocomposites of enhanced adsorption and photocatalytic features (Kang et al. 2020, Li, He, et al. 2020a, Mousavi and Janjani 2020, Rono et al. 2020). g-C_3N_4 is a non-toxic, environmentally friendly, inexpensive, easily available and effective material, but the pristine g-C_3N_4 has a band gap of ~2.7 eV which can absorb a small region of visible light. Furthermore, it exhibits rather high charge carriers recombination when used in the photocatalytic processes, which yields nearly low performances (Jourshabani, Shariatinia, and Badiei 2018b, 2017c). Therefore, it is commonly modified using different heteroatom dopants, metals, and semiconductors to expand its light absorption area and optimize its physicochemical and structural characteristics. CNTs reveal high specific surface areas, large adsorption capacities and extraordinary stability at rather high pressures (Jana and Singh 2018, Salehi and Shariatinia 2016a, 2016b, Salehi, Shariatinia, and Sadeghi 2019). Many studies have hitherto been done on using these nanomaterials in water treatment processes, and some examples are given here.

Two treated henequen and jute natural fibers were used as precursors to produce activated carbon cloths, which were applied in water treatment (Nieto-Delgado, Partida-Gutierrez, and Rangel-Mendez 2019). Both physical and chemical activation approaches were employed using steam/carbon dioxide and zinc chloride, respectively. The activated carbon cloths were explored in water purification processes by adsorption of cadmium ions, methylene blue, and phenol. The porous structures of the activated carbon cloths were changed by varying the activation technique indicating the chemical activation developed microporous cloths, but the physical activation generated little meso- and microporous cloths. The surface areas of the cloths were in the range of 480–1200 m^2/g, and their surfaces were neutral to acidic containing acidic functionalities in the range of 0.03–0.35 meq/g. It was exhibited that the surface chemistry and pore size distributions of the activated carbon cloths greatly influenced their adsorption capabilities. The microporous materials could adsorb micro-pollutants like cadmium ions and phenol, whereas mesoporous cloths more adsorbed the methylene blue. The activated carbon materials containing more acid groups had a greater sorption performance for the cadmium ions than the pristine cloths. Accordingly, henequen and jute were known as appropriate materials for the green synthesis of activated carbon cloths instead of using chemicals/synthetic compounds. In another study, TiO_2/activated carbon photocatalyst was developed hydrothermally and used in the photocatalytic degradation of aflatoxin B1 upon the UV–Vis light illumination (Sun et al. 2019).

Effects of parameters influencing the aflatoxin B1 photodegradation efficacy, like light source, pH, and catalyst amount, were examined. The TiO_2/activated carbon nanocomposite illustrated a superior aflatoxin B1 degradation performance of 98% compared to that of the pure TiO_2 (76%) due to its higher visible light absorption and greater surface area. The aflatoxin B1 degradation data was fitted to the pseudo-first-order kinetic model, and it was indicated that the hydroxyl radicals and holes had significant effects on the aflatoxin B1 photodegradation. It was noted that the TiO_2/activated carbon nanocomposite had a great sorption capability and photocatalytic property that were valuable for the water treatment process. Activated carbon powders were employed dynamic microfiltration membranes to mitigate the fouling throughout cross-flow pretreatment of seawater (see Figure 1.1) (Soesanto et al. 2019). Changing the mechanism

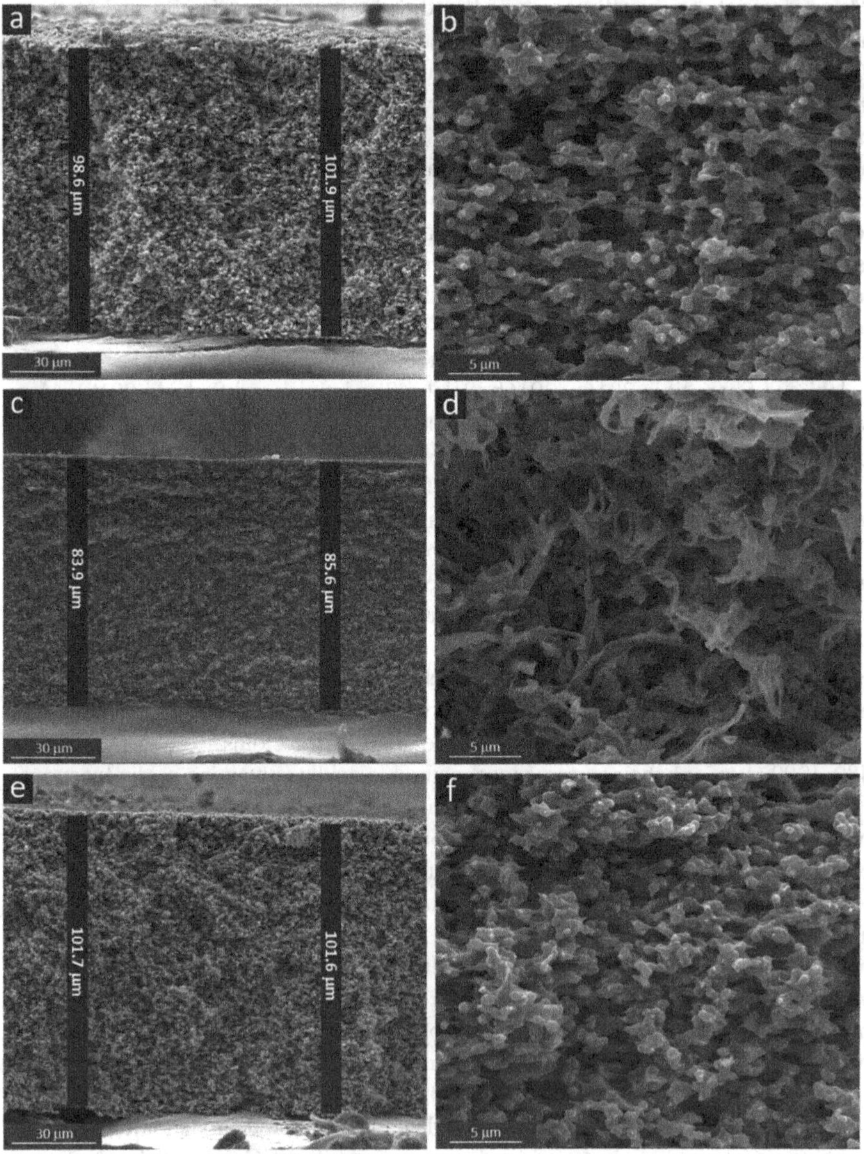

FIGURE 1.1 The SEM photographs depicting cross-sections of (a, b) pristine PVDF membrane, (c, d) used PVDF membrane without activated carbon and (e, f) used PVDF membrane containing activated carbon. Reprinted with permission from (Soesanto et al. 2019).

of fouling and path of water molecules improved the pseudo-steady-state rejection and flux of cross-flow filtration by 29 and 53%, respectively, in comparison to those of microfiltration without using activated carbon powdered membranes. Furthermore, the Fenton oxidation process was used in the membrane cross-flow microfiltration, confirming initial filtration flux recovery of up to 28% if it was only utilized, but 52% if it was used along with the alkaline purification technique.

Milkvetch species was used as a precursor for the synthesis of activated carbon that was mixed with multiwall CNT and employed as an adsorbent for the removal of humic acid (Noorimotlagh et al. 2020). The highest humic acid sorption of 22.57 mg/g was measured at pH = 3, but it reached equilibrium in 30 min using a higher adsorbent amount, 0.4–0.5 g/L, the removal efficacy was enhanced from 90.3 to 97.6%. The adsorption kinetics was fitted, $R^2 = 0.9991$, to the pseudo-second-order model and the Langmuir isotherm model, $R^2 = 0.9967$, confirming monolayer sorption of humic acid on the CNT/activated carbon, which indicated the highest capacity equal to 73.29 mg/g. The thermodynamic data exhibited that the humic acid sorption was happened spontaneously by a physical endothermic process. Thus, the CNT/activated carbon was a favorable adsorbent for the adsorption of the humic acid aqueous contaminant. Some nanocomposites of CNT/ZnO nanospheres, CNT/Z, and reduced graphene oxide, GO/Z, containing diverse ratios (10 and 30 wt%) of CNT or GO were prepared by the hydrothermal method (Khairy, Naguib, and Mohamed 2020). These materials were used as heterogeneous catalysts for the photocatalytic degradation of 10 ppm 4-nitrophenol through the Fenton method under visible light illumination at λ>420 nm and 160 W, ultrasonic irradiation at 20 kHz and 60 W plus simultaneous irradiation of both visible light and ultrasound. The 4-nitrophenol degradation by the photo-Fenton mechanism was accomplished in 15 min over the CNT (30)/Z (70) photocatalyst at 25°C and pH = 4 using catalyst amount = 1.75 g/L in the presence of 40 mM H_2O_2 (Figure 1.2). The 4-nitrophenol sonophotocatalytic degradation rate constant was 0.3 min^{-1}, greater than those of photocatalytic and sonocatalytic methods, i.e., 0.06 and 0.15 min^{-1}, respectively. The 4-nitrophenol degradation rate constants were enhanced at pH 11, indicating more $^{\bullet}OH$ radicals are released in the alkaline condition. In another study, a sandwich-shaped support coated by single-walled carbon nanotubes (SWCNTs) was fabricated to prepare antifouling and high-performance thin film composite membrane for the forward osmosis technology (Deng et al. 2020). The sandwich-shaped support was achieved by depositing polydopamine-loaded SWCNTs onto both surfaces of microfiltration polyethersulfone (PES) membrane, the back surface an antifouling layer that enhanced the prevention of foulant sorption, hydrophilicity, and rejection of bovine serum albumin, ~98.1%. The surface characteristics and filtration performances of the CNT-PES membranes are given in Table 1.2. Furthermore, Table 1.3 displays the properties of the sandwich-like CNT-PES-CNT supports. The thin film composite forward osmosis membrane displayed exceptional permselectivity with 35.7 $L/m^2.h$ water flux and 1.42 $g/m^2.h$ reverse salt flux while adsorbed 1 M NaCl draw solution. Dynamic antifouling tests confirmed that the modified thin film composite membrane had an efficient antifouling property with 19.0% average fouling throughout the cross-flow test and 8.4% throughout the bovine serum albumin sorption process, which were much lower compared with those of the unmodified thin film composite membrane (36.1% and 15.4%).

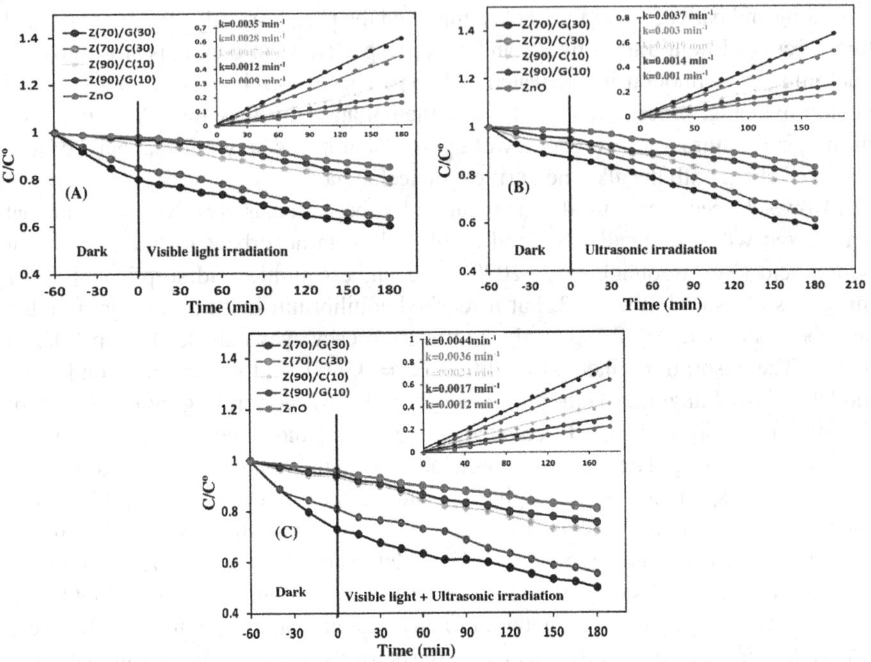

FIGURE 1.2 Visible light photocatalytic (A), sonocatalytic (B), and sonophotocatalytic (C) degradation of 4-nitrophenol on various catalysts with their first-order kinetic curves as insets at the conditions of 1 g/L catalyst, 10 ppm of 4-nitrophenol, and 25°C. Reprinted with permission from (Khairy, Naguib, and Mohamed 2020).

Magnetic functionalized fullerene nanocomposite was synthesized using sustainable, thermal, and catalytic decomposition of poly (ethylene terephthalate) waste bottles and ferrocene catalyst and magnetite (Elessawy et al. 2020). The growth mechanism was described for the formation of magnetic fullerene nanocomposite, which was utilized for the sorption of antibiotic ciprofloxacin. Several sorption factors were examined, like temperature, ciprofloxacin concentration, and time, optimized by the response surface method. Furthermore, the antibiotic effect was investigated on the *E. coli* DH5, an organism in the environment, confirming that this process solves the issue of antibiotic-resistant bacteria. In another work, the surface of commercial Hombikat TiO_2 was modified by C_{60} fullerene nanoparticles in tetrahydrofuran, nC_{60}-THF, and C_{60} $(OH)_{24}$ full erenol nanoparticles, FNP (Djordjevic et al. 2018). The photocatalytic activities of TiO_2, FNP, nC_{60}-THF, FNP/TiO_2 and nC_{60}-THF/TiO_2 were studied to investigate mesotrione degradation under sunlight radiation. The addition of the FNP into the FNP/TiO_2 nanocomposite enhanced its negative charge and the TiO_2 catalytic activity. Besides, the TiO_2/FNP exhibited smaller band gap energy than those of the nC_{60}-THF/TiO_2 and TiO_2. Also, the FNP/TiO_2 revealed the maximum photocatalytic activity. The influences of different amounts of $KBrO_3$ and H_2O_2 electron acceptors and scavengers were explored onto the mesotrione removal kinetics in aqueous solution without/with FNP and TiO_2 under sunlight radiation. The mesotrione degradation rate was enhanced in the presence of the electron acceptors in

TABLE 1.2
Surface Properties and Filtration Performances of the CNT-PES Membranes[a]

Samples[c]	PDA-SWCNTs Deposition Amount (g/m²)	Roughness (nm)	Water Contact Angle (°)	Zeta Potential (mV, pH = 7)	PWP[b] (Lm^{-2} h^{-1} bar^{-1})	BSA Rejection (%)
CNT-PES-0	0	254 ± 72	64.8 ± 2.6	−77.1 ± 1.2	20,890.3 ± 1110.6	0
CNT-PES-1	0.16	302 ± 56	43.0 ± 2.8	−67.5 ± 1.0	10,882.6 ± 1607.5	18.0 ± 0.4
CNT-PES-2	0.48	106 ± 18	38.1 ± 0.7	−63.5 ± 2.9	1237.4 ± 722.2	80.9 ± 0.3
CNT-PES-3	0.80	27.6 ± 8	32.1 ± 3.3	−46.7 ± 3.3	334.1 ± 78.5	97.9 ± 0.3
CNT-PES-4	1.12	50.9 ± 9	39.0 ± 1.3	−42.8 ± 1.1	288.2 ± 75.3	98.1 ± 0.2

[a] All characterizations were performed on the CNTs back layer (the backside of CNT-PES membrane).
[b] PWP: Pure water permeability.
[c] CNT-PES: PES support with vacuum-filtrated CNTs back layer.

TABLE 1.3
Properties of the Sandwich-Like CNT-PES-CNT Supports[a]

Samples[c]	Roughness (nm)	Water Contact Angle (°)	Zeta Potential[b] (mV)	PWP (Lm^{-2} h^{-1} bar^{-1})	BSA Rejection (%)
PES	40.7 ± 1.6	62.4 ± 2.2	−58.3 ± 1.1	20,890.3 ± 1110.6	0
PES-CNT	12.3 ± 1.6	41.2 ± 2.5	−38.9 ± 1.6	550.6 ± 31.7	0
CNT-PES	27.6 ± 8.0	32.1 ± 3.3	−46.7 ± 3.3	334.1 ± 78.5	97.9 ± 0.3
CNT-PES-CNT	14.5 ± 4.9	33.0 ± 0.2	−36.3 ± 1.3	249.7 ± 18.8	98.0 ± 0.1

[a] All characterizations were performed on the top surface of PES or the CNT's top layers.
[b] Zeta potentials were determined at pH = 7.
[c] PES-CNT: PES support with spray-coated CNTs top layer; CNT-PES: PES support with vacuum-filtrated CNTs back layer; CNT-PES-CNT: PES support with two CNTs layers deposited on both sides.

comparison to the value measured when only O_2 was used. Moreover, the initial mesotrione degradation was perhaps happened by the hydroxyl radicals, but the reaction mostly proceeded through holes after 60 min light radiation. The highest effective nanomatcrials for the mesotrione mineralization and degradation were TiO_2/7 mm $KBrO_3$ and TiO_2/7 mm $KBrO_3$/40 mL FNP, respectively. C_{60}/O-C_3N_4 hybrid nanocomposites with enhanced photocatalytic activities were synthesized through microwave technique (Li, Jiang, et al. 2020b). In the porous structures of C_{60}/O-C_3N_4 hybrid nanocomposites, oxygen atoms are doped, C_{60} was formed inh-situ, and covalent bonds were simultaneously created. Also, C_{60} was only formed in the C_{60}/O-C_3N_4 hybrids synthesized at temperatures >170°C during reaction times >10 min. The X-ray photoelectron spectra confirmed that C60 was formed in-situ upon microwave radiation and the doped oxygen atoms were in the form of –OH groups. The C_{60}/O-C_3N_4 nanocomposite had a porous hierarchical nanostructure and revealed a greater surface area (104.5 m^2/g) than that of the g-C_3N_4 (53.4 m^2/g). C–N chemical bond was formed between the C_{60} and the O-doped g-C_3N_4 and acted as an electron transporter linkage. The greatest photocatalytic performance was measured for the C_{60}/O-C_3N_4 nanocomposite with the highest UV–Vis diffuse reflectance spectra absorption.

Nanocomposite of GO and Fe_3O_4 was fabricated and used for the removal of total organic carbon (TOC), in which the GO caused flocculation and coagulation, and the Fe_3O_4 enhanced the sedimentation ratio (Parsa, Pourfakhar, and Baghdadi 2020). The Box–Behnken method was used in the Design-Expert software to design all experiments considering significant factors affecting the removal of turbidity and TOC, including $FeCl_3$ and GO concentrations plus pH. The optimized conditions achieved between aspH and $FeCl_3$, and GO concentrations were equal to 7.5 and 36, and 18 mg/L, respectively. The adsorption data were tried to fit with some isotherms such as Langmuir, UT, Temkin, Radke–Prausnitz, Freundlich, Redlich–Peterson, and Dubinin–Radushkevich, and it was found that the data were most fitted to the Radke–Prausnitz model. Also, the highest sorption capacity (1608 mg/g) was correlated to the Langmuir isotherm. Thus, the GO/Fe_3O_4 nanocomposite was very effective in turbidity and TOC removal through the flocculation/coagulation method. Recently, a porous three-dimensional, 3D, Ag-AgBr/$BiVO_4$/graphene aerogel (GA) was synthesized as a Z-scheme photocatalyst through the hydrothermal process (Lin, Xie, et al. 2020b). The scanning electron microscopy (SEM) and TEM images of the synthesized materials are provided in Figure 1.3. This photocatalyst could effectively separate electron and hole thereby assisted the charge transport, which enhanced the water treatment performance due to its porous macroscopic 3D structure as well as the existence of GA as a highly conductive material on which the $BiVO_4$ nanocrystals and the Ag-Ag Brnano particles were extremely dispersed to form a heterojunction leading to effective visible light absorption and photo-created charge carriers' separation, see Figures 1.4a–e. The Ag-AgBr/$BiVO_4$/GA displayed outstanding photocatalytic activity in the methyl orange degradation upon visible light radiation, with 93.92% degradation in 24 min, plus substantial disinfection property against *S. aureus* and *E. coli* indicating 100% bactericidal efficacy in 24 min. Besides, when chitosan was chelated to the Ag-AgBr/$BiVO_4$/GA, brilliant recycling stability was achieved in methyl orange degradation after eight cycles (6.3% weight loss), demonstrating 91.84% degradation efficacy. A nanofiltration membrane was developed based on phosphorylated chitosan functionalized by a suitable quantity of GO nanosheets via the covalent linkages (Song et al. 2020). The membrane surface negative charge, roughness and hydrophilicity were −56.4 mV, 18.7 nm, and 41.9°, respectively. This nanofiltration membrane exhibited high performance for the separation of salt and anionic dye. Also, the surface porosity of the membrane was improved by 16.8%. The nanofiltration membrane illustrated greater permeate fluxes than those of the unmodified membrane for the aqueous solutions containing ponceau S, xylenol orange, direct black 38,

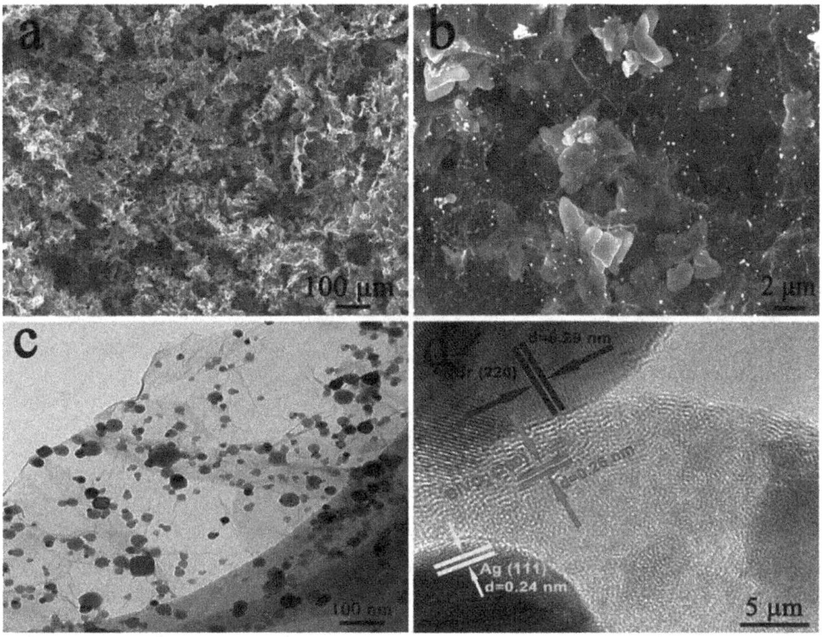

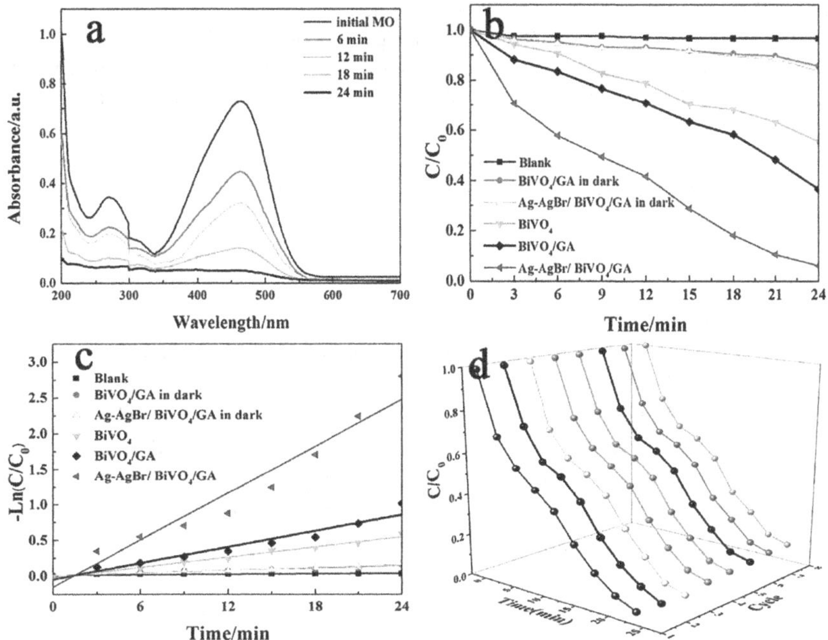

FIGURE 1.3 A (a, b) The SEM images of Ag-AgBr/$BiVO_4$/GA at the different magnifications (c) TEM image of Ag-AgBr/$BiVO_4$/GA), and (d) high-resolution transmission electron microscopy (HRTEM) image of Ag-AgBr/$BiVO_4$/GA.

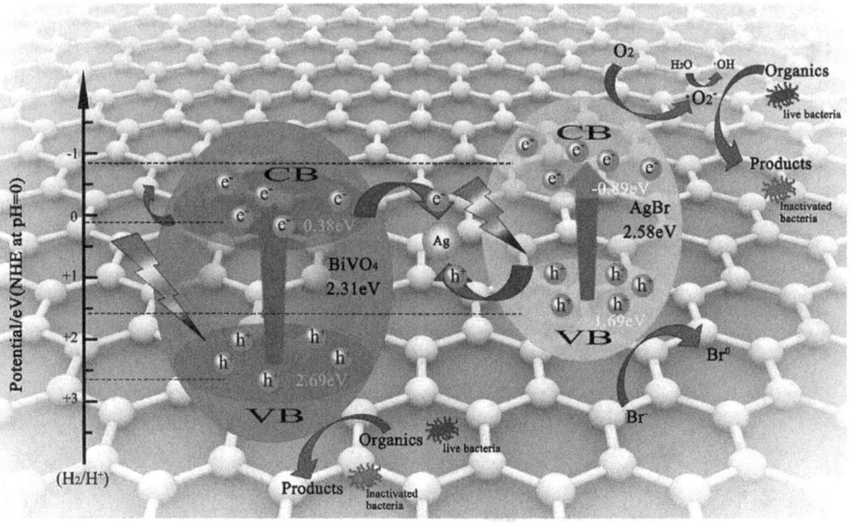

FIGURE 1.4 (a) The absorption spectra of methyl orange aqueous solution in the presence of the Ag-AgBr/$BiVO_4$/GA. (b) Photocatalytic activities of the catalysts in the methyl orange degradation under the visible light radiation and the absorptive curve of Ag-AgBr/$BiVO_4$/GA in darkness. (c). Kinetic curves of the photocatalytic degradation over the as-prepared catalysts. (d) Recycling experiment of methyl orange degradation over the Ag-AgBr/$BiVO_4$/GA under the visible light radiation. Reprinted with permission from (Lin, Xie, et al. 2020b). (e) The suggested mechanism for photocatalytic methyl orange degradation using the Ag-AgBr/$BiVO_4$/GA composite under visible light radiation. Reprinted with permission from (Lin, Xie, et al. 2020b).

Na_2SO_4 and NaCl that were 75.1, 76.3, 58.8, 78.8 and 83.1%, respectively, under steady-state condition. Furthermore, The nanofiltration membrane revealed somewhat greater removal of anionic dyes but considerably larger removal of salt. Besides, the optimized membrane had a superior antifouling feature with an exceptional flux drop (8.67%) and particularly irreversible fouling (0.17%) for the direct black 38.

CQDs were quickly synthesized in only 10 min by starch pyrolysis under microwave radiation and indicated a fluorescence emission peak at 526 nm upon excitation at 390 nm (Mahmoud, Fekry, and Abdelfattah 2020). A bionano-adsorbent was developed by incorporating the CQDs into poly (anthranilic acid-formaldehyde-phthalic acid), called PAFP, to achieve the PAFP/CQDs, which had 28.79 m^2/g surface area, see Figure 1.5a and b. The PAFP/CQDs nano-adsorbent had 30–90 mg/L sorbent capacities, and exhibited maximum U (VI) removal amounts using the changes in the range of 95.5–98.0%. The adsorption data obeyed the Freundlich isotherm, and they were fitted to the pseudo-second-order model. The PAFP/CQDs was an outstanding bionano-adsorbent for removing U (VI) from seawater and wastewater with 96.0 and 97.3% efficiencies, respectively. Also, the PAFP/CQDs illustrated exceptional reusability for multiple effective usage for U (VI) recovery from diverse water solutions. In another work, S, N-doped carbon quantum dots, S, N-CQD, were used to develop hybrid nanoflowers with hexamethylenetetramine functionalized ZnOby the hydrothermal method (1.. 6) (Qu et al. 2020). The S, N-CQD/ZnOhad a considerably high photocatalytic capacity under irradiation of near-infrared and visible lights so that 72.8% degradation of malachite green happened in 180 min upon near-infrared light radiation. Furthermore, nearly 85.8 and 92.9% ciprofloxacin was degraded using S, N-CQD/ZnO under natural sunlight (in 50 min) and simulated sunlight (in 20 min), respectively. It was found that the functionalized surface, high electron transport, and up-conversion luminescence effect of the S, N-CQD, and the

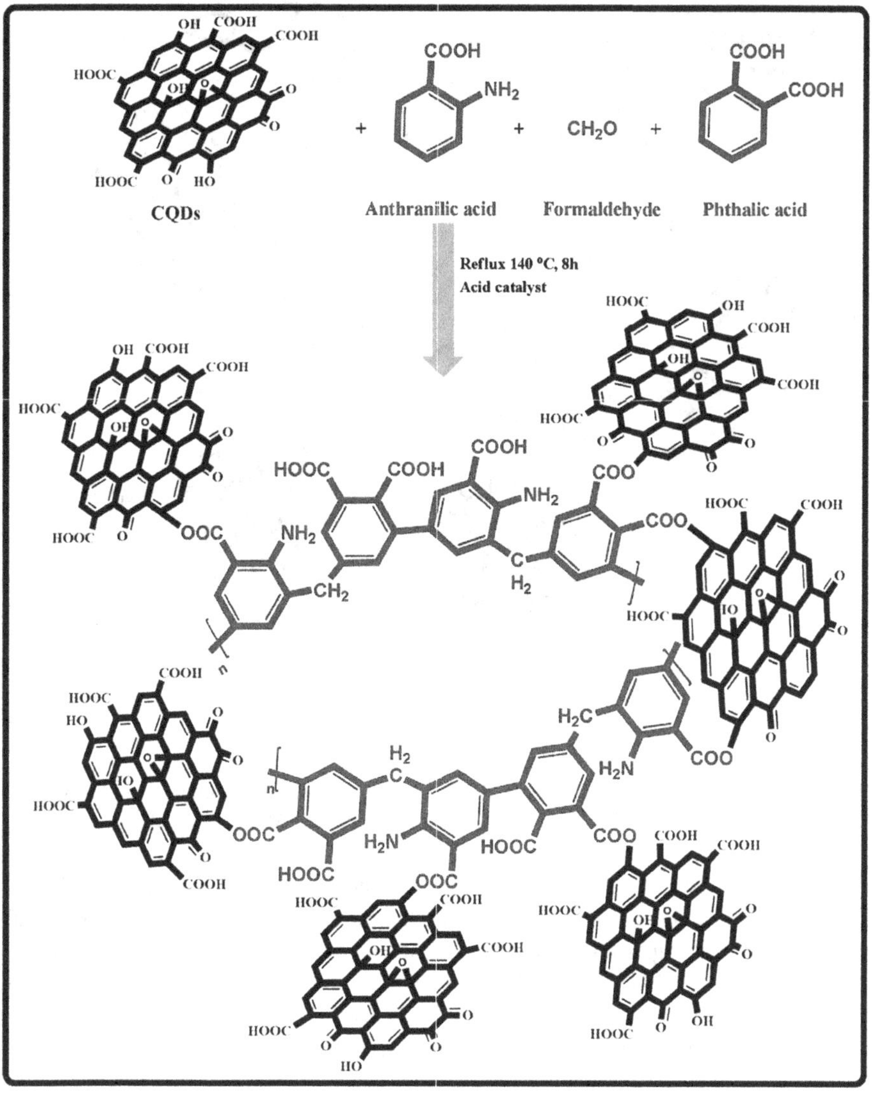

(a)

FIGURE 1.5 (a) Synthesis of CQDs@PAFP. Reprinted with permission from (Mahmoud, Fekry, and Abdelfattah 2020). *(Continued)*

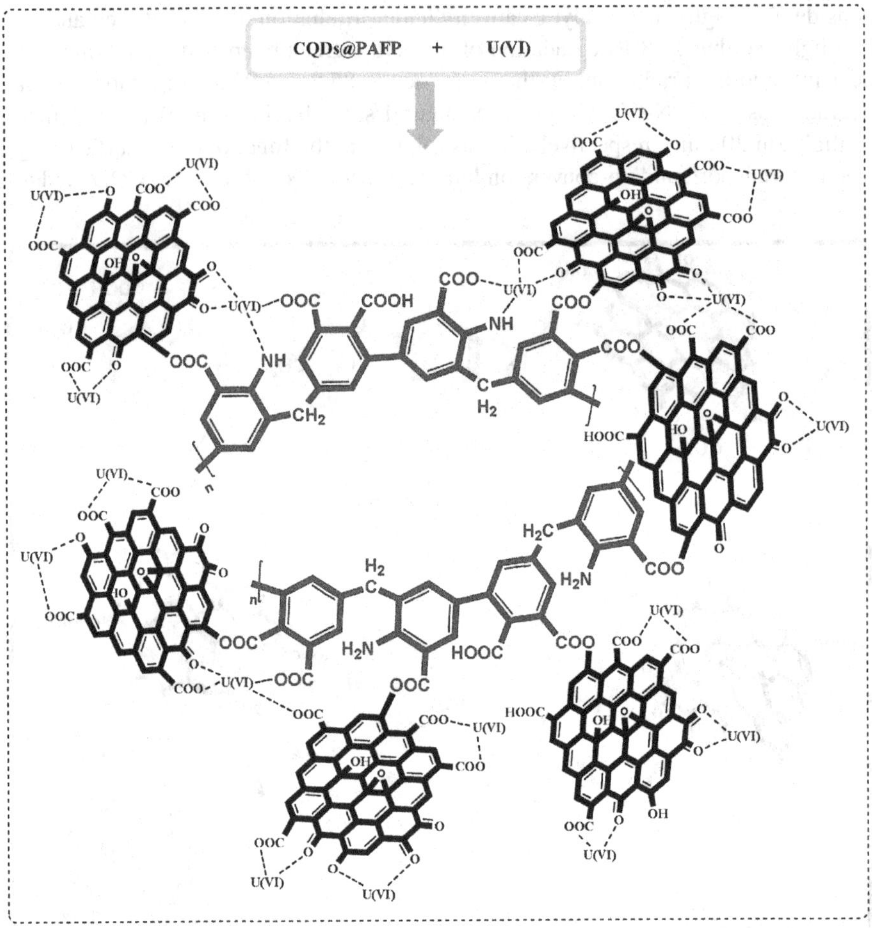

(b)

FIGURE 1.5 (Continued) (b) Proposed mechanism of U(VI) adsorption onto the CQDs@ PAFP. Reprinted with permission from (Mahmoud, Fekry, and Abdelfattah 2020).

extremely reactive ZnO nanoflowers greatly boosted the photocatalytic capability of the S, N-CQD/ZnO (Figure 1.6). Moreover, the radical trapping tests proved that all hydroxyl radicals, superoxide radicals, and holes greatly affected the photodegradation performance. The S, N-CQD/ZnO was used in photocatalytic degradation of antibiotics in real water samples and exhibited >60% efficacy under sunlight radiation for 120 min. Thus, the S, N-CQD/ZnO could be regarded as a relatively inexpensive photocatalyst for application in industrial water treatment processes. Thin film hollow fiber nanocomposite membranes were fabricated using Na^+ loaded carbon quantum dots, CQDs-Na, added to the polyamide and applied in the desalination of brackish water (Gai, Zhao, and Chung 2019). Such CQDs-Na containing thin film nanocomposite membranes revealed thinner polyamide film, greater effective

FIGURE 1.6 (a–b) SEM images of the ZnO/N,S-CQDs$_{0.54}$, TEM image of (c) N,S-CQDs and (d) ZnO/N,S-CQDs$_{0.54}$ (insets are the corresponding TEM images of ZnO/N,SCQDs$_{0.54}$). Reprinted with permission from (Qu et al. 2020).

surface area and additional hydrophilic oxygen comprising functionalities on the polyamide film among other conventional composite membranes. In fact, the spaces between the polyamide chains were bigger as the CQDs-Na particles were located in these places. Consequently, adding 1 wt.% of the CQDs-Nato to the polyamide film improved the pure water permeability by 47.1% (from 1.74 to 2.56 L/m^2.h.bar) with maintaining the 97.7% NaCl rejection. Also, stabilizing the hollow fiber thin film nanocomposite membranes loaded by 1 wt.% of the CQDs-Na at 23 bar increased the salt rejection to 98.6% and the pure water permeability to 4.27 L/m^2.h.bar in identical conditions as a result of the membrane deformation upon employing great hydraulic pressure. If 2000 ppm of aqueous NaCl solution feed was used, the optimum rejection and water flux at 15 bar were 98.6 ± 0.35% and 57.65 ± 3.26 L/m^2.h, respectively. Thus, the CQDs-Na containing hollow fiber membranes were auspicious materials for the desalination of brackish water.

Fe_2O_3 modified g-C_3N_4 was synthesized by an economical and simple process using melamine and hematite iron oxide and used in simultaneous adsorption and oxidation of As (III) cations to As (V) under irradiation of both UV and visible lights (Kim et al. 2020). The g-C_3N_4 existing in the Fe_2O_3/g-C_3N_4 oxidized the As (III) to As (V) cations, and the As (V) ions were adsorbed onto the amorphous Fe_2O_3; thus, the arsenic ions were removed from water. The As (III) removal efficiencies were 33%, 41% and 50% under darkness, visible and UV lights, respectively, indicating the Fe_2O_3/g-C_3N_4 had both of the adsorption and photocatalytic oxidation abilities.

In another work, oxygen and cobalt doped g-C_3N_4 with little nitrogen vacancies was achieved as a metal to ligand (cobalt to oxygen) charge transfer by a non-toxic and low-cost thermal polymerization technique (Yu et al. 2020). The cobalt ions were doped into the g-C_3N_4 framework as Co^{2+} cations coordinated to the N atoms in the aromatic rings, but they did not form cobalt oxide particles. The presence of cobalt cations preserved the sorption capacity of the g-C_3N_4 and acted as separation centers which assisted the interfacial transport of electrons. Hence, the oxygen and cobalt doped g-C_3N_4 exhibited an exceptional photocatalytic potential under the visible light illumination for the oxidation of endocrine disruptor bisphenol A. Mesoporous g-C_3N_4 (MCN), photocatalyst was incorporated into polyvinylidene fluoride (PVDF) membrane via an immersion precipitation phase inversion method (Yang, Ding, et al. 2020a). Figure 1.7 resents the SEM images of the surfaces and cross-sections of the fabricated PVDF membranes. The PVDF/MCN_{80} membrane displayed boosted anti-fouling and hydrophilicity characteristics. When the MCN amount was enhanced, the contact angles of membranes declined from 68.33 to 57.12, but their flux recovery rates were enlarged by 28% compared with that of the bare PVDF. The PVDF/MCN membranes had self-cleaning, photocatalytic, and antimicrobial properties. The PVDF/MCN_{80} membrane was employed in the degradation of cefotaxime under sunlight radiation (see Figure 1.8), indicating its degradation rate was 97.4% even after five cycles of usage. Furthermore, the PVDF/MCN_{80} membrane revealed 3 log inactivation of *E. coli* by the visible light radiation for 4 h (Figure 1.9). The reactive oxygen species and photo-created h^+ were the main species in the degradation of organic materials and inactivation of bacteria. Thus, the PVDF/MCN_{80} hybrid membrane could be used in the purification of real wastewaters under sunlight irradiation.

1.4 APPLICATIONS OF BORON NITRIDE, METAL–ORGANIC FRAMEWORKS (MOFS), AND ZEOLITES IN WATER TREATMENT

Currently, boron nitride, MOFs and zeolites are broadly utilized in water treatment as adsorbents, membrane fillers and photocatalysts' supports. MOFs are inorganic coordination polymers obtained through covalent bonds formed between metal cations and organic linkers, but zeolites are crystalline aluminosilicates or phospho-aluminosilicates existing as natural minerals or chemically synthesized in the laboratory, biocompatibility, biodegradability, non/low toxicity, and diverse porous structures (Yaripour et al. 2015, Shariatinia and Bagherpour 2018, Siemens, Dynes, and Chang 2020, Xu et al. 2020). Boron nitride, MOFs, and zeolites are synthesized via several approaches and illuminate valuable properties like great sorption capacities, high specific surface areas, variable pore volumes/diameters.

Boron nitride nanosheets with a hierarchical structure and urchin-like morphology (Figures 1.10 and 1.11a–g) was synthesized using analogous core-shell precursor containing boron and catalytic thermochemical vapor deposition method with both vapor/liquid/solid plus vapor/solid growth mechanisms, see Figure 1.11h (Wang et al. 2018). The boron nitride nanosheets exhibited high capacities 92.85 and

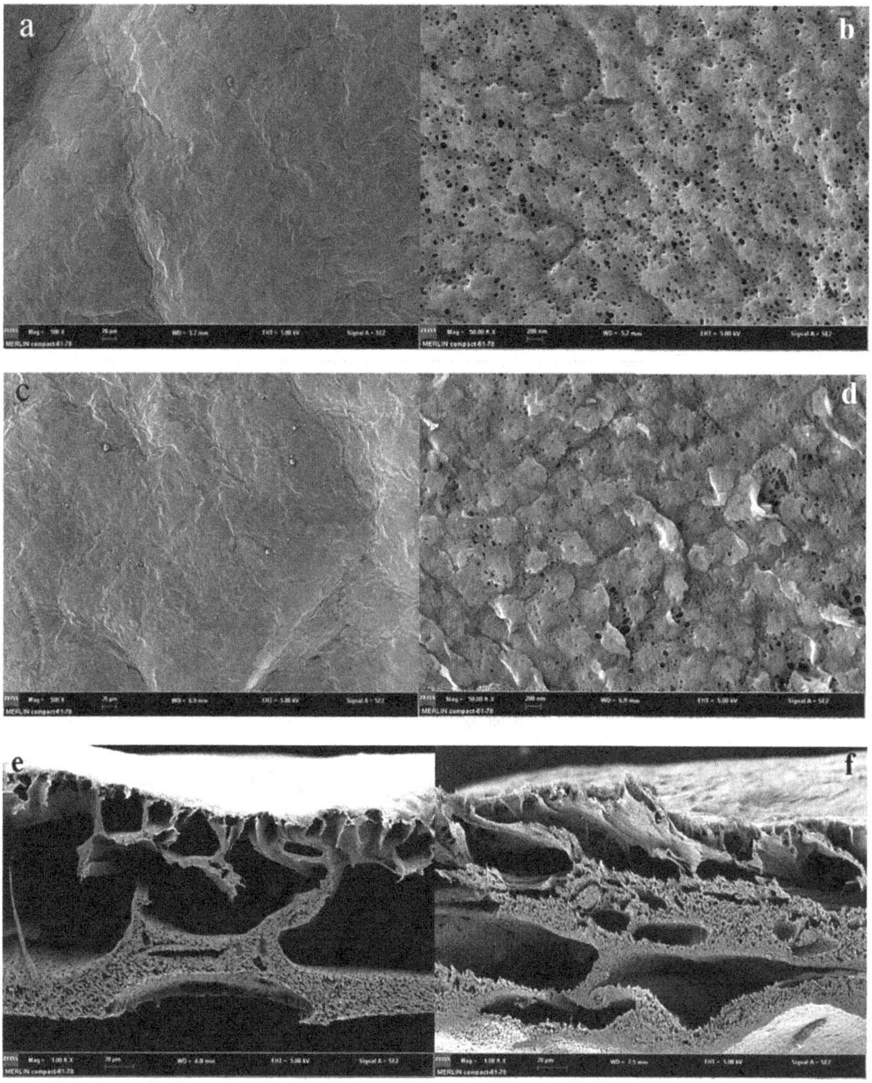

FIGURE 1.7 The SEM images of the surface of the pristine PVDF membrane magnified (a) 500× and (b) 50,000×. The SEM images of the surface of the PVDF/MCN_{80} membrane magnified (c) 500× and (d) 50,000×. The cross-section SEM images of the (e) pristine PVDF and (f) PVDF/MCN_{80} membranes. Reprinted with permission from (Yang, Ding, et al. 2020a).

115.07 mg/g for the removal of $Cu^{2+}$and Pb^{2+} cations from water, respectively. These outstanding adsorption performances were primarily attributed to numerous lattice defects, edge active sites, larger interplanar space, and exceptional structural features, which were advantageous for accommodation/adsorption of the Cu^{2+} and Pb^{2+} heavy metal cations and water purification. In recent research, boron nitride/SnO_2 nanocomposite was synthesized as a photocatalyst with a high surface area and

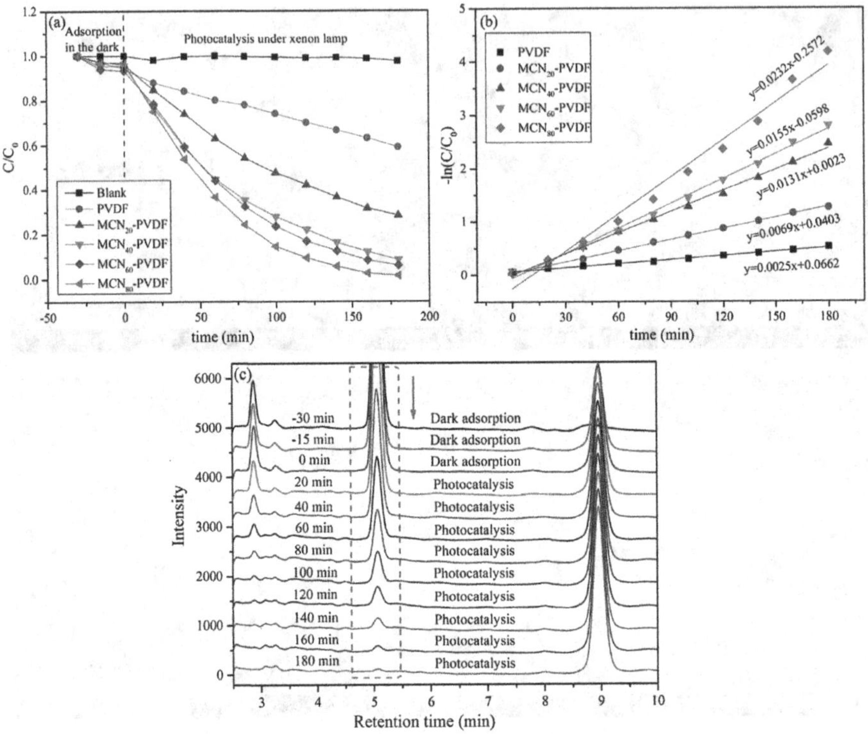

FIGURE 1.8 (a) Photodegradation of cefotaxime (2 mg/L) under the visible light, (b) photodegradation kinetics of cefotaxime degradation by PVDF/MCN$_x$, (c) chromatograms of cefotaxime degradation by PVDF/MCN$_{80}$ with diverse radiation times. Reprinted with permission from (Yang, Ding, et al. 2020a).

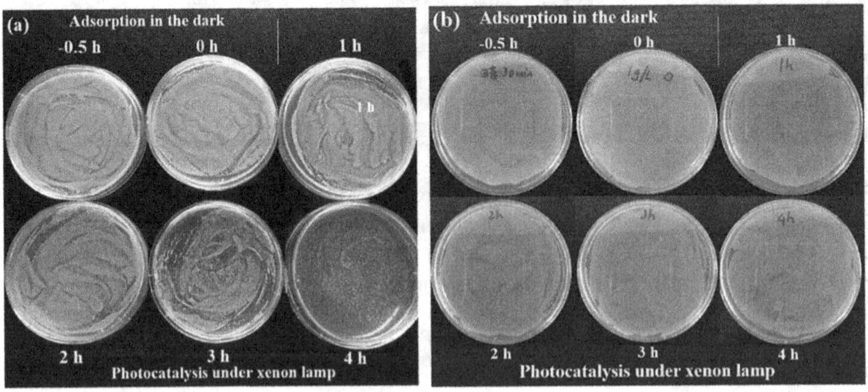

FIGURE 1.9 (a) Antimicrobial activity of PVDF/MCN$_{80}$ membrane against *E. coli* at varied times, (b) parallel antimicrobial tests without membranes. Reprinted with permission from (Yang, Ding, et al. 2020a).

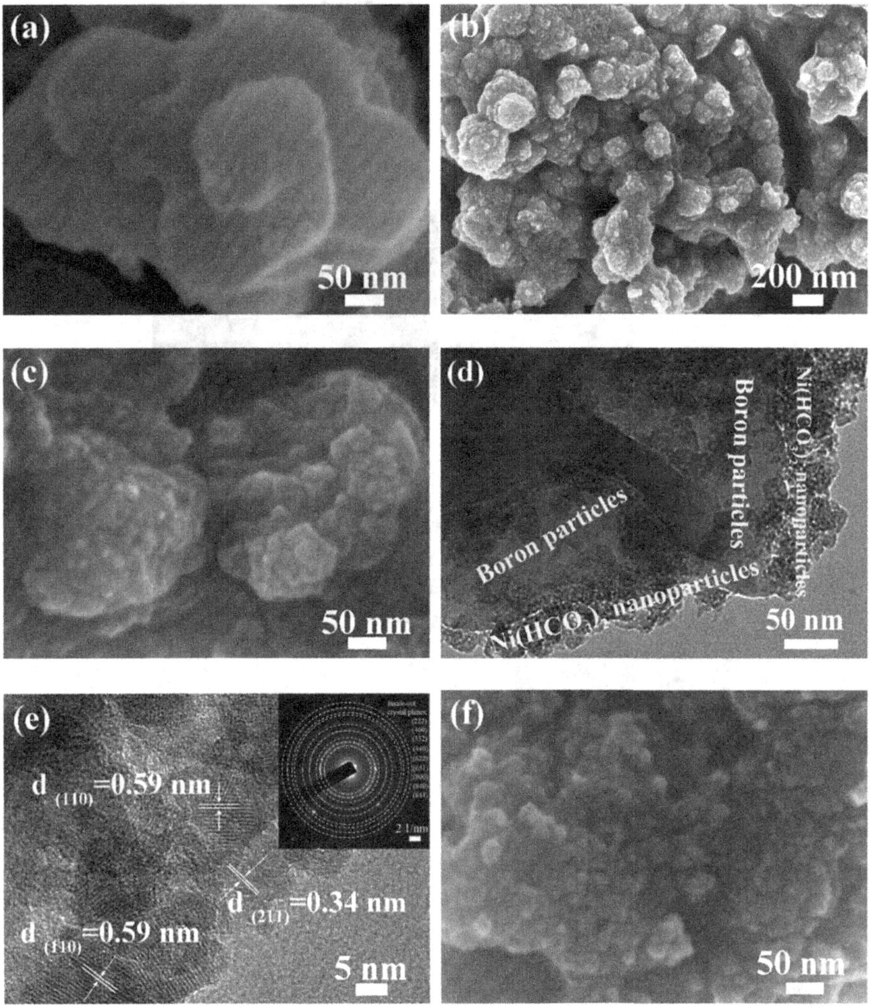

FIGURE 1.10 FESEM image of the initial boron powders (a); FESEM image at low magnification (b), high magnification (c) and TEM image (d) of boron-containing precursor obtained at 60°C; HRTEM image of $Ni(HCO_3)_2$ nanoparticles and corresponding selected area electron diffraction (SAED) pattern (e); FESEM image of the pretreated precursor at 330°Cin air atmosphere (f). Reprinted with permission from (Wang et al. 2018).

a tetragonal crystalline nanosheet structure grown in (102) crystallographic plane direction (Singh, Singh, et al. 2020a). The boron nitride/SnO_2 nanocomposite produced hydroxyl radicals on their active sites during the photocatalytic degradation of organic methyl orange dye and colorless salicylic acid contaminants, which were degraded by ~92% in 7 min and ~82% in 40 min, respectively, confirming this nanocomposite was suitable for the industrial water treatment. A thin film nanocomposite polyamide membrane loaded with boron nitride nanotubes was fabricated. It was

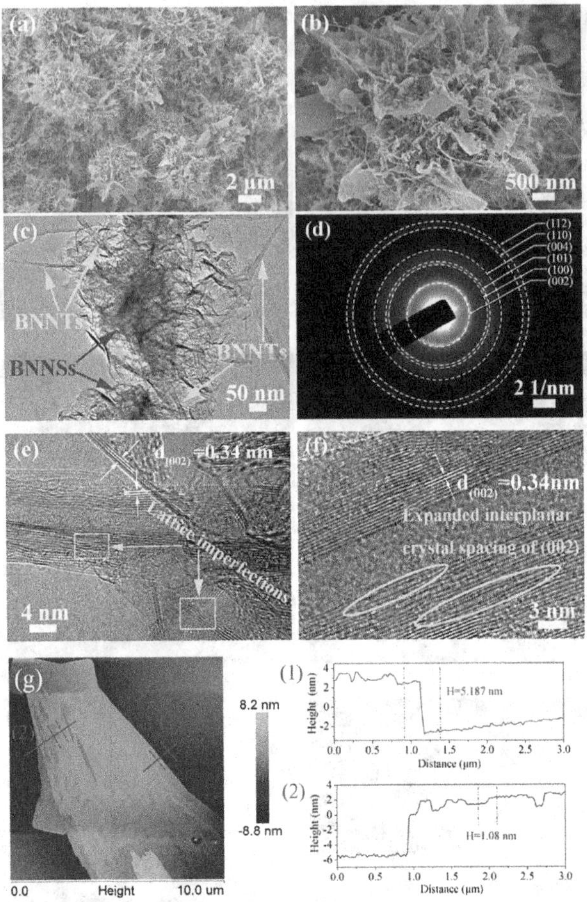

(a)

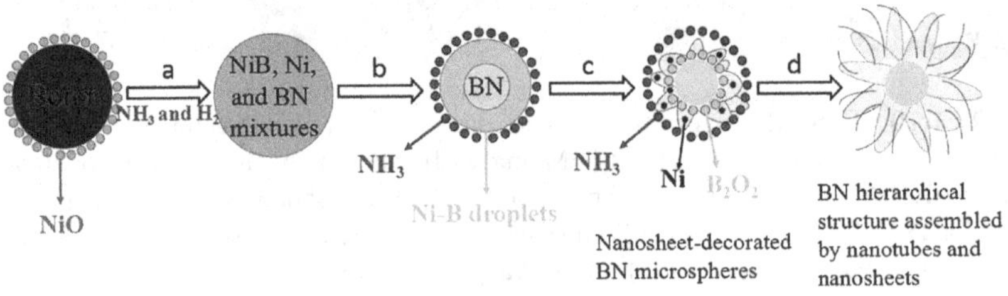

(b)

FIGURE 1.11 The FE-SEM images at (a) low and (b)high magnifications, (c) the TEM image and corresponding SAED pattern of (d) BN hierarchical structure; (f) the HRTEM image of the binding domain between boron nitride nanosheets (BNNSs) and boron nitride nanotubes (BNNTs) (e) and single BNNTs areas; (g) the atomic force microscopy (AFM) images and height profiles displaying typical size and thickness of the BNNSs.Reprinted with permission from (Wang et al. 2018); (h) A scheme indicating the synthesis pathway of the boron nitride hierarchical structure assembled with nanosheets and nanotubes. Reprinted with permission from (Wang et al. 2018).

found that only adding 0.02 wt% boron nitride nanotubes caused a four-fold enhancement to the permeance of pure water without decreasing the rejection of humic acid, methylene blue or divalent salts in comparison to those of the bare polyamide membrane (Casanova et al. 2020). However, incorporation of >0.02 wt% boron nitride nanotubes resulted in their agglomeration and total performance loss. The membranes composed of 0.02 wt% boron nitride nanotubes exhibited 4.5 $L/m^2.h.bar$ pure water permeance plus >80% and >90% rejection of $CaCl_2$ and $MgSO_4$, respectively. The fouling experiments using humic acid displayed >95% flux recovery ratio and ~50% smaller flux loss throughout the fouling tests than the bare polyamide membrane. Such results are significantly improved compared with those of thin film nanocomposite membranes containing CNTs and commercial polyamide membranes only using a very low amount of boron nitride nanotubes, which were highly appealing for water treatment processes because nanofiltration membranes demonstrate high organic fouling.

Two porous NH_2-UiO-66 (Zr) and UiO-66 (Zr) MOFs were synthesized, which had particles sizes of 3.56 and 7.56 nm and surface areas of 985 and 1420 m^2/g, respectively (Zango et al. 2020). These materials were used as adsorbents for the removal of chrysene and anthracene from water. It was shown that the NH_2-UiO-66 (Zr) and UiO-66 (Zr) could adsorb 96.4 and 98.6% of anthracene was in 25 min but 95.7 and 97.9% of chrysene in 30 min, respectively. The molecular docking data modeled the surface interactions among these MOFs, and the contaminants confirming the computed binding energies were close to those achieved from the experiments. In another work, gold nanoparticles (GNPs) and silver nanoparticles (SNPs) were grown situ onto Ce/Tb-doped Y-benzene tricaroboxylate MOFs, i.e., MOF-76 and the photocatalytic activities of the nanocomposites were evaluated for the p-nitrophenol reduction within water (Singh, Kukkar, et al. 2020b). The big surface areas of the nanocomposites meant they could effectively absorb the light. Thus, the electron transport was expedited between their conduction and valance bands which promoted the p-nitrophenol reduction. Table 1.4 represents the performances of the nanocatalysts applied in the p-nitrophenol catalytic reduction. The GNPs@MOF-76 (Ce) (5 mg/mL, 10 mL) revealed the highest catalytic performance of 96.3% with a 0.33 min^{-1}rate constant. Additionally, the GNPs@MOF-76 (Ce) nanocomposite

TABLE 1.4

Performance of the Nanocatalysts Used for the Catalytic Reduction of p-Nitrophenol

S. No.	Nanocatalysts	Induction Time (s)	$K'_{app}(min^{-1})$	$t_{1/2}$ (min)	Reduction %
1	GNPs	5	-7.3×10^{-2}	9.65	79.4
2	SNPs	5	-3.5×10^{-2}	19.97	75.1
3	MOF-76 (1)	120	-4.8×10^{-4}	1,447.66	1.90
4	MOF-76 (2)	120	-1.82×10^{-3}	380.8	3.9
5	MOF-76 (1a)	5	−0.33	2.1	96.3
6	MOF-76 (1b)	5	-1.6×10^{-1}	4.33	85.3
7	MOF-76 (2a)	5	-1.74×10^{-1}	3.96	82.6
8	MOF-76 (2b)	5	-1.8×10^{-1}	3.84	83.5

demonstrated appropriate reusability indicating 77.45% photocatalytic efficacy after three cycles. Also, this nanocomposite displayed greater efficiency than that of the P25 TiO_2 commercial nanophotocatalyst in the p-nitrophenol reduction verifying this modified MOF could be efficiently utilized in the industrial contaminant removal. Recently, a continuous crystalline film of zeolite imidazole framework-8, ZIF-8, was deposited onto the flexible polysulfone (PS) membrane surface functionalized with polydopamine (PDA) by the immersion approach (Wang, Hou, et al. 2020c). The ZIF-8 coated film considerably increased the membrane roughness from 17.2 to 154.0 nm relative to that of the polydopamine/polysulfone material (Figure 1.12). Also, the deposited ZIF-8 film diminished the contact angle of the polydopamine/polysulfone membrane surface from 45.4 to 38.6°. In addition, this membrane displayed high microbicidal activity of 99% against *E. coli* that was much greater than that of the polydopamine/polysulfone membrane, i.e., 64% (Figure 1.13). The composite ZIF-8/polydopamine/polysulfone membrane exhibited 9.6 L/m^2.h large water flux and 3.8 g/m^2.h small solute reverse flux by means of $MgCl_2$ (1 M) and deionized water as draw feed solutions, respectively, in the forward osmosis process, see Figure 1.14.

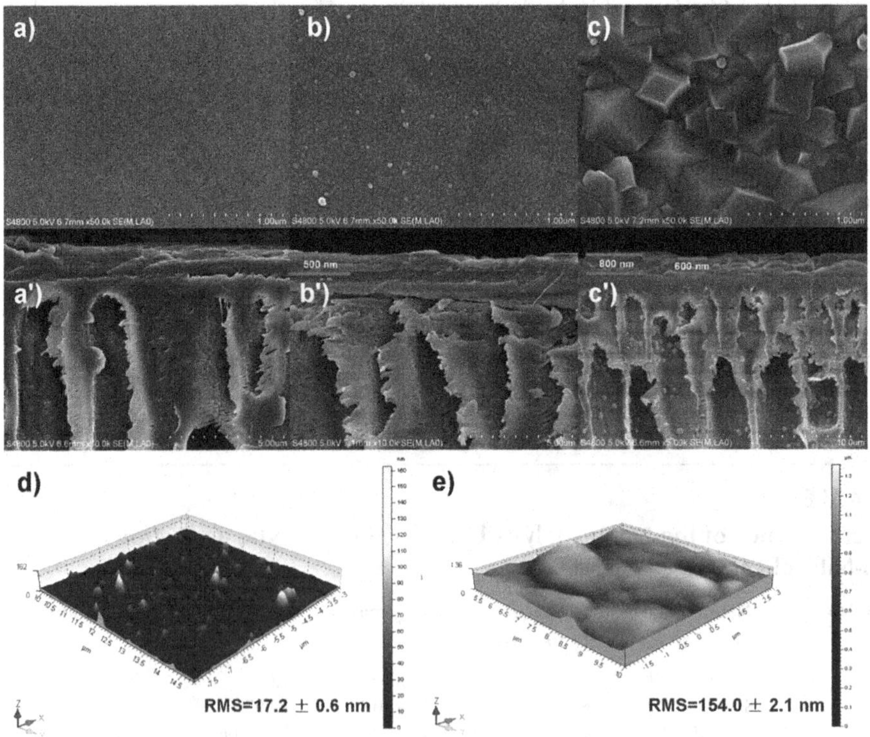

FIGURE 1.12 Morphological and roughness characterization of the composite membranes. (a, a′) PS substrate; (b, b′) PDA/PS membrane; (c, c′) ZIF-8/PDA/PS composite membrane. (d, e) AFM images and surface roughness of PDA/PS and ZIF-8/PDA/PS membrane. Reprinted with permission from (Wang, Hou, et al. 2020c).

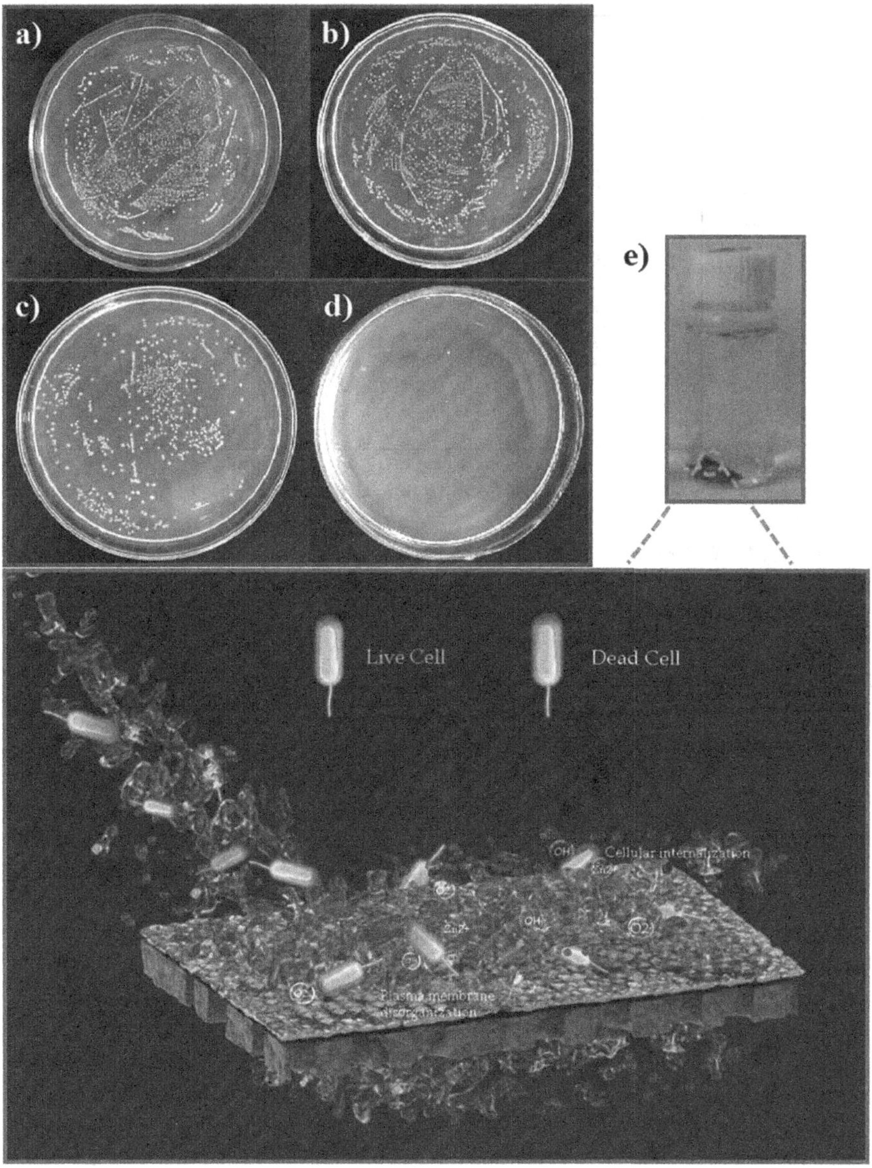

FIGURE 1.13 Images of agar plates showing bacterial colonies from (a) control sample (without membrane); (b) PS substrate; (c) PDA/PS membrane; (d) ZIF-8/PDA/PS membrane. (e) Schematic antibacterial mechanism of the ZIF-8/PDA/PS membrane. Reprinted with permission from (Wang, Hou, et al. 2020c).

In order to remediate the groundwater contaminated metal ions, a permeable reactive barrier can be applied using inexpensive reactive solutions, which simultaneously eliminates pollutants. For this purpose, diverse zeolites and nanoscale zero-valent iron, nZVI, composites were examined as adsorbents for the removal of cadmium, Cd,

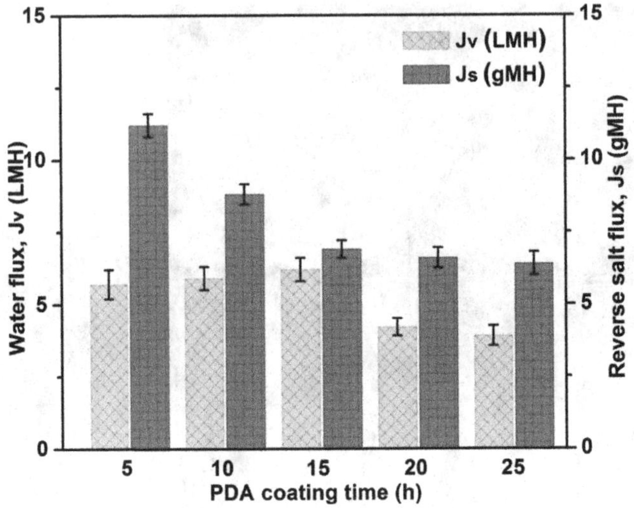

FIGURE 1.14 Water flux (J_v) and reverse salt flux (J_s) of the PDA/PS membrane in forward osmosis process with different PDA deposition time. Reprinted with permission from (Wang, Hou, et al. 2020c).

cations (Tasharrofi et al. 2020). The best zeolite/nZVI composite adsorbent revealed 20.6 g/kg removal efficacy. Additionally, the effects of ionic strength, temperature, pH, and the existence of several metal cations were explored in the performance of Cd removal by the thermodynamic and kinetic tests. Accordingly, it was found that the sodium zeolite, NaZ, afforded numerous specific sites for the ion exchange and the deposition and stabilization of then zero-valent iron (ZVI) and prevented its agglomeration and leakage under severe experimental conditions. Hence, the nZVI/NaZ composite could remove the Cd cations through sorption, confirming it was suitable material for application in the permeable reactive barriers in groundwater remediation. Recently, a graphene/Zeolite/$Bi_8La_{10}O_{27}$, called G/Z/$Bi_8La_{10}O_{27}$, nanocomposite was synthesized by the microwave method and employed as a photocatalyst in photodegradation of texbrite, methylene blue, and rhodamine B upon visible light radiation (Areerob, Cho, and Oh 2017). It was exhibited that the G/Z/$Bi_8La_{10}O_{27}$ nanocomposite had a more durable photodegradation activity upon the visible light illumination than that of the $Bi_8La_{10}O_{27}$ approving the G/Z/$Bi_8La_{10}O_{27}$ was a promising candidate with a high photocatalytic capacity for degradation of pollutant dyes existing in wastewaters. Also, the comprehensive mechanisms of photocatalytic degradation reactions were explored (Figure 1.15). A RO polyamide membrane was developed through an interfacial polymerization reaction between trimesoyl chloride and m-phenylenediamine and was filled by mineral hydrophilic clinoptilolite zeolite incorporated into the polymer-matrix (Safarpour et al. 2017). The influence of treatment with glow discharge plasma using diverse pressures of the plasma gas was examined onto the chemical and physical characteristics of the clinoptilolite plus the membranes. The Fourier transform infrared, FT-IR, spectra, and SEM of the treated and untreated clinoptilolite established that its surface features were varied, and Si/OH/Al bonds were formed upon applying

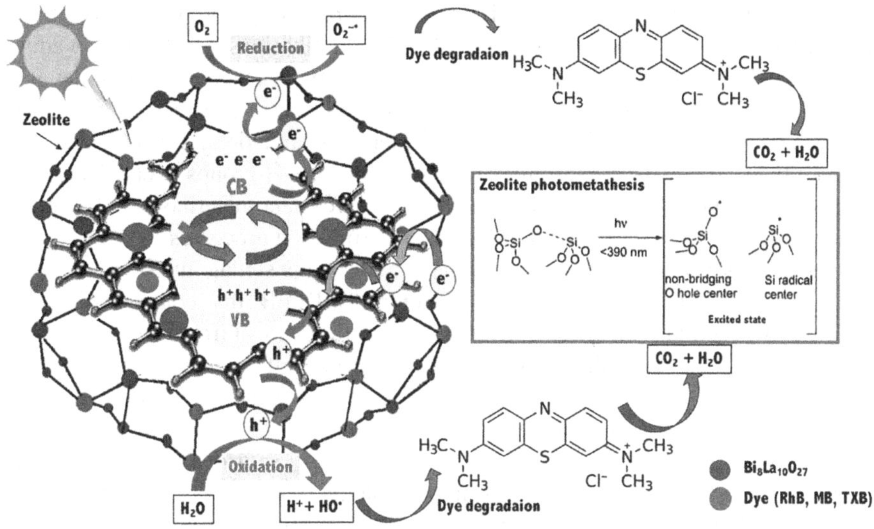

$$Bi_8La_{10}O_{27}\text{-}rGO + hv \rightarrow Bi_8La_{10}O_{27}(h^+) + rGO(e^-) \quad (1.8)$$

$$OH_2 + h^+ \rightarrow OH^\bullet \quad (1.9)$$

$$O_2 + e^- \rightarrow O_2^{\bullet-} \quad (1.10)$$

$$O_2^{\bullet-} + H_2O \rightarrow OH_2^\bullet + OH^- \quad (1.11)$$

$$e^- + OH_2^\bullet + H^+ \rightarrow H_2O_2 \quad (1.12)$$

$$e^- + H_2O_2 \rightarrow OH^\bullet + OH^- \quad (1.13)$$

$$OH_2^\bullet + \text{dye} \rightarrow \text{degraded product} \quad (1.14)$$

FIGURE 1.15 Proposed mechanism of charge separation and photocatalytic process over the graphene/Zeolite/$Bi_8La_{10}O_{27}$ photocatalyst under the visible light radiation. Reprinted with permission from (Areerob, Cho, and Oh 2017).

the plasma. Adding hydrophilic clinoptilolite into the polyamide film declined its surface roughness. Furthermore, the modified membrane revealed a higher water contact angle and enhanced hydrophilicity membranes as verified by the data achieved from the permeation experiments. The membrane loaded by 0.01 wt.% of the clinoptilolite and treated by 1.0 Torr of oxygen gas plasma displayed fouling recovery of 88% and maximum water flux of 39%, greater than those of the bare polyamide membrane.

1.5 APPLICATIONS OF MESOPOROUS SILICA NANOMATERIALS IN WATER TREATMENT

Mesoporous silica, MCM, and SBA nanomaterials are commonly applied in water treatment (Ebrahimpour, Hassaninejad-Darzi, and Zavvar Mousavi 2020). These materials (particularly the nanoparticles) show extraordinary features such as hierarchical mesoporous structures, large pore diameters, and pore volumes, high specific surface areas of >1000 m^2/g, uniform size distributions, economical in addition to facile synthesis and surface functionalization methods (Shariatinia and Esmaeilzadeh 2019, Han, Chen, et al. 2020b). Indeed, the surfaces of such materials are easily functionalized through plentiful inorganic and organic compounds to improve their properties because they have plentiful Si-OH groups over their walls/surfaces (Liu, Luo, et al. 2020c). The mesoporous silica nanomaterials are very interesting substances for employment in the water purification processes.

Mesoporous silica nanoparticles with large pores and a stellate morphology were synthesized, functionalized with desferrioxamine B, DFoB, that could particularly chelate to iron cations and specifically used in the iron removal from biological solutions (Duenas-Ramirez et al. 2020). The DfoBas iron-specific chelate was covalently grafted to the mesoporous silica nanoparticles in a controlled way and revealed a great grafting degree of 730 nmol/mg or 0.85 ligand/nm^2. Also, the extremely chelating silica with the stellate morphology was capable of capturing iron (III) cations stabilized by nitrilotriacetic acid, NTA, in physiological pH solution via a rapid rate of <30 min when the DFoB/FeNTA ratio was 1/0.85, the nanoparticles of DFoB/stellate silica exhibited efficiency and capturing ability of 78% and 480 nmol Fe^{3+}/mg SiO_2, respectively. The selectivity of the DFoB/stellate silica nanomaterial was investigated through performing the removal reaction in two diverse solutions containing several metal cations, including a solution composed of different equimolar metal ions plus a Barth's buffer solution simulating the composition of the brain. In these two environments, the DFoB/stellate silica as a chelating agent had a high iron selectivity compared to other cations with diverse (e.g., K^+, Na^+) and identical (e.g., Al^{3+}) valences. Accordingly, such a nanosystem was very useful for treating aqueous solutions contaminated with metal cations using an extremely chelating agent. Recently, mesoporous silica was self-assembled to graphene/Cu_2O nanocomposite using nonionic Pluronic® P123 surfactant. Its performance was explored in the photodegradation of some water-soluble organic dyes, including texbrite BAC-L, texbrite NFW-L, texbrite BA-L, and safranine O within water upon visible light radiation (Nguyen and Oh 2018). The effects of parameters such as catalyst amount and pH were studied, and the photocatalytic degradations of dyes were explored by acquiring their UV–Vis spectra. The graphene/Cu_2O/mesoporous SiO_2 illustrated exceptional photocatalytic capacity due to the high sorption ability of the mesoporous SiO_2 with big surface area, pore volume and pore diameter plus the catalytic degradation influence of Cu_2O. The graphene/Cu_2O/mesoporous SiO_2 photocatalyst was highly stable. The optimum sample with the outstanding photodegradation performance was developed by combining mesoporous SiO_2 and 10wt.% of graphene/Cu_2O composite. In another study, porous hollow nanoparticles were achieved and used in RO membrane of high performance to modify its structure resembling the skin layer (Yan et al. 2019). For this purpose, nanospheres of amine-functionalized hollow mesoporous silica were synthesized by a hard template technique, exhibiting average sizes in the range of 110–130 nm. Amine functionalized silica nanospheres were incorporated into the organic phase, and after that, they were added to the skin layer resembling RO membrane throughout the interfacial polymerization reaction. The amine-functionalized silica nanosphere improved the hydrophilicity of the RO membrane. The amine-functionalized silica nanosphere enhanced the flux of the RO membrane by 41.5% and somewhat decreased the rejection. Also, the amine-functionalized silica nanosphere was synthesized without hollow cores and added to the RO membrane to examine its hollow cores' influence on the performance of the skin layer membrane. It was found that the hollow cores within the amine-functionalized silica nanospheres improved the skin layer RO membrane's performance, indicating good rejection as well as great flux.

MCM-41/rice husk composite was developed by the hydrothermal process in which the rice husk was used as the source of silica and a substrate on which the MCM-41 was deposited (Sohrabnezhad and Daraie Mooshangaie 2019). AgBr/Ag nanoparticles were introduced in-situ onto the MCM-41/rice husk. The cetyltrimethylammonium bromide was applied as a source of bromide ions and a template to synthesize the AgBr and MCM-41 nanoparticles, respectively. The nanocomposite was used to degrade Eriochrome Black-T pollutant dye in water upon visible light radiation. Table 1.5 demonstrates the photocatalytic performances of the AgBr nanoparticles, rice husk (RH)/MCM-41 and Ag/AgBr-RH/MCM-41 nanocomposites upon the visible light radiation. It was established that the MCM-41 particles were formed with average particle sizes of 150–200 nm onto the rice husk surface, and the AgBr/Ag nanoparticles were deposited, indicating average diameters of 2–30 nm. The photocatalytic Eriochrome Black-T degradation performance was found to be 99.04% when 1.48% of the AgBr was loaded onto the nanocomposite. The highest Eriochrome Black-T sorption happened at pH2, whereas its decomposition was greater at pH7. In another study, a microspherical core-shell composite of mesoporous MCM-41@mTiO_2 was achieved via combined hydrothermal and sol-gel methods, which exhibited 316.8 m^2/g specific surface area, 0.42 cm^3/g pore volume plus two diverse pore sizes of 2.6 and 11.0 nm (Wei et al. 2016). The MCM-41@ mTiO_2 composite was utilized in the photocatalytic degradation of 2-sec-butyl-4,6-dinitrophenol in water upon the UV–Vis light illumination. The photocatalytic degradation data proved that this composite had a greater performance under

TABLE 1.5

The Photocatalytic Performances of the AgBr Nanoparticles, Rice Husk (RH)/ MCM-41and Ag/AgBr-RH/MCM-41 Nanocomposites under Visible Light, $C_{0(\text{Eriochrome Black-T})}$ = 25 ppm, the Photocatalyst Dosage: 0.4 g/L, at 30 min in Dark and 45 min under Visible Light

Sample	Adsorbed Dye[a] (%)	Degraded Dye[b] (%)	Removed Dye[c] (%)
AgBr NPs (pH = 7)	10.0	48.48	58.48
RH/MCM-41 (pH = 7)	74.75	3.41	78.16
Ag/AgBr-RH/MCM-41 (pH = 7)	45.30	53.74	99.04
Ag/AgBr-RH/MCM-41 (pH = 2)	75.14	20.45	95.58
Ag/AgBr-RH/MCM-41 (pH = 9)	35.70	28.50	64.20

[a] Determined with UV–vis after 30 min of stirring in dark.

[b] Determined with UV–vis after 45 min of irradiation with Vis light.

[c] The total amount of dye removed after 75 min of reaction.

identical conditions than those of commercial Degussa P25 and anatase TiO_2. Hence, the MCM-41@mTiO_2 photocatalyst was introduced as a promising material for the photodegradation of dangerous contaminants existing within wastewater. Monodispersed nanoparticles of MCM-48 with spherical morphology and a three-dimensional mesoporous cubic structure were achieved and added as a nano-filler into the organic/aqueous phase to develop some thin film nanocomposite membranes via the interfacial polymerization reaction between trimesoyl chloride and m-phenylenediamine (Liu et al. 2016). It was shown that the MCM-48 nanoparticles were dispersed in the polyamide film and clipped between the polysulfone support and the polyamide film. The water flux was evaluated using NaCl solution (2000 ppm) at the pressure of 16 bar. Besides, the water flux was steadily enhanced from 24 to 40 L/m^2.h by increasing the MCM-48 amount within the organic solution but it did not considerably affect the salt rejection, >95%. However, the water flux was improved from 24 to 68 L/m^2.h while the salt rejection was diminished to 97–80% by increasing the MCM-48 quantity within the aqueous solution. A lower amount of the MCM-48 was added to the aqueous solution compared with its amount introduced into the organic solution to achieve a similar performance. Both types of the thin film nanocomposite membranes displayed improved long-lasting stability confirming the MCM-48 was very stable within these membranes and dispersed either in the aqueous or the organic phase.

A hybrid of SBA-15 mesoporous silica functionalized with 3-aminopropyltrimethoxysilane (NH_2-SBA-15) and polyvinylpyrrolidone, PVP, was fabricated for the adsorption of Cu (II), Ni (II) and Pb (II) divalent heavy metal cations (Betiha et al. 2020). In fact, the rice husk was utilized as the silica source and a Schiff base was formed between the amine groups existing over the SBA-15 surface and the PVP. It was shown that the NH_2-SBA-15 illustrated medium adsorption of heavy metal cations. When the PVP was grafted to the NH_2-SBA-15, it showed great adsorption of heavy metal cations, with sorption data that were greatly fitted to the Langmuir model. Table 1.6 presents the Langmuir, Freundlich, and Temkin adsorption data for the sorption of Pb (II), Cu (II) and Ni(II) onto the PVP–SBA-15 at room temperature. The adsorption tests revealed that the SBA-15/PVP, NH_2-SBA-15, and SBA-15 had equilibrium sorption capacities of 72, 128 and 175 mg/g for the Ni(II), Cu(II) and Pb(II), respectively. Table 1.7 indicates the sorption of various heavy metal cations in a real wastewater sample onto the PVP–SBA-15 at 25°C and pH = 6.66. The

TABLE 1.6
Langmuir, Freundlich, and Temkin Adsorption Parameters for the Adsorption of Pb(II), Cu(II) and Ni(II) on PVP–SBA-15 at Room Temperature

Metals	Langmuir				Freundlich			Temkin		
	q_m (mg/g)	b (L/mg)	R^2	R_L	K_f	n	R^2	A (l/g)	B (J/mol)	R^2
Pb(II)	175.439	0.0067	0.996	0.597	2.143	1.37	0.964	0.252	14.78	0.914
Cu(II)	128.205	0.0070	0.997	0.599	1.581	1.363	0.973	0.242	10.96	0.923
Ni(II)	72.464	0.0090	0.998	0.503	1.38	1.437	0.976	0.26	7.94	0.901

TABLE 1.7
Adsorption of Heavy Metal from Real Wastewater Sample over PVP–SBA-15 at 25°C and pH = 6.66

Heavy Metals	Concentration mg/L		
	Before Adsorption	After Adsorption	Removal (%)
Cr (VI)	0.5	0.001	99.8
Cd (II)	0.2	nil	100
Pb (II)	1	nil	100
Hg(II)	0.2	nil	100
Ag(I)	0.5	0.006	98.8
Cu (II)	1.5	nil	100
Ni (II)	1	0.004	99.6
Sn(IV)	2	0.01	99.5
As(III)	2	nil	100
B(III)	1	0.08	92

TABLE 1.8
Thermodynamic Parameters for Adsorption of Heavy Metals on the PVP–SBA-15 at Diverse Temperatures

Metals	ΔH^0 (kJ/mol)	ΔS^0 ($J.mol^{-1}$ K^{-1})	ΔG^0 (KJ/mol)				R^2
			293 K	303 K	313 K	323 K	
Pb(II)	20.997	121.334	−14.55	−15.76	−16.98	−18.19	0.9986
Cu(II)	34.832	160.127	−12.08	−13.68	−15.28	−16.88	0.9994
Ni(II)	47.074	194.930	−10.03	−11.98	−13.93	−15.88	0.9998

kinetic experiments displayed that the sorption of heavy metal ions obeyed the pseudo-first-order rate law. The thermodynamic parameters for the sorption of heavy metals on the PVP–SBA-15 at diverse temperatures are gathered in Table 1.8. The ΔS°, ΔH° and ΔG° values established that the endothermic and spontaneous sorption of heavy metal ions happened onto the SBA-15/PVP. Therefore, the SBA-15/PVP was a suitable adsorbent for removing heavy metal cations from contaminated water. In another research, SBA-15/CdS nanomaterial was achieved through the solvothermal technique, which revealed nearly ordered mesoporous architecture, high crystallinity, and specific surface area (Wang, Shao, et al. 2020b). It was indicated that the SBA-15/CdS composite practically did not decrease the light absorption capacity of CdS, while the composite photocapacity was highly improved. The SBA-15/CdS displayed high photocatalytic salicylic acid degradation upon the visible light illumination. The greatest salicylic acid (10 mg/L) degradation efficacy was 84.93% upon 6 h light radiation using 30 mg of the photocatalyst (0.75 g/L). The activation entropy, activation enthalpy, and activation energy values of the photocatalytic reaction were −281.00 J/mol K, 3.13 kJ/mol, and 2.90 kJ/mol, respectively. The mechanism of the photocatalytic degradation was examined, and it was established that the superoxide ($^{\bullet}O_2^-$) radicals were the most reactive species, the h^+ and e^- also affected the reaction, but the $^{\bullet}OH$ radicals had a small contribution to the photodegradation process. Indeed, the SBA-15 protected the CdS nanomaterial so that the SBA-15/CdS nanocomposite exhibited greater photocatalytic property and photoluminescence intensity than those of the pure CdS nanoparticles. To purify the petroleum industry discharges at all stages from drilling/exploration to transport, polydopamine, pDA, modified mesoporous SBA-15 silica was utilized as the filler into the composite ultrafiltration polyetherimide membranes (Kaleekkal et al. 2018). Figure 1.16 illustrates the SEM and the TEM images of the fabricated membranes. These membranes were extremely hydrophilic and revealed high porosity as they contained interconnected pores and a thin skin layer. The upper surfaces on such membranes were highly rough so that they indicated oleophobic properties in water.

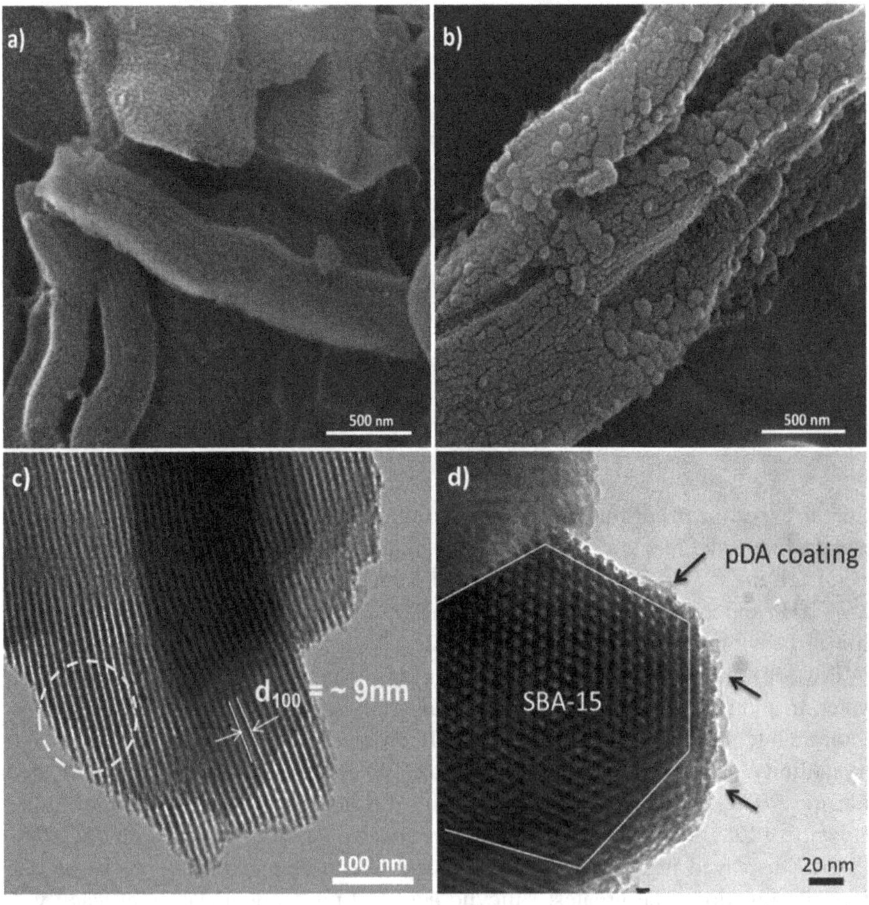

FIGURE 1.16 High-resolution SEM images of pristine SBA-15 (a) and pda-SBA-15 (b). HR-TEM Images of pda-SBA-15 at higher magnifications showing the lattice structure and polydopamine deposition (c and d). Reprinted with permission from (Kaleekkal et al. 2018).

These membranes had enhanced water/emulsion flux; however, they maintained >99.8% of oil rejection using a synthetic emulsion of surfactant/water/motor oil. The membranes were applied for the oil removal from water. It was found that the flux drop was only <15% for up to 9 h by three alternating backwashes using the permeate oil/grease content of <10 ppm that was quite a lot lower than the allowed discharge quantities. Furthermore, the composite membranes displayed high microbicidal activities against Gram-negative *Pseudomonas aeruginosa* and Gram-positive *Bacillus subtilis* microorganisms. Hence, the polyetherimide membranes composed of the polydopamine modified SBA-15 were valuable materials for separating oil from water.

1.6 APPLICATIONS OF METALS AND METAL OXIDES/SULFIDES/SELENIDES NANOPARTICLES/QUANTUM DOTS IN WATER TREATMENT

Metals (like Ag, Au, Pt, Fe, and Cu nanoparticles), metallic alloys (such as Ag-Au, Co-Ag, Fe-Pd and Fe-Ni nanoparticles), metal oxides, or mixed metal oxides (e.g., ZnO, TiO_2, Fe_3O_4, $MgFe_2O_4$, $MnFe_2O_4$, $LaFeO_3$, and $ZnAl_2O_4$), metal sulfides (ZnS, CdS, CuS, FeS, CoS, and NiS_2) and metal selenides (CdSe, ZnSe) semiconductors are extensively used in water treatment reactions because they display extremely valuable properties (Jalali-Moghadam and Shariatinia 2018a, 2018b, Zolfaghari-Isavandi and Shariatinia 2018a, 2018b, Din et al. 2019, Trawiński and Skibiński 2019, Zhang, Zhang, et al. 2020a). Metals and metallic alloys are electrically and thermally conductive materials that display outstanding catalytic and photocatalytic properties. Also, they indicate characteristic surface plasmon resonance peaks in the UV–Vis spectra as well as emission bands in the photoluminescence spectra (Ahmad et al. 2020). Metal oxides, mixed metal oxides, metal sulfides, and metal selenides are semiconductor materials that exhibit exceptional photocatalytic and catalytic features, as well as photoluminescence emission spectra. Some of the metals and metallic alloys like Ag and Ag-Au, plus semiconductors such as ZnO nanoparticles, reveal bactericidal and antiviral properties against different microorganisms as a result of their abilities to attach the bacterial/viral cells and create reactive oxygen species, ROS, which unbalance the protein activity and release metal cations that cause cytotoxicity to the bacterial and viral cells and ultimately cell death (Shariatinia and Sardsahra 2016, Zolfaghari et al. 2019, Rafique et al. 2020, Danwittayakul, Songngam, and Sukkasi 2020, Shah, Fiaz, et al. 2020b, Shariatinia and Zolfaghari-Isavandi 2020).

Sand coated by Fe/Pd nanoparticles, called Fe/Pd-S, was applied to remove tetracycline from water (Ravikumar et al. 2020). The optimal removal of tetracycline, i.e., 99.78 ± 0.05%, was measured by batch tests using 20 μg/mL tetracycline concentration, 1000 mg of the IS-Fe/Pd weight during 180 min interaction. The mechanism of tetracycline removal was the combined adsorption and degradation onto the Fe/Pd-S nanoparticles. Also, the cell viability assays by the *Pseudomonas* and *Bacillus* microorganisms as well as *Chlorellasp.* algae proved that the tetracycline toxicity was significantly decreased upon its treatment by the Fe/Pd-S nanoparticles. Additionally, the effects of changing tetracycline concentration, flow rate and bed height were explored on the removal performance of the Fe/Pd-S nanoparticles. Under the optimized conditions, including 1 mL/min flow rate, 10 cm bed height, and 20 μg/mL tetracycline concentration, 433 ± 13 μg/mg removal efficiency was achieved within the column reactor. Using tetracycline spiked natural water solutions, 395 ± 15 and 350 ± 18 μg/mg capacities were measured for tap water and lake, respectively. In another work, gold nanoparticles, AuNPs, were phytosynthesized using *Elaeisguineensis* plant leaves extract, indicating desirable morphology and size by adjusting the synthesis conditions such as temperature and extract quantity (Ahmad et al. 2020). Also, the photocatalytic activity of the gold nanoparticles with diverse shapes and sizes was examined in the methylene blue degradation upon the visible light radiation. It was found that the temperature of reaction and the amount of the *Elaeisguineensis* leaf extract had great influences on the shapes and sizes and the photocatalytic capacities of the AuNPs. The AuNPs synthesized at 70°C displayed aspherical morphology and an average nanoparticle size of 16.26 ± 5.84 nm with the maximum photodegradation of 92.55% after 60 min. The greatest photocatalytic property of the AuNPs was associated with higher sorption of methylene blue molecules onto the AuNPs in the water that highly assisted the photodegradation reaction. As well, the mechanism of methylene blue decomposition was described. A nanocomposite membrane was fabricated using cellulose acetate, graphene, and Cu nanorods and/or Ag nanoparticles that simultaneously exhibited enhanced membrane anti-biofouling and desalination (El-Gendi et al. 2018). For this purpose, a 25 wt.% mixture of the cellulose acetate polymer in acetone composed of diverse amounts of graphene and Ag-Cu was cast at 18°C. The separation of saline (10 and 5 g/L of NaCl) solutions was evaluated by the developed membrane in the RO process. The Ag and Cu NPs were released from the membrane as observed through batch test using the atomic absorption spectra. The influences of Cu–Ag NPs and graphene were explored on the anti-biofouling property of membrane against Gram-negative and Gram-positive bacteria (Figure 1.17). The membrane was immersed into suspensions containing bacteria to perform the anti-biofouling property. The antibacterial capacities were determined by measuring the inhibition zones, and the formation of biofilms was observed in their SEM images. The reference cellulose acetate membrane did not show any bactericidal potency as bacteria were grown onto its surface, but the cellulose acetate/graphene/Cu–Ag NPs membrane exhibited bactericidal characteristics. Furthermore, the membrane illustrated highly improved salt rejection and water flux in the permeability experiments.

ZnO nanoparticles co-doped with Cr and Cowere were synthesized and used to remove antibiotics and dyes (Li et al. 2018). The optimum Cr:Co:Zn ratio was 6:4:100, which afforded the highest sorption performances in removing tetracycline hydrochloride and methyl orange, i.e., 874.46 and 1057.90 mg/g, respectively. Both the tetracycline hydrochloride and methyl orange sorption data were fitted to the Langmuir isotherm and the pseudosecond order kinetic. The thermodynamics tests demonstrated that the sorption processes of both tetracycline hydrochloride and methyl orange onto the Cr, Coco-doped ZnO NPs happened as endothermic and

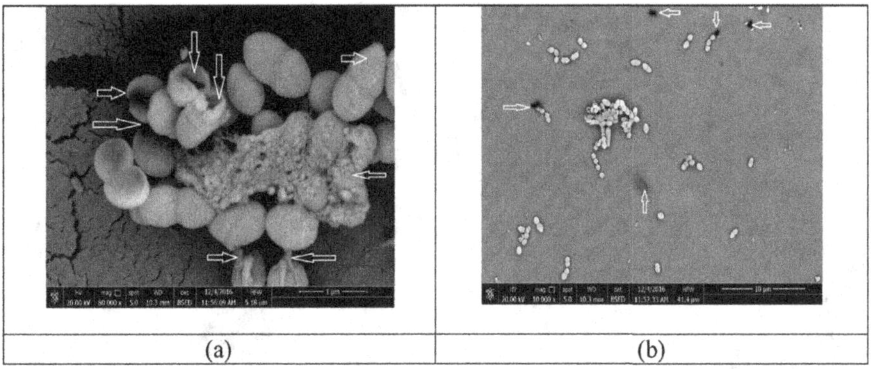

FIGURE 1.17 (a) Cell membrane of *Enterococcus facium* attached to the membrane containing 0.01 mg of Cu being severely destroyed and the cytoplasm flowing out, (b) accumulation of Cu NPs onto the bacterial cells (black cells in front of arrows). Reprinted with permission from (El-Gendi et al. 2018).

spontaneous phenomena. It was exhibited that the Cr,Coco-doped ZnO had more positive charges on its surface and displayed greater specific surface area and extra crystalline defects by the entrance of the Cr^{3+} and Co^{3+} cations substituted by the Zn^{2+}cations within the ZnO crystalline lattice, which improved its sorption capacity. Furthermore, the Cr, Coco-doped ZnO NPs illustrated exceptional removal of Congo Red, Direct Red, Methyl Blue, and Evans Blue. The adsorbents' reusability was examined to evade additional contamination confirming this nanomaterial was a valuable candidate for application in the wastewater treatment processes. Recently, copper selenide (CuSe) nanostructures were utilized as photocatalysts to degrade malachite green and methylene blue upon visible light radiation (Nouri et al. 2020). The photocatalytic activity of the CuSe nanostructures was greatly changed by varying the Se/Cu molar ratio in the range of 0.6–1.5 achieved under the ambient environments. The SEM micrographs displayed that when the Se/Cu <1, the CuSe nanostructures had a particulate shape, but when the Se/Cu ≥1, they revealed a combination of nanodiscs and nanoparticles as hexagonal and spherical shapes, respectively. The X-ray diffraction (XRD) patterns demonstrated that varying the Se/Cu ratio produced diverse phases of copper selenide nanostructures such as Cu_3Se_2, CuSe, and $Cu_{1.8}Se$. Also, the bandgaps of the copper selenide nanostructures were not considerably changed by changing the Se/Cu ratio. The photocatalytic capacity of the copper selenide nanostructures in the decomposition of malachite green and methylene blue under visible light radiation exhibited that when the Se/Cu >1 was used, the photocatalysts indicated a superior performance compared to those having the Se/Cu ≤1. Moreover, the optimal Se/Cu ratio was determined to be 1.4, which illustrated the utmost photocatalytic activity. In another study, nanosheet aggregates of Cu_2ZnSnS_4 were synthesized in the kesterite phase with a controlled shape and great purity by the inexpensive solvothermal process and used to fabricate membrane with salt blocking capability for application in solar desalination (Zhang et al. 2019).

FIGURE 1.18 The SEM images of the Cu_2ZnSnS_4 nanosheets prepared using diverse precursor concentrations: (a) C1, (b) C2, and (c) C3. (d) Partial enlarged SEM image in Fig. 18(c). Reprinted with permission from (Zhang et al. 2019).

Figure 1.18 exhibits the SEM images of the Cu_2ZnSnS_4 nanosheets synthesized using diverse concentrations of precursor solutions. The membrane containing Cu_2ZnSnS_4 nanosheets was employed in a solar steam creation system with great solar absorption, high stability, and little thermal energy loss, which showed an extraordinary water evaporation rate (1.54 kg/m^{2}h) and 78.85% solar steam conversion performance. The artificial seawater desalinated via evaporation exhibited a higher quality than the distilled water, see Figure 1.19. Besides, almost 100% of the organic dyes were decomposed using this technology. Therefore, the Cu_2ZnSnS_4 membrane was proposed as a promising nanomaterial that could be utilized to desalinate seawater and for re-sanitization of chemical wastewaters and urban domestic water. Iron sulfide nanoparticles (FeS NPs) were stabilized using carboxymethyl cellulose and used in the removal of toxic hexavalent chromium, Cr(VI), heavy metal cations from polluted groundwater as well as saturated soil (Wang et al. 2019). It was exhibited that the FeS NPs had 1046.1 mg/g adsorption capacity for the aqueous Cr(VI) cations through a mechanism involving sorption, reduction, and co-precipitation. Removal of Cr(VI) from the water was strongly changed by pH, and dissolved oxygen was a rival to Cr(VI) for sorption onto the adsorbent. Column experiments confirmed that the effluent's Cr(VI) concentration was <0.005 mg/L after eluting 45 pore volume of FeS NPs. The Cr(VI) was declined from 4.58 mg/L in original Cr(VI) polluted soil to 46.8–80.7 μg/L in the treated soil (from surface to bottom). Consequently, the FeS NPs were effective in remediating Cr(VI) polluted groundwater as well as saturated soil.

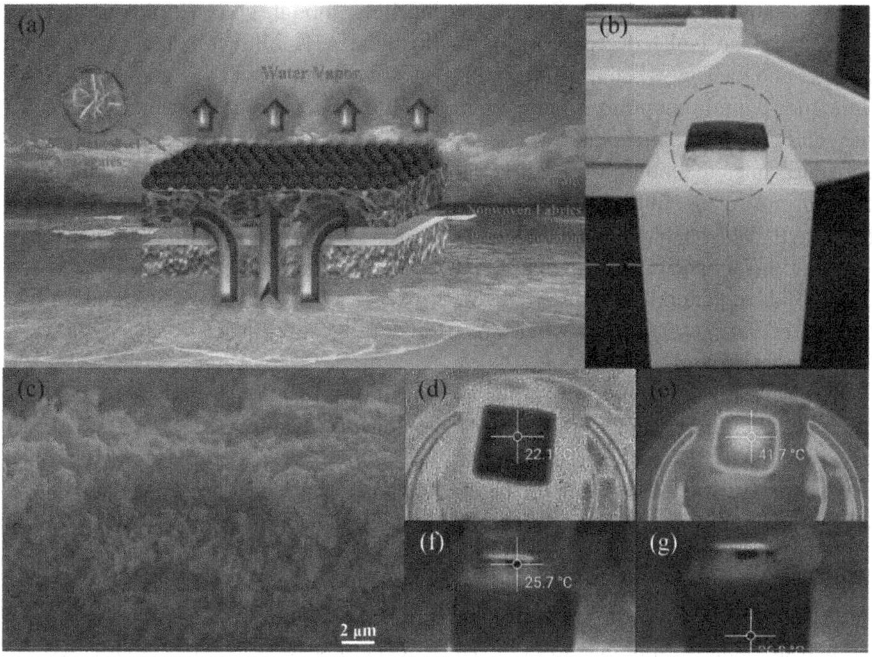

FIGURE 1.19 (a) Schematic design of the Cu_2ZnSnS_4 nanosheets membrane-based solar desalination device. (b) Digital photograph of the solar desalination device. (c) Cross-sectional SEM pattern of the Cu_2ZnSnS_4 nanosheets C3 membrane. (d)–(g) IR images of the solar desalination device while the water evaporation process is over time. Reprinted with permission from (Zhang et al. 2019).

1.7 APPLICATIONS OF CLAYS AND LAYERED DOUBLE HYDROXIDE (LDH) NANOPARTICLES IN WATER TREATMENT

Clays and LDHs are eminent, naturally existing minerals that are usually applied in water purification processes due to their brilliant properties, such as high adhesiveness and hydrophilic nature. Furthermore, they are synthesized and functionalized in various sizes and morphologies by inexpensive and simple procedures (Toledo-Jaldin et al. 2018, Rahimi-Aghdam et al. 2019, Rahman et al. 2019, Shivaraju et al. 2019, Zhong, Qiu, and Chen 2020). The clays and LDHs reveal high surface areas, exceptional adsorption capacities, plus great colloidal and adhesive characteristics. Indeed, the outstanding crystalline structures of clays and LDHs permit the expansion of their interlayer spaces to intercalate with and accommodate several materials with maintaining their octahedral and tetrahedral crystalline configurations. The protonated/neutral hydrophilic pollutants intercalated/adsorbed into the interlayer spaces can then be released through replacement with other cations presenting within the solution (Zhong, Qiu, and Chen 2020). The properties of clays and LDHs can also be modified via their functionalization using various inorganic/organic materials.

Clays are layered, fine-grained, and natural soil materials containing hydrated aluminum phyllosilicates (Shivaraju et al. 2019). They are categorized into four groups – smectite, kaolinite, illite, and chlorite – though chlorites are sometimes classified as an isolated group in the phyllosilicates. Kaolinites contain halloysite, dickite, and nacrite, whereas illites comprise glauconite, which is a green clay sand. Smectites contain montmorillonites, pyrophyllite, talc, saponite, vermiculite, hectorite, sauconite, saponite, nontronite, and bentonite but chlorites comprise chamosite, cookeite, amesite, and nimite (Mlih et al. 2020). These mineral clays are comprised of aluminosilicates in every layer and display remarkable cation exchange features because they have abundant cations like Na^+, K^+, Ca^{2+}, Mg^{2+}, and water molecules in their interlayer spaces (Kausar et al. 2018). LDHs are naturally available layered and ionic solids of general formula $[M_{1-n}{}^{2+}M_n{}^{3+}(OH)_2]^{n+}$ where M^{2+} and M^{3+} stand for two cations of various valences which are octahedrally coordinated to the hydroxide, OH^-, anions (Yang, Zhang, et al. 2020c).

The bentonite clay was utilized as a natural, low-cost commercial adsorbent for the removal of carbendazim pollutants from aqueous solution (Rizzi et al. 2020). It was found that 80% of the carbendazim was quickly removed after 15 min when 1 g of sorbent was used, indicating the bentonite had 1.5 mg/g sorption capacity. Also, if the contact time was extended to 60 min, about 100% of carbendazim was removed. The removal process was sustainable because the adsorbent and the contaminant were both recovered, confirming this process could lower the expenses related to its industrial scale-up. For this purpose, some adsorption-desorption cycles were performed in ethanol using one sorbent. The effects of several factors that affected the removal phenomenon were investigated, including the clay and pollutant amounts, temperature, pH and addition of salts to the solution. It was shown that the sorption obeyed the pseudosecond order kinetic model. Also, the adsorption data well followed the Temkin and Freundlich isotherms signifying the carbendazim was adsorbed physically and chemically in a heterogeneous manner onto the clay. The initial influx tests confirmed that when 2 g of sorbent was used, 100% of the carbendazim contaminant was removed within 2 s. In another effort, the bentonite clay was applied as the support for the deposition of Nb_2O_5 semiconductor, which formed pellets immobilized onto glass slides and employed for photocatalytic decomposition of reactive blue 19 dye (Hass Caetano Lacerda et al. 2020). The photodegradation experiments employing the immobilized photocatalyst were performed in a non-continuous refrigerated reactor so that a 125 W mercury vapor lamp was utilized to irradiate the substrate. It was found that the bentonite-supported $Nb_2O_{5z,}$ which formed pellets, had superior photocatalytic activity for the degradation of the reactive blue 19 dye with 98% efficiency within 2 h. However, the Nb_2O_5 immobilized onto the glass slide only indicated 78% photocatalytic degradation after 5 h. Furthermore, the photocatalytic decomposition reactions were accomplished without using H_2O_2. It was revealed that 90% photocatalytic degradation was achieved within 2 h, showing that it was not necessary to add the H_2O_2 into the reaction medium. Recently, extremely hydrophilic thin film nanocomposite membranes were fabricated using electrospun nanofibrous poly (vinylidene fluoride), PVDF, loaded by the bentonite nanoclay (Shah, Cho, et al. 2020a). The membranes were developed through interfacial polymerization of a selective film of polyamide polymer and employed in the forward osmosis process. The nanoclay bentonite tuned the important characteristics of the nanofibrous PVDF

TABLE 1.9
Intrinsic Characteristics of the Fabricated Membranes

Membranes	A (L/m^2 h bar)	B (L/m^2 h)	B/A (kPa)	Rejection (%)	S Values (μm)	Pore Size (nm)	Thickness (μm)
PVDF18_0	2.256 ± 0.14	0.323 ± 0.02	14.3	80.7	527.9	767.9	61.5 ± 2.6
PVDF18_0.25	2.484 ± 0.09	0.340 ± 0.01	13.6	81.4	412.6	876.5	67.5 ± 2.0
PVDF18_0.5	2.586 ± 0.24	0.373 ± 0.04	14.4	80.3	350.3	952.5	71.2 ± 2.7
PVDF18_1.0	2.686 ± 0.23	0.412 ± 0.03	15.3	79.6	268.2	997	78.5 ± 2.0
PVDF18_2.0	2.818 ± 0.10	0.495 ± 0.02	17.7	77.2	187.9	1173.3	82.7 ± 3.3

thin film nanocomposite membranes like mechanical features, a fraction of β-phase/polar phase and hydrophilicity. Table 1.9 displays the intrinsic characteristics of the prepared membranes. The bentonite/PVDF significantly improved the efficacy of the forward osmosis process. The thin film nanocomposite membranes exhibited greater water permeability, mechanical strength, and hydrophilicity than those of the pure electrospun nanofibrous PVDF thin film membrane because of increasing the hydrophilicity/β-phase fraction through the synergistic influence of bentonite nanoclay and electrospinning. Besides, the thin film nanocomposite membrane containing the maximum nanoclay amount of 2.0 wt.% displayed a very large water flux in the forward osmosis (40.64 L/m^2.h) without significant enhancement of the reverse solute flux when NaCl (1 M) was used as the draw solution.

In order to decrease the amount of diclofenac sodium within aqueous solutions, biochar derived from the *Syagruscoronata* was applied as a support for the deposition of the MgAl LDH, which was then employed in the removal of the diclofenac sodium from water (de Souza dos Santos et al. 2020). The sorption batch tests established >82% removal of diclofenac sodium (200 mg/L). The equilibrium and kinetic experiments proved that the sorption was happened through the chemisorption mechanism with the highest adsorption capacities of 168.04 and 138.83 mg/g measured at 60 and 30°C, respectively. Thermodynamic studies revealed that the sorption of the diclofenac sodium occurred spontaneously as an exothermic process that changed the adsorbent structure. Therefore, the biochar/MgAl LDH nanocomposite was considered a valuable candidate for removing diclofenac sodium contaminants from aqueous solutions. ZnCuFeCrLDHs were synthesized by the plasma treatment method and applied as photocatalysts for the photodegradation of methyl orange pollutants upon the visible light radiation, indicating outstanding photocatalytic performances, see Figure 1.20 (Tao et al. 2020). In fact, the plasma treatment process greatly affected the photocatalytic activities of the LDHs upon the visible light

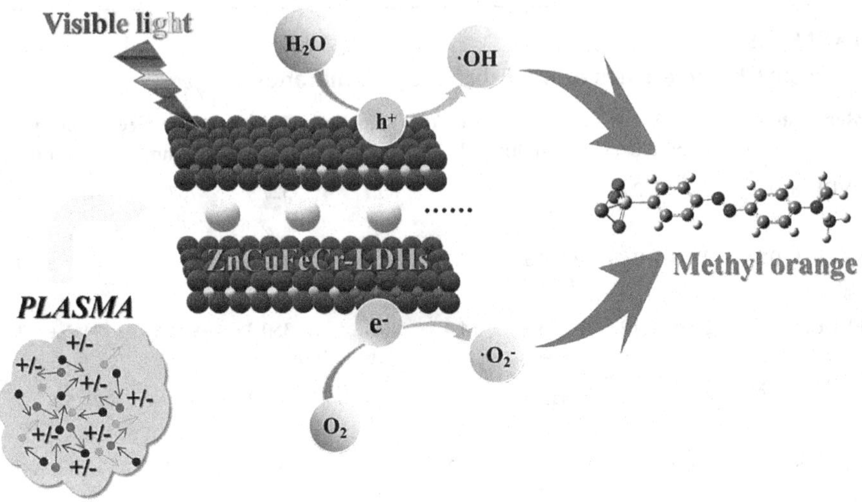

FIGURE 1.20 Suggested mechanism of the photocatalytic degradation of methyl orange. Reprinted with permission from (Tao et al. 2020).

radiation. It illustrated that it was superior to the common co-precipitation technique. Application of the plasma treatment more tightly arranged of the laminar atoms which formed more stable crystalline structures, enlarged the specific surface areas, produced extra oxygen vacancies, and declined recombination of the photo-generated electrons and holes throughout the photocatalytic reaction. Moreover, the mechanisms of synthesis and the photocatalytic reactions were described. Also, the influence of the synthesis time onto the photodegradation performance was studied and the optimum synthesis parameters were introduced. The highest methyl orange (20 mg/L, 1000 mL) degradation rate was 96.8% in 42 min using 1 g/L of the photocatalyst under the visible light illumination when employing 30.9 kV discharge voltage for the synthesis of the photocatalyst. Recently, the internal concentration polarization was decreased using LDHs/polysulfone nanocomposite substrates used as the membranes in the forward osmosis process (Lu et al. 2019). These substrates demonstrated greater porosities plus interconnected three-dimensional macro-porous networks due to corrosion of the LDHs through the application of HCl solution, which likely created connections between the finger-resembling pores within the substrates. As a result, high water flux was measured for the forward osmosis membrane developed by the interfacial polymerization process. The HCl-treated nanocomposite membrane exhibited the largest water flux of 47.2 $L/m^2.h.bar$ using deionized water and NaCl (1 M) as feed and draw solutions, respectively. Furthermore, the water permeability coefficient was 7.29 $L/m^2.h.bar$ by employing the RO process at 5 bar pressure and using deionized water as the feed solution. Such results were greatly superior compared to those of the commercially used forward osmosis membrane. Also, the HCl-treated nanocomposite membrane solved the permeability/selectivity problem because enhanced selectivity was attained between NaCl and water. Consequently, such an HCl-treated nanocomposite membrane was an efficient substrate for the osmosis membranes.

1.8 APPLICATIONS OF POLYMER NANOCOMPOSITES IN WATER TREATMENT

Polymer nanocomposites fabricated using various kinds of natural and synthetic polymers such as chitosan, polyamide, polydopamine, poly (vinylidene fluoride), polypyrrole, and polyaniline are utilized in different water treatment processes (Ghosh, Kouamé, et al. 2015a, 2015b, Ghosh, Remita, and Basu 2018, Floresyona et al. 2017, Dongre et al. 2019, Idumah et al. 2019, Yuan et al. 2020) because they exhibit outstanding properties (Indherjith et al. 2019, Shariatinia 2018, 2019, 2020, Ravikumar and Udayakumar 2020). Also, polymer nanocomposites can be incorporated with lots of diverse nanomaterials as nanofillers with high adsorption and photocatalytic activities, which can simultaneously enhance the mechanical and structural stability of the nanocomposites (Shariatinia et al. 2015, Hussein et al. 2019, Shariatinia and Barzegari 2019, Shariatinia and Mazloom-Jalali 2019, Dewangan et al. 2020).

Poly (maleic anhydride)-graft-poly (vinyl alcohol), PMA-g-PVA, was used as a comb polymer to functionalize the magnetic, Fe_3O_4, nanoparticles that were employed as effective polymer nanocomposite adsorbents of heavy metal cations indicating reversible, stimuli-sensitive and controlled sorption-desorption property (Liu, Guan, et al. 2020d). Figure 1.21 depicts the SEM and TEM images of the Fe_3O_4, Fe_3O_4/PVA and Fe_3O_4/PMA-g-PVA materials. This adsorbent exhibited that the sorption capacity data for the adsorption of Ni^{2+}, Ag^+, Pb^{2+}, Co^{2+} and Cd^{2+} were well fitted to the pseudo-second-order kinetic (so that the correlation coefficients were >0.98) and Langmuir isotherm model. The adsorption capacities were measured to be 0.8194, 0.8634, 0.8429, 0.7682, and 0.6720 mmol/g for the Ni^{2+}, Ag^+, Pb^{2+} Co^{2+}, and Cd^{2+} cations, respectively, that was five times larger compared to those of the magnetic nanoparticles functionalized by linear polymer because the topology of the latter was

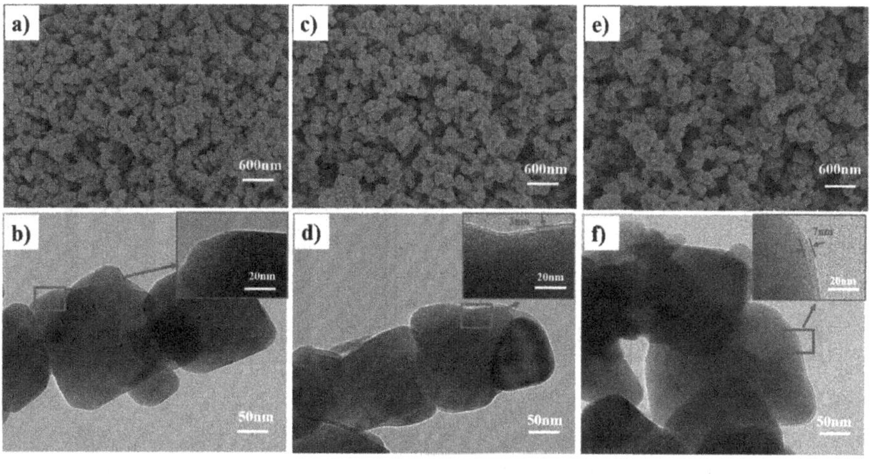

FIGURE 1.21 The SEM images of (a) Fe_3O_4, (c) Fe_3O_4/PVA and (e) Fe_3O_4/PMA-g-PVA, and TEM images of (b) Fe_3O_4, (d) Fe_3O_4/PVA and (f) Fe_3O_4/PMA-g-PVA. Reprinted with permission from (Liu, Guan, et al. 2020d).

not improved. It was shown that the removal amounts of the magnetic nanoparticles functionalized by the comb polymer were considerably enhanced from 17.6, 21.1, 18.64, 15.68 and 16.3% to 89.2, 92.15, 90.16, 84.52 and 83.56%, respectively, for the sorption of theNi^{2+}, Ag^{+}, Cd^{2+}, Co^{2+}, and Pb^{2+} cations when the pH of solution was increased from 4 to 7, confirming this sorbent had a pH-sensitive property, see Figure 1.22a, b. Thus, the adsorption mechanism was described by the pH-triggered variations in the electrostatic interactions plus molecular shapes. Such adsorbent was highly sustainable and reusable because almost >80% removal performance was

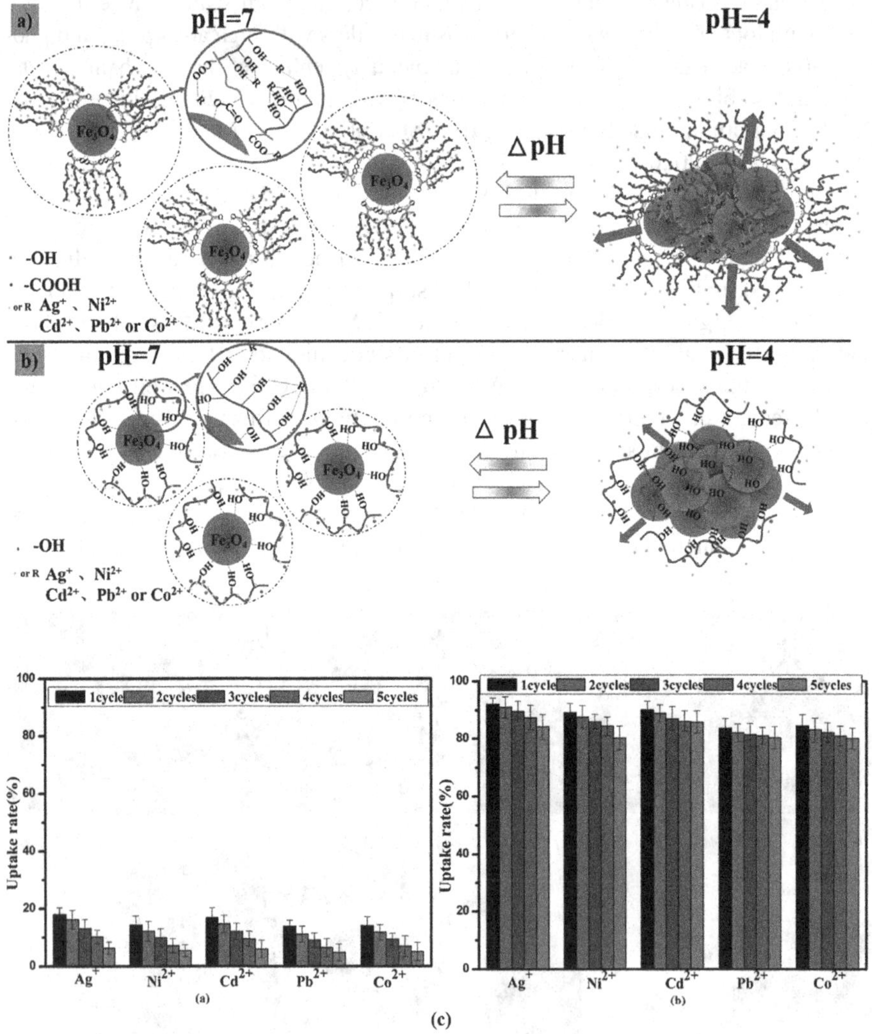

FIGURE 1.22 Mechanism of adsorption of heavy metal cations onto the (a) Fe_3O_4/PMA-g-PVA and (b) Fe_3O_4/PVA. (c) Uptake rates of the regenerated (a) Fe_3O_4/PVA and (b) Fe_3O_4/PMA-g-PVA after five successive cycles of adsorption and desorption processes. Reprinted with permission from (Liu, Guan, et al. 2020d).

achieved after five successive cycles of absorption and desorption experiments; see Figure 1.22c. Consequently, changes in the molecular topology of the polymer structure greatly enhanced the pH-responsive sorption capacity/removal rate of heavy metal cations existing within wastewaters. Recently, a polymer nanocomposite was developed using cross-linked glycerol dimethacrylate and the MCM-41 mesoporous silica that was applied to remove the methylene blue as a model pollutant dye from aqueous solution (Cherifi et al. 2020). The synthesis of glycerol dimethacrylate was initially done through the ring-opening of glycidyl methacrylate by means of methacrylic anhydride that was performed via a green solvent-free catalytic reaction at ambient temperature. The nanocomposites were achieved through in-situ polymerization of the glycerol dimethacrylate within the external and internal MCM-41 channels/surfaces and employed in the methylene blue uptake from water. The effects of adsorbent dosage, polymer amount, dye concentration, and contact time were examined and described by the sorption capacity and affinity. It was indicated that the sorption of methylene blue was greater when enhancing the poly (glycerol dimethacrylate) amount, so that the optimal nanocomposite displayed the highest sorption capacity of the methylene blue. The sorption of methylene blue onto the optimal nanocomposite obeyed the pseudo second-order kinetic and the Langmuir isotherm model. The greatest quantity of the methylene blue adsorbed onto the optimal nanocomposite was 111.11 mg/g.

Polyaniline is one of the most important conductive polymers indicating exceptional features like excellent electrochemical characteristics, high stability, electrical conductivity, and facile, inexpensive synthesis/polymerization reaction. Hence, a polymer nanocomposite was developed based on the polyaniline and PbS (lead sulfide) quantum dots used as a photocatalyst for the photodegradation of rhodamine 6G dye upon the visible light radiation (Chhabra et al. 2020). This polyaniline-PbS nanocomposite was achieved via chemical oxidation polymerization of aniline solution to which the PbS quantum dots were simultaneously added. The nanocomposite catalyst exhibited a higher conductivity. It displayed ~87% performance when used for the dye photodegradation after 50 min. The photodegradation rate was 5.03 mmol/g.h and the quantum efficiency was 7.98×10^{-6} molecule/photon. As the polyaniline-PbS photocatalyst had a higher charge transport capability, it effectively degraded diverse concentrations of the dye molecules. It was found that the electron/hole pairs produced by the visible light radiation over the polyaniline/PbS nanocomposite caused effective degradation and oxidation of the rhodamine 6G. In another work, polyaniline functionalized $LaNiSbWO_4$/graphene oxide nanocomposite was synthesized through the sonochemical technique and applied in the photocatalytic degradation of gallic acid and Safranin-O organic compounds upon the visible light illumination with $\lambda > 420$ nm (Oh et al. 2020). Figures 1.23 and 1.24a–d display the SEM and TEM images of the synthesized materials. Besides the great photocatalytic performance of the $LaNiSbWO_4$/graphene oxide/polyaniline nanocomposite, the existence of free electrons more enhanced its visible light photocatalytic activity. Figure 1.24e illustrates the dye degradation mechanism over the $LaNiSbWO_4$/graphene oxide/polyaniline nanocomposite. The optimal $LaNiSbWO_4$/graphene oxide/polyaniline nanocomposite illustrated the highest photocatalytic performance, which was around three times greater than that of the $LaNiSbWO_4$. The $LaNiSbWO_4$/

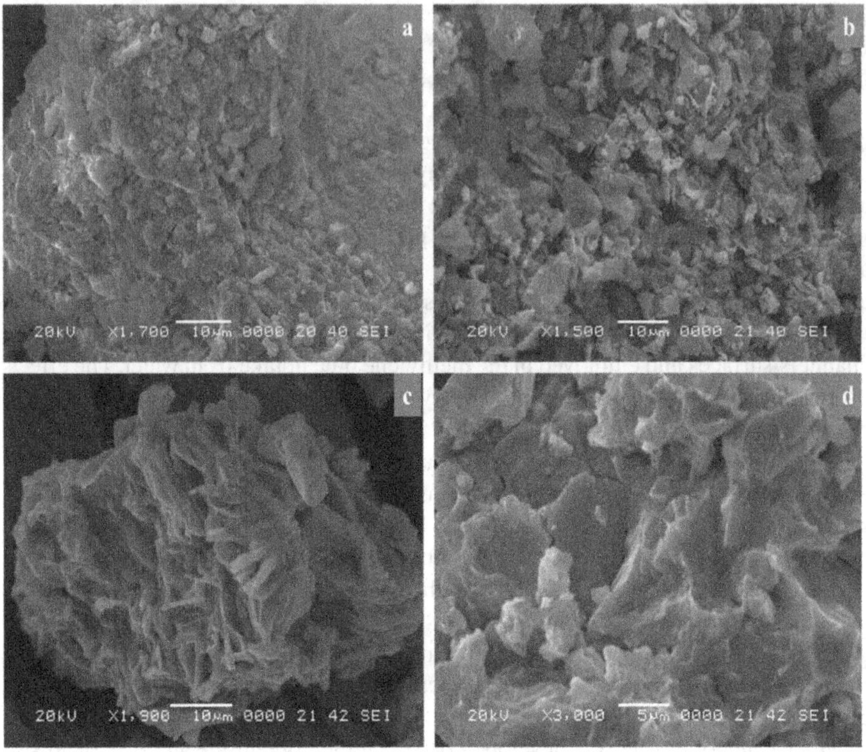

FIGURE 1.23 The SEM images of (a) $LaNiSbWO_4$ (LNSW), (b) $LaNiSbWO_4$/graphene oxide (LNSWG), and (c, d) $LaNiSbWO_4$/graphene oxide/polyaniline (NSWGP) nanocomposite synthesized via the sonochemical method. Reprinted with permission from (Oh et al. 2020).

graphene oxide/polyaniline photocatalyst revealed 84% and 92% photodegradation of the Safranin-O and gallic acid, respectively, in 180 min under identical conditions. The nanocomposite's higher photocatalytic capability was associated with the synergistic influences of $LaNiSbWO_4$/graphene oxide and polyaniline that stimulated the separation of the photo-created electrons and holes.

Several thin film nanocomposite membranes were fabricated for the RO process, in which the p-aminophenol functionalized graphene oxide (m-GO) as the filler nanomaterial was incorporated into the polyamide skin layers by the interfacial polymerization reaction (Zhang, Ruan, et al. 2020d). The bare membrane denoted P0, but the four RO membranes containing 0.001, 0.002, 0.003 and 0.005 wt.% amounts of GO were named G1, G2, G3 and G4, respectively. Similarly, the four RO membranes containing 0.001, 0.002, 0.003, and 0.005 wt.% quantities of m-GO were called M1, M2, M3, and M4, respectively. Figure 1.25 reveals the surface and cross-sectional SEM images of the P0 and RO G2–G4 and M2–M4 nanocomposite membranes. The

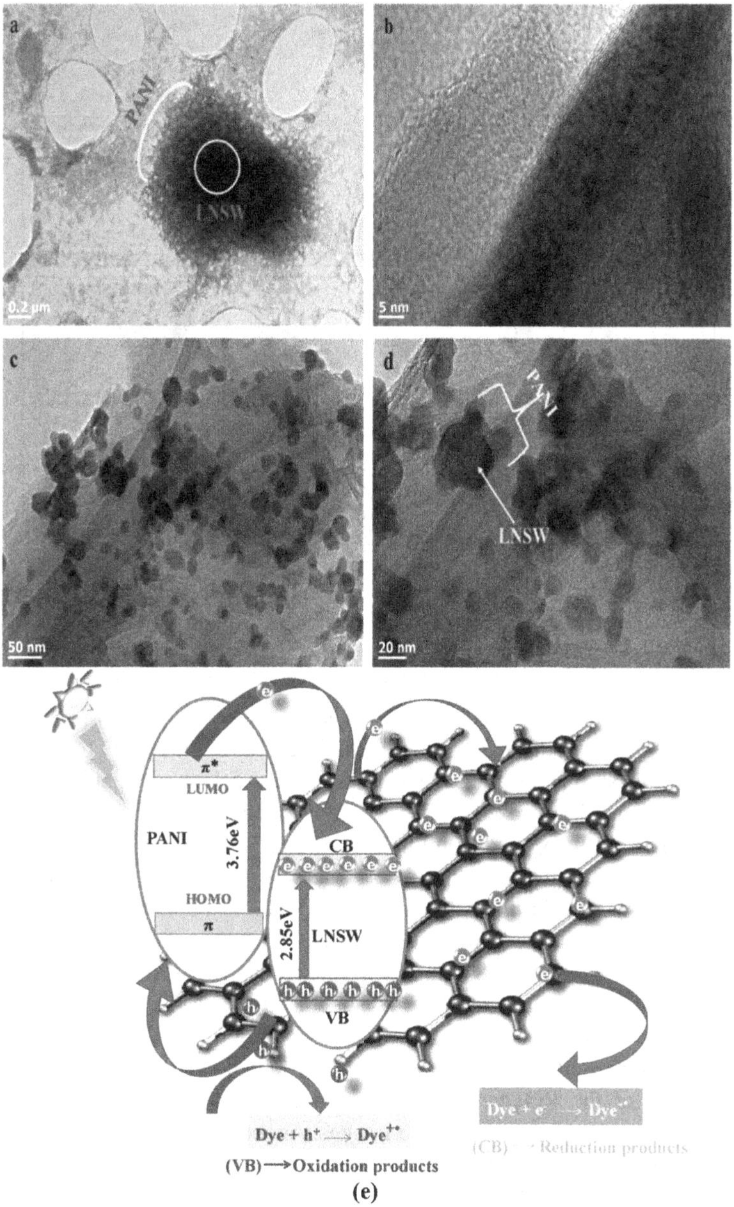

FIGURE 1.24 The TEM images of (a) LNSW, (b) LNSWG and (c, d) LNSWGP. (e) Dye degradation mechanism over the NSWGP nanocomposite. The band gap energies of PANI = 3.76 eV, $LaNiSbWO_4$ (LNSW) = 2.85 and LNSWGP = 1.75 eV. Reprinted with permission from (Oh et al. 2020).

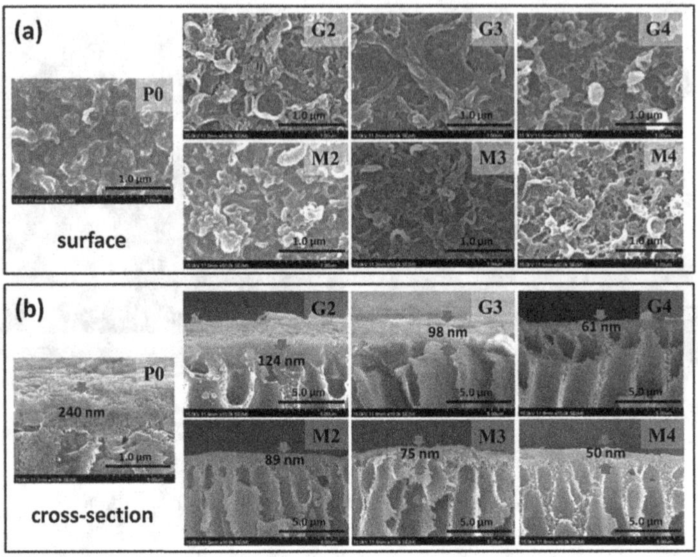

FIGURE 1.25 The (a) surface and (b) cross-sectional SEM images of the reverse osmosis P0, G2–G4 and M2–M4 nanocomposite membranes. Reprinted with permission from (Zhang, Ruan, et al. 2020d).

m-GO filler nanomaterial added to the skin layer reduced its hydrophilicity as the water contact angle was dropped from 69.6 to 48.2°. Also, the m-GO filler nanomaterial decreased the skin layer thickness from 240 nm to 50 nm compared to that of the bare polyamide membrane used in the RO. Consequently, the thin film nanocomposite RO membrane at the optimal conditions showed 23.6 L/m^2.h water flux and 99.7% NaCl rejection, indicating an extraordinary increase (24.5%) in the water flux than that of the bare RO membrane. Additionally, the statistical data on the dead/live fluorescent images demonstrated that m-GO loaded thin film nanocomposite RO membrane exhibited 95.26 and 96.78% antibacterial activities against *S. aureus* and *E. coli* bacteria when 0.005 wt.% of m-GO was added, reflecting these values were much greater compared with those measured for the GO-containing RO membrane (90.43, 90.64%) and the bare RO membrane (2.48, 4.95%). Table 1.10 illustrates the antibacterial activities of the RO membranes against the *E. coli* and *S. aureus* bacteria achieved by analyzing the fluorescence images. Accordingly, them-GO incorporated thin film nanocomposite RO membrane had a high performance for the separation with appropriate bactericidal activities. Recently, a thin film nanocomposite membrane was developed for application in the forward osmosis process and its adsorption performance and structural characteristics were investigated (Saeedi-Jurkuyeh et al. 2020). The nanocomposite membrane was utilized in the removal of heavy metal cations from both the industrial and artificial wastewaters. The forward osmosis membrane had a support layer produced through adding diverse amounts, i.e., 0–8 wt.%, of 1-methy-l,2-pyrrolidone, polysulfone and polyethylene glycol 400 by the phase inversion procedure. Also, the membrane had an active layer created by adding varied quantities, 0–0.012 wt.%, of 1,3,5-benzene trichloride, 1,3-phenylenediamine,

TABLE 1.10
The Antibacterial Activities of the Reverse Osmosis Membranes against the *E. coli* and *S. aureus* Bacteria Achieved by Analyzing the Fluorescence Images

Sample	Total Number of Bacteria		Number of Live Bacteria		Bacterial Killing Ratio (%)	
	E. coli	*S. aureus*	*E. coli*	*S. aureus*	*E. coli*	*S. aureus*
P0	323 ± 3	202 ± 5	307 ± 4	197 ± 5	4.9	2.5
G1	299 ± 2	199 ± 3	211 ± 7	170 ± 3	29.4	14.6
G2	318 ± 3	181 ± 6	104 ± 3	134 ± 2	67.3	25.9
G3	301 ± 1	211 ± 2	47 ± 5	87 ± 4	84.4	58.7
G4	299 ± 2	209 ± 1	28 ± 2	20 ± 3	90.6	90.4
M1	301 ± 2	205 ± 2	142 ± 8	182 ± 5	52.8	11.2
M2	295 ± 1	197 ± 2	54 ± 3	99 ± 6	81.7	49.8
M3	305 ± 1	205 ± 2	14 ± 4	34 ± 2	95.4	83.4
M4	311 ± 2	211 ± 1	10 ± 2	10 ± 1	96.8	95.3

and graphene oxide to the polyamide layer through the interfacial polymerization process. The forward osmosis membrane displayed greater porosity, hydrophilicity, water flux, salt rejection, water permeability, and less specific reverse salt flux, reverse salt flux, and inner concentration polarization than the commonly utilized thin film nanocomposite membrane. The water flux of the thin film nanocomposite membrane, i.e., 34.3 L/m^2.h, was enhanced by 129% and 174% relative to those of the commercial forward osmosis and thin film nanocomposite membranes, i.e., 15 and 12.5 L/m^2.h, respectively. The Cr^{3+}, Cd^{2+}, and Pb^{2+}rejection rates by the thin film nanocomposite forward osmosis membrane were 98.3, 99.7, and 99.9%, respectively. The flux recovery was 96% and >14.5 h breakthrough time was achieved for these heavy metal cations approving this forward osmosis membrane had an exceptional structure, separation capacity, and forward osmosis performance.

1.9 CONCLUSION

The advancements achieved in the synthesis and development of numerous nanomaterials have been highly beneficial for various water treatment technologies such as adsorption, photocatalytic degradation, and separation using nanocomposite membranes. These advanced nanomaterials include graphitic carbon nitride (g-C_3N_4), carbon nanotubes (CNTs), boron nitride, fullerenes, graphene nanosheets, CQDs, MOFs, zeolites, mesoporous silica, MCM and SBA derivatives, metals and metal oxides/sulfides/selenides nanoparticles/quantum dots, clay nanomaterials, LDHs, and polymer nanocomposites. Therefore, wastewater purification and seawater reuse/desalination will be simply performed using such valuable nanomaterials. Most of the nanomaterials reveal exceptional characteristics like non/low toxicity, biocompatibility, biodegradability, and high surface areas. Also, some of them indicate micro- and/or meso-porous structures and bioavailability. These features endowed them with a wide range of extraordinary opportunities for application in several water treatment processes. Though lots of studies have so far been done in the water treatment area, it is still required to find more effective materials with greater safety using more cost-effective methods. For instance, some serious challenges are encountered when applying different water treatment processes, such as remaining pollutants in the environment after their adsorption onto the adsorbents, nearly rapid inactivation of photocatalysts utilized at large scales for degradation of pollutants, usage of harmful and/or costly oxidation processes and antibacterial agents – such as those using H_2O_2, ozone, and chlorine – as well as relatively fast and irreversible fouling, and low to moderate mechanical strengths of membranes. These problems can be solved to a large extent by choosing efficient nanomaterials and controlling their size, morphology, charge, porous structure, hydrophilicity, surface chemistry, synthesis methods and costs, biocompatibility, and toxicity/safety. Finally, the most effective nanomaterials and methods will be very useful for industrial water purification purposes and require deployment of highly stable, efficient, and inexpensive nanomaterials for the systemic analyses on the long-lasting treatments of real seawater, brackish water, or wastewater samples. It is believed that using high technology and advanced nanomaterials will lead to potable water treatment devices and decrease the costs of producing freshwater.

ACKNOWLEDGMENTS

The financial support of this work by Research Office of Amirkabir University of Technology (Tehran Polytechnic) is gratefully acknowledged.

CONFLICTS OF INTEREST

The author declares that there are not any personal or financial conflicts of interests.

REFERENCES

Abdurahman, Mohamed Hussein, and Ahmad Zuhairi Abdullah. 2020. Mechanism and reaction kinetic of hybrid ozonation-ultrasonication treatment for intensified degradation of emerging organic contaminants in water: A critical review. *Chemical Engineering and Processing-Process Intensification* 154:108047. https://doi.org/10.1016/j.cep.2020.108047.

Afzal, Muhammad Junaid, Erum Pervaiz, Sarah Farrukh, Tahir Ahmed, Zhang Bingxue, and Minghui Yang. 2019. Highly integrated nanocomposites of RGO/TiO_2 nanotubes for enhanced removal of microbes from water. *Environmental Technology* 40 (19):2567–2576. https://doi.org/10.1080/09593330.2018.1447021.

Ahmad, Tausif, Mohamad Azmi Bustam, Muhammad Zulfiqar, Muhammad Moniruzzaman, Alamin Idris, Jibran Iqbal, Hafiz Muhammad Anwaar Asghar, and Sami Ullah. 2020. Controllable phytosynthesis of gold nanoparticles and investigation of their size and morphology-dependent photocatalytic activity under visible light. *Journal of Photochemistry and Photobiology A: Chemistry* 392:112429. https://doi.org/10.1016/j.jphotochem.2020.112429.

Alijani, Hassan, Mostafa Hossein Beyki, Yousef Fazli, and Zahra Shariatinia. 2017a. A nanohybrid of mixed ferrite—polyaniline derivative copolymer for efficient adsorption of lead ions: Design of experiment for optimal condition, kinetic and isotherm study. *Desalination and Water Treatment* 66:338–345.

Alijani, Hassan, Mostafa Hossein Beyki, Zahra Shariatinia, Mehrnoosh Bayat, and Farzaneh Shemirani. 2014. A new approach for one step synthesis of magnetic carbon nanotubes/diatomite earth composite by chemical vapor deposition method: Application for removal of lead ions. *Chemical Engineering Journal* 253:456–463. https://doi.org/10.1016/j.cej.2014.05.021.

Alijani, Hassan, and Zahra Shariatinia. 2017. Effective aqueous arsenic removal using zero valent iron doped MWCNT synthesized by in situ CVD method using natural α-Fe_2O_3 as a precursor. *Chemosphere* 171:502–511. https://doi.org/10.1016/j.chemosphere.2016.12.106.

Alijani, Hassan, Zahra Shariatinia, and Abdolreza Aroujalian. 2017b. Enhanced hexavalent chromium adsorption from aqueous solutions onto amine functionalized MWCNTs synthesized using iron (0)/egg-shell as an efficient catalyst. *Desalination and Water Treatment* 59:168–180.

Alijani, Hassan, Zahra Shariatinia, and Abdolreza Aroujalian Mashhadi. 2015. Water assisted synthesis of MWCNTs over natural magnetic rock: An effective magnetic adsorbent with enhanced mercury(II) adsorption property. *Chemical Engineering Journal* 281:468–481. https://doi.org/10.1016/j.cej.2015.07.025.

Álvarez, Miguel A., M. Ruidíaz-Martínez, G. Cruz-Quesada, M. Victoria López-Ramón, José Rivera-Utrilla, Manuel Sánchez-Polo, and Antonio J. Mota. 2020. Removal of parabens from water by UV-driven advanced oxidation processes. *Chemical Engineering Journal* 379:122334. https://doi.org/10.1016/j.cej.2019.122334.

Areerob, Yonrapach, Kwang-Youn Cho, and Won-Chun Oh. 2017. Microwave assisted synthesis of graphene-Bi8La10O27-Zeolite nanocomposite with efficient photocatalytic activity towards organic dye degradation. *Journal of Photochemistry and Photobiology A: Chemistry* 340:157–169. https://doi.org/10.1016/j.jphotochem.2017.03.018.

Bahamon, D., and Lourdes. F. Vega. 2019. Molecular simulations of phenol and ibuprofen removal from water using multilayered graphene oxide membranes. *Molecular Physics* 117 (23–24):3703–3714. https://doi.org/10.1080/00268976.2019.1662129.

Bargozideh, Samin, Mahboubeh Tasviri, and Mohammad Kianifar. 2020. Construction of novel magnetic $BiFeO_3/MoS_2$ composite for enhanced visible-light photocatalytic performance towards purification of dye pollutants. *International Journal of Environmental Analytical Chemistry*:1–15. https://doi.org/10.1080/03067319.2020.1811259.

Basso, Alex, Andrei Pavei Battisti, Regina de Fátima Peralta Muniz Moreira, and Humberto Jorge José. 2020. Photocatalytic effect of addition of TiO_2 to acrylic-based paint for passive toluene degradation. *Environmental Technology* 41 (12):1568–1579. https://doi.org/10.1080/09593330.2018.1542034.

Behzadifard, Zahra, Zahra Shariatinia, and Milad Jourshabani. 2018. Novel visible light driven $CuO/SmFeO_3$ nanocomposite photocatalysts with enhanced photocatalytic activities for degradation of organic pollutants. *Journal of Molecular Liquids* 262:533–548. https://doi.org/10.1016/j.molliq.2018.04.126.

Bekbölet, Miray 1997. Photocatalytic bactericidal activity of TiO_2 in aqueous suspensions of *E.* Coli. *Water Science and Technology* 35 (11):95–100. https://doi.org/10.1016/S0273-1223(97)00241-2.

Bekbölet, Miray, and Claudia V. Araz. 1996. Inactivation of *E.* coli by photocatalytic oxidation. *Chemosphere* 32 (5):959–965. https://doi.org/10.1016/0045-6535(95)00359-2.

Betiha, Mohamed A., Yasser Mohamed Moustafa, Mohamed Fathy El-Shahat, and Errakhi Rafik. 2020. Polyvinylpyrrolidone-Aminopropyl-SBA-15 Schiff Base hybrid for efficient removal of divalent heavy metal cations from wastewater. *Journal of Hazardous Materials* 397:122675. https://doi.org/10.1016/j.jhazmat.2020.122675.

Bodzek, Michał, Krystyna Konieczny, and Anna Kwiecińska-Mydlak. 2020. Nanotechnology in water and wastewater treatment. Graphene – the nanomaterial for next generation of semipermeable membranes. *Critical Reviews in Environmental Science and Technology* 50 (15):1515–1579. https://doi.org/10.1080/10643389.2019.1664258.

Çağlar Yılmaz, Hatice, Emrah Akgeyik, Salma Bougarrani, Mohammed El Azzouzi, and Sema Erdemoğlu. 2020. Photocatalytic degradation of amoxicillin using Co-doped TiO_2 synthesized by reflux method and monitoring of degradation products by LC–MS/MS. *Journal of Dispersion Science and Technology* 41 (3):414–425. https://doi.org/10.1080/01932691.2019.1583576.

Casanova, Serena, Tian-Yin Liu, Yong-Min J. Chew, Andrew Livingston, and Davide Mattia. 2020. High flux thin-film nanocomposites with embedded boron nitride nanotubes for nanofiltration. *Journal of Membrane Science* 597:117749. https://doi.org/10.1016/j.memsci.2019.117749.

Chen, Dongjie, Yanling Cheng, Nan Zhou, Paul Chen, Yunpu Wang, Kun Li, Shuhao Huo, Pengfei Cheng, Peng Peng, Renchuang Zhang, Lu Wang, Hui Liu, Yuhuan Liu, and Roger Ruan. 2020a. Photocatalytic degradation of organic pollutants using TiO_2-based photocatalysts: A review. *Journal of Cleaner Production* 268:121725. https://doi.org/10.1016/j.jclepro.2020.121725.

Chen, Fei, Lian-Lian Liu, Ying-Jie Zhang, Jing-Hang Wu, Gui-Xiang Huang, Qi Yang, Jie-Jie Chen, and Han-Qing Yu. 2020b. Enhanced full solar spectrum photocatalysis by nitrogen-doped graphene quantum dots decorated BiO_{2-x} nanosheets: Ultrafast charge transfer and molecular oxygen activation. *Applied Catalysis B: Environmental* 277:119218. https://doi.org/10.1016/j.apcatb.2020.119218.

Chen, Yize, Yoshifumi Nakazawa, Yoshihiko Matsui, Nobutaka Shirasaki, and Taku Matsushita. 2020c. Sulfate ion in raw water affects performance of high-basicity PACl coagulants produced by $Al(OH)_3$ dissolution and base-titration: Removal of SPAC particles by coagulation-flocculation, sedimentation, and sand filtration. *Water Research* 183:116093. https://doi.org/10.1016/j.watres.2020.116093.

Cheng, Dongle, Huu Hao Ngo, Wenshan Guo, Soon Wang Chang, Dinh Duc Nguyen, Yiwen Liu, Xinbo Zhang, Xue Shan, and Yi Liu. 2020. Contribution of antibiotics to the fate of antibiotic resistance genes in anaerobic treatment processes of swine wastewater: A review. *Bioresource Technology* 299:122654. https://doi.org/10.1016/j.biortech.2019.122654.

Cherifi, Zakaria, Bouhadjar Boukoussa, Adel Mokhtar, Mohammed Hachemaoui, Fatima Zohra Zeggai, Aniss Zaoui, Khaldoun Bachari, and Rachid Meghabar. 2020. Preparation of new nanocomposite poly(GDMA)/mesoporous silica and its adsorption behavior towards cationic dye. *Reactive and Functional Polymers* 153:104611. https://doi.org/10.1016/j.reactfunctpolym.2020.104611.

Chhabra, Varun A., Rajnish Kaur, Manrajvir S. Walia, Ki-Hyun Kim, and Akash Deep. 2020. PANI/PbS QD nanocomposite structure for visible light driven photocatalytic degradation of rhodamine 6G. *Environmental Research* 186:109615. https://doi.org/10.1016/j.envres.2020.109615.

Combatt, Maria Paulina Mendoza, William Caires Silva Amorim, Erick Matheus da Silveira Brito, Allan F. Cupertino, Regina Célia Santos Mendonça, and Heverton Augusto Pereira. 2020. Design of parallel plate electrocoagulation reactors supplied by photovoltaic system applied to water treatment. *Computers and Electronics in Agriculture* 177:105676. https://doi.org/10.1016/j.compag.2020.105676.

Danwittayakul, Supamas, Supachai Songngam, and Sittha Sukkasi. 2020. Enhanced solar water disinfection using ZnO supported photocatalysts. *Environmental Technology* 41 (3):349–356. https://doi.org/10.1080/09593330.2018.1498921.

Deng, Luyao, Qun Wang, Xiaochan An, Zhuangzhi Li, and Yunxia Hu. 2020. Towards enhanced antifouling and flux performances of thin-film composite forward osmosis membrane via constructing a sandwich-like carbon nanotubes-coated support. *Desalination* 479:114311. https://doi.org/10.1016/j.desal.2020.114311.

Dewangan, Ranjana, Ayesha Hashmi, Anupama Asthana, Ajaya K. Singh, and Md Abu Bin Hasan Susan. 2020. Degradation of methylene blue and methyl violet using graphene oxide/NiO/β-cyclodextrin nanocomposites as photocatalyst. *International Journal of Environmental Analytical Chemistry*:1–20. https://doi.org/10.1080/03067319.2020.1802443.

Din, Muhammad Imran, Amna Ghulam Nabi, Zaib Hussain, Muhammad Arshad, Azeem Intisar, Ahsan Sharif, Ejaz Ahmed, Hafiz Arslan Mehmood, and Muhammad Latif Mirza. 2019. Innovative seizure of metal/metal oxide nanoparticles in water purification: A critical review of potential risks. *Critical Reviews in Analytical Chemistry* 49 (6):534–541. https://doi.org/10.1080/10408347.2018.1564647.

Djordjevic, Aleksandar, Daniela Šojić Merkulov, Marina Lazarević, Ivana Borišev, Igor Medić, Vladimir Pavlović, Bojan Miljević, and Biljana Abramović. 2018. Enhancement of nano titanium dioxide coatings by fullerene and polyhydroxy fullerene in the photocatalytic degradation of the herbicide mesotrione. *Chemosphere* 196:145–152. https://doi.org/10.1016/j.chemosphere.2017.12.160.

Dongre, Rajendra S., Kishor Kumar Sadasivuni, Kalim Deshmukh, Akansha Mehta, Soumen Basu, Jostna S. Meshram, Mariam Al Ali Al-Maadeed, and Alamgir Karim. 2019. Natural polymer based composite membranes for water purification: a review. *Polymer-Plastics Technology and Materials* 58 (12):1295–1310. https://doi.org/10.1080/25740881.2018.1563116.

dos Santos, Grazielle de Souza Emanuelle, Alessandra Honjo Ide, José Leandro Silva Duarte, Gordon McKay, Antonio Osimar Sousa Silva, and Lucas Meili. 2020. Adsorption of anti-inflammatory drug diclofenac by MgAl/layered double hydroxide supported on Syagrus coronata biochar. *Powder Technology* 364:229–240. https://doi.org/10.1016/j.powtec.2020.01.083.

Duenas-Ramirez, Paula, Caroline Bertagnolli, Roxane Müller, Kevin Sartori, Anne Boos, Mourad Elhabiri, Sylvie Bégin-Colin, and Damien Mertz. 2020. Highly chelating stellate mesoporous silica nanoparticles for specific iron removal from biological media. *Journal of Colloid and Interface Science* 579:140–151. https://doi.org/10.1016/j.jcis.2020.06.013.

Ebrahimpour, Mehdi, Seyed Karim Hassaninejad-Darzi, and Hassan Zavvar Mousavi. 2020. Adsorption of ternary toxic crystal violet, malachite green and methylene blue onto synthesised SBA-15 mesoporous nanoparticles. *International Journal of Environmental Analytical Chemistry*:1–24. https://doi.org/10.1080/03067319.2020.1757085.

Elami, Dariush, and Kambiz Seyyedi. 2020. Removing of carmoisine dye pollutant from contaminated waters by photocatalytic method using a thin film fixed bed reactor. *Journal of Environmental Science and Health, Part A* 55 (2):193–208. https://doi.org/10.1080/10934529.2019.1673089.

Elessawy, Noha A., Mohamed Elnouby, Marwa Hassan Gouda, Hesham A. Hamad, Nahla A. Taha, Mohamed Gouda, and Mohamed S. Mohy Eldin. 2020. Ciprofloxacin removal using magnetic fullerene nanocomposite obtained from sustainable PET bottle wastes: Adsorption process optimization, kinetics, isotherm, regeneration and recycling studies. *Chemosphere* 239:124728. https://doi.org/10.1016/j.chemosphere.2019.124728.

El-Gendi, Ayman, Farag A. Samhan, Nahla Ismail, and Lara A. Nezam El-Dein. 2018. Synergistic role of Ag nanoparticles and Cu nanorods dispersed on graphene on membrane desalination and biofouling. *Journal of Industrial and Engineering Chemistry* 65:127–136. https://doi.org/10.1016/j.jiec.2018.04.021.

Feng, Shuting, Tian Chen, Zhichao Liu, Jianhui Shi, Xiuping Yue, and Yuzhen Li. 2020. Z-scheme CdS/CQDs/g-C_3N_4 composites with visible-near-infrared light response for efficient photocatalytic organic pollutant degradation. *Science of The Total Environment* 704:135404. https://doi.org/10.1016/j.scitotenv.2019.135404.

Fernandes, André, Patrycja Makoś, Zhaohui Wang, and Grzegorz Boczkaj. 2020. Synergistic effect of TiO_2 photocatalytic advanced oxidation processes in the treatment of refinery effluents. *Chemical Engineering Journal* 391:123488. https://doi.org/10.1016/j.cej.2019.123488.

Floresyona, Dita, Fabrice Goubard, Pierre-Henri Aubert, Isabelle Lampre, Jérémie Mathurin, Alexandre Dazzi, Srabanti Ghosh, Patricia Beaunier, François Brisset, and Samy Remita. 2017. Highly active poly (3-hexylthiophene) nanostructures for photocatalysis under solar light. *Applied Catalysis B: Environmental* 209:23–32.

Gai, Wenxiao, Die Ling Zhao, and Tai-Shung Chung. 2019. Thin film nanocomposite hollow fiber membranes comprising Na+-functionalized carbon quantum dots for brackish water desalination. *Water Research* 154:54–61. https://doi.org/10.1016/j.watres.2019.01.043.

Ganta, Deepak, Carlos Guzman, Keith Combrink, and Mario Fuentes. 2020. Adsorption and removal of thymol from water using a Zeolite imidazolate framework-8 nanomaterial. *Analytical Letters*:1–12. https://doi.org/10.1080/00032719.2020.1774601.

Ghosh, Srabanti, Natalie A. Kouamé, Laurence Ramos, Samy Remita, Alexandre Dazzi, Ariane Deniset-Besseau, Patricia Beaunier, Fabrice Goubard, Pierre-Henri Aubert, and Hynd Remita. 2015a. Conducting polymer nanostructures for photocatalysis under visible light. *Nature Materials* 14 (5):505–511. https://doi.org/10.1038/nmat4220.

Ghosh, Srabanti, Natalie Amoin Kouamé, Samy Remita, Laurence Ramos, Fabrice Goubard, Pierre-Henri Aubert, Alexandre Dazzi, Ariane Deniset-Besseau, and Hynd Remita. 2015b. Visible-light active conducting polymer nanostructures with superior photocatalytic activity. *Scientific Reports* 5 (1):18002. https://doi.org/10.1038/srep18002.

Ghosh, Srabanti, Hynd Remita, and Rajendra N Basu. 2018. Visible-light-induced reduction of Cr (VI) by PDPB-ZnO nanohybrids and its photo-electrochemical response. *Applied Catalysis B: Environmental* 239:362–372.

Guo, Youhong, Hengyi Lu, Fei Zhao, Xingyi Zhou, Wen Shi, and Guihua Yu. 2020. Biomass-derived hybrid hydrogel evaporators for cost-effective solar water purification. *Advanced Materials* 32 (11):1907061. https://doi.org/10.1002/adma.201907061.

Han, Meina, Xiaoguang Duan, Guoliang Cao, Shishu Zhu, and Shih-Hsin Ho. 2020a. Graphitic nitride-catalyzed advanced oxidation processes (AOPs) for landfill leachate treatment: A mini review. *Process Safety and Environmental Protection* 139:230–240. https://doi.org/10.1016/j.psep.2020.04.046.

Han, Ya, Jinjin Chen, Xingxing Gu, and Jianrong Chen. 2020b. Adsorption of multi-bivalent heavy metal ions in aqueous solution onto aminopropyl-functionalized MCM-48 preparation by co-condensation. *Separation Science and Technology*:1–11. https://doi.org/10.1080/01496395.2020.1799009.

Huang, Min-Yue, Yan Chen, Xi Yan, Xiao-Jing Guo, Lei Dong, and Wan-Zhong Lang. 2020. Two-dimensional Montmorillonite membranes with efficient water filtration. *Journal of Membrane Science* 614:118540. https://doi.org/10.1016/j.memsci.2020.118540.

Hussein, Mahmoud A., Reda M. El-Shishtawy, Khalid A. Alamry, Abdullah M. Asiri, and Saleh A. Mohamed. 2019. Efficient water disinfection using hybrid polyaniline/graphene/carbon nanotube nanocomposites. *Environmental Technology* 40 (21):2813–2824. https://doi.org/10.1080/09593330.2018.1466921.

Idumah, Christopher Igwe, Azman Hassan, James Ogbu, J. U. Ndem, and Iheoma Chigoziri Nwuzor. 2019. Recently emerging advancements in halloysite nanotubes polymer nanocomposites. *Composite Interfaces* 26 (9):751–824. https://doi.org/10.1080/09276440.2018.1534475.

Indherjith, Sakthinathan, Soundararajan Karthikeyan, J. Helen Ratna Monica, and Karthik Krishna Kumar. 2019. Graphene oxide & reduced graphene oxide polysulfone nanocomposite pellets: An alternative adsorbent of antibiotic pollutant-ciprofloxacin. *Separation Science and Technology* 54 (5):667–674. https://doi.org/10.1080/01496395.2018.1518986.

Ioannidi, Alexandra, Paula Oulego, Sergio Collado, Athanasia Petala, Victor Arniella, Zacharias Frontistis, George N. Angelopoulos, Mario Diaz, and Dionissios Mantzavinos. 2020. Persulfate activation by modified red mud for the oxidation of antibiotic sulfamethoxazole in water. *Journal of Environmental Management* 270:110820. https://doi.org/10.1016/j.jenvman.2020.110820.

Jalali-Moghadam, Elnaz, and Zahra Shariatinia. 2018a. Al^{3+} doping into TiO_2 photoanodes improved the performances of amine anchored CdS quantum dot sensitized solar cells. *Materials Research Bulletin* 98:121–132.

Jalali-Moghadam, Elnaz, and Zahra Shariatinia. 2018b. Quantum dot sensitized solar cells fabricated by means of a novel inorganic spinel nanoparticle. *Applied Surface Science* 441:1–11.

Jana, Malay, and Raj N. Singh. 2018. Progress in CVD synthesis of layered hexagonal boron nitride with tunable properties and their applications. *International Materials Reviews* 63 (3):162–203. https://doi.org/10.1080/09506608.2017.1322833.

Jourshabani, Milad, Byeong-Kyu Lee, and Zahra Shariatinia. 2020. From traditional strategies to Z-scheme configuration in graphitic carbon nitride photocatalysts: Recent progress and future challenges. *Applied Catalysis B: Environmental* 276:119157. https://doi.org/10.1016/j.apcatb.2020.119157.

Jourshabani, Milad, Zahra Shariatinia, Gopal Achari, Cooper H. Langford, and Alireza Badiei. 2018. Facile synthesis of NiS_2 nanoparticles ingrained in a sulfur-doped carbon nitride framework with enhanced visible light photocatalytic activity: two functional roles of thiourea. *Journal of Materials Chemistry A* 6 (27):13448–13466. https://doi.org/10.1039/C8TA03068E.

Jourshabani, Milad, Zahra Shariatinia, and Alireza Badiei. 2017a. Controllable synthesis of mesoporous sulfur-doped carbon nitride materials for enhanced visible light photocatalytic degradation. *Langmuir* 33 (28):7062–7078. https://doi.org/10.1021/acs.langmuir.7b01767.

Jourshabani, Milad, Zahra Shariatinia, and Alireza Badiei. 2017b. Facile one-pot synthesis of cerium oxide/sulfur-doped graphitic carbon nitride (g-C_3N_4) as efficient nanophotocatalysts under visible light irradiation. *Journal of Colloid and Interface Science* 507:59–73. https://doi.org/10.1016/j.jcis.2017.07.106.

Jourshabani, Milad, Zahra Shariatinia, and Alireza Badiei. 2017c. In situ fabrication of SnO_2/S-doped g-C_3N_4 nanocomposites and improved visible light driven photodegradation of methylene blue. *Journal of Molecular Liquids* 248:688–702.

Jourshabani, Milad, Zahra Shariatinia, and Alireza Badiei. 2017d. Sulfur-doped mesoporous carbon nitride decorated with Cu particles for efficient photocatalytic degradation under visible-light irradiation. *The Journal of Physical Chemistry C* 121 (35):19239–19253. https://doi.org/10.1021/acs.jpcc.7b05556.

Jourshabani, Milad, Zahra Shariatinia, and Alireza Badiei. 2018a. High efficiency visible-light-driven $Fe_2O_{3-x}S_x$/S-doped g-C_3N_4 heterojunction photocatalysts: Direct Z-scheme mechanism. *Journal of Materials Science & Technology* 34 (9):1511–1525. https://doi.org/10.1016/j.jmst.2017.12.020.

Jourshabani, Milad, Zahra Shariatinia, and Alireza Badiei. 2018b. Synthesis and characterization of novel Sm_2O_3/S-doped g-C_3N_4 nanocomposites with enhanced photocatalytic activities under visible light irradiation. *Applied Surface Science* 427:375–387.

Kaleekkal, Noel Jacob, Ramakrishnan Radhakrishnan, Vishnu Sunil, Geethanzali Kamalanathan, Arijit Sengupta, and Ranil Wickramasinghe. 2018. Performance evaluation of novel nanostructured modified mesoporous silica/polyetherimide composite membranes for the treatment of oil/water emulsion. *Separation and Purification Technology* 205:32–47. https://doi.org/10.1016/j.seppur.2018.05.007.

Kang, Guo-dong, and Yi-ming Cao. 2012. Development of antifouling reverse osmosis membranes for water treatment: A review. *Water Research* 46 (3):584–600. https://doi.org/10.1016/j.watres.2011.11.041.

Kang, Song Gun, Tong Ho Choe, Chol Ung Ryom, and Myong Chol Ri. 2020. Research on synthesis and photocatalytic activity of $ZnFe_2O_4/Ag/g\text{-}C_3N_4$ nanosheets composites. *Composite Interfaces*:1–13. https://doi.org/10.1080/09276440.2020.1747347.

Kausar, Abida, Munawar Iqbal, Anum Javed, Kiran Aftab, Zill-i-Huma Nazli, Haq Nawaz Bhatti, and Shazia Nouren. 2018. Dyes adsorption using clay and modified clay: A review. *Journal of Molecular Liquids* 256:395–407. https://doi.org/10.1016/j.molliq.2018.02.034.

Kaveh, Reyhaneh, Zahra Shariatinia, and Ahmad Arefazar. 2016. Improvement of polyacrylonitrile ultrafiltration membranes' properties using decane-functionalized reduced graphene oxide nanoparticles. *Water Science and Technology: Water Supply* 16 (5):1378–1387.

Khairy, Mohamed, Eman M. Naguib, and Mohamed Mokhtar Mohamed. 2020. Enhancement of photocatalytic and sonophotocatalytic degradation of 4-nitrophenol by ZnO/graphene oxide and ZnO/carbon nanotube nanocomposites. *Journal of Photochemistry and Photobiology A: Chemistry* 396:112507. https://doi.org/10.1016/j.jphotochem.2020.112507.

Kim, Jong-Gook, Hye-Bin Kim, Jeong-Hwan Choi, and Kitae Baek. 2020. Bifunctional iron-modified graphitic carbon nitride ($g\text{-}C_3N_4$) for simultaneous oxidation and adsorption of arsenic. *Environmental Research* 188:109832. https://doi.org/10.1016/j.envres.2020.109832.

Kommu, Anitha, and Jayant K. Singh. 2020. A review on graphene-based materials for removal of toxic pollutants from wastewater. *Soft Materials*:1–26. https://doi.org/10.1080/1539445X.2020.1739710.

Kozari, Argyri, Andreana Paloglou, and Dimitra Voutsa. 2020. Formation potential of emerging disinfection by-products during ozonation and chlorination of sewage effluents. *Science of The Total Environment* 700:134449. https://doi.org/10.1016/j.scitotenv.2019.134449.

Kumar, Suneel, Ajay Kumar, Ashish Kumar, and Venkata Krishnan. 2020. Nanoscale zinc oxide based heterojunctions as visible light active photocatalysts for hydrogen energy and environmental remediation. *Catalysis Reviews* 62 (3):346–405. https://doi.org/10.1080/01614940.2019.1684649.

Lacerda, Hass Caetano Elenice, Francielli Casanova Monteiro, Juliana Regina Kloss, and Sérgio Toshio Fujiwara. 2020. Bentonite clay modified with Nb_2O_5: An efficient and reused photocatalyst for the degradation of reactive textile dye. *Journal of Photochemistry and Photobiology A: Chemistry* 388:112084. https://doi.org/10.1016/j.jphotochem.2019.112084.

Lata, S., and Vikas. 2019. Externally predictive quantum-mechanical models for the adsorption of aromatic organic compounds by graphene-oxide nanomaterials. *SAR and QSAR in Environmental Research* 30 (12):847–863. https://doi.org/10.1080/1062936X.2019.1666164.

Lee, Kyun Ho, Haedong Kim, Jung Won KuK, Jae Dong Chung, Sungsu Park, and Eilhann E. Kwon. 2020. Micro-bubble flow simulation of dissolved air flotation process for water treatment using computational fluid dynamics technique. *Environmental Pollution* 256:112050. https://doi.org/10.1016/j.envpol.2019.01.011.

Li, Jie, Shi He, Wei Dai, Tian Wu, Rui Li, Hongyan Qi, Junhui Tao, Lin Zhang, Yu Zhang, Qing Huang, and Yonghong Tian. 2020a. Facile synthesis of porous boron nitride-supported α-Fe_2O_3 nanoparticles for enhanced regeneration performance. *Materials Technology*:1–7. https://doi.org/10.1080/10667857.2020.1794277.

Li, Juan, Jin Jiang, Su-Yan Pang, Yi Yang, Shaofang Sun, Lihong Wang, and Panxin Wang. 2020b. Transformation of X-ray contrast media by conventional and advanced oxidation processes during water treatment: Efficiency, oxidation intermediates, and formation of iodinated byproducts. *Water Research* 185:116234. https://doi.org/10.1016/j.watres.2020.116234.

Li, Zhenjiang, Yongkai Sun, Jing Xing, Yucheng Xing, and Alan Meng. 2018. One step synthesis of Co/Cr-codoped ZnO nanoparticle with superb adsorption properties for various anionic organic pollutants and its regeneration. *Journal of Hazardous Materials* 352:204–214. https://doi.org/10.1016/j.jhazmat.2018.03.049.

Lian, Jianjun, Fanjie Zhou, Bo Chen, Mei Yang, Shisheng Wang, Zailiang Liu, and Siping Niu. 2020a. Enhanced adsorption of molybdenum(VI) onto drinking water treatment residues modified by thermal treatment and acid activation. *Journal of Cleaner Production* 244:118719. https://doi.org/10.1016/j.jclepro.2019.118719.

Liang, Jiahao, Qinghong Wang, Qing X. Li, Liangyan Jiang, Jiawen Kong, Ming Ke, Muhammad Arslan, Mohamed Gamal El-Din, and Chunmao Chen. 2020b. Aerobic sludge granulation in shale gas flowback water treatment: Assessment of the bacterial community dynamics and modeling of bioreactor performance using artificial neural network. *Bioresource Technology* 313:123687. https://doi.org/10.1016/j.biortech.2020.123687.

Lin, Jui-Yen, Minsoo Kim, Dan Li, Hyunook Kim, and Chin-pao Huang. 2020a. The removal of phosphate by thermally treated red mud from water: The effect of surface chemistry on phosphate immobilization. *Chemosphere* 247:125867. https://doi.org/10.1016/j.chemosphere.2020.125867.

Lin, Li, Qin Xie, Mingjing Zhang, Chunxiu Liu, Yunsong Zhang, Guangtu Wang, Ping Zou, Jun Zeng, Hui Chen, and Maojun Zhao. 2020b. Construction of Z-scheme Ag-AgBr/$BiVO_4$/graphene aerogel with enhanced photocatalytic degradation and antibacterial activities. *Colloids and Surfaces A: Physicochemical and Engineering Aspects* 601:124978. https://doi.org/10.1016/j.colsurfa.2020.124978.

Liu, Botao, Azmatullah Khan, Ki-Hyun Kim, Deepak Kukkar, and Ming Zhang. 2020a. The adsorptive removal of lead ions in aquatic media: Performance comparison between advanced functional materials and conventional materials. *Critical Reviews in Environmental Science and Technology* 50 (23):2441–2483. 10.1080/10643389.2019.1694820.

Liu, Jingyi, Wenyong Hu, Maogui Sun, Ouyang Xiong, Haibin Yu, Haopeng Feng, Xuan Wu, Lin Tang, and Yaoyu Zhou. 2019. Enhancement of Fenton processes at initial circumneutral pH for the degradation of norfloxacin with Fe@Fe_2O_3 core-shell nanomaterials. *Environmental Technology* 40 (27):3632–3640. https://doi.org/10.1080/09593330.2018.1483972.

Liu, Lu, Guiru Zhu, Zhaofeng Liu, and Congjie Gao. 2016. Effect of MCM-48 nanoparticles on the performance of thin film nanocomposite membranes for reverse osmosis application. *Desalination* 394:72–82. https://doi.org/10.1016/j.desal.2016.04.028.

Liu, Shujuan, Changrui Jiang, Yunpeng Gao, Jiaxin He, Kexin Li, and Yan Feng. 2020b. One-step microwave synthesis of covalently bonded OC_3N_4/C60 with enhanced photocatalytic properties. *Materials Research Bulletin* 122:110668. https://doi.org/10.1016/j.materresbull.2019.110668.

Liu, Shuqiong, Jianqiang Luo, Jianguo Ma, Jianqiang Li, Song Li, Lina Meng, and Shujuan Liu. 2020c. Removal of uranium from aqueous solutions using amine-functionalized magnetic platelet large-pore SBA-15. *Journal of Nuclear Science and Technology*:1–11. https://doi.org/10.1080/00223131.2020.1796838.

Liu, Xiao, Jianan Guan, Guanghong Lai, Qian Xu, Xiabing Bai, Ziming Wang, and Suping Cui. 2020d. Stimuli-responsive adsorption behavior toward heavy metal ions based on comb polymer functionalized magnetic nanoparticles. *Journal of Cleaner Production* 253:119915. https://doi.org/10.1016/j.jclepro.2019.119915.

Lu, Peng, Wenjun Li, Sheng Yang, Yayu Wei, Zhongguo Zhang, and Yanshuo Li. 2019. Layered double hydroxides (LDHs) as novel macropore-templates: The importance of porous structures for forward osmosis desalination. *Journal of Membrane Science* 585:175–183. https://doi.org/10.1016/j.memsci.2019.05.045.

Mahmoud, Mohamed E., Nesma A. Fekry, and Amir M. Abdelfattah. 2020. Removal of uranium (VI) from water by the action of microwave-rapid green synthesized carbon quantum dots from starch-water system and supported onto polymeric matrix. *Journal of Hazardous Materials* 397:122770. https://doi.org/10.1016/j.jhazmat.2020.122770.

Mansor, Nur Adawiyah, and Kheng Soo Tay. 2020. Potential toxic effects of chlorination and UV/chlorination in the treatment of hydrochlorothiazide in the water. *Science of The Total Environment* 714:136745. https://doi.org/10.1016/j.scitotenv.2020.136745.

Matsunaga, Tadashi, and Mina Okochi. 1995. TiO_2-Mediated photochemical disinfection of *E.* coli using optical fibers. *Environmental Science & Technology* 29 (2):501–505. https://doi.org/10.1021/es00002a028.

Matsunaga, Tadashi, Ryozo Tomoda, Toshiaki Nakajima, Noriyuki Nakamura, and Tamotsu Komine. 1988. Continuous-sterilization system that uses photosemiconductor powders. *Applied and Environmental Microbiology* 54 (6):1330–1333.

Matsunaga, Tadashi, Ryozo Tomoda, Toshiaki Nakajima, and Hitoshi Wake. 1985. Photoelectrochemical sterilization of microbial cells by semiconductor powders. *FEMS Microbiology Letters* 29 (1):211–214.

Mazhar, Mohd Aamir, Nadeem A. Khan, Sirajuddin Ahmed, Afzal Husain Khan, Azhar Hussain, Rahisuddin, Fazlollah Changani, Mahmood Yousefi, Shahin Ahmadi, and Viola Vambol. 2020. Chlorination disinfection by-products in municipal drinking water – A review. *Journal of Cleaner Production* 273:123159. https://doi.org/10.1016/j.jclepro.2020.123159.

Mirzaeifard, Zahra, Zahra Shariatinia, Milad Jourshabani, and Seyed Mahmood Rezaei Darvishi. 2020. ZnO photocatalyst revisited: Effective photocatalytic degradation of emerging contaminants using S-doped ZnO nanoparticles under visible light radiation. *Industrial & Engineering Chemistry Research.* https://doi.org/10.1021/acs.iecr.0c03192.

Mlih, Rawan, Franciszek Bydalek, Erwin Klumpp, Nader Yaghi, Roland Bol, and Jannis Wenk. 2020. Light-expanded clay aggregate (LECA) as a substrate in constructed wetlands: A review. *Ecological Engineering* 148:105783. https://doi.org/10.1016/j.ecoleng.2020.105783.

Monteoliva-García, Antonio, Jaime Martín-Pascual, Maria del Mar Muñío, and José Manuel Poyatos. 2020. Effects of carrier addition on water quality and pharmaceutical removal capacity of a membrane bioreactor – advanced oxidation process combined treatment. *Science of The Total Environment* 708:135104. https://doi.org/10.1016/j.scitotenv.2019.135104.

Morcali, Mehmet Hakan, and Asli Baysal. 2019. The miniaturised process for lead removal from water samples using novel bioconjugated sorbents. *International Journal of Environmental Analytical Chemistry* 99 (14):1397–1414. https://doi.org/10.1080/03067319.2019.1622695.

Mousavi, Seyyed Alireza, and Hosna Janjani. 2020. Antibiotics adsorption from aqueous solutions using carbon nanotubes: A systematic review. *Toxin Reviews* 39 (2):87–98. https://doi.org/10.1080/15569543.2018.1483405.

Nagajyothi, Patnamsetty Chidanandha, Surya Veerendra Prabhakar Vattikuti, Kamakshaiah Charyulu Devarayapalli, Kisoo Yoo, Jaesool Shim, and Thupakula Venkata Madhukar Sreekanth. 2019. Green synthesis: Photocatalytic degradation of textile dyes using metal and metal oxide nanoparticles-latest trends and advancements. *Critical Reviews in Environmental Science and Technology*:1–107. https://doi.org/10.1080/10643389.2019.1705103.

Nasrollahzadeh, Mahmoud, Mohaddeseh Sajjadi, Siavash Iravani, and Rajender S. Varma. 2021. Carbon-based sustainable nanomaterials for water treatment: State-of-art and future perspectives. *Chemosphere* 263:128005. https://doi.org/10.1016/j.chemosphere.2020.128005.

Nguyen, Dinh Cung Tien, and Won-Chun Oh. 2018. Ternary self-assembly method of mesoporous silica and Cu_2O combined graphene composite by nonionic surfactant and photocatalytic degradation of cationic-anionic dye pollutants. *Separation and Purification Technology* 190:77–89. https://doi.org/10.1016/j.seppur.2017.08.054.

Nieto-Delgado, Cesar, Dulce Partida-Gutierrez, and J. Rene Rangel-Mendez. 2019. Preparation of activated carbon cloths from renewable natural fabrics and their performance during the adsorption of model organic and inorganic pollutants in water. *Journal of Cleaner Production* 213:650–658. https://doi.org/10.1016/j.jclepro.2018.12.184.

Noorimotlagh, Zahra, Maryam Ravanbakhsh, Mohammad Reza Valizadeh, Behrouz Bayati, George Z. Kyzas, Mehdi Ahmadi, Nadereh Rahbar, and Neemat Jaafarzadeh. 2020. Optimization and genetic programming modeling of humic acid adsorption onto prepared activated carbon and modified by multi-wall carbon nanotubes. *Polyhedron* 179:114354. https://doi.org/10.1016/j.poly.2020.114354.

Nouri, Morteza, Ramin Yousefi, Nasser Zare-Dehnavi, and Farid Jamali-Sheini. 2020. Tuning crystal phase and morphology of copper selenide nanostructures and their visible-light photocatalytic applications to degrade organic pollutants. *Colloids and Surfaces A: Physicochemical and Engineering Aspects* 586:124196. https://doi.org/10.1016/j.colsurfa.2019.124196.

Novakovic, Mladenka, Goran Strbac, Maja Petrovic, Dragana Strbac, and Ivana Mihajlovic. 2020. Decomposition of pharmaceutical micropollutant – diclofenac by photocatalytic nanopowder mixtures in aqueous media: effect of optimization parameters, identification of intermediates and economic considerations. *Journal of Environmental Science and Health, Part A* 55 (4):483–497. https://doi.org/10.1080/10934529.2019.1701895.

Oh, Won-Chun, Kamrun Nahar Fatema, Yin Liu, Chang Sung Lim, Kwang Youn Cho, Chong-Hun Jung, and Md Rokon Ud Dowla Biswas. 2020. Sonochemical synthesis of quaternary $LaNiSbWO_4$-G-PANI polymer nanocomposite for photocatalytic degradation of Safranin-O and gallic acid under visible light irradiation. *Journal of Photochemistry and Photobiology A: Chemistry* 394:112484. https://doi.org/10.1016/j.jphotochem.2020.112484.

Parsa, Mohammad Mohammadi, Hossein Pourfakhar, and Majid Baghdadi. 2020. Application of graphene oxide nanosheets in the coagulation-flocculation process for removal of Total Organic Carbon (TOC) from surface water. *Journal of Water Process Engineering* 37:101367. https://doi.org/10.1016/j.jwpe.2020.101367.

Pendergast, MaryTheresa M., and Eric M. V. Hoek. 2011. A review of water treatment membrane nanotechnologies. *Energy & Environmental Science* 4 (6):1946–1971. https://doi.org/10.1039/C0EE00541J.

Qu, Yanning, Xiaojian Xu, Renliang Huang, Wei Qi, Rongxin Su, and Zhimin He. 2020. Enhanced photocatalytic degradation of antibiotics in water over functionalized N,S-doped carbon quantum dots embedded ZnO nanoflowers under sunlight irradiation. *Chemical Engineering Journal* 382:123016. https://doi.org/10.1016/j.cej.2019.123016.

Rafique, Muhammad, Rabbia Tahir, Syed Sajid Ali Gillani, Muhammad Bilal Tahir, Muhammad Shakil, Tahir Iqbal, and Muhammad O. Abdellahi. 2020. Plant-mediated green synthesis of zinc oxide nanoparticles from Syzygium Cumini for seed germination and wastewater purification. *International Journal of Environmental Analytical Chemistry*:1–16. https://doi.org/10.1080/03067319.2020.1715379.

Rahimi-Aghdam, Taher, Zahra Shariatinia, Minna Hakkarainen, and Vahid Haddadi-Asl. 2019. Polyacrylonitrile/N,P co-doped graphene quantum dots-layered double hydroxide nanocomposite: Flame retardant property, thermal stability and fire hazard. *European Polymer Journal* 120:109256. https://doi.org/10.1016/j.eurpolymj.2019.109256.

Rahimi-Aghdam, Taher, Zahra Shariatinia, Minna Hakkarainen, and Vahid Haddadi-Asl. 2020. Nitrogen and phosphorous doped graphene quantum dots: Excellent flame retardants and smoke suppressants for polyacrylonitrile nanocomposites. *Journal of Hazardous Materials* 381:121013. https://doi.org/10.1016/j.jhazmat.2019.121013.

Rahman, Mir Tamzid, Tomohito Kameda, Takao Miura, Shogo Kumagai, and Toshiaki Yoshioka. 2019. Application of Mg–Al layered double hydroxide for treating acidic mine wastewater: A novel approach to sludge reduction. *Chemistry and Ecology* 35 (2):128–142. https://doi.org/10.1080/02757540.2018.1534964.

Ravikumar, K., and J. Udayakumar. 2020. Preparation and characterisation of green clay-polymer nanocomposite for heavy metals removal. *Chemistry and Ecology* 36 (3):270–291. https://doi.org/10.1080/02757540.2020.1723559.

Ravikumar, Konda Venkata Giri, Hemamalathi Kubendiran, Rajat Gupta, Ashutosh Gupta, Pankaj Sharma, Sruthi Ann Alex, Chandrasekaran Natarajan, Bhaskar Das, and Amitava Mukherjee. 2020. In-situ coating of Fe/Pd nanoparticles on sand and its application for removal of tetracycline from aqueous solution. *Journal of Water Process Engineering* 36:101400. https://doi.org/10.1016/j.jwpe.2020.101400.

Remanan, Sanjay, Nagarajan Padmavathy, Sabyasachi Ghosh, Subhadip Mondal, Suryasarathi Bose, and Narayan Ch Das. 2020. Porous graphene-based membranes: Preparation and properties of a unique two-dimensional nanomaterial membrane for water purification. *Separation & Purification Reviews*:1–21. https://doi.org/10.1080/15422119.2020.1725048.

Reza, Md Sumon, Cheong Sing Yun, Shammya Afroze, Nikdalila Radenahmad, Muhammad S. Abu Bakar, Rahman Saidur, Juntakan Taweekun, and Abul K. Azad. 2020. Preparation of activated carbon from biomass and its' applications in water and gas purification, a review. *Arab Journal of Basic and Applied Sciences* 27 (1):208–238. https://doi.org/10.1080/25765299.2020.1766799.

Rizzi, Vito, Jennifer Gubitosa, Paola Fini, Roberto Romita, Angela Agostiano, Sergio Nuzzo, and Pinalysa Cosma. 2020. Commercial bentonite clay as low-cost and recyclable natural adsorbent for the Carbendazim removal/recover from water: Overview on the adsorption process and preliminary photodegradation considerations. *Colloids and Surfaces A: Physicochemical and Engineering Aspects* 602:125060. https://doi.org/10.1016/j.colsurfa.2020.125060.

Rono, Nicholas, Joshua K. Kibet, Bice S. Martincigh, and Vincent O. Nyamori. 2020. A review of the current status of graphitic carbon nitride. *Critical Reviews in Solid State and Materials Sciences*:1–29. https://doi.org/10.1080/10408436.2019.1709414.

Saeedi-Jurkuyeh, Alireza, Ahmad Jonidi Jafari, Roshanak Rezaei Kalantary, and Ali Esrafili. 2020. A novel synthetic thin-film nanocomposite forward osmosis membrane modified by graphene oxide and polyethylene glycol for heavy metals removal from aqueous solutions. *Reactive and Functional Polymers* 146:104397. https://doi.org/10.1016/j.reactfunctpolym.2019.104397.

Safarpour, Mahdie, Vahid Vatanpour, Alireza Khataee, Hamed Zarrabi, Peyman Gholami, and Mohammad Ehsan Yekavalangi. 2017. High flux and fouling resistant reverse osmosis membrane modified with plasma treated natural zeolite. *Desalination* 411:89–100. https://doi.org/10.1016/j.desal.2017.02.012.

Saffari, Reyhaneh, Zahra Shariatinia, and Milad Jourshabani. 2020. Synthesis and photocatalytic degradation activities of phosphorus containing ZnO microparticles under visible light irradiation for water treatment applications. *Environmental Pollution* 259:113902. https://doi.org/10.1016/j.envpol.2019.113902.

Saito, Toshiyuki, Tatsuo Iwase, Jun Horie, and Toshio Morioka. 1992. Mode of photocatalytic bactericidal action of powdered semiconductor TiO_2 on mutans streptococci. *Journal of Photochemistry and Photobiology B: Biology* 14 (4):369–379. https://doi.org/10.1016/1011-1344(92)85115-B.

Salehi, Masoumeh, and Zahra Shariatinia. 2016a. An optimization of MnO_2 amount in CNT-MnO_2 nanocomposite as a high rate cathode catalyst for the rechargeable Li-O_2 batteries. *Electrochimica Acta* 188:428–440. https://doi.org/10.1016/j.electacta.2015.12.016.

Salehi, Masoumeh, and Zahra Shariatinia. 2016b. Synthesis of star-like MnO_2-CeO_2/CNT composite as an efficient cathode catalyst applied in lithium-oxygen batteries. *Electrochimica Acta* 222:821–829. https://doi.org/10.1016/j.electacta.2016.11.043.

Salehi, Masoumeh, Zahra Shariatinia, and Abbas Sadeghi. 2019. Application of RGO/CNT nanocomposite as cathode material in lithium-air battery. *Journal of Electroanalytical Chemistry* 832:165–173. https://doi.org/10.1016/j.jelechem.2018.10.053.

Savino, Angelo, and Giuseppe Angeli. 1985. Photodynamic inactivation of *E.* coli by immobilized or coated dyes on insoluble supports. *Water Research* 19 (12):1465–1469. https://doi.org/10.1016/0043-1354(85)90390-2.

Shah, Aatif Ali, Young Hoon Cho, Seung-Eun Nam, Ahrumi Park, You-In Park, and Hosik Park. 2020a. High performance thin-film nanocomposite forward osmosis membrane based on PVDF/bentonite nanofiber support. *Journal of Industrial and Engineering Chemistry* 86:90–99. https://doi.org/10.1016/j.jiec.2020.02.016.

Shah, Jafar Hussain, Mohammad Fiaz, Muhammad Athar, Jafar Ali, Mahnoor Rubab, Rashid Mehmood, Syed Umair Ullah Jamil, and Ridha Djellabi. 2020b. Facile synthesis of N/B-double-doped Mn_2O_3 and WO_3 nanoparticles for dye degradation under visible light. *Environmental Technology* 41 (18):2372–2381. https://doi.org/10.1080/09593330.2019.1567604.

Shariatinia, Zahra. 2018. Carboxymethyl chitosan: Properties and biomedical applications. *International Journal of Biological Macromolecules* 120:1406–1419. https://doi.org/10.1016/j.ijbiomac.2018.09.131.

Shariatinia, Zahra. 2019. Chapter 2 – Pharmaceutical applications of natural polysaccharides. In *Natural Polysaccharides in Drug Delivery and Biomedical Applications*, edited by Md Saquib Hasnain and Amit Kumar Nayak, 15–57. Cambridge, MA: Academic Press.

Shariatinia, Zahra. 2020. Biopolymeric Nanocomposites in Drug Delivery. In *Advanced Biopolymeric Systems for Drug Delivery*, edited by Amit Kumar Nayak and Md Saquib Hasnain, 233–290. Cham: Springer International Publishing.

Shariatinia, Zahra, and Ali Bagherpour. 2018. Synthesis of zeolite NaY and its nanocomposites with chitosan as adsorbents for lead(II) removal from aqueous solution. *Powder Technology* 338:744–763. https://doi.org/10.1016/j.powtec.2018.07.082.

Shariatinia, Zahra, and Azam Barzegari. 2019. 22 – Polysaccharide hydrogel films/membranes for transdermal delivery of therapeutics. In *Polysaccharide Carriers for Drug Delivery*, edited by Sabyasachi Maiti and Sougata Jana, 639–684. Woodhead Publishing, Sawston, UK.

Shariatinia, Zahra, and Alireza Esmaeilzadeh. 2019. Hybrid silica aerogel nanocomposite adsorbents designed for Cd(II) removal from aqueous solution. *Water Environment Research* 91 (12):1624–1637. https://doi.org/10.1002/wer.1162.

Shariatinia, Zahra, and Azin Mazloom-Jalali. 2019. Chitosan nanocomposite drug delivery systems designed for the ifosfamide anticancer drug using molecular dynamics simulations. *Journal of Molecular Liquids* 273:346–367. https://doi.org/10.1016/j.molliq.2018.10.047.

Shariatinia, Zahra, Zahra Nikfar, Khodayar Gholivand, and Sara Abolghasemi Tarei. 2015. Antibacterial activities of novel nanocomposite biofilms of chitosan/phosphoramide/Ag NPs. *Polymer Composites* 36 (3):454–466. https://doi.org/10.1002/pc.22960.

Shariatinia, Zahra, and Farzaneh Bagherpour Sardsahra. 2016. Synthesis and characterization of novel spinel Zn1.114La1.264Al0.5O4.271 nanoparticles. *Journal of Alloys and Compounds* 686:384–393. https://doi.org/10.1016/j.jallcom.2016.06.061.

Shariatinia, Zahra, and Zahra Zolfaghari-Isavandi. 2020. Application of $Zn_xLa_yFe_zO_4$ spinel nanomaterial in quantum dot sensitized solar cells. *Optik* 212:164682. https://doi.org/10.1016/j.ijleo.2020.164682.

Shen, Guoqiang, Lun Pan, Rongrong Zhang, Shangcong Sun, Fang Hou, Xiangwen Zhang, and Ji-Jun Zou. 2020. Low-spin-state hematite with superior adsorption of anionic contaminations for water purification. *Advanced Materials* 32 (11):1905988. https://doi.org/10.1002/adma.201905988.

Shivaraju, H. Puttaiah, Henok Egumbo, P. Madhusudan, K. M. Anil Kumar, and G. Midhun. 2019. Preparation of affordable and multifunctional clay-based ceramic filter matrix for treatment of drinking water. *Environmental Technology* 40 (13):1633–1643. https://doi.org/10.1080/09593330.2018.1430853.

Siemens, Ashley M., James J. Dynes, and Wonjae Chang. 2020. Sodium adsorption by reusable zeolite adsorbents: Integrated adsorption cycles for salinised groundwater treatment. *Environmental Technology*:1–12. https://doi.org/10.1080/09593330.2020.1721567.

Singh, Bikramjeet, Kulwinder Singh, Manjeet Kumar, Shagun Thakur, and Akshay Kumar. 2020a. Insights of preferred growth, elemental and morphological properties of BN/SnO_2 composite for photocatalytic applications towards organic pollutants. *Chemical Physics* 531:110659. https://doi.org/10.1016/j.chemphys.2019.110659.

Singh, Karanveer, Deepak Kukkar, Ravinder Singh, Preeti Kukkar, Nardev Bajaj, Jagpreet Singh, Mohit Rawat, Akshay Kumar, and Ki-Hyun Kim. 2020b. In situ green synthesis of Au/Ag nanostructures on a metal-organic framework surface for photocatalytic reduction of p-nitrophenol. *Journal of Industrial and Engineering Chemistry* 81:196–205. https://doi.org/10.1016/j.jiec.2019.09.008.

Soesanto, Jansen Fajar, Kuo-Jen Hwang, Chiao-Wei Cheng, Hung-Yuan Tsai, Allen Huang, Chien-Hua Chen, Tung-Wen Cheng, and Kuo-Lun Tung. 2019. Fenton oxidation-based cleaning technology for powdered activated carbon-precoated dynamic membranes used in microfiltration seawater pretreatment systems. *Journal of Membrane Science* 591:117298. https://doi.org/10.1016/j.memsci.2019.117298.

Sohrabnezhad, Shabnam, and Sima Daraie Mooshangaie. 2019. In situ fabrication of n-type Ag/AgBr nanoparticles in MCM-41 with rice husk (RH/MCM-41) composite for the removal of Eriochrome Black-T. *Materials Science and Engineering: B* 240:16–22. https://doi.org/10.1016/j.mseb.2019.01.007.

Soltanabadi, Yousef, Milad Jourshabani, and Zahra Shariatinia. 2018. Synthesis of novel CuO/$LaFeO_3$ nanocomposite photocatalysts with superior Fenton-like and visible light photocatalytic activities for degradation of aqueous organic contaminants. *Separation and Purification Technology* 202:227–241. https://doi.org/10.1016/j.seppur.2018.03.019.

Som, Ipsita, Mouni Roy, and Rajnarayan Saha. 2020. Advances in nanomaterial-based water treatment approaches for photocatalytic degradation of water pollutants. *ChemCatChem* 12 (13):3409–3433. https://doi.org/10.1002/cctc.201902081.

Son, Dong-Jin, Woo-Yeol Kim, Bo-Rim Jung, Duk Chang, and Ki-Ho Hong. 2020. Pilot-scale anoxic/aerobic biofilter system combined with chemical precipitation for tertiary treatment of wastewater. *Journal of Water Process Engineering* 35:101224. https://doi.org/10.1016/j.jwpe.2020.101224.

Song, Yuefei, Yueke Sun, Mingzhen Chen, Penglin Huang, Tiemei Li, Xiaozhuan Zhang, and Kai Jiang. 2020. Efficient removal and fouling-resistant of anionic dyes by nanofiltration membrane with phosphorylated chitosan modified graphene oxide nanosheets incorporated selective layer. *Journal of Water Process Engineering* 34:101086. https://doi.org/10.1016/j.jwpe.2019.101086.

Sun, Jiayu, Bihan Li, Qi Wang, Ping Zhang, Yulei Zhang, Li Gao, and Xiaochen Li. 2020. Preparation of phosphorus-doped tungsten trioxide nanomaterials and their photocatalytic performances. *Environmental Technology*:1–11. https://doi.org/10.1080/09593330.2020.1745292.

Sun, Shumin, Ran Zhao, Yanli Xie, and Yong Liu. 2019. Photocatalytic degradation of aflatoxin B1 by activated carbon supported TiO_2 catalyst. *Food Control* 100:183–188. https://doi.org/10.1016/j.foodcont.2019.01.014.

Tahir, Muhammad Bilal, Sohail Farman, Muhammad Rafique, Muhammad Shakil, Muhammad Isa Khan, Mohsin Ijaz, Iqra Mubeen, Maria Ashraf, and Khalid Nadeem Riaz. 2019a. Photocatalytic performance of hybrid WO_3/TiO_2 nanomaterials for the degradation of methylene blue under visible light irradiation. *International Journal of Environmental Analytical Chemistry*:1–13. https://doi.org/10.1080/03067319.2019.1685093.

Tahir, Muhammad Bilal, Tasmia Nawaz, Ghulam Nabi, M. Sagir, M. Isa Khan, and Nafisa Malik. 2020. Role of nanophotocatalysts for the treatment of hazardous organic and inorganic pollutants in wastewater. *International Journal of Environmental Analytical Chemistry*:1–25. https://doi.org/10.1080/03067319.2020.1723570.

Tahir, Muhammad Bilal, Sunaina Tufail, Adeel Ahmad, M. Rafique, Tahir Iqbal, Muhammad Abrar, Tasmia Nawaz, Muhammad Yaqoob Khan, and Mohsin Ijaz. 2019b. Semiconductor nanomaterials for the detoxification of dyes in real wastewater under visible-light photocatalysis. *International Journal of Environmental Analytical Chemistry*:1–15. https://doi.org/10.1080/03067319.2019.1686494.

Tao, Xumei, Chen Yang, Liang Huang, and Shuyong Shang. 2020. Novel plasma assisted preparation of ZnCuFeCr layered double hydroxides with improved photocatalytic performance of methyl orange degradation. *Applied Surface Science* 507:145053. https://doi.org/10.1016/j.apsusc.2019.145053.

Tasharrofi, Saeideh, Zahra Rouzitalab, Davood Mohammady Maklavany, Ali Esmaeili, Mohammadmehdi Rabieezadeh, Mojtaba Askarieh, Alimorad Rashidi, and Hossein Taghdisian. 2020. Adsorption of cadmium using modified zeolite-supported nanoscale zero-valent iron composites as a reactive material for PRBs. *Science of The Total Environment* 736:139570. https://doi.org/10.1016/j.scitotenv.2020.139570.

Toledo-Jaldin, Helen Paola, Alien Blanco-Flores, Víctor Sánchez-Mendieta, and Osnieski Martín-Hernández. 2018. Influence of the chain length of surfactant in the modification of zeolites and clays: Removal of atrazine from water solutions. *Environmental Technology* 39 (20):2679–2690. https://doi.org/10.1080/09593330.2017.1365097.

Trawiński, Jakub, and Robert Skibiński. 2019. Multivariate comparison of photocatalytic properties of thirteen nanostructured metal oxides for water purification. *Journal of Environmental Science and Health, Part A* 54 (9):851–864. https://doi.org/10.1080/10934529.2019.1598169.

Vatanparast, Morteza, and Zahra Shariatinia. 2018a. AlN and AlP doped graphene quantum dots as novel drug delivery systems for 5-fluorouracil drug: Theoretical studies. *Journal of Fluorine Chemistry* 211:81–93. https://doi.org/10.1016/j.jfluchem.2018.04.003.

Vatanparast, Morteza, and Zahra Shariatinia. 2018b. Computational studies on the doped graphene quantum dots as potential carriers in drug delivery systems for isoniazid drug. *Structural Chemistry* 29 (5):1427–1448. https://doi.org/10.1007/s11224-018-1129-x.

Vatanparast, Morteza, and Zahra Shariatinia. 2019a. Hexagonal boron nitride nanosheet as novel drug delivery system for anticancer drugs: Insights from DFT calculations and molecular dynamics simulations. *Journal of Molecular Graphics and Modelling* 89:50–59. https://doi.org/10.1016/j.jmgm.2019.02.012.

Vatanparast, Morteza, and Zahra Shariatinia. 2019b. Revealing the role of different nitrogen functionalities in the drug delivery performance of graphene quantum dots: A combined density functional theory and molecular dynamics approach. *Journal of Materials Chemistry B* 7 (40):6156–6171. https://doi.org/10.1039/C9TB00971J.

Wang, Haitao, Xueyue Mi, Yi Li, and Sihui Zhan. 2020a. 3D graphene-based macrostructures for water treatment. *Advanced Materials* 32 (3):1806843. https://doi.org/10.1002/adma.201806843.

Wang, Heng, Weimin Wang, Hao Wang, Fan Zhang, Yawei Li, and Zhengyi Fu. 2018. Urchin-like boron nitride hierarchical structure assembled by nanotubes-nanosheets for effective removal of heavy metal ions. *Ceramics International* 44 (11):12216–12224. https://doi.org/10.1016/j.ceramint.2018.04.003.

Wang, Junhong, Xianzhao Shao, Junhai Liu, Xiaohui Ji, Jianqi Ma, and Guanghui Tian. 2020b. Fabrication of CdS-SBA-15 nanomaterials and their photocatalytic activity for degradation of salicylic acid under visible light. *Ecotoxicology and Environmental Safety* 190:110139. https://doi.org/10.1016/j.ecoenv.2019.110139.

Wang, Tao, Yuanyuan Liu, Jiajia Wang, Xizhi Wang, Bin Liu, and Yingxu Wang. 2019. In-situ remediation of hexavalent chromium contaminated groundwater and saturated soil using stabilized iron sulfide nanoparticles. *Journal of Environmental Management* 231:679–686. https://doi.org/10.1016/j.jenvman.2018.10.085.

Wang, Xin-ping, Jingwei Hou, Fu-shan Chen, and Xiang-min Meng. 2020c. In-situ growth of metal-organic framework film on a polydopamine-modified flexible substrate for antibacterial and forward osmosis membranes. *Separation and Purification Technology* 236:116239. https://doi.org/10.1016/j.seppur.2019.116239.

Watts, Richard J., Sungho Kong, Margaret P. Orr, Glenn C. Miller, and Berch E. Henry. 1995. Photocatalytic inactivation of coliform bacteria and viruses in secondary wastewater effluent. *Water Research* 29 (1):95–100. https://doi.org/10.1016/0043-1354(94)E0122-M.

Wei, Chang, Wen Yuan Lin, Zulkarnain Zainal, Nathan E. Williams, Kai Zhu, Andrew P. Kruzic, Russell L. Smith, and Krishnan Rajeshwar. 1994. Bactericidal Activity of TiO_2 Photocatalyst in Aqueous Media: Toward a Solar-Assisted Water Disinfection System. *Environmental Science & Technology* 28 (5):934–938. https://doi.org/10.1021/es00054a027.

Wei, Xiao-Na, Hui-Long Wang, Zhen-Duo Li, Zhi-Qiang Huang, Hui-Ping Qi, and Wen-Feng Jiang. 2016. Fabrication of the novel core-shell MCM-41@mTiO_2 composite microspheres with large specific surface area for enhanced photocatalytic degradation of dinitro butyl phenol (DNBP). *Applied Surface Science* 372:108–115. https://doi.org/10.1016/j.apsusc.2016.03.047.

Xiong, Hong-bin, Ya-ni Ma, and Tian-xin Liu. 2020a. Purification-analysis of urban rivers by combining graphene photocatalysis with sewage treatment improvement based on the MIKE11 model. *Environmental Technology*:1–10. https://doi.org/10.1080/09593330.2020.1797897.

Xiong, Jie, Xibao Li, Juntong Huang, Xiaoming Gao, Zhi Chen, Jiyou Liu, Hai Li, Bangbang Kang, Wenqing Yao, and Yongfa Zhu. 2020b. CN/rGO@BPQDs high-low junctions with stretching spatial charge separation ability for photocatalytic degradation and H_2O_2 production. *Applied Catalysis B: Environmental* 266:118602. https://doi.org/10.1016/j.apcatb.2020.118602.

Xu, Shengjie, Feng Li, Baowei Su, Michael Z. Hu, Xueli Gao, and Congjie Gao. 2019. Novel graphene quantum dots (GQDs)-incorporated thin film composite (TFC) membranes for forward osmosis (FO) desalination. *Desalination* 451:219–230. https://doi.org/10.1016/j.desal.2018.04.004.

Xu, Xiu-Dian, Yu Liang, Jun-Feng Li, Lei Zhou, Li-Zhuang Chen, and Fang-Ming Wang. 2020. A stable zinc(II)-organic framework as rapid and multi-responsive luminescent sensor for metal ions in water. *Journal of Coordination Chemistry* 73 (5):867–876. https://doi.org/10.1080/00958972.2020.1746772.

Yan, Wentao, Mengqi Shi, Zhi Wang, Yong Zhou, Lifen Liu, Song Zhao, Yanli Ji, Jixiao Wang, and Congjie Gao. 2019. Amino-modified hollow mesoporous silica nanospheres-incorporated reverse osmosis membrane with high performance. *Journal of Membrane Science* 581:168–177. https://doi.org/10.1016/j.memsci.2019.03.042.

Yang, Fan, Guoyu Ding, Jin Wang, Zihan Liang, Boru Gao, Mengmeng Dou, Ce Xu, and Shuming Li. 2020a. Self-cleaning, antimicrobial, and antifouling membrane via integrating mesoporous graphitic carbon nitride into polyvinylidene fluoride. *Journal of Membrane Science* 606:118146. https://doi.org/10.1016/j.memsci.2020.118146.

Yang, Yang, Xin Li, Chengyun Zhou, Weiping Xiong, Guangming Zeng, Danlian Huang, Chen Zhang, Wenjun Wang, Biao Song, Xiang Tang, Xiaopei Li, and Hai Guo. 2020b. Recent advances in application of graphitic carbon nitride-based catalysts for degrading organic contaminants in water through advanced oxidation processes beyond photocatalysis: A critical review. *Water Research* 184:116200. https://doi.org/10.1016/j.watres.2020.116200.

Yang, Zhong-zhu, Chang Zhang, Guang-ming Zeng, Xiao-fei Tan, Hou Wang, Dan-lian Huang, Kai-hua Yang, Jing-jing Wei, Chi Ma, and Kai Nie. 2020c. Design and engineering of layered double hydroxide based catalysts for water depollution by advanced oxidation processes: A review. *Journal of Materials Chemistry A* 8 (8):4141–4173. https://doi.org/10.1039/C9TA13522G.

Yaripour, Fereydoon, Zahra Shariatinia, Saeed Sahebdelfar, and Akbar Irandoukht. 2015. Effect of boron incorporation on the structure, products selectivities and lifetime of H-ZSM-5 nanocatalyst designed for application in methanol-to-olefins (MTO) reaction. *Microporous and Mesoporous Materials* 203:41–53. https://doi.org/10.1016/j.micromeso.2014.10.024.

Yu, Yalin, Shangdi Wu, Jiayu Gu, Rui Liu, Zhe Wang, Huan Chen, and Fang Jiang. 2020. Visible-light photocatalytic degradation of bisphenol A using cobalt-to-oxygen doped graphitic carbon nitride with nitrogen vacancies via metal-to-ligand charge transfer. *Journal of Hazardous Materials* 384:121247. https://doi.org/10.1016/j.jhazmat.2019.121247.

Yuan, Xiaojiao, Marek P Kobylanski, Zhenpeng Cui, Jian Li, Patricia Beaunier, Diana Dragoe, Christophe Colbeau-Justin, Adriana Zaleska-Medynska, and Hynd Remita. 2020. Highly active composite TiO_2-polypyrrole nanostructures for water and air depollution under visible light irradiation. *Journal of Environmental Chemical Engineering* 8 (5):104178.

Zango, Zakariyya Uba, Nonni Soraya Sambudi, Khairulazhar Jumbri, Noor Hana Hanif Abu Bakar, Nor Ain Fathihah Abdullah, El-Sayed Moussa Negim, and Bahruddin Saad. 2020. Experimental and molecular docking model studies for the adsorption of polycyclic aromatic hydrocarbons onto UiO-66(Zr) and NH2-UiO-66(Zr) metal-organic frameworks. *Chemical Engineering Science* 220:115608. https://doi.org/10.1016/j.ces.2020.115608.

Zhang, Yang, Bo-Tao Zhang, Yanguo Teng, Juanjuan Zhao, Lulu Kuang, and Xiaojie Sun. 2020a. Carbon nanofibers supported Co/Ag bimetallic nanoparticles for heterogeneous activation of peroxymonosulfate and efficient oxidation of amoxicillin. *Journal of Hazardous Materials* 400:123290. https://doi.org/10.1016/j.jhazmat.2020.123290.

Zhang, Jin, Yawei Yang, Jianqiu Zhao, Zhonghua Dai, Weiguo Liu, Chaobo Chen, Song Gao, D. A. Golosov, S. M. Zavadski, and S. N. Melnikov. 2019. Shape tailored Cu_2ZnSnS_4 nanosheet aggregates for high efficiency solar desalination. *Materials Research Bulletin* 118:110529. https://doi.org/10.1016/j.materresbull.2019.110529.

Zhang, Mengxue, Avanthi Deshani Igalavithana, Liheng Xu, Binoy Sarkar, Deyi Hou, Ming Zhang, Amit Bhatnagar, Won Chul Cho, and Yong Sik Ok. 2020b. Engineered/designer hierarchical porous carbon materials for organic pollutant removal from water and wastewater: A critical review. *Critical Reviews in Environmental Science and Technology*:1–34. https://doi.org/10.1080/10643389.2020.1780102.

Zhang, Qingchun, Lei Jiang, Jun Wang, Yongfa Zhu, Yujuan Pu, and Weidong Dai. 2020c. Photocatalytic degradation of tetracycline antibiotics using three-dimensional network structure perylene diimide supramolecular organic photocatalyst under visible-light irradiation. *Applied Catalysis B: Environmental* 277:119122. https://doi.org/10.1016/j.apcatb.2020.119122.

Zhang, Yan, Huimin Ruan, Changmeng Guo, Junbin Liao, Jiangnan Shen, and Congjie Gao. 2020d. Thin-film nanocomposite reverse osmosis membranes with enhanced antibacterial resistance by incorporating p-aminophenol-modified graphene oxide. *Separation and Purification Technology* 234:116017. https://doi.org/10.1016/j.seppur.2019.116017.

Zhao, Die Ling, and Tai-Shung Chung. 2018. Applications of carbon quantum dots (CQDs) in membrane technologies: A review. *Water Research* 147:43–49. https://doi.org/10.1016/j.watres.2018.09.040.

Zhong, Pei, Xinhong Qiu, and Jinyi Chen. 2020. Removal of bisphenol A using Mg-Al-layer double hydroxide and Mg-Al calined layer double hydroxide. *Separation Science and Technology* 55 (3):501–512. https://doi.org/10.1080/01496395.2019.1577265.

Zhou, Liang, Zhihang Liu, Zhipeng Guan, Baozhu Tian, Lingzhi Wang, Yi Zhou, Yanbo Zhou, Juying Lei, Jinlong Zhang, and Yongdi Liu. 2020. 0D/2D plasmonic $Cu_{2-x}S/g\text{-}C_3N_4$ nanosheets harnessing UV-vis-NIR broad spectrum for photocatalytic degradation of antibiotic pollutant. *Applied Catalysis B: Environmental* 263:118326. https://doi.org/10.1016/j.apcatb.2019.118326.

Zolfaghari-Isavandi, Zahra, Ehsan Hassanabadi, Didac Pitarch-Tena, Seog Joon Yoon, Zahra Shariatinia, Jao van de Lagemaat, Joseph M. Luther, and Iván Mora-Seró. 2019. Operation mechanism of perovskite quantum dot solar cells probed by impedance spectroscopy. *ACS Energy Letters* 4 (1):251–258. https://doi.org/10.1021/acsenergylett.8b02157.

Zolfaghari-Isavandi, Zahra, and Zahra Shariatinia. 2018a. Enhanced efficiency of quantum dot sensitized solar cells using Cu_2O/TiO_2 nanocomposite photoanodes. *Journal of Alloys and Compounds* 737:99–112. https://doi.org/10.1016/j.jallcom.2017.12.036.

Zolfaghari-Isavandi, Zahra, and Zahra Shariatinia. 2018b. Fabrication of CdS quantum dot sensitized solar cells using nitrogen functionalized $CNTs/TiO_2$ nanocomposites. *Diamond and Related Materials* 81:1–15. https://doi.org/10.1016/j.diamond.2017.11.004.

Zyoud, Ahed, Maysaa Ateeq, Muath H. Helal, Samer H. Zyoud, and Hikmat S. Hilal. 2019. Photocatalytic degradation of phenazopyridine contaminant in soil with direct solar light. *Environmental Technology* 40 (22):2928–2939. https://doi.org/10.1080/09593330.2018.1459873.

2 Enhancement in Degradation of Antibiotics Using Photocatalytic Semiconductors under Visible Light Irradiation

A Review

Rupali Setia, Harshita Chawla, and Seema Garg
Amity University, Noida, India

CONTENTS

2.1 Introduction 69
2.2 General Mechanism of Degradation of Antibiotics by Visible Light-Active Photocatalysts 71
2.3 Advancements in Photocatalytic Degradation of Antibiotics 72
  2.3.1 Doping Visible Light-Active Semiconductors with Metals and Non-Metals 72
  2.3.2 Heterojunction Formation 74
  2.3.3 Using Visible Light-Active Photocatalysts 75
  2.3.4 Modification of Interface Using Co-Catalysts 76
  2.3.5 Creation of Oxygen Vacancies 78
  2.3.6 SPR-Enhanced Photocatalysts 80
2.4 Conclusion and Future Aspects 84
References 86

2.1 INTRODUCTION

Over the past two decades, water pollution is becoming a major problem for human civilization. The need for clean water is constantly demanding new water treatment methods. We need clean water for most of our basic necessities like personal hygiene,

DOI: 10.1201/9781003118749-2

cooking, and other activities. Clean water is crucial for the survival of the human race. Contaminants, which get discharged from agricultural industries, pharmaceutical industries, and human waste, continue to pollute our water bodies (Verlicchi et al., 2010). Thus, it is the need of the hour to develop new sustainable methods for water treatment (Li and Shi, 2016).

Multiple ways have been constructed for wastewater treatment, including the physical method, biodegradation method, chemical water treatment method, and sludge method. Physical methods include processes such as screening, sedimentation, and skimming. The biological method involves using biological organisms such as microorganisms to break down contaminants present in wastewater. Chemical methods involve using chemicals such as chlorine for the cleansing of wastewater. Sludge treatment includes a liquid-solid separation process, for which the motive is to make the solid phase contain a minimum amount of moisture, and the liquid part carry a minimum amount of solid particles.

A sustainable and economical method for the degradation of organic matter by the use of visible light is known as photocatalysis. Photocatalysis is a method by which chemical reactions can be initiated with the help of visible light. As suggested by Einstein, sunlight contains packets of energy called quanta. This energy can be utilized for the degradation of organic matter. A catalyst that works in the existence of visible light is called a photocatalyst. When the energy supplied by the visible light source is more than the band gap of the catalyst, the electrons and holes get excited and reach the catalyst's surface. Electrons are responsible for starting reductive pathways. During this, O_2 is converted into $^*O_2^-$ radical. This free radical can then further lead to more chemical reactions.

In contrast, the holes present on the top platform of the photocatalyst lead to the reduction of H_2O or the organic contaminants and lead to the formation of *OH. These photocatalytic species thus formed, i.e.,$^*O_2^-$ and $^*OH$, lead to a chain reaction of the degradation of more contaminants (Liu et al., 2013). Since the past few years, titanium dioxide has been used for the photodegradation of organic waste matter (Veréb et al., 2014; Wang et al., 2014b; Gu et al., 2017; Alapi et al., 2018; Yadav et al., 2019). However, problems are being faced due to its wide band gap of 3.2 eV (Yang et al., 2010; Su et al., 2011; Wang et al., 2012; Asahi et al., 2014).

Different ways have been studied to narrow down the band gap of titanium dioxide, which include doping it with metals and non-metals (Jiexiang et al., 2015; Di et al., 2017; Yadav et al., 2021). Apparently, in order to use visible light as much as possible, compounds of bismuth, bismuthoxyhalides, i.e., BiOX (X = I, Br, Cl), have been studied as they are a new group of visible light active photocatalysts (Di et al., 2015; Yu et al., 2017). These compounds have a smaller band gap, which can be a great boon in photocatalysis as the material can capture more sunlight to degrade the pollutant faster. They have a layered structure, so the electron-hole recombination possibility is less. This property makes it a better photocatalyst. BiOI has caught the interest of various researchers because of its tetragonal matlockite geometry displayed by $[Bi_2O_2]^{2+}$ slabs, which are separated in an alternative way by two slabs of iodine atoms. It has been observed that BiOI and BiOBr have a band gap of 1.7–1.9 and 2.8 eV, respectively (Garg et al., 2018; Yadav et al., 2019). Thus, the electrons in the valence band will require less energy to jump from the valence band to the conduction band. Thus, more reaction can occur by using minimum light energy (Batt et al., 2006; Li and Shi, 2016; Sápi et al., 2021).

A world without pharmaceuticals cannot be imagined. They have become a major part of our lives. When they enter the human body, certain parts are metabolized in the body, whereas some parts get discharged in the form of human waste which reaches water bodies (Azimi and Nezamzadeh-Ejhieh, 2015; Wang et al., 2017; Yang et al., 2017). Unfortunately, most of these pharmaceuticals are uncooperative and common methods of wastewater treatment do not work on them (de Witte et al., 2011; Hu et al., 2020). Antibiotics are a group of drugs that have antibacterial properties. Some commonly used antibiotics are sulfamethazine, diclofenac, norfloxacin, carbamazepine, sulfamethoxazole, and ciprofloxacin. Ciprofloxacin is a powerful antibiotic that is commonly used in households. This drug reaches water bodies via human waste. Since this drug is very stable and microorganisms cannot degrade it, it leads to bioaccumulation in the food cycle and proves to be a threat to human life (Jiang et al., 2019). Photocatalysis is a great method to degrade antibiotics present in wastewater (Yang et al., 2021). It converts antibiotics into photoactive species ($^*O_2^-$ and $^*OH$), further degrading increasing numbers of antibiotic molecules (Ming Yan et al., 2016; Yan et al., 2017). Thus, photocatalysis is efficient, cost-effective, and the most suitable method for the erosion of antibiotics (Wang et al., 2017).

2.2 GENERAL MECHANISM OF DEGRADATION OF ANTIBIOTICS BY VISIBLE LIGHT-ACTIVE PHOTOCATALYSTS

The following steps are followed for the degradation of antibiotics:

i. When light beams of higher energy than that of the band gap of the photocatalytic compound fall onto the photocatalytic compound, there is a separation of electrons and holes. Electrons possess reducing properties, and the holes have oxidizing properties. Both of them reach the conduction band by crossing the valence band. This is the top surface of the photocatalyst. This process is known as migration.
ii. During migration, some of the electrons and holes recombine and produce energy. This energy is called thermal energy, which is released as heat.
iii. On reaching the surface of the catalyst, reduction begins by a pathway called the reduction pathway. During this, O_2 is converted into $^*O_2^-$ radical by the electrons. This free radical can then further lead to more chemical reactions.
iv. Meanwhile, the holes of the photocatalytic substance help reduce H_2O or the antibiotics and lead to the formation of $^*OH$ (Li and Shi, 2016).

The general mechanism is displayed in Figure 2.1 as follows:

These photocatalytic species so formed, i.e., $^*O_2^-$ and $^*OH$ lead to the chain reaction for the degradation of more antibiotic molecules. All these steps affect the overall photocatalytic efficiency and the rate of degradation of the photocatalyst. Thus, the goal is band-gap modification of the photocatalyst because the narrower the band, the less the energy required, lowering the rates of recombination of holes and electrons, and producing more photocatalytic species for increased degradation of

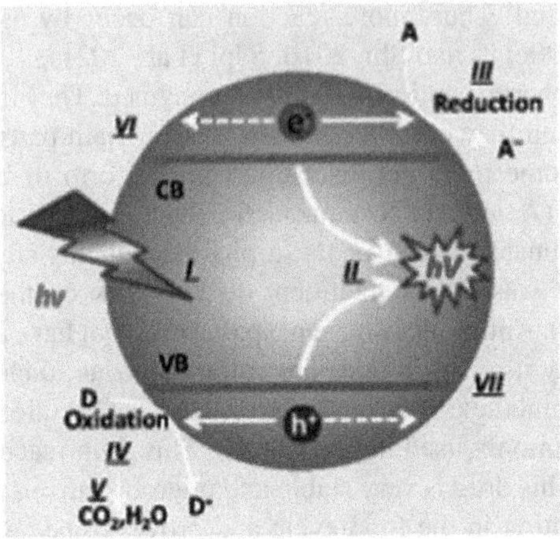

FIGURE 2.1 Basic mechanism of photodegradation of antibiotics by the process of photocatalysis ref. (Li and Shi, 2016. Reprinted with permission from Elsevier.)

antibiotics (Li and Shi, 2016). We can make optimum use of visible light and photocatalyst by reducing the band gap of photocatalyst by applying certain material inside the conduction and valence band that can trap photoexcited holes and electrons and prevent their recombination (Topkaya et al., 2014; Wang et al., 2014a).

2.3 ADVANCEMENTS IN PHOTOCATALYTIC DEGRADATION OF ANTIBIOTICS

Over the past few decades, researchers have been focusing on new ways to treat wastewater containing antibiotics (Wang et al., 2014a; Ghosh et al., 2015; Huang et al., 2017; Hunge et al., 2017, 2018). As discussed earlier, antibiotics are highly stable, and conventional methods of wastewater treatment doesn't degrade them. Here follows other ways to treat antibiotic-containing wastewater, using visible light active semiconductors with metals and non-metals, heterojunction formation, using visible light active photocatalysts, modification of interface using co-catalysts, creation of oxygen vacancies, and surface plasmon resonance (SPR) enhanced photocatalysts. These are discussed next.

2.3.1 Doping Visible Light-Active Semiconductors with Metals and Non-Metals

Visible light active semiconductors such as titanium dioxide face problems of having a large band gap. As a result, its photocatalytic activity is confined only under sunlight. Doping titanium dioxide with metals and metal ions on the surface can help narrow down its band gap, i.e., increasing its photocatalytic efficiency (Ni et al., 2007). When a semiconductor is doped with a metal ion, the metal ion acts as a photoinduced charge trap, i.e., it can trap electrons and holes and thus control the recombination rate (Choi et al., 1994). However, problems can occur due to excess electrons (or holes) and the low number of recombination centers, which can cause a reduction in photocatalytic activity. Henceforth, metal or non-metal are always doped at the exterior of visible light active photocatalysts. The following chemical reactions explain impurity level formation and charge entrapping (Choi et al., 1994; Li and Shi, 2016).

Impurity level formation:

$$M^{n+} + hv \rightarrow M^{(n+1)} + e_{CB}^{-} \tag{2.1}$$

$$M^{n+} + hv \rightarrow M^{(n-1)} + h_{VB}^{+} \tag{2.2}$$

Charge entrapping:

$$M^{n+} + e_{CB}^{-} \rightarrow M^{(n-1)+} \tag{2.3}$$

$$M^{n+} + hv_{VB}^{+} \rightarrow M^{(n+1)^{+}} \tag{2.4}$$

Here M^{n+} and M represents doped metal ions and a metal and conduction band (CB), while the valence band (VB) symbolizes the conduction band and valence band, respectively (Duan et al., 2012).

Vignesh et al. (2014) prepared TiO_2 nanoparticles by modifying them with zinc phthalocyanine (Znpc–TiO_2) using the chemical impregnation method (Kim et al., 2004; Vignesh et al., 2014) so as to increase the photocatalytic power of titanium dioxide in the presence of visible light. Photocatalytic activity of the synthesized Znpc–TiO_2 nanoparticles was checked by degradation of erythromycin (a drug with similar properties to penicillin) under visible light irradiation by the process of photocatalysis. The results revealed that Znpc–TiO_2 nanoparticles possess a larger surface area than TiO_2 nanoparticles. Also, Znpc–TiO_2 nanoparticles have higher photocatalytic activity (74.21%) than TiO_2 nanoparticles (31.57%). Also, it was found that this compound was stable and reusable for several processes. Non-metal ions act as better dopants because they are less capable of capturing photoinduced charges. Hence, non-metal ions help in inclining the photocatalytic efficiency of a photocatalyst. Although, inducing metal ions can also help increase the photocatalytic power of a photocatalyst up to a certain order (Chen and Chu, 2012).

Carbon doped titanium dioxide (C-TiO_2) was developed, which was used for the photodegradation of norfloxacin. Norfloxacin is an antibiotic employed to heal common as well as complex urinary tract infections. C-TiO_2 catalyst was prepared by carbonization of TiO_2. In order to draw a contrast with TiO_2, the photocatalytic activities of both the photocatalysts were checked in parallel. It was reported that 78% of norfloxacin was degraded by C-TiO_2 within 70 minutes. The decay curve showed pseudo-first-order kinetics. On the contrary, TiO_2 (pure) was able to use 25% of norfloxacin under similar conditions (Vignesh et al., 2014).

Thus, we can say that a doping photocatalyst with metal and non-metal ions is very beneficial as it not only showed better photocatalytic activity, but also higher stability (Shi et al., 2020).

2.3.2 Heterojunction Formation

The formation of catalysts by heterojunction is a great way of developing photocatalysts with enhanced properties (Li et al., 2021b). This involves combining two semiconductors with similar band gap energy (Liu et al., 2014). The advantage it offers is that it enables the separation of photoinduced charges and thus boosts photocatalytic efficiency. There are three main heterojunction systems Type I, Type II, and Type III (Moniz et al., 2015; Murillo-Sierra et al., 2021). The most common out of the three types is known as the Type II heterojunction system. It is proven to be the best among the three types because it prevents the regrouping of electrons with holes, thus hiking the photocatalytic efficiency of the photocatalyst (Li and Shi, 2016). The basic mechanism of Type II catalysis is depicted in Figure 2.2. As we can see, semiconductor 1 emits photoexcited electrons, which reach semiconductor 2, which possess negatively charged VB. Due to the positively charged VB of semiconductor 1, photoexcited holes will move from semiconductor 2 to semiconductor 1. Henceforth, there is a complete separation of photoinduced charges, resulting in enhanced photocatalytic activity (Li and Shi, 2016).

Guo et al. (2013) synthesized a Co_3O_4 catalyzed peroxymonosulfate system for the degradation of amoxicillin. Amoxicillin is one of the most commonly used drugs used in the treatment of humans and animals. It is used to heal infections occurred due to microorganisms. The motive of their study was to understand how the parameters like temperature, catalyst dosage, pH, and oxidant concentration affect the rate of amoxicillin degradation. The ideal conditions were reported to be as: pH = 6.0, reaction temperature = 60°C, Co_3O_4 dosage = 0.06 g, oxone concentration = 0.01 mol L^{-1}, and reaction time = 45 minutes (Ibrahim et al., 2017).

In 2019, Peyman Gholami et al. prepared biochar (BC)-incorporated Zn-Co layered double-hydroxide (LDH) nanostructure for the photodegradation of gemifloxacin (GMF). According to WHO, the presence of gemifloxacin in the form of quinoline antibiotics is toxic and causes the deaths of humans and animals. It also causes antibiotic resistance and genotoxicity (Acisli et al., 2017; Darvishi Cheshmeh Soltani

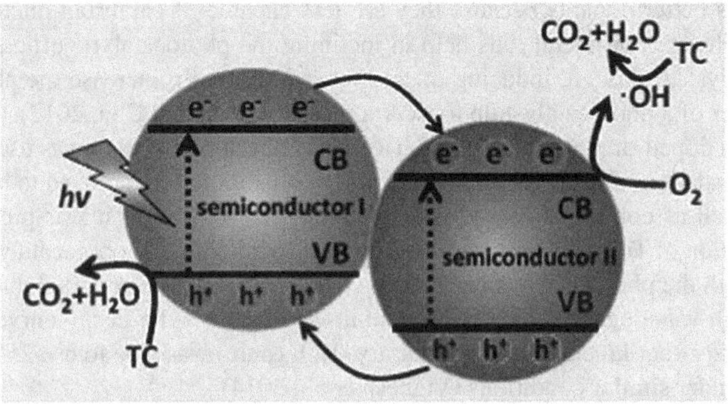

FIGURE 2.2 Demonstration of Type II heterogeneous semiconductor mechanism. (Li and Shi, 2016. Reprinted with permission from Elsevier.)

and Mashayekhi, 2018). Advanced oxidation processes convert harmful and toxic factory wastes into something less harmful for the environment (Gholami et al., 2019). Zn-Co-LDH@BC nanocomposite displayed great photocatalytic activity and efficiency caused because of excellent separation of photogenerated carriers. The method used for synthesizing the nanocomposite was the hydrothermal method. It was reported that Zn-Co-LDH photocatalyst was able to degrade 92.7% of GMF by photocatalysis. The factors affecting the photocatalytic efficiency of the catalyst were: solute concentration and the photocatalyst dosage. It was observed that when the reactor carrying Zn-Co-LDH @BC nanostructures was oxygenated, there was a hike in the photodegradation of Gemifloxacin (Wang et al., 2014b; Xue et al., 2014; Simon et al., 2018; Gholami et al., 2019).

2.3.3 Using Visible Light-Active Photocatalysts

Visible light active photocatalysts have attained immense interest from many researchers. Their capabilities of being highly stable and showing great photocatalytic efficiency are the two main vital reasons for their popularity. Titanium dioxide is the most common visible light-active photocatalyst. Its activity is poor because it has a wide band gap. Due to this, its photocatalytic activity is restricted only within visible light limits Yadav et al., 2020). Different ways have been formed so that its photocatalytic efficiency can be hiked, but there is also an urge to produce novel visible light active photocatalytic agents (Veréb et al., 2014). Some researchers are investigating a new class of photocatalyst, which is made from compounds of bismuth.

Durán–Álvarez et al. (2020) prepared different kinds of bismuth oxoiodides (BiOI, $Bi_4O_5I_2$, $Bi_7O_9I_3$ and Bi_5O_7I) by solvothermal method. The precursors used for the synthesis were Bi $(NO_3)_3.5H_2O$ and KI (Bárdos et al., 2020). They obtained BiOI in the form of nanocrystals and microspheres (MS) (Li et al., 2014; Durán-Álvarez et al., 2020). The decrease in iodine loading was made by changing the following parameters: temperature, solvent, and heating time (Chen et al., 2018). They used the deposition-precipitation method for the deposition of gold nanoparticles on the $Bi_xO_yI_z$ materials. Due to a decrease in the amounts of iodine, the physical properties, electrochemical, and optical properties of the bismuth oxyhalide materials changed. Photocatalytic activity of the synthesized bismuth oxyhalides was proceeded by degradation of oxytetracycline. Oxytetracycline is an antibiotic that belongs to the tetracycline family. This drug is used as a veterinary drug in Europe and the USA. It is being reported that 2500 tons of tetracycline are used each year to treat animals (Halling-Sørensen et al., 2003). The problem arises from the fact that the hydrochloric salt of oxytetracycline is highly polar and thus, results in water pollution. Thus, the goal of their project was to photocatalytically degrade oxytetracycline in tap water. The photocatalytic activity was checked by photodegradation of oxytetracycline (250 mL of 30 gL^{-1}) under visible light (Veréb et al., 2014). $Bi_xO_yI_z$ materials proved to be great catalysts and the amount of iodine present in the material influenced the catalytic performance of the material. The photocatalytic reaction was conducted for 5 hours each time. BiOI had the lowest degradation rate and it degraded 51% of oxytetracycline. Bi_5O_7I degraded 46% of oxytetracycline and proved to have a poor photocatalytic efficiency after 5 hours of irradiation. $Bi_4O_5I_2$, $Bi_7O_9I_3$

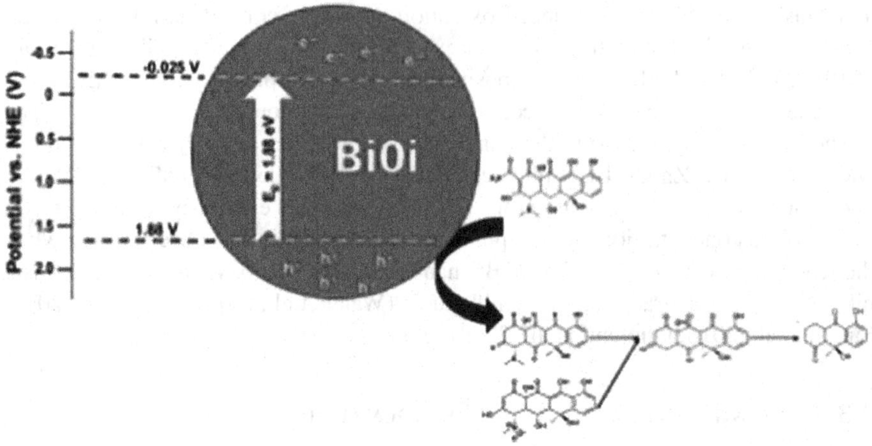

FIGURE 2.3 Mechanism of photodegradation of oxytetracycline using BiOI. (Durán-Álvarez et al., 2020. Reprinted with permission from Elsevier.)

displayed similar degradation rates of oxytetracycline, around 75% (Wen et al., 2017.) BiOI microspheres had the highest degradation rate of 84% for the degradation of oxytetracycline by supplying visible light for a time duration of 5 hours. The degradation rate of Au/BiOI was found to be lower than that of pure BiOI microspheres. The reason was that the gold nanoparticles could have possibly blocked the reaction sites of the BiOI mesospheres, thereby causing a fall in the photocatalytic efficiency (Durán-Álvarez et al., 2020).

Figure 2.3 shows the mechanism of the degradation of oxytetracycline by using BiOI as a photocatalyst. BiOI adsorbs visible light falling on it. Due to this, electrons and holes get excited and jump from the valence band (VB) to the CB. BiOI has1.88 eV as the band gap energy. Some energy is also released in heat because of the regrouping of holes with electrons in the CB. The electrons start a reductive pathway and convert O_2 into $^*O_2^-$, a photoactive species which will start a chain reaction. On the other hand, the holes create an oxidative pathway and convert water molecules or the antibiotic (oxytetracycline) into OH^*. These photoactive species ($^*O_2^-$ and OH^*) start up a chain reaction and thus degrade more and more molecules of oxytetracycline. (Hu et al., 2020; Verma and Haritash, 2020).

2.3.4 Modification of Interface Using Co-Catalysts

Catalysts used for photocatalysis have caused issues recently due to numerous problems, which include low photocatalytic efficiency. Commonly used photocatalysts such as TiO_2, ZnO_2, Fe_2O_3, WO_3, and CdS are struggling with the same problems. Titanium dioxide is the most common material employed for photocatalysis due to its great stability, resistance from photo-corrosion, great oxidative properties, and high chemical stability.

Oros-Ruiz et al. (2013) deposited Ag, Ni, and Cu nanoparticles on a TiO_2-P25 catalyst and studied how it affected its photocatalytic power (Sirtori et al., 2010). The weight of metal loading was kept under 0.5wt.%. The average size of the nanoparticle was kept at 5 nm. They characterized the obtained materials by XRD, EDS, ICP UV-vis, TPR, and TEM methods. They observed that the deposition of the nanoparticles helped in improving the photocatalytic efficiency of the TiO_2-P25 catalyst. Its photocatalytic efficiency was tested on 2,4-diamino-5-(3,4,5-trimethoxybenzoyl) pyrimidine, commonly known as trimethoprim. Trimethoprim is an antibiotic generally used along with sulfamethoxazole to heal many types of infections, including ear and urinary tract infections. Its single dose is excreted by 40–60% in the form of urine by humans (Siemens et al., 2008). Thus, this drug leads to water pollution. The team conducted the first study on the interaction of metallic nanoparticles on the surface of the titanium dioxide photocatalyst. They prepared Ni/TiO_2-P25, Au/ TiO_2-P25, and Cu/ TiO_2-P25 using deposition-precipitation with urea (DPU) in the dark. Due to urea, pH reached the value of 7.3. Ag/TiO_2-P25 was prepared using deposition-precipitation using NaOH as it was not getting deposited with the help of urea. Photodegradation of trimethoprim of concentration 40 ppm was conducted in the presence of UV light irradiation with the source as UV lamp (primary emission = 254 nm and 2.2 mW/cm^2). It was conducted under pHv value of 6. It was observed that UV light was able to degrade 20% of trimethoprim, and TiO_2-P25 was able to degrade 54% of trimethoprim in a time duration of 300 minutes. Cu/TiO_2-P25, Ni/ TiO_2-P25, Au/TiO_2-P25 and Ag/TiO_2-P25 were able to degrade 72%, 64%, 81%, and 78% of the antibiotic under the same set of conditions, respectively. Figure 2.4 shows the UV-visible spectra of the synthesized photocatalysts (Oros-Ruiz et al., 2013).

The table shown below (Table 2.1) summarizes the rate constant, particle size, band gap energy, photocatalyst, and metal loading.

It can be concluded that the photocatalytic property of TiO_2 was thus modified by the addition of metal nanoparticles on its surface. The photocatalytic activity of the photocatalysts was found by the trimethoprim photodegradation. The photocatalytic

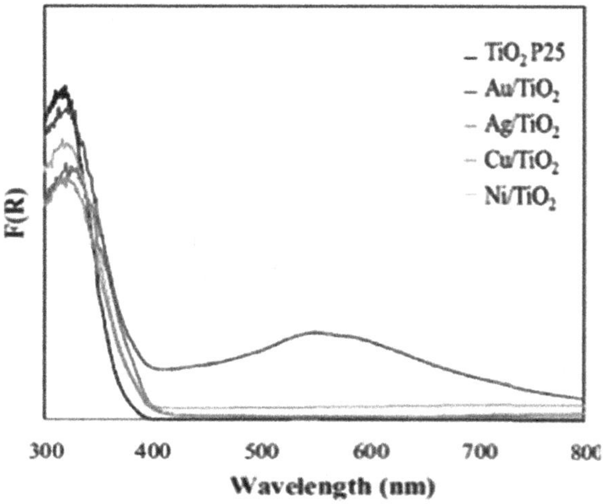

FIGURE 2.4 UV-visible spectra of synthesized samples ref. (Oros-Ruiz et al., 2013. Reprinted with permission from Elsevier.)

TABLE 2.1
Characteristic Details of the Synthesized Samples for the Photodegradation of Trimethoprim

Photocatalyst	E_g (eV)	Metal Loading (wt.%)	Size of Particle (nm)	Surface Area (m^2/g)	Rate Constant k (min^{-1})
UV light	–	–	–	–	0.64×10^{-3}
TiO_2-P25	3.1	–		57	02.4×10^{-3}
Au/TiO_2-P25	2.8	0.48	2.0 + 0.3	52	05.2×10^{-3}
Ag/TiO_2-P25	2.9	0.49	1.8 + 0.26	53	05.2×10^{-3}
Cu/TiO_2-P25	2.9	0.47	2.1 + 0.15	57	04.5×10^{-3}
Ni/TiO_2-P25	2.9	0.45	1.8 + 0.16	52	03.0×10^{-3}

Source: Taken from S. Oros-Ruiz, et al., Photocatalytic degradation of trimethoprim by metallic nanoparticles supported on TiO_2-P25, (J. Hazard. Mater. 2013. Reprinted with permission from Elsevier.)

efficiencies of the samples are in this order: Ni<Cu< Ag ≈ Au. Electron affinity and electronegativity are two main factors that influence the degradation of trimethoprim because they collect electrons, which helps create more OH^* radicals, which in turn causes oxidation of the drug trimethoprim (Oros-Ruiz et al., 2013).

2.3.5 Creation of Oxygen Vacancies

Photocatalysis is a great method for the degradation of antibiotics with the help of light. This method of degradation is renewable, clean, and energy-efficient (Elward and Chakraborty, 2013). The necessary condition is that the redox potential of the antibiotic should be lower or equal to that of band gap energy of the photocatalytic agent (Iwase et al., 2011; Zhou et al., 2014; Li et al., 2021a; Shao et al., 2021). Heterogeneous catalysts can overcome this problem to a certain extent by degradation of antibiotics under visible light by photocatalysis. It has also been proven that heterogeneous catalysts offer better electron-hole separation.

Oxygen vacancies act as electron mediators, which means that they help in increasing the production of photoinduced charges, which are electrons and holes (Iwase et al., 2011; Dong et al., 2013; Ahmed et al., 2015; Jo and Sivakumar Natarajan, 2015). These photo holes and photoelectrons are accountable for redox reactions. Thus, photoexcited charges transfer from the valence to the CB, where they start redox reactions. Electrons are responsible for reductive pathway and covert O_2 into $^*O_2^-$, a photoactive species which will start a chain reaction. On the other hand, the holes start up an oxidative pathway and convert water molecules or the antibiotic into OH^*. These photoactive species ($^*O_2^-$ and $OH\cdot$) start up a chain reaction and thus degrade more and more antibiotic molecules.

The Z-scheme is a great method to improve heterogeneous catalysts, aiding the photocatalytic efficiency of catalysts, and shows high charge separation (Ding et al., 2017). Also, it helps the catalysts to show strong redox ability. In 2017, Jie Ding et al. developed a Z-scheme $BiO_{1-x}Br/BiO_2CO_3$ photocatalyst with oxygen vacancy

TABLE 2.2
The Detailed Experimental Parameters[a]

Sample Name	Molar Ratio of Precursors ($NaBr/Bi(NO_3)_3$)	Reaction Duration (hours)
BRC-1	0.5	6
BRC-2	0.5	3
BRC-3	1	6
BRC-4	1	3
BRC-5	0	6

[a] The amount of $Bi(NO_3)_3$ is 1 mmol, reaction temperature is 180° C and the solvent is mannitol aqueous solution (25 mL, 0.1 M)Taken from Ding, et al. (2017). "Z-scheme BiO1-xBr/Bi2O2CO3 photocatalyst with rich oxygen vacancy as electron mediator for highly efficient degradation of antibiotics." *Applied Catalysis B: Environmental* **205**: 281–291 (Reprinted with permission from Elsevier).

using facile-time dependent solvothermal method for photodegradation of antibiotics (CIP, 4MMA, etc.) by irradiation of visible light. They synthesized multiple $BiO_{1-x}Br/BiO_2CO_3$ photocatalyst samples by changing the following parameters: molar ratio and reaction time, and using NaBr and $Bi(NO_3)_3$ as precursors. This is shown in Table 2.2 as above.

Photocatalytic activities were checked by photodegradation of BPA (20 mg/L), CIP (40 mg/L) and 4 MMA (50 mg/L) by irradiation of visible light. The process involved dissolving photocatalyst (0.001 g) in antibiotic solution (40 mL), followed by placing it in the dark for 2 hours to allow desorption-adsorption to occur. Then, the solution was subjected to visible light in order to photodegradation to occur. Figure 2.5a shows a graph between CIP concentration (C_t/C_0) versus time of irradiation for all the samples. It shows that BRC-1 had the highest photodegradation rate as compared to all other samples. It also showed that photocatalyst like BiOBr/ BiO_2CO_3 hetero-catalyst has high rate of photocatalysis. It was demonstrated that synthesized catalysts followed pseudo-order of reaction as shown in Figure 2.5b. The equation of pseudo-order reaction is as follows,

$$\ln\left(\frac{C_o}{C_t}\right) = kt \tag{2.5}$$

It was found that $BiO_{1-x}Br/BiO_2CO_3$ nanocomposite with rich oxygen vacancy showed higher photocatalytic activity than non-oxygen vacancy containing BiOBr/ BiO_2CO_3 catalyst, $BiO_{1-x}Br$, and BiO_2CO_3. Only $BiO_{1-x}Br/BiO_2CO_3$ could completely degrade the antibiotics found by TOC (total organic carbon) removal analysis (Jo and Sivakumar Natarajan, 2015; Ding et al., 2017). The oxygen vacancy was the reason for the high photocatalytic activity, working as an electron mediator for the formation of photoactive species ($^*O_2^-$& OH^*), and helping to maximize the redox reaction rate (Linic et al., 2011; Ding et al., 2017).

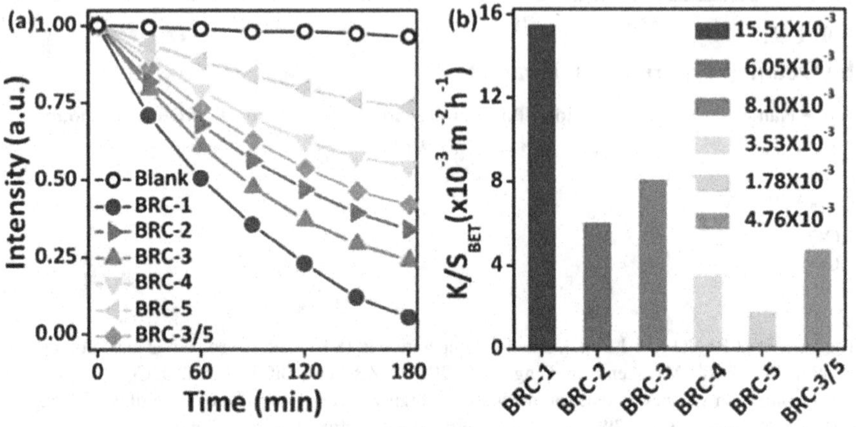

FIGURE 2.5 (a) Photodegradation of CIP by samples and (b) pseudo-order rate constants of the samples ref. (Ding et al., 2017. Reprinted with permission from Elsevier.)

2.3.6 SPR-Enhanced Photocatalysts

SPR-enhanced photocatalysts like Au and Ag are gaining much popularity over the past few decades. Due to their upgraded photocatalytic efficiency, they are being recognized as a new class of visible light photocatalysts. SPR is a phenomenon in which the frequency of the incident radiation matches with the collective oscillation of the electrons present in the CB of an atom of metal. Due to this collective oscillation of free charges, there is the development of plasmons. A surface plasmon is formed because of the oscillation of free charges on the surface of metal atoms. The surface plasmon has great interactive properties with visible light, which lays the principle of SPR (Li and Shi, 2016). Figure 2.6 shows the mechanism of photogenerated transfer of charge in an SPR-enhanced system.

When an incident light possesses photons of frequency equal to that of the total oscillation of electrons of the metal atom strike on the outer boundary of the metal atom, the electrons transfer from the valence band of atom to the conduction band of

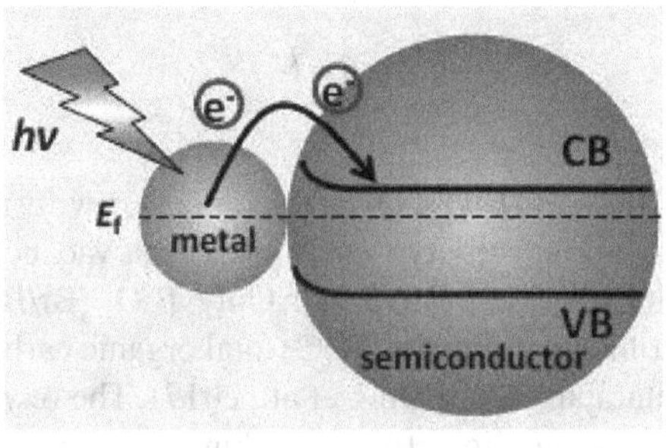

FIGURE 2.6 Schematic representation of photogenerated charge transfer in a plasmonic photocatalyst system (Li and Shi, 2016. Reprinted with permission from Elsevier.)

the photocatalyst. This will increase the number of electrons on the conduction band, which is the photocatalytic part of the photocatalyst (Li and Shi, 2016; Zhou et al., 2017b). Gold and silver are the two elements that show plasmon resonance (Li et al., 2008; Zhou et al., 2017a).

Ren et al. (2020) prepared novel Ag/g-C_3N_4 nanosheets. They emphasized sphotodegradation of three types of tetracycline antibiotics (TCs), which are oxytetracycline (OTC), tetracycline (TC), and chlortetracycline (CTC) using novel Ag/g-C_3N_4 nanosheets (Liu et al., 2016; Sarmah et al., 2006) These three types of tetracyclines belong to the tetracycline family, used globally for treatment and growth purposes. It has been reported that a high amount of tetracycline antibiotics is expelled by humans in the form of urine, which reaches water bodies and leads to the development of antibiotic-resistant genes and bacteria (Zhang et al., 2012). In pig and chicken manure, the amounts of TC, OTC, and CTC were found to be 3–25 mg/kg in Beijing (Hollmann et al., 2014; Mousavi and Habibi-Yangjeh, 2018). g-C_3N_4 is a robust semiconductor with high thermal stability and a band gap of 2.7 eV, making it an ideal semiconductor for photocatalysis in the presence of visible light (Fu et al., 2019a, 2019b). Its high rate of repairing electrons with holes is one of the reasons why there was a need to develop an upgraded form of this catalyst. Thus, modifying g-C_3N_4 with Ag is a great step because this shows SPR. It also reduces the rate of recombination of photoexcited electrons with holes (Chi et al., 2016; Lu et al., 2017; Asadzadeh-Khaneghah et al., 2018; Xu et al., 2018; Sharma et al., 2019; Li et al., 2019; Qin et al., 2019; Ren et al., 2020). In their work, nanosheets of Ag/g-C_3N_4were prepared. The technique involved thermo-polymerization of urea as the precursor followed by photo-deposition of evenly dispersed Ag nanoparticles. The factors which influenced the photocatalytic activity of nanosheets were pH and the amount of Ag loading. The surface, optical, component, and electrochemical properties were checked using XRD, XPS, TEM, BET, FTIR, UV-vis DRS, EIS, and I-T responsive curve. The photocatalytic activity was conducted by degradation of the three TCs at 25°C. The process involved dissolving 0.1 g of catalyst into 100 mL TCs solution (20 mg/L) and ultrasonicate it for thirty minutes. This was followed by adjusting its pH to 7 and mixing the solution on a magnetic stirrer for thirty minutes in the dark at room temperature. This process allowed desorption-adsorption processes to happen. The visible light source was a Xe lamp (300 nm), and a filter of 420 nm was used. 5 mL of the solution was collected out after definite time sets, and a filter of 0.45 μm was used to filtrate the catalyst from the solution. The amounts of TCs present were investigated with the help of liquid chromatography. Figure 2.7a shows that in the absence of light, the amount of TC decreased slightly due to the adsorption of TC on the photocatalyst. Figure 2.7b depicts the photocatalytic performance of Ag/g-C_3N_4 nanosheets over regular g-C_3N_4, Ag(12%)/g-$C_3N_{4,}$ Ag(8%)/g-C_3N_4 and Ag(4%)/g-C_3N_4 were able to degrade 35%, 77%, 83% and 70% of TC respectively within 120 minutes of exposure to visible light. It was found that the maximum degradation rate was of Ag(8%)/g-$C_3N_{4,}$ and there was increasing order of photocatalytic activities with metal loading. Similar results were observed for photodegradation of oxytetracycline and chlortetracycline displayed in Figure 2.7c–f. Ag(8%)/g-C_3N_4 was able to degrade 85% and 81% of CTC and OTC, respectively.

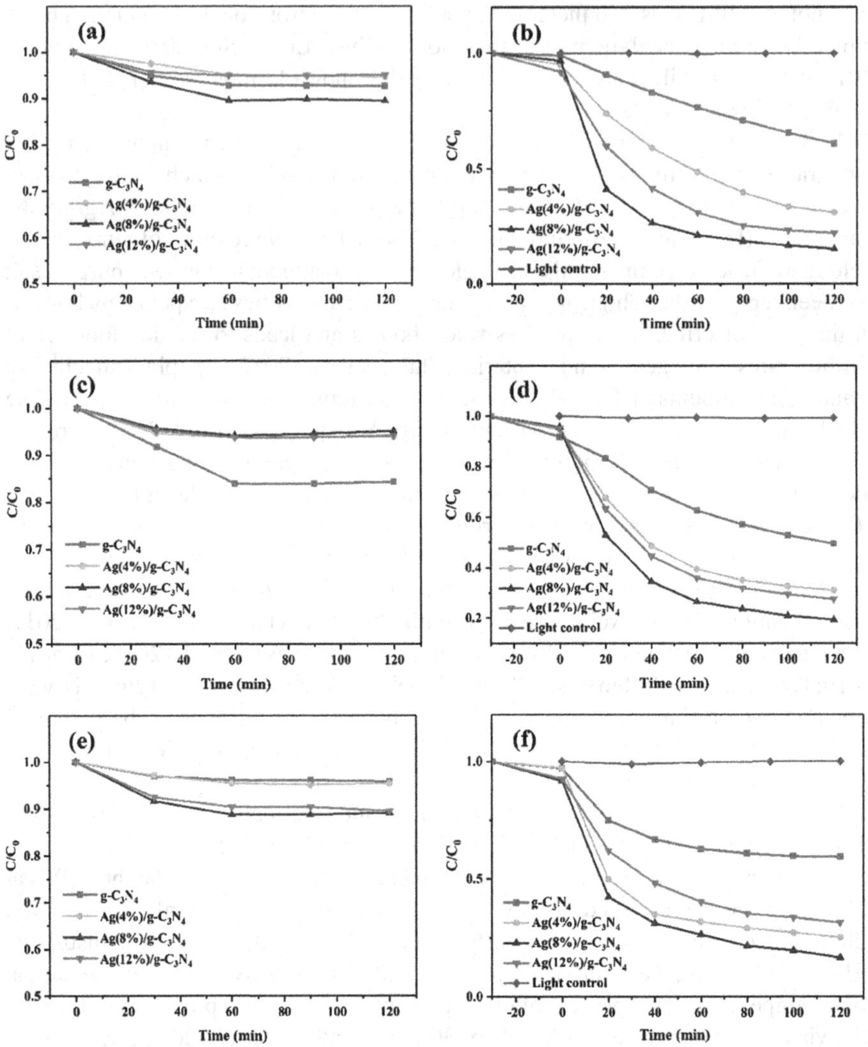

FIGURE 2.7 Degradation performance of tetracycline in the dark (a), degradation performance of tetracycline (b), degradation performance of oxytetracycline in the dark (c), degradation performance of oxytetracycline (d), degradation performance of chlortetracycline in the dark (e), degradation performance of chlortetracycline (f) with synthesized catalysts (1 mg/L) in the presence of visible light. (Guo et al., 2013; Moniz et al., 2015; Ren et al., 2020. Reprinted with permission from Elsevier.)

The influence of pH was studied in a series of experiments carried on Ag(8%)/g-C_3N_4 on tetracyclines (20 mg/L). The pH was gradually hiked from 3.00 to 11.00, and so was the degradation rate. At pH 11, the degradation rate was maximum, proving that the basic medium promotes photodegradation. At pH < 4, Ag(8%)/g-C_3N_4 surfaces acted as being positively charged, and at a pH of more than 4, the top layer of Ag(8%)/g-C_3N_4 acted as being negatively charged. Figure 2.8d displays pH_{pzc},

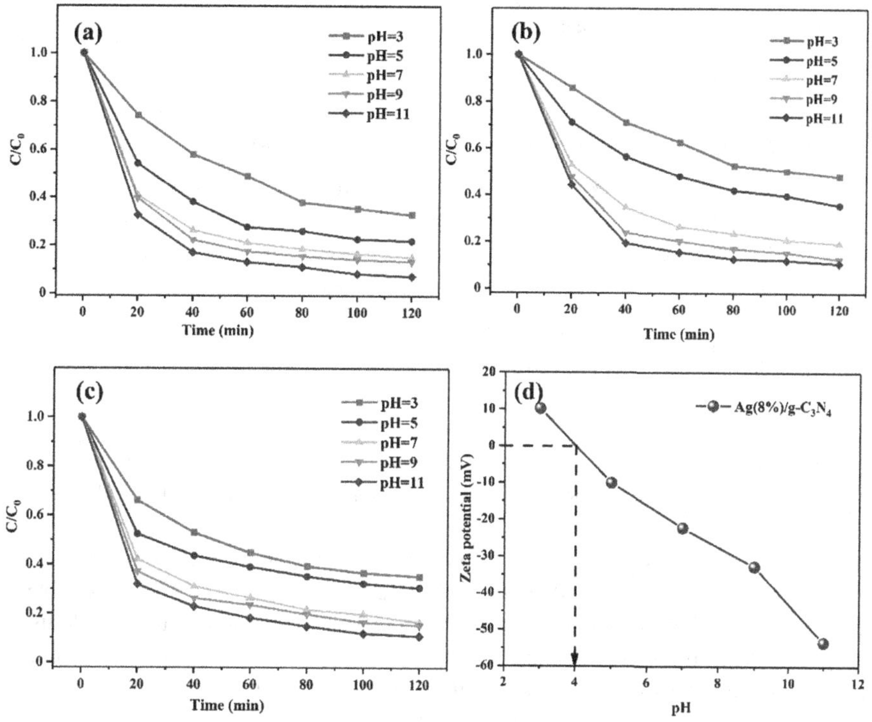

FIGURE 2.8 Effect of pH on photodegradation of TC (original concentration = 20 mg/L) (a), on photodegradation of OTC (original concentration = 20 mg/L) (b), photodegradation of CTC (original concentration = 20 mg/L) (c), in presence of Ag(8%)/g-C3N4 (1 mg/L); zeta-potential plot of Ag(8%)/g-C3N4 (d). (Ren et al., 2020. Reprinted with permission from Elsevier.)

point of zero charge, of Ag (8%)/g-C_3N_4, which was reported as 4. At pH = 3, Ag(8%)/g-C_3N_4 was acting as cations which faced electrostatic repulsion with oxygen carrying-positive charged groups, which acts as a barrier for adsorption of antibiotics by the photocatalyst. After increasing the pH value to 5, the electrostatic attraction between tetracyclines (positively charged) and the Ag (8%)/g-C_3N_4 (negatively charged) allows tetracyclines to be adsorbed over the catalyst, resulting in improving its catalytic efficiency. On the other hand, at pH values greater than or equal to 7, the tetracycline molecules possess a negative charge, hence causing electrostatic repulsion between tetracycline molecules. At the same time, the Ag (8%)/g-C_3N_4 catalyst can decrease in the adsorption of tetracyclines. Apparently, with the hike of pH, a greater number of OH^- can cause chemical reactions with photogenerated holes and $^*OH$ (hydroxyl radicals). Thus, $^*OH$ radicals can, in turn, break down more molecules of tetracyclines. Consequently, the degradation strength of a catalyst to degrade tetracyclines was sequentially enhanced with the increase of pH by shifting from 7 to 11. The following chemical reaction shows the reaction of photogenerated holes to develop photoactive species, i.e., $^*OH$ radical (Ren et al., 2020):

$$h^+ + OH^- \rightarrow {}^* OH \tag{2.6}$$

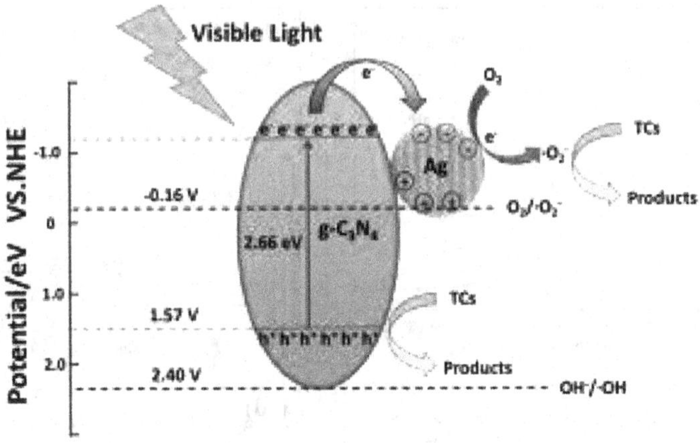

FIGURE 2.9 Schematic representation of photodegradation of TCs by Ag/g-C3N4 photocatalyst under visible light ref. (Ren et al., 2020. Reprinted with permission from Elsevier.)

Based on reactive species studies, a possible mechanism of photodegradation of tetracyclines by Ag/g-C_3N_4 is depicted in Figure 2.9. According to a study, the band gap of g-C_3N_4 (2.70 eV) is similar to UV-vis (2.66 eV). After absorbing visible light energy, electrons transfer from the valence to the conduction band. This leaves vacant spots known as holes (h^+), which are left in the valence band of the photocatalyst. The electrons, which are photoexcited, transfer to the conduction band of g-C_3N_4 will jump to the Ag nanoparticles because Ag has low fermi energy level. The electrons begin the reduction process, converting O_2 into $^*O_2^-$, a photoactive species which will start a chain reaction. The oxidation potential of $OH^-/^*OH$ is 2.40 V, greater than that of the valence band of g-C_3N_4 = 1.57 V; hence, the holes present in the valence band of g-C_3N_4 are not able to react with ^-OH to form hydroxyl radical. It was demonstrated by a radical trapping experiment. The photodegradation of TCs was due to the oxidation of photogenerated holes present on the valence band.$^*O_2^-$ photoactive species was developed by the oxidation of molecular oxygen by the photogenerated electrons. Figure 2.9 displays the mechanism of degradation of TCs by Ag/g-C_3N_4 photocatalyst. This pathway of degradation of TCs was developed using GC-MS. Thus, Ag/g-$C_3N_4$is an excellent photocatalyst with high stability, great photocatalytic performance, and can be used for more future applications (Ren et al., 2020).

2.4 CONCLUSION AND FUTURE ASPECTS

Antibiotics pollute water bodies and contaminate them such that antibiotic-resistant bacteria and microorganisms are produced. This is problematic for aquatic animals as well as humans. Conventional water treatment methods do not work on these antibiotics; thus, photocatalysis has a certain edge over these classical water treatment methods. Since photocatalysis occurs with the help of natural light, thus this method is clean, green, and produces no harmful by-products. Thus, in this chapter, we have discussed ways to improvise and increase the photocatalytic properties of visible light-active photocatalysts. We have discussed six main ways to treat wastewater containing antibiotics. These methods are doping visible light active semiconductors with metals and non-metals, heterojunction formation, using visible light active photocatalysts, modification of interface using co-catalysts, creation of oxygen vacancies, and SPR-enhanced photocatalysts. All these methods included the usage of a visible light-active photocatalyst. Two main ways were adopted to enhance the photocatalytic strength of these photocatalytic agents: to either decrease the band gap, or improve the separation of charge between photoexcited holes and electrons. It was depicted in this chapter that photocatalysis is a new way of treating wastewater containing antibiotics.

Current Scenario

The goal is to improvise the overall photocatalytic performance of visible light active photocatalysts. To date, the two main challenges faced by researchers are the swift transfer of charge and the sustainable separation of charge in photocatalysts. Compounds of bismuth, such as bismuth oxyhalides, have emerged as a new and effective category of visible light active photocatalysts. This is because bismuth oxyhalides have a layered structure. BiOI has attracted the interest of various researchers because of its tetragonal matlockite geometry displayed by $[Bi_2O_2]^{2+}$ slabs, which are separated in an alternative way by two slabs of iodine atoms (Garg et al., 2018; Tabatabai-Yazdi et al., 2021; Yadav et al., 2021). Hence, the electrons take less energy to migrate from the valence to the conduction band. This makes them efficient photocatalysts for the photodegradation of antibiotics. 2D graphene sheets are also attracting the interest of various researchers due to their vast surface area and fast charge-carrying speeds. Studies on photocatalysts based on graphene for antibiotics-degradation have already begun (Sharma et al., 2019). The reasons applicable for these studies are that migration of photoinduced charges is faster in graphene sheets, and so is charge separation (Sharma et al., 2019; Chawla et al., 2021).

Future Challenges

To date, much has been studied regarded bismuth oxyhalides, and efforts are being made to increase their photocatalytic activity. However, incoming challenges need to be handled by future researchers to design efficient, green, and economical photocatalysts.

- The very first challenge is that BiOX is not yet fully studied. There is much scope left in this compound, which includes doping it with certain metals and non-metals and heterojunction catalysis.
- Bismuth ferrite, a multiferroic photocatalyst, has much scope with its magnetic properties. Its applications are yet to be studied.
- Even by a slight change of pH and moving toward the alkaline region, the photocatalytic activity of bismuth oxyhalides improves significantly. Therefore, new routes for the preparation of bismuth oxyhalides must be studied.
- More studies should entail the recyclability of bismuth ferrite and bismuth oxyhalides after completing one round of photodegradation of antibiotics and other organic pollutants (Sharma et al., 2019).

REFERENCES

Acisli, O., Khataee, A., Soltani, R. D. C., & Karaca, S. (2017). Ultrasound-assisted Fenton process using siderite nanoparticles prepared via planetary ball milling for removal of reactive yellow 81 in aqueous phase. *Ultrasonics Sonochemistry*, *35*, 210–218. https://doi.org/10.1016/j.ultsonch.2016.09.020

Ahmed, M. G., Kandiel, T. A., Ahmed, A. Y., Kretschmer, I., Rashwan, F., & Bahnemann, D. (2015). Enhanced photoelectrochemical water oxidation on nanostructured hematite photoanodes via p-CaFe2O4/n-Fe_2O_3 heterojunction formation. *The Journal of Physical Chemistry C*, *119*(11), 5864–5871. https://doi.org/10.1021/jp512804p

Alapi, T., Farkas, J., Náfrádi, M., Hlogyik, T., Bartus, C., Schrantz, K., & Hernadi, K. (2018). Comparison of Advanced Oxidation Processes in the decomposition of diuron and monuron – efficiency, intermediates, Electric Energy per Order and the effect of various matrices. *Environmental Science: Water Research & Technology*, *4*. https://doi.org/10.1039/C8EW00202A

Asadzadeh-Khaneghah, S., Habibi-Yangjeh, A., & Seifzadeh, D. (2018). Graphitic carbon nitride nanosheets coupled with carbon dots and BiOI nanoparticles: Boosting visible-light-driven photocatalytic activity. *Journal of the Taiwan Institute of Chemical Engineers*, *87*, 98–111. https://doi.org/10.1016/j.jtice.2018.03.017

Asahi, R., Morikawa, T., Irie, H., & Ohwaki, T. (2014). Nitrogen-doped titanium dioxide as visible-light-sensitive photocatalyst: Designs, developments, and prospects. *Chemical Reviews*, *114*(19), 9824–9852. https://doi.org/10.1021/cr5000738

Azimi, S., & Nezamzadeh-Ejhieh, A. (2015). Enhanced activity of clinoptilolite-supported hybridized PbS–CdS semiconductors for the photocatalytic degradation of a mixture of tetracycline and cephalexin aqueous solution. *Journal of Molecular Catalysis A: Chemical*, *408*, 152–160. https://doi.org/10.1016/j.molcata.2015.07.017

Bárdos, E., Márta, V., Baia, L., Todea, M., Kovács, G., Baán, K., Garg, S., Pap, Z., & Hernadi, K. (2020). Hydrothermal crystallization of bismuth oxybromide (BiOBr) in the presence of different shape controlling agents. *Applied Surface Science*, *518*, 146184. https://doi.org/10.1016/j.apsusc.2020.146184

Batt, A., Bruce, I., & Aga, D. (2006). Evaluating the vulnerability of surface waters to antibiotic contamination from varying wastewater treatment plant discharges. *Environmental Pollution (Barking, Essex : 1987)*, *142*, 295–302. https://doi.org/10.1016/j.envpol.2005.10.010

Chawla, H., Chandra, A., Ingole, P. P., & Garg, S. (2021). Recent advancements in enhancement of photocatalytic activity using bismuth-based metal oxides Bi_2MO_6 (M=W, Mo, Cr) for environmental remediation and clean energy production. *Journal of Industrial and Engineering Chemistry*, *95*, 1–15. https://doi.org/10.1016/j.jiec.2020.12.028

Chen, M., & Chu, W. (2012). Degradation of antibiotic norfloxacin in aqueous solution by visible-light-mediated C-TiO_2 photocatalysis. *Journal of Hazardous Materials*, *219–220*, 183—189. https://doi.org/10.1016/j.jhazmat.2012.03.074

Chen, Y., Zhu, G., Hojamberdiev, M., Gao, J., Zhu, R., Wang, C., Wei, X., & Liu, P. (2018). Three-dimensional Ag_2O/Bi_5O_7I p-n heterojunction photocatalyst harnessing UV-vis-NIR broad spectrum for photodegradation of organic pollutants. *Journal of Hazardous Materials*, *344*, 42—54. https://doi.org/10.1016/j.jhazmat.2017.10.015

Chi, S., Ji, C., Sun, S., Jiang, H., Qu, R., & Sun, C. (2016). Magnetically separated meso-g-C3N4/Fe_3O_4: Bifuctional composites for removal of arsenite by simultaneous visible-light catalysis and adsorption. *Industrial & Engineering Chemistry Research*, *55*(46), 12060–12067. https://doi.org/10.1021/acs.iecr.6b02178

Choi, W., Termin, A., & Hoffmann, M. R. (1994). The role of metal ion dopants in quantum-sized TiO_2: Correlation between photoreactivity and charge carrier recombination dynamics. *The Journal of Physical Chemistry*, *98*(51), 13669–13679. https://doi.org/10.1021/j100102a038

de Witte, B., van Langenhove, H., Demeestere, K., & Dewulf, J. (2011). Advanced oxidation of pharmaceuticals: Chemical analysis and biological assessment of degradation products. *Critical Reviews in Environmental Science and Technology*, *41*(3), 215–242. https://doi.org/10.1080/10643380902728698

Di, J., Xia, J., Ji, M., Yin, S., Li, H., Xu, H., Zhang, Q., & Li, H. (2015). Controllable synthesis of Bi4O5Br2 ultrathin nanosheets for photocatalytic removal of ciprofloxacin and mechanism insight. *Journal of Materials Chemistry A*, *3*(29), 15108–15118. https://doi.org/10.1039/C5TA02388B

Di, J., Xia, J., Li, H., Guo, S., & Dai, S. (2017). Bismuth oxyhalide layered materials for energy and environmental applications. *Nano Energy*, *41*, 172–192. https://doi.org/10.1016/j.nanoen.2017.09.008

Ding, J., Dai, Z., Qin, F., Zhao, H., Zhao, S., & Chen, R. (2017). Z-scheme BiO1-xBr/$Bi_2O_2C_O3$ photocatalyst with rich oxygen vacancy as electron mediator for highly efficient degradation of antibiotics. *Applied Catalysis B: Environmental*, *205*, 281–291. https://doi.org/10.1016/j.apcatb.2016.12.018

Dong, F., Zhao, Z., Xiong, T., Ni, Z., Zhang, W., Sun, Y., & Ho, W.-K. (2013). In situ construction of g-C3N4/g-C3N4 metal-free heterojunction for enhanced visible-light photocatalysis. *ACS Applied Materials & Interfaces*, *5*(21), 11392–11401. https://doi.org/10.1021/am403653a

Duan, M., Li, J., Li, M., Zhang, Z., & Wang, C. (2012). Pt(II) porphyrin modified TiO_2 composites as photocatalysts for efficient 4-NP degradation. *Applied Surface Science*, *258*, 5499–5504. https://doi.org/10.1016/j.apsusc.2012.02.069

Durán-Álvarez, J. C., Martínez-Avelar, C., González-Cervantes, E., Gutiérrez-Márquez, R. A., Rodríguez-Varela, M., Varela, A. S., Castillón, F., & Zanella, R. (2020). Degradation and mineralization of oxytetracycline in pure and tap water under visible light irradiation using bismuth oxyiodides and the effect of depositing Au nanoparticles. *Journal of Photochemistry and Photobiology A: Chemistry*, *388*, 112163. https://doi.org/10.1016/j.jphotochem.2019.112163

Elward, J. M., & Chakraborty, A. (2013). Effect of dot size on exciton binding energy and electron–hole recombination probability in CdSe quantum dots. *Journal of Chemical Theory and Computation*, *9*(10), 4351–4359. https://doi.org/10.1021/ct400485s

Fu, Y., Qin, L., Huang, D., Zeng, G., Lai, C., Li, B., He, J., Yi, H., Zhang, M., Cheng, M., & Wen, X. (2019a). Chitosan functionalized activated coke for Au nanoparticles anchoring: Green synthesis and catalytic activities in hydrogenation of nitrophenols and azo dyes. *Applied Catalysis B: Environmental*, *255*, 117740. https://doi.org/10.1016/j.apcatb.2019.05.042

Fu, Y., Xu, P., Huang, D., Zeng, G., Lai, C., Qin, L., Li, B., He, J., Yi, H., Cheng, M., & Zhang, C. (2019b). Au nanoparticles decorated on activated coke via a facile preparation for efficient catalytic reduction of nitrophenols and azo dyes. *Applied Surface Science*, *473*, 578–588. https://doi.org/10.1016/j.apsusc.2018.12.207

Garg, S., Yadav, M., Chandra, A., Gahlawat, S., Ingole, P. P., Pap, Z., & Hernadi, K. (2018). Plant leaf extracts as photocatalytic activity tailoring agents for BiOCl towards environmental remediation. *Ecotoxicology and Environmental Safety*, *165*, 357–366. https://doi.org/10.1016/j.ecoenv.2018.09.024

Gholami, P., Khataee, A., Soltani, R. D. C., Dinpazhoh, L., & Bhatnagar, A. (2019). Photocatalytic degradation of gemifloxacin antibiotic using Zn-Co-LDH@biochar nanocomposite. *Journal of Hazardous Materials*, *382*, 121070. https://www.unboundmedicine.com/medline/citation/31470301/Photocatalytic_degradation_of_gemifloxacin_antibiotic_using_Zn-Co-LDH@biochar_nanocomposite.

Ghosh, S., Kouamé, N. A., Ramos, L., Remita, S., Dazzi, A., Deniset-Besseau, A., Beaunier, P., Goubard, F., Aubert, P.-H., & Remita, H. (2015). Conducting polymer nanostructures for photocatalysis under visible light. *Nature Materials*, *14*(5), 505–511. https://doi.org/10.1038/nmat4220

Gu, Y., Zhao, L., Yang, M., Xiong, Y., Wu, Z., Zhou, M., & Yan, J. (2017). Preparation and characterization of highly photocatalytic active hierarchical BiOX (X=Cl, Br, I) micro-flowers for rhodamine B degradation with kinetic modelling studies. *Journal of Central South University*, *24*(4), 754–765. https://doi.org/10.1007/s11771-017-3477-x

Guo, W., Su, S., Yi, C., & Ma, Z. (2013). Degradation of antibiotics amoxicillin by Co3O4-catalyzed peroxymonosulfate system. *Environmental Progress & Sustainable Energy*, *32*. https://doi.org/10.1002/ep.10633

Halling-Sørensen, B., Lykkeberg, A., Ingerslev, F., Blackwell, P., & Tjørnelund, J. (2003). Characterisation of the abiotic degradation pathways of oxytetracyclines in soil interstitial water using LC-MS-MS. *Chemosphere*, *50*(10), 1331–1342. https://doi.org/10.1016/s0045-6535(02)00766-x

Hollmann, D., Karnahl, M., Tschierlei, S., Kailasam, K., Schneider, M., Radnik, J., Grabow, K., Bentrup, U., Junge, H., Beller, M., Lochbrunner, S., Thomas, A., & Brückner, A. (2014). Structure–activity relationships in bulk polymeric and Sol–Gel-derived carbon nitrides during photocatalytic hydrogen production. *Chemistry of Materials*, *26*(4), 1727–1733. https://doi.org/10.1021/cm500034p

Hu, X., Hu, X., Peng, Q., Zhou, L., Tan, X., Jiang, L., Tang, C., Wang, H., Liu, S., Wang, Y., & Ning, Z. (2020). Mechanisms underlying the photocatalytic degradation pathway of ciprofloxacin with heterogeneous TiO_2. *Chemical Engineering Journal*, *380*, 122366. https://doi.org/10.1016/j.cej.2019.122366

Huang, H., Cao, R., Shixin, Y., Xu, K., Hao, W., Wang, Y., Dong, F., Zhang, T., & Zhang, Y. (2017). Single-unit-cell layer established Bi 2 WO 6 3D hierarchical architectures: Efficient adsorption, photocatalysis and dye-sensitized photoelectrochemical performance. *Applied Catalysis B: Environmental*, *219* https://doi.org/10.1016/j.apcatb.2017.07.084

Hunge, Y., Mahadik, M., Bulakhe, R., Yadav, S., Shim, J.-J., Moholkar, A., & Bhosale, C. (2017). Oxidative degradation of benzoic acid using spray deposited WO3/TiO_2 thin films. *Journal of Materials Science: Materials in Electronics*, *28*. https://doi.org/10.1007/s10854-017-7740-6

Hunge, Y. M., Yadav, A. A., Mahadik, M. A., Mathe, V., & Bhosale, C. H. (2018). A highly efficient visible-light responsive sprayed WO3/FTO photoanode for photoelectrocatalytic degradation of brilliant blue. *Journal of The Taiwan Institute of Chemical Engineers*, *85*, 273–281.

Ibrahim, F. A., Al-Ghobashy, M. A., Abd El-Rahman, M. K., & Abo-Elmagd, I. F. (2017). Optimization and in line potentiometric monitoring of enhanced photocatalytic degradation kinetics of gemifloxacin using TiO_2 nanoparticles/H_2O_2. *Environmental Science and Pollution Research International*, *24*(30), 23880—23892. https://doi.org/10.1007/s11356-017-0045-8

Iwase, A., Ng, Y. H., Ishiguro, Y., Kudo, A., & Amal, R. (2011). Reduced graphene oxide as a solid-state electron mediator in Z-scheme photocatalytic water splitting under visible light. *Journal of the American Chemical Society*, *133*(29), 11054–11057. https://doi.org/10.1021/ja203296z

Jiang, Y., Zhang, M., Xin, Y., Chai, C., & Chen, Q. (2019). Construction of immobilized CuS/TiO_2 nanobeltsheterojunctionphotocatalyst for photocatalytic degradation of enrofloxacin: Synthesis, characterization, influencing factors and mechanism insight. *Journal of Chemical Technology & Biotechnology*, *94*(7), 2219–2228. https://doi.org/10.1002/jctb.6006

Jiexiang, X., Di, J., Li, H., Hui, X., Li, H., & Guo, S. (2015). Ionic liquid-induced strategy for carbon quantum dots/BiOX (X= Br, Cl) hybrid nanosheets with superior visible light-driven photocatalysis. *Applied Catalysis B: Environmental*, *181*. https://doi.org/10.1016/j.apcatb.2015.07.035

Jo, W.-K., & Sivakumar Natarajan, T. (2015). Facile synthesis of novel redox-mediator-free direct Z-scheme CaIn2S4 marigold-flower-like/TiO_2 photocatalysts with superior photocatalytic efficiency. *ACS Applied Materials & Interfaces*, *7*(31), 17138–17154. https://doi.org/10.1021/acsami.5b03935

Kim, Y.-H., Heinze, T. M., Beger, R., Pothuluri, J. V., & Cerniglia, C. E. (2004). A kinetic study on the degradation of erythromycin A in aqueous solution. *International Journal of Pharmaceutics*, *271*(1), 63–76. https://doi.org/10.1016/j.ijpharm.2003.10.023

Li, D., & Shi, W. (2016). Recent developments in visible-light photocatalytic degradation of antibiotics. *Chinese Journal of Catalysis*, *37*(6), 792–799. https://doi.org/10.1016/S1872-2067(15)61054-3

Li, D., Wang, H., Tang, H., Yang, X., & Liu, Q. (2019). Remarkable enhancement in solar oxygen evolution from MoSe2/Ag3PO4 heterojunction photocatalyst via in situ constructing interfacial contact. *ACS Sustainable Chemistry & Engineering*, *7*(9), 8466–8474. https://doi.org/10.1021/acssuschemeng.9b00252

Li, J., Liu, L., Liang, Q., Zhou, M., Yao, C., Xu, S., & Li, Z. (2021a). Core-shell ZIF-8@MIL-68(In) derived ZnO nanoparticles-embedded In2O3 hollow tubular with oxygen vacancy for photocatalytic degradation of antibiotic pollutant. *Journal of Hazardous Materials*, 125395. https://doi.org/10.1016/j.jhazmat.2021.125395

Li, J., Yu, Y., & Zhang, L. (2014). Bismuth oxyhalide nanomaterials: Layered structures meet photocatalysis. *Nanoscale*, *6*(15), 8473–8488. https://doi.org/10.1039/C4NR02553A

Li, K., Yediler, A., Yang, M., Schulte-Hostede, S., & Wong, M. H. (2008). Ozonation of oxytetracycline and toxicological assessment of its oxidation by-products. *Chemosphere*, *72*(3), 473–478. https://doi.org/10.1016/j.chemosphere.2008.02.008

Li, S., Wang, C., Liu, Y., Xue, B., Jiang, W., Liu, Y., Mo, L., & Chen, X. (2021b). Photocatalytic degradation of antibiotics using a novel Ag/Ag2S/Bi2MoO6 plasmonic p-n heterojunction photocatalyst: Mineralization activity, degradation pathways and boosted charge separation mechanism. *Chemical Engineering Journal*, *415*, 128991. https://doi.org/10.1016/j.cej.2021.128991

Linic, S., Christopher, P., & Ingram, D. B. (2011). Plasmonic-metal nanostructures for efficient conversion of solar to chemical energy. *Nature Materials*, *10*(12), 911–921. https://doi.org/10.1038/nmat3151

Liu, H., Dong, X., Li, G., Su, X., & Zhu, Z. (2013). Synthesis of C, Ag co-modified TiO_2 photocatalyst and its application in waste water purification. *Applied Surface Science*, *271*, 276–283. https://doi.org/10.1016/j.apsusc.2013.01.181

Liu, Y., He, X., Fu, Y., & Dionysiou, D. D. (2016). Kinetics and mechanism investigation on the destruction of oxytetracycline by UV-254nm activation of persulfate. *Journal of Hazardous Materials*, *305*, 229—239. https://doi.org/10.1016/j.jhazmat.2015.11.043

Liu, Z., Niu, J., Feng, P., Sui, Y., & Zhu, Y. (2014). One-pot synthesis of Bi24O31Br10/Bi4V2O11 heterostructures and their photocatalytic properties. *RSC Advances*, *4*(82), 43399–43405. https://doi.org/10.1039/C4RA04815F

Lu, D., Wang, H., Zhao, X., Kondamareddy, K. K., Ding, J., Li, C., & Fang, P. (2017). Highly efficient visible-light-induced photoactivity of Z-scheme g-C3N4/Ag/MoS2 ternary photocatalysts for organic pollutant degradation and production of hydrogen. *ACS Sustainable Chemistry & Engineering*, *5*(2), 1436–1445. https://doi.org/10.1021/acssuschemeng.6b02010

Moniz, S. J. A., Shevlin, S. A., Martin, D. J., Guo, Z.-X., & Tang, J. (2015). Visible-light driven heterojunction photocatalysts for water splitting: A critical review. *Energy & Environmental Science*, *8*(3), 731–759. https://doi.org/10.1039/C4EE03271C

Mousavi, M., & Habibi-Yangjeh, A. (2018). Magnetically recoverable highly efficient visible-light-active g-C3N4/Fe_3O_4/Ag_2WO_4/AgBr nanocomposites for photocatalytic degradations of environmental pollutants. *Advanced Powder Technology*, *29*(1), 94–105. https://doi.org/10.1016/j.apt.2017.10.016

Murillo-Sierra, J. C., Hernández-Ramírez, A., Zhao, Z.-Y., Martínez-Hernández, A., & Gracia-Pinilla, M. A. (2021). Construction of direct Z-scheme WO_3/ZnS heterojunction to enhance the photocatalytic degradation of tetracycline antibiotic. *Journal of Environmental Chemical Engineering*, *9*(2), 105111. https://doi.org/10.1016/j.jece.2021.105111

Ni, M., Leung, M. K. H., Leung, D. Y. C., & Sumathy, K. (2007). A review and recent developments in photocatalytic water-splitting using TiO_2 for hydrogen production. *Renewable and Sustainable Energy Reviews*, *11*(3), 401–425. https://doi.org/10.1016/j.rser.2005.01.009

Oros-Ruiz, S., Zanella, R., & Prado, B. (2013). Photocatalytic degradation of trimethoprim by metallic nanoparticles supported on TiO_2-P25. *Journal of Hazardous Materials*, *263*, 28–35. https://doi.org/10.1016/j.jhazmat.2013.04.010

Qin, L., Zeng, Z., Zeng, G., Lai, C., Duan, A., Xiao, R., Huang, D., Fu, Y., Yi, H., Li, B., Liu, X., Liu, S., Zhang, M., & Jiang, D. (2019). Cooperative catalytic performance of bimetallic Ni-Au nanocatalyst for highly efficient hydrogenation of nitroaromatics and corresponding mechanism insight. *Applied Catalysis B: Environmental*, *259*, 118035. https://doi.org/10.1016/j.apcatb.2019.118035

Ren, Z., Chen, F., Wen, K., & Lu, J. (2020). Enhanced photocatalytic activity for tetracyclines degradation with Ag modified g-C3N4 composite under visible light. *Journal of Photochemistry and Photobiology A: Chemistry*, *389*, 112217. https://doi.org/10.1016/j.jphotochem.2019.112217

Sápi, A., Mutyala, S., Garg, S., Yadav, M., Gómez-Pérez, J. F., Czirok, F., Sándor, Z., Hernadi, K., Farkas, F., Kovačič, S., Kukovecz, Á., & Kónya, Z. (2021). Size controlled Pt over mesoporousNiOnanocomposite catalysts: Thermal catalysis vs. photocatalysis. *Journal of Porous Materials*. https://doi.org/10.1007/s10934-020-00978-x

Sarmah, A., Meyer, M., & Boxall, A. (2006). A Global Perspective on the Use, Sales, Exposure Pathways, Occurrence, Fate and Effects of Veterinary Antibiotics (VAs) in the Environment. *Chemosphere*, *65*, 725–759. https://doi.org/10.1016/j.chemosphere.2006.03.026

Shao, L., Cheng, S., Yang, Z., Xia, X., & Liu, Y. (2021). Nickel aluminum layered double hydroxide nanosheets grown on oxygen vacancy-rich TiO_2 nanobelts for enhanced photodegradation of an antibiotic. *Journal of Photochemistry and Photobiology A: Chemistry*, *411*, 113209. https://doi.org/10.1016/j.jphotochem.2021.113209

Sharma, K., Dutta, V., Sharma, S., Raizada, P., Hosseini-Bandegharaei, A., Thakur, P., & Singh, P. (2019). Recent advances in enhanced photocatalytic activity of bismuth oxyhalides for efficient photocatalysis of organic pollutants in water: A review. *Journal of Industrial and Engineering Chemistry*, *78*, 1–20. https://doi.org/10.1016/j.jiec.2019.06.022

Shi, W., Li, M., Huang, X., Ren, H., Guo, F., Tang, Y., & Lu, C. (2020). Construction of CuBi2O4/Bi2MoO6 p-n heterojunction with nanosheets-on-microrods structure for improved photocatalytic activity towards broad-spectrum antibiotics degradation. *Chemical Engineering Journal*, *394*, 125009. https://doi.org/10.1016/j.cej.2020.125009

Siemens, J., Huschek, G., Siebe, C., & Kaupenjohann, M. (2008). Concentrations and mobility of human pharmaceuticals in the world's largest wastewater irrigation system, Mexico City-Mezquital Valley. *Water Research*, *42*(8–9), 2124–2134. https://doi.org/10.1016/j.watres.2007.11.019

Simon, G., Gyulavári, T., Hernadi, K., Molnár, M., Pap, Z., Veréb, G., Schrantz, K., Náfrádi, M., & Alapi, T. (2018). Photocatalytic ozonation of monuron over suspended and immobilized TiO_2 –study of transformation, mineralization and economic feasibility. *Journal of Photochemistry and Photobiology A: Chemistry*, *356*. https://doi.org/10.1016/j.jphotochem.2018.01.025

Sirtori, C., Agüera, A., Gernjak, W., & Malato, S. (2010). Effect of water-matrix composition on Trimethoprim solar photodegradation kinetics and pathways. *Water Research*, *44*(9), 2735—2744. https://doi.org/10.1016/j.watres.2010.02.006

Soltani, R. D. S., & Mashayekhi, M. (2018). Decomposition of ibuprofen in water via an electrochemical process with nano-sized carbon black-coated carbon cloth as oxygen-permeable cathode integrated with ultrasound. *Chemosphere*, *194*, 471—480. https://doi.org/10.1016/j.chemosphere.2017.12.033

Su, M., He, C., Sharma, V., Asi, M., Xia, D., Li, X., Huiqi, D., & Xiong, Y. (2011). Mesoporous zinc ferrite: Synthesis, characterization, and photocatalytic activity with H_2O_2/visible light. *Journal of Hazardous Materials*, *211–212*, 95–103. https://doi.org/10.1016/j.jhazmat.2011.10.006

Tabatabai-Yazdi, F.-S., EbrahimianPirbazari, A., Esmaeili Khalil Saraei, F., & Gilani, N. (2021). Construction of graphene based photocatalysts for photocatalytic degradation of organic pollutant and modeling using artificial intelligence techniques. *Physica B: Condensed Matter*, *608*, 412869. https://doi.org/10.1016/j.physb.2021.412869

Topkaya, E., Konyar, M., Yatmaz, H. C., & Öztürk, K. (2014). Pure ZnO and composite ZnO/TiO_2 catalyst plates: A comparative study for the degradation of azo dye, pesticide and antibiotic in aqueous solutions. *Journal of Colloid and Interface Science*, *430*, 6–11. https://doi.org/10.1016/j.jcis.2014.05.022

Veréb, G., Ambrus, Z., Pap, Z., Mogyorosi, K., Dombi, A., & Hernadi, K. (2014). Immobilization of crystallized photocatalysts on ceramic paper by titanium(IV) ethoxide and photocatalytic decomposition of phenol. *Reaction Kinetics, Mechanisms and Catalysis*, *113*. https://doi.org/10.1007/s11144-014-0734-y

Verlicchi, P., Galletti, A., Petrovic, M., & Barcelo, D. (2010). Hospital effluents as a source of emerging pollutants: An overview of micropollutants and sustainable treatment options. *Journal of Hydrology*, *389*, 416–428. https://doi.org/10.1016/j.jhydrol.2010.06.005

Verma, M., & Haritash, A. K. (2020). Photocatalytic degradation of Amoxicillin in pharmaceutical wastewater: A potential tool to manage residual antibiotics. *Environmental Technology & Innovation*, *20*, 101072. https://doi.org/10.1016/j.eti.2020.101072

Vignesh, K., Rajarajan, M., & Suganthi, A. (2014). Photocatalytic degradation of erythromycin under visible light by zinc phthalocyanine-modified titania nanoparticles. *Materials Science in Semiconductor Processing*, *23*, 98–103. https://doi.org/10.1016/j.mssp.2014.02.050

Wang, H., Zhang, L., Chen, Z., Hu, J., Li, S., Wang, Z., Liu, J., & Wang, X. (2014a). Semiconductor heterojunction photocatalysts: Design, construction, and photocatalytic performances. *Chemical Society Reviews*, *43*(15), 5234–5244. https://doi.org/10.1039/C4CS00126E

Wang, J.-P., Yang, H.-C., & Hsieh, C.-T. (2012). Visible-light photodegradation of dye on co-doped titania nanotubes prepared by hydrothermal synthesis. *International Journal of Photoenergy*, *2012*, 206534. https://doi.org/10.1155/2012/206534

Wang, K., Zhang, G., Li, J., Li, Y., & Wu, X. (2017). 0D/2D Z-scheme heterojunctions of bismuth tantalate quantum dots/Ultrathin g-C3N4 nanosheets for highly efficient visible light photocatalytic degradation of antibiotics. *ACS Applied Materials & Interfaces*, *9*. https://doi.org/10.1021/acsami.7b14275

Wang, T., Guidong, Y., Liu, J., Yang, B., Ding, S., Yan, Z., & Xiao, T. (2014b). Orthogonal synthesis, structural characteristics, and enhanced visible-light photocatalysis of mesoporous Fe_2O_3/TiO_2 heterostructured microspheres. *Applied Surface Science*, *311*, 314–323. https://doi.org/10.1016/j.apsusc.2014.05.060

Wen, X.-J., Niu, C.-G., Zhang, L., & Zeng, G.-M. (2017). Fabrication of SnO_2 nanopaticles/BiOI n–p heterostructure for wider spectrum visible-light photocatalytic degradation of antibiotic oxytetracycline hydrochloride. *ACS Sustainable Chemistry & Engineering*, *5*(6), 5134–5147. https://doi.org/10.1021/acssuschemeng.7b00501

Xu, M., Chen, Y., Qin, J., Feng, Y., Li, W., Chen, W., Zhu, J., Li, H., & Bian, Z. (2018). Unveiling the role of defects on oxygen activation and photodegradation of organic pollutants. *Environmental Science & Technology*, *52*(23), 13879–13886. https://doi.org/10.1021/acs.est.8b03558

Xue, C., Xia, J., Wang, T., Zhao, S., Yang, G., Yang, B., Dai, Y., & Yang, G. (2014). A facile and efficient solvothermal fabrication of three-dimensionally hierarchical BiOBr microspheres with exceptional photocatalytic activity. *Materials Letters*, *133*, 274–277. https://doi.org/10.1016/j.matlet.2014.07.016

Yadav, M., Garg, S., Chandra, A., Gläser, R., & Hernadi, K. (2020). Green BiOI impregnated 2-dimensional cylindrical carbon block: A promising solution for environmental remediation and easy recovery of the photocatalyst. *Separation and Purification Technology*, *240*, 116628. https://doi.org/10.1016/j.seppur.2020.116628

Yadav, M., Garg, S., Chandra, A., & Hernadi, K. (2019). Fabrication of leaf extract mediated bismuth oxybromide/oxyiodide (BiOBrxI1−x) photocatalysts with tunable band gap and enhanced optical absorption for degradation of organic pollutants. *Journal of Colloid and Interface Science*, *555*, 304–314. https://doi.org/10.1016/j.jcis.2019.07.090

Yadav, M., Garg, S., Chandra, A., Ingole, P. P., Bardos, E., & Hernadi, K. (2021). Quercetin-mediated 3-D hierarchical BiOI-Q and BiOI-Q-Ag nanostructures with enhanced photodegradation efficiency. *Journal of Alloys and Compounds*, *856*, 156812. https://doi.org/10.1016/j.jallcom.2020.156812

Yan, M, Hua, Y. Q., Zhu, F. F., Gu, W., Jiang, J. H., Shen, H. Q., & Shi, W. D. (2017). Fabrication of nitrogen doped graphene quantum dots-BiOI/MnNb2O6 p-n junction photocatalysts with enhanced visible light efficiency in photocatalytic degradation of antibiotics. *Applied Catalysis B: Environmental*, *202*, 518–527. https://doi.org/10.1016/j.apcatb.2016.09.039

Yan, Ming, Zhu, F., Gu, W., Sun, L., Shi, W., & Hua, Y. (2016). Construction of nitrogen-doped graphene quantum dots-BiVO4/g-C3N4 Z-scheme photocatalyst and enhanced photocatalytic degradation of antibiotics under visible light. *RSC Advances*, *6*(66), 61162–61174. https://doi.org/10.1039/C6RA07589D

Yang, C., Cheng, J. H., Chen, Y. C., & Hu, Y. Y. (2017). CdS nanoparticles immobilized on porous carbon polyhedrons derived from a metal-organic framework with enhanced visible light photocatalytic activity for antibiotic degradation. *Applied Surface Science*, *420*, 252–259. https://doi.org/10.1016/j.apsusc.2017.05.102

Yang, L., Luo, S., Li, Y., Xiao, Y., Kang, Q., & Cai, Q. (2010). High efficient photocatalytic degradation of p-nitrophenol on a unique Cu_2O/TiO_2 p-n heterojunction network catalyst. *Environmental Science & Technology*, *44*(19), 7641–7646. https://doi.org/10.1021/es101711k

Yang, Z., Li, L., Yu, H., Liu, M., Chi, Y., Sha, J., & Xu, S. (2021). Facile synthesis of highly crystalline g-C3N4 nanosheets with remarkable visible light photocatalytic activity for antibiotics removal. *Chemosphere*, *271*, 129503. https://doi.org/10.1016/j.chemosphere.2020.129503

Yu, H., Huang, H., Xu, K., Hao, W., Guo, Y., Wang, S., Shen, X., Pan, S., & Zhang, Y. (2017). Liquid-phase exfoliation into monolayered BiOBr nanosheets for photocatalytic oxidation and reduction. *ACS Sustainable Chemistry & Engineering*, *5*(11), 10499–10508. https://doi.org/10.1021/acssuschemeng.7b02508

Zhang, F., Li, Y., Yang, M., & Li, W. (2012). Content of heavy metals in animal feeds and manures from farms of different scales in northeast China. *International Journal of Environmental Research and Public Health*, *9*(8), 2658–2668. https://doi.org/10.3390/ijerph9082658

Zhou, C., Lai, C., Huang, D., Zeng, G., Zhang, C., Cheng, M., Hu, L., Wan, J., Xiong, W., Wen, M., Wen, X., & Qin, L. (2017a). Highly porous carbon nitride by supramolecular preassembly of monomers for photocatalytic removal of sulfamethazine under visible light driven. *Applied Catalysis B: Environmental*, *220*. https://doi.org/10.1016/j.apcatb.2017.08.055

Zhou, M., Hou, Z., Zhang, L., Liu, Y., Gao, Q., & Chen, X. (2017b). n/n junctioned g-C3N4 for enhanced photocatalytic H_2 generation. *Sustainable Energy & Fuels*, *1*(2), 317–323. https://doi.org/10.1039/C6SE00004E

Zhou, P., Yu, J., & Jaroniec, M. (2014). All-Solid-State Z-scheme photocatalytic systems. *Advanced Materials*, *26*(29), 4920–4935. https://doi.org/10.1002/adma.201400288

3 Graphitic Carbon Nitride-Based Nanomaterials as Photocatalysts for Organic Pollutant Degradation

Anise Akhundi and Aziz Habibi-Yangjeh
University of Mohaghegh Ardabili, Ardabil, Iran

CONTENTS

3.1 Introduction ... 93
3.2 Pure g-C_3N_4 ... 96
3.2.1 Electronic Structure of g-C_3N_4 ... 96
3.2.2 Photocatalytic Applications of g-C_3N_4 ... 96
3.2.3 Modified g-C_3N_4 and Its Derivatives ... 98
3.3 g-C_3N_4-Based Nanocomposites ... 99
3.3.1 Synthesis and Different Types of g-C_3N_4-Based Nanocomposites ... 99
3.3.2 Basic Principles of Photocatalytic Degradation Reaction ... 103
3.4 g-C_3N_4-Based Composites for Degradation Reactions ... 103
3.4.1 Type II Heterojunction ... 103
3.4.2 Z-Scheme Heterojunction ... 105
3.4.3 Effect of Environmental Parameters ... 109
3.5 Conclusions and Future Perspectives ... 109
Acknowledgment ... 110
References ... 110

3.1 INTRODUCTION

With the fast growth of industrialization, the world is experiencing an increasing discharge of organic pollutants in water sources. As the wastewater treatment by degradation of these pollutants under natural conditions is difficult or even impossible (Li et al. 2018), the use of semiconductor nanomaterials in photocatalysis technology is

DOI: 10.1201/9781003118749-3

becoming more and more popular as a promising approach to address water pollution (Bahnemann 2004; Ahmed et al. 2017). Over the past years, semiconductor-mediated photocatalysis has attracted a great deal of attention due to its plus points, including stability of the utilized materials, cost-effectiveness, and high capability in the complete mineralization of organic pollutants. TiO_2 and TiO_2-based photocatalysts are extensively applied in pollutants degradation, as well as other photocatalytic reactions owing to their low cost, appropriate stability, and non-toxicity (Guo et al. 2020; Huang et al. 2020; Ma et al. 2020; Wu et al., 2020a, 2020b). However, the wide band gap energy (E_g) of TiO_2 makes it insensitive to visible light, and its photocatalytic activation occurs only within the UV range, which occupies ca. 5% of the solar spectrum (Ghoreishian et al., 2021; Wu et al., 2021). Therefore, the design and fabrication of novel visible-light-driven photocatalysts are desirable.

In 2009, Wang et al. (2009) used graphitic carbon nitride (g-C_3N_4) photocatalyst for water splitting reaction to release H_2 and O_2. Since then, this conjugated semiconductor has attracted researchers' attention because of its unique properties such as exceptional physical and chemical stability, facile preparation, appealing electronic and optical properties, low cost, and moderate band gap (2.7 eV) (Zhu et al. 2014; Gong et al. 2015a, 2015b; Dong et al. 2016; Kumar, Karthikeyan, and Lee 2018). This polymeric photocatalyst is fabricated by simple thermal condensation of inexpensive and abundant nitrogen precursors such as melamine, urea, thiourea, cyanamide, dicyandiamide, and their mixture (Figure 3.1).

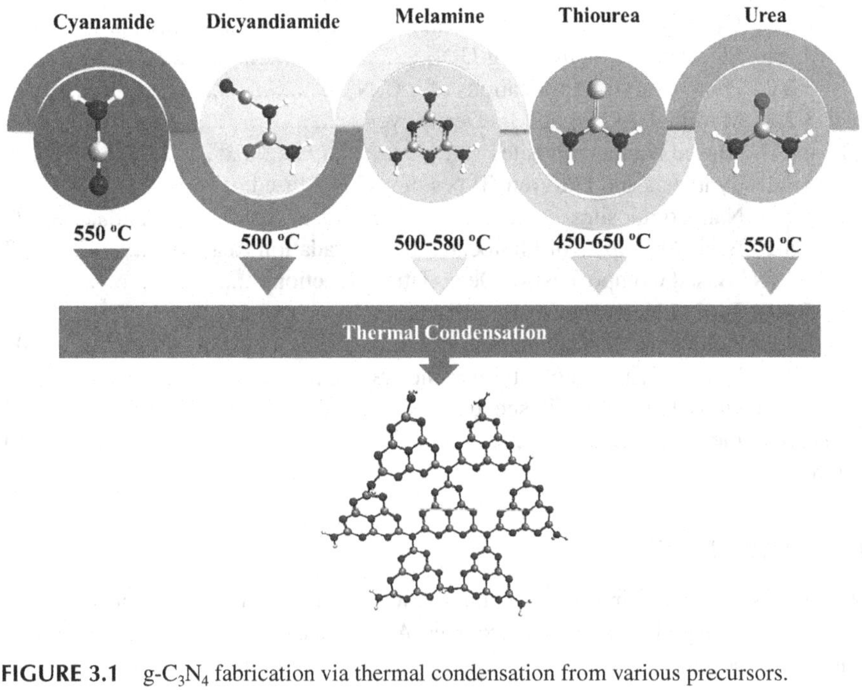

FIGURE 3.1 g-C_3N_4 fabrication via thermal condensation from various precursors.

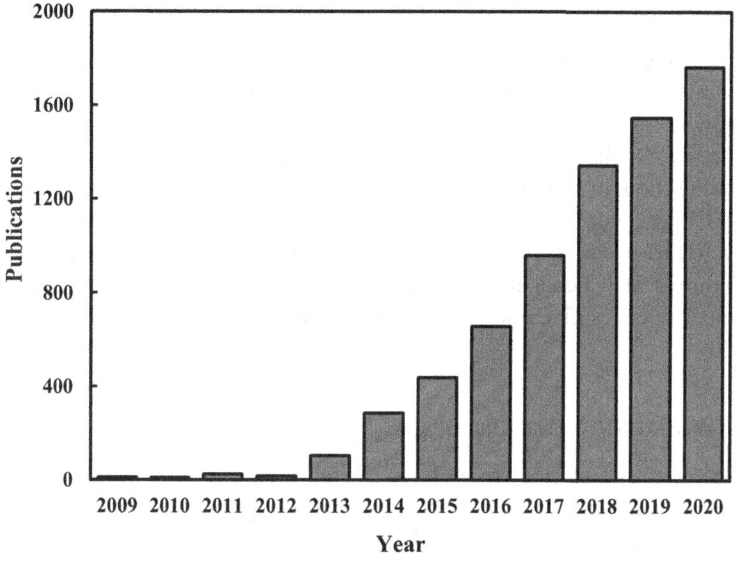

FIGURE 3.2 Representation of the number of publications containing the keyword "g-C_3N_4" published from 2009 to October 2020. The data was obtained from Scopus.

In recent years, the increased number of published papers containing the keyword "g-C_3N_4" is strong evidence that it is a hot research material (Figure 3.2). These publications have emerged in various fields, including bioimaging (Zhang et al., 2020a; Zheng et al. 2020), sensors (Wang et al. 2019); Kumar et al., 2020; Vilian et al. 2020; Yuan et al. 2020 biofuel cells (Sarkar et al. 2019), supercapacitors (Safaei et al. 2018; Ghaemmaghami and Mohammadi 2019; Shi et al. 2021), lithium-ion batteries (Zuo et al. 2020), organic synthesis (Cao et al., 2020), CO_2 reduction (R. Liu et al., 2020), water splitting (Nasir et al., 2019; Zhang et al., 2020b; Malik and Tomer 2021), pollutants degradation (Acharya and Parida 2020; Kumar et al., 2020; Zhang et al., 2020c; Tan et al., 2021) and so on.

Nevertheless, pure g-C_3N_4 has some demerits, namely high charge carriers recombination, insufficient surface area, and poor visible-light absorption, which restricted its photocatalytic performance. Many efforts have been devoted to addressing these drawbacks, such as doping metal and non-metal elements (Jiang et al. 2017; Vu et al. 2018; Huang et al. 2019; Wei et al. 2020), semiconductor integration (Ong et al. 2016; Raziq et al. 2020), defect engineering (Jiang et al. 2020; Huang et al. 2021), surface modification (Patnaik, Martha, and Parida 2016; Du et al. 2020), morphology modification (Mahvelati-Shamsabadi and Lee 2020) etc.

This chapter is devoted to g-C_3N_4-based nanostructures and gives an outline for their modification strategies for photocatalytic applications. Specific attention will be paid to different kinds of g-C_3N_4-based nanocomposites. It was also will discuss applying nanocomposites in the degradation of various organic pollutants. Finally, challenges and opportunities for improving the performance of photocatalytic systems are discussed.

3.2 PURE g-C_3N_4

Among various allotropes of carbon nitrides, g-C_3N_4 is known as the most stable allotrope. The 2D architecture of g-C_3N_4 is similar to graphite, in which some of the carbon atoms have substituted by nitrogen and make a planar structure with sp^2 hybridization. S-triazine, along with tri-s-triazine rings in an individual sheet of g-C_3N_4 were thought of as tectonic units. However, according to density functional theory (DFT) calculations, the tri-s-triazine-based structure (heptazine) has higher stability and tri-s-triazine units are considered as the main blocks in g-C_3N_4 structure (Figure 3.3) (Zheng et al. 2012). This conjugated structure and its unique electronic properties give rise to researchers' interest in the semiconducting materials field.

3.2.1 Electronic Structure of g-C_3N_4

The study of the electronic structure of g-C_3N_4 is based on both theoretical (DFT calculations) and experimental (UV–vis spectroscopy) findings (Dong et al. 2014). Wang et al. (2009) leveraged DFT calculations to investigate the reduction and oxidation capability of g-C_3N_4 conduction and valance bands. Evidently, by investigation of the HOMO-LUMO band gap during the condensation path to g-C_3N_4 formation, polymeric melon has a direct band gap of 2.6 eV, which decreases to 2.1 eV to form g-C_3N_4. Their findings revealed that the valance band (VB) mainly consists of nitrogen Pz orbitals, while conduction band (CB) is predominantly driven by Pz orbitals of carbon (Figure 3.4).

3.2.2 Photocatalytic Applications of g-C_3N_4

Compared to TiO_2, pure g-C_3N_4 represented higher photocatalytic performance under visible light when dealing with organic pollutants and dyes due to its moderate 2.7 eV

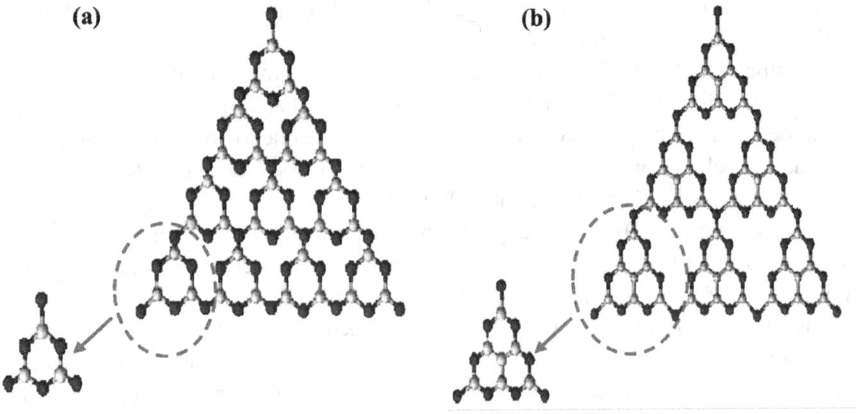

FIGURE 3.3 Tectons in a single layer of g-C_3N_4: (a) s-triazine and (b) tri-s-triazine. (Adapted from Mousavi et al. 2018.)

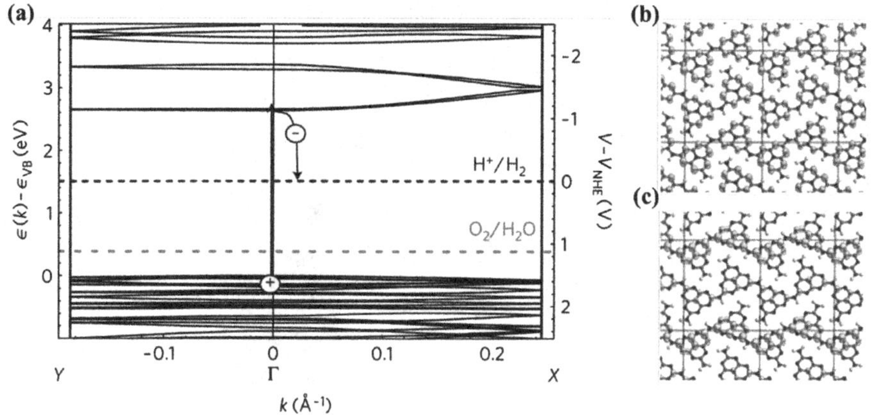

FIGURE 3.4 (a) DFT band structure, (b) VB Kohn–Sham orbitals, and (c) CB of melon. Gray, blue, and white atoms are representing carbon, nitrogen, and hydrogen, respectively. (Adapted from Wang et al., 2009.)

band gap, which is analogous to the absorption edge of 460 nm. In order to estimate the photocatalyst ability in reduction and oxidation reactions, knowing about the band structures is of great importance. The rate of charge carriers' generation and their separation are highly dependent on the position of band energies. For this purpose, the CB and VB potentials of semiconductors can be determined using the following equations:

$$E_{CB} = \chi - E^{e} - 0.5E_{g} \tag{3.1}$$

$$E_{VB} = E_{CB} + E_{g} \tag{3.2}$$

$$\chi = \left[x(A)^{a} x(B)^{b} x(C)^{c} \right]^{1/(a+b+c)} \tag{3.3}$$

In which E_{CB} and E_{VB} depict the CB and VB edge potentials, respectively. χ is the semiconductor absolute electronegativity; E^{e} illustrates the energy of free electrons on the hydrogen scale, and E_{g} stands for the band gap energy of the photocatalyst (Bi 2014; Li et al., 2015a). According to Equations (3.1–3.3), the VB, CB, and χ values for g-C_3N_4 are calculated to be +1.57, −1.13, and 4.73 eV, respectively. Suitable energy levels enable g-C_3N_4 to absorb visible-light wavelengths and be used in a wide range of photocatalytic reactions, including CO_2 reduction, water splitting, organic transformation, and so on. Figure 3.5 represents practical and potential applications of g-C_3N_4-based materials. Despite the outstanding properties of g-C_3N_4, its photocatalytic performance in practical implementations is limited because of (i) its low efficiency in the absorption of visible light (only blue light), (ii) high charge recombination rate, which causes the restricted formation of radical species during photocatalytic processes, and (iii) low specific surface area (<10 m^2g^{-1}), which results in limited reactive sites and light-harvesting capabilities (Mamba and Mishra 2016).

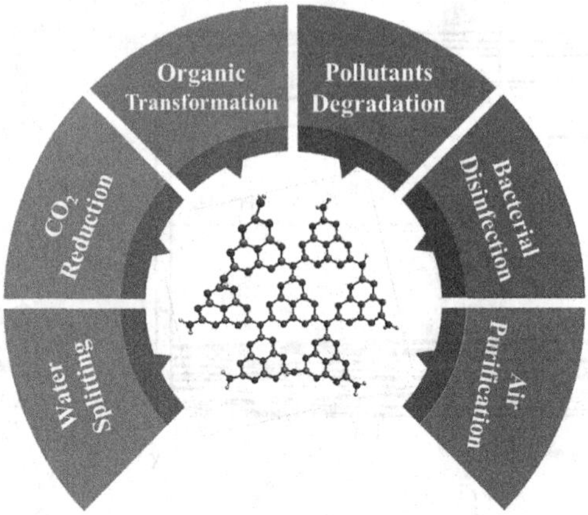

FIGURE 3.5 Photocatalytic utilizations of g-C_3N_4-based materials.

3.2.3 Modified g-C_3N_4 and Its Derivatives

Considering the above-mentioned drawbacks, modification of g-C_3N_4 structure is crucially important. Several approaches have been applied to address these limitations, including (1) modification of band gap, which promotes light-absorption ability, (2) morphology engineering, including design and fabrication of various nanostructures such as quantum dots, nanotubes, nanowires, and ultrathin nanosheets, which can boost the specific surface area, and (3) strategies for dealing with charge carriers recombination such as doping with non-metal and metal atoms, and construction of g-C_3N_4-based nanocomposites. Figure 3.6 summarized the applied methods for modification of g-C_3N_4.

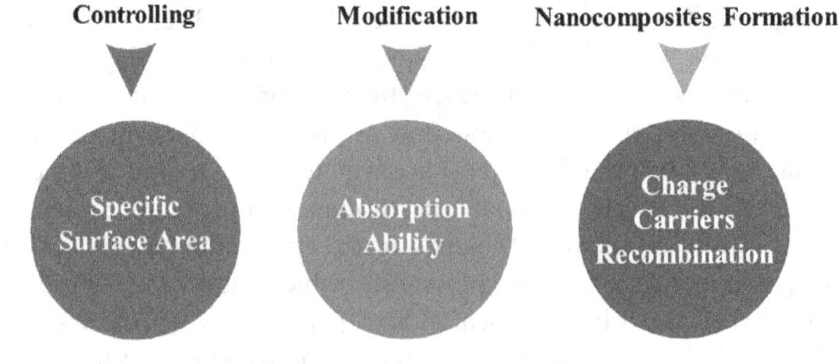

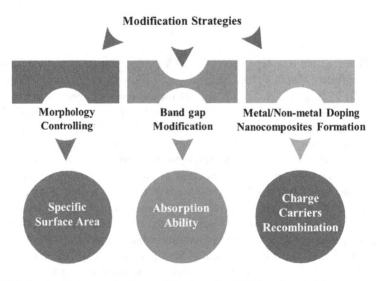

FIGURE 3.6 Applied strategies for modification of g-C_3N_4-based materials.

3.3 g-C_3N_4-BASED NANOCOMPOSITES

According to the typical photocatalytic mechanism, when a semiconductor is irradiated with photons, which have equal or bigger energy than the semiconductor band gap, electrons excitation will happen, and they move to the CB; consequently, holes will be produced in the VB. Then, the photoinduced electrons and holes can transport to the semiconductor surface to contribute to the oxidation and reduction reactions through the production of different reactive species. However, during this pathway, they may also recombine and produce heat, decreasing photocatalytic efficiency. In order to overcome this issue, constructing a nanocomposite is a well-planned strategy, in which the effective separation of charge carriers and faster transferring of them to the surface of the photocatalyst take place, which inhibits the recombination and improves the photocatalytic performance. In this regard, several semiconductors, such as ZnO (Kumaresan et al. 2020), TiO_2 (Deng et al. 2020; Gündo, Park, and Öztürk 2020), Ag_3PO_4 (Shi et al. 2020; Tang et al. 2020), AgBr (Chen et al. 2020a), AgCl, Ag_3VO_4 (Zhu et al. 2015), Ag_2CrO_4 (Che et al. 2018), In_2O_3 (He et al. 2020; Cao et al. 2014), SnO_2 (Zhu et al. 2019), MoS_2 (Wu et al. 2020c; Zhang et al. 2020d) and CdS (Pan et al. 2021) have been widely used for g-C_3N_4-based nanocomposites fabrication.

3.3.1 Synthesis and Different Types of g-C_3N_4-Based Nanocomposites

Due to the extraordinary stability of g-C3N4, various approaches have been applied to synthesize g-C_3N_4-based nanocomposites, such as solvothermal/hydrothermal, calcination, sol-gel, microwave, and so forth. Hydro(solvo) thermal synthesis method is usually carried out using an autoclave under specific pressure and temperature. This technique is beneficial for controlling the crystallinity and morphology of nanomaterials. Furthermore, the process is facile and environmentally friendly. For example, Chou et al. prepared BiO_xI_y/g-C_3N_4 composites via the hydrothermal method and evaluated their effective performance by photocatalytic degradation of crystal violet (Chou et al. 2016). In another study, Lu et al. (2017b) synthesized C-TiO_2/g-C_3N_4 nanocomposite for the decomposition of MO.

So far, the utilization of diverse procedures for synthesizing g-C_3N_4-based nanomaterials has led to various morphologies (Figure 3.7) (Lu et al. 2014; Tahir et al. 2014; Guo et al. 2016; Wu et al. 2017). According to the literature, the common method for preparing pure g-C_3N_4 is the calcination of nitrogen-rich precursors. In addition, g-C_3N_4-based nanocomposites are obtained by calcination of the mixture of the specified precursors with prefabricated catalysts at controlled temperature. A great case in point is the preparation of ZrO_2/g-C_3N_4 nanocomposite via calcination method by Ismael et al., which was employed for photocatalytic splitting of water under irradiation of visible light (Ismael, Wu, and Wark 2019). In another work, Ismael and Wu used a simple calcination approach to prepare $LaFeO_3$/g-C_3N_4 nanocomposite and tested for RhB and 4-CP degradation (Ismael and Wu 2019). Despite the plus points of the calcination method, such as its simplicity and rapidness, some drawbacks limit its application. This technique involves the mixing of precursors only in solid phases, and as a result, the spread of nanoparticles on the surface of carbon nitride as a support is commonly nonhomogeneous. The sol-gel method has

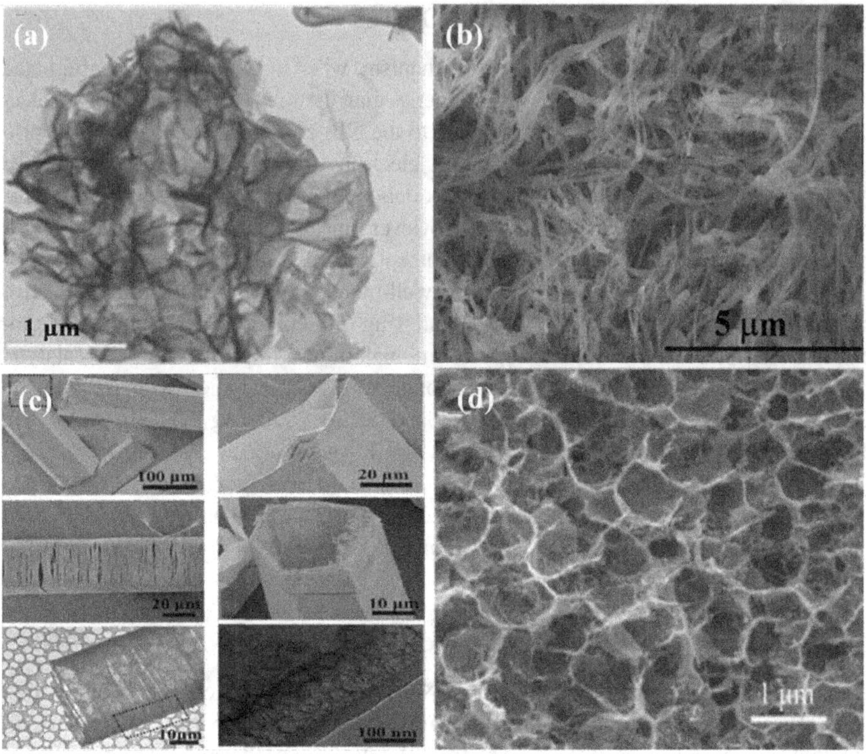

FIGURE 3.7 SEM and TEM images of g-C_3N_4 with different morphologies (a) nanosheets, (b) nanowires, (c) nanotubes, and (d) carbon/3D g-C_3N_4. (Adapted from Lu et al. 2014; Tahir et al. 2014; Guo et al. 2016; Wu et al., 2017.)

attracted a great deal of attention because of the low amount of consumed energy during the process, suitable homogeneity, appropriate crystallinity, and higher surface area of the products (Patil, Aruna, and Ekambaram 1997; Mukasyan and Dinka 2007). For example, g-C_3N_4/TiO_2 nanocomposites were prepared via the sol-gel method to degrade dyes (Li et al. 2016a). Furthermore, several g-C_3N_4-based nanocomposites, such as $AgNbO_3$/g-C_3N_4 (Chen et al. 2019a) and MoS_2/g-C_3N_4 (Monga et al. 2020, were prepared using the microwave method.

The formation of binary and ternary nanocomposites through making the junction between two or three semiconductors is one of the most promising approaches to successfully separating photogenerated charge carriers. As a result of heterojunction formation, electrons and holes can transfer from one semiconductor to another with different energy levels. This method has gained a great deal of attention from many researchers due to the synergistic effect between the components, resulting in improved charge separation, a broader range of light absorption, and finally, improved performance in photocatalytic reactions. Many studies were carried out to form g-C_3N_4-based composites. A combination of transition metal oxides, mostly active under UV light irradiation, with g-C3N4 has been reported as a strategy for making them practically useful in visible regions. Evidently, several semiconductors with a broad band gap, including ZnO, TiO_2, SnO_2 (Yin et al. 2014), and CeO_2 (Huang et al., 2013a) were applied to synthesis g-C_3N_4-based composites for boosting their response to the visible region. Moreover, the combination of g-C_3N_4 with many narrow band gap semiconductors such as Cu_2O (Peng et al. 2014), MoO_3 (He et al. 2014), WO_3 (Doan et al. 2014), and In_2O_3 (Cao et al. 2014) has also been reported with many activities.

Generally, based on photocatalytic mechanisms, the formed heterojunctions can be categorized into three types of type I (Figure 3.8a), type II (Figure 3.8b), and Z-scheme (Figure 3.8c, d). Type II and Z-scheme mechanisms have attracted more attention from researchers (Huang et al. 2018). As shown in Figure 3.8a, in type I, the separation of electrons and holes is inefficient; because both of them are transferring toward one of the semiconductors, and the recombination of charge carriers is highly probable. Then, this configuration will not considerably affect the photocatalytic ability. In type II heterojunction systems, photogenerated holes migrate from VB of one semiconductor to the other. On the other hand, CB electrons move in the opposite direction, which accumulates electrons and holes in different semiconductors, effectively diminishing charges recombination. Fabrication of this type of heterojunction is a straightforward way of boosting photoactivity. Therefore, the design of these systems is highly desired, and most of the reported g-C_3N_4-based composites have type II configuration.

Nevertheless, due to the migration of electrons and holes to more positive CB and more negative VB, respectively, photogenerated charge carriers in type II systems have lowered redox abilities. Furthermore, the existence of electrostatic repulsion between similar charges may bring difficulties in the migration of electrons and holes in the CB and VB of the semiconductors (Low et al. 2017; Jourshabani, Lee, and Shariatinia 2020). To overcome these pitfalls, Z-scheme heterojunction is proposed as a novel

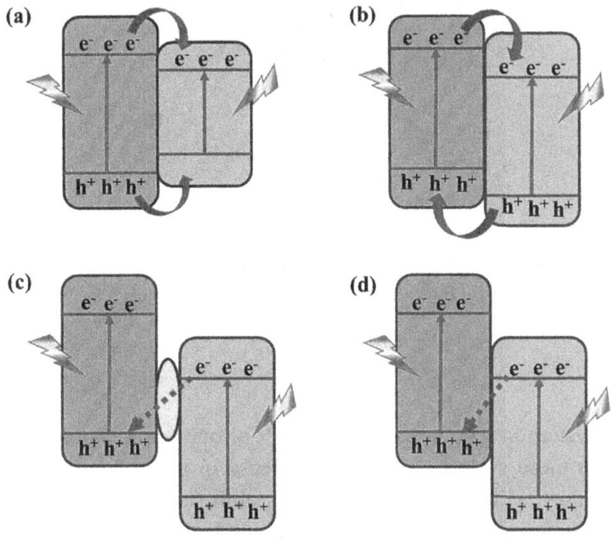

FIGURE 3.8 (a) Type I, (b) type II, (c) indirect Z-scheme, and (d) direct Z-scheme heterojunctions.

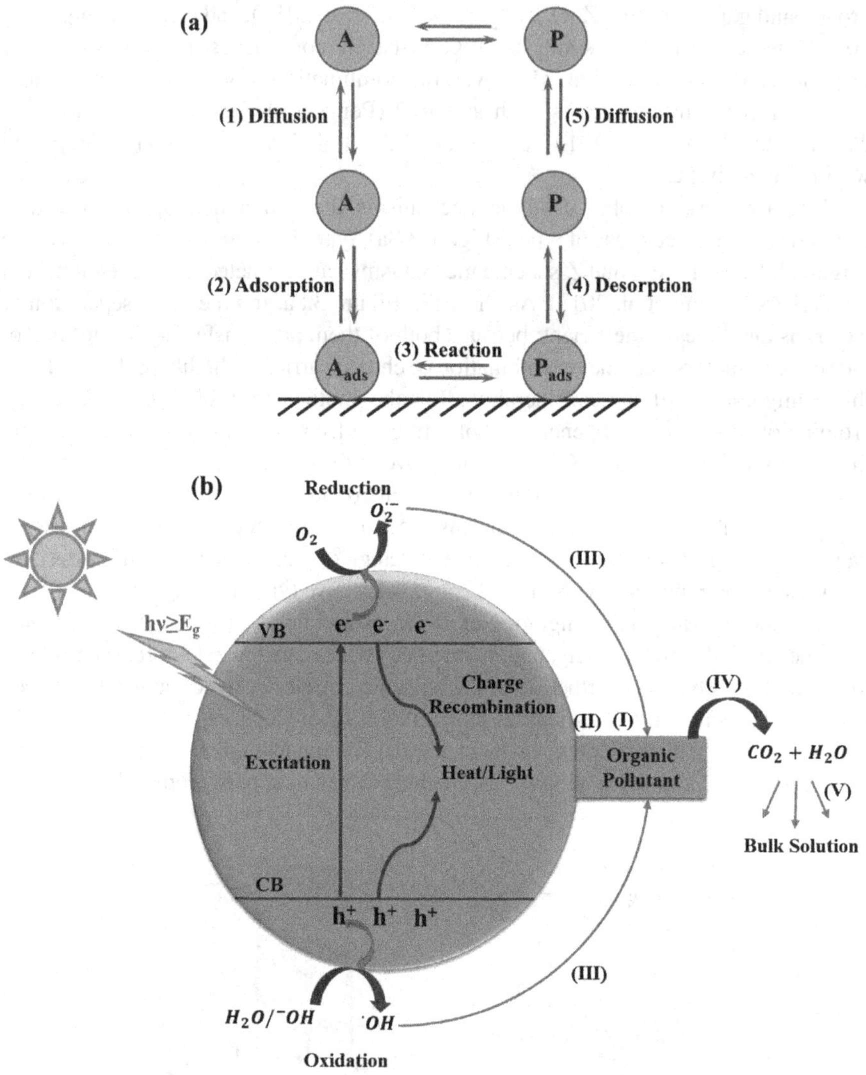

FIGURE 3.9 Key stages in (a) reaction of a pollutant (A) over the surface of a solid photocatalyst, and (b) photocatalytic degradation of organic contaminants.

heterojunction system in which the energy levels of one semiconductor are lower than the other one. In these systems, despite a decrease in the number of photogenerated charge carriers, they will indicate higher redox abilities due to the recombination of electrons of the semiconductor with lower CB and holes on the other one (Ismael 2020).

3.3.2 Basic Principles of Photocatalytic Degradation Reaction

Photodegradation of pollutants in aqueous solutions is relatively a complex procedure and involves plenty of pathways. Generally, pollutants decomposition in these processes can be sum up in five stages. At first, organic pollutants move from the bulk solution to the semiconductor surface (i). Then, the adsorption of the pollutants onto the surface of the photocatalyst takes place (ii). In the third step, oxidizing species degrade the organic compounds (iii). This is followed by the desorption of products from the semiconductor surface (iv). Finally, the diffusion of the generated products from the semiconductor surface to the bulk solution occurs (v) (Willner and Eichen 1987; Herrmann 1999; Pirkanniemi 2002). Figure 3.9a, b exhibits the key steps in the reaction of a pollutant (A) over the surface of a solid photocatalyst and a detailed description of photocatalytic degradation of pollutants.

3.4 g-C_3N_4-BASED COMPOSITES FOR DEGRADATION REACTIONS

Environmental pollution, namely, pollution in water resources, as a result of industrialization, is the main problem for humanity, hindering sustainability and damaging the survival of the natural environment (Gong et al. 2018; Xue et al. 2018; He et al. 2019a). Therefore, applying different effective strategies to deal with water pollutants is urgently necessary. In this context, the photocatalytic process is the most promising, low-cost, and efficient method for mineralizing polluted wastewater (Chen et al. 2020b). Construction of different heterojunctions containing g-C_3N_4 is extensively used for the decomposition of organic contaminants such as rhodamine B (RhB), methyl orange (MO), methylene blue (MB), and antibiotics. Figure 3.10a–c shows the molecular structure of the commonly used dyes for photodegradation study. With its suitable band position under the light irradiation, O_2 molecules can be reduced to $^{\bullet}O_2^-$ via photogenerated electrons, while holes oxidize ^{-}OH to $^{\bullet}OH$. Accordingly, the produced active species accelerate the mineralization of organic pollutants to CO_2 and H_2O (Teixeira et al., 2018). In this context, a variety of carbon nitride-based nanocomposites were used. Lin et al. employed SiO_2/g-C_3N_4 core-shell nanospheres for the degradation of RhB. The efficiency of degradation reached 94.3% after 150 min irradiation (Lin et al., 2015). Li et al. (2016a) prepared g-C_3N_4/TiO_2 composites with the sol-gel method for photocatalytic degradation of MB. The degradation efficiency reached 92% after 360 min irradiation (). Recently Li et al. (2020d) employed a g-C_3N_4/$BiVO_4$ core-shell structure for degradation of MO. The prepared Z-scheme heterojunction decomposed 94.2% of MO during 50 min irradiation of visible light.

3.4.1 Type II Heterojunction

Generally, according to the charge transport pathways, most of the g-C_3N_4-based nanomaterials, which have been reported in the previous studies, are type II

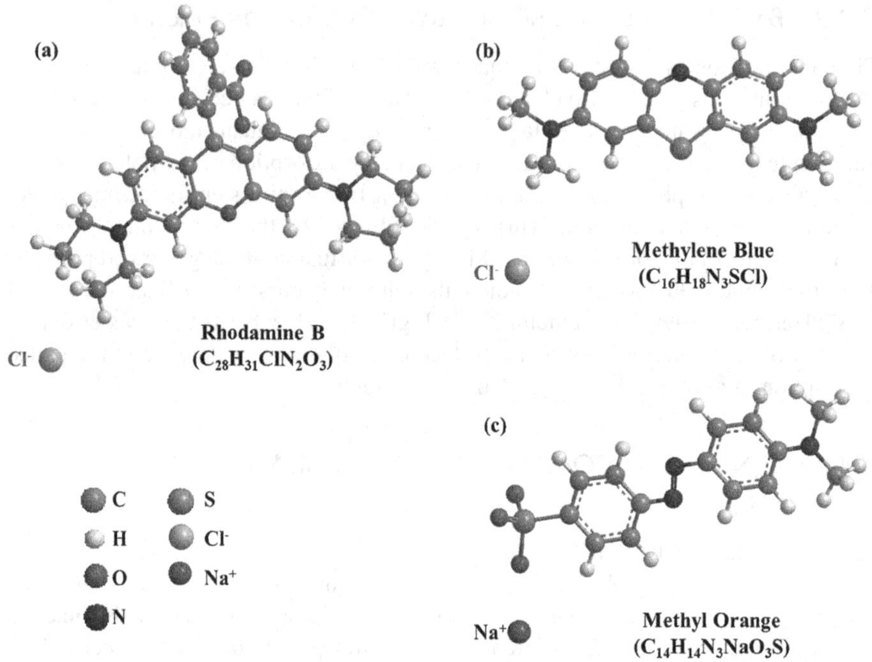

FIGURE 3.10 Molecular structure of three commonly used dyes for photodegradation study: (a) rhodamine B, (b) methylene blue, and (c) methyl orange.

heterojunction systems. Several semiconductors such as TiO_2 (Fang et al. 2020; Lin, Yu, and Huang 2020), CdS (Li et al. 2020b), MoS_2 (Chen et al. 2019b) and ZnO (Jin et al. 2020) have been used for construction of type II heterojunction with g-C_3N_4, that substantially accelerate the separation of photoinduced charge carriers (Li et al. 2020c). For instance, Liu et al. synthesized N-doped $KTiNbO_5$/g-C_3N_4 heterojunction via a simple calcination method (Liu et al. 2018). According to Figure 3.11a–c, both of the components, as well as the nanocomposite, have a layered structure. PL and time-resolved PL techniques were applied to evaluate the recombination of the electron/hole pairs. According to these findings, the emission intensity of the nanocomposite notably decreased and the lifespan of electron/hole pairs increased because of their efficient separation (Figure 3.11d, e). The fabricated samples were employed for photocatalytic decontamination of bisphenol A and rhodamine B contaminants. The high photocatalytic performance is accredited with ordered layered heterojunction containing a type II mechanism for charge transfer.

In a different study, the heterojunctions of porous g-C_3N_4 with Sb-doped SnO_2, which were fabricated by Huang et al., exhibited improved performance in photocatalytic oxidation of isopropanol due to the formation of type II heterojunction between the components (Yang et al. 2018). Table 3.1 summarizes recent research

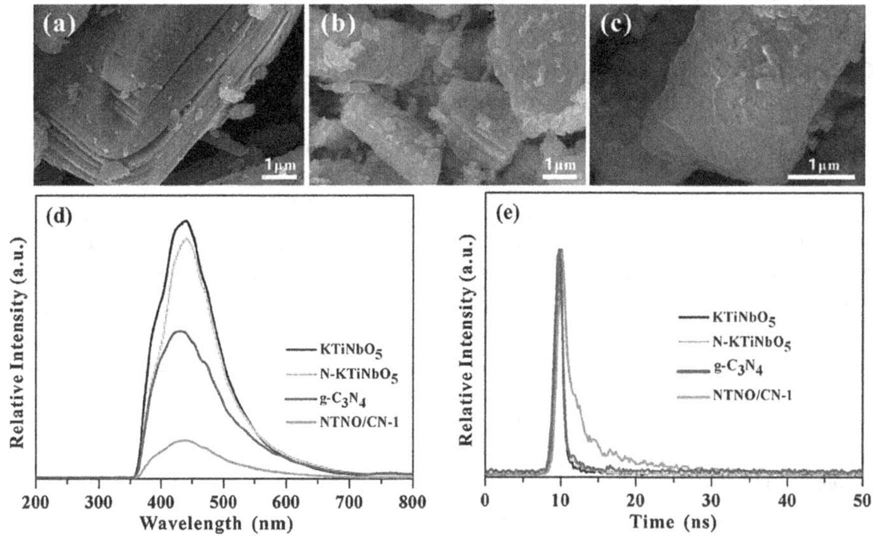

FIGURE 3.11 SEM images of (a) $KTiNbO_5$, (b) N-$KTiNbO_5$ and (c) NTNO/CN-1, (d) PL spectra and (e) TR-PL decay spectra of selected samples. (Adapted from Liu et al. 2018.)

about the application of g-C_3N_4-based nanocomposites with type II configuration in the photocatalytic degradation of organic pollutants.

3.4.2 Z-Scheme Heterojunction

In the photocatalytic systems with Z-scheme pathway of electron-hole pairs, the redox ability of photocatalyst and separation of charge carriers is improved since each component of the composite conducts just one oxidation or reduction reaction (Ghosh 2020). g-C_3N_4-based Z-scheme photocatalytic structures are classified into three groups: Z-scheme systems with a metal mediator between g-C_3N_4 and the combined semiconductor; Z-scheme systems with a carbon conductor between g-C_3N_4 and the semiconductor; and direct Z-scheme systems.

In Z-scheme photocatalytic systems with a metal mediator, the noble metals act as an electron-transfer bridge between the components. Lu et al. (2017a) reported g-C_3N_4/Ag/MoS_2 plasmonic photocatalyst for degradation of RhB. As can be seen in Figure 3.12, the separation of charges takes place via the Z-scheme mechanism.

In another study, Cu_2O/Cu/g-C_3N_4 composites were prepared through a single-step reduction strategy (Bao and Chen 2017). The metallic Cu had a charge separation mediator role, and the improved photocatalytic ability for degradation of MO was assigned to the Z-scheme charge separation mechanism.

Studies revealed that carbonaceous materials such as carbon dots and graphene oxide can operate as electron-transmission bridges and assist the separation of

TABLE 3.1
Representative Type II g-C_3N_4-Based Heterojunctions for Photocatalytic Decontamination of Organic Pollutants

Photocatalysts	Pollutant	Light Source	Improvement (relative to g-C_3N_4)	References
g-C_3N_4/MoS_2	RhB	300 W Xe	~5.6	Li et al. (2016b)
Sb-doped SnO_2/g-C_3N_4	IPA	300 W Xe	~9.1	Yang et al. (2018)
$BiVO_4$/g-C_3N_4	RhB	300 W Xe ($\lambda > 420$ nm)	~4.3	Sun et al. (2019)
Ni–Mn–LDH/g-C_3N_4	RhB	Hg ($\lambda > 420$ nm)	~24.2	Shakeel et al. (2019)
SnS_2/g-C_3N_4	RhB	300 W Xe ($400 < \lambda < 780$)	~4	Zhang et al. (2015)
Ti^{3+}-doped TiO_2/g-C_3N_4	MB	30 W, LED	~7.6	Li et al. (2015b)
$SnNb_2O_6$/g-C_3N_4	MB	500 W tungsten ($\lambda > 420$ nm)	~3.9	Zhang et al. (2016)
WO_3/g-C_3N_4	MB	300 W Xe ($\lambda > 400$ nm)	~2.9	Huang et al., (2013b)
g-C_3N_4/N-$KTiNbO_5$	RhB	300 W Xe	~2	Liu et al. (2018)
g-C_3N_4/$Bi_4O_5I_2$	RhB	Visible light ($\lambda > 420$ nm)	~1.9	Tian et al. (2016)
g-C_3N_4/CeO_2	MB	300 W Xe	~3.1	Huang et al. (2013a)
g-C_3N_4/BiOCl	4-CP	300 W Xe	~5.3	Wang et al. (2018)
g-C_3N_4/$Bi_2O_2CO_3$	RhB	500 W Xe	~3	Huang et al. (2016)

charge carriers. Accordingly, several Z-scheme photocatalytic systems with a carbon mediator between g-C_3N_4 and various semiconductors have been reported. Compared to noble metal mediators, in these systems, the ohmic resistance for electron transfer is low. Lu et al. (2019) reported g-C_3N_4/RGO/WO_3 photocatalyst, in which reduced graphene oxide (RGO) acts as an electron mediator. The photocatalyst was prepared using the photoreduction method and employed for the degradation of ciprofloxacin (CIP). The PL technique, along with photoelectrochemical measurements, demonstrated that RGO effectively increased charge carriers' separation and transfer. Bao and Chen (2018) reported the synthesis of BiOBr/RGO/protonated g-C_3N_4 for decomposition of RhB. On account of the effective charge separation, 10% BiOBr/RGO/pg-C_3N_4 mineralized 88% of RhB to CO_2 and H_2O.

In the case of direct Z-scheme mechanism, g-C_3N_4 has heterojunction with a semiconductor containing appropriate band edges, in which electrons move from the CB of one semiconductor toward the VB of the other one. This contact has a significant influence on the photocatalytic performance of the formed heterojunction. Therefore, besides band alignments, considerations about the geometrical

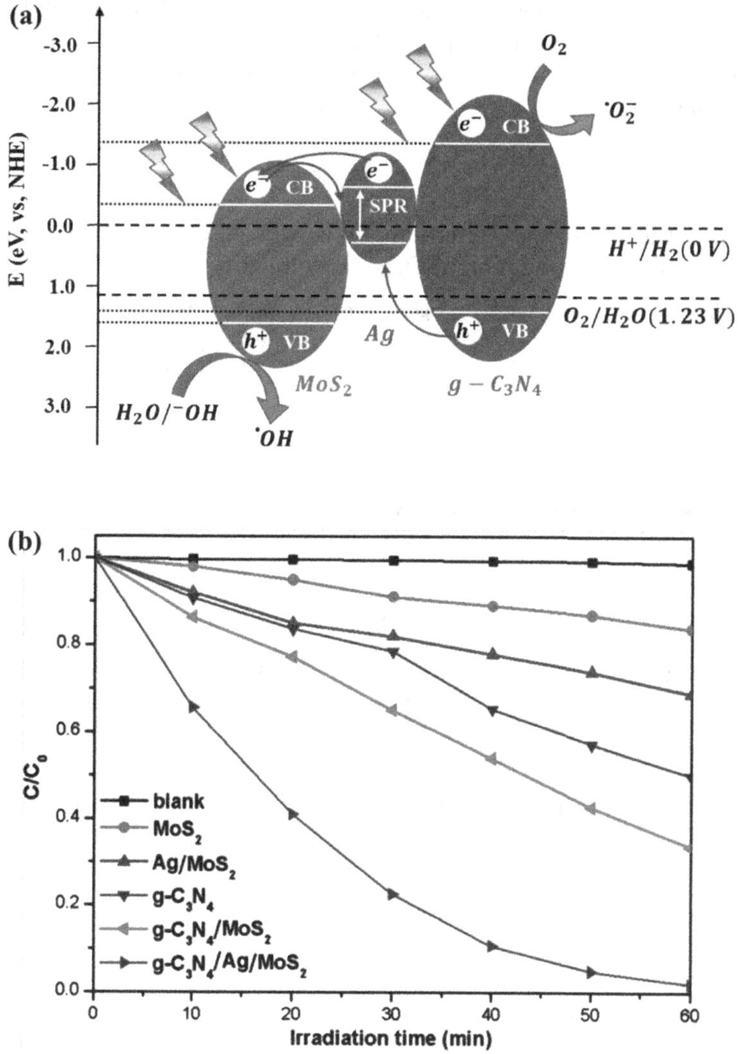

FIGURE 3.12 (a) Proposed mechanism for charges separation in the g-C_3N_4/Ag/MoS_2 nanocomposites, (b) Photocatalytic degradation of RhB over different samples. (Adapted from Lu et al. 2017a.)

configurations should also be taken into account. Zhang et al. (2018) fabricated direct Z-scheme g-C_3N_4/$Ag_2VO_2PO_4$ photocatalyst via the hydrothermal procedure. The prepared photocatalyst was utilized to degrade MO and the g-C_3N_4/$Ag_2VO_2PO_4$-0.4 exhibited the best photoactivity, which was increased about 4- and 3-folds relative to the $Ag_2VO_2PO_4$ and g-C_3N_4, respectively. In another study, Dadigala et al. (2019) synthesized direct Z-scheme V_2O_5 nanorods/g-C_3N_4 nanosheets (VONRs/CNNs) for degradation of Congo red (CR) under sunlight.

The photoactivity of the optimized composite (4-VONRs/CNNs) for CR degradation raised 18.69 and 9.33-fold compared with the g-C_3N_4 and g-C_3N_4 nanosheets, respectively. Representative examples of g-C_3N_4-based Z-scheme photocatalysts can be found in Table 3.2.

TABLE 3.2
A Number of g-C_3N_4-Based Heterojunctions with Z-Scheme Configuration for Photocatalytic Degradation of Organic Pollutants

Photocatalysts	Pollutant	Light Source	Improvement (relative to g-C_3N_4)	References
g-C_3N_4/BiOBr/Au	RhB	300 W Xe (λ = 380 nm)	~20	Bai et al. (2016)
g-C_3N_4/Ag/MoS_2	RhB	300 W Xe (λ > 420 nm)	~20	Lu et al. (2017a)
Cu_2O/Cu/g-C_3N_4	MO	300 W Xe (λ > 420 nm)	~4	Bao and Chen (2017)
g-C_3N_4/Au/Bi_2WO_6	RhB	300 W Xe (λ > 400 nm)	~1.62	Li et al. (2020a)
g-C_3N_4/RGO/WO_3	CIP	300 W Xe (λ > 420 nm)	~1.3	Lu et al. (2019)
BiOBr/RGO/pg-C_3N_4	RhB	300 W Xe (λ > 420 nm)	~3	Bao and Chen (2018)
BiOBr/CDs/g-C_3N_4	CIP	300 W Xe (with different cutoff filters)	~3.9	Zhang et al. (2019)
g-C_3N_4/$Ag_2VO_2PO_4$	MO	1000 W halogen (λ > 420 nm)	~3	Zhang et al. (2018)
V_2O_5/g-C_3N_4	CR	Sunlight, 100,000 lx	~18.7	Dadigala et al. (2019)
Bi_2O_3/g-C_3N_4	TC	250 W Xe (λ > 420 nm)	~5.3	Hong et al. (2018)
Co_3O_4/g-C_3N_4	TC	500 W Xe (λ > 420 nm)	~3.3	Wu et al. (2018)
g-C_3N_4/MnO_2	RhB	300 W Xe	~9	Xia et al. (2018)
BiOI/g-C_3N_{4*}	Phenol	60-W LED (λ > 400 nm)	~3	He et al. (2019b)
$NiTiO_3$/g-C_3N_4	RhB	30 W visible LED	~3.13	Huang et al. (2017)
g-C_3N_4/Ag_2WO_4	MO	300 W Xe	~2	Zhu et al. (2017)
MIL-88A/g-C_3N_4	RhB	1000 W iodine Tungsten	~4.7	Shao et al. (2019)
$CuInS_2$/g-C_3N_4	TC	300 W Xe (λ > 420 nm)	~15	Guo et al. (2019)
g-C_3N_4/$BiVO_4$	MO	250 W Xe (λ > 420 nm)	~4.9	Li et al. (2020d)
SnS_2/g-C_3N_4	RhB	300 W Xe (λ > 400 nm)	~50.3	Song et al. (2019)
Bi_2WO_6/g-C_3N_4	RhB	300 W Xe (λ > 420 nm)	~1.46	Guo et al. (2018)
WO_3/g-C_3N_4	RhB	300 W Xe (λ > 420 nm)	~2.1	Chai et al. (2018)

3.4.3 Effect of Environmental Parameters

Many pieces of research exhibited that the photocatalytic performance of g-C_3N_4-based nanomaterials highly relies on various experimental conditions (Xu et al. 2018). The solution pH has a crucially important role on organic pollutants photodegradation via influencing the surface charge of the photocatalysts and pollutants and subsequently on the adsorption of the contaminant molecules on the catalyst surface. When the charge of the photocatalyst surface is opposite of the pollutant's charge, the adsorption of contaminants on the surface of the catalyst will be considerable, and therefore the photocatalytic performance will be improved. On the other hand, pH has an impact on the generation of reactive oxygen species. For example, in higher pH, the formation of $^{\bullet}OH$ is thermodynamically pleasing (Raj et al. 2012). Dissolved oxygen also has significant importance on the production of reactive oxygen species. The reaction between electrons and dissolved oxygen leads to the generation of $^{\bullet}O_2^-$, which can promote degradation of pollutants by its oxidation ability and at the same time by consuming electrons, which prevents recombination of charge carriers (Yan, Li, and Zou 2009). Besides, reaction temperature plays an important role in photocatalytic processes. The redox reactions will be accelerated in high temperatures. Furthermore, the presence of radical scavengers in the solution can affect the degradation rate. Therefore, the formulation and construction of g-C_3N_4-based nanocomposites to achieve an optimized photocatalytic performance, investigation of solution pH, controlling dissolved oxygen, reaction temperature, and the presence of radical scavengers are all of great importance in obtaining desired photocatalytic ability (Zheng, Shen, and Shuai 2017).

3.5 CONCLUSIONS AND FUTURE PERSPECTIVES

All in all, semiconductor photocatalysis is a thoughtful approach to address the environmental crisis. Hence, it is appealing to researchers to design low-cost and sustainable materials in this field. g-C_3N_4, as an earth-abundant photocatalyst, has demonstrated vast potential because of its exceptional characteristics such as great thermal and chemical stability, suitable band gap, and simplicity of preparation procedure. Consequently, g-C_3N_4 has some attractive features in meeting energy and environmental concerns. Nevertheless, some downsides, such as restricted visible-light utilization, limited specific surface area, and high-speed recombination of electrons and holes, have a negative effect on the photocatalytic ability of pure g-C_3N_4. These pitfalls can be controlled by making some improvements through doping elements in g-C_3N_4, morphology modifications, and fabrication of g-C_3N_4-based nanocomposites. However, it is difficult for the modifications such as doping elements or making changes in the morphology of the nanostructure to come across all these problems simultaneously. For instance, elemental doping can provide a narrower band gap and extend the light absorption capacity, while this strategy cannot overcome the low redox ability. This chapter mainly concentrated on g-C_3N_4-based nanocomposites, especially type II and Z-scheme heterojunctions, which can accelerate photogenerated charges separation and strengthens the redox ability in photocatalytic systems. In type II systems, close contact of g-C_3N_4 to other semiconductors with higher or lower band energies can provide a facile pathway for transferring the produced charges. The Z-scheme systems can be divided into three groups of g-C_3N_4/noble metal/semiconductor, g-C_3N_4/carbonaceous materials/semiconductor, and direct g-C_3N_4/semiconductor. These systems can not only enhance the charges separation ability, but also fulfill the redox capability of the systems.

Unlike plenty of works on the construction of g-C_3N_4-based nanocomposites and achieved improvements in their utilization for the degradation of pollutants, there are still some challenges in exploring environmentally friendly synthesis routes and designing efficient photocatalysts for industrial applications, which should be considered. For practical applications, it is necessary to explore a green synthetic pathway to produce g-C_3N_4-based photocatalysts with sufficient stability on a large scale. Moreover, since the visible-light absorbance of g-C_3N_4 is mainly limited to the blue-violet region, the efficiency in harvesting sunlight is low. Computation studies based on DFT, along with a deep investigation of the charges transport mechanism, could provide a roadmap toward designing novel nanocomposites with desirable performance. Hence, extensive research is still necessary to achieve the desired efficiency for the degradation reaction, which is economically acceptable for large-scale utilization.

ACKNOWLEDGMENT

This work is financially supported by the University of Mohaghegh Ardabili and the authors would like to give thanks for this commitment.

REFERENCES

Acharya, R., and K. Parida. 2020. A Review on TiO_2/g-C_3N_4 Visible-Light-Responsive Photocatalysts for Sustainable Energy Generation and Environmental Remediation. *Journal of Environmental Chemical Engineering* 8 (4): 103896.

Ahmed, M. B., J. L. Zhou, H. H. Ngo, W. Guo, N. S. Thomaidis, and J. Xu. 2017. Progress in the Biological and Chemical Treatment Technologies for Emerging Contaminant Removal from Wastewater: A Critical Review. *Journal of Hazardous Materials* 323: 274–298.

Bahnemann, D. 2004. Photocatalytic Water Treatment: Solar Energy Applications. *Solar Energy* 77 (5): 445–459.

Bai, Y., T. Chen, P. Wang, L. Wang, L. Ye, X. Shi, and W. Bai. 2016. Size-Dependent Role of Gold in g-C_3N_4/BiOBr/Au System for Photocatalytic CO_2 Reduction and Dye Degradation. *Solar Energy Materials and Solar Cells* 157: 406–414.

Bao, Y., and K. Chen. 2017. A Novel Z-Scheme Visible Light Driven Cu_2O/Cu/g-C_3N_4 Photocatalyst Using Metallic Copper as a Charge Transfer Mediator. *Molecular Catalysis* 432: 187–195.

Bao, Y., and K. Chen. 2018. Novel Z-Scheme BiOBr/Reduced Graphene Oxide/Protonated g-C_3N_4 Photocatalyst: Synthesis, Characterization, Visible Light Photocatalytic Activity and Mechanism. *Applied Surface Science* 437: 51–61.

Bi, N. 2014. Novel Bi_2S_3/$Bi_2O_2CO_3$ Heterojunction Photocatalysts with Enhanced Visible Light Responsive Activity and Wastewater Treatment. *Journal of Materials Chemistry A* 2: 4208–4216.

Cao, Q., B. Kumru, M. Antonietti, and B. V. K. J. Schmidt. 2020. Graphitic Carbon Nitride and Polymers: A Mutual Combination for Advanced Properties. *Materials Horizons* 7 (3): 762–786.

Cao, S., X. Liu, Y. Yuan, Z. Zhang, and Y. Liao. 2014. Solar-to-Fuels Conversion over In_2O_3/g-C_3N_4 Hybrid Photocatalysts. *Applied Catalysis B, Environmental* 147: 940–946.

Chai, B., C. Liu, J. Yan, Z. Ren, and Z. Wang. 2018. In-Situ Synthesis of WO_3 Nanoplates Anchored on g-C_3N_4 Z-Scheme Photocatalysts for Significantly Enhanced Photocatalytic Activity. *Applied Surface Science* 448: 1–8.

Che, Y., B. Lu, Q. Qi, H. Chang, J. Zhai, K. Wang, and Z. Liu. 2018. Bio-Inspired Z-Scheme g-C_3N_4/Ag_2CrO_4 for Efficient Visible-Light Photocatalytic Hydrogen Generation. *Scientific Reports* 8: 16504.

Chen, M., C. Guo, S. Hou, J. Lv, Y. Zhang, H. Zhang, and J. Xu. 2020a. A Novel Z-Scheme AgBr/P-g-C_3N_4 Heterojunction Photocatalyst: Excellent Photocatalytic Performance and Photocatalytic Mechanism for Ephedrine Degradation. *Applied Catalysis B: Environmental* 266: 118614.

Chen, P., P. Xing, Z. Chen, X. Hu, H. Lin, L. Zhao, and Y. He. 2019a. In-Situ Synthesis of $AgNbO_3$/g-C_3N_4 Photocatalyst via Microwave Heating Method for Efficiently Photocatalytic H_2 Generation. *Journal of Colloid and Interface Science* 534: 163–171.

Chen, T., D. Yin, F. Zhao, K. K. Kyu, B. Liu, D. Chen, K. Huang, L. L. Deng, and L. Li. 2019b. Fabrication of 2D Heterojunction Photocatalyst Co-g-C_3N_4/MoS_2 with Enhanced Solar-Light-Driven Photocatalytic Activity. *New Journal of Chemistry* 43: 463–473.

Chen, Z., S. Zhang, Y. Liu, N. Saleh, S. Omar, S. Wang, and X. Wang. 2020b. Synthesis and Fabrication of G-C_3N_4-Based Materials and Their Application in Elimination of Pollutants. *Science of the Total Environment* 731: 139054.

Chou, S. Y., C. C. Chen, Y. M. Dai, J. H. Lin, and W. W. Lee. 2016. Novel Synthesis of Bismuth Oxyiodide/Graphitic Carbon Nitride Nanocomposites with Enhanced Visible-Light Photocatalytic Activity. *RSC Advances* 6 (40): 33478–33491.

Dadigala, R., R. Bandi, B. R. Gangapuram, A. Dasari, H. H. Belay, and V. Guttena. 2019. Fabrication of Novel 1D/2D V_2O_5/g-C_3N_4 Composites as Z-Scheme Photocatalysts for CR Degradation and Cr (VI) Reduction under Sunlight Irradiation. *Journal of Environmental Chemical Engineering* 7: 102822.

Deng, H., X. Wang, L. Wang, Z. Li, P. Liang, and J. Ou. 2020. Enhanced Photocatalytic Reduction of Aqueous Re (VII) in Ambient Air by Amorphous TiO_2/g-C_3N_4 Photocatalysts: Implications for Tc (VII) Elimination. *Chemical Engineering Journal* 401: 125977.

Doan, A. T., X. D. Nguyen Thi, P. H. Nguyen, V. N. Nguyen Thi, S. J. Kim, and V. Vo. 2014. Graphitic G-C_3N_4-WO_3 Composite: Synthesis and Photocatalytic Properties. *Bulletin of the Korean Chemical Society* 35 (6): 1794–1798.

Dong, G., Y. Zhang, Q. Pan, and J. Qiu. 2014. A Fantastic Graphitic Carbon Nitride (g-C_3N_4) Material: Electronic Structure, Photocatalytic and Photoelectronic Properties. *Journal of Photochemistry & Photobiology, C: Photochemistry Reviews* 20: 33–50.

Dong, Y., Q. Wang, H. Wu, Y. Chen, C. H. Lu, Y. Chi, and H. H. Yang. 2016. Graphitic Carbon Nitride Materials: Sensing, Imaging and Therapy. *Small* 12 (39): 5376–5393.

Du, C., X. Lan, G. An, Q. Li, and G. Bai. 2020. Direct Surface Modi Fi Cation of Graphitic C_3N_4 with Porous Organic Polymer and Silver Nanoparticles for Promoting CO_2 Conversion. *ACS Sustainable Chemistry & Engineering* 8: 7051–7058.

Fang, Y., W. Huang, S. Yang, X. Zhou, C. Ge, Q. Gao, Y. Fang, and S. Zhang. 2020. Facile Synthesis of Anatase/Rutile TiO_2/g-C_3N_4 Multi-Heterostructure for Efficient Photocatalytic Overall Water Splitting. *International Journal of Hydrogen Energy* 45 (35): 17378–17387.

Ghaemmaghami, M., and R. Mohammadi. 2019. Carbon Nitride as a New Way to Facilitate the next Generation of Carbon-Based Supercapacitors. *Sustainable Energy and Fuels* 3 (9): 2176–2204.

Ghoreishian, S. M., K. S. Ranjith, H. Lee, B. Park, M. Norouzi, S. Z. Nikoo, W. S. Kim, Y. K. Han, and Y. S. Huh. 2021. Tuning the Phase Composition of 1D TiO_2 by Fe/Sn Co-Doping Strategy for Enhanced Visible-Light-Driven Photocatalytic and Photoelectrochemical Performances. *Journal of Alloys and Compounds* 851: 156826.

Ghosh, S. (Editor). 2020. *Heterostructured Photocatalysts for Solar Energy Conversion*. 1st Ed. Berkeley, CA: Elsevier.

Gong, X., D. Huang, Y. Liu, G. Zeng, and R. Wang. 2018. Pyrolysis and Reutilization of Plant Residues after Phytoremediation of Heavy Metals Contaminated Sediments: For Heavy Metals Stabilization and Dye Adsorption. *Bioresource Technology* 253: 64–71.

Gong, Y., M. Li, H. Li, and Y. Wang. 2015a. Graphitic Carbon Nitride Polymers: Promising Catalysts or Catalyst Supports for Heterogeneous Oxidation and Hydrogenation. *Green Chemistry* 17 (2): 715–736.

Gong, Y., M. Li, and Y. Wang. 2015b. Carbon Nitride in Energy Conversion and Storage: Recent Advances and Future Prospects. *ChemSusChem* 8 (6): 931–946.

Gündo, P., J. Park, and A. Öztürk. 2020. Preparation and Photocatalytic Activity of G-C_3N_4/TiO_2 Heterojunctions under Solar Light Illumination. *Ceramics International* 46: 21431–21438.

Guo, F., W. Shi, M. Li, Y. Shi, and H. Wen. 2019. 2D/2D Z-Scheme Heterojunction of $CuInS_2$/g-C_3N_4 for Enhanced Visible-Light-Driven Photocatalytic Activity towards the Degradation of Tetracycline. *Separation and Purification Technology* 210: 608–615.

Guo, S., Z. Deng, M. Li, B. Jiang, C. Tian, Q. Pan, and H. Fu. 2016. Phosphorus-Doped Carbon Nitride Tubes with a Layered Micro-Nanostructure for Enhanced Visible-Light Photocatalytic Hydrogen Evolution. *Angewandte Chemie* 128: 1862–1866.

Guo, W., K. Fan, J. Zhang, and C. Xu. 2018. 2D/2D Z-Scheme Bi_2WO_6/Porous-g-C_3N_4 with Synergy of Adsorption and Visible-Light-Driven Photodegradation. *Applied Surface Science* 447: 125–134.

Guo, W., J. Zou, B. Guo, J. Xiong, C. Liu, Z. Xie, and L. Wu. 2020. Pd Nanoclusters/TiO_2(B) Nanosheets with Surface Defects toward Rapid Photocatalytic Dehalogenation of Polyhalogenated Biphenyls under Visible Light. *Applied Catalysis B: Environmental* 277: 119255.

He, D., C. Zhang, G. Zeng, Y. Yang, and D. Huang. 2019a. A Multifunctional Platform by Controlling of Carbon Nitride in the Core-Shell Structure: From Design to Construction, and Catalysis Applications. *Applied Catalysis B: Environmental* 258: 117957.

He, J., P. Lv, J. Zhu, and H. Li. 2020. Selective CO_2 Reduction to HCOOH on a Pt/In_2O_3/g-C_3N_4 Multifunctional Visible-Photocatalyst. *RSC Advances* 10: 22460–22467.

He, R., K. Cheng, Z. Wei, S. Zhang, and D. Xu. 2019b. Room-Temperature in Situ Fabrication and Enhanced Photocatalytic Activity of Direct Z-Scheme BiOI/g-C_3N_4 Photocatalyst. *Applied Surface Science* 465 (August 2018): 964–972.

He, Y., L. Zhang, X. Wang, Y. Wu, H. Lin, L. Zhao, W. Weng, H. Wan, and M. Fan. 2014. Enhanced Photodegradation Activity of Methyl Orange over Z-Scheme Type MoO_3-g-C_3N_4 Composite under Visible Light Irradiation. *RSC Advances* 4 (26): 13610–13619.

Herrmann, J. 1999. Heterogeneous Photocatalysis : Fundamentals and Applications to the Removal of Various Types of Aqueous Pollutants. *Catalysis Today* 53: 115–129.

Hong, Y., C. Li, B. Yin, D. Li, Z. Zhang, and B. Mao. 2018. Promoting Visible-Light-Induced Photocatalytic Degradation of Tetracycline by an Efficient and Stable Beta-Bi_2O_3@g-C_3N_4 Core/Shell Nanocomposite. *Chemical Engineering Journal Journal* 338: 137–146.

Huang, D., X. Yan, M. Yan, G. Zeng, C. Zhou, J. Wan, M. Cheng, and W. Xue. 2018. Graphitic Carbon Nitride-Based Heterojunction Photoactive Nanocomposites: Applications and Mechanism Insight: Review-Article. *ACS Applied Materials and Interfaces* 10 (25): 21035–21055.

Huang, H., J. Feng, S. Zhang, H. Zhang, X. Wang, T. Yu, C. Chen, et al. 2020. Molecular-Level Understanding of the Deactivation Pathways during Methanol Photo-Reforming on Pt-Decorated TiO_2. *Applied Catalysis B: Environmental* 272: 118980.

Huang, H., K. Xiao, N. Tian, X. Du, and Y. Zhang. 2016. Dual Visible-Light Active Components Containing Self-Doped $Bi_2O_2CO_3$/g-C_3N_4 2D-2D Heterojunction with Enhanced Visible-Light-Driven Photocatalytic Activity. *Colloids and Surfaces A: Physicochemical and Engineering Aspects* 511: 64–72.

Huang, J., D. Li, R. Li, Q. Zhang, T. Chen, and H. Liu. 2019. An Efficient Metal-Free Phosphorus and Oxygen Co-Doped g-C_3N_4 Photocatalyst with Enhanced Visible Light Photocatalytic Activity for the Degradation of Fluoroquinolone Antibiotics. *Chemical Engineering Journal* 374: 242–253.

Huang, L., Y. Li, H. Xu, Y. Xu, J. Xia, K. Wang, H. Li, and X. Cheng. 2013a. Synthesis and Characterization of CeO_2/g-C_3N4 Composites with Enhanced Visible-Light Photocatatalytic Activity Liying. *RSC Advances* 3 (44): 22269–22279.

Huang, L., H. Xu, Y. Li, H. Li, X. Cheng, J. Xia, Y. Xu, and G. Cai. 2013b. Visible-Light-Induced WO_3/g-C_3N_4 Composites with Enhanced Photocatalytic Activity. *Dalton Transactions* 42 (24): 8606–8616.

Huang, S., F. Ge, J. Yan, H. Li, X. Zhu, Y. Xu, and H. Xu. 2021. Synthesis of Carbon Nitride in Moist Environments: A Defect Engineering Strategy toward Superior Photocatalytic Hydrogen Evolution Reaction. *Journal of Energy Chemistry* 54: 403–413.

Huang, Z., X. Zeng, K. Li, S. Gao, Q. Wang, and J. Lu. 2017. Z-Scheme $NiTiO_3$/G-C_3N_4 Heterojunctions with Enhanced Photoelectrochemical and Photocatalytic Performances under Visible LED Light Irradiation. *ACS Applied Materials & Interfaces* 9: 41120–41125.

Ismael, M. 2020. A Review on Graphitic Carbon Nitride (g-C_3N_4) Based Nanocomposites : Synthesis, Categories, and Their Application in Photocatalysis. *Journal of Alloys and Compounds* 846: 156446.

Ismael, M., and Y. Wu. 2019. A Facile Synthesis Method for Fabrication of $LaFeO_3$/g-C_3N_4 Nanocomposite as Efficient Visible-Light-Driven Photocatalyst for Photodegradation of RhB and 4-CP. *New Journal of Chemistry* 43 (35): 13783–13793.

Ismael, M., Y. Wu, and M. Wark. 2019. Photocatalytic Activity of ZrO_2 Composites with Graphitic Carbon Nitride for Hydrogen Production under Visible Light. *New Journal of Chemistry* 43 (11): 4455–4462.

Jiang, J., X. Wang, Y. Liu, Y. Ma, T. Li, Y. Lin, T. Xie, and S. Dong. 2020. Photo-Fenton Degradation of Emerging Pollutants over Fe-POM Nanoparticle/Porous and Ultrathin g-C_3N_4 Nanosheet with Rich Nitrogen Defect Degradation Mechanism, Pathways, and Products Toxicity Assessment. *Applied Catalysis B: Environmental* 278: 119349.

Jiang, L., X. Yuan, Y. Pan, J. Liang, G. Zeng, Z. Wu, and H. Wang. 2017. Doping of Graphitic Carbon Nitride for Photocatalysis: A Reveiw. *Applied Catalysis B: Environmental* 217: 388–406.

Jin, C., W. Li, Y. Chen, R. Li, J. Huo, Q. He, and Y. Wang. 2020. Efficient Photocatalytic Degradation and Adsorption of Tetracycline over Type-II Heterojunctions Consisting of ZnO Nanorods and K-Doped Exfoliated Nanosheets. *Industrial and Engineering Chemistry Research* 59 (7): 2860–2873.

Jourshabani, M., B. K. Lee, and Z. Shariatinia. 2020. From Traditional Strategies to Z-Scheme Configuration in Graphitic Carbon Nitride Photocatalysts: Recent Progress and Future Challenges. *Applied Catalysis B: Environmental* 276 (April): 119157.

Kumar, A., P. Raizada, P. Singh, R. V. Saini, A. K. Saini, and A. Hosseini-Bandegharaei. 2020. Perspective and Status of Polymeric Graphitic Carbon Nitride Based Z-Scheme Photocatalytic Systems for Sustainable Photocatalytic Water Purification. *Chemical Engineering Journal* 391: 123496.

Kumar, S., S. Karthikeyan, and A. Lee. 2018. G-C_3N_4-Based Nanomaterials for Visible Light-Driven Photocatalysis. *Catalysts* 8: 74.

Kumaresan, N., M. M. Angelin, M. Sarathbavan, K. Ramamurthi, K. Sethuraman, and R. R. Babu. 2020. Synergetic Effect of G-C_3N_4/ZnO Binary Nanocomposites Heterojunction on Improving Charge Carrier Separation through 2D/1D Nanostructures for e Ff Ective Photocatalytic Activity under the Sunlight Irradiation. *Separation and Purification Technology* 244: 116356.

Li, C., Z. Sun, Y. Xue, G. Yao, and S. Zheng. 2016a. A Facile Synthesis of G-C_3N_4/TiO_2 Hybrid Photocatalysts by Sol-Gel Method and Its Enhanced Photodegradation towards Methylene Blue under Visible Light. *Advanced Powder Technology* 27 (2): 330–337.

Li, J., E. Liu, Y. Ma, X. Hu, J. Wan, L. Sun, and J. Fan. 2016b. Synthesis of MoS_2/g-C_3N_4 Nanosheets as 2D Heterojunction Photocatalysts with Enhanced Visible Light Activity. *Applied Surface Science* 364: 694–702.

Li, J., X. Wang, G. Zhao, C. Chen, Z. Chai, A. Alsaedi, T. Hayat, and X. Wang. 2018. Metal-Organic Framework-Based Materials: Superior Adsorbents for the Capture of Toxic and Radioactive Metal Ions. *Chemical Society Reviews* 47 (7): 2322–2356.

Li, J., W. Zhao, Y. Guo, Z. Wei, M. Han, and H. He. 2015a. Facile Synthesis and High Activity of Novel $BiVO_4$/$FeVO_4$ Heterojunction Photocatalyst for Degradation of Metronidazole. *Applied Surface Science* 351: 270–279.

Li, K., S. Gao, Q. Wang, H. Xu, Z. Wang, B. Huang, Y. Dai, and J. Lu. 2015b. In-Situ-Reduced Synthesis of Ti^{3+} Self-Doped TiO_2/g-C_3N_4 Heterojunctions with High Photocatalytic Performance under LED Light Irradiation. *ACS Applied Materials and Interfaces* 7 (17): 9023–9030.

Li, Q., M. Lu, W. Wang, W. Zhao, G. Chen, and H. Shi. 2020a. Fabrication of 2D/2D g-C_3N_4/Au/Bi_2WO_6 Z-Scheme Photocatalyst with Enhanced Visible-Light-Driven Photocatalytic Activity. *Applied Surface Science* 508 (September 2019): 144182.

Li, X., M. Edelmannová, P. Huo, and K. Kočí. 2020b. Fabrication of Highly Stable CdS/g-C_3N_4 Composite for Enhanced Photocatalytic Degradation of RhB and Reduction of CO_2. *Journal of Materials Science* 55 (8): 3299–3313.

Li, Y., M. Zhou, B. Cheng, and Y. Shao. 2020c. Recent Advances in G-C_3N_4-Based Heterojunction Photocatalysts. *Journal of Materials Science & Technology* 56: 1–17.

Li, Z., C. Jin, M. Wang, J. Kang, Z. Wu, D. Yang, and T. Zhu. 2020d. Novel Rugby-like g-C_3N_4/$BiVO_4$ Core/Shell Z-Scheme Composites Prepared via Low-Temperature Hydrothermal Method for Enhanced Photocatalytic Performance. *Separation and Purification Technology* 232: 115937.

Lin, B., C. Xue, X. Yan, G. Yang, G. Yang, and B. Yang. 2015. Facile Fabrication of Novel SiO_2/g-C_3N_4 Core – Shell Nanosphere Photocatalysts with Enhanced Visible Light Activity. *Applied Surface Science* 357: 346–355.

Lin, Z., B. Yu, and J. Huang. 2020. Cellulose-Derived Hierarchical g-C_3N_4/TiO_2-Nanotube Heterostructured Composites with Enhanced Visible-Light Photocatalytic Performance. *Langmuir* 36 (21): 5967–5978.

Liu, C., H. Zhu, Y. Zhu, P. Dong, H. Hou, Q. Xu, X. Chen, X. Xi, and W. Hou. 2018. Ordered Layered N-Doped $KTiNbO_5$/g-C_3N_4 Heterojunction with Enhanced Visible Light Photocatalytic Activity. *Applied Catalysis B: Environmental* 228: 54–63.

Liu, R., Z. Chen, Y. Yao, Y. Li, W. A. Cheema, and D. Wang. 2020. Recent Advancements in G-C_3N_4-Based Photocatalysts for Photocatalytic CO_2 Reduction: A Mini Review. *RSC Advances* 10: 29408–29418.

Low, J., C. Jiang, B. Cheng, S. Wageh, and A. A. Al-Ghamdi. 2017. A Review of Direct Z-Scheme Photocatalysts. *Small Methods* 1(5): 1700080.

Lu, D., H. Wang, X. Zhao, K. K. Kondamareddy, J. Ding, C. Li, and P. Fang. 2017a. Highly Efficient Visible-Light-Induced Photoactivity of Z-Scheme G-C_3N_4/Ag/MoS_2 Ternary Photocatalysts for Organic Pollutant Degradation and Production of Hydrogen. *ACS Sustainable Chemistry & Engineering* 5: 1436–1445.

Lu, N., P. Wang, Y. Su, H. Yu, N. Liu, and X. Quan. 2019. Construction of Z-Scheme g-C_3N_4/RGO/WO_3 with in Situ Photoreduced Graphene Oxide as Electron Mediator for Effi Cient Photocatalytic Degradation of Ciprofloxacin. *Chemosphere* 215: 444–453.

Lu, X., K. Xu, P. Chen, K. Jia, S. Liu, and C. Wu. 2014. Facile One Step Method Realizing Scalable Production of G-C_3N_4 Nanosheets and Study of Their Photocatalytic H_2 Evolution Activity. *Journal of Materials Chemistry A* 2: 18924–18928.

Lu, Z., L. Zeng, W. Song, Z. Qin, D. Zeng, and C. Xie. 2017b. In Situ Synthesis of C-TiO_2/g-C_3N_4 Heterojunction Nanocomposite as Highly Visible Light Active Photocatalyst Originated from Effective Interfacial Charge Transfer. *Applied Catalysis B: Environmental* 202: 489–499.

Ma, Y., Q. Tang, W. Y. Sun, Z. Y. Yao, W. Zhu, T. Li, and J. Wang. 2020. Assembling Ultrafine TiO_2 Nanoparticles on UiO-66 Octahedrons to Promote Selective Photocatalytic Conversion of CO_2 to CH_4 at a Low Concentration. *Applied Catalysis B: Environmental* 270: 118856.

Mahvelati-Shamsabadi, T., and B. Lee. 2020. Design of Ag/g-C_3N_4 on TiO_2 Nanotree Arrays via Ultrasonic-Assisted Spin Coating as an Efficient Photoanode for Solar Water Oxidation : Morphology Modification and Junction Improvement. *Catalysis Today* 358: 412–421.

Malik, R., and V. K. Tomer. 2021. State-of-the-Art Review of Morphological Advancements in Graphitic Carbon Nitride (g-CN) for Sustainable Hydrogen Production. *Renewable and Sustainable Energy Reviews* 135: 110235.

Mamba, G., and A. K. Mishra. 2016. Graphitic Carbon Nitride (g-C_3N_4) Nanocomposites: A New and Exciting Generation of Visible Light Driven Photocatalysts for Environmental Pollution Remediation. *Applied Catalysis B: Environmental* 198: 347–377.

Monga, D., D. Ilager, N. P. Shetti, S. Basu, and T. M. Aminabhavi. 2020. 2D/2d Heterojunction of MoS_2/g-C_3N_4 Nanoflowers for Enhanced Visible-Light-Driven Photocatalytic and Electrochemical Degradation of Organic Pollutants. *Journal of Environmental Management* 274: 111208.

Mousavi, M., A. Habibi-Yangjeh, and S. Rahim Pouran. 2018. Review on magnetically separable graphitic carbon nitride-based nanocomposites as promising visible-light-driven photocatalysts. *Journal of Materials Science: Materials in Electronics* 29: 1719–1747.

Mukasyan, A. S., and P. Dinka. 2007. Novel Approaches to Solution-Combustion Synthesis of Nanomaterials. *International Journal of Self-Propagating High-Temperature Synthesis* 16 (1): 23–35.

Nasir, M. S., G. Yang, I. Ayub, S. Wang, L. Wang, X. Wang, W. Yan, S. Peng, and S. Ramakarishna. 2019. Recent Development in Graphitic Carbon Nitride Based Photocatalysis for Hydrogen Generation. *Applied Catalysis B: Environmental* 257 (June): 117855.

Ong, W.-J., L.-L. Tan, Y. H. Ng, S.-T. Yong, and S.-P. Chai. 2016. Graphitic Carbon Nitride (g-C_3N_4)-Based Photocatalysts for Artificial Photosynthesis and Environmental Remediation: Are We a Step Closer to Achieving Sustainability? *Chemical Reviews* 116 (12): 7159–7329.

Pan, J., P. Wang, P. Wang, Q. Yu, J. Wang, and C. Song. 2021. The Photocatalytic Overall Water Splitting Hydrogen Production of G-C_3N_4/CdS Hollow Core – Shell Heterojunction via the HER/OER Matching of Pt/MnO_X. *Chemical Engineering Journal Journal* 405: 126622.

Patil, K. C., S. T. Aruna, and S. Ekambaram. 1997. Combustion Synthesis. *Current Opinion in Solid State and Materials Science* 2 (2): 158–165.

Patnaik, S., S. Martha, and K. M. Parida. 2016. An Overview of the Structural, Textural and Morphological Modulations of g-C_3N_4 towards Photocatalytic Hydrogen Production. *RSC Advances* 6 (52): 46929–46951.

Peng, B., S. Zhang, S. Yang, H. Wang, H. Yu, S. Zhang, and F. Peng. 2014. Synthesis and Characterization of G-C_3N_4/Cu_2O Composite Catalyst with Enhanced Photocatalytic Activity under Visible Light Irradiation. *Materials Research Bulletin* 56: 19–24.

Pirkanniemi, K. 2002. Heterogeneous Water Phase Catalysis as an Environmental Application: A Review. *Chemosphere* 48: 1047–1060.

Raj, R., H. Ozaki, T. Okada, S. Taniguchi, and R. Takanami. 2012. Factors Influencing UV Photodecomposition of Perfluorooctanoic Acid in Water. *Chemical Engineering Journal* 180: 197–203.

Raziq, F., A. Hayat, M. Humayun, S. Kumar, B. Mane, M. B. Faheem, A. Ali, et al. 2020. Photocatalytic Solar Fuel Production and Environmental Remediation through Experimental and DFT Based Research on CdSe-QDs-Coupled P. *Applied Catalysis B: Environmental* 270: 118867.

Safaei, J., N. A. Mohamed, M. F. Mohamad Noh, M. F. Soh, N. A. Ludin, M. A. Ibrahim, W. N. Roslam Wan Isahak, and M. A. Mat Teridi. 2018. Graphitic Carbon Nitride (g-C_3N_4) Electrodes for Energy Conversion and Storage: A Review on Photoelectrochemical Water Splitting, Solar Cells and Supercapacitors. *Journal of Materials Chemistry A* 6 (45): 22346–22380.

Sarkar, S., S. S. Sumukh, K. Roy, N. Kamboj, T. Purkait, M. Das, and R. S. Dey. 2019. Facile One Step Synthesis of Cu-g-C_3N_4 Electrocatalyst Realized Oxygen Reduction Reaction with Excellent Methanol Crossover Impact and Durability. *Journal of Colloid and Interface Science* 558: 182–189.

Shakeel, M., M. Arif, G. Yasin, B. Li, and H. D. Khan. 2019. Layered by Layered Ni-Mn-LDH/g-C_3N_4 Nanohybrid for Multi-Purpose Photo/Electrocatalysis: Morphology Controlled Strategy for Effective Charge Carriers Separation. *Applied Catalysis B: Environmental* 242 (October 2018): 485–498.

Shao, Z., D. Zhang, H. Li, C. Su, X. Pu, and Y. Geng. 2019. Fabrication of MIL-88A/g-C_3N_4 Direct Z-Scheme Heterojunction with Enhanced Visible-Light Photocatalytic Activity. *Separation and Purification Technology* 220: 16–24.

Shi, W., C. Liu, M. Li, X. Lin, F. Guo, and J. Shi. 2020. Fabrication of Ternary Ag_3PO_4/$Co_3(PO_4)_2$/g-C_3N_4 Heterostructure with Following Type II and Z-Scheme Dual Pathways for Enhanced Visible-Light Photocatalytic Activity Weilong. *Journal of Hazardous Materials* 389: 121907.

Shi, Y., P. Chen, J. Chen, D. Chen, H. Shu, H. Jiang, and X. Luo. 2021. Hollow Prussian Blue Analogue/g-C_3N_4 Nanobox for All-Solid-State Asymmetric Supercapacitor. *Chemical Engineering Journal* 404: 126284.

Song, Y., J. Gu, K. Xia, J. Yi, H. Chen, and X. She. 2019. Construction of 2D SnS_2/g-C_3N_4 Z-Scheme Composite with Superior Visible-Light Photocatalytic Performance. *Applied Surface Science* 467–468: 56–64.

Sun, Z., Z. Yu, Y. Liu, C. Shi, M. Zhu, and A. Wang. 2019. Construction of 2D/2D $BiVO_4$/g-C_3N_4 Nanosheet Heterostructures with Improved Photocatalytic Activity. *Journal of Colloid and Interface Science* 533: 251–258.

Tahir, M., C. Cao, N. Mahmood, F. K. Butt, A. Mahmood, F. Idrees, S. Hussain, M. Tanveer, and I. Aslam. 2014. Multifunctional G-C_3N_4 Nano Fibers: A Template-Free Fabrication and Enhanced Optical, Electrochemical, and Photocatalyst Properties. *ACS Applied Materials & Interfaces* 6: 1258–1265.

Tan, J., Z. Li, J. Li, J. Wu, X. Yao, and T. Zhang. 2021. Graphitic Carbon Nitride-Based Materials in Activating Persulfate for Aqueous Organic Pollutants Degradation: A Review on Materials Design and Mechanisms. *Chemosphere* 262: 127675.

Tang, M., Y. Ao, C. Wang, and P. Wang. 2020. Facile Synthesis of Dual Z-Scheme g-C_3N_4/Ag_3PO_4/AgI Composite Photocatalysts with Enhanced Performance for the Degradation of a Typical Neonicotinoid Pesticide. *Applied Catalysis B: Environmental* 268: 118395.

Teixeira, I. F., E. C. M. Barbosa, S. C. E. Tsang, and P. H. C. Camargo. 2018. Carbon Nitrides and Metal Nanoparticles: From Controlled Synthesis to Design Principles for Improved Photocatalysis. *Chemical Society Reviews* 47 (20): 7783–7817.

Tian, N., Y. Zhang, C. Liu, S. Yu, M. Li, and H. Huang. 2016. G-C_3N_4/$Bi_4O_5I_2$ 2D-2D Heterojunctional Nanosheets with Enhanced Visible-Light Photocatalytic Activity. *RSC Advances* 6 (13): 10895–10903.

Vilian, A. T. E., S. Young, M. Rethinasabapathy, R. Umapathi, S. Hwang, C. Woo, B. Park, Y. Suk, and Y. Han. 2020. Improved Conductivity of Flower-like $MnWO_4$ on Defect Engineered Graphitic Carbon Nitride as an Efficient Electrocatalyst for Ultrasensitive Sensing of Chloramphenicol. *Journal of Hazardous Materials* 399: 122868.

Vu, M., M. Sakar, C. Nguyen, and T. Do. 2018. Chemically Bonded Ni Cocatalyst onto the S Doped G-C_3N_4 Nanosheets and Their Synergistic Enhancement in H_2 Production under Sunlight Irradiation. *ACS Sustainable Chemistry & Engineering* 6: 4194–4203.

Wang, Q., W. Wang, L. Zhong, D. Liu, X. Cao, and F. Cui. 2018. Oxygen Vacancy-Rich 2D/2D BiOCl-g-C_3N_4 Ultrathin Heterostructure Nanosheets for Enhanced Visible-Light-Driven Photocatalytic Activity in Environmental Remediation. *Applied Catalysis B: Environmental* 220 (May 2017): 290–302.

Wang, X., K. Maeda, A. Thomas, K. Takanabe, G. Xin, J. M. Carlsson, K. Domen, and M. Antonietti. 2009. A Metal-Free Polymeric Photocatalyst for Hydrogen Production from Water under Visible Light. *Nature Materials* 8 (1): 76–80.

Wang, Y., R. Zhang, Z. Zhang, J. Cao, and T. Ma. 2019. Host–Guest Recognition on 2D Graphitic Carbon Nitride for Nanosensing. *Advanced Materials Interfaces* 6: 1901429.

Wei, B., W. Wang, J. Sun, Q. Mei, Z. An, H. Cao, D. Han, and J. Xie. 2020. Insight into the Effect of Boron Doping on Electronic Structure, Photocatalytic and Adsorption Performance of g-C_3N_4 by First-Principles Study. *Applied Surface Science* 511: 145549.

Willner, I., and Y. Eichen. 1987. TiOz and CIS Colloids Stabilized by 8-Cyclodextrins: Tailored Semiconductor-Receptor Systems as a Means to Control Interfacial Electron-Transfer Processes. *Journal of the American Chemical Society* 109: 6862–6863.

Wu, H., T. Inaba, Z. M. Wang, and T. Endo. 2020a. Photocatalytic TiO_2@CS-Embedded Cellulose Nanofiber Mixed Matrix Membrane. *Applied Catalysis B: Environmental* 276: 119111.

Wu, H., C. Li, H. Che, H. Hu, W. Hu, C. Liu, J. Ai, and H. Dong. 2018. Decoration of Mesoporous Co_3O_4 Nanospheres Assembled by Monocrystal Nanodots on g-C_3N_4 to Construct Z-Scheme System for Improving Photocatalytic Performance. *Applied Surface Science* 440: 308–319.

Wu, J., Y. Zhang, J. Zhou, K. Wang, Y. Z. Zheng, and X. Tao. 2020b. Uniformly Assembling N-Type Metal Oxide Nanostructures (TiO_2 Nanoparticles and SnO_2 Nanowires) onto P Doped g-C_3N_4 Nanosheets for Efficient Photocatalytic Water Splitting. *Applied Catalysis B: Environmental* 278: 119301.

Wu, S., X. Li, Y. Tian, Y. Lin, and Y. H. Hu. 2021. Excellent Photocatalytic Degradation of Tetracycline over Black Anatase-TiO_2 under Visible Light. *Chemical Engineering Journal* 406: 126747.

Wu, X., S. Li, B. Wang, J. Liu, and M. Yu. 2017. From Biomass Chitin to Mesoporous Nanosheets Assembled Loofa Sponge-like N-Doped Carbon/g-C_3N_4 3D Network Architectures as Ultralow-Cost Bifunctional Oxygen Catalysts. *Microporous and Mesoporous Materials* 240: 216–226.

Wu, Z., X. He, Y. Xue, X. Yang, Y. Li, Q. Li, and B. Yu. 2020c. Cyclodextrins Grafted MoS_2/g-C_3N_4 as High-Performance Photocatalysts for the Removal of Glyphosate and Cr (VI) from Simulated Agricultural Runof. *Chemical Engineering Journal* 399: 125747.

Xia, P., B. Zhu, B. Cheng, J. Yu, and J. Xu. 2018. 2D/2D G-C_3N_4/MnO_2 Nanocomposite as a Direct Z-Scheme Photocatalyst for Enhanced Photocatalytic Activity. *ACS Sustainable Chemistry & Engineering* 6: 965–973.

Xu, B., M. B. Ahmed, J. L. Zhou, A. Altaee, G. Xu, and M. Wu. 2018. Graphitic Carbon Nitride Based Nanocomposites for the Photocatalysis of Organic Contaminants under Visible Irradiation: Progress, Limitations and Future Directions. *Science of the Total Environment* 633: 546–559.

Xue, W., D. Huang, G. Zeng, J. Wan, and C. Zhang. 2018. Nanoscale Zero-Valent Iron Coated with Rhamnolipid as an Effective Stabilizer for Immobilization of Cd and Pb in River Sediments. *Journal of Hazardous Materials* 341: 381–389.

Yan, S. C., Z. S. Li, and Z. G. Zou. 2009. Photodegradation Performance of G-C_3N_4 Fabricated by Directly Heating Melamine. *Langmuir* 25 (11): 11269–11273.

Yang, L., J. Huang, L. Shi, L. Cao, H. Liu, Y. Liu, Y. Li, H. Song, Y. Jie, and J. Ye. 2018. Sb Doped SnO_2-Decorated Porous g-C3N4 Nanosheet Heterostructures with Enhanced Photocatalytic Activities under Visible Light Irradiation. *Applied Catalysis B: Environmental* 221: 670–680.

Yin, R., Q. Luo, D. Wang, H. Sun, Y. Li, X. Li, and J. An. 2014. SnO_2/g-C_3N_4 Photocatalyst with Enhanced Visible-Light Photocatalytic Activity. *Journal of Materials Science* 49 (17): 6067–6073.

Yuan, Q., L. Li, Y. Tang, and X. Zhang. 2020. A Facile Pt-Doped g-C_3N_4 Photocatalytic Biosensor for Visual Detection of Superoxide Dismutase in Serum Samples. *Sensors & Actuators: B. Chemical* 318: 128238.

Zhang, H., D. Zheng, Z. Cai, Z. Song, Y. Xu, R. Chen, C. Lin, and L. Guo. 2020a. Graphitic Carbon Nitride Nanomaterials for Multicolor Light-Emitting Diodes and Bioimaging. *ACS Applied Nano Materials* 3: 6798–6805.

Zhang, J. H., M. J. Wei, Z. W. Wei, M. Pan, and C. Y. Su. 2020b. Ultrathin Graphitic Carbon Nitride Nanosheets for Photocatalytic Hydrogen Evolution. *ACS Applied Nano Materials* 3 (2): 1010–1018.

Zhang, M., C. Lai, B. Li, D. Huang, G. Zeng, P. Xu, L. Qin, et al. 2019. Rational Design 2D/2D BiOBr/CDs/g-C_3N_4 Z-Scheme Heterojunction Photocatalyst with Carbon Dots as Solid-State Electron Mediators for Enhanced Visible and NIR Photocatalytic Activity: Kinetics, Intermediates, and Mechanism Insight. *Journal of Catalysis* 369: 469–481.

Zhang, T., X. Shao, D. Zhang, X. Pu, Y. Tang, J. Yin, B. Ge, and W. Li. 2018. Synthesis of Direct Z-Scheme g-C_3N_4/$Ag_2VO_2PO_4$ Photocatalysts with Enhanced Visible Light Photocatalytic Activity. *Separation and Purification Technology* 195: 332–338.

Zhang, X., X. Yuan, L. Jiang, J. Zhang, H. Yu, H. Wang, and G. Zeng. 2020c. Powerful Combination of 2D G-C_3N_4 and 2D Nanomaterials for Photocatalysis: Recent Advances. *Chemical Engineering Journal* 390: 124475.

Zhang, Z., J. Huang, M. Zhang, Q. Yuan, and B. Dong. 2015. Ultrathin Hexagonal SnS_2 Nanosheets Coupled with g-C_3N_4 Nanosheets as 2D/2D Heterojunction Photocatalysts toward High Photocatalytic Activity. *Applied Catalysis B: Environmental* 163: 298–305.

Zhang, Z., D. Jiang, D. Li, M. He, and M. Chen. 2016. Construction of $SnNb_2O_6$ Nanosheet/g-C_3N_4 Nanosheet Two-Dimensional Heterostructures with Improved Photocatalytic Activity: Synergistic Effect and Mechanism Insight. *Applied Catalysis B: Environmental* 183: 113–123.

Zhang, Z., C. Liu, Z. Dong, Y. Dai, G. Xiong, Y. Liu, Y. Wang, Y. Wang, and Y. Liu. 2020d. Synthesis of Flower-like MoS_2/g-C_3N_4 Nanosheet Heterojunctions with Enhanced Photocatalytic Reduction Activity of Uranium (VI). *Applied Surface Science* 520: 146352.

Zheng, H., Z. Zhao, J. B. Phan, H. Ning, Q. Huang, R. Wang, J. Zhang, and W. Chen. 2020. Highly Efficient Metal-Free Two-Dimensional Luminescent Melem Nanosheets for Bioimaging. *ACS Applied Materials & Interfaces* 12: 2145–2151.

Zheng, Q., H. Shen, and D. Shuai. 2017. Emerging Investigators Series: Advances and Challenges of Graphitic Carbon Nitride as a Visible-Light-Responsive Photocatalyst for Sustainable Water Purification. *Environmental Science: Water Research and Technology* 3 (6): 982–1001.

Zheng, Y., J. Liu, J. Liang, M. Jaroniec, and S. Z. Qiao. 2012. Graphitic Carbon Nitride Materials: Controllable Synthesis and Applications in Fuel Cells and Photocatalysis. *Energy and Environmental Science* 5 (5): 6717–6731.

Zhu, B., P. Xia, Y. Li, W. Ho, and J. Yu. 2017. Fabrication and Photocatalytic Activity Enhanced Mechanism of Direct Z-Scheme g-C_3N_4/Ag_2WO_4 Photocatalyst. *Applied Surface Science* 391: 175–183.

Zhu, J., P. Xiao, H. Li, and S. A. C. Carabineiro. 2014. Graphitic Carbon Nitride: Synthesis, Properties, and Applications in Catalysis. *ACS Applied Materials & Interfaces* 6 (19): 16449–16465.

Zhu, K., Y. Lv, J. Liu, W. Wang, C. Wang, S. Li, P. Wang, M. Zhang, A. Meng, and Z. Li. 2019. Facile Fabrication of G-C_3N_4/SnO_2 Composites and Ball Milling Treatment for Enhanced Photocatalytic Performance. *Journal of Alloys and Compounds* 802: 13–18.

Zhu, T., Y. Song, H. Ji, Y. Xu, Y. Song, J. Xia, S. Yin, et al. 2015. Synthesis of G-C_3N_4/Ag_3VO_4 Composites with Enhanced Photocatalytic Activity under Visible Light Irradiation. *Chemical Engineering Journal* 271: 96–105.

Zuo, Y., X. Xu, C. Zhang, J. Li, R. Du, X. Wang, X. Han, et al. 2020. SnS_2/g-C_3N_4/Graphite Nanocomposites as Durable Lithium-Ion Battery Anode with High Pseudocapacitance Contribution. *Electrochimica Acta* 349: 136369.

4 2D Materials for Wastewater Treatments

Nishanth Thomas, Amit Goswami, Kris O'Dowd, Gerard McGranaghan, and Suresh C. Pillai
Institute of Technology Sligo, Sligo, Ireland

CONTENTS

4.1 Introduction 121
4.2 Graphene and Graphene Oxide in Water Purification 123
  4.2.1 Graphene 123
  4.2.2 Graphene Oxide 127
4.3 MXenes 135
4.4 Transition Metal Dichalcogenides 139
4.5 Materials for Photocatalytic Water Treatment 144
  4.5.1 Antibacterial Action of Graphene Family Nanomaterials 144
  4.5.2 Photocatalytic Degradation and Disinfection 144
  4.5.3 Graphene and *g*-C_3N_4 145
  4.5.4 Iron Oxide 146
  4.5.5 Manganese Oxide 147
  4.5.6 Metal Oxyhalides 147
  4.5.7 Boron Nitride Nanosheets 148
4.6 Future Advances 149
4.7 Conclusion 149
References 150

4.1 INTRODUCTION

The scarcity of pure drinking water is one of the most considerable challenges faced by humanity (Elimelech and Phillip 2011). Sustainable technologies for treating wastewater and fabricating membranes for desalination are two important high-throughput areas of research (Van der Bruggen and Vandecasteele 2002). Degradation of contaminants by advanced oxidation processes and separation of pollutants by passage through filtration membranes are the major strategies for purifying contaminated water (Comninellis et al. 2008; Pendergast and Hoek 2011). Water filtration methods vary depending on the types of membranes employed for purification. The filtration methods can be classified as reverse osmosis, nanofiltration, ultrafiltration, microfiltration and particle filtration depending on the pore size of membranes used (Figure 4.1) (Lee, Elam, and Darling 2016).

DOI: 10.1201/9781003118749-4

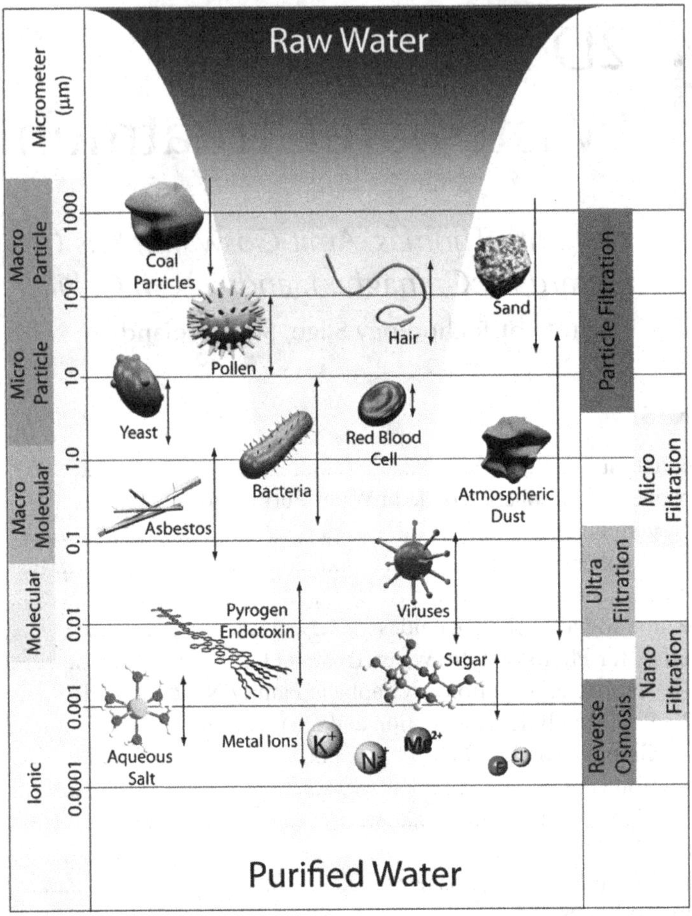

FIGURE 4.1 A schematic illustration of various water filtration methods employed according to the differences in the pore size of membranes used. Reproduced with permission from Lee, Elam, and Darling (2016). Copyright (2016), The Royal Society of Chemistry.

The polymeric materials used in fabricating conventional water filtration membranes have the drawbacks of poor chemical resistance and lower water permeability rates. (B. Van der Bruggen, Mänttäri, and Nyström 2008). With the advent of graphene, 2D materials attained a great deal of attention in separation and purification technologies (Geim and Grigorieva 2013; Xu et al. 2013; Dubertret, Heine, and Terrones 2015). 2D material membranes have excellent thermal and chemical stability, and they can achieve higher water permeation rates owing to their unique atomic thickness. (Mi 2014; Joshi et al. 2014; Liu, Jin, and Xu 2016). Various classes of 2D materials, such as the graphene family of nanomaterials (Xu et al. 2013), transition metal dichalcogenides (TMDs) (Geim and Grigorieva 2013), MXenes (Li, Li, and Van der Bruggen 2020a), zeolites (Cui et al. 2008), metal-organic frameworks (MOFs) (Kadhom and Deng 2018), etc. are explored as building blocks for water filtration membranes. 2D material-based membranes are generally fabricated as nanosheets or laminar membranes. The nanosheet architecture consists of a monolayer or a few layers of 2D materials having uniform pore sizes. A laminar membrane is made by assembling layers of 2D material nanosheets, and it allows the selective passage of molecules through their interlayer spacing.

This chapter discusses various classes of 2D material membranes, their fabrication strategies, and compares their advantages with the conventional membrane materials used in separation techniques. One separate section is devoted to discussing 2D materials explored for the photocatalytic method of water treatment. 2D photocatalytic materials have a high surface area, superior charge separation ability and excellent light-harvesting capabilities (Liu et al. 2019; Xie et al. 2020). Their unique properties make them attractive for photocatalytic degradation of pollutants and disinfection of different microorganisms.

4.2 GRAPHENE AND GRAPHENE OXIDE IN WATER PURIFICATION

Despite a significant adoption of polymeric membranes in recent years, the trade-offs between salt rejection and water flux are still large, and the issues of fouling and chemical sludge formation persist. The accumulation of sludge in the membrane pores leads to higher pressures for a constant water flux (Roy et al. 2020). However, carbon-based materials such as graphene, graphene oxides (GO), and CNTs possess characteristics desirable in a membrane used for desalination purposes, such as high mechanical and chemical robustness, high specific area, high anti-fouling, and antimicrobial properties, along with low cytotoxicity toward mammalian cells (Geim and Novoselov 2009; Dervin, Dionysiou, and Pillai 2016; Ihsanullah 2019; Firouzjaei et al. 2020). Unlike conventional thin-film composite (TFC) and thin-film nanocomposite (TFN) membranes based on polyamides, carbon-based materials show remarkable water transport properties with permeability 2–3 orders of magnitudes higher than the former (Aqra and Ramanathan 2020). The past decade has seen significant improvements in the membrane fabrication technology of these carbon-based nanomaterials in the primary rejection layer (PRL), surface located nanocomposite (SLN), TFN or mixed matrix substrate (MMS) (Yang, Ma, and Tang 2018b; Roy et al. 2020). Based on variables such as functional groups attached, and pore or channel dimensions, membranes with varying characteristics can be obtained (Figure 4.2) (Yang, Ma, and Tang 2018b)

4.2.1 Graphene

Once considered impossible to remain stable in its isolated form due to thermodynamic instabilities (Landau 1937), the discovery of graphene by Novoselov and Geim (Novoselov et al. 2005) led to a whole new range of possibilities in multiple sectors (Schwierz 2010; Brownson, Kampouris, and Banks 2011; Yang et al. 2018a). The sp^2 hybridized carbon atoms form a hexagonal planar honeycomb lattice (Figure 4.3) with the thickness of a graphene monolayer of about 0.3 nm (Yang, Ma, and Tang 2018b). The high specific area, hydrophilic nature, astonishing mechanical and chemical robustness make graphene an extremely favorable candidate to be used in desalination membranes.

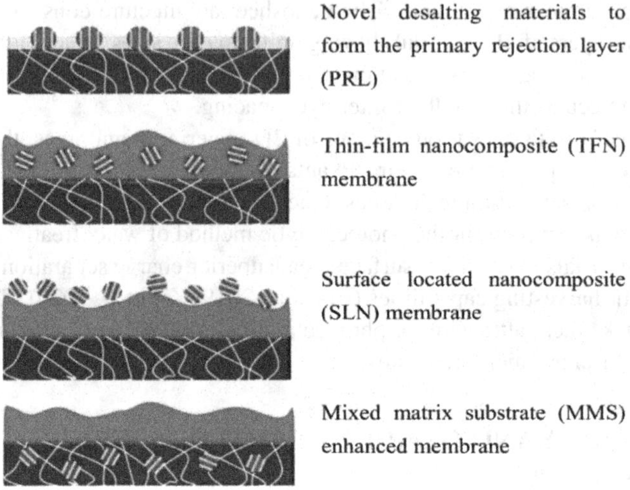

FIGURE 4.2 Various types of membranes based on the location of desalting material. Here the balls with red stripes represent the incorporated desalting material. Reproduced with permission from Yang, Ma, and Tang (2018b). Copyright (2017), Elsevier.

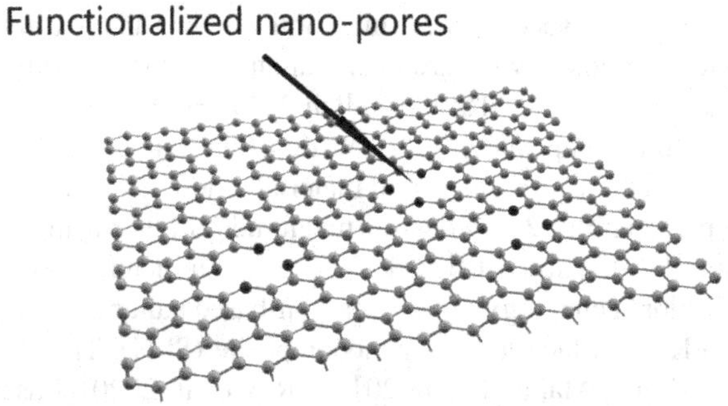

FIGURE 4.3 Functionalized nanoporous graphene (NPG).

Generally adopted methods to produce graphene are chemical vapor deposition (Zhang, Zhang, and Zhou 2013b), epitaxial growth (Xu et al. 2017b), and chemical and mechanical exfoliation (Novoselov et al. 2005; Yang et al. 2019). In its intact form, graphene is impervious to molecules as small as helium (Bunch et al. 2008); therefore, various methods like laser, plasma etching (Surwade et al. 2015), ion beam (Bell et al. 2009) and electron beam (Garaj et al. 2010) are utilized to create nanopores (Figure 4.3) on the surface of the graphene monolayer. The porous surface thus produced is known as nanoporous graphene (NPG). The number of defects on NPG can be increased by a longer exposure time to an external energy source, as observed in case of exposure to oxygen plasma (Surwade et al. 2015).

The membrane produced (Surwade et al. 2015) had near 100% salt rejection along with high water flux of 10^6 g $m^{-2}s^{-1}$ at 40°C with external pressure difference. In a novel method, O'Hern et al. (2014) utilized a two-step method to produce nanopores by first generating isolated defects in graphene lattice by ion bombardment and later enlarging them using a chemical etching treatment. The method adopted allowed the creation of pores with high density (over 10^{12} cm^{-2}) and a diameter of 0.40 nm ± 0.24 nm.

During the transfer and growth process, various undesired effects like tears are introduced into the NPG, which result in molecular leakage leading to impeding desalination performance. Therefore, various leakage sealing processes like the one presented in Figure 4.4 have been adopted by researchers in order to minimize the effect of such defects on desalination performance (O'Hern et al. 2015; Cheng et al. 2020).

The atomic thin porous graphene layer results in ultrafast convective water flow compared to the slow diffusion flow found in relatively thicker polyamide TFC membranes. Molecular dynamic simulation studies show a higher permeability by 2–3 orders of magnitude compared with conventional RO membranes. Likewise, salt rejection efficiency is superior to other membranes such as hexagonal boron nitride (Nguyen and Beskok 2020). The water flux can be enhanced further by attaching various hydrophilic functional groups to the pore edges. By attaching the hydrophilic hydroxyl group to pore edges, Cohen-Tanugi and Grossman (2012) saw a near

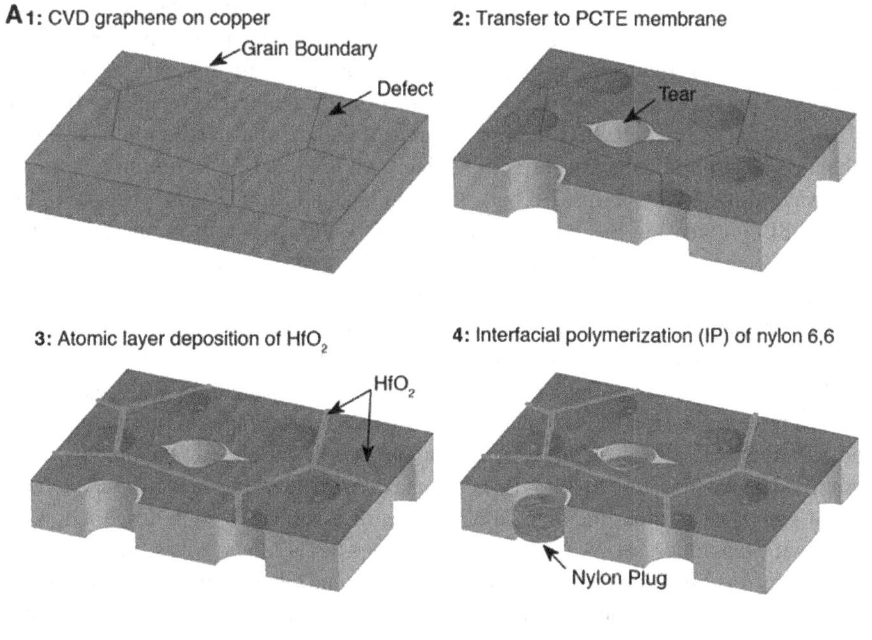

FIGURE 4.4 Defect sealing of nanoporous graphene. Reprinted with permission from O'Hern et al. (2015). Copyright (2015) American Chemical Society.

doubling of the water flux. Similarly, pore functionalization with various groups or ions enables selectivity for different ions. For example, pore functionalization with negatively charged nitrogen and fluorine ions allows the passage of cations like Li^+, Na^+ and K^+ whereas functionalization with positively charged hydrogen ions allows passage of Cl^- and Br^- (Sint, Wang, and Král 2008). Water permeation and ion selectivity thus can be altered based on pore size and pore functionalization with various groups like carboxyl, hydroxyl, and amides. A reduction in pore size will enhance the salt rejection but decrease the water flux. NPG membranes show extreme mechanical robustness withstanding pressures as high as 570 bar (Cohen-Tanugi and Grossman 2014), nearly ten times more than pressures associated with conventional RO membranes. An increase in porosity may enable the NPG membrane to withstand even higher pressures. However, the performance is critically related to the base substrate on which graphene is attached. When the substrate is incorporated with graphene, NF, or RO membranes with enhanced chemical and mechanical stability can be obtained. By modifying PES (poly-ether-sulphone) using polydopamine crosslinked graphene, Ndlwana et al. (2020) showed high hydrophilicity, tensile strength, and salt rejection up to 99.95%. Furthermore, taking into account the excellent RO performance of single-layer NPG membrane, molecular dynamic (MD) simulation studies based on multilayer graphene demonstrate that their desalination properties are similar to single-layer graphene (SLG) (Cohen-Tanugi, Lin, and Grossman 2016). However, since the mass production of SLG is relatively more difficult than MLG (Shao, Zhao, and Qu 2020), these MLG layers will be of interest for subsequent studies. The performance can be altered based on the membrane thickness and other configurational variables.

Various additional treatments and external factors can significantly affect the desalination performance of graphene membranes. Charged nanoporous graphene membranes show reduced pressure drops compared to uncharged membranes, although with small compromises in salt rejection efficiency (Nguyen and Beskok 2020). Similarly, stacked SLG membrane treated under solar irradiation has shown enhanced adsorption capacity for cations like Na^+, Pb^{+2} and Fe^{+3} with an extremely high water flux of 167.1 L m^{-2} h^{-1} bar^{-1} (Zhang, Hu, and Zhou 2020b). The sunlight-treated SLG consisted of larger channels for water transport with defect-free nanosheets as compared with pristine SLG. A recent study based on using an external electric field to tailor the permeability and salt rejection showed a 59.6% enhancement in water flux compared to a case of no electric field with 100% salt rejection (Mortazavi, Moosavi, and Nouri-Borujerdi 2020). The oscillation frequency and amplitude of such electric field are decisive in the final desalination performance.

The above discussion shows the ability of graphene membranes in high permeation and salt rejection. However, most of the studies are based on MD simulations, and practical data is scarce due to difficulties associated with the experimental characterization of graphene. In other methods, the potential of graphene in solar desalination by interfacial evaporation shows high evaporation rates (~1.42 kg m^{-2} h^{-1}) and high solar thermal conversion efficiency (~92.55%) (Chen, Li, and Chen 2020d; Wu et al. 2020).

Furthermore, graphene and its composites have also been adopted in the removal of various pesticides from wastewater (Naushad 2018). These pesticides are widely adopted for improving crop quality. However, due to a heterocyclic ring

TABLE 4.1
Water Flux and Permeability of Graphene Membranes in Water Purification

Experimental/ Simulation	Pressure	Pore Size	Water Flux	Salt Rejection	References
Experimental	0.17 bar	5-μm	3.6×10^6 L m^{-2} h^{-1}	~100% for KCl at 40° C	Surwade et al. (2015)
Simulation	1000–2000 bar	1.5 to 62 $Å^2$	10–100 L/ cm^2 / day/MPa (416.66–4166.6 L m^{-2} h^{-1} bar^{-1}), 2×RO membrane for saltwater	greater than 90% for saltwater	Cohen-Tanugi and Grossman (2012)
Experimental	–	0.22 μm	167.1 L m^{-2} h^{-1} bar^{-1}	99.8% for Pb^{+2}, 50%–80% for ionic salts	Lei Zhang, Hu, and Zhou (2020b)
Simulation	1000–3000 ba	12.6–19.08 Å	Up to 59.6% higher with an external electric field compared to no electric field	100% for KCl and NaCl	Mortazavi, Moosavi, and Nouri-Borujerdi (2020)
Experimental	4–10 bar	0.369–2.93 nm	21.1–164.3 L m^{-2} h^{-1}	99.85% for NaCl and 99.95% for $MgSO_4$	Ndlwana, Motsa, and Mamba (2020)

arrangement, these pesticides are not biodegradable under normal conditions, allowing them to stay in the environment for longer durations (Rajapaksha et al. 2018). Also, rather than staying localized, they tend to move in the environment, resulting in their wide presence in surface water. By using graphene and its composites, these pesticides are eliminated via an adsorption mechanism. Compared to pristine graphene, its composites like silica nanoparticles coated with graphene show near 100 times more adsorption capability for pesticide like aromatic phenanthrene (Yang, Chen, and Zhu 2015). The pesticide replaced the hydrogen-bonded water molecules on the graphene-coated silica nanoparticles. Also, it was observed that the contaminant could even be desorbed from such adsorbent, making it reusable. Other studies have been reported for the efficient removal of pesticides like Ametryn (Zhang et al. 2015), Chlorpyrifos (Gupta et al. 2012), Malathion (Liu et al. 2013) and others.

Here Table 4.1 summarizes water flux and rejection for various salts, metals, and dyes obtained using graphene membranes.

4.2.2 Graphene Oxide

Apart from graphene, many of its derivatives have been used in desalination, and water purification applications, of which GO and reduced GO make up the widest category. Generally, oxygen-related groups are incorporated into raw graphite by various oxidation methods like Brodie (1860), Staudenmaier (1898), Hummer and Offeman (1958), or the modified Hummers method. GO fabrication methods are relatively easier when compared to fabrication methods for graphene materials, having properties similar to pristine graphene (Dervin, Dionysiou, and Pillai 2016). Many epoxides, hydroxyl, and carboxyl groups present in GO nanosheets obtained by the above methods make them water attractive. By increasing the oxidation temperature, a higher number of epoxy groups and lesser hydroxyl groups are obtained, which is favorable to enhancing the water flux (Zhaozan Xu et al. 2020; Chen et al. 2020). When stacked together, these GO nanosheets make a graphene oxide framework (GOF), showing high mechanical strength due to hydrogen bonding between the nanosheets (Sun et al. 2013b), making them favorable for pressure-driven flow in terms of mechanical stability. However, a 2017 MD simulation-based study by Chen et al. (2017) showed GO membranes to be unsuitable for pressure-driven flow, showing the formation of hydration ions by water and salts which could pass through the gaps, even when the hydration radius was larger than the width of the gap (Figure 4.5). This occurred due to the breakage of hydrogen bonds of the outer hydration shell under external pressure. On the contrary, a high desalination performance can be achieved in concentration-driven diffusion flow with GO membranes (Sun et al. 2016).

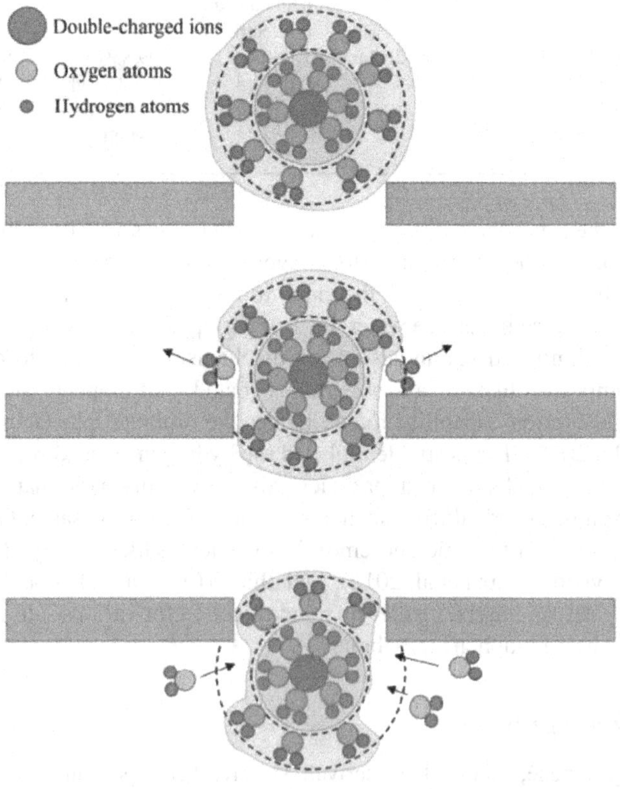

FIGURE 4.5 Ions with hydration radius greater than the nano-slit width passing through the slit. Reprinted with permission from Chen et al. (2017) Copyright (2017) American Chemical Society.

The salt rejection rate for such GO membranes is given as (Li, Zhu, and Zhu 2019)-

$$R(\%) = \frac{(C_f - C_p)\%}{100} \tag{4.1}$$

where C_f and C_p are feed and salt, permeate concentrations in ppm, respectively.

And the water flux is given by (Li, Zhu, and Zhu 2019)-

$$\text{Water flux} = \frac{Q}{(A \times \delta t)} \tag{4.2}$$

where Q is the permeate quantity *(L)*, A is film area *(m^2)* and δt is processing time (*s*).

The presence of oxygen-containing groups in GO membranes ensures layer spacing and empty spaces between non-oxidized regions forming nano-capillary-like channels (Pacilé et al. 2011). Water flow occurs by continuous adsorption of water molecules at the hydrophilic edges and then diffusion along the channels formed between adjacent hydrophobic nanosheets in the interlayer. This way, the oxidized regions are responsible for high ion rejection, whereas the non-oxidized regions result in frictionless water flow with high permeability (Zhang et al. 2020a). Large changes in salt rejection capacity and water permeability can be made by varying the interlayer spacing (*d*-spacing) and GOF thickness. The interlayer distance for pristine GO is narrow in general (effective pore size 0.3 nm; Qian, Zhou, and Huang 2018) and hence puts limitations over selectivity and long-term operation. The trade-off between ion selectivity and water flux found in pristine GO membranes thus is a limitation for their use in desalination and purification applications.

On the other hand, due to laminated layers rich in oxygen functional groups, water tends to enter between the layers, increasing the interlayer spacing (swelling) and allowing larger ions to pass, resulting in reduced ion separation ability (Zhang and Chung 2017). Therefore, control over the interlayer spacing is critical in GO membranes and is a limiting factor for their extended use in water purification and desalination applications. Chen et al. (2017) achieved the stabilized spacing by using K^+, Na^+, Ca^+, Li^+, Mg^{+2} ions by soaking the GOMs in KCl solution (Figure 4.6).

Various methodologies have addressed these limitations associated with d-spacing to vary and control the interlayer spacing, resulting in variation in permeability, salt rejection, and ion selectivity. For example, the interlayer spacing can be increased by inserting nanofibers (A. Huang and Feng 2018), nanoparticles or polymeric materials (Kim et al. 2017; Dervin, Dionysiou, and Pillai 2016) between layers. The process is known as intercalation and is adopted widely to increase the water flux owing to a larger interlayer spacing. In a recent work, Xu et al. (2020) fabricated nanosheets of coordinated tannic acid-functionalized GO with Fe^{3+} ions (Fe/GO-TAx). The resultant membrane showed higher rejection for organic dyes (up to 99% removal) with more interlayer spacing compared with pristine GO membranes resulting in permeability of 61.2 L m^{-2} h^{-1} bar^{-1}. This permeability is near six times higher than obtained for a pristine GO membrane.

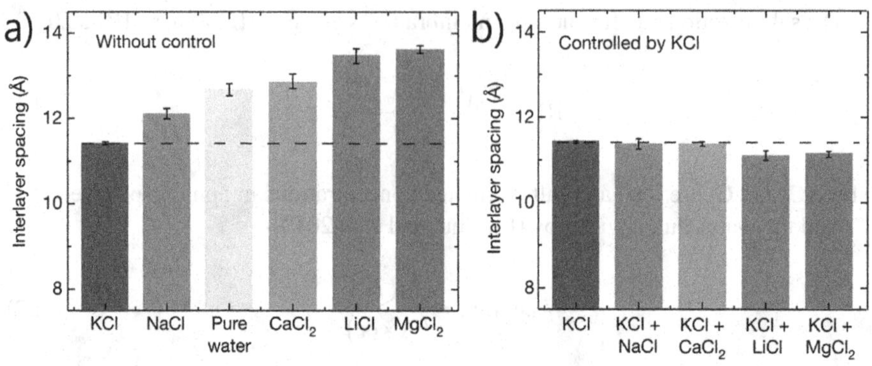

FIGURE 4.6 Interlayer spacing of GO membranes. (a) Immersed in water and different salt solutions. (b) Soaked in KCl before immersion in same salt solutions. Reprinted with permission from Chen et al. (2017). Copyright (2017) Macmillan Publishers Limited.

Similarly, reducing the interlayer spacing will result in better salt rejection, compromising the permeability. The chemical reduction of pristine GO membranes eliminates the oxygen-containing groups, which ensures layer spacing. This results in the thickness reduction of the GOF. The water permeation can now be improved by additional hydrophilic layers like polydopamine. This way, ultrathin rGO membranes have been shown to result in astonishing water flux alongside high selectivity and salt rejection (Li et al. 2013; Liu, Wang, and Zhang 2015). Such a synergistic combination of reduced layer spacing and hydrophilic layering has shown a remarkable water flux of 36.6 L m^{-2} h^{-1} with salt rejection of 92% (Yang et al. 2017). The possibility of tailored interlayer spacing has resulted in another recent work, demonstrating the prevention of salt accumulation even in high saline solution in solar thermal-based desalination application (Zhuang et al. 2020). Furthermore, the high anti-fouling properties and stability of such membranes have been demonstrated in experimental works.

The use of GO can be found in any of the four layers named PRL, SLN, TFN, or MMS. When utilized as the PRL, a layer-by-layer assembly by spin coating, drop-casting, or vacuum filtration are the most utilized fabrication techniques. In pioneering works, Nair et al. (2012) used a layer-by-layer spray or spin coating method. The PRL membrane thus obtained was permeable to water vapor with permeability at least '1010' times faster than helium. The extremely high permeability to water was attributed to a low friction flow in the channels, as discussed before. A free-standing GO membrane was fabricated in another study (Sun et al. 2013b) by a drop-casting method which showed excellent mechanical strength, complete blocking, and hindrance to copper salts and organic contaminants and quick permeation to sodium salts.

When incorporated into the polyamide rejection layer, an interfacial polymerization method is generally used for membrane fabrication. The GO nano-plates are dispersed into an aqueous solution of MPD or organic TMC, and interfacial polymerization results in the final membrane. Water flow occurs in the channels formed between the nano-plates (Figure 4.7). The resulting membrane shows enhanced hydrophilicity, permeability, and high chemical resistance compared to

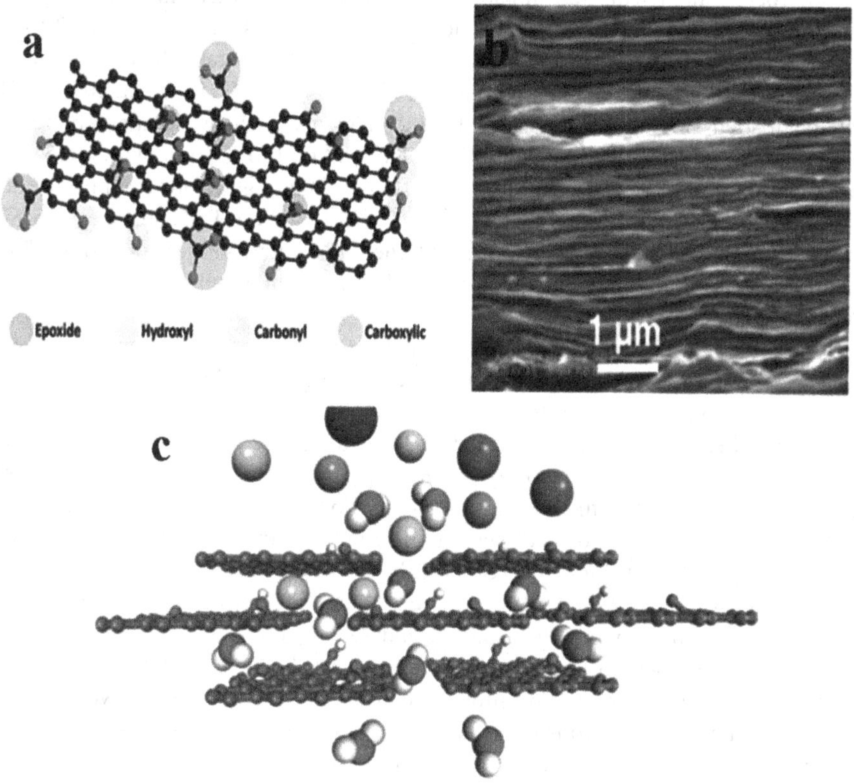

FIGURE 4.7 A schematic illustration showing (a) chemically functionalized graphene oxide, (b) a scanning electron microscopy (SEM) image of graphene oxide-based framework, and (c) mechanism representing the passage of water molecules through the GO-based membrane. Reproduced with permission from Yang, Ma, and Tang (2018b). Copyright (2017), Elsevier.

TFC membranes (Chae et al. 2015). The enhancement in hydrophilicity is due to the intrinsic hydrophilic nature of GO nanosheets. Furthermore, the formation of hydrogen bonds or polyester bonds between GO and polyamide impedes undesired chemical interactions which cause chlorination or oxidation (Kim et al. 2013; Zhang, Su et al. 2013a). By using the above interfacial polymerization method to produce polyimide-based TFN membranes, a water flux of 31.80 L m^{-2} h^{-1} and salt rejection of 98.8%, were obtained, along with remarkable anti-biofouling properties toward *E. coli* and *S. aureus* (Hamdy and Taher 2020).

Similarly, instead of GO incorporation into the PRL or TFN, the nano-plates can be located at the surface of the PRL. The main motive of this approach lies in enhancing the anti-fouling or antibacterial properties of the overall layer. Moreover, this approach can provide additional benefits like increased mechanical stability of the membrane owing to stable bonds between sheets (Joshi et al. 2014). Antimicrobial properties of GO-based materials are well established and vary in the sequence of graphite oxide < graphite < reduced GO < GO (Liu et al. 2011). GO can be loaded on the PRL surface by mechanisms such as covalent bonding (Perreault, Tousley, and Elimelech 2013), electrostatic force (Choi et al. 2013), or coordination bonds (Y. Gao, Hu, and Mi 2014) without significant compromise to the intrinsic permeability and salt rejection capacity of the PRL surface. A 65% inactivation in the bacterial growth was obtained on a conventional polyamide TFC membrane modified with GO nano-plates by covalent bonding in a study by Perreault et al. (2013).

Similarly, a 2020 study shows enhanced chlorine resistance, hydrophilicity and salt rejection by incorporating GO nanosheets with grafted PAMAM (polyamidoamine) on conventional RO membrane (Vatanpour and Sanadgol 2020). Unlike conventional techniques where the coating leads to reduced water flux, SLN based GO does not affect the permeability of PRL significantly. Additionally, as the surface in direct contact with the saline water is now GO, the smoother and more hydrophilic surface is more resistant to fouling. Further, several studies have reported using GO to form MMS (Ganesh, Isloor, and Ismail 2013; Yu et al. 2013; Xu et al. 2014; Zinadini et al. 2014) where GO can be beneficial in enhancing the permeability, antimicrobial performance, and salt rejection of the base substrate. This is an indirect way of membrane performance enhancement but can have significant improvements over conventional membranes. Such type of a forward osmosis membrane was developed by Park and co-workers (2015) by interfacial polymerization on a substrate incorporated with GO. This membrane showed a significant enhancement in forward osmosis water flux with improved membrane selectivity. Similarly, Mukherjee et al. (2016) incorporated GO nanosheets into polysulphone matrix with a focus to eliminate heavy metal ions. Experiments showed steady-state permeate flux of near 30 L m^{-2} h^{-1} with rejection between 90%–96% for Pb, Cu, Cd, and Cr.

The strategy adopted for the fabrication of GO membranes has a great impact on its desalination performance. Even the same membrane preparation method can result in different membrane characteristics if different schemes during fabrication are adopted. For instance, in GO-based membranes, high salt rejection is achieved only by thick membranes. However, it is inevitable to think that the water flux will be reduced for thicker membranes. Hence, the trade-offs between rejection and permeability remain an issue for GO membranes. In an effort to overcome this issue, a self-assembly method for GO nanosheets is adopted. Observations showed that by controlling the deposition rate of single-layered GO, the self-assembly of these layers could overcome the trade-offs between water flux and selectivity (Xu et al. 2017a). For lower deposition rates, a more thermodynamically favored structure was obtained (Figure 4.8), which resulted in permeability 2.5–4 times higher, and salt rejection 1.8–4 times higher than the structure obtained by faster deposition rates. Similar conclusions have been drawn in a recent study where the

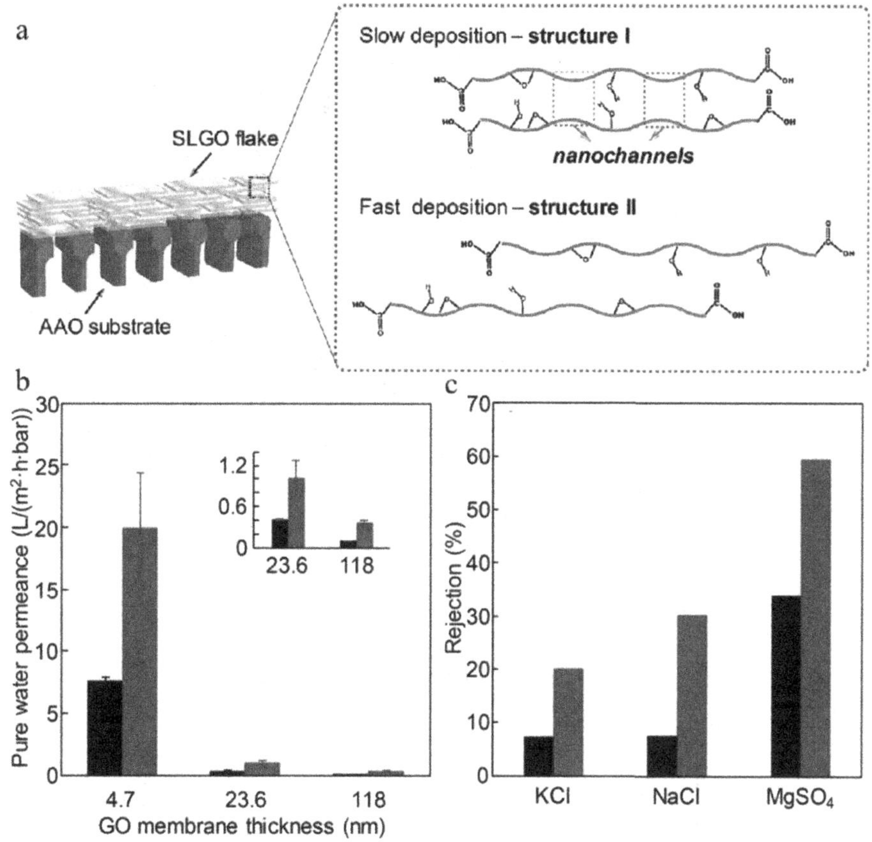

FIGURE 4.8 Effect of deposition rate: (a) Structure 1 obtained with a slow deposition rate is thermodynamically more stable than structure II obtained by a fast deposition rate. (b) Effect of membrane thickness on permeability (black bars indicate fast deposition whereas red bars indicate slow deposition). (c) Rejection of a 4.9 nm-thick membrane. Reprinted with permission from Xu et al. (2017a). Copyright (2017) American Chemical Society.

authors have suggested that in the self-assembly of GO sheets, the stacking of oxidized regions in adjacent GO sheets and matching between non-oxidized regions can produce more interlayer nanochannels which can significantly enhance water permeation in thicker membranes (Lei Zhang et al. 2020a). Similar to graphene, GO-based membranes have been utilized in solar desalination with improved evaporation rate (1.27 kg m^{-2} h^{-1}) and conversion efficiency (79%) under highly saline environments (X. Guo et al. 2020b; Zhuang et al. 2020). Further, similar to graphene and its composites, GO and rGO can adsorb and eliminate many types of pesticides such as chlorpyrifos, endosulfan, and malathion (Maliyekkal et al. 2013) from wastewater.

TABLE 4.2
Graphene Oxide-Based Membranes

Substrate	Type	Experimental/ Simulation	Pressure	Water Flux	Salt Rejection	References
PES	PRL	Experimental	Up to 5 bar	61.2 $L\ m^{-2}\ h^{-1}\ bar^{-1}$	99% for organic dyes	D. Xu et al. (2020a)
TiO2/Polycarbonate	PRL	Experimental	1 bar	7 $L\ m^{-2}\ h^{-1}\ bar^{-1}$	100% for 10 ppm methyl orange dye	C. Xu et al. (2013a)
PAN nanofiber	PRL	Experimental	1 bar	2 $L\ m^{-2}\ h^{-1}\ bar^{-1}$	100% rejection of Congo red and 56.7% for Na_2SO_4	Jianqiang Wang et al. (2016)
PDA/Al2O3	PRL	Experimental	–	48.4 $L\ m^{-2}\ h^{-1}$	99.7% for 3.5 wt % NaCl at 90° C	K. Xu et al. (2016)
Polyamide	TFN	Experimental	15.5 bar	16.6 ± 0.2 $L\ m^{-2}\ h^{-1}$	>99% for 2000 ppm NaCl	Chae et al. (2015)
Polyimide	TFN	Experimental	20 bar	31.80 $L\ m^{-2}\ h^{-1}$	98.8% for 2000 ppm NaCl solution	Hamdy and Taher (2020)
PAMAM	SLN	Experimental	15 bar	>30 $L\ m^{-2}\ h^{-1}$	>95% for NaCl, ~91% for $NaNO_3$	Vatanpour and Sanadgol (2020)
Polysulfone	MMS	Experimental	10 bar	19.77 $L\ m^{-2}\ h^{-1}$	98.7% for NaCl	Park et al. (2015)
Polysulfone	MMS	Experimental	2.76–6.90 bar	30 $L\ m^{-2}\ h^{-1}$	90%–96% for Pb, Cu, Cd, and Cr	Mukherjee, Bhunia, and De (2016)

Various types of GO membranes and their water purification performance are summarized in Table 4.2.

4.3 MXenes

Researchers are continuously developing advanced materials for improving water purification technologies. In 2011 Yury Gogotsi's group at Drexel University, USA reported synthesizing a new class of two-dimensional materials called MXenes (Naguib et al. 2011, 2012, 2014). MXenes are transition metal carbides, nitrides, and carbonitrides with a general formula of $M_{n+1}X_nT_x$. Here M stands for an early transition metal element, such as titanium. X could be nitrogen and/or oxygen, and T represents the surface functionalities attached during synthesis (e.g., oxygen, fluorine, and hydroxyl groups) (Karahan et al. 2020). MXenes are generally synthesized by a bottom-up approach (i.e., chemical vapor deposition) or a top-down approach (i.e., chemical etching of MAX phases) (Ihsanullah 2020). Figure 4.9 represents a systematic approach to synthesizing MXenes starting from MAX phases (Naguib et al. 2012). HF treatment causes the etching of the MAX phases, and later sonication results in the exfoliated MXene sheets. The nomenclature of MXenes represents the loss of "A layers" from the MAX phases, and it emphasizes the two-dimensional structure similar to that of graphene.

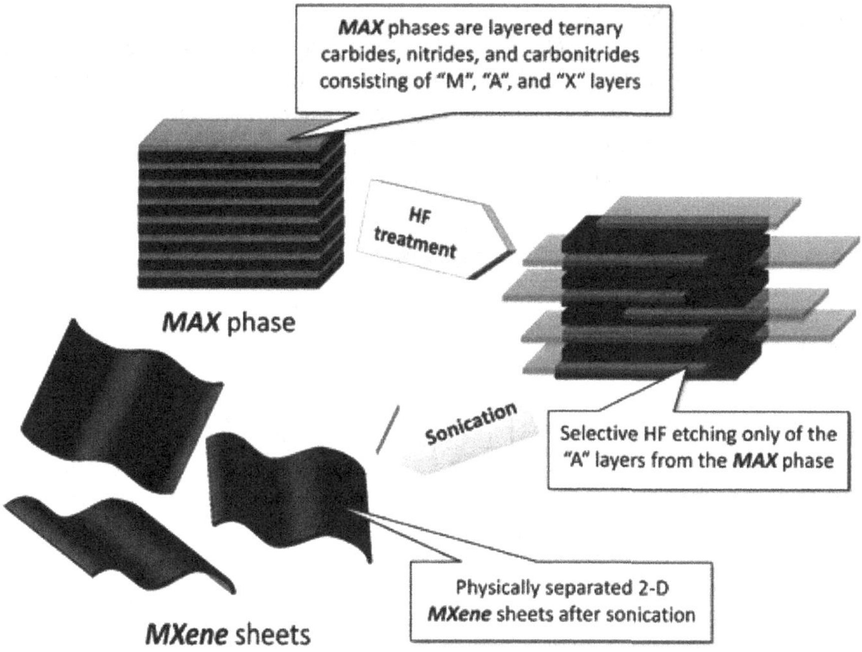

FIGURE 4.9 Schematic representation of the synthesis of the MXene sheets from MAX phases. Reproduced with permission from Naguib et al. (2012). Copyright (2012), American Chemical Society.

$Ti_3C_2T_x$ is one of the most widely studied members of the MXene family (Jingtao Wang et al. 2018; Lipatov et al. 2018; Pandey et al. 2018; Xiong et al. 2018). Flexibility, electrical conductivity and hydrophilic surfaces make MXenes an excellent candidate for ion separation applications (Al-Hamadani et al. 2020; J. Chen et al. 2020b; Ding et al. 2020; Ihsanullah 2020; Karahan et al. 2020; Rasool et al. 2019). The $Ti_3C_2T_x$ MXene sheets had an interlayer spacing of approximately 6 Å (Ren et al. 2015) and a negative surface charge of −29 mV (zeta potential) at pH 7 (Ying et al. 2015). Since the MXenes have a conductivity similar to that of metals (2400 S/cm) (Ling et al. 2014), their potential for ion separation can be altered by applying external voltage. Gogotsi and co-workers compared the selective cation permeation ability of $Ti_3C_2T_x$ MXenes with the graphene oxide membranes (Ren et al. 2015). The study reported that MXenes perform better than graphene oxide membranes in the selective separation of cations having a higher number of charges. In a similar study, a lamellar membrane was fabricated by stacking 2D $Ti_3C_2T_x$ MXenes on anodic aluminum oxide substrate (Ding et al. 2017). The positively charged colloidal $Fe(OH)_3$ solution was intercalated into the negatively charged MXene nanosheets, expanding the nanochannels present in the MXenes. The membrane exhibited water permeation greater than 1000 L m^{-2} h^{-1} bar^{-1}. All molecules larger than 2.5 nm were filtered out, with a rejection rate of over 90%.

Uncontrolled release of antibiotics into the water bodies results in the attainment of antibiotic resistance for the existing bacteria and generation of new classes of superbugs (Wright 2000; Alanis 2005; Laxminarayan et al. 2013; Adegoke et al. 2017). A titanium carbide ($Ti_3C_2T_x$) lamellar nanosheet with a large aspect ratio was studied to demonstrate the separation of different types of antibiotics (Li et al. 2020). The regular pore structure of the lamellar membrane was attributed to the exceptional antibiotic separation ability and higher flux rate achieved. The separation performance of tetracycline antibiotics following the change in thickness of the

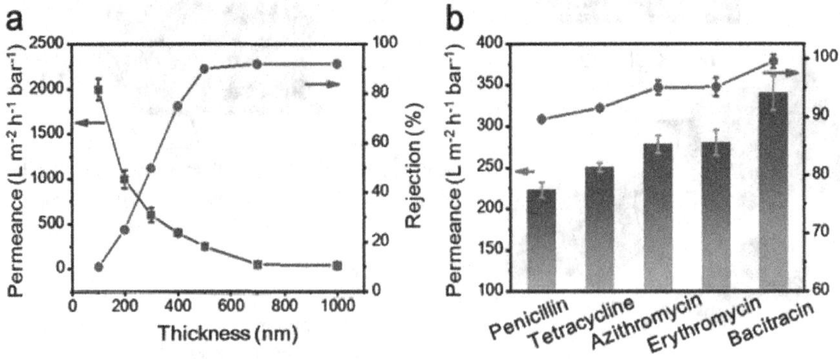

FIGURE 4.10 (a) The change in rejection percentage of tetracycline antibiotic to the thickness of the membrane. (b) A comparison of the rejection performance of different antibiotics studied. Reproduced with permission from Li et al. (2020). Copyright (2020), Wiley-VCH Verlag GmbH.

membrane is shown in Figure 4.10a. Membrane thickness above 500 nm could reach 90% rejection of tetracycline antibiotics. The penicillin molecule with a size of 1.4 nm × 0.7 nm achieved 89.5% rejection with 223.1 L m^{-2} h^{-1} bar^{-1} water flux (Figure 4.10b). As the molecular size of antibiotics increases, the system attains a higher flux rate with an increased rejection of antibiotic molecules, which can be attributed to the size effect of the solute molecules. The studied antibiotics depicted a similar rejection performance with both solvents: water and ethanol.

MXene based membranes have been reported to separate methylene blue and methyl orange dye with excellent water flux (Kim, Yu, and Yoon 2020; Zhang et al. 2020). Graphene oxide/MXene composite lamellar membranes have shown ultrafast

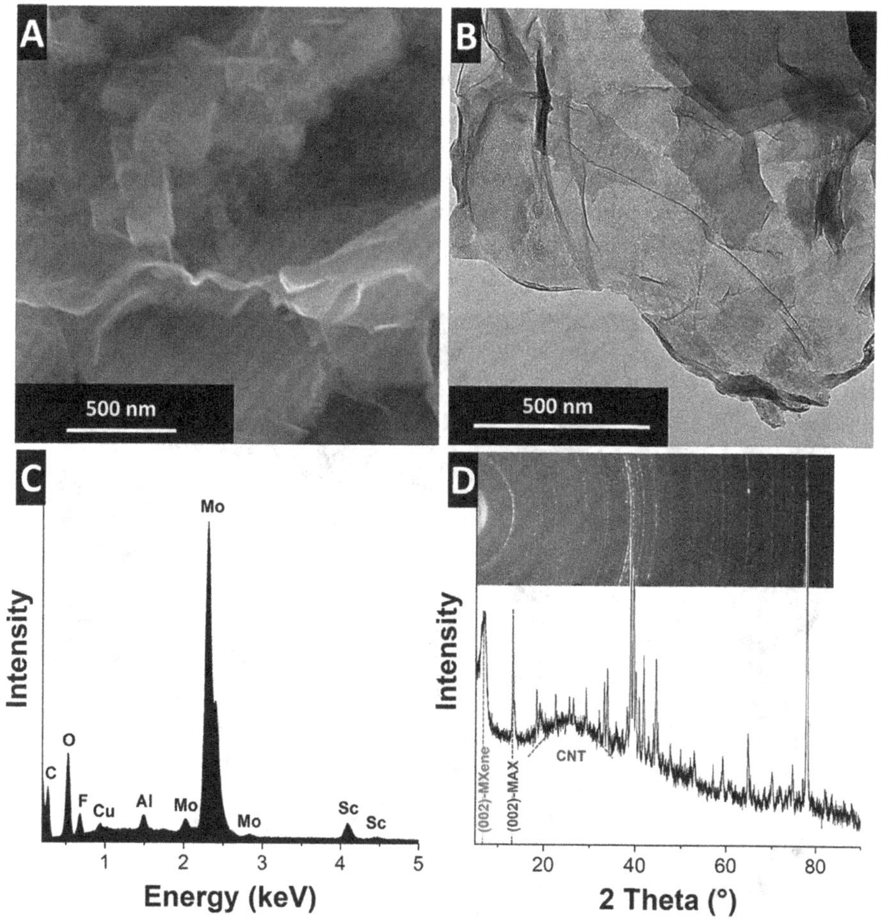

FIGURE 4.11 Characterization of the $Mo_{1.33}C$-MXene used for the preparation of electrode material. (a) SEM image; (b) TEM image; (c) EDS spectrum; and (d) XRD spectrum with 2D scattering pattern. Reproduced with permission from Srimuk et al. (2018). Copyright (2020), American Chemical Society.

organic solvent and water permeation (Liu et al. 2020; Wei et al. 2019). In a recent study, a molybdenum carbide MXene based electrode was fabricated to demonstrate the removal of cations and anions via capacitive deionization (Srimuk et al. 2018). The binder-free electrode material was prepared by entangling $Mo_{1.33}C$-MXene with carbon nanotubes. The material effectively showed desalination both in the seawater, and brackish water and the electrode attained a desalination performance of 15 mg/g in 600 mM NaCl solution. Figure 4.11 represents various characterization methods of the $Mo_{1.33}C$-MXene used as electrode material. The wrinkle layer morphology of MXene was observed via SEM image, and the TEM image depicts the exfoliated 2D MXene layers. The $Mo_{1.33}C$-MXene was prepared by HF etching on the precursor material $(Mo_{2/3}Sc_{1/3})_2AlC$. Here the smaller peaks of Al and Sc in the energy-dispersive X-ray (EDS) spectra designate effective removal of these elements during the HF acid treatment. In the X-ray diffractogram, the (002) reflection at 6.6° 2θ value is the signature peak of MXene.

Desalination of seawater to achieve the purity level of drinking water is an area of intense research focus (Sholl and Lively 2016). It is challenging to utilize 2D membranes for seawater desalination because they show a poor rejection rate for monovalent ions such as Na^+ (Han, Xu, and Gao 2013; Hirunpinyopas et al. 2017). Also, the 2D membranes tend to swell in an aqueous solution, which decreases the ion sieving performance of the membranes (Abraham et al. 2017). In a recent study, a non-swelling lamellar MXene membrane was developed by intercalating Al^{3+} ions

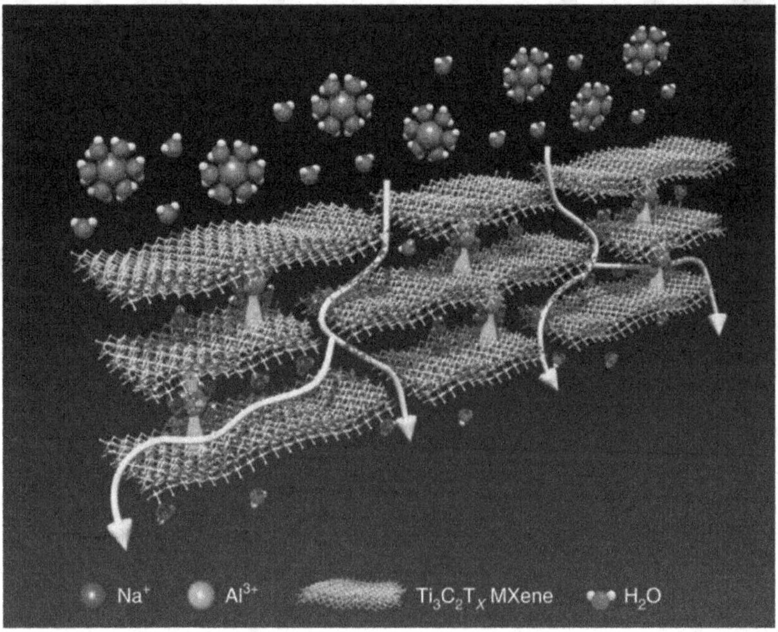

FIGURE 4.12 Schematic illustration representing the intercalation of Al^{3+} ions between $Ti_3C_2T_x$ MXenes nanosheets. Hydrated Na^+ ions are filtered out, and water molecules are passed through the membrane. Reproduced with permission from Ding et al. (2020). Copyright (2020), Springer Nature Limited.

TABLE 4.3
Membrane Materials Fabricated Using 2D MXenes and Their Pollutant Removal Efficiency

Material Used for Membrane	Pollutant/ Compound Removed	Removal Efficiency	Water Flux	References
Lamellar membrane of $Ti_3C_2T_x$ MXene	Molecules with size larger than 2.5 nm	Rejection rate over 90 %	Greater than 1000 L m^{-2} h^{-1} bar^{-1}	Ding et al. (2017)
$Ti_3C_2T_x$ lamellar nanosheets with regular pore structure	Antibiotics e.g., tetracycline	89.5% rejection	223.1 L m^{-2} h^{-1} bar^{-1}	Z. K. Li et al. (2020b)
$Mo_{1.33}C$-MXene	Cations and anions from saline water	15 mg/g from 600 mM NaCl solution via capacitive deionization	–	
Al^{3+} intercalated $Ti_3C_2T_x$ MXene	NaCl	96.5% NaCl rejection	2.8 L m^{-2} h^{-1}	Ding et al. (2020)

into the 2D $Ti_3C_2T_x$ MXene nanolayers (Figure 4.12) (Ding et al. 2020). The strong interaction between the Al^{3+} ions and the oxygen functional groups on the MXene surface was responsible for the non-swelling stability of the membranes. The membrane exhibited 96.5% NaCl rejection capacity with higher water fluxes of 2.8 L m^{-2} h^{-1} and the membrane was stable in the aqueous solution up to 400 hours.

Here Table 4.3 summarizes the different examples of MXene based membrane materials and their efficiency in separating different classes of pollutants from water.

4.4 TRANSITION METAL DICHALCOGENIDES

TMDs are inorganic analogs of graphene with the chemical formula MX_2 (Dickinson and Pauling 1923; Radisavljevic et al. 2011; Chhowalla et al. 2013). M represents transition metals such as Ti, Mo, W, etc., and X represents the chalcogenide elements such as S, Se, and Te (Manzeli et al. 2017). Among the different TMDs, MoS_2 is the most widely studied material for applications related to water purification (Hasija et al. 2020). In MoS_2, each Mo metal atom is sandwiched between two sulfur atoms, and they form a layered arrangement with an interlayer distance of approximately 6.5 Å (Figure 4.13) (Radisavljevic et al. 2011).

The bonds between the metal and chalcogen atoms are covalent, and the layers are connected by weak van der Waals interactions (Wei et al. 2018). Therefore, the TMDs can be exfoliated to separate layers by mechanical exfoliation (Cui et al. 2015), liquid-phase exfoliation (Knirsch et al. 2015), etc. A popular methodology for large-scale synthesis of TMD nanosheets is the liquid-phase exfoliation technique using ultrasonication or intercalation process (Smith et al. 2011). Choosing a suitable solvent (e.g., *N*-methylpyrrolidone, acetone) with surface energy similar to that of

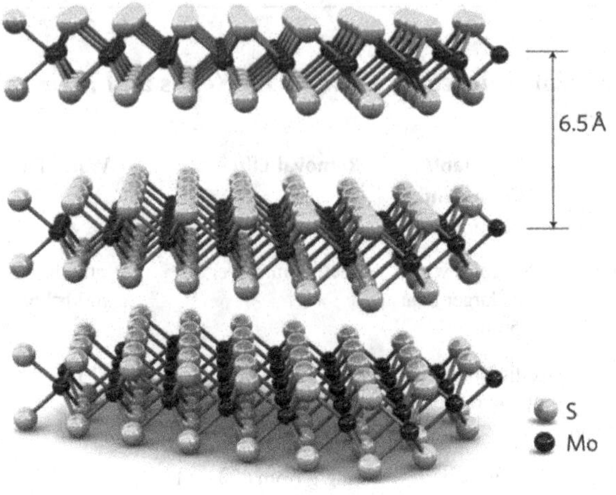

FIGURE 4.13 3D representation of the single layers of MoS_2 nanosheets. Reproduced with permission from Radisavljevic et al. (2011). Copyright (2011), Macmillan Publishers Limited.

TMDs is essential for the proper dispersion of 2D TMD layers in the solvent (Smith et al. 2011; Carey et al. 2015).

In 2013, a laminar membrane of MoS_2 was reported by assembling the chemically exfoliated atom thick MoS_2 nanosheets (L. Sun, Huang, and Peng 2013a). The membrane achieved a water permeation flux, 3–5 times that of graphene oxide membranes, rejecting 89% of Evans blue (1.2 × 3.1 nm in size) molecules. Studies performed by applying external pressure proved that the nanochannels were intact up to the pressure of 1 MPa. The laminar MoS_2 membranes were also stable in harsh chemical conditions such as strong acid and strong alkaline solutions. With the current advancements in electron beam lithography, nanopores with precise diameters of 1–10 nm can be achieved on the MoS_2 nanosheet membranes (K. Liu et al. 2014). MD simulation studies performed on single-layer MoS_2 nanosheets revealed that the pore areas ranging from 20 to 60 $Å^2$ are capable of achieving an ion rejection up to 88% (Heiranian, Farimani, and Aluru 2015). MD simulation studies on the pores with only Mo atoms on the edges concluded that those pores could reach 70% higher water flux rates compared to similar graphene nanopores. Similarly, theoretical studies on $MoSe_2$, $MoTe_2$, WS_2, and WSe_2 summarized that the key role in desalination efficiency is played by the transition metal element rather than the chalcogen atom.

MoS_2 based materials are explored for removing pollutants by adsorption technique because the contaminants can be desorbed from the MoS_2 material by altering the solution polarity using the solvents like tetrahydrofuran and acetone (Q. Huang et al. 2017). A polyacrylamide-MoS_2 based nano-fibrous membrane was effective in adsorbing rhodamine B dye up to 77.70 mg per one gram of the catalyst (Lu et al. 2018). After 30 repetitions of the adsorption/desorption process, 97.7% of the rhodamine B was removed. The desorption of dye material from the membrane was achieved by changing the solvent polarity by impregnating the membrane in acetone.

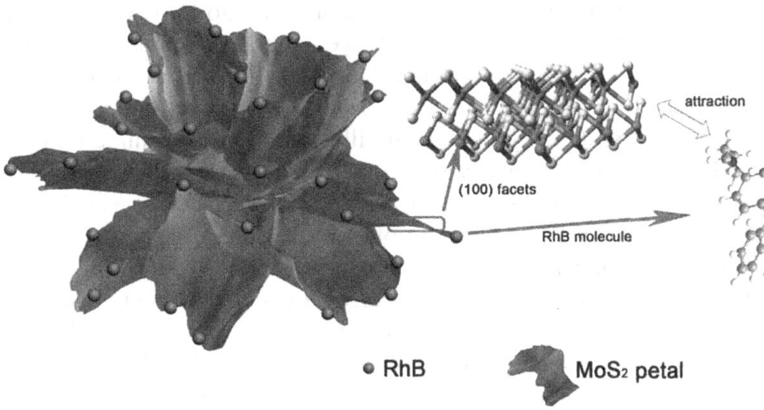

FIGURE 4.14 Schematic illustration of rhodamine b molecules loaded on the flower-like MoS_2. Reproduced with permission from Q. Huang et al. (2017). Copyright (2017), American Chemical Society.

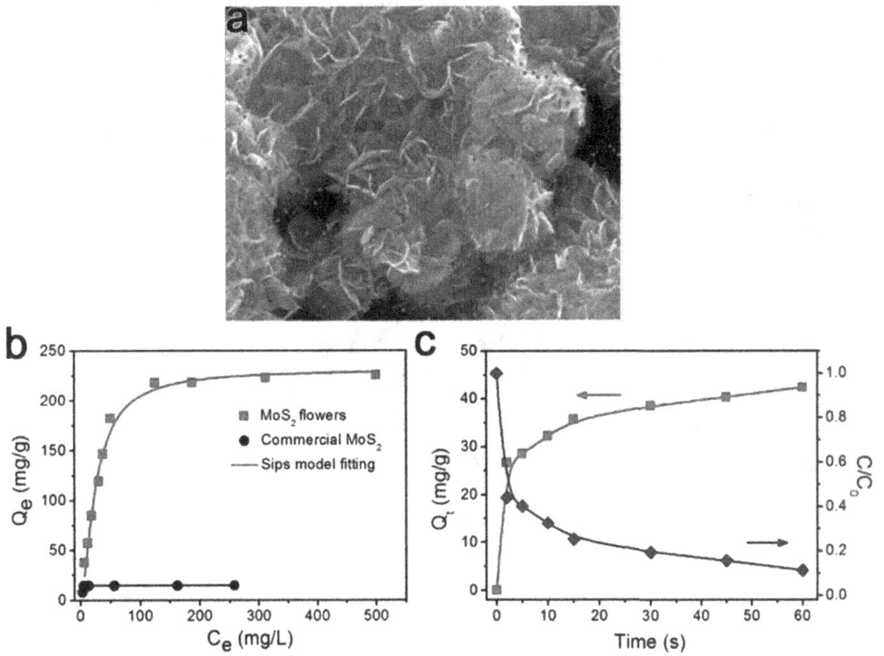

FIGURE 4.15 (a) EDS mapping showing the distribution of rhodamine b dye molecules on the flower-like MoS_2. The orange color dots represent the element nitrogen. (b) Sips model fitting, which concludes multilayer adsorption of dye molecules. (c) graphical representation of progressive adsorption of dye molecules with 42.4 mg/g loading speed. Reproduced with permission from (Q. Huang et al. 2017). Copyright (2017), American Chemical Society.

In a similar study, flower-like MoS_2 was demonstrated to have a switchable accumulation/release property of rhodamine b dye in tetrahydrofuran (THF) solvent (Q. Huang et al. 2017). The polar covalent bonds (Mo-S) at the edges of MoS_2 were responsible for attracting the zwitterionic rhodamine b molecules into the MoS_2. Figure 4.14 represents a schematic illustration of rhodamine b molecules adsorbed on a MoS_2 nanomaterial. Nitrogen element signal obtained from the energy-dispersive X-ray (EDS) analysis concluded the selective adsorption of dye molecules on the petals of the MoS_2 nanoflowers (Figure 4.15a). The kinetics profile of adsorption (Figure 4.15b) of rhodamine b was studied by the Sips model, and the fitted curves indicated multilayer adsorption with correlation coefficient (R^2) = 0.984 and fitted value of heterogeneity (n_s) = 1.5. Figure 4.15c demonstrates the progressive adsorption of dye molecules on the flower-like MoS_2 with a 42.4 mg/g loading speed and efficient rhodamine b removal (90%) within 60 seconds.

The fabric coloring industry heavily depends on different coloring dyes and inorganic salts such as Na_2SO_4 and NaCl (Li et al. 2014; Yu et al. 2021). Since the polyethersulfone matrix membrane material is hydrophobic in nature, they are modified with inorganic 2D materials to enhance their hydrophilicity and water flux (Zinadini et al. 2014; Safarpour, Vatanpour, and Khataee 2016; Abdel-Karim et al. 2018). MoS_2 nanosheet material was chemically modified by incorporating a zwitterionic compound into the polyethersulfone matrix of the membrane (Yu et al. 2021). Because of the large pore size offered by the 2D material integrated nanofiltration membrane, this system was demonstrated for the fractionalization of dyes and inorganic salts from dye industry wastewater. High dye retention (reactive black) of 98.2%, and

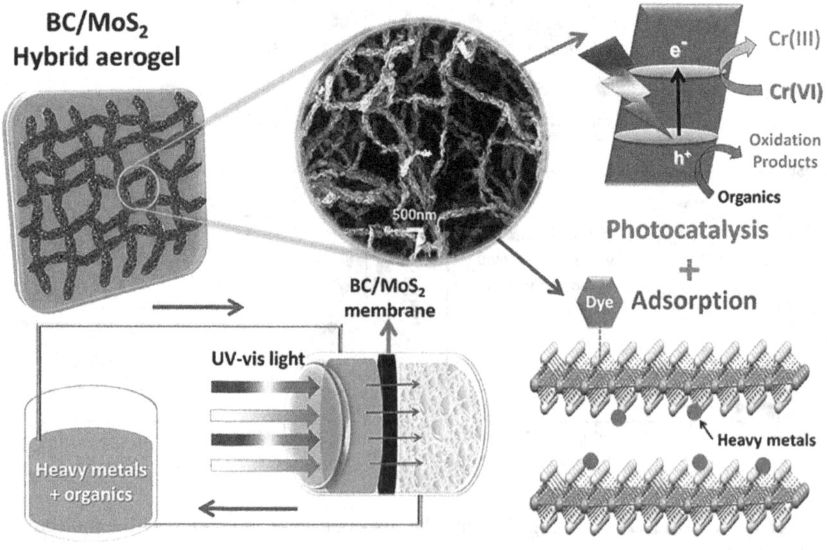

FIGURE 4.16 Schematic illustration of the highly porous architecture of the bacterial cellulose-MoS_2 membranes and the *in-flow* removal of contaminants *via* adsorption and photocatalytic strategies. Reproduced with permission from Ferreira-Neto et al. (2020). Copyright (2020), American Chemical Society.

TABLE 4.4
Membrane Materials Fabricated Using 2D MoS_2 and Their Pollutant Removal Efficiency

Material Used for Membrane	Pollutant/ Compound Removed	Removal Efficiency	Water Flux	References
Atom thick MoS_2 nanosheets	Evans blue dye	89% rejection	245 L h^{-1} m^{-2} bar^{-1}	Sun, Huang, and Peng (2013a)
Single-layer MoS_2 nanosheets (molecular dynamic simulation studies)	Ions present in saline water	88% of ions	70% greater than that of graphene nanopores	Heiranian, Farimani, and Aluru (2015)
Polyacrylamide-MoS_2 membrane	rhodamine B (Adsorption based)	77.70 mg/g with 97.7% efficiency	–	Y. Lu et al. (2018)
Zwitterionic compound modified MoS_2 polyethersulfone matrix membrane	reactive black	98.2% dye retention	108.3 L m^{-2} h^{-1} at 0.6 MPa	Safarpour, Vatanpour, and Khataee (2016)
Bacterial nanocellulose-MoS_2 hybrid aerogel membrane	Cr (VI) heavy metal removal	88 % Cr (VI) removal within 120 minutes	–	Ferreira-Neto et al. (2020)

selective permeation of salts (NaCl, 1.1% rejection) was achieved. The separated dye was used for the coloring process, and inorganic saltwater was reused in the same industry, thus demonstrating an environmentally friendly sustainable model.

In a recent study fabricated a bacterial nanocellulose-MoS_2 hybrid aerogel membrane and demonstrated high contaminant removal by adsorption-cum-photocatalytic mechanism (Ferreira-Neto et al. 2020). The hydrothermal process and supercritical drying synthesis methodology resulted in the precise coating of MoS_2 nanostructures on the bacterial cellulose nanofibrils and achieved the porous hybrid architecture. The removal performance of contaminants varied based on their size, and the smaller molecules such as CrO_4^{2-} were effectively removed compared to larger dye molecules. The membrane showed 88% Cr (VI) heavy metal removal within 120 minutes (K_{obs} = 0.0012 min^{-1}). Figure 4.16 represents a schematic diagram of the nanoporous structure of the hybrid membrane and the *in-flow* photocatalytic degradation mechanism of contaminants.

Here Table 4.4 summarizes the various examples of MoS_2 based membrane materials and their efficiency in separating different pollutants from water.

4.5 MATERIALS FOR PHOTOCATALYTIC WATER TREATMENT

4.5.1 Antibacterial Action of Graphene Family Nanomaterials

The antibacterial properties of graphene family nanomaterials (GFN) come from a variety of methods including: membrane stress, oxidative stress, electron transfer, photothermal reaction, interference of protein-protein interaction, and coverage and isolation. Membrane stress is caused due to the physical stress exerted by sharp edges on GFN, causing damage to the bacteria membrane. Oxidative stress results from the formation of reactive oxygen species (ROS) that can decrease the activity of enzymes, the fragmentation of DNA and lipid peroxidation (Han et al. 2019). Electron transfer results from an electron moving from the membrane of bacteria to the GFN, causing damage to the membrane. Photothermal reaction is a process in which the GFN absorb light and transfer the energy; this energy can be exerted as heat up to 100°C to inactivate the bacteria. Interference of protein-protein is caused by the graphene blocking contact of proteins leading to biological dysfunction and even cell death. Coverage and isolation are whereby GFN sheets completely surround a bacteria, preventing it from interacting with its surrounding environment (Hu et al. 2010; Liu et al. 2011). A modified graphene oxide with sodium anthraquinone-2-sulfonate was found to be capable of eliminating 99% of *E. coli* when exposed to visible light for 160 minutes (Zhang et al. 2020).

4.5.2 Photocatalytic Degradation and Disinfection

Photocatalytic disinfection works because of the absorption of photons with higher energy than the band gap of the employed photocatalyst. When this occurs, electrons will move to the conduction band from the valence band as they have become excited (An and Yu 2011). When the electron leaves the valence band, it will leave a hole; the electron and the hole are referred to as charge carriers. Either charge carrier can result in the formation of a ROS. A hydroxyl radical is formed due to the hole in the valence band causing oxidation, and a superoxide radical anion is formed in the

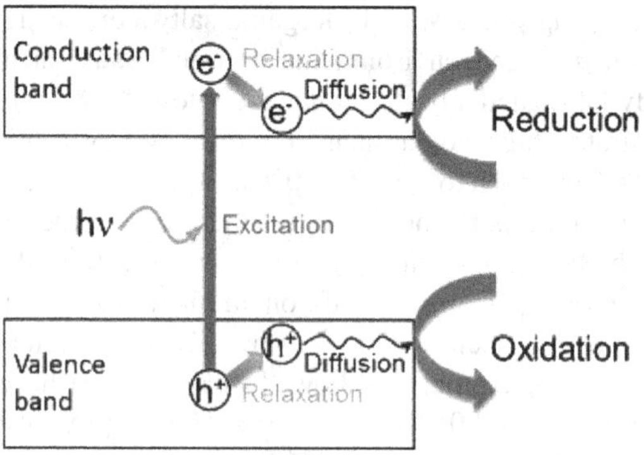

FIGURE 4.17 Schematic illustration of oxidation and reduction processes during photocatalysis. Reproduced with permission from Nakata and Fujishima (2012). Copyright (2012), Elsevier.

conduction band due to reduction (Figure 4.17) (Nakata and Fujishima 2012). It is these ROS that are used for photocatalytic disinfection (Hu et al. 2020). The ROS have very little selectivity and, as a result, will break down a wide variety of pollutants (Lu et al. 2016). Photo-Fenton and electro-Fenton are both widely used methods of photocatalytic disinfection that have seen widespread use over the years (Affam, Chaudhuri, and Kutty 2018; Babuponnusami and Muthukumar 2012; Changotra, Rajput, and Dhir 2019). The use of an iron catalyst 2D material in Fenton reactions can potentially lead to advanced and effective catalysts. Heterogenous photocatalysts use a support structure to adhere to a photocatalyst that uses advanced oxidation processes to create ROS and degrade pollutants (Oluwole, Omotola, and Olatunji 2020).

Photocatalytic degradation requires a significant amount of surface area to be effective when eliminating organic compounds. A 2D membrane that can allow liquid to make contact with all of the molecules and can allow large amounts of the catalyst to interact with the pollutant (Wang et al. 2020).

4.5.3 Graphene and g-C_3N_4

Graphene has multiple derivatives: graphene itself is a single sheet of graphite, while graphene oxide is an oxidized form of graphene, and reduced graphene oxide is a thermally reduced form of GO (Mohammed et al. 2020). Graphitic carbon nitride (g-C_3N_4) is a 2D structure that connects tri-s-triazine units with amino groups. As a result, layers form weak van der Waals forces. The nitrogen atoms present allow the compound to be a semiconductor (Shaolin Zhang et al. 2017). g-C_3N_4 has a bandgap of between 2.7 to 2.8e V (Saha et al. 2020). The integration of metal ions into the

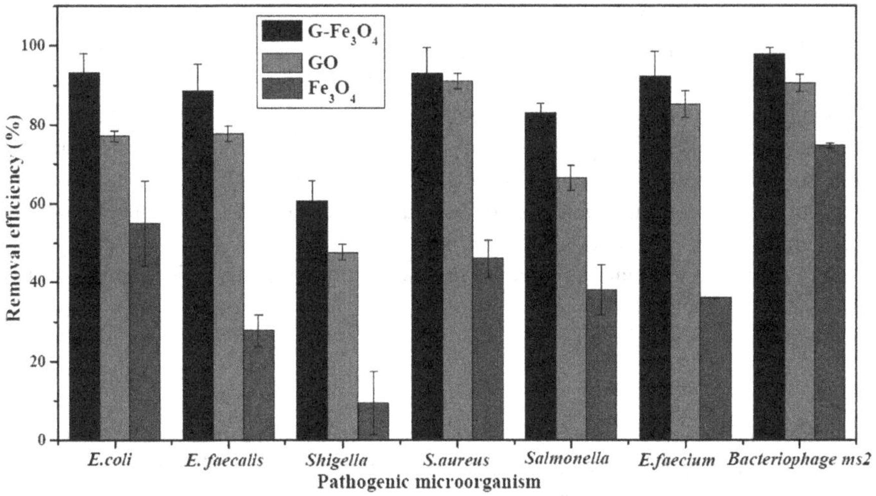

FIGURE 4.18 Removal of bacteria by G-Fe_3O_4, GO, and Fe_3O_4. Reproduced with permission from Zhan et al. (2015) Copyright (2015), American Chemical Society.

TABLE 4.5
Graphene Composites and Bacterial Disinfection

Composite	Bacteria	Concentration	% Elimination	References
G-TiO2	*E. coli*	10^6 cfu/mL	90.50	Cao et al. (2013)
G-ZnO	*E. coli*	10^5 cfu/mL	99.99	Kavitha et al. (2012)
G-Fe_3O_4	*E. coli*	10^8 cfu/mL	93.09	Zhan et al. (2015)
G-Fe_3O_4	*S. aureus*	10^2 cfu/mL	92.79	Zhan et al. (2015)
G-Fe_3O_4	*E. faecalis*	10^2 cfu/mL	88.52	Zhan et al. (2015)
G-Fe_3O_4	*E. faecium*	10^2 cfu/mL	92.26	Zhan et al. (2015)
G-Fe_3O_4	*Shigella*	10^2 cfu/mL	60.49	Zhan et al. (2015)
G-Fe_3O_4	*Salmonella*	10^2 cfu/mL	82.85	Zhan et al. (2015)
Ag-GO	*E. coli*	10^4 cfu/mL	95.60	Y. Q. Chen et al. (2020c)
AQS-GO	*E. coli*	10^8 cfu/mL	99.00	L Zhang et al. (2020a)

g-C_3N_4 greatly enhanced the photodegradation of organic pollutants in water (Gao et al. 2014). The formation of a nanoporous *g*-C_3N_4 showed an increase in photodegradation of 3.4 times higher than the standard *g*-C_3N_4. The use of graphene-based materials along with the metal oxides such as TiO_2, ZnO (Gao, Ng, and Sun 2013), and Fe_3O_4 (Singh et al. 2020) have been studied to enhance the photocatalytic activity of metal oxides. TiO_2 quantum dots were shown to be capable of inactivating *E. coli* by 91% at 60 μg mL^{-1} (Ahmed et al. 2019). Graphene oxide and silver composite was used for the disinfection of *E. coli*; this was capable of sterilization using only 7.0 μg mL^{-1} of the GO-Ag (Chen et al. 2020). *g*-C_3N_4 combined with graphene oxide quantum dots were capable of 99.6% inactivation of *E.coli* (Xu et al. 2020a). A graphene Fe_3O_4 composite was compared to graphene and Fe_3O_4 in its elimination of a range of Gram-positive and Gram-negative bacteria, which can be seen in Figure 4.18 (Zhan et al. 2015). With electro-Fenton being used for possible wastewater treatment, the development of cathodes using graphene has shown considerable promise (Mousset et al. 2016). Using pristine graphene, one study was able to show that these 2D materials could degrade water pollutants. This degradation was greatly enhanced when the materials were layered to form a foam. A novel study used waste material from polyethylene-terephthalate (PET) bottles to produce GO, which was used for adsorption; however, this could also be used as a green source for the photocatalytic treatment of water (Elessawy et al. 2020).

Here Table 4.5 summarizes different graphene composites used for photocatalytic deactivation of bacteria.

4.5.4 Iron Oxide

Iron oxide nanomaterials are beginning to see an increase in interest due to several specific properties like their "high surface-area-to-volume ratio", having good biocompatibility, magnetic properties, a modifiable surface and their nano-size (P. A. Xu et al. 2012). These properties make them suitable for a range of technologies. Their magnetism makes them suitable for adsorption of heavy metals and allows for easier

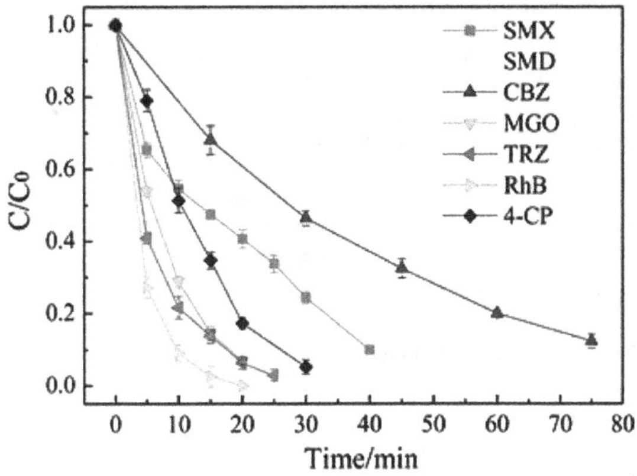

FIGURE 4.19 Degradation of sulfamethoxazole (SMX), sulfacetamide (SMD), carbamazepine (CBZ), malachite green (MGO), tartrazine (TRZ), rhodamine B (RhB) and 4-chlorophenol (4-CP) by MnO_2 Fe_2O_3 hybrid. Reproduced with permission from Guo et al. (2020a) Copyright (2020), Elsevier.

removal after adsorption from the water (Pang et al. 2011). Iron oxide materials are also suitable for removing dyes and aromatic compounds from water (Saharan et al. 2014). A bandgap of 2.2 eV makes them suitable for visible light photocatalysis (Akhavan and Azimirad 2009). The combination of γ-Fe_2O_3 and graphene oxide for the removal of methylene blue was found to have 90.60% photodegradation (Rehman et al. 2020).

4.5.5 Manganese Oxide

Manganese oxide has been found to have excellent catalytic activity in wastewater treatment, with dye removal seen at neutral pH. This is due to the electrostatic attraction between the dye molecules and manganese oxide (Chen and He 2008). A MnO_2 Fe_2O_3 hybrid was used to degrade multiple pollutants in water with considerable success (Guo et al. 2020a). When used to degrade malachite green, rhodamine B, sulfacetamide, 4-chlorophenol, tartrazine, carbamazepine, and sulfamethoxazole degradation efficiencies of 93.10%, 100%, 91.50%, 94.80%, 97.10%, 87.60% and 90% respectively were found. See Figure 4.19

4.5.6 Metal Oxyhalides

Metal oxyhalides such as metal-doped bismuth oxyhalides have been used as alternative catalysts for the photo-Fenton process. Using the bismuth oxyhalide to degrade the tartrazine dye produced a 91% degradation, 95% decolorization, and TOC reduction of 59% (Tekin, Ersöz, and Atalay 2018). Photo-catalytically active metal oxide can have a range of bandgaps depending on the metal present. When three elements were examined using a base bismuth oxyhalide, bandgaps of 1.85, 2.76, and 3.44 were found for iodine, bromine, and chlorine, respectively. Bromine was found to be

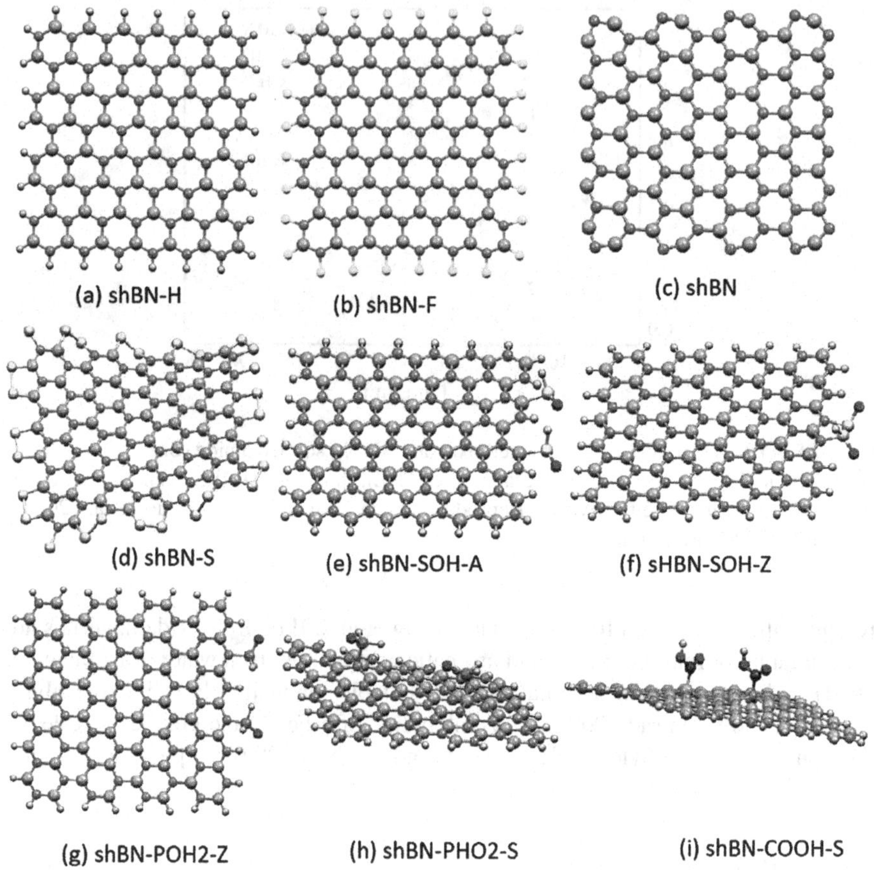

FIGURE 4.20 Modified boron nitride nanosheet quantum dots with different elements: colors represent the elements: gray = carbon, orange = phosphorus, red = oxygen, pink = boron, blue = nitrogen, silver = silver, green = fluorine, and yellow = sulfur. Reproduced with permission from Abdelsalam et al. (2020). Copyright (2020), The Royal Society of Chemistry.

most photocatalytic when used to degrade rhodamine B dye, iodine showed no photocatalytic activity at all, and chlorine showed the greatest activity when degrading isopropanol (H. An et al. 2008). A separate study analyzing BiOCl, BiOBr, and BiOI showed total organic carbon removal of 52%, 44%, and 31%, respectively (Li et al. 2020). This variability would require a range of configurations to ensure the correct treatment of wastewater.

4.5.7 Boron Nitride Nanosheets

Boron nitride (BN) nanosheets are 2D materials that are formed when BN powder is exfoliated (Ikram et al. 2020). These have a wide bandgap (~5.5eV), making them much more suitable for co-catalysts (Gao, Yao, and Meng 2020). When used on methyl blue, a degradation percentage of 99% was achieved. Moreover, when used as quantum dots, they were found to be excellent adsorbents for the hexahydrate elements Co, Fe, Cr, Zn, Cd and Cu (Abdelsalam et al. 2020). This study also showed that these nanosheets could retain their structural stability with some elements when modified, but could lead to deformation with others (Figure 4.20). A BN nanosheet doped with AgI was capable of degrading 96.17% of rhodamine B within 70 minutes (Choi, Reddy, and Kim 2015). When a BN/Bi_2MoO_6 was used to degrade iohexol using only 3.5 wt%, a degradation rate of 92% was obtained after 150 min (He et al. 2020). A separate study using a graphene-like BN for the degradation of rhodamine B showed 95.1% degradation after 20 minutes (Song et al. 2020).

The unique set of properties of 2D materials makes them highly applicable to the treatment of wastewater. These properties include antibacterial uses of the graphene family of materials and photocatalytic activity of metal oxides and graphitic carbon nitride (Szunerits and Boukherroub 2016; Luo, Liu, and Wang 2016). Various 2D materials show excellent adsorption ability to heavy metals and other inorganic contaminants, which are otherwise impossible or difficult to break down (Liu et al. 2019).

4.6 FUTURE ADVANCES

Currently, there are various techniques available for the exfoliation of different nanolayers from bulk material. However, precisely controlling the thickness of individual layers and altering their surface properties is still a challenging task. Therefore, new exfoliation methodologies need to be explored in order to achieve nano layers with uniform thickness. The stability of 2D material membranes is considerably higher compared to conventional polymeric membranes (Van der Bruggen, Mänttäri, and Nyström 2008). Nevertheless, 2D material membranes undergo deformation and swelling upon long-term exposure to solvents (Abraham et al. 2017). Hence, future research should concentrate on developing materials having long-term stability in different classes of solvents. Synthetic methodologies should be able to control the stacking distance between the layers in laminar membranes. Only the advanced fabrication techniques can discriminate and separate the contaminants having a similar size. According to current understanding in the field, size exclusion and chemical interaction are attributed as the method for separation in 2D material membranes. Proper theoretical understanding of the mass transport methodologies within a 2D material is an immediate research direction for theoretical material scientists.

4.7 CONCLUSION

Two-dimensional (2D) materials are emerging as excellent systems for the fabrication of nanosheet and lamellar membranes for applications in separation. The unique atomic thickness of 2D materials endows them with higher water permeation rates. Incorporating a functionalized graphene family of materials into the thin-film composite membranes alters these membranes' selectivity and water permeation flux. The distinctive antimicrobial properties of graphene-based materials make them exciting candidates for applications in the disinfection of microorganisms. The non-swelling membranes fabricated by incorporating Al^{3+} ions into the nanolayers of $Ti_3C_2T_x$ MXene resulted in a 96.5% rejection of NaCl molecules. It throws light on the potential of MXene for applications in desalination purposes. The nanocomposite membranes fabricated by MXenes and TMDs have a great future in preparing separation membranes for large-scale applications. In the area of photo-catalytically active 2D materials, the major research is concentrated on techniques for efficient light absorption and excellent charge separation. Further research is needed to gain insight into the mechanism of photocatalytic inactivation of microorganisms by 2D materials. It is expected that developing more accessible routes to the synthesis and functionalization of 2D materials will bring about more sustainable solutions for water treatment technologies.

REFERENCES

Abdel-Karim, A., S. Leaper, M. Alberto, A. Vijayaraghavan, X. Fan, S. M. Holmes, E. R. Souaya, M. I. Badawy, and P. Gorgojo. 2018. "High Fl Ux and Fouling Resistant Fl at Sheet Polyethersulfone Membranes Incorporated with Graphene Oxide for Ultra Fi Ltration Applications" 334 (August 2017): 789–799. https://doi.org/10.1016/j.cej.2017.10.069.

Abdelsalam, H, W O Younis, V A Saroka, N H Teleb, S Yunoki, and Q Zhang. 2020. "Interaction of Hydrated Metals with Chemically Modified Hexagonal Boron Nitride Quantum Dots: Wastewater Treatment and Water Splitting." *Physical Chemistry Chemical Physics* 22 (4): 2566–2579. https://doi.org/10.1039/c9cp06823f.

Abraham, J, K S Vasu, C D Williams, K Gopinadhan, Y Su, C T Cherian, J Dix, et al. 2017. "Tunable Sieving of Ions Using Graphene Oxide Membranes." *Nature Nanotechnology* 12 (6): 546–550. https://doi.org/10.1038/nnano.2017.21.

Adegoke, A A, A C Faleye, G Singh, and T A Stenström. 2017. "Antibiotic Resistant Superbugs: Assessment of the Interrelationship of Occurrence in Clinical Settings and Environmental Niches." *Molecules* 22 (1). https://doi.org/10.3390/molecules22010029.

Affam, A C, M Chaudhuri, and S R M Kutty. 2018. "Comparison of Five Advanced Oxidation Processes for Degradation of Pesticide in Aqueous Solution." *Bulletin of Chemical Reaction Engineering and Catalysis* 13 (1): 179–186. https://doi.org/10.9767/bcrec.13.1.1394.179-186.

Ahmed, F, C Awada, S A Ansari, A Aljaafari, and A Alshoaibi. 2019. "Photocatalytic Inactivation of Escherichia Coli under UV Light Irradiation Using Large Surface Area Anatase TiO2 Quantum Dots." *Royal Society Open Science* 6 (12): 10. https://doi.org/10.1098/rsos.191444.

Akhavan, O, and R Azimirad. 2009. "Photocatalytic Property of Fe_2O_3 Nanograin Chains Coated by TiO_2 Nanolayer in Visible Light Irradiation." *Applied Catalysis A: General* 369 (1): 77–82. https://doi.org/10.1016/j.apcata.2009.09.001.

Alanis, A J 2005. "Resistance to Antibiotics: Are We in the Post-Antibiotic Era?" *Archives of Medical Research* 36 (6): 697–705. https://doi.org/10.1016/j.arcmed.2005.06.009.

Al-Hamadani, Y A J, B M Jun, M Yoon, N Taheri-Qazvini, S A Snyder, M Jang, J Heo, and Y Yoon. 2020. "Applications of MXene-Based Membranes in Water Purification: A Review." *Chemosphere* 254: 126821. https://doi.org/10.1016/j.chemosphere.2020.126821.

An, H, Y Du, T Wang, C Wang, W Hao, and J Zhang. 2008. "Photocatalytic Properties of BiOX (X = Cl, Br, and I)." *Rare Metals* 27 (3): 243–250. https://doi.org/10.1016/S1001-0521(08)60123-0.

An, X Q, and J C Yu. 2011. "Graphene-Based Photocatalytic Composites." *RSC Advances* 1 (8): 1426–1434. https://doi.org/10.1039/c1ra00382h.

Aqra, M W, and A A Ramanathan. 2020. "Graphene and Related 2D Materials for Desalination: A Review of Recent Patents." *Jordan Journal of Physics* 13 (3): 233–242. https://doi.org/10.47011/13.3.7.

Babuponnusami, A, and K Muthukumar. 2012. "Advanced Oxidation of Phenol: A Comparison between Fenton, Electro-Fenton, Sono-Electro-Fenton and Photo-Electro-Fenton Processes." *Chemical Engineering Journal* 183: 1–9. https://doi.org/10.1016/j.cej.2011.12.010.

Bell, D C, M C Lemme, L A Stern, J R Williams, and C M Marcus. 2009. "Precision Cutting and Patterning of Graphene with Helium Ions." *Nanotechnology* 20 (45): 455301. https://doi.org/10.1088/0957-4484/20/45/455301.

Brodie, B C. 1860. "Sur Le Poids Atomique Du Graphite." *Annales de Chimie et de Physique* 59 (466): e472.

Brownson, D A C, D K Kampouris, and C E Banks. 2011. "An Overview of Graphene in Energy Production and Storage Applications." *Journal of Power Sources* 196 (11): 4873–4885. https://doi.org/10.1016/j.jpowsour.2011.02.022.

Bunch, J S, S S Verbridge, J S Alden, A M Van Der Zande, J M Parpia, H G Craighead, and P L McEuen. 2008. "Impermeable Atomic Membranes from Graphene Sheets." *Nano Letters* 8 (8): 2458–2462. https://doi.org/10.1021/nl801457b.

Cao, B, S Cao, P Dong, J Gao, and J Wang. 2013. "High Antibacterial Activity of Ultrafine TiO_2/Graphene Sheets Nanocomposites under Visible Light Irradiation." *Materials Letters* 93: 349–352. https://doi.org/10.1016/j.matlet.2012.11.136.

Carey, B J, T Daeneke, E P Nguyen, Y Wang, O Jian Zhen, S Zhuiykov, and K Kalantar-Zadeh. 2015. "Two Solvent Grinding Sonication Method for the Synthesis of Two-Dimensional Tungsten Disulphide Flakes." *Chemical Communications* 51 (18): 3770–3773. https://doi.org/10.1039/c4cc08399g.

Chae, H R, J Lee, C H Lee, I C Kim, and P K Park. 2015. "Graphene Oxide-Embedded Thin-Film Composite Reverse Osmosis Membrane with High Flux, Anti-Biofouling, and Chlorine Resistance." *Journal of Membrane Science* 483: 128–135. https://doi.org/10.1016/j.memsci.2015.02.045.

Changotra, R, H Rajput, and A Dhir. 2019. "Treatment of Real Pharmaceutical Wastewater Using Combined Approach of Fenton Applications and Aerobic Biological Treatment." *Journal of Photochemistry and Photobiology A-Chemistry* 376: 175–184. https://doi.org/10.1016/j.jphotochem.2019.02.029.

Chen, B, H Jiang, X Liu, and X Hu. 2017a. "Molecular Insight into Water Desalination Across Multilayer Graphene Oxide Membranes." *ACS Applied Materials and Interfaces* 9 (27). https://doi.org/10.1021/acsami.7b05307.

Chen, C, L Jia, J Li, L Zhang, L Liang, E Chen, Z Kong, X Wang, W Zhang, and J W Shen. 2020a. "Understanding the Effect of Hydroxyl/Epoxy Group on Water Desalination through Lamellar Graphene Oxide Membranes via Molecular Dynamics Simulation." *Desalination* 491 (March): 114560. https://doi.org/10.1016/j.desal.2020.114560.

Chen, H, and J He. 2008. "Facile Synthesis of Monodisperse Manganese Oxide Nanostructures and Their Application in Water Treatment." *The Journal of Physical Chemistry C* 112 (45): 17540–45. https://doi.org/10.1021/jp806160g.

Chen, J, Q Huang, H Huang, L Mao, M Liu, X Zhang, and Yen Wei. 2020b. "Recent Progress and Advances in the Environmental Applications of MXene Related Materials." *Nanoscale* 12 (6): 3574–3592. https://doi.org/10.1039/c9nr08542d.

Chen, L, G Shi, J Shen, B Peng, B Zhang, Y Wang, F Bian, et al. 2017b. "Ion Sieving in Graphene Oxide Membranes via Cationic Control of Interlayer Spacing." *Nature* 550 (7676): 1–4. https://doi.org/10.1038/nature24044.

Chen, Y Q, W Wu, Z Q Xu, C Jiang, S Han, J Ruan, and Y Wang. 2020c. "Photothermal-Assisted Antibacterial Application of Graphene Oxide-Ag Nanocomposites against Clinically Isolated Multi-Drug Resistant Escherichia Coli." *Royal Society Open Science* 7 (7): 13. https://doi.org/10.1098/rsos.192019.

Chen, Z, Q Li, and X Chen. 2020d. "Porous Graphene/Polyimide Membrane with a Three-Dimensional Architecture for Rapid and Efficient Solar Desalination via Interfacial Evaporation." *ACS Sustainable Chemistry and Engineering* 8 (36). https://doi.org/10.1021/acssuschemeng.0c05306.

Cheng, P, M M Kelly, N K Moehring, W Ko, A P Li, J C Idrobo, M S H Boutilier, and P R Kidambi. 2020. "Facile Size-Selective Defect Sealing in Large-Area Atomically Thin Graphene Membranes for Sub-Nanometer Scale Separations." *Nano Letters* 20 (8). https://doi.org/10.1021/acs.nanolett.0c01934.

Chhowalla, M, H S Shin, G Eda, L J Li, K P Loh, and H Zhang. 2013. "The Chemistry of Two-Dimensional Layered Transition Metal Dichalcogenide Nanosheets." *Nature Chemistry* 5 (4): 263–275. https://doi.org/10.1038/nchem.1589.

Choi, J, D Amaranatha Reddy, and T K Kim. 2015. "Enhanced Photocatalytic Activity and Anti-Photocorrosion of AgI Nanostructures by Coupling with Graphene-Analogue Boron Nitride Nanosheets." *Ceramics International* 41 (10, Part A): 13793–13803. https://doi.org/10.1016/j.ceramint.2015.08.062.

Choi, W, J Choi, J Bang, and J H Lee. 2013. "Layer-by-Layer Assembly of Graphene Oxide Nanosheets on Polyamide Membranes for Durable Reverse-Osmosis Applications." *ACS Applied Materials and Interfaces* 5 (23): 12510–12519. https://doi.org/10.1021/am403790s.

Cohen-Tanugi, D, and J C Grossman. 2012. "Water Desalination across Nanoporous Graphene." *Nano Letters* 12 (7): 3602–3608. https://doi.org/10.1021/nl3012853.

Cohen-Tanugi, D, and J C Grossman. 2014. "Mechanical Strength of Nanoporous Graphene as a Desalination Membrane." *Nano Letters* 14 (11): 6171–6178. https://doi.org/10.1021/nl502399y.

Cohen-Tanugi, D, C L Li, and J C Grossman. 2016. "Multilayer Nanoporous Graphene Membranes for Water Desalination." *Nano Letters* 16 (2). https://doi.org/10.1021/acs.nanolett.5b04089.

Comninellis, C, A Kapalka, S Malato, S A Parsons, I Poulios, and D Mantzavinos. 2008. "Advanced Oxidation Processes for Water Treatment: Advances and Trends for R&D." *Journal of Chemical Technology and Biotechnology* 83 (6): 769–776. https://doi.org/10.1002/jctb.1873.

Cui, J, X Zhang, H Liu, S Liu, and K L Yeung. 2008. "Preparation and Application of Zeolite/Ceramic Microfiltration Membranes for Treatment of Oil Contaminated Water." *Journal of Membrane Science* 325 (1): 420–426. https://doi.org/10.1016/j.memsci.2008.08.015.

Cui, X, G H Lee, Y D Kim, G Arefe, P Y Huang, C H Lee, D A Chenet, et al. 2015. "Multi-Terminal Transport Measurements of MoS2 Using a van Der Waals Heterostructure Device Platform." *Nature Nanotechnology* 10 (6): 534–540. https://doi.org/10.1038/nnano.2015.70.

Dervin, S, D D Dionysiou, and Suresh C Pillai. 2016. "2D Nanostructures for Water Purification: Graphene and Beyond." *Nanoscale* https://doi.org/10.1039/c6nr04508a.

Dickinson, R G, and L Pauling. 1923. "The Crystal Structure of Molybdenite." *Journal of the American Chemical Society* 45 (6): 1466–1471. https://doi.org/10.1021/ja01659a020.

Ding, L, L Li, Y Liu, Y Wu, Z Lu, J Deng, Y Wei, J Caro, and H Wang. 2020. "Effective Ion Sieving with Ti3C2Tx MXene Membranes for Production of Drinking Water from Seawater." *Nature Sustainability* 3 (4): 296–302. https://doi.org/10.1038/s41893-020-0474-0.

Ding, L, Y Wei, Y Wang, H Chen, J Caro, and H Wang. 2017. "A Two-Dimensional Lamellar Membrane: MXene Nanosheet Stacks." *Angewandte Chemie - International Edition* 56 (7): 1825–1829. https://doi.org/10.1002/anie.201609306.

Dubertret, B, T Heine, and M Terrones. 2015. "The Rise of Two-Dimensional Materials." *Accounts of Chemical Research* 48 (1): 1–2. https://doi.org/10.1021/ar5004434.

Elessawy, N A, M H Gouda, S M Ali, M Salerno, and M S M Eldin. 2020. "Effective Elimination of Contaminant Antibiotics Using High-Surface-Area Magnetic-Functionalized Graphene Nanocomposites Developed from Plastic Waste." *Materials* 13 (7): 17. https://doi.org/10.3390/ma13071517.

Elimelech, M, and W A Phillip. 2011. "The Future of Seawater Desalination: Energy, Technology, and the Environment." *Science* 333 (6043): 712–717. https://doi.org/10.1126/science.1200488.

Ferreira-Neto, E P, S Ullah, T C A Da Silva, R R Domeneguetti, A P Perissinotto, F S De Vicente, U P Rodrigues-Filho, and S J L Ribeiro. 2020. "Bacterial Nanocellulose/MoS_2 Hybrid Aerogels as Bifunctional Adsorbent/Photocatalyst Membranes for in-Flow Water Decontamination." *ACS Applied Materials and Interfaces* 12 (37): 41627–41643. https://doi.org/10.1021/acsami.0c14137.

Firouzjaei, M D, S Fatemeh Seyedpour, S A Aktij, M Giagnorio, N Bazrafshan, A Mollahosseini, F Samadi, et al. 2020. "Recent Advances in Functionalized Polymer Membranes for Biofouling Control and Mitigation in Forward Osmosis." *Journal of Membrane Science* 596: 117604. https://doi.org/10.1016/j.memsci.2019.117604.

Ganesh, B M, A M Isloor, and A F Ismail. 2013. "Enhanced Hydrophilicity and Salt Rejection Study of Graphene Oxide-Polysulfone Mixed Matrix Membrane." *Desalination* 313: 199–207. https://doi.org/10.1016/j.desal.2012.11.037.

Gao, H, S Yan, J Wang, and Z Zou. 2014. "Ion Coordination Significantly Enhances the Photocatalytic Activity of Graphitic-Phase Carbon Nitride." *Dalton Transactions* 43 (22): 8178–8183. https://doi.org/10.1039/C3DT53224K.

Gao, P, K Ng, and D D Sun. 2013. "Sulfonated Graphene Oxide–ZnO–Ag Photocatalyst for Fast Photodegradation and Disinfection under Visible Light." *Journal of Hazardous Materials* 262: 826–835. https://doi.org/10.1016/j.jhazmat.2013.09.055.

Gao, X Y, Y Yao, and X C Meng. 2020. "Recent Development on BN-Based Photocatalysis: A Review." *Materials Science in Semiconductor Processing* 120: 13. https://doi.org/10.1016/j.mssp.2020.105256.

Gao, Y, M Hu, and B Mi. 2014. "Membrane Surface Modification with TiO_2-Graphene Oxide for Enhanced Photocatalytic Performance." *Journal of Membrane Science* 455: 349–356. https://doi.org/10.1016/j.memsci.2014.01.011.

Garaj, S, W Hubbard, A Reina, J Kong, D Branton, and J A Golovchenko. 2010. "Graphene as a Subnanometre Trans-Electrode Membrane." *Nature* 467 (7312): 190–193. https://doi.org/10.1038/nature09379.

Geim, A K, and I V Grigorieva. 2013. "Van Der Waals Heterostructures." *Nature* 499 (7459): 419–425. https://doi.org/10.1038/nature12385.

Geim, A K, and K S Novoselov. 2009. "The Rise of Graphene." In *Nanoscience and Technology: A Collection of Reviews from Nature Journals*, 11–19. World Scientific. https://doi.org/10.1142/9789814287005_0002.

Guo, R N, Y Y Wang, J J Li, X W Cheng, and D D Dionysiou. 2020a. "Sulfamethoxazole Degradation by Visible Light Assisted Peroxymonosulfate Process Based on Nanohybrid Manganese Dioxide Incorporating Ferric Oxide." *Applied Catalysis B-Environmental* 278: 13. https://doi.org/10.1016/j.apcatb.2020.119297.

Guo, X, H Gao, S Wang, L Yin, and Y Dai. 2020b. "Scalable, Flexible and Reusable Graphene Oxide-Functionalized Electrospun Nanofibrous Membrane for Solar Photothermal Desalination." *Desalination* 488 https://doi.org/10.1016/j.desal.2020.114535.

Gupta, S S, T S Sreeprasad, S M Maliyekkal, S K Das, and T Pradeep. 2012. "Graphene from Sugar and Its Application in Water Purification." *ACS Applied Materials and Interfaces* 4 (8): 4156–4163. https://doi.org/10.1021/am300889u.

Hamdy, G, and A Taher. 2020. "Enhanced Chlorine-Resistant and Low Biofouling Reverse Osmosis Polyimide-Graphene Oxide Thin Film Nanocomposite Membranes for Water Desalination." *Polymer Engineering and Science* 60 (10). https://doi.org/10.1002/pen.25495.

Han, W, Z N Wu, Y Li, and Y Y Wang. 2019. "Graphene Family Nanomaterials (GFNs)-Promising Materials for Antimicrobial Coating and Film: A Review." *Chemical Engineering Journal* 358: 1022–1037. https://doi.org/10.1016/j.cej.2018.10.106.

Han, Y, Z Xu, and C Gao. 2013. "Ultrathin Graphene Nanofiltration Membrane for Water Purification." *Advanced Functional Materials* 23 (29): 3693–3700. https://doi.org/10.1002/adfm.201202601.

Hasija, V, P Raizada, V K Thakur, A A Parwaz Khan, A M Asiri, and P Singh. 2020. "An Overview of Strategies for Enhancement in Photocatalytic Oxidative Ability of MoS2for Water Purification." *Journal of Environmental Chemical Engineering* 8 (5): 104307. https://doi.org/10.1016/j.jece.2020.104307.

He, H, W Wang, C Xu, S Yang, C Sun, X Wang, Y Yao, et al. 2020. "Highly Efficient Degradation of Iohexol on a Heterostructured Graphene-Analogue Boron Nitride Coupled Bi2MoO6 Photocatalyst under Simulated Sunlight." *Science of The Total Environment* 730: 139100. https://doi.org/10.1016/j.scitotenv.2020.139100.

Heiranian, M, A B Farimani, and N R Aluru. 2015. "Water Desalination with a Single-Layer MoS 2 Nanopore." *Nature Communications* 6: 1–6. https://doi.org/10.1038/ncomms9616.

Hirunpinyopas, W, E Prestat, S D Worrall, S J Haigh, R A W Dryfe, and M A Bissett. 2017. "Desalination and Nanofiltration through Functionalized Laminar MoS2 Membranes." *ACS Nano* 11 (11): 11082–11090. https://doi.org/10.1021/acsnano.7b05124.

Hu, W, C Peng, W Luo, M Lv, X Li, D Li, Q Huang, and C Fan. 2010. "Graphene-Based Antibacterial Paper." *ACS Nano* 4 (7): 4317–4323. https://doi.org/10.1021/nn101097v.

Hu, X Y, X K Zeng, Y Liu, J Lu, S Yuan, Y C Yin, J Hu, D T McCarthy, and X W Zhang. 2020. "Nano-Layer Based 1T-Rich MoS2/g-C3N4 Co-Catalyst System for Enhanced Photocatalytic and Photoelectrochemical Activity." *Applied Catalysis B-Environmental* 268: 9. https://doi.org/10.1016/j.apcatb.2019.118466.

Huang, A, and B Feng. 2018. "Synthesis of Novel Graphene Oxide-Polyimide Hollow Fiber Membranes for Seawater Desalination." *Journal of Membrane Science* 548 https://doi.org/10.1016/j.memsci.2017.11.016.

Huang, Q, Y Fang, J Shi, Y Liang, Y Zhu, and G Xu. 2017. "Flower-Like Molybdenum Disulfide for Polarity-Triggered Accumulation/Release of Small Molecules." *ACS Applied Materials and Interfaces* 9 (41): 36431–36437. https://doi.org/10.1021/acsami.7b11940.

Hummers, W S, and R E Offeman. 1958. "Preparation of Graphitic Oxide." *Journal of the American Chemical Society* 80 (6): 1339. https://doi.org/10.1021/ja01539a017.

Ihsanullah. 2019. "Carbon Nanotube Membranes for Water Purification: Developments, Challenges, and Prospects for the Future." *Separation and Purification Technology* 209 (September): 307–337. https://doi.org/10.1016/j.seppur.2018.07.043.

Ihsanullah, I. 2020. "MXenes (Two-Dimensional Metal Carbides) as Emerging Nanomaterials for Water Purification: Progress, Challenges and Prospects." *Chemical Engineering Journal* 388 (December 2019): 124340. https://doi.org/10.1016/j.cej.2020.124340.

Ikram, M, J Hassan, M Imran, J Haider, A Ul-Hamid, I Shahzadi, A Raza, U Qumar, and S Ali. 2020. "2D Chemically Exfoliated Hexagonal Boron Nitride (HBN) Nanosheets Doped with Ni: Synthesis, Properties and Catalytic Application for the Treatment of Industrial Wastewater." *Applied Nanoscience* 10 (9): 3525–3528. https://doi.org/10.1007/s13204-020-01439-2.

Joshi, R K, P Carbone, F C Wang, V G Kravets, Y Su, I V Grigorieva, H A Wu, A K Geim, and R R Nair. 2014. "Precise and Ultrafast Molecular Sieving through Graphene Oxide Membranes." *Science* 343 (6172): 752–754. https://doi.org/10.1126/science.1245711.

Kadhom, M, and B Deng. 2018. "Metal-Organic Frameworks (MOFs) in Water Filtration Membranes for Desalination and Other Applications." *Applied Materials Today* 11: 219–230. https://doi.org/10.1016/j.apmt.2018.02.008.

Karahan, H E, K Goh, C Zhang, E Yang, C Yıldırım, C Y Chuah, M. Göktuğ Ahunbay, et al. 2020. "MXene Materials for Designing Advanced Separation Membranes." *Advanced Materials* 32 (29): 1–23. https://doi.org/10.1002/adma.201906697.

Kavitha, T, A I Gopalan, K-P Lee, and S-Y Park. 2012. "Glucose Sensing, Photocatalytic and Antibacterial Properties of Graphene–ZnO Nanoparticle Hybrids." *Carbon* 50 (8): 2994–3000. https://doi.org/10.1016/j.carbon.2012.02.082.

Kim, S, X Lin, R Ou, H Liu, X Zhang, G P Simon, C D Easton, and H Wang. 2017. "Highly Crosslinked, Chlorine Tolerant Polymer Network Entwined Graphene Oxide Membrane for Water Desalination." *Journal of Materials Chemistry A* 5 (4). https://doi.org/10.1039/c6ta07350f.

Kim, S, M Yu, and Y Yoon. 2020. "Fouling and Retention Mechanisms of Selected Cationic and Anionic Dyes in a Ti3C2Tx MXene-Ultrafiltration Hybrid System." *ACS Applied Materials and Interfaces* 12 (14): 16557–16565. https://doi.org/10.1021/acsami.0c02454.

Kim, S G, D H Hyeon, J H Chun, B H Chun, and S H Kim. 2013. "Novel Thin Nanocomposite RO Membranes for Chlorine Resistance." *Desalination and Water Treatment* 51 (31–33): 6338–6345. https://doi.org/10.1080/19443994.2013.780994.

Knirsch, K C, N C Berner, H C Nerl, C S Cucinotta, Z Gholamvand, N McEvoy, Z Wang, et al. 2015. "Basal-Plane Functionalization of Chemically Exfoliated Molybdenum Disulfide by Diazonium Salts." *ACS Nano* 9 (6): 6018–6030. https://doi.org/10.1021/acsnano.5b00965.

Landau, L D. 1937. "On the Theory of Phase Changes II." *Physikalische Zeitschrift der Soviet Union* 11 (545): 26–35.

Laxminarayan, R, A Duse, C Wattal, A K M Zaidi, H F L Wertheim, N Sumpradit, E Vlieghe, et al. 2013. "Antibiotic Resistance-the Need for Global Solutions." *The Lancet Infectious Diseases* 13 (12): 1057–1098. https://doi.org/10.1016/S1473-3099(13)70318-9.

Lee, A, J W Elam, and S B Darling. 2016. "Membrane Materials for Water Purification: Design, Development, and Application." *Environmental Science: Water Research and Technology* 2 (1): 17–42. https://doi.org/10.1039/c5ew00159e.

Li, H, Z Song, X Zhang, Y Huang, S Li, Y Mao, H J Ploehn, Y Bao, and M Yu. 2013. "Ultrathin, Molecular-Sieving Graphene Oxide Membranes for Selective Hydrogen Separation." *Science* 342 (6154): 95–98. https://doi.org/10.1126/science.1236686.

Li, J, X Li, and B Van der Bruggen. 2020a. "An MXene-Based Membrane for Molecular Separation." *Environmental Science: Nano* 7 (5): 1289–1304. https://doi.org/10.1039/c9en01478k.

Li, X, Y Chen, X Hu, Y Zhang, and L Hu. 2014. "Desalination of Dye Solution Utilizing PVA/PVDF Hollow Fi Ber Composite Membrane Modi Fi Ed with TiO_2 Nanoparticles." *Journal of Membrane Science* 471: 118–129. https://doi.org/10.1016/j.memsci.2014.08.018.

Li, X, B Zhu, and J Zhu. 2019. "Graphene Oxide Based Materials for Desalination." *Carbon* 146: 320–328. https://doi.org/10.1016/j.carbon.2019.02.007.

Li, Z K, Y Wei, X Gao, L Ding, Z Lu, J Deng, X Yang, J Caro, and H Wang. 2020b. "Antibiotics Separation with MXene Membranes Based on Regularly Stacked High-Aspect-Ratio Nanosheets." *Angewandte Chemie - International Edition* 59 (24): 9751–9756. https://doi.org/10.1002/anie.202002935.

Li, Z S, G H Huang, K Liu, X K Tang, Q Peng, J Huang, M L Ao, and G F Zhang. 2020c. "Hierarchical BiOX (X=Cl, Br, I) Microrods Derived from Bismuth-MOFs: In Situ Synthesis, Photocatalytic Activity and Mechanism." *Journal of Cleaner Production* 272: 11. https://doi.org/10.1016/j.jclepro.2020.122892.

Ling, Z, C E Ren, M Q Zhao, J Yang, J M Giammarco, J Qiu, M W Barsoum, and Y Gogotsi. 2014. "Flexible and Conductive MXene Films and Nanocomposites with High Capacitance." *Proceedings of the National Academy of Sciences of the United States of America* 111 (47): 16676–16681. https://doi.org/10.1073/pnas.1414215111.

Lipatov, A, H Lu, M Alhabeb, B Anasori, A Gruverman, Y Gogotsi, and A Sinitskii. 2018. "Elastic Properties of 2D Ti3C2Tx MXene Monolayers and Bilayers." *Science Advances* 4 (6): 1–8. https://doi.org/10.1126/sciadv.aat0491.

Liu, C, Q Wang, F Jia, and S Song. 2019. "Adsorption of Heavy Metals on Molybdenum Disulfide in Water: A Critical Review." *Journal of Molecular Liquids* 292: 111390. https://doi.org/10.1016/j.molliq.2019.111390.

Liu, G, W Jin, and N Xu. 2016. "Two-Dimensional-Material Membranes: A New Family of High-Performance Separation Membranes." *Angewandte Chemie - International Edition* 55 (43): 13384–13397. https://doi.org/10.1002/anie.201600438.

Liu, H, H Wang, and X Zhang. 2015. "Facile Fabrication of Freestanding Ultrathin Reduced Graphene Oxide Membranes for Water Purification." *Advanced Materials* 27 (2): 249–254. https://doi.org/10.1002/adma.201404054.

Liu, K, J Feng, A Kis, and A Radenovic. 2014. "Atomically Thin Molybdenum Disulfide Nanopores with High Sensitivity for Dna Translocation." *ACS Nano* 8 (3): 2504–2511. https://doi.org/10.1021/nn406102h.

Liu, S, T H Zeng, M Hofmann, E Burcombe, J Wei, R Jiang, J Kong, and Y Chen. 2011. "Antibacterial Activity of Graphite, Graphite Oxide, Graphene Oxide, and Reduced Graphene Oxide: Membrane and Oxidative Stress." *ACS Nano* 5 (9): 6971–6980. https://doi.org/10.1021/nn202451x.

Liu, T, X Liu, N Graham, W Yu, and K Sun. 2020. "Two-Dimensional MXene Incorporated Graphene Oxide Composite Membrane with Enhanced Water Purification Performance." *Journal of Membrane Science* 593 (August 2019): 117431. https://doi.org/10.1016/j.memsci.2019.117431.

Liu, X, H Zhang, Y Ma, X Wu, L Meng, Y Guo, G Yu, and Y Liu. 2013. "Graphene-Coated Silica as a Highly Efficient Sorbent for Residual Organophosphorus Pesticides in Water." *Journal of Materials Chemistry A* 1 (5): 1875–1884. https://doi.org/10.1039/c2ta00173j.

Liu, Y, X Zeng, X Hu, J Hu, and X Zhang. 2019. "Two-Dimensional Nanomaterials for Photocatalytic Water Disinfection: Recent Progress and Future Challenges." *Journal of Chemical Technology and Biotechnology* 94 (1): 22–37. https://doi.org/10.1002/jctb.5779.

Lu, H, J Wang, M Stoller, T Wang, Y Bao, and H Hao. 2016. "An Overview of Nanomaterials for Water and Wastewater Treatment." *Advances in Materials Science and Engineering* 2016: 4964828. https://doi.org/10.1155/2016/4964828.

Lu, Y, Y Fang, X Xiao, S Qi, C Huan, Y Zhan, H Cheng, and G Xu. 2018. "Petal-like Molybdenum Disulfide Loaded Nanofibers Membrane with Superhydrophilic Property for Dye Adsorption" 553 (May): 210–217. https://doi.org/10.1016/j.colsurfa.2018.05.056.

Luo, B, G Liu, and L Wang. 2016. "Recent Advances in 2D Materials for Photocatalysis." *Nanoscale* 8 (13): 6904–6920. https://doi.org/10.1039/c6nr00546b.

Maliyekkal, S M, T S Sreeprasad, D Krishnan, S Kouser, A K Mishra, U V Waghmare, and T Pradeep. 2013. "Graphene: A Reusable Substrate for Unprecedented Adsorption of Pesticides." *Small* 9 (2): 273–283. https://doi.org/10.1002/smll.201201125.

Manzeli, S, D Ovchinnikov, D Pasquier, O V Yazyev, and A Kis. 2017. "2D Transition Metal Dichalcogenides." *Nature Reviews Materials* 2. https://doi.org/10.1038/natrevmats.2017.33.

Mi, B. 2014. "Graphene Oxide Membranes for Ionic and Molecular Sieving." *Science* 343 (6172): 740–742. https://doi.org/10.1126/science.1250247.

Mohammed, H, A Kumar, E Bekyarova, Y Al-Hadeethi, X X Zhang, M G Chen, M S Ansari, A Cochis, and L Rimondini. 2020. "Antimicrobial Mechanisms and Effectiveness of Graphene and Graphene-Functionalized Biomaterials: A Scope Review." *Frontiers in Bioengineering and Biotechnology* 8: 22. https://doi.org/10.3389/fbioe.2020.00465.

Mortazavi, V, A Moosavi, and A Nouri-Borujerdi. 2020. "Enhancing Water Desalination in Graphene-Based Membranes via an Oscillating Electric Field." *Desalination* 495. https://doi.org/10.1016/j.desal.2020.114672.

Mousset, E, Z Wang, J Hammaker, and O Lefebvre. 2016. "Physico-Chemical Properties of Pristine Graphene and Its Performance as Electrode Material for Electro-Fenton Treatment of Wastewater." *Electrochimica Acta* 214: 217–230. https://doi.org/10.1016/j.electacta.2016.08.002.

Mukherjee, R, P Bhunia, and S De. 2016. "Impact of Graphene Oxide on Removal of Heavy Metals Using Mixed Matrix Membrane." *Chemical Engineering Journal* 292: 284–297. https://doi.org/10.1016/j.cej.2016.02.015.

Naguib, M, M Kurtoglu, V Presser, J Lu, J Niu, M Heon, L Hultman, Y Gogotsi, and M W Barsoum. 2011. "Two-Dimensional Nanocrystals Produced by Exfoliation of Ti 3AlC 2." *Advanced Materials* 23 (37): 4248–4253. https://doi.org/10.1002/adma.201102306.

Naguib, M, O Mashtalir, J Carle, V Presser, J Lu, L Hultman, Y Gogotsi, and M W Barsoum. 2012. "Two-Dimensional Transition Metal Carbides." *ACS Nano* 6 (2): 1322–1331. https://doi.org/10.1021/nn204153h.

Naguib, M, V N Mochalin, M W Barsoum, and Y Gogotsi. 2014. "25th Anniversary Article: MXenes: A New Family of Two-Dimensional Materials." *Advanced Materials* 26 (7): 992–1005. https://doi.org/10.1002/adma.201304138.

Nair, R R, H A Wu, P N Jayaram, I V Grigorieva, and A K Geim. 2012. "Unimpeded Permeation of Water through Helium-Leak-Tight Graphene-Based Membranes." *Science* 335 (6067): 442–444. https://doi.org/10.1126/science.1211694.

Nakata, K, and A Fujishima. 2012. "TiO_2 Photocatalysis: Design and Applications." *Journal of Photochemistry and Photobiology C: Photochemistry Reviews* 13 (3): 169–189. https://doi.org/10.1016/j.jphotochemrev.2012.06.001.

Naushad, M. 2018. *A New Generation Material Graphene: Applications in Water Technology. A New Generation Material Graphene: Applications in Water Technology*. Springer International Publishing. https://doi.org/10.1007/978-3-319-75484-0.

Ndlwana, L, M M Motsa, and B B Mamba. 2020. "A Unique Method for Dopamine-Cross-Linked Graphene Nanoplatelets within Polyethersulfone Membranes (GNP-PDA/PES) for Enhanced Mechanochemical Resistance during NF and RO Desalination." *European Polymer Journal* 136 (July): 109889. https://doi.org/10.1016/j.eurpolymj.2020.109889.

Nguyen, C T, and A Beskok. 2020. "Water Desalination Performance of H-BN and Optimized Charged Graphene Membranes." *Microfluidics and Nanofluidics* 24 (5). https://doi.org/10.1007/s10404-020-02340-8.

Novoselov, K S, D Jiang, F Schedin, T J Booth, V V Khotkevich, S V Morozov, and A K Geim. 2005. "Two-Dimensional Atomic Crystals." *Proceedings of the National Academy of Sciences of the United States of America* 102 (30): 10451–10453. https://doi.org/10.1073/pnas.0502848102.

O'Hern, S C, M S H Boutilier, J-C Idrobo, Y Song, J Kong, T Laoui, M Atieh, and R Karnik. 2014. "Selective Ionic Transport through Tunable Subnanometer Pores in Single-Layer Graphene Membranes." *Nano Letters* 14 (3): 1234–1241.

O'Hern, S C, D Jang, S Bose, J C Idrobo, Y Song, T Laoui, J Kong, and R Karnik. 2015. "Nanofiltration across Defect-Sealed Nanoporous Monolayer Graphene." *Nano Letters* 15 (5). https://doi.org/10.1021/acs.nanolett.5b00456.

Oluwole, A O, E O Omotola, and O S Olatunji. 2020. "Pharmaceuticals and Personal Care Products in Water and Wastewater: A Review of Treatment Processes and Use of Photocatalyst Immobilized on Functionalized Carbon in AOP Degradation." *BMC Chemistry* 14 (1): 29. https://doi.org/10.1186/s13065-020-00714-1.

Pacilé, D, J C Meyer, A Fraile Rodríguez, M Papagno, C Gómez-Navarro, R S Sundaram, M Burghard, K Kern, C Carbone, and U Kaiser. 2011. "Electronic Properties and Atomic Structure of Graphene Oxide Membranes." *Carbon* 49 (3): 966–972. https://doi.org/10.1016/j.carbon.2010.09.063.

Pandey, R P, K Rasool, V E Madhavan, B Aïssa, Y Gogotsi, and K A Mahmoud. 2018. "Ultrahigh-Flux and Fouling-Resistant Membranes Based on Layered Silver/MXene (Ti3C2T:X) Nanosheets." *Journal of Materials Chemistry A* 6 (8): 3522–3533. https://doi.org/10.1039/c7ta10888e.

Pang, Y, G Zeng, L Tang, Y Zhang, Y Liu, X Lei, Z Li, J Zhang, Z Liu, and Y Xiong. 2011. "Preparation and Application of Stability Enhanced Magnetic Nanoparticles for Rapid Removal of Cr(VI)." *Chemical Engineering Journal* 175: 222–227. https://doi.org/10.1016/j.cej.2011.09.098.

Park, M J, S Phuntsho, T He, G M Nisola, L D Tijing, X-M Li, G Chen, W-J Chung, and H K Shon. 2015. "Graphene Oxide Incorporated Polysulfone Substrate for the Fabrication of Flat-Sheet Thin-Film Composite Forward Osmosis Membranes." *Journal of Membrane Science* 493: 496–507.

Pendergast, M M, and E M V Hoek. 2011. "A Review of Water Treatment Membrane Nanotechnologies." *Energy and Environmental Science* 4 (6): 1946–1971. https://doi.org/10.1039/c0ee00541j.

Perreault, F, M E Tousley, and M Elimelech. 2013. "Thin-Film Composite Polyamide Membranes Functionalized with Biocidal Graphene Oxide Nanosheets." *Environmental Science and Technology Letters* 1 (1): 71–76. https://doi.org/10.1021/ez4001356.

Qian, Y, C Zhou, and A Huang. 2018. "Cross-Linking Modification with Diamine Monomers to Enhance Desalination Performance of Graphene Oxide Membranes." *Carbon* 136: 28–37. https://doi.org/10.1016/j.carbon.2018.04.062.

Radisavljevic, B, A Radenovic, J Brivio, V Giacometti, and A Kis. 2011. "Single-Layer MoS2 Transistors." *Nature Nanotechnology* 6 (3): 147–150. https://doi.org/10.1038/nnano.2010.279.

Rajapaksha, P P, A Power, S Chandra, and J Chapman. 2018. "Graphene, Electrospun Membranes and Granular Activated Carbon for Eliminating Heavy Metals, Pesticides and Bacteria in Water and Wastewater Treatment Processes." *Analyst* 143 (23): 5629–5645. https://doi.org/10.1039/c8an00922h.

Rasool, K, R P Pandey, P Abdul Rasheed, S Buczek, Y Gogotsi, and K A Mahmoud. 2019. "Water Treatment and Environmental Remediation Applications of Two-Dimensional Metal Carbides (MXenes)." *Materials Today* 30 (November): 80–102. https://doi.org/10.1016/j.mattod.2019.05.017.

Rehman, A, A Daud, M F Warsi, I Shakir, P O Agboola, M I Sarwar, and S Zulfiqar. 2020. "Nanostructured Maghemite and Magnetite and Their Nanocomposites with Graphene Oxide for Photocatalytic Degradation of Methylene Blue." *Materials Chemistry and Physics* 256: 123752. https://doi.org/10.1016/j.matchemphys.2020.123752.

Ren, C E, K B Hatzell, M Alhabeb, Z Ling, K A Mahmoud, and Y Gogotsi. 2015. "Charge- and Size-Selective Ion Sieving Through Ti3C2Tx MXene Membranes." *Journal of Physical Chemistry Letters* 6 (20): 4026–4031. https://doi.org/10.1021/acs.jpclett.5b01895.

Roy, K, A Mukherjee, N R Maddela, S Chakraborty, B Shen, M Li, D Du, Y Peng, F Lu, and L C Garciá Cruzatty. 2020. "Outlook on the Bottleneck of Carbon Nanotube in Desalination and Membrane-Based Water Treatment-A Review." *Journal of Environmental Chemical Engineering* 8 (1): 103572. https://doi.org/10.1016/j.jece.2019.103572.

Safarpour, M, V Vatanpour, and A Khataee. 2016. "Preparation and Characterization of Graphene Oxide/TiO_2 Blended PES Nano Fi Ltration Membrane with Improved Antifouling and Separation Performance." *DES* 393: 65–78. https://doi.org/10.1016/j.desal.2015.07.003.

Saha, D, M C Visconti, M M Desipio, and R Thorpe. 2020. "Inactivation of Antibiotic Resistance Gene by Ternary Nanocomposites of Carbon Nitride, Reduced Graphene Oxide and Iron Oxide under Visible Light." *Chemical Engineering Journal* 382: 10. https://doi.org/10.1016/j.cej.2019.122857.

Saharan, P, G R Chaudhary, S K Mehta, and A Umar. 2014. "Removal of Water Contaminants by Iron Oxide Nanomaterials." *Journal of Nanoscience and Nanotechnology* 14 (1): 627–643. https://doi.org/10.1166/jnn.2014.9053.

Schwierz, F. 2010. "Graphene Transistors." *Nature Nanotechnology* 5 (7): 487.

Shao, C, Y Zhao, and L Qu. 2020. "Tunable Graphene Systems for Water Desalination." *ChemNanoMat* https://doi.org/10.1002/cnma.202000041.

Sholl, D S, and R P Lively. 2016. "Comment." *Nature* 532 (435): 6–9.

Singh, P, P Shandilya, P Raizada, A Sudhaik, A Rahmani-Sani, and A Hosseini-Bandegharaei. 2020. "Review on Various Strategies for Enhancing Photocatalytic Activity of Graphene Based Nanocomposites for Water Purification." *Arabian Journal of Chemistry* 13 (1): 3498–3520. https://doi.org/10.1016/j.arabjc.2018.12.001.

Sint, K, B Wang, and P Král. 2008. "Selective Ion Passage through Functionalized Graphene Nanopores." *Journal of the American Chemical Society* 130 (49): 16448–49. https://doi.org/10.1021/ja804409f.

Smith, R J, P J King, M Lotya, C Wirtz, U Khan, S De, A O'Neill, et al. 2011. "Large-Scale Exfoliation of Inorganic Layered Compounds in Aqueous Surfactant Solutions." *Advanced Materials* 23 (34): 3944–3948. https://doi.org/10.1002/adma.201102584.

Song, L L, H L Jia, H L Zhang, and J L Cao. 2020. "Graphene-like h-BN/CdS 2D/3D Heterostructure Composite as an Efficient Photocatalyst for Rapid Removing Rhodamine B and Cr(VI) in Water." *Ceramics International* 46 (15): 24674–24681. https://doi.org/10.1016/j.ceramint.2020.06.257.

Srimuk, P, J Halim, J Lee, Q Tao, J Rosen, and V Presser. 2018. "Two-Dimensional Molybdenum Carbide (MXene) with Divacancy Ordering for Brackish and Seawater Desalination via Cation and Anion Intercalation." *ACS Sustainable Chemistry and Engineering* 6 (3): 3739–3747. https://doi.org/10.1021/acssuschemeng.7b04095.

Staudenmaier, L 1898. "Method for Representing the Graphits\a Ure." *Reports of the German Chemical Society* 31 (2): 1481–1487.

Sun, L, H Huang, and X Peng. 2013a. "Laminar MoS2 Membranes for Molecule Separation." *Chemical Communications* 49 (91): 10718–10720. https://doi.org/10.1039/c3cc46136j.

Sun, P, R Ma, H Deng, Z Song, Z Zhen, K Wang, T Sasaki, Z Xu, and H Zhu. 2016. "Intrinsic High Water/Ion Selectivity of Graphene Oxide Lamellar Membranes in Concentration Gradient-Driven Diffusion." *Chemical Science* 7 (12): 6988–6994. https://doi.org/10.1039/C6SC02865A.

Sun, P, M Zhu, K Wang, M Zhong, J Wei, D Wu, Z Xu, and H Zhu. 2013b. "Selective Ion Penetration of Graphene Oxide Membranes." *ACS Nano* 7 (1): 428–437. https://doi.org/10.1021/nn304471w.

Surwade, S P, S N Smirnov, I V Vlassiouk, R R Unocic, G M Veith, S Dai, and S M Mahurin. 2015. "Water Desalination Using Nanoporous Single-Layer Graphene." *Nature Nanotechnology* 10 (5): 459–464. https://doi.org/10.1038/nnano.2015.37.

Szunerits, S, and R Boukherroub. 2016. "Antibacterial Activity of Graphene-Based Materials." *Journal of Materials Chemistry B* 4 (43): 6892–6912. https://doi.org/10.1039/c6tb01647b.

Tekin, G, G Ersöz, and S Atalay. 2018. "Visible Light Assisted Fenton Oxidation of Tartrazine Using Metal Doped Bismuth Oxyhalides as Novel Photocatalysts." *Journal of Environmental Management* 228: 441–450. https://doi.org/10.1016/j.jenvman.2018.08.099.

Van der Bruggen, B, M Mänttäri, and M Nyström. 2008. "Drawbacks of Applying Nanofiltration and How to Avoid Them: A Review." *Separation and Purification Technology* 63 (2): 251–263. https://doi.org/10.1016/j.seppur.2008.05.010.

Van der Bruggen, B, and C Vandecasteele. 2002. "Distillation vs. Membrane Filtration: Overview of Process Evolutions in Seawater Desalination." *Desalination* 143 (3): 207–218. https://doi.org/10.1016/S0011-9164(02)00259-X.

Vatanpour, V, and A Sanadgol. 2020. "Surface Modification of Reverse Osmosis Membranes by Grafting of Polyamidoamine Dendrimer Containing Graphene Oxide Nanosheets for Desalination Improvement." *Desalination* 491 (January): 114442. https://doi.org/10.1016/j.desal.2020.114442.

Wang, H, X Liu, P Niu, S Wang, J Shi, and L Li. 2020. "Porous Two-Dimensional Materials for Photocatalytic and Electrocatalytic Applications." *Matter* 2 (6): 1377–1413. https://doi.org/10.1016/j.matt.2020.04.002.

Wang, J, P Chen, B Shi, W Guo, M Jaroniec, and S Z Qiao. 2018. "A Regularly Channeled Lamellar Membrane for Unparalleled Water and Organics Permeation." *Angewandte Chemie – International Edition* 57 (23): 6814–6818. https://doi.org/10.1002/anie.201801094.

Wang, J, P Zhang, B Liang, Y Liu, T Xu, L Wang, B Cao, and K Pan. 2016. "Graphene Oxide as an Effective Barrier on a Porous Nanofibrous Membrane for Water Treatment." *ACS Applied Materials and Interfaces* 8 (9): 6211–6218. https://doi.org/10.1021/acsami.5b12723.

Wei, S, Y Xie, Y Xing, L Wang, H Ye, X Xiong, S Wang, and K Han. 2019. "Two-Dimensional Graphene Oxide/MXene Composite Lamellar Membranes for Efficient Solvent Permeation and Molecular Separation." *Journal of Membrane Science* 582 (March): 414–422. https://doi.org/10.1016/j.memsci.2019.03.085.

Wei, Z, Bo L, C Xia, Y Cui, J He, J-B Xia, and J Li. 2018. "Various Structures of 2D Transition-Metal Dichalcogenides and Their Applications." *Small Methods* 2 (11): 1800094. https://doi.org/10.1002/smtd.201800094.

Wright, G D 2000. "Resisting Resistance: New Chemical Strategies for Battling Superbugs." *Chemistry and Biology* 7 (6): 127–132. https://doi.org/10.1016/S1074-5521(00)00126-5.

Wu, S, B Gong, H Yang, Y Tian, C Xu, X Guo, G Xiong, et al. 2020. "Plasma-Made Graphene Nanostructures with Molecularly Dispersed F and Na Sites for Solar Desalination of Oil-Contaminated Seawater with Complete In-Water and In-Air Oil Rejection." *ACS Applied Materials and Interfaces* 12 (34). https://doi.org/10.1021/acsami.0c07921.

Xie, Z, Y P Peng, L Yu, C Xing, M Qiu, J Hu, and H Zhang. 2020. "Solar-Inspired Water Purification Based on Emerging 2D Materials: Status and Challenges." *Solar RRL* 4 (3). https://doi.org/10.1002/solr.201900400.

Xiong, D, X Li, Z Bai, and S Lu. 2018. "Recent Advances in Layered Ti3C2Tx MXene for Electrochemical Energy Storage." *Small* 14 (17): 1–29. https://doi.org/10.1002/smll.201703419.

Xu, C, A Cui, Y Xu, and X Fu. 2013a. "Graphene Oxide-TiO_2 Composite Filtration Membranes and Their Potential Application for Water Purification." *Carbon* 62: 465–471. https://doi.org/10.1016/j.carbon.2013.06.035.

Xu, D, H Liang, X Zhu, L Yang, X Luo, Y Guo, Y Liu, L Bai, G Li, and X Tang. 2020a. "Metal-Polyphenol Dual Crosslinked Graphene Oxide Membrane for Desalination of Textile Wastewater." *Desalination* 487. https://doi.org/10.1016/j.desal.2020.114503.

Xu, J, J Huang, Z P Wang, and Y F Zhu. 2020b. "Enhanced Visible-Light Photocatalytic Degradation and Disinfection Performance of Oxidized Nanoporous g-C3N4 via Decoration with Graphene Oxide Quantum Dots." *Chinese Journal of Catalysis* 41 (3): 474–484. https://doi.org/10.1016/s1872-2067(19)63501-1.

Xu, K, B Feng, C Zhou, and A Huang. 2016. "Synthesis of Highly Stable Graphene Oxide Membranes on Polydopamine Functionalized Supports for Seawater Desalination." *Chemical Engineering Science* 146: 159–165. https://doi.org/10.1016/j.ces.2016.03.003.

Xu, M, T Liang, M Shi, and H Chen. 2013b. "Graphene-like Two-Dimensional Materials." *Chemical Reviews* 113 (5): 3766–3798. https://doi.org/10.1021/cr300263a.

Xu, P A, G M Zeng, D L Huang, C L Feng, S Hu, M H Zhao, C Lai, et al. 2012. "Use of Iron Oxide Nanomaterials in Wastewater Treatment: A Review." *Science of the Total Environment* 424: 1–10. https://doi.org/10.1016/j.scitotenv.2012.02.023.

Xu, W L, C Fang, F Zhou, Z Song, Q Liu, R Qiao, and M Yu. 2017a. "Self-Assembly: A Facile Way of Forming Ultrathin, High-Performance Graphene Oxide Membranes for Water Purification." *Nano Letters* 17 (5): 2928–2933. https://doi.org/10.1021/acs.nanolett.7b00148.

Xu, X, Z Zhang, J Dong, D Yi, J Niu, M Wu, L Lin, et al. 2017b. "Ultrafast Epitaxial Growth of Metre-Sized Single-Crystal Graphene on Industrial Cu Foil." *Science Bulletin* 62 (15): 1074–1080. https://doi.org/10.1016/j.scib.2017.07.005.

Xu, Z, X Yan, Z Du, J Li, and F Cheng. 2020c. "Effect of Oxygenic Groups on Desalination Performance Improvement of Graphene Oxide-Based Membrane in Membrane Distillation." *Separation and Purification Technology* 251. https://doi.org/10.1016/j.seppur.2020.117304.

Xu, Z, J Zhang, M Shan, Y Li, B Li, J Niu, B Zhou, and X Qian. 2014. "Organosilane-Functionalized Graphene Oxide for Enhanced Antifouling and Mechanical Properties of Polyvinylidene Fluoride Ultrafiltration Membranes." *Journal of Membrane Science* 458: 1–13. https://doi.org/10.1016/j.memsci.2014.01.050.

Yang, E, C M Kim, J H Song, H Ki, M H Ham, and I S Kim. 2017. "Enhanced Desalination Performance of Forward Osmosis Membranes Based on Reduced Graphene Oxide Laminates Coated with Hydrophilic Polydopamine." *Carbon* 117: 293–300. https://doi.org/10.1016/j.carbon.2017.03.005.

Yang, K, B Chen, and L Zhu. 2015. "Graphene-Coated Materials Using Silica Particles as a Framework for Highly Efficient Removal of Aromatic Pollutants in Water." *Scientific Reports* 5 (February): 1–12. https://doi.org/10.1038/srep11641.

Yang, Y, H Hou, G Zou, W Shi, H Shuai, J Li, and X Ji. 2019. "Electrochemical Exfoliation of Graphene-like Two-Dimensional Nanomaterials." *Nanoscale* 11 (1): 16–33. https://doi.org/10.1039/c8nr08227h.

Yang, Y, R Zhao, T Zhang, K Zhao, P Xiao, Y Ma, P M Ajayan, G Shi, and Y Chen. 2018a. "Graphene-Based Standalone Solar Energy Converter for Water Desalination and Purification." *ACS Nano* 12 (1): 829–835. https://doi.org/10.1021/acsnano.7b08196.

Yang, Z, X H Ma, and C Y Tang. 2018b. "Recent Development of Novel Membranes for Desalination." *Desalination* 434 (May 2017): 37–59. https://doi.org/10.1016/j.desal.2017.11.046.

Ying, Y, Y Liu, X Wang, Y Mao, W Cao, P Hu, and X Peng. 2015. "Two-Dimensional Titanium Carbide for Efficiently Reductive Removal of Highly Toxic Chromium(VI) from Water." *ACS Applied Materials and Interfaces* 7 (3): 1795–1803. https://doi.org/10.1021/am5074722.

Yu, L, Y Zhang, B Zhang, J Liu, H Zhang, and C Song. 2013. "Preparation and Characterization of HPEI-GO/PES Ultrafiltration Membrane with Antifouling and Antibacterial Properties." *Journal of Membrane Science* 447: 452–462.

Yu, S, C Gao, H Su, and M Liu. 2021. "Nanofiltration Used for Desalination and Concentration in Dye Production" 140 (2001): 3–6.

Zhan, S, D Zhu, S Ma, W Yu, Y Jia, Y Li, H Yu, and Z Shen. 2015. "Highly Efficient Removal of Pathogenic Bacteria with Magnetic Graphene Composite." *ACS Applied Materials & Interfaces* 7 (7): 4290–4298. https://doi.org/10.1021/am508682s.

Zhang, C, R Z Zhang, Y Q Ma, W B Guan, X L Wu, X Liu, H Li, Y L Du, and C P Pan. 2015. "Preparation of Cellulose/Graphene Composite and Its Applications for Triazine Pesticides Adsorption from Water." *ACS Sustainable Chemistry and Engineering* 3 (3): 396–405. https://doi.org/10.1021/sc500738k.

Zhang, L, P P Chen, Y Xu, W Y Nie, and Y F Zhou. 2020a. "Enhanced Photo-Induced Antibacterial Application of Graphene Oxide Modified by Sodium Anthraquinone-2-Sulfonate under Visible Light." *Applied Catalysis B-Environmental* 265: 10. https://doi.org/10.1016/j.apcatb.2019.118572.

Zhang, L, X Hu, and Q Zhou. 2020b. "Sunlight-Assisted Tailoring of Surface Nanostructures on Single-Layer Graphene Nanosheets for Highly Efficient Cation Capture and High-Flux Desalination." *Carbon* 161. https://doi.org/10.1016/j.carbon.2020.01.112.

Zhang, L, W Li, M Zhang, and S Chen. 2020c. "Self-Assembly of Graphene Oxide Sheets: The Key Step toward Highly Efficient Desalination." *Nanoscale* 12 (40): 20749–20758. https://doi.org/10.1039/d0nr05548d.

Zhang, S, N T Hang, Z Zhang, H Yue, and W Yang. 2017. "Preparation of G-C3N4/Graphene Composite for Detecting NO2 at Room Temperature." *Nanomaterials* 7 (1). https://doi.org/10.3390/nano7010012.

Zhang, S, S Liao, F Qi, R Liu, T Xiao, J Hu, K Li, R Wang, and Y Min. 2020d. "Direct Deposition of Two-Dimensional MXene Nanosheets on Commercially Available Filter for Fast and Efficient Dye Removal." *Journal of Hazardous Materials* 384 (July 2019): 121367. https://doi.org/10.1016/j.jhazmat.2019.121367.

Zhang, Y, and T S Chung. 2017. "Graphene Oxide Membranes for Nanofiltration." *Current Opinion in Chemical Engineering* 16: 9–15. https://doi.org/10.1016/j.coche.2017.03.002.

Zhang, Y, Y Su, J Peng, X Zhao, J Liu, J Zhao, and Z Jiang. 2013a. "Composite Nanofiltration Membranes Prepared by Interfacial Polymerization with Natural Material Tannic Acid and Trimesoyl Chloride." *Journal of Membrane Science* 429: 235–242. https://doi.org/10.1016/j.memsci.2012.11.059.

Zhang, Y, L Zhang, and C Zhou. 2013b. "Review of Chemical Vapor Deposition of Graphene and Related Applications." *Accounts of Chemical Research* 46 (10): 2329–2339. https://doi.org/10.1021/ar300203n.

Zhuang, P, H Fu, N Xu, B Li, J Xu, and L Zhou. 2020. "Free-Standing Reduced Graphene Oxide (RGO) Membrane for Salt-Rejecting Solar Desalination via Size Effect." *Nanophotonics* 9 (15). https://doi.org/10.1515/nanoph-2020-0396.

Zinadini, S, A A Zinatizadeh, M Rahimi, V Vatanpour, and H Zangeneh. 2014. "Preparation of a Novel Antifouling Mixed Matrix PES Membrane by Embedding Graphene Oxide Nanoplates." *Journal of Membrane Science* 453: 292–301. https://doi.org/10.1016/j.memsci.2013.10.070.

5 TiO_2-Based Nanomaterial for Pollutant Removal

M. Dhaneesha
Kannur University, Kannur, India
Shahanas Beegam
University of Calicut, Kerala, India
P. Periyat
Kannur University, Kannur, India

CONTENTS

5.1 Introduction 164
5.2 TiO_2 Photocatalyst 165
5.3 Electronic Process in TiO_2 and Photocatalysis Mechanism 165
5.4 Modifications of TiO_2 Photocatalyst for Efficient Removal of Water Pollutants 167
5.4.1 Doping 167
5.4.2 Doping with Metals 168
5.4.3 Non-Metal Doping 168
5.4.4 Surface Chemical Modifications 170
5.4.5 Combined Effect of Doping and Coupling 171
5.5 Factors Affecting Pollutant Removal by Photocatalysis 171
5.5.1 pH of the Reaction Medium 171
5.5.2 Crystal Phase 172
5.5.3 Percentage of (001) Facets of Anatase Phase 172
5.5.4 Oxygen Concentration 173
5.5.5 Catalyst Loading 173
5.5.6 Crystal Size/Surface Area of the Catalyst 173
5.5.7 Reaction Temperature 173
5.5.8 Amount of Dopant 173
5.5.9 Calcination Temperature of the Catalyst 174
5.6 Achieving Selective Photocatalysis in TiO_2 for Pollutant Removal 174
5.7 Conclusion 175
References 176

DOI: 10.1201/9781003118749-5

5.1 INTRODUCTION

Nanoscience and technology is a very rapidly growing interdisciplinary area of research. Nanomaterials are single or multi-phased polycrystalline particles of organic or inorganic materials with a size of 1–100 nm possess unique optical, electrical, magnetic, and other properties. They received great interest because of these unique properties (Rao and Cheetham 2001). A wide variety of nanomaterials such as metal, semiconductor, biological, polymeric are available for various applications. Among these, semiconductor nanomaterials are still in the research stage and have promising environmental applications in water treatment and remediation (W. Wu, Chang Zhong Jiang, and Roy 2015).

Water shortage and contamination of available drinking water by organic pollutants is an overwhelming problem worldwide (Alharbi et al. 2018). In the last decade, remediation of industrial hazardous waste contamination in water and related problems becomes a high national and international priority. High toxic pollutants, which are difficult to remove by existing biological technologies, require some novel techniques to transform them into harmless and eco-friendly compounds like carbon dioxide, water, and other inorganic species. The prime condition is that the technique should be inexpensive and green without secondary pollution (Lee and Park 2013).

The latest promising approach for water pollution control and water purification uses an advanced oxidation process (AOP) (Lazar, Varghese, and Nair 2012). The key factor of AOPs includes homogeneous and heterogeneous photocatalytic activity based on UV/visible light irradiation, ozonation, ultrasound, Fenton's reagent, electrolysis, and wet air oxidation (Comninellis et al. 2008). AOPs can be defined as an aqueous oxidation method that depends on the intermediacy of reactive oxygen species (ROS) such OH· radicals, which lead to the destruction of the target pollutants. The reactive hydroxyl radicals are the strongest oxidizing species that can virtually oxidize any organic/inorganic compound that may present as contaminants in the water bodies (Elsalamony 2016). The formed hydroxyl radicals quickly and non-selectively fragment the contaminants into inorganic molecules. Usually, the OH· Radicals are generated using primary oxidants (e.g., hydrogen peroxide, ozone, or oxygen, etc.) and/or energy sources like UV or catalysts. AOPs can decrease the total amount of organic pollutants very effectively and significantly bring down the total organic carbon (TOC) and chemical oxygen demand (COD) when applied under proper conditions, which make it a promising water treatment technique of the 21st century (Gągol, Przyjazny, and Boczkaj 2018).

Semiconductor photocatalysis is a well-established research area where semiconductor metal oxides nanomaterials are used to degrade organic pollutants in the presence of solar energy. Different types of semiconductor photocatalyst nanomaterials such as TiO_2, ZnO, and other metal oxides (such as V_2O_5, W_2O_3, MO_3, and CeO_2) are used for this purpose (Wang et al. 2013). Out of these photosensitive semiconductor nanomaterials, TiO_2 is considered an ideal and most attractive semiconductor because of its low cost, high chemical stability, availability, and highly oxidizing nature. The underlying principle of water disinfection and purification using TiO_2 is that it utilizes solar energy for water oxidation for the production of reactive oxygen (ROS) species, which are toxic to water contaminating microbes and degrade organic pollutants (M. J. Wu et al. 2014).

5.2 TiO_2 PHOTOCATALYST

TiO_2 occurs in different forms such as rutile, anatase, brookite, and TiO_2 B form. It is well known that rutile is the most stable form of TiO_2. TiO_2 is an n-type semiconductor having a band gap of between 3 and 3.2 eV. Anatase and rutile phases of TiO_2 are the most studied polymorphs, which have more useful for practical applications (Grätzel 2005). Properties of anatase and rutile are summarized in Table 5.1.

Generally, photocatalysis is a redox reaction process. The electrons in the valence band get excited to the conduction band by creating positive holes in the valence band of semiconducting materials by irradiating light with suitable energy. TiO_2 is considered an ideal catalyst since it has been used from ancient times as a white pigment, and thus, is considered safe for humans and the environment. Commercially, the application of TiO_2 as a potential photocatalyst for pollutant removal is dependent on several factors, which include: reactions take place at room temperature; TiO_2 completely mineralizes the organic pollutants into non-toxic substances; and is inexpensive, reusable, and can be used as a support on various substrates.

5.3 ELECTRONIC PROCESS IN TIO_2 AND PHOTOCATALYSIS MECHANISM

TiO_2 is an n-type semiconductor. In TiO_2, the titanium ions are in an octahedral environment, generally have a Ti^{4+}oxidation state with a $3d^{\circ}$electronic configuration. The valence band of TiO_2 consists of 2p orbitals of oxygen, whereas the conduction band is made up of 3d orbitals of titanium (Asahi et al. 2000). Irradiation of UV light on TiO_2 surface excites electrons from the filled valence band to the empty conduction band, creating electron-hole pairs as shown in Figure 5.1. The photogenerated electron-hole pairs undergo recombination and loss the extra energy through non-radioactively or radioactively as light or heat as shown in Equations 5.1 and 5.2.

$$TiO_2 + h\nu \rightarrow h^+{}_{\nu B} + e^-{}_{CB} \quad (5.1)$$

$$e^-{}_{CB} +\_\ h^+{}_{\nu B} \rightarrow \text{Energy} \quad (5.2)$$

TABLE 5.1
Structural and Physical Properties of Anatase and Rutile (Gopal, Moberly Chan, and De Jonghe 1997)

Properties	Anatase	Rutile
Molecular weight (g/mol)	79.88	79.88
Melting point (°C)	1825	1843
Specific gravity	3.9	4.0
Light absorption (nm)	<390	<415
Mohr's Hardness	5.5	6.5–7.0
Refractive index	2.55	2.75
Dielectric constant	48	114
Crystal structure	Tetragonal	Tetragonal
Lattice constants (Å)	a = b = 3.78, c = 9.52	a = b = 4.59, c = 2.96
Density (g/cm³)	3.79	4.13

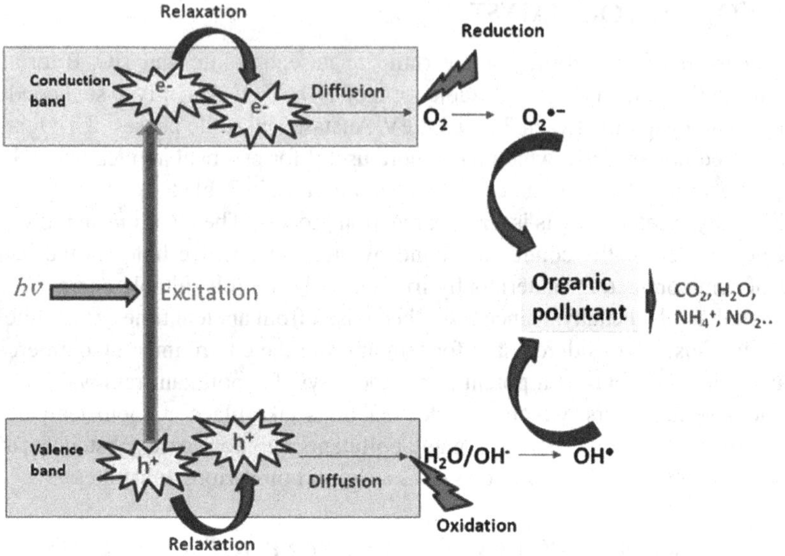

FIGURE 5.1 Band gap structure of UV active anatase TiO_2.

This will scale down the overall efficiency of the light-induced process. Various strategies have been adopted to reduce the recombination rate, including heterojunction coupling, doping with ions, and nano-sized crystalline particles. The excited e^- present in the conduction band is now in a pure 3d state. As a result of the transition probability of e^- to the valence band decreases because of dissimilar parity, leading to a decrease in the probability of e^-/h^+ recombination (S. Banerjee et al. 2006). These charge carriers can easily migrate to the surface of the TiO_2 catalyst and trigger secondary reactions with the molecules adsorbed on its surface or with the surroundings of the charged particles.

For example, the excited electrons present in the conduction band can reduce the oxygen molecule to produce superoxide ($\cdot O_2^-$) radicals, which may again react with H^+ ion to produce hydroperoxy radicals ($\cdot OOH$) and these two ROS can oxidize the organic pollutants. At the same time, the positive holes present in the valence band can oxidize the water or hydroxyl molecule and produce hydroxyl radicals ($\cdot OH$), which again oxidizes the organic pollutants (Figure 5.1; Equations 5.3 and 5.4) (Swagata Banerjee et al. 2014). The ROS such as superoxide, hydroxyl, or hydroperoxide radicals can then oxidize pollutants producing CO_2, H_2O, and mineral salts (Equation 5.5) (Hoffman, Carraway, and Hoffmann 1994).

$$h^+{}_{vB} + H_2O \rightarrow OH^\bullet + H^+ \tag{5.3}$$

$$e^-{}_{CB} + O_2 \rightarrow O_2^\bullet \rightarrow OOH^\bullet \tag{5.4}$$

$$OH^\bullet / O_2^\bullet / OOH^\bullet + \text{Pollutants} \rightarrow H_2O + CO_2 \tag{5.5}$$

The primary condition for a semiconductor for organic compound degradation is that the oxidation potential for converting OH^-/H_2O to hydroxyl radical lies above the position of TiO_2 valence band (3.2 eV). The band potential of a semiconductor is controlled by pH of the reaction medium. The reaction of OH^- to hydroxyl molecule is favored at the basic condition, and that from H_2O is favored at the acidic range. At neutral pH also the oxidation potential of hydroxyl radical lies above the valence band position of TiO_2. Thus the formation of hydroxyl radical is thermodynamically favored at the entire pH range. The oxidation potential of several organic compounds lies above the valence band of TiO_2 and the direct oxidation of these pollutants by TiO_2 valence band holes is thermodynamically possible.

The anatase phase is commonly selected for photocatalytic applications due to its better photoactivity. The better photoactivity of anatase phase is due to its lower rate of electron-hole recombination, the higher number of hydroxyl radicals present on the surface of TiO_2, higher surface area, and lower cost (Sclafani, Palmisano, and Davì 1991). Solar radiation contains only 5% of UV, limiting the use of anatase TiO_2 for practical applications. Various chemical modifications have been proposed to modify TiO_2 for harvesting solar radiations to raise the photocatalytic efficiency of TiO_2.

5.4 MODIFICATIONS OF TIO_2 PHOTOCATALYST FOR EFFICIENT REMOVAL OF WATER POLLUTANTS

Modifying electronic band structure using different techniques such as doping, surface chemical modifications, and coupling with a narrow band gap semiconductor of TiO_2 will help to overcome the low efficiency of TiO_2 nanomaterials in the visible range and ultimately helps for efficient removal of pollutants.

5.4.1 Doping

Doping is the process of adding impurities in very small concentrations into the crystal structure of TiO_2 without giving rise to new crystallographic phases, forms, or structures. Generally, doping of TiO_2 has been taken as a common approach for bandgap engineering of the material, i.e., a decrease in the bandgap or introduction of intra-bandgap states, which results in the absorption of more visible light (Gupta and Tripathi 2011). Different types of doping techniques comprise self-doping, doping with metal/non-metal ions, and co-doping.

In self-doping the property of the semiconductor, TiO_2 is transformed without the introduction of any metals or non-metals. Justicia et al. (2005) developed Self-doped TiO_2 thin films for effective visible light photocatalysis for pollutant removal. Here oxygen vacancies induced by doping are good electron trappers, and are located just below the conduction band of TiO_2. The surface promoted electrons after visible light absorption are employed in the photocatalytic oxidation of NO. Hamdy, Amrollahi, and Mul (2012) reported the one-step Ti^{3+} rich blue-colored TiO_2 synthesis, active under visible light treatment, but the material was not very stable. In contrast, Etacheri et al. (1993) developed a dopant-free antatase TiO_2 photocatalyst stable up to 900°C.

5.4.2 Doping with Metals

Modification on the surface of TiO_2 nanoparticles seems to be easier and more useful than the modification of bulk TiO_2 (Justicia et al. 2005). Modifying the pure photocatalyst with various metal ions, especially d-block metals, leads to the insertion of impurity energy levels between the parent conduction band and valence bands. As a result, electrons could be promoted from the dopant d-band to the conduction band or from the valence band to the dopant d-band (Zou et al. 2001). The resulting dopant energy level and photocatalytic activity of semiconductors strongly depend on different factors such as the dopant concentration, chemical nature, and structural and chemical environment. In the case of dopant concentration, at lower amounts, the dopant may act as a trapping site by introducing localized states that can improve charge separation and finally increase the photocatalytic efficiency of TiO_2. Higher dopant concentrationsan promote the recombination of charge carriers (Joshi et al. 2009). Hence, an optimum concentration of dopant is typically in the order of one or less than one percent. Nagaveni, Hegde, and Madras (2004) synthesized anatase TiO_2 nanoparticles doped with W, Ce, V, Zr, Fe, and Cu ion using the solution combustion method and observed the amount of dopant ion present was limited to a narrow range of concentrations.

Ben Chobba et al. (2019) investigated the Gd-doped TiO_2 nanoparticles for methylene blue (MB) dye degradation and antimicrobial activities toward *Stenotrophomonasm altophilia* and *Micrococcus luteus*. Data showed that doping with Gd enhances photocatalytic activity and, consequently, antimicrobial activity. Khairy and Zakaria (2014) studied the effect of Cu and Zn doping of TiO_2 nanoparticles prepared by the sol-gel method for their photocatalytic activities toward degradation of organic dye methyl orange (MO) and Cu doped TiO_2 nanoparticles exhibited good activity.

5.4.3 Non-Metal Doping

Unlike non-metal anions, metal cations frequently give well-localized d-band deep in the band gap of TiO_2 and are likely to form recombination centers of charge carriers (Mills, Hodgen, and Lee 2005). Therefore, non-metal doping gives a promising way to absorb maximum light absorption for the photocatalytic applications of TiO_2. Numbers of studies are already done based on the non-metal doping on TiO_2 under visible light-responsive photocatalysis. According to the report of Asahi et al. (2001), N-doping shifts the absorption edge of TiO_2 into the visible region (<500 nm) through the replacement of oxygen in the anatase crystal lattice by N-atom, and hence electrons from the 2p orbital of N can be excited to 3d orbital of Ti. Oxygen deficiency also boosted the absorption of visible light of TiO_2 and Ihara (2003) reported nitrogen doped in the oxygen vacancy site and blocked the tendency for reoxidation. Qian et al. (2014) reported a low-temperature N-doping method for the synthesis of highly efficient n-doped TiO_2-hybridized graphene for water treatment. Photodegradation data indicated that the MB dye decomposition rate of N-doped TiO_2/graphene photocatalyst is double as fast as commercial catalyst Degussa P25. Han et al. (2011) reported an innovative light-activated S-doped TiO_2 nanofilms synthesized using the sol-gel method based on the self-assembly technique. The photocatalytic degradation of hepatotoxin microcystin-LR C was very high for nanofilms calcined at 350°, when compared with nanofilms calcined at 400°C and 500°C, under visible light irradiation. Bento, Correa, and Pillis (2019) studied the photocatalytic activity of S-doped and undoped TiO_2 films grown at 400°C using metallorganic chemical vapor deposition (MOCVD) technique for water treatment under visible light irradiation, and they found that S-doped TiO_2 films are very promising photocatalysts for water treatment.

All these reports concluded that the lattice oxygen atoms were replaced by non-metal atoms such as N, S, C, F, etc., producing an isolated narrow band between the valence and conduction band of TiO_2 and the resulting narrowed band gap.

Figure 5.2 shows that introducing an extrinsic electronic level of 2p (N or C) or 3p (S) orbitals of dopant with potential energy greater than that of O2p orbital causes an excitation of electrons to the conduction band using visible radiation, in the case of N or S doped TiO_2.

During co-doping, the wavelength of absorption and photocatalytic activity of the co-doped sample differs from those doped individually (Bingham and Daoud 2011). Recently, non-metals have been co-doped with metals and other non-metals to attain visible light photocatalysis more effectively. This recommends that more photons are being effectively utilizes by the co-doped catalyst. Co-doping of TiO_2 can be used as an effective and successful way to improve the charge separation. TiO_2 co-doped with Eu^{3+} and Fe^{3+} in optimal concentration significantly improved the photocatalytic activity compared with undoped TiO_2. Several reports are available based on metal-non-metal co-doped TiO_2 exhibiting efficient photocatalytic activity (P. Yang et al. 2002). Yao et al. (2018) reported a one-step *in situ* hydrothermal synthesis of N/Ti^{3+} co-doping multiphasic TiO_2/BiOBr heterojunctions for the enhanced optical absorption and degradation of different organic dyes such as MB, Rhodamine B, and *p*-Nitrophenol (Sayed et al. 2018).

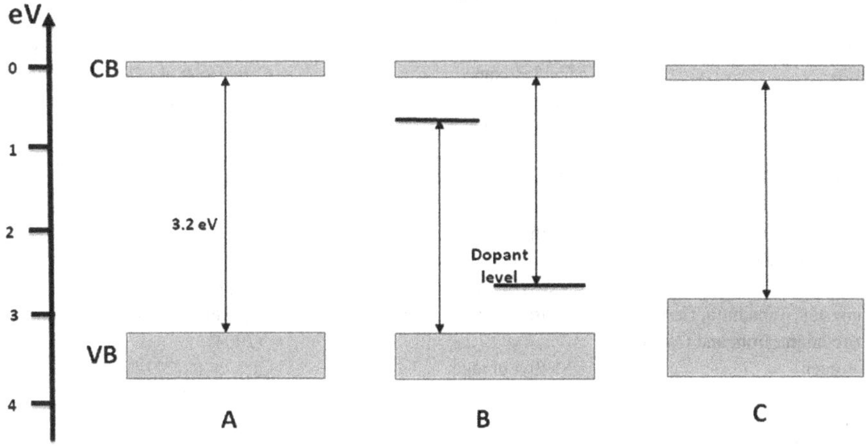

FIGURE 5.2 Schemes showing the different changes that happen to the band gap of anatase TiO_2 on doping with various non-metals: (A) band gap of normal TiO_2; (B) doped TiO_2 with localized dopant levels near the valance band (VB) and the conduction band (CB); and (C) band gap narrowing resulting from broadening of the VB.

5.4.4 Surface Chemical Modifications

Sensitization is the process of photocurrent production with light energy less than that of the semiconductor band gap, and the materials which absorb light are known as sensitizers (Grätzel 2005). Examples of sensitizers for TiO_2 materials include metals, organic dyes, and inorganic semiconductors with narrow band gaps. The efficiency of the sensitized TiO_2 mainly depends on the light interaction efficiency of the sensitizer and the sensitizer's efficiency for charge transfer. Different surface chemical modifications of TiO_2 to improve the photocatalytic pollutant removal include coupling of TiO_2 by narrow bandgap semiconductor, dye sensitization, sensitization by metal nanoparticles, and capping with other semiconductor or metal nanoparticles (Daghrir, Drogui, and Robert 2013; Table 5.2).

Dye sensitization generally occurs through excitation of the sensitizer adsorbed via physisorption or chemisorption on the TiO_2 surface, followed by charge transfer to the semiconductor. This sensitization shifts the wavelength of absorption of TiO_2 to the visible region, an important criterion to operate under natural sunlight for a photocatalyst.

Sensitization or coupling of TiO_2 by narrow bandgap semiconductors is another important TiO_2 modification method to enhance the optical absorption properties in the visible region using various groups. Narrow band gap semiconductors are widely used as sensitizers to improve the photocatalytic performance of TiO_2. In the case of

TABLE 5.2
The Different Metal/Non-Metal Ion Doped TiO_2 as Photocatalyst for Water Detoxification

Doped Metal/Non-Metal Ions	Pollutant	Reference
Nitrogen	Methylene blue, Gaseous acetaldehyde	Asahi et al. (2001)
Sulfur	Methyl orange	Bento, Correa, and Pillis (2019)
Gold	Cyclohexane	Carneiro et al. (2011)
Gadolinium	Methylene blue dye, *Stenotrophomonas maltophilia, Micrococcus luteus*	Ben Chobba et al. (2019)
Sulfur	Hepatotoxin microcystin-LR (MC-LR)	Han et al. (2011)
Nitrogen	Methyl orange	Joshi et al. (2009)
Copper and Zinc	Methyl orange	Khairy and Zakaria (2014)
Zinc	Methyl orange	Liao, Badour, and Liao (2008)
Tungsten, Vanadium, Cerium, zirconium, Iron, and Copper	4-nitrophenol	Nagaveni, Hegde, and Madras (2004)
Nitrogen	Methyl orange	Qian et al. (2014)
Manganese and Cobalt	Enoxacin	Sayed et al. (2018)
Copper	Rhodamine B	Xin et al. (2008)
Iron and Europium	Chloroform	Yang et al. (2002)
Gold	Cyclohexanone	Hernández-Alonso et al. (2009), Carneiro et al. (2011)
Molybdenum oxide (MoO)	Cyclohexane	Ciambelli et al. (2008)
Low temperature synthesized TiO_2	Methyl orange and Methylene blue	M. A. Lazar and Daoud (2012)

coupled semiconductors, illumination of one semiconductor makes a response in the other at their interface. Hoyer and Könenkamp (1995) reported that comparing the absorption and photocurrent spectra of coupled TiO_2/PbS system, only PbS clusters with sizes lesser than 25 A° contribute toward sensitization. Fitzmaurice, Frei, and Rabani (1995) found that excitation TiO_2 semiconductor nanoparticles sensitized using AgI results in stable electron-hole pairs and electrons can easily migrate from AgI to TiO_2. Bi_2S_3/TiO_2, WO_3/TiO_2, CdS/TiO_2 SnO_2/TiO_2, etc. are the other few examples for the photocatalytic system enhanced *via* heterojunction coupling to improve photocatalytic efficiency.

Capping is the process of enhancing emissive properties by coating one semiconductor or metal nanomaterial on the surface of another semiconductor or metal nanoparticle core with a different band gap. In core-shell geometry (Justicia et al., 2005). However, the charge separation mechanism in the capped and coupled semiconductor systems is similar, while the charge collection and interfacial charge transfer differ. In the case of coupled semiconductors, the two semiconductor systems are in touch with each other, and both electrons and holes are available for redox processes on different particle surfaces. Whereas in capped semiconductor systems, only one charge carrier will be available at the surface area, and the opposite charge will transfer to the inner semiconductor system, consequently e the selectivity of the interfacial transfer and enhancing the oxidation or reduction reaction (Rajeshwar, de Tacconi, and Chenthamarakshan 2001; Justicia et al. 2005). The Au-capped TiO_2 nanomaterials improve the efficiency of interfacial charge transfer and enhance the efficiency of removing thiocyanate by oxidation (Dawson and Kamat 2001). Thomas et al. (2019) reported an Ultrafast organic dye removal technique using oleic acid as a capping agent for the synthesis of TiO_2 nanocrystallites prepared *by* precipitation method and found that nanocrystalline TiO_2 removed more than 95% MB dye in 10 minutes.

5.4.5 Combined Effect of Doping and Coupling

Doping introduces an additional state in TiO_2 band gap and increases the visible light response, improving photocatalytic pollutant removal efficiency. Also, coupling enhances the visible light activity by forming a heterojunction between TiO_2 and another small bandgap semiconductor.

5.5 FACTORS AFFECTING POLLUTANT REMOVAL BY PHOTOCATALYSIS

The following factors affect the photocatalytic degradation of pollutants in wastewater.

5.5.1 pH of the Reaction Medium

The pH of the reaction medium influences the surface charge, particle size, and the band edge position of TiO_2 (Saquib et al. 2008). As a result of protonation or deprotonation of hydroxyl groups in surface, TiO_2 shows +Ve charged surface at pH lower than isoelectric point (pH = 6.628) and –Ve charged surface above this pH. The buildup of –Ve charge at the surface improves the transfer of photogenerated holes due to an upward band bending.

5.5.2 Crystal Phase

Among different crystalline phases of TiO_2, anatase (3.2 eV, 387 nm) is generally considered the photocatalytically active phase, which is extensively studied. The brookite phase is very hard to get in pure form; thus, it is the least studied compared to the other two. When compared to anatase rutile (band gap = 3.0 eV (418 nm)), it seems to be a less active photocatalyst (Wold 1993), and as a result, most of the studies on TiO_2 photocatalysts is on the anatase phase. The reason for the high activity is due to the high surface area, the difference in the structural energy between two-phase types, the slightly upward movement in the Fermi level, less oxygen adsorption capacity and high hydroxylation amount in the anatase phase (Tanaka, Capule, and Hisanaga 1991).

In both rutile and anatase phases, the position of the valence band is deep, and the resulting positively charged holes show good oxidative power. Although, in the case of the anatase phase the conduction band is close to the –ve position when compared to the rutile phase. Consequently, the reducing power is high to anatase phase when compared to rutile phase of. Also, the crystal phase of anatase formed at lower temperatures exhibits higher surface area than the rutile phase. Thus, the anatase phase shows high photocatalytic efficiency than the rutile phase.

Commercial catalyst, Degussa P25 comprises of 20% rutile and 80% anatase with a Brunauer–Emmett–Teller (BET) surface area of 55 ± 15 m^2g^{-1} and crystallite sizes of 30 nm in 0.1 μm diameter aggregates and is produced by high-temperature sintering of $TiCl_4$ in the presence of hydrogen and oxygen (Tahiri, Serpone, and Le Van Mao 1996). The enhanced photocatalytic activity of Degussa P25 is due to the ability of excited electrons generated in the anatase TiO_2 particles to be transferred to the rutile particles, thus minimizing charge recombination (Hurum et al. 2005). Recently, Degussa P25 TiO_2 has set the standard for photocatalysis in environmental pollutant removal applications.

5.5.3 Percentage of (001) Facets of Anatase Phase

Recent literature attributes that different facets of anatase TiO_2 show different chemical activity, a property determined by the surface energy (ɣ) of the facet (Tian et al. 2014). The surface energy of different facets of anatase TiO_2 follows the order $\gamma_{(001)}$ (0.90 J/m^2) > $\gamma_{(100)}$ (0.53 J/m^2) > $\gamma_{(101)}$ (0.44 J/m^2), that means the surface energy of (001) facet of anatase TiO_2 is almost double than that of the (101) facet (Nian and Teng 2006). Selloni (2008) found that in an anatase crystal, the reactivity of (001) facets are higher than that of (101) facets, which means that highly energetic and largely sized (001) facet exhibit higher photocatalytic activity. However, the lowest stability of (001) facet makes its preparation difficult. Generally, the anatase TiO_2 nanobelt grows in a very low surface energy (101) direction. Only a small percentage of (001) facet is exposed on the top of the nanobelt, which clarifies the reason behind the very low photocatalytic activity of TiO_2 nanobelts. By adjusting the synthesis procedure as well as the crystallization process, anatase TiO_2 nanobelts grown along the (101) direction with two dominant (001) facets can be synthesized (Selloni 2008). Yang et al. (2008) synthesized uniform anatase crystal (47% of 001 facets) using hydrofluoric acid as a morphology controlling agent. From the above-mentioned literature, it can be concluded that exposing the most active facets of TiO_2 nanostructure is the main problem to be overcome for obtaining highly efficient TiO_2 photocatalysts for pollutant removal.

5.5.4 Oxygen Concentration

Dioxygen inhibits the recombination of the electron/hole pair by trapping the photogenerated electrons and is an important factor determining reaction rate. The rate of organic pollutant degradation is proportional to dissolved oxygen concentration until the saturation point (Mills, Davies, and Worsley 1993).

5.5.5 Catalyst Loading

The organic pollutant degradation rate is linearly related to the amount of TiO_2 present in the sample, finally leveling off to a steady value (Serpone and Salinaro 1999). When excess TiO_2 is present, which leads to the aggregation of TiO_2r, which in turn reduces the accessibility of reaction sites. Increased solution opacity and scattering of light will also decrease the reaction rates.

5.5.6 Crystal Size/Surface Area of the Catalyst

The surface area of the catalyst determines rates of electron/hole recombination because it affects the number of molecules that can adsorb to the surface of the photocatalyst (Fox and Dulay 1993). In general, nanometer-sized TiO_2 has been found to be very effective (P25 particle size of 25–35 nm); however, some micrometer-sized photocatalysts catalysts are reasonably active.

5.5.7 Reaction Temperature

Usually, photochemical reactions are not sensitive to temperature, whereas photodegradation is slightly sensitive to temperature because of different steps in the TiO_2 degradations (adsorption, desorption, surface migration, rearrangements). Activation energies needed for photodegradation are in the 5–15 kJ/mol range (Serpone and Salinaro 1999).

5.5.8 Amount of Dopant

The amount of dopant on TiO_2 should be optimum for the degradation of organic pollutants in wastewater. An increase in dopant concentration beyond this limit will negatively affect the photocatalyst activity (Liao, Badour, and Liao 2008). An extra amount of dopant on the TiO_2 surface will block the UV irradiation falls on the TiO_2 surface and thus hinder the interfacial electron-hole transfer, resulting in lesser photoactivity (Li et al. 2007). Xin et al. (2008) found that excess dopant present in the reaction produces additional oxygen vacancies, which becomes the centers of recombination of photoinduced electrons and holes and extra Cu_2O mask the TiO_2 surface, which leads to lowering the photodegradation efficiency of the catalyst.

5.5.9 Calcination Temperature of the Catalyst

Photocatalytic activity is generally increased with I calcination temperature (Li et al. 2007). It reaches its maximum because of the complete crystallization of the anatase phase, and the further increase will decrease the photocatalytic activity due to the conversion of anatase to rutile form. As a result, it is very difficult to predict the photocatalytic activities from the physicochemical properties of TiO_2 nanomaterials. Optimal conditions are sought by taking into account all these considerations, which may vary from case to case.

5.6 ACHIEVING SELECTIVE PHOTOCATALYSIS IN TIO_2 FOR POLLUTANT REMOVAL

Nanocrystalline TiO_2 is receiving increasing attention because of its quick degradation from a broad range of contaminants. However, its non-selectivity/poor selectivity may limit its application to differentiate highly toxic contaminants and organic contaminants of low toxicity. Generally, less toxic substances will be in high concentration, and more toxic ones will be less concentrated in wastewaters. Furthermore, lower toxic contaminants can be easily degraded by biological means, and higher harmful one is non-biodegradable. Therefore, controlling the selectivity of the TiO_2 photocatalyst is an important goal in the frontiers of TiO_2 research and development.

TiO_2 based selective photocatalysis can be categorized into selective degradation and selective formation. Selective degradation is the process of selectively degrading one of the contaminants without any concentration change in the other molecules from a mixture. In contrast, in the case of selective formation, the desired product is formed through selective photocatalysis. Both of these are worked under the same principle. The difference is that in the case of selective formation, the wanted product will be desorbed from the catalytic surface without undergoing further degradation (Scheme 5.1) (Karpova, Preis, and Kallas 2007).

Non-selectivity can convert to selectivity by simply changing the physical and chemical properties of the catalyst proceed via the following routes:

- Changing the pH of the reaction medium
- Using substrates with different polarity
- With commercial TiO_2 at different calcination temperature
- Coating with a layer of polymer on TiO_2 surface
- Through anodization process to produce TiO_2 with appropriate pore size
- Changing the energy of the photons.

TiO_2 doped with metal oxide can also exhibit selective photocatalysis of pollutant removal. For example, TiO_2 doped with molybdenum oxide (MoO) could selectively degrade cyclohexane into benzene, whereas undoped TiO_2 degrade cyclohexane into

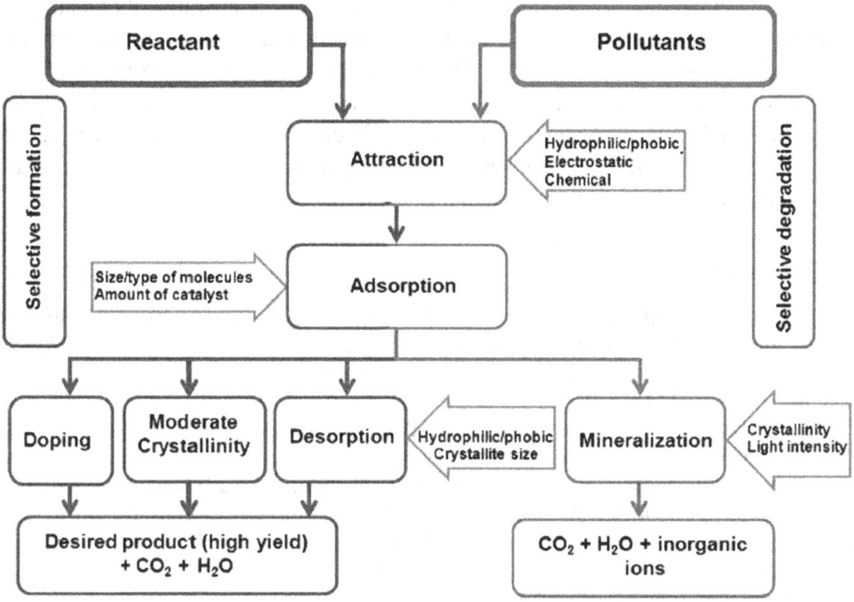

SCHEME 5.1 Origin of selectivity in TiO_2 mediated photocatalytic organic transformations and in photocatalytic degradation reactions. Reproduced with permission from M. A. Lazar and Daoud (2013).

carbon dioxide and water non-selectively (Ciambelli et al. 2008). Similar observations were found in TiO_2 coated tungsten oxide (WO_3), iridium doped and iridium-palladium co-doped P25 catalyst for the selective photocatalytic degradation of benzyl alcohol to benzaldehyde (Feng et al. 2011). Similarly (Tang, Zhang, and Xu 2012) reported that metal ions (Cu^{2+}, Fe^{2+}, Co^{2+}, Ni^{2+}, and Mn^{2+}) and doped titanate nanotubes convert benzylic and allylic alcohols into their corresponding aldehyde. Modification of anatase TiO_2 with Na_2CO_3 and NH_4OH at low temperature exhibits selectivity in the degradation of dyes mainly due to the factors such as moderate crystallinity, smallest size, and surface charge of the catalyst (M. A. Lazar and Daoud 2012). The negative effect of TiO_2-selective photocatalysis through doping has also been reported in various studies. For example, gold doped TiO_2 and zirconium oxide (ZrO_2) doped TiO_2 delayed the formation of cyclohexanone from cyclohexane (Hernández-Alonso et al. 2009; Carneiro et al. 2011).

5.7 CONCLUSION

Presently significant research studies have been carried out on TiO_2 semiconductors nanomaterial because of various possibilities of applications. TiO_2 nanomaterials mediated photocatalysis has shown great potential in the degradation of complex organic pollutants, which is very promising for wastewater treatment. A major limitation of TiO_2 nanomaterials is their inability to absorb in the visible range. To improve the efficiency of TiO_2 nanomaterials in the visible range, modification of electronic band structure of TiO_2 using different strategies. Generally, photocatalytic degradation by TiO_2 follows a free radical mechanism and follows complete mineralization of pollutant non-selectively. This non-selectivity can convert to selectivity by simply changing the catalyst's physical and chemical properties, helping to remove the highly toxic contaminants and producing useful compounds from an effluent. Hence, improving the selectivity of TiO_2 photocatalysis is an important target in recent water treatment photocatalysis.

REFERENCES

Alharbi, Omar M.L., Al Arsh Basheer, Rafat A. Khattab, and Imran Ali. 2018. "Health and Environmental Effects of Persistent Organic Pollutants." *Journal of Molecular Liquids* 263 (August): 442–453. doi:10.1016/j.molliq.2018.05.029.

Asahi, Ryoji, Takeshi Morikawa, Takeshi Ohwaki, K Aoki, and Yasunori Taga. 2001. "Visible-Light Photocatalysis in Nitrogen-Doped Titanium Oxides." *Science (New York, N.Y.)* 293 (5528): 269–271. doi:10.1126/science.1061051.

Asahi, R, Y Taga, W Mannstadt, and A J Freeman. 2000. "Electronic and Optical Properties of Anatase TiO 2." *Physical Review B* 61 (11): 7459–7465. doi:10.1103/PhysRevB.61.7459.

Banerjee, Sangam, Judy Gopal, Pandurangan Muraleedharan, Ashok K. Tyagi, and Baldev Raj. 2006. "Physics and Chemistry of Photocatalytic Titanium Dioxide: Visualization of Bactericidal Activity Using Atomic Force Microscopy." *Current Science* 90: 1378.

Banerjee, Swagata, Suresh C. Pillai, Polycarpos Falaras, Kevin E. O'Shea, John A. Byrne, and Dionysios D. Dionysiou. 2014. "New Insights into the Mechanism of Visible Light Photocatalysis." *The Journal of Physical Chemistry Letters* 5 (15): 2543–2554. doi:10.1021/jz501030x.

Bento, Rodrigo T., Olandir V. Correa, and Marina F. Pillis. 2019. "Photocatalytic Activity of Undoped and Sulfur-Doped TiO_2 Films Grown by MOCVD for Water Treatment under Visible Light." *Journal of the European Ceramic Society.* doi:10.1016/j.jeurceramsoc.2019.02.046.

Bingham, Sonny, and Walid A. Daoud. 2011. "Recent Advances in Making Nano-Sized TiO_2 Visible-Light Active through Rare-Earth Metal Doping." *J. Mater. Chem.* 21 (7): 2041–2050. doi:10.1039/C0JM02271C.

Carneiro, Joana T., Tom J. Savenije, Jacob A. Moulijn, and Guido Mul. 2011. "The Effect of Au on TiO_2 Catalyzed Selective Photocatalytic Oxidation of Cyclohexane." *Journal of Photochemistry and Photobiology A: Chemistry* 217 (2–3): 326–332. doi:10.1016/j.jphotochem.2010.10.027.

Chobba, Marwa Ben, Mouna Messaoud, Maduka L. Weththimuni, Jamel Bouaziz, Maurizio Licchelli, Filomena De Leo, and Clara Urzì. 2019. "Preparation and Characterization of Photocatalytic Gd-Doped TiO_2 Nanoparticles for Water Treatment." *Environmental Science and Pollution Research* 26 (32): 32734–32745. doi:10.1007/s11356-019-04680-7.

Ciambelli, P., D. Sannino, V. Palma, and V. Vaiano. 2008. "The Effect of Sulphate Doping on Nanosized TiO_2 and $MoOx/TiO_2$ Catalysts in Cyclohexane Photooxidative Dehydrogenation." *International Journal of Photoenergy* 2008: 1–8. doi:10.1155/2008/258631.

Comninellis, Christos, Agnieszka Kapalka, Sixto Malato, Simon A. Parsons, Ioannis Poulios, and Dionissios Mantzavinos. 2008. "Advanced Oxidation Processes for Water Treatment: Advances and Trends for R&D." *Journal of Chemical Technology & Biotechnology* 83 (6): 769–776. doi:10.1002/jctb.1873.

Daghrir, Rimeh, Patrick Drogui, and Didier Robert. 2013. "Modified TiO_2 for Environmental Photocatalytic Applications: A Review." *Industrial & Engineering Chemistry Research* 52 (10): 3581–3599. doi:10.1021/ie303468t.

Dawson, Amy, and Prashant V. Kamat. 2001. "Semiconductor–Metal Nanocomposites. Photoinduced Fusion and Photocatalysis of Gold-Capped TiO_2 (TiO_2 /Gold) Nanoparticles." *The Journal of Physical Chemistry B* 105 (5): 960–966. doi:10.1021/jp0033263.

Elsalamony, Radwa. 2016. "Advances in Photo-Catalytic Materials for Environmental Applications." *Research & Reviews: Journal of Material Sciences* 04 (02): 26–50. doi:10.4172/2321-6212.1000145.

Etacheri, Vinodkumar, Michael K. Seery, Marye Anne Fox, Maria T. Dulay. (1993) Heterogeneous Photocatalysis. *Chemical Reviews*. https://doi.org/10.1021/cr00017a016; Steven J. Hinder, and Suresh C. Pillai. 2011. "Oxygen Rich Titania: A Dopant Free, High Temperature Stable, and Visible-Light Active Anatase Photocatalyst." *Advanced Functional Materials* 21 (19): 3744–3752. doi:10.1002/adfm.201100301.

Feng, Wei, Guangjun Wu, Landong Li, and Naijia Guan. 2011. "Solvent-Free Selective Photocatalytic Oxidation of Benzyl Alcohol over Modified TiO2." *Green Chemistry*. doi:10.1039/c1gc15595d.

Fitzmaurice, Donald, Heinz Frei, and Joseph Rabani. 1995. "Time-Resolved Optical Study on the Charge Carrier Dynamics in a TiO_2/AgI Sandwich Colloid." *The Journal of Physical Chemistry* 99 (22): 9176–9181. doi:10.1021/j100022a034.

Fox, Marye Anne, and Maria T. Dulay. 1993. "Heterogeneous Photocatalysis." *Chemical Reviews* 93 (1): 341–357. doi:10.1021/cr00017a016.

Gągol, Michał, Andrzej Przyjazny, and Grzegorz Boczkaj. 2018. "Wastewater Treatment by Means of Advanced Oxidation Processes Based on Cavitation – A Review." *Chemical Engineering Journal* 338 (April): 599–627. doi:10.1016/j.cej.2018.01.049.

Gopal, Madan, Warren J. Moberly Chan, and Lutgard C. De Jonghe. 1997. "Room Temperature Synthesis of Crystalline Metal Oxides." *Journal of Materials Science* 32: 6001–6008. doi:10.1023/A:1018671212890.

Grätzel, Michael. 2005. "Dye-Sensitized Solid-State Heterojunction Solar Cells." *MRS Bulletin* 30 (1): 23–27. doi:10.1557/mrs2005.4.

Gupta, Shipra Mital, and Manoj Tripathi. 2011. "A Review of TiO_2 Nanoparticles." *Chinese Science Bulletin* 56 (16): 1639–1657. doi:10.1007/s11434-011-4476-1.

Hamdy, Mohamed S., Rezvaneh Amrollahi, and Guido Mul. 2012. "Surface Ti 3+ -Containing (Blue) Titania: A Unique Photocatalyst with High Activity and Selectivity in Visible Light-Stimulated Selective Oxidation." *ACS Catalysis* 2 (12): 2641–2647. doi:10.1021/cs300593d.

Han, Changseok, Miguel Pelaez, Vlassis Likodimos, Athanassios G. Kontos, Polycarpos Falaras, Kevin O'Shea, and Dionysios D. Dionysiou. 2011. "Innovative Visible Light-Activated Sulfur Doped TiO_2 Films for Water Treatment." *Applied Catalysis B: Environmental* 107 (1–2): 77–87. doi:10.1016/j.apcatb.2011.06.039.

Hernández-Alonso, María D., Ana R. Almeida, Jacob A. Moulijn, and Guido Mul. 2009. "Identification of the Role of Surface Acidity in the Deactivation of TiO_2 in the Selective Photo-Oxidation of Cyclohexane." *Catalysis Today*. doi:10.1016/j.cattod.2008.09.025.

Hoffman, Amy J., Elizabeth R. Carraway, and Michael R. Hoffmann. 1994. "Photocatalytic Production of H_2O_2 and Organic Peroxides on Quantum-Sized Semiconductor Colloids." *Environmental Science & Technology* 28 (5): 776–785. doi:10.1021/es00054a006.

Hoyer, P., and R. Könenkamp. 1995. "Photoconduction in Porous TiO_2 Sensitized by PbS Quantum Dots." *Applied Physics Letters* 66 (3): 349–351. doi:10.1063/1.114209.

Hurum, Deanna C., Kimberly A. Gray, Tijana Rajh, and Marion C. Thurnauer. 2005. "Recombination Pathways in the Degussa P25 Formulation of TiO 2 : Surface versus Lattice Mechanisms." *The Journal of Physical Chemistry B* 109 (2): 977–980. doi:10.1021/jp045395d.

Ihara, Tatsuhiko 2003. "Visible-Light-Active Titanium Oxide Photocatalyst Realized by an Oxygen-Deficient Structure and by Nitrogen Doping." *Applied Catalysis B: Environmental* 42 (4): 403–409. doi:10.1016/S0926-3373(02)00269-2.

Joshi, Meenal M., Nitin K. Labhsetwar, Priti A. Mangrulkar, Saumitra N. Tijare, Sanjay P. Kamble, and Sadhana S. Rayalu. 2009. "Visible Light Induced Photoreduction of Methyl Orange by N-Doped Mesoporous Titania." *Applied Catalysis A: General* 357 (1): 26–33. doi:10.1016/j.apcata.2008.12.030.

Justicia, Isaac, Gabriel Garcia, Luis Vázquez, Jóse Santiso, Pablo Ordejón, Giovanni A. Battiston, Rosalba Gerbasi, and Albert Figueras. 2005. "Self-Doped Titanium Oxide Thin Films for Efficient Visible Light Photocatalysis." *Sensors and Actuators B: Chemical* 109 (1): 52–56. doi:10.1016/j.snb.2005.03.021.

Karpova, T., S. Preis, and J. Kallas. 2007. "Selective Photocatalytic Oxidation of Steroid Estrogens in Water Treatment: Urea as Co-Pollutant." *Journal of Hazardous Materials* 146 (3): 465–471. doi:10.1016/j.jhazmat.2007.04.047.

Khairy, Mohamed, and Wan Zakaria. 2014. "Effect of Metal-Doping of TiO_2 Nanoparticles on Their Photocatalytic Activities toward Removal of Organic Dyes." *Egyptian Journal of Petroleum.* doi:10.1016/j.ejpe.2014.09.010.

Lazar, Manoj A., and Walid A. Daoud. 2012. "Selective Adsorption and Photocatalysis of Low-Temperature Base-Modified Anatase Nanocrystals." *RSC Advances* 2 (2): 447–452. doi:10.1039/C1RA00539A.

Lazar, Manoj A., and Walid A. Daoud.2013. "Achieving Selectivity in TiO2-Based Photocatalysis." *RSC Advances* 3 (13): 4130. doi:10.1039/c2ra22665k.

Lazar, Manoj A., Shaji Varghese, and Santhosh Nair. 2012. "Photocatalytic Water Treatment by Titanium Dioxide: Recent Updates." *Catalysts* 2 (4): 572–601. doi:10.3390/catal2040572.

Lee, Seul Yi, and Soo Jin Park. 2013. "TiO_2 Photocatalyst for Water Treatment Applications." *Journal of Industrial and Engineering Chemistry.* doi:10.1016/j.jiec.2013.07.012.

Li, Yuexiang, Shaoqin Peng, Fengyi Jiang, Gongxuan Lu, and Shuben Li. 2007. "Effect of Doping TiO_2 with Alkaline-Earth Metal Ions on Its Photocatalytic Activity." *Journal of the Serbian Chemical Society* 72 (4): 393–402. doi:10.2298/JSC0704393L.

Liao, Dongliang, Chadi A. Badour, and Baoqiang Liao. 2008. "Preparation of Nanosized TiO2/ZnO Composite Catalyst and Its Photocatalytic Activity for Degradation of Methyl Orange." *Journal of Photochemistry and Photobiology A: Chemistry* 194 (1): 11–19. doi:10.1016/j.jphotochem.2007.07.008.

Mills, Andrew, Richard H. Davies, and David Worsley. 1993. "Water Purification by Semiconductor Photocatalysis." *Chemical Society Reviews* 22 (6): 417. doi:10.1039/cs9932200417.

Mills, Andrew, Stephanie Hodgen, and Soo Keun Lee. 2005. "Self-Cleaning Titania Films: An Overview of Direct, Lateral and Remote Photo-Oxidation Processes." *Research on Chemical Intermediates* 31 (4–6): 295–308. doi:10.1163/1568567053956644.

Nagaveni, K., M. S. Hegde, and Giridhar Madras. 2004. "Structure and Photocatalytic Activity of Ti 1- x M x O 2±δ (M = W, V, Ce, Zr, Fe, and Cu) Synthesized by Solution Combustion Method." *The Journal of Physical Chemistry B* 108 (52): 20204–20212. doi:10.1021/jp047917v.

Nian, Jun-Nan, and Hsisheng Teng. 2006. "Hydrothermal Synthesis of Single-Crystalline Anatase TiO 2 Nanorods with Nanotubes as the Precursor." *The Journal of Physical Chemistry B* 110 (9): 4193–4198. doi:10.1021/jp0567321.

Qian, Wen, P. Alex Greaney, Simon Fowler, Sheng-Kuei Chiu, Andrea M. Goforth, and Jun Jiao. 2014. "Low-Temperature Nitrogen Doping in Ammonia Solution for Production of N-Doped TiO_2-Hybridized Graphene as a Highly Efficient Photocatalyst for Water Treatment." *ACS Sustainable Chemistry & Engineering* 2 (7): 1802–1810. doi:10.1021/sc5001176.

Rajeshwar, Krishnan, Norma R. de Tacconi, and C. R. Chenthamarakshan. 2001. "Semiconductor-Based Composite Materials: Preparation, Properties, and Performance." *Chemistry of Materials* 13 (9): 2765–2782. doi:10.1021/cm010254z.

Rao, C. N.R., and A. K. Cheetham. 2001. "Science and Technology of Nanomaterials: Current Status and Future Prospects." *Journal of Materials Chemistry* 11: 2887. doi:10.1039/b105058n.

Saquib, M., M. Abu Tariq, M. Faisal, and M. Muneer. 2008. "Photocatalytic Degradation of Two Selected Dye Derivatives in Aqueous Suspensions of Titanium Dioxide." *Desalination* 219 (1–3): 301–311. doi:10.1016/j.desal.2007.06.006.

Sayed, Murtaza, Aranda Arooj, Noor S. Shah, Javed Ali Khan, Luqman Ali Shah, Faiza Rehman, Hamidreza Arandiyan, Asad M. Khan, and Abdur Rahman Khan. 2018. "Narrowing the Band Gap of TiO_2 by Co-Doping with Mn2+ and Co2+ for Efficient Photocatalytic Degradation of Enoxacin and Its Additional Peroxidase like Activity: A Mechanistic Approach." *Journal of Molecular Liquids.* doi:10.1016/j.molliq.2018.09.102.

Sclafani, Antonino, Leonardo Palmisano, and Eugenio Davì. 1991. "Photocatalytic Degradaton of Phenol in Aqueous Polycrystalline TiO_2 Dispersions: The Influence of Fe3+, Fe2+ and Ag+ on the Reaction Rate." *Journal of Photochemistry and Photobiology A: Chemistry* 56 (1): 113–123. doi:10.1016/1010-6030(91)80011-6.

Selloni, Annabella. 2008. "Anatase Shows Its Reactive Side." *Nature Materials* 7 (8): 613–615. doi:10.1038/nmat2241.

Serpone, Nick, and Angela Salinaro. 1999. "Terminology, Relative Photonic Efficiencies and Quantum Yields in Heterogeneous Photocatalysis. Part I: Suggested Protocol." *Pure and Applied Chemistry* 71 (2): 303–320. doi:10.1351/pac199971020303.

Tahiri, Halima, Nick Serpone, and Raymond Le Van Mao. 1996. "Application of Concept of Relative Photonic Efficiencies and Surface Characterization of a New Titania Photocatalyst Designed for Environmental Remediation." *Journal of Photochemistry and Photobiology A: Chemistry.* doi:10.1016/1010-6030(95)04195-8.

Tanaka, Keiichi, Mario F.V. Capule, and Teruaki Hisanaga. 1991. "Effect of Crystallinity of TiO_2 on Its Photocatalytic Action." *Chemical Physics Letters* 187 (1–2): 73–76. doi:10.1016/0009-2614(91)90486-S.

Tang, Zi Rong, Yanhui Zhang, and Yi Jun Xu. 2012. "Tuning the Optical Property and Photocatalytic Performance of Titanate Nanotube toward Selective Oxidation of Alcohols under Ambient Conditions." *ACS Applied Materials and Interfaces.* doi:10.1021/am3001852.

Thomas, Jasmine, T. K. Harsha, P. K. Anitha, and Nygil Thomas. 2019. "Ultrafast Dye Removal Using Capped TiO_2 Nanocrystallites." *In AIP Conference Proceedings.* doi:10.1063/1.5130359.

Tian, Jian, Zhenhuan Zhao, Anil Kumar, Robert I. Boughton, and Hong Liu. 2014. "Recent Progress in Design, Synthesis, and Applications of One-Dimensional TiO_2 Nanostructured Surface Heterostructures: A Review." *Chem. Soc. Rev.* 43 (20): 6920–6937. doi:10.1039/C4CS00180J.

Wang, Yajun, Qisheng Wang, Xueying Zhan, Fengmei Wang, Muhammad Safdar, and Jun He. 2013. "Visible Light Driven Type II Heterostructures and Their Enhanced Photocatalysis Properties: A Review." *Nanoscale* 5 (18): 8326. doi:10.1039/c3nr01577g.

Wold, Aaron. 1993. "Photocatalytic Properties of Titanium Dioxide (TiO2)." *Chemistry of Materials* 5 (3): 280–283. doi:10.1021/cm00027a008.

Wu, Ming J., Tadeusz Bak, Patrick J. O'Doherty, Michelle C. Moffitt, Janusz Nowotny, Trevor D. Bailey, and Cindy Kersaitis. 2014. "Photocatalysis of Titanium Dioxide for Water Disinfection: Challenges and Future Perspectives." *International Journal of Photochemistry* 2014: 1–9. doi:10.1155/2014/973484.

Wu, Wei, Chang Zhong Jiang, Chang Zhong Jiang, and Vellaisamy A. L. Roy. 2015. "Recent Progress in Magnetic Iron Oxide–Semiconductor Composite Nanomaterials as Promising Photocatalysts." *Nanoscale* 7 (1): 38–58. doi:10.1039/C4NR04244A.

Xin, Baifu, Peng Wang, Dandan Ding, Jia Liu, Zhiyu Ren, and Honggang Fu. 2008. "Effect of Surface Species on Cu-TiO_2 Photocatalytic Activity." *Applied Surface Science* 254 (9): 2569–2574. doi:10.1016/j.apsusc.2007.09.002.

Yang, Hua Gui, Cheng Hua Sun, Shi Zhang Qiao, Jin Zou, Gang Liu, Sean Campbell Smith, Hui Ming Cheng, and Gao Qing Lu. 2008. "Anatase TiO_2 Single Crystals with a Large Percentage of Reactive Facets." *Nature* 453 (7195): 638–641. doi:10.1038/nature06964.

Yang, Ping, Cheng Lu, Nanping Hua, and Yukou Du. 2002. "Titanium Dioxide Nanoparticles Co-Doped with Fe3+ and Eu3+ Ions for Photocatalysis." *Materials Letters* 57 (4): 794–801. doi:10.1016/S0167-577X(02)00875-3.

Yao, Yuan, Mingxuan Sun, Xiaojiao Yuan, Yuanhua Zhu, Xiaojing Lin, and Sambandam Anandan. 2018. "One-Step Hydrothermal Synthesis of N/Ti3+ Co-Doping Multiphasic TiO2/BiOBr Heterojunctions towards Enhanced Sonocatalytic Performance." *Ultrasonics Sonochemistry* 49 (December): 69–78. doi:10.1016/j.ultsonch.2018.07.025.

Zou, Zhigang, Jinhua Ye, Kazuhiro Sayama, and Hironori Arakawa. 2001. "Direct Splitting of Water under Visible Light Irradiation with an Oxide Semiconductor Photocatalyst." *Nature* 414 (6864): 625–627. doi:10.1038/414625a.

6 Nanomaterials for the Removal of Heavy Metals from Water

Subhadeep Biswas, Mohammad Danish, and Anjali Pal
Indian Institute of Technology Kharagpur, India

CONTENTS

6.1 Introduction 182
6.2 Different Nanoparticle-Based Technologies for Metal Removal 184
6.2.1 Adsorption 184
6.2.1.1 Adsorption Isotherm, Kinetics, and Thermodynamic Parameters 185
6.2.2 Coprecipitation 187
6.2.3 Membrane Technologies 188
6.2.4 Removal of Metals by Reduction 189
6.3 Removal of Different Heavy Metals (Or Metalloids) 190
6.3.1 Arsenic Removal 190
6.3.2 Cadmium Removal 192
6.3.3 Chromium Removal 193
6.3.4 Lead Removal 195
6.3.5 Mercury Removal 197
6.3.6 Copper Removal 198
6.3.7 Selective Removal of Heavy Metals 200
6.4 Different Characterization Analysis of Nanoparticles and Removal Mechanism Studies 201
6.4.1 Fourier Transform Infrared Spectroscopy (FTIR) 201
6.4.2 Scanning and Transmission Electron Microscopic (SEM and TEM) Analysis 202
6.4.3 X-ray Diffraction (XRD) Analysis 203
6.4.4 X-ray Photoelectron Spectroscopy (XPS) Analysis 204
6.4.5 Other Characterization Techniques 205
6.5 Regeneration and Reusability 205
6.6 Future Scope and Challenges 207
Acknowledgment 208
References 208

DOI: 10.1201/9781003118749-6

6.1 INTRODUCTION

Since the last century, the rapid progress of human society and lifestyle has been noticed with the advancement of science and technology. The development of newer technologies and ideas has consequently aided the growth of industrialization. As industrial production has gradually increased over time, so has the discharge of waste materials left after production. Several industrial units discharge various kinds of waste products into the environment. Natural water bodies often act as a receiver of these waste products and consequently get polluted. Heavy metals and emerging contaminants are some of the major waste products that pollute the environment. Heavy metals constitute a class of recalcitrant water pollutants that are non-biodegradable, and they possess high toxicity even when present at low concentrations. They are bioaccumulative in nature, and most of them are carcinogenic and mutagenic. Some of the heavy metals are industry-specific. For example, chromium is mostly found in the effluents of the tannery industry. Chromium in the natural environment exists in two oxidation states (+3 and +6), and out of the two forms, hexavalent chromium is more dangerous in terms of health aspects. Hexavalent chromium (Cr(VI)) has been enlisted as highly carcinogenic and mutagenic (Bertagnolli et al., 2014). As Cr(VI) is soluble in water throughout a wide range of pH, so it is very difficult to immobilize it. Apart from the tannery industry effluents, Cr(VI) is also prevalent in the discharges of the electroplating industry. Namasivayam and Sureshkumar (2008) reported the presence of hexavalent chromium in the wastewater of the electroplating industry at a concentration of 9.9 mg/L, which is much higher in the Indian context than the acceptable dose (0.05 mg/L) (Bai & Abraham, 2002). Besides this, the effluents of the electroplating and metal plating industry, cadmium-nickel battery industry, and pigment industry can be a common source of cadmium (Pal & Pal, 2017). On the other hand, lead paint, lead piping, lead batteries are identified as potential causes of lead contamination in the environment (Singh et al., 2012).

It is already highlighted that the major sources of heavy metals in the water are industrial units, but it is important to note that there are several geogenic sources also. The occurrence of arsenic (As) in the water bodies is mostly geogenic, and it is due to the leaching of arsenic-bearing rocks. Mukherjee et al. (2008) stated that throughout the whole world, the number of geographical provinces where the source of As in the hydrologic system is purely geogenic is at least 20. It is estimated that more than 100 million people are at risk due to the As contamination in drinking water. During the acid mine drainage, a huge quantity of heavy metals is discharged from the rocks to the environment. Whatever may be the source, it is well established that heavy metals should be eliminated from water bodies to save the aquatic ecology and human health. Several heavy metal remediation techniques such as adsorption, coprecipitation, membrane technologies, etc. have emerged out. Nanoparticle research during the last few decades has added new technical advantages to the existing metal removal processes. Nanoparticles are referred to as those particles with at least one dimension in the range of 1–100 nm. The optical properties of metal nanoparticles are related to their size, and they have different colors owing to the difference in absorption in the visible region (Khan et al., 2019). The nanoparticles may be categorized into 0D, 1D, 2D, and 3D depending on their overall size and shape (Tiwari et al., 2012). Due to the extremely small size, the properties of nanoparticles often largely differ from those of the bulk particles of the same material. One of the most relevant examples in this context is that gold (Au) in general behaves as an inert material and does not possess any catalytic activity. However, Haruta's research has proved that gold nanoparticles can successfully act as a novel catalyst for CO oxidation (Haruta et al., 1987). In the current world, nanoparticles have been utilized in various fields. From the report of (Cho et al. 2008), it is observed that the specially designed nanoparticles for cancer drug delivery have overcome many of the existing challenges of the conventional drug delivery system. Hamidi et al. (2008) highlighted the importance of the hydrogel nanoparticle system in the modern world. Among various nanoparticles, zinc oxide is a common one. It is known for its application as a gas sensor, biosensor, cosmetics, etc. It has unique optical and anti-microbial properties. Sabir et al. (2014) described its significance and importance in the agriculture field. Rastogi et al. (2019) reported applying silicon nanoparticles for plant growth, as silicon is a metalloid and is required by plants as an important element. The results showed that the nanoparticles behaved differently in comparison to their bulk counterpart.

As the size of these particles is extremely small, the specific surface area is very large, and that is why they possess various distinguished properties. The mechanical properties exhibited by the nanoparticles are different from those of the parent material, and thus the nanoparticles can act as a surface modifier of different materials (Guo et al., 2014). Moreover, the rising of nanotechnology has given birth to various new generation smart materials, which are promising materials for various environmental purposes. Metal-organic frameworks (MOFs), MXenes, nano zero-valent iron (nZVI), graphitic carbon nitride (g-C_3N_4), nanocellulose, etc. (Ngwabebhoh & Yildiz, 2019; Wu et al., 2019; Ihsanullah, 2020) are counted as some of the prominent examples of new-age nanomaterials which find immense use in wastewater treatment as well as other environmental remediation purposes. MOFs are synthesized through coordination between metal precursors and organic ligands (Wu et al., 2019). Owing to their excellent porosity, they are one of the notable adsorbents. MXenes nanomaterials are commonly represented as $M_{n+1}X_nT_x$ (n = 1–3), M denotes early transition metals such as Cr, Mo, V, etc., X stands for carbon and/or nitrogen, and T for surface termination groups such as oxygen, fluorine, etc. (Ihsanullah, 2020). g-C_3N_4 has been recognized as one of the most stable allotropes of C_3N_4, possessing a controllable structure (Wu et al., 2019), superb thermal and chemical stability. Moreover, due to its environmentally benign nature, it is currently applied for various environmental remediation processes. Besides these latest developed materials, carbon nanotubes also serve as a promising new-age alternative of activated carbon for wastewater treatment. In their review, Sharma et al. (2009) reported the high efficiency of carbon nanotubes toward heavy metal adsorption. In another study, Rao et al. (2007) also discussed the efficiency of surface oxidized CNTs for divalent heavy metal adsorption (such as Cd^{2+}, Ni^{2+}, Pb^{2+}, etc.).

In the present chapter, heavy metal remediation from water bodies using nanoparticles has been reviewed. Most of the works covered here are of recent times. Firstly, different nanoparticle-mediated heavy metal removal techniques have been discussed in short. After that, various reported studies on the abatement of different heavy metals viz., cadmium, arsenic, chromium, lead, copper, and mercury have been

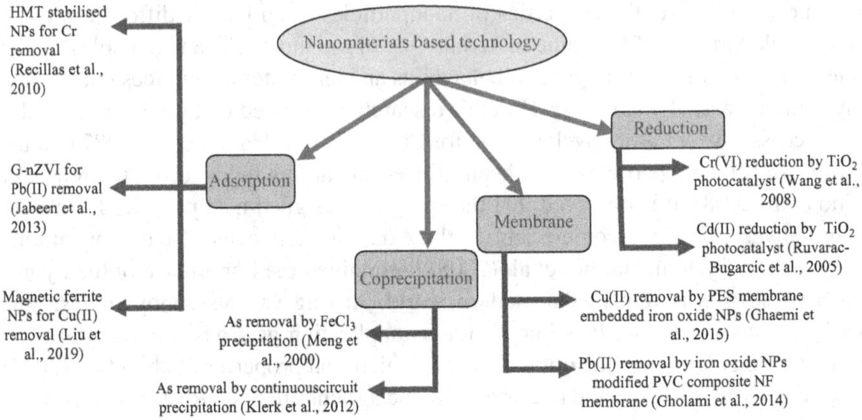

FIGURE 6.1 Schematic diagram of various nanoparticles mediated techniques for different heavy metal.

highlighted. A schematic (Figure 6.1) has been presented to highlight different nanoparticle-mediated techniques for heavy metal removal. Adsorption, coprecipitation, membrane technology, reductive removal with suitable examples have been presented in the scheme. Advanced characterization techniques such as scanning electron microscopy (SEM), X-ray diffraction (XRD), X-ray photoelectron spectroscopy (XPS), which are commonly employed for getting better insight into the mechanism of the removal process, have been presented. Finally, the possibility of regeneration and reusability of various nanocomposites after one cycle of heavy metal remediation has been covered, along with the challenges and future scopes.

6.2 DIFFERENT NANOPARTICLE-BASED TECHNOLOGIES FOR METAL REMOVAL

6.2.1 Adsorption

Adsorption is one of the simplest physicochemical technologies often used for wastewater remediation purposes. Since old times activated carbon served as the most versatile and widely used adsorbent for water and wastewater treatment. There are various reports available in the literature regarding the use of activated carbon for the adsorptive removal of various organic and inorganic pollutants. However, due to the high cost (Sheshdeh et al., 2014) and less regeneration capacity, there arose a huge urge among the researchers to find a suitable alternative. Various low-cost adsorbents have already been experimented out. In this regard, nanoparticles added some significant contributions to the adsorption field. Owing to the high surface-area-to-volume ratio, they often serve as the most suitable sorbent. They also often act as a surface modifier in adsorption studies (Badruddoza et al., 2013; Sheshdeh et al., 2014). Vojoudi et al. (2017) applied newly synthesized magnetic nanomaterials for the fast and enhanced removal of lead (Pb), mercury (Hg), and palladium (Pd) ions. Mahdavi et al. (2013) reported using TiO_2, MgO, and Al_2O_3 nanoparticles for adsorption of heavy metal ions (Cd^{2+}, Cu^{2+}, Ni^{2+}, and Pb^{2+}) from aqueous media.

6.2.1.1 Adsorption Isotherm, Kinetics, and Thermodynamic Parameters

The experimental data are often fitted to various isotherm and kinetic models to better understand the adsorptive removal of heavy metals. Isotherm studies deal with the variation of the adsorption capacities of the material. The experiments are conducted at equilibrium conditions keeping either the dose or the initial concentration fixed. Two-parameter isotherm models such as the Langmuir, Freundlich are most commonly used. The nonlinear form of the Langmuir model is often expressed as Equation (6.1)

$$q_e = \frac{QbC_e}{1 + bC_e} \tag{6.1}$$

where q_e is the amount of heavy metal adsorbed per unit weight of adsorbent at equilibrium, Q is the saturated monolayer adsorption capacity, b is the binding energy of the adsorption system, and C_e is the equilibrium concentration of the heavy metal in the water. Turk and Alp (2014) found that As removal was best fitted to the Langmuir isotherm model. The fitting of the experimental data is shown in Figure 6.2. It can be observed that the plot of q_e vs. C_e, is linear for As(V), while it is non-linear for As(III). However, the plot of $1/q_e$ vs. $1/C_e$ is linear for both.

The equation of the Freundlich model can be expressed by Equation (6.2):

$$q_e = K_f C_e^{\frac{1}{n}} \tag{6.2}$$

where K_f is the Freundlich constant and n is the adsorption intensity.

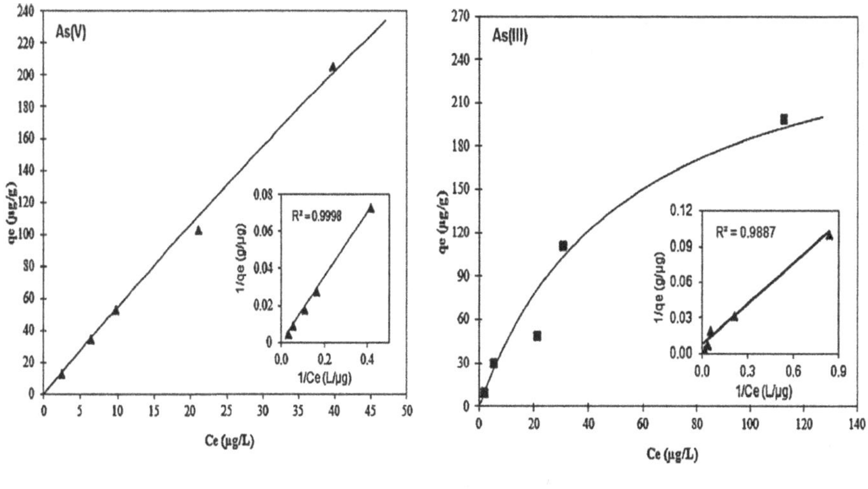

FIGURE 6.2 Equilibrium isotherm for As(V) and As(III) adsorbed at pH 9 and 25°C on M-FeHT. (Reprinted with permission from (Turk & Alp, 2014).)

The Dubinin Radushkevich (D-R) isotherm model is also used to calculate the energy of the adsorption process. The D-R isotherm is expressed as Equation (6.3) (Lunge et al., 2014):

$$\ln q_e = \ln Q_m - k_{ads}\,\varepsilon^2 \tag{6.3}$$

where

$$\varepsilon = RT\ln\left(1+\frac{1}{C_e}\right) \tag{6.4}$$

In the expression of the D-R isotherm model, Q_m is the theoretical adsorption capacity, k_{ads} is the constant related to the adsorption energy, ε, as shown in Equation (6.4) is the Polyani potential.

Apart from these two-parameter models, three-parameter isotherm models such as the Redlich-Peterson model have also been used by researchers. The Redlich-Peterson model can be expressed as Equation (6.5) (Cui et al., 2012):

$$q_e = \frac{AC_e}{1+BC_e^g} \tag{6.5}$$

where A, B, g are the Redlich-Peterson constants of adsorption.

In the work of Cui et al. (2012), the experimental data of removal of arsenic were best fitted by the Redlich-Peterson isotherm model.

Figure 6.3 shows the curve fitting of experimental data of the work done by Lin et al. (2018) to various isotherm models.

Kinetic studies offer insight into the rate of the adsorption process. Commonly used kinetic models for the adsorption studies are the pseudo-first, pseudo-second, intra-particle diffusion, and Elovich model. The pseudo-first model represents the physisorption process, while the pseudo-second indicates the chemisorption phenomenon. Equations of the pseudo-first, pseudo-second, and Elovich models are shown as Equations (6.6) to (6.8), respectively (Nayak & Pal, 2017):

$$\log(q_e - q_t) = \log q_e - \left(\frac{k_1}{2.303}\right)t \tag{6.6}$$

$$\frac{t}{q_t} = \frac{1}{k_2 q_e^2} + \left(\frac{1}{q_e}\right)t \tag{6.7}$$

$$q_t = \frac{1}{\beta}\ln(\alpha\beta) + \frac{1}{\beta}\ln t \tag{6.8}$$

where q_t is the amount of heavy metal adsorption capacity (mg/g) at time t, q_e is the adsorption capacity at equilibrium, α (mg/g) is the initial adsorption rate, and β (mg/g) allied to the extent of surface coverage and activation energy. Most of the reported studies in the literature reveal that pseudo-second-order fits best with the experimental data.

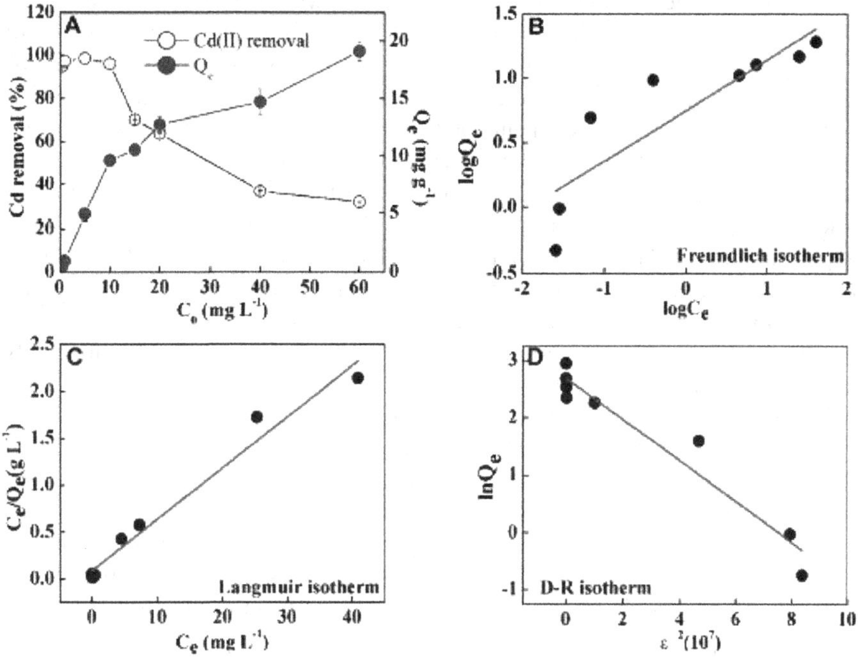

FIGURE 6.3 Effect of initial concentration of Cd (II) on adsorption and the adsorption isotherm model of Freundlich (B), Langmuir (C) and D-R (D) models. (Reprinted with permission from (Lin et al., 2018).)

From the thermodynamic parameters, an idea regarding the spontaneity of the process is obtained. Gibbs free energy change, enthalpy change, and entropy change of the process are calculated by Equations (6.9 and 6.10)

$$\ln K_e = \left(\frac{\Delta S}{R}\right) - \left(\frac{\Delta H}{RT}\right) \tag{6.9}$$

$$\Delta G = -RT\ln K_e \tag{6.10}$$

where ΔG stands for Gibb's free energy, ΔS stands for change in entropy, ΔH change in enthalpy, R the universal gas constant, and T is the temperature in K. Negative value of ΔG indicates that the process is the spontaneous and positive value of ΔH means that the adsorption process is endothermic in nature.

6.2.2 Coprecipitation

Heavy metal alleviation using precipitation is another easy physico-chemical process. This technique mainly involves the formation of the metal hydroxides under alkaline conditions and precipitation of the same. Due to its easiness and low cost to treat a large volume of wastewater, the technique is widely accepted. In the case of arsenic remediation, coprecipitation with ferric chloride is one of the most useful techniques. Meng et al. (2000) studied the interference by various ions such as sulfate, carbonate, and silicate on the removal of As by $FeCl_3$ precipitation. It was observed that among the three mentioned above, the silicate ion showed maximum interference on As(III) removal by coprecipitation. The other two ions showed minimum interference. Klerk et al. (2012) reported a method for arsenic removal by continuous circuit coprecipitation combined with lime neutralization. In the case of simple chemical precipitation, one major drawback is the production of bulk sludge having a large quantity of water. The introduction of nanoparticles in coagulation precipitation undoubtedly provided an attractive solution to the sludge bulking. Owing to the presence of enormous surface area, easy availability commercially, possession of various functional groups, and the specialty to form relatively thicker sludge, the nanoparticles are becoming one of the most promising choices for precipitating metal ions (Pacheco et al., 2006). However, the reported studies for heavy metal removal using the nanoparticle-mediated precipitation technique are comparatively fewer than other techniques.

Cai et al. (2010) synthesized polyacrylic acid-stabilized calcium carbonate nanoparticles and explored the same for removing various heavy metals from water. The study reported a high removal efficiency, and the main mechanism mentioned behind the removal was precipitation.

6.2.3 Membrane Technologies

Membrane technology is counted as one of the most advanced water treatment methods. For alleviating heavy metal pollution, membrane-based technologies are often considered superior and promising concerning conventional technologies. Membrane technologies are commonly classified as microfiltration (MF), ultrafiltration (UF), nanofiltration (NF), and reverse osmosis (RO). Among all these methods, the NF and RO techniques are applied in most cases for heavy metal removal. The separation mechanism is mainly based on molecular sieving, Donnan exclusion, and solution-diffusion principles. Nanoparticles are often explored for the preparation of membranes. Metal oxide nanoparticles are used in many cases for optimizing the performance of ceramic membranes. However, in most studies, iron oxide nanoparticles have been covered (Daraei et al., 2012).

Especially for Cu(II) removal, iron oxide-modified nanomaterials have been reported to play a very important role in the membrane separation process. Polyethersulfone (PES) membrane embedded with Fe_3O_4 nanoparticles has shown promising results toward Cu removal (Ghaemi et al., 2015). It was concluded that the NH group of the aniline was mainly responsible for enhanced Cu(II) adsorption. Preparation of nanocomposite membrane with γ-Al_2O_3 nanoparticles was reported by Ghaemi (2016) for Cu(II) removal from water. The mixed matrix membrane showed higher water permeation in comparison to the pristine membrane due to the addition of a small number of nanoparticles ($\leq$ 1wt%). This is an obvious indication of the improvement of porosity and hydrophilicity. Moreover, Cu(II) removal efficiency was also greatly improved due to the presence of nanoparticles. The membrane test revealed that adsorption on alumina nanoparticles is the main factor for Cu removal. Gholami et al. (2014) reported preparing a polyvinyl chloride-based nanocomposite-nanofiltration membrane and its further modification with iron oxide nanoparticles. Finally, the newly modified membrane was successfully applied for lead removal from water. The inclusion of iron oxide nanoparticles increased the rejection rate significantly because the nanoparticles themselves acted as the additional adsorbent in the removal process.

Mukherjee et al. (2016) showed that incorporating graphene oxide nanoparticles in the preparation of mixed matrix membranes enhanced the porosity, hydrophilicity, and negative charge on the surface of the membrane. Tests were conducted to remove various heavy metals like Pb, Cu, Cd, Cr. The aforementioned heavy metal ions' rejection rate was above 90%. Through the sharing of lone pair of electrons, the oxygen of the graphene oxide material helped bind the metal ions causing high removal efficiency. Moreover, the impregnation of the nanographene oxide on the membrane support helped obtain stability, prevent leaching, and promote easy regeneration.

6.2.4 Removal of Metals by Reduction

Apart from the three techniques mentioned earlier, adopted for removing many heavy metals, another method for removing especially hexavalent chromium is to first reduce it to trivalent form and then removing it. Hexavalent chromium is soluble throughout a wide pH range, while Cr(III) gets precipitated at higher pH. There are several reported methods on nanoparticle-mediated catalytic reduction of hexavalent chromium and thereafter its elimination from water bodies. Ku and Jung (2001) prepared TiO_2 photocatalyst and under the irradiation of UV light Cr(VI) was converted to Cr(III). At lower pH, the reduction efficiency was better in comparison to that at alkaline pH. Wang et al. (2008) also conducted a photocatalytic reduction of Cr(VI) by different TiO_2 photocatalysts. The TiO_2 was subjected to precalcination at different temperatures to obtain different surface areas and photocatalytic activity. The reduction reaction followed a pseudo-first-order reaction.

Apart from Cr(VI) removal, Cd(II) removal from water bodies has also been reported by the photoreduction principle. Skubal et al. (2002) removed Cd(II) from wastewater by TiO_2 photocatalyst through reduction. Nanosized TiO_2 colloids were prepared through controlled hydrolysis of $TiCl_4$ followed by modification of the surface by thiolactic acid. Photocatalytic reduction of Cd(II) by TiO_2 nanoparticles have also been reported by Ruvarac-Bugarcic et al. (2005)

Liu et al. (2013) applied chitosan bead-supported zero-valent iron nanoparticles as a permeable reactive barrier to eliminate heavy metal ions from electroplating wastewater. The reduction of Cr(VI) to Cr(III) was confirmed by an XPS study. For other metals like Pb(II) and Cu(II) removal, the mechanism was predominantly reduction followed by auxiliary adsorption. In Table 6.1, various reductive removal methods of heavy metals have been summarized.

TABLE 6.1
Reductive Removal of Some Heavy Metals by Nanomaterials

Heavy Metal Reduced	Nanomaterial	Process	Reaction Mechanism Model	References
Cr	TiO_2	Photocatalytic reduction	Langmuir-Hinshelwood, Pseudo first	Ku and Jung (2001)
Cr	TiO_2	Photocatalytic reduction	Pseudo first	Wang et al. (2008))
Cd	Thiolactic acid-modified TiO_2 nanoparticles	Adsorption and reduction		Skubal et al. (2002)
Cd	Amino acid-modified TiO_2	Photocatalytic reduction		Ruvarac-Bugarcic et al. (2005)
Cr, Cd, Pb, Cu	Chitosan bead supported zvi	Reduction and auxiliary adsorption		Liu et al. (2013)
Cu, Cr, Cd, Zn, Pb	$CuCo_2S_4$ modified $MoSe_2/BiVO_4$ nanocomposite	Photocatalytic reduction		Shahzad et al. (2019)
Cr	Sulfate modified titania	Photocatalytic reduction		Mohapatra et al. (2005)
Cr	Starch stabilized Fe^0 nanoparticles	Reduction	Pseudo-first	Alidokht et al. (2011)

6.3 REMOVAL OF DIFFERENT HEAVY METALS (OR METALLOIDS)

6.3.1 Arsenic Removal

Arsenic (As), a ubiquitous element, is present either as As(III) or As(V) in both surface and subsurface water bodies in various parts of the world. Owing to its well-known toxicity, As removal from water is a prime need from a health point of view. Considering health safety, the World Health Organization (WHO) reduced the permissible limit of As in drinking water from 50 to 10 μg/L (Mukherjee et al., 2008). Among the two forms of arsenic, As(III) removal is relatively more difficult since it exists as a neutral form in the pH range of 4–9. Moreover, in terms of toxicity, it is nearly 60 times more toxic than As(V) (Ratnaike, 2003). There are various processes reported so far regarding As removal. They include adsorption, coprecipitation, ion exchange, membrane separation, etc. Among the various available options for As remediation, the adsorption process is the most simple, easy, and versatile. Due to various technical and operational problems, often adsorption process does not fit suitable for a practical purpose. However, nanotechnologies have emerged as a promising method for As removal. Among various applications of metal oxide nanoparticle-based adsorbents, the use of iron or iron oxide-based nanosorbent has been reported most. Ascorbic acid-coated superparamagnetic

Fe_3O_4 nanoparticles having high surface area have been utilized by Feng et al. (2012) for As removal. Due to having a diameter less than 10 nm, the nanoparticles possess a high specific surface area (179 m^2/g), leading to high adsorption capacity. From the Langmuir model, the maximum adsorption capacity for As (V) and As (III) has been found as 16.56 mg/g and 46.06 mg/g, respectively. Yu et al. (2013) reported the synthesis of magnetic composites of cellulose@iron oxide nanoparticles and applied the same for As removal. One of the distinct advantages of the iron oxide nanoparticles is the easy separation utilizing an external magnetic field. Turk and Alp (Turk & Alp, 2014) also reported the utilization of Fe-hydrotalcite supported magnetite nanoparticles for both As(III) and As(V) removal. The spent adsorbent has been recovered using magnetic separation. Lunge et al. (2014) synthesized magnetic iron oxide nanoparticles using tea waste as the support template. The newly prepared magnetic nanoadsorbent showed super magnetic properties and was successfully deployed for the As adsorption purpose from the water medium. The maximum capacity of the adsorbent reached up to 188.69 mg/g for As(III) and 153.8 mg/g for As(V), while the mean sorption energy calculated from the D-R isotherm model was 64.27 kJ/mole, which indicated that the sorption process is physicochemical. The sorption followed the pseudo-second order kinetic model best. Although the use of iron oxide nanoparticles for removal of As has been mostly reported, the use of other nanomaterials for As remediation is also investigated by some researchers. Cui et al. (2012) used zirconium oxide (ZrO_2) nanoparticles for the same purpose. Due to unique properties such as the high specific area, large mesoporous volume, and the presence of high-affinity hydroxyl groups, exceptional performance regarding the adsorption of both As(III) and As(V) was observed. The inner-sphere complex formation mechanism was reported as the responsible factor for As removal. Moreover, the amorphous ZrO_2, prepared by the hydrothermal process, has a high pore volume, a loose structure that facilitates the adsorption process. Copper oxide nanoparticles were also explored by Martinson and Reddy (2009) for As removal from groundwater. The as-prepared CuO nanoparticles were 12–18 nm in diameter, and the adsorption process was very fast (within minutes). The maximum adsorption capacity of the nanoadsorbent toward As removal was observed to be 26.9 and 22.6 mg/g for As(III) and As(V), respectively. While examining the interference caused by the coexisting anions, it was found that phosphate anions at a concentration > 0.2 mM caused a reduction in adsorption efficiency. On the other hand, the presence of sulfate and silicate showed no appreciable interference. Goswami et al. (2012) synthesized copper oxide nanoparticles by thermal refluxing and the prepared nanomaterial was used for As remediation purposes. From the detailed thermodynamic and kinetic study, it was observed that the process was endothermic in nature and it followed the pseudo-second-order kinetic model best. Unlike other studies, here it was noticed that the presence of both phosphate and sulfate adversely affects the adsorption process. However, the presence of phosphate caused more hindrance to the removal of As in comparison to sulfate. The presence of sulfate anions up to 280 mg/L caused a $<10\%$ reduction in the removal efficiency, while phosphate anions at a concentration of 0.2 mM reduced the capacity by 20%. The study concluded that the affinity of As toward the potential adsorption site was higher than the competing anions.

6.3.2 Cadmium Removal

Cadmium (Cd) is also one of the most toxic heavy metal ions often detected in water bodies. Consumption of Cd causes cell death, kidney, and lung damage. WHO has prescribed a maximum limit of 0.005 mg/L of cadmium in drinking water. USEPA classifies it as a B1 carcinogen. Cd accumulates in the human body with a half-life of about 15–20 years (Sethi et al., 2006). The source of Cd is mostly anthropogenic. Chen et al. synthesized Fe_3O_4 sulfonated magnetic nanoparticle for the removal of Cd(II) and Pb(II) from water bodies (Chen et al., 2017). The nanoadsorbents were synthesized by sulfo-functionalization of mercaptopropyl trimethoxysilane (MPTS) coated Fe_3O_4 nanoparticles by free radical polymerization. Shellac-coated iron oxide nanoparticles were employed by Gong et al. (2012) for the Cd(II) remediation purpose. Fourier-transform infrared spectroscopy (FTIR) analysis revealed a reaction occurring between the carboxylic groups present on the surface of the shellac modified iron oxide nanoparticles and Cd(II) ions as the main mechanism behind the adsorptive removal of cadmium.

As mentioned in our previous section, besides adsorption, nanoparticles play important roles in other technologies also. For example, the utilization of nanoparticle-based membrane for Cd(II) removal by ultrafiltration has been reported. Jawor and Hoek (2010) evaluated the chemical properties, metal-binding kinetics, capacity, and reversibility while studying the Cd(II) removal by ultrafiltration through a zeolite-polymer nanoparticle-based membrane. Skubal et al. (2002) synthesized colloids of TiO_2 of nanosize dimensions by the hydrolysis of $TiCl_4$, thereby modifying the surface with a bidental chelating agent. The as-prepared nano colloids were deployed for the Cd(II) removal from simulated wastewater at a concentration of 65 ppm. The technique was innovative as it involved removing cadmium by conversion of +2 state to elemental Cd. Other techniques for Cd remediation involve precipitation with a suitable sol-gel. Pacheco et al. (2006) prepared Al-Si and Si-Al nanoparticles and utilized the same for Cd removal at an initial concentration of 140 ppm by precipitating the sludge formed.

Huang and Keller (2015) synthesized EDTA functionalized magnetic nanoparticles for the adsorptive elimination of lead and cadmium. The removal efficiency of the adsorbent was quite appreciable in a wide range of pH (3–10) with an easy regeneration using 1% HCl without any significant reduction in efficiency even after several cycles of reuse. Wan et al. (2018) used biochar-supported hydrated manganese oxide for adsorption of lead and cadmium. The composite nanoparticle-based adsorbent was highly efficient in Cd(II) removal and in overcoming all difficulties faced by either biochar or manganese oxide singly. The hydrated manganese oxide provides a suitable site for capturing the metal ions via the formation of an inner-sphere complex. Kataria and Garg (2018) prepared Fe_3O_4 loaded sawdust carbon and EDTA modified Fe_3O_4 loaded sawdust carbon by biogenic green synthesis method and applied it for the removal of Cd(II) from water. The preparation has been shown in Figure 6.4. After being thoroughly washed, the sawdust and $Fe(NO_3)_3.9H_2O$ were completely mixed at 90°C, and ammonia was added dropwise up to pH~10. Sawdust solution act as the green reducing solvent. It was interesting to observe that, at a lower pH, the adsorbent showed not such a good performance due to the protonation of

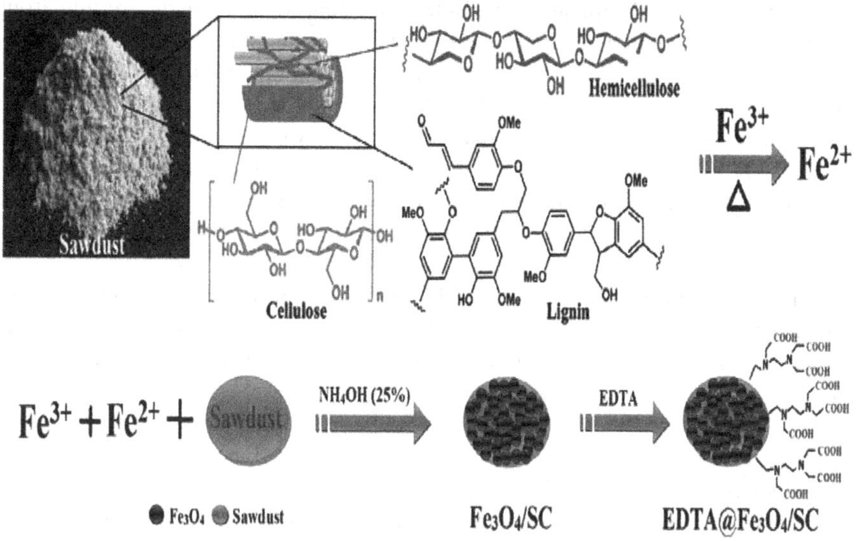

FIGURE 6.4 Scheme for synthesis of Fe_3O_4/SC and EDTA@Fe_3O_4/SC ncs. (Reprinted with permission from (Kataria & Garg, 2018).)

most of its adsorption sites. However, upon increasing the pH up to a certain level, the removal efficiency increased appreciably. Koju et al. (2018) utilized alumina nanoparticles for Cd remediation purposes. Experimental results concluded that the adsorption occurred through the heterogeneous adsorption sites and chemisorption was the predominant mechanism. Chowdhury and Yanful (2013) studied the kinetics of Cd(II) adsorption by mixed maghemite-magnetite nanoparticles. Adsorption occurred partly due to diffusion and partly due to electrostatic interaction. The experimental data suggested that the process followed the pseudo-second-order kinetic model best. Lin et al. (2018) applied biosynthesized iron oxide nanoparticles for Cd(II) adsorption purposes. Chitosan-supported zero-valent iron was successfully deployed by Ahmadi et al. (2017) for the same purpose. Tabesh et al. (2018) reported the assignment of γ-Al_2O_3 prepared by the sol-gel method for Cd(II) removal. Besides iron oxide, other nanoparticles like TiO_2, CeO_2 have also been used. Contreras et al. (2012) found that although other nanoparticles showed some good results in eliminating Cd(II) from water, the removal efficiency of iron oxide nanoparticles was the highest.

6.3.3 Chromium Removal

Chromium is a redox-active element. It is already mentioned in the previous section that it usually exists as either Cr(III) or Cr(VI) in the environment. Hexavalent chromium is highly carcinogenic, and it exists either in the form of dichromate ($Cr_2O_7^{2-}$), hydrochromate ($HCrO_4^-$), or chromate (CrO_4^{2-}) depending on the solution pH (Park & Jang, 2002). Due to negative charge by Cr(VI) anionic species, they are poorly adsorbed by the negatively charged soil particles. Hence Cr(VI) removal from aqueous media is of significant importance, and consequently, there are numerous reports on it. Recillas et al. (2010) have synthesized hexamethylenetetramine (HMT) stabilized CeO_2 nanoparticles (6.5 nm mean size) and applied the same for Cr(VI) adsorption. All the experiments were conducted at pH~7 (same as that of drinking water or natural water). After 4 h of reaction, the removal percentages were obtained as 96.5%, 67.8%, 50.6%, corresponding to the initial Cr(VI) concentrations of 0.6, 37.5, 80 mg/L and the maximum adsorption capacities reached 1.88, 83.33 and 121.95 mg/g, respectively. The experimental data fitted well to the Freundlich adsorption isotherm.

Geng et al. (2009) prepared chitosan-Fe^0 nanoparticles (chitosan-Fe^0) using non-toxic and biodegradable chitosan as a stabilizer. The study revealed that chitosan-Fe^0 had the potential to become an effective agent for *in situ* subsurface environmental remediation. The overall mechanism behind Cr(VI) removal was a two-step process. Firstly, Cr(VI) got physically adsorbed onto the chitosan iron nanocomposite and then reduced to Cr(III). The reduction of Cr(VI) to Cr(III) followed pseudo-first-order kinetics. The rate constants increased with the increase in temperature and iron loading while it showed a declining trend with the rise in initial Cr(VI) concentration and pH. It was concluded that the system's acidity played a major role in reducing Cr(VI) to Cr(III). From the thermodynamic study, the apparent activation energy was obtained as 33 kJ/mol, indicating a chemically controlled reaction.

Jabeen et al. (2011) synthesized graphene-supported iron nanoparticles through impregnation via sodium borohydride reduction of graphene oxide. The material was used to remove Cr(VI) from aqueous solutions. This newly synthesized iron nanoparticle decorated graphene was found to show an enhanced magnetic property, larger surface area, and higher Cr(VI) adsorption capacity compared to bare iron nanoparticles. Results showed that the maximum adsorption capacity for bare iron oxide nanoparticles (nZVI) and graphene-supported iron oxide nanoparticles (G-nZVI) were 148 and 162 mg/g, respectively, and the reduction of hexavalent chromium to trivalent chromium is higher for G-nZVI than for nZVI.

Hu et al. (2005) synthesized nanoscale maghemite and evaluated the same as adsorbent of Cr(VI). During the adsorption study, the effects of influencing factors such as pH, temperature, initial Cr(VI) concentration, and coexisting common ions for Cr(VI) adsorption were studied. It was found that the adsorption reached equilibrium within 15 min and was independent of the initial Cr concentration. The maximum adsorption was obtained at pH 2.5. The experimental data fitted well to the Freundlich isotherm. Interference study showed that the competition from common coexisting ions such as Na^+, Ca^{2+}, Mg^{2+}, Cu^{2+}, Ni^{2+}, NO_3^-, and Cl^- was negligible. Maghemite nanoparticles prepared by the coprecipitation method were used for Cr(VI) remediation (Jiang et al., 2013). The study showed that removing Cr(VI) by the maghemite nanoparticles followed a pseudo-second-order and the intra-particle diffusion kinetic model. It proved that Cr(VI) adsorption onto the maghemite occurred via two distinct phases: the diffusion-controlled by external surface, followed by intra-particle diffusion. Equilibrium data fitted well to the Langmuir and Langmuir–Freundlich (L–F) models, and it was concluded that the adsorption of Cr(VI) was spontaneous and highly favorable. The heterogeneity index was found as 0.55, which implied heterogeneous monolayer adsorption. The adsorptive removal was better at a lower pH value, and the maximum elimination was obtained at a pH of 4. The maghemite nanoparticles' uptake of Cr(VI) was quite rapid and effective under various environmental conditions. Moreover, after treatment, the effluent obtained was at the drinking water standard.

By the thermo-decomposition method, Zhu et al. (2012) synthesized double core-shell-structured magnetic graphene nanocomposites by combining iron oxide and amorphous silica and deployed the same for the hexavalent chromium remediation purpose. Adsorption of Cr(VI) was very fast, and almost the whole process was completed within 5 min. Like other reported studies here, the adsorption was also better achieved at a lower pH, and the experimental data followed the pseudo-second-order kinetic model. The large magnetization of the nanocomposite (96.3 emu/g) helped in quick separation from the liquid media.

This study highlighted that the recycling process (using a permanent magnet) of both the magnetic graphene nanocomposites (MGNC) adsorbent and the adsorbed Cr(VI) was more energetically and economically sustainable. The significantly reduced treatment time required to remove the Cr(VI) and the applicability in treating the solutions with low pH made MGNCs a promising candidate for the efficient removal of heavy metals from the real wastewater.

Ma et al. (2012) reported the preparation of ethylenediamine-reduced graphene oxide (ED-DMF-RGO) nanosheets by refluxing GO with ED and applying it for chromium removal from water. The nanocomposite was improved while using dimethylformamide as the solvent (Zhang et al., 2013a). Hexavalent chromium was reduced to Cr(III) state. The removal efficiency of ED-DMF-RGO was higher (75.6%) than that (59.4%) of ED-RGO. In comparison to the traditional adsorbent, i.e., activated carbon, the removal capacity of ED-DMF-RGO (92.15 mg/g) was much higher (27 times) (Selvi et al., 2001). Wu et al. (2013) prepared CTAB modified graphene by Hummer's method for exploring it as an adsorbent of Cr(VI). The adsorption process was quite fast, and the equilibrium was attained within 40 min.

In another work, it is reported that zero-valent iron (ZVI) nanoparticles having a specific surface area of 35 m^2/g have been successfully explored for the reduction followed by immobilization of Cr(VI) (Cao & Zhang, 2006). Xu and Zhao (2007) stabilized ZVI using carboxymethyl cellulose and applied the same material for Cr(VI) reductive immobilization purposes. In the batch study, 90% removal was achieved when the dose of ZVI was 0.12 g/L.

6.3.4 Lead Removal

Lead (Pb) is another deleterious heavy metal and has serious health implications. Pb is common in the discharge of battery and paint industries. Various studies have been reported regarding the removal of Pb by nanoparticles. Zhang et al. (2011) fabricated three nanocomposite adsorbents (denoted as ZrP–Cl, ZrP–S, and ZrP–N) by encapsulating zirconium phosphate (ZrP) nanoparticles into three macroporous polystyrene resins with various surface groups, i.e., $-CH_2Cl$, $-SO_3^-$, and $-CH_2N^+(CH_3)_3$, respectively for Pb removal from water. The effect of the functional groups on nano-ZrP dispersion and the effect of ZrP immobilization on the mechanical strength of the resulting nanocomposites were investigated. It was found that the presence of the charged functional groups ($-SO_3^-$ and $-CH_2N^+(CH_3)_3$) was more favorable than the neutral $-CH_2Cl$ group to improve nano-ZrP dispersion (i.e., to achieve smaller ZrP nanoparticles). It was also revealed that ZrP–N and ZrP–S were more efficient than ZrP–Cl for Pb removal. Results showed that ZrP–S exhibited a higher preference toward lead ion at high calcium levels as compared to ZrP–N, as a result of the potential Donnan membrane effect. The results also shed some light on the generation of environmental nanocomposites with high capacity and excellent mechanical strength because it was found that nano-ZrP immobilization would enhance the compressive strength and improve the wear performance of the resulting nanocomposites with the ZrP loadings up to 5 wt%.

A ZVI nanoparticle graphene composite (G-nZVI) was prepared via sodium borohydride reduction of graphene oxide and iron chloride under an argon atmosphere (Jabeen et al., 2013). The adsorption experiments showed maximum Pb(II) adsorption capacity for the G-nZVI composite with 6 wt% graphene oxide loading. The adsorption of Pb(II) ions followed a pseudo-second-order kinetic model, and adsorption isotherms could be described using the Freundlich equations. It was also found that G-nZVI had great promise as an efficient adsorbent for lead immobilization from water, as it exhibited stability, reducing power, and a large surface area. The added advantage was that the material was magnetically separable.

Recillas et al. (2011) reported the synthesis of nanoparticle (NP) suspensions of CeO_2, Fe_3O_4, and TiO_2 and tested for Pb removal in water cleaning processes. The study showed that the results obtained were promising for lead removal via adsorption. In this study, the adsorption capacity obtained for the NPs was: 189 mg Pb/g NPs CeO_2, 83 mg Pb/g NPs Fe_3O_4, and 159 mg Pb/g NP TiO_2. The toxicity of the synthesized NPs, NPs after lead adsorption, and the supernatant after NPs' separation was also examined. To study the interaction with living organisms and prevent future environmental damages, Germination test in Tomato (*Lycopersicom esculentum*), Lettuce (*Lactuca sativa*), Cucumber (*Cucumis sativus*) seeds, and the Microtox® assay, based on the use of bioluminescent marine bacterium, *Photobacterium phosphoreum/Vibrio fischeri*, have been conducted. The CeO_2 NPs showed a high level of Pb removal although presented high phytotoxicity and the TiO_2 NPs inhibited the lead toxicity against the marine bacterium. It was also observed that the media used to stabilize the NPs (tetramethylammonium hydroxide and hexamethylenetetramine) presented a significant reduction in the germination index. The study also revealed that TiO_2 and Fe_3O_4 NPs did not exhibit any toxicity, and they could be used as adsorbents for Pb(II) removal.

Rahimi et al. (2015) synthesized needle-like lepidocrocite nanoparticles with a diameter of 10–30 nm and a length of 100–300 nm, and goethite nanospheres with a diameter of 10–60 nm, and finally examined their potentiality as adsorbents for Pb removal from solutions. It was found that the most influential factor in the batch adsorption experiments by nanoparticles was the adsorbent dose and the type of adsorbent. The optimum condition for Pb removal by nanoparticles was obtained as C_0 = 5 mg/L, pH = 5.2 for goethite as the more efficient adsorbent. The maximum experimental adsorption capacity of Pb^{2+} for goethite was ~1.64 times higher than lepidocrocite. Experimental data best fitted with Langmuir and Sips isotherm model. The type of nanosorbent was the most important factor to obtain maximum Pb(II) removal.

Wang et al. (2007a) reported the preparation and application of manganese oxide-coated carbon nanotubes (MnO_2/CNTs) for Pb(II) removal from aqueous solution. Langmuir isotherm model fitted well to the experimental data and the maximum adsorption capacity was reported as 78.74 mg/g. It was found that the adsorption efficiency decreased with the decrease in pH value. Multiwalled CNTs were applied by Xu et al. (2008) for the lead removal purpose. Graphene nanosheets and SiO_2 modified graphene nanocomposites, which were further converted to graphene oxide, have also been reported as the potential adsorber of lead from water bodies (Huang et al., 2011; Hao et al., 2012). He et al. (2011) utilized GO coupled with chitosan (CS) for lead adsorption purposes, while Fan et al. (2013) used magnetic GO-CS for the same purpose. Madadrang et al. (2012) showed that incorporating various functional groups to the GO can enhance lead removal capacity significantly.

Pristine and acid-modified multiwalled CNTs were prepared by soaking in the HNO_3 solution for different durations (Wang et al., 2007b). After that, the CNTs were annealed in an argon flow at a temperature of 800°C for 2 h. Thus, the modified CNTs were applied for Pb(II) adsorption and it was found that the maximum adsorption capacities of the acidified, annealed, and pristine multiwalled CNTs were 91, 22.5, and 7.2 mg/g, respectively.

The exploitation of ZVI or Fe(0)) for *in situ* remedial treatment has been prolonged for different kinds of contaminants (Ponder et al., 2000). The ZVI removes contaminants present in aqueous media employing reduction. Ferrogel-supported ZVI has been assigned for the removal of Cr(VI) and Pb(II) by reducing to Cr(III) and Pb(0).

6.3.5 Mercury Removal

Mercury is another notable toxic heavy metal often detected in water bodies. It is very dangerous even if it is present at a low concentration. Deleterious effects due to mercury include damage to the nervous, immune, and digestive systems. Hence Hg(II) remediation constitutes an important topic of present-day environmental clean-up programs. Like the previously mentioned heavy metals, there are several reports on Hg(II) removal via nanoparticle-based technology. Parham et al. (2012) applied 2-mercaptobenzothioazole-modified magnetic iron oxide nanoparticles to remove mercury from water samples. Within a short period, the modified magnetic iron oxide nanoparticles could remove 98.6% Hg(II) with an initial concentration of 50 ng/mL. Moreover, a variation of pH and the presence of an electrolyte (e.g., NaCl) did not show any considerable effect on the removal efficiency. The maximum adsorption capacity was found as 590 µg/g. Wang et al. (2020) synthesized a covalent organic framework that supported Ag nanoparticles and applied the same for Hg(II) adsorption from acidic wastewater. The composite showed excellent performance (removal rate of 99%), high selectivity, stability, and reusability. From various characterization studies and DFT calculations, it was found that during the adsorption process the organic framework supplied electrons to Hg(II) via Ag nanoparticles, which led to the formation of an amalgam.

Fakhri (2015) applied copper oxide nanoparticles for the removal of mercury from aqueous media. The adsorption process was optimized using response surface methodology, and the reactions were conducted based on the Box–Behnken design. The influence of adsorbent dose, pH, and temperature on the mercury removal was studied, and the interactions among the independent parameters were evaluated using ANOVA. The optimum pH, adsorbent dose, and temperature were reported as 9, 0.05 g, and 278 K, respectively.

6.3.6 Copper Removal

Although copper is not often counted as a severely toxic heavy metal like those discussed earlier, its presence in water bodies beyond a certain level is undesirable, and hence its removal is necessary. Suganthi and Ramani (2017) modified the surface of iron oxide nanoparticles using iron precursor and bacterial supernatant (*S. thermolineatus*) and applied the same for Cu(II) adsorption from the effluent of pigment industry. Thus, synthesized surface-modified iron oxide nanoparticles showed removal efficiency up to 85% ± 4% from the effluent of the pigment industry. Liu et al. (2019) applied magnetic ferrite nanoparticles for Cu adsorption from copper-ammonia wastewater. Under optimized conditions, the maximum adsorption capacity of the nanoadsorbent reached up to 124.8 mg/g. The experimental data showed that the adsorption process followed pseudo-second-order kinetics and the Langmuir isotherm model. Neeraj et al. (2016) carried out copper sequestration using chitosan-coated magnetic nanoparticles. From the thermodynamic parameters, it was concluded that the process was energetically feasible, spontaneous, and exothermic. Moreover, the predominant mechanism behind the removal of Cu(II) ions was physisorption.

Apart from adsorption, as already mentioned, nanoparticles also can improve the removal efficiency of the membrane process. Ghaemi (2016) used ɣ-alumina nanoparticles in the nanocomposite membrane for the improved removal of Cu(II) ions. The mixed matrix membrane was prepared using PES and alumina nanoparticles in different amounts by the phase inversion method. Water permeation, porosity, and the hydrophilicity of the mixed membrane were increased compared with the pristine membrane, and consequently, Cu removal also improved. In Table 6.2, various nanoadsorbents with their performance for removing various metal ions by nanomaterials have been tabulated.

TABLE 6.2
Removal of Various Heavy Metals by Nanomaterials

Metal Removed	Nanomaterial Description	Favorable Condition	Maximum Adsorption Capacity (mg of metal/g of nanomaterial)/ Efficiency	References
As	Fe-hydrotalcite supported magnetite nanoparticle	pH 9 and equilibrium time of 15 min	1281.3 μg/g for As(V) and 121.4 μg/g for As(III)	Turk and Alp (2014)
As	Tea waste supported magnetic iron oxide nanoparticles	pH 7, dose 1 g/L	188.69 mg/g for As(III) and 153.8 mg/g for As(V)	Lunge et al. (2014)
As	Amorphous zirconium oxide nanoparticles	pH 7 for As(III) and pH 2 for As(V)	83 mg/g for As(III) and 95 mg/g for As(V)	Cui et al. (2012)

(Continued)

TABLE 6.2 (Continued)
Removal of Various Heavy Metals by Nanomaterials

Cd	Biosynthesized iron oxide nanoparticles	pH 8.07, dose 2.5 g/L, temp 45°C, ionic strength 0.07 mol/L	3.94 mg/g	Lin et al. (2018)
As	Superparamagnetic Fe_3O_4 nanoparticles	Acidic pH	16.56 mg/g for As(V) and 46.06 mg/g for As(III)	Feng et al. (2012)
As	Cupric oxide nanoparticles	pH 9.3 for As(III)	26.9 mg/g for As(III) and 22.6 mg/g for As(V)	Martinson and Reddy (2009)
Pb and Cd	EDTA functionalized magnetic nanoparticle	pH 3–10	79.4 mg/g for Cd and 100.2 mg/g for Pb	Huang and Keller (2015)
Pb and Cd	Biochar-supported hydrated manganese oxide	Near neutral pH	67.9 mg/g for Pb and 22.3 mg/g for Cd	Wan et al. (2018)
Cd	Fe_3O_4 loaded sawdust carbon	pH 6, dose 2 g/L, equilibrium time 30 min	63.3 mg/g	Kataria and Garg (2018)
Cd	Alumina nanoparticles	_	0.68 mg/g	Koju et al. (2018)
Cr(VI)	Hexamethylenetetramine (HMT) stabilized CeO_2 nanoparticles	Neutral condition	121.95 mg/g	Recillas et al. (2010)
Cr(VI)	Nanoscale zero-valent iron decorated graphene(G-nZVI)	Acidic condition	162 mg/g	Jabeen et al. (2011)
Cr(VI)	Maghemite nanoparticles	Acidic condition	97.3%	Hu et al. (2005)
Cr(VI)	Maghemite nanoparticles	Acidic as well as neutral condition	_	Jiang et al. (2013)
Cr(VI)	Magnetic graphene nanocomposites	Acidic condition	_	Zhu et al. (2012)
Cr(VI)	Ethylenediamine dimethyl formamide reduced graphene oxide (ED-DMF-RGO) composite	_	92.15 mg/g	Zhang et al. (2013a)
Pb(II)	TiO_2	_	159 mg/g	Recillas et al. (2011)
Pb(II)	Manganese oxide-coated carbon nanotubes (MnO_2/CNTs)	Neutral condition	78.74 mg/g	Wang et al. (2007a)

(Continued)

TABLE 6.2 (Continued)
Removal of Various Heavy Metals by Nanomaterials

Metal Removed	Nanomaterial Description	Favorable Condition	Maximum Adsorption Capacity (mg of metal/g of nanomaterial)/ Efficiency	References
Pb(II)	Water-dispersible magnetic chitosan/ graphene oxide composite	pH 5	50.23 mg/g	Fan et al. (2013)
Pb(II)	Acidified, annealed and pristine multiwalled CNTs	–	91 mg/g	Wang et al. (2007b)

6.3.7 Selective Removal of Heavy Metals

In most practical scenarios, various heavy metal ions are present in the wastewater. Therefore, it is necessary to develop materials for the selective removal of heavy metals from water. During the last decade, various nanoparticles-aided methods have been documented as promising techniques for the selective abatement of heavy metals. Fato et al. (2019) tested the ultrafine magnetite nanoparticles for the removal of four types of heavy metals (lead, cadmium, copper, and nickel) from river water. From the mixtures of four metal ions, the removal efficiency toward lead was highest (86%) and least for nickel (54%). Hu et al. (2006) prepared maghemite nanoparticles by sol-gel method and applied the same for selective adsorption of heavy metals like Cr, Cu, and Ni from wastewater. The adsorption process was highly dependent on pH, which mainly controls the selective adsorption process. The pH for the Cr, Cu, and Ni adsorption experiments was kept as 2.5, 6.5, and 8.5, respectively. Experimental data fitted well with the Langmuir isotherm model and the maghemite nanoparticles showed excellent reusability after adsorption. Removal of Cr and Cu occurs via electrostatic attraction and ion exchange mechanism, while the removal of Ni was solely due to the electrostatic attraction. Li et al. (2003) tested nitric acid-treated multiwalled carbon nanotubes for the adsorption of Pb^{2+}, Cu^{2+}, and Cd^{2+}. It was noticed that the affinity of the nanoadsorbent was best at Pb^{2+} removal, followed by Cu^{2+} and Cd^{2+}. The maximum adsorption capacity for the lead was found as 97.08 mg/g, whereas it was 24.49 mg/g and 10.86 mg/g for Cu^{2+} and Cd^{2+}, respectively. Sun et al. (2018) reported the preparation of novel nanocomposite by doping hybrid Mg and Zr oxides onto the negatively charged polystyrene and employed the same for the selective adsorption of Cu from highly acidic wastewater. The newly prepared adsorbent showed nearly 20 times more selectivity toward Cu removal and was able to treat 4650 kg of wastewater per kg of the adsorbent. Xiang-Feng et al. (2014) pre-treated silica fumes with nitric acid and modified the surface of the silica nanoparticles by various modifying agents such as (3-mercaptopropyl) triethoxysilane (MPTES) and (3-aminopropyl) trithoxysilane (APTES) for the selective adsorption of heavy metals like Pb, Cu, Hg, Cd, and Zn from aqueous solution. It was observed that the silica modified with MPTES resulted in the maximum removal of Cu and Pb, while the silica modified with APTES resulted in the removal of Pb and Hg.

Youssef and Malhat (2014) prepared TiO_2 nanowires by hydrothermal methods and applied the same for the adsorption of heavy metals viz., Pb, Cu, Fe, Cd, and Zn. It was noticed that the maximum removal occurred in the case of Pb (97.06%) while the least was noticed in the case of Zn (35.18%). Chen et al. (2011) reported excellent selective removal of Cr(VI) species using MCM-41 magnetic nanoadsorbents. Magnetic iron oxide nanoparticles were entrapped within the matrix of MCM-41 to form the nanocomposite. The material showed promising Cr(VI) removal efficiency in distilled water, mountain, and tap water samples. Although the calcium ions present in the natural water system inhibited the adsorption of Cr(VI), to some extent it was still comparable with other reported studies. The maximum adsorption capacity reached up to 100 mg/g. Carboxymethyl-β-cyclodextrin modified Fe_3O_4 nanoparticles were deployed for the selective adsorption of Pb, Cd, and Ni from water (Badruddoza et al., 2013). The maximum uptake for Pb was reported to be 64.5 mg/g, while that for Cd and Ni was 27.7 and 13.2 mg/g, respectively. Experimental data fitted well with the Langmuir isotherm model and pseudo-second-order kinetic model. The affinity of the nanoadsorbent toward different metal ion species was in the order of Pb>>Cd>Ni. The high affinity toward lead can be explained by hard acid and soft base theory. Here hard is referred to as the species that possesses the low size and high charge, while soft refers to the species which has a relatively large size and low charge.

Acid-modified CNTs were synthesized by Stafiej and Pyrzynska (2007), and after filtering through a 0.45 μm filter, it was applied for heavy metal adsorption purposes. It was observed that the affinity for different metal species was in the order of Cu(II) > Pb(II) > Co(II) > Zn(II) >Mn(II).

6.4 DIFFERENT CHARACTERIZATION ANALYSIS OF NANOPARTICLES AND REMOVAL MECHANISM STUDIES

6.4.1 Fourier Transform Infrared Spectroscopy (FTIR)

Fourier transform infrared spectroscopy (FTIR) was carried out to determine the functional groups present in the nanocomposite. From the sharp peak at 2360 cm^{-1} Kataria and Garg (2018) concluded that it was due to the bond vibration mode of iron oxide. Peaks due to the stretching vibration of the FeO group are also noticed at 687, 563, and 416 cm^{-1}. Rahimi et al. (2015) denoted the peaks at 466 and 746 cm^{-1}, which were due to the out-of-plane bending vibration of Fe-OH. In another report, the peaks corresponding to Fe-O stretching were reported at 539.2 and 526.8 cm^{-1} (Jiang et al., 2013). In most of the FTIR spectra, a broad peak was often noticed in the region of 3500 cm^{-1}, which was due to the presence of OH groups. Apart from this peak, a peak at 1600 cm^{-1} was also assigned to the stretching vibration of OH. Peaks at 2863 and 2774 cm^{-1} are commonly associated with the stretching vibration of the C-H bond. Peaks at 1716 and 1561 cm^{-1} are mostly due to the C = O bond owing to the presence of lignin, hemicellulose, etc.

From the FTIR spectra, Koju et al. (2018) concluded that the substitution of OH groups occurred on the alumina nanocomposite during the Cd removal process. Peaks due to alumina were found at ~500–750 cm^{-1} (due to AlO_6) and around 800 cm^{-1} (due to AlO_4). Goswami et al. (2012) reported the peak at 517 cm^{-1} corresponding to Cu-O. While investigating the mechanism behind the As removal by ZrO_2 nanoparticles, Cui et al. (2012) studied the FTIR spectra. After arsenic removal, a new peak appeared at 621 cm^{-1}, which was assigned to the stretching vibration of uncomplexed As(III)-O bond.

6.4.2 Scanning and Transmission Electron Microscopic (SEM and TEM) Analysis

Morphological analysis of the nanomaterials was carried out using SEM and transmission electron microscopy (TEM) study. In the study of As removal by Cui et al. (2012), the FESEM images of the amorphous ZrO_2 indicated that those nanoparticles possessed a highly porous structure, and the aggregation was highly beneficial for As removal. On the other hand, results from the TEM study conformed to the XRD pattern. The material was poorly crystalline and had a non-uniform shape. It was found that the average size of the nanoparticles was ~5 nm. Turk and Alp (2014) also conducted a TEM study to better understand the size and nucleation of the Fe-hydrotalcite nanoparticles. It was revealed that the magnetic nanoparticles were approximately spherical in appearance with an average diameter of ~50 nm. The TEM image has been shown in Figure 6.5. From the SEM images of Fe_3O_4 and sulfonated magnetic nanoparticles (Fe_3O_4-SO_3HMNPs), the size of the nanoparticles was found to be ~ 80 nm (Chen et al., 2017). Al-Qahatani (2017) carried out the SEM analysis of the zero-valent silver nanoparticles in the context of Cd remediation. Electron microscopic studies were conducted to get an insight into particle growth and size.

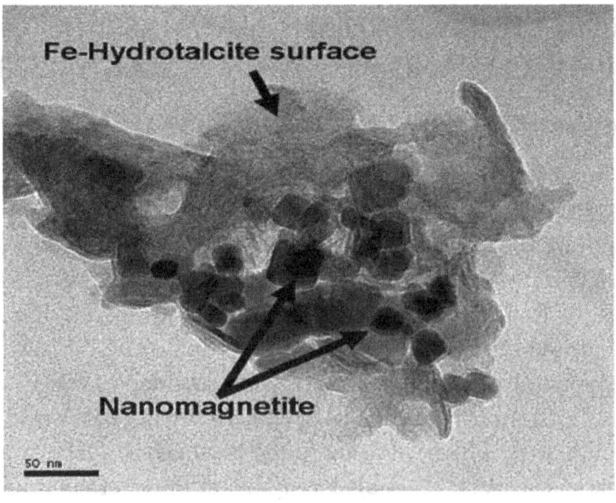

FIGURE 6.5 TEM images of NM and FeHT. (Reprinted with permission from (Turk & Alp, 2014).)

Agglomeration among the nanoparticles during the adsorption-desorption could also be explained using TEM images (Recillas et al., 2010).

From the micrographs of the iron oxide prepared by tangerine peel extract, the effect of the concentration of the peel extract on the iron oxide preparation was revealed (Ehrampoush et al., 2015). Tangerine peel extract was varied in the range of 2–10%. In the SEM images of *Phanerochaete chrysosporium*-loaded TiO_2 nanoparticles (PTNs), it was clear that the nanoparticles were surrounded by *P. chrysosporium* hypae with a favorable network for Cd removal. From the TEM image, the coating of the nanoparticles was displayed (Chen et al., 2013).

6.4.3 X-Ray Diffraction (XRD) Analysis

XRD studies are carried out to discover the crystallinity of the metal nanoparticles involved. To gain insight into the crystallinity of the structure of the sawdust, Kataria and Garg (2018) conducted an XRD analysis of the sawdust, Fe_3O_4-loaded sawdust, and EDTA-modified Fe_3O_4 loaded sawdust. It was interesting to observe that before magnetization, broad diffraction peaks were identified at 22° and 26.22° for the sawdust, which indicated the presence of lignocellulose and amorphous carbon. After the inclusion of Fe_3O_4, new diffraction peaks occur at 30.2°, 35.24°, 43.2°, 57.3°, and 67.7°, which confirmed the presence of Fe_3O_4 nanoparticles. But after EDTA modification, no new peak in the XRD spectrum was obtained. The XRD pattern has been shown in Figure 6.6. In the work of Cd removal by glycerol-modified alumina nanoparticles, XRD patterns of the synthesized and modified alumina were recorded before Cd adsorption (Koju et al., 2018). In both the XRD spectra, sharp peaks were recorded. Moreover, the sharp peaks in the glycerol-modified alumina revealed that a phase change of γ-alumina to α-alumina occurred.

From the peaks of the XRD spectra of Cu(II) oxide nanoparticles used for As remediation, the presence of single-phase monoclinic Cu(II) oxide was confirmed (Goswami et al., 2012). Additional peaks at 16.2° and 32.4° indicated the presence of $Cu_2Cl(OH)_2$.

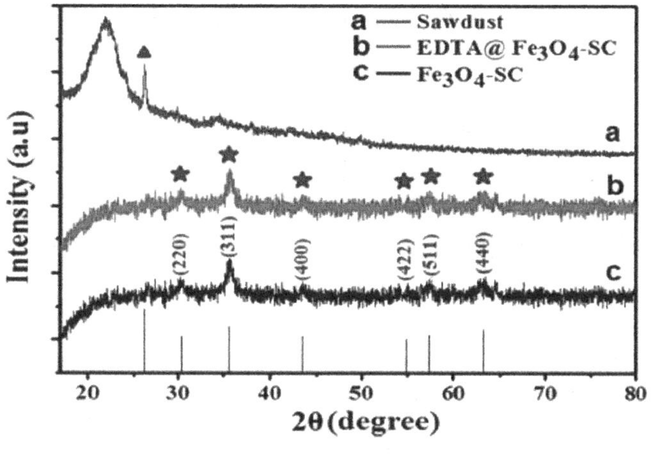

FIGURE 6.6 XRD pattern of raw sawdust, Fe_3O_4/SC ncs and EDTA@ Fe_3O_4/SC ncs. (Reprinted with permission from (Kataria & Garg, 2018).)

6.4.4 X-ray Photoelectron Spectroscopy (XPS) Analysis

XPS technique helps to determine the oxidation states of the metal in the nanocomposite. It also helps in determining the interaction between the adsorbent and the heavy metal. Wan et al. (2018) explained the adsorption of Pb(II) and Cd(II) by biochar-supported hydrated manganese oxide nanoparticles with the help of XPS analysis. The spectra of Pb $4f_{7/2}$ were compared in the case of Pb removal, and $Pb(NO_3)_2$ was used as the reference for comparison of the peak shift in the XPS spectra. In XPS spectra of both $Pb(NO_3)_2$ and biochar-loaded Pb (BC-Pb), the peak appeared almost at the same place (~139 eV). However, in the case of manganese oxide-loaded biochar after Pb adsorption (HMO-Pb), the peak shifted toward the left by 1°. From this, it is implied that the interaction between Pb and HMO is stronger than the outer sphere complexation mechanism between Pb and the oxygen group of biochar.

Regarding the composition of the biosynthesized iron oxide, Lin et al. (2018) confirmed the presence of iron from the peaks at 710.6 and 724.3 eV (corresponding to the binding energies of $Fe2p_{3/2}$ and $Fe2p_{1/2}$), respectively. The absence of any additional satellite peak around 719 eV confirms the phase purity of Fe_3O_4 in the iron oxide nanoparticles matrix. However, after the reaction a peak appeared at 719 eV, which may be due to the oxidation of Fe_3O_4. Moreover, the peaks at 530.1 eV may be due to the lattice oxygen of Fe_3O_4.

Geng et al. (2009) explained, with the help of XPS analysis, the Cr(VI) removal by chitosan-supported ZVI. From the XPS spectra, the Cr $2p_{3/2}$ and Cr $2p_{1/2}$ peaks at 577.1 eV and 586.9 eV are well distinguished. The spin-orbit splitting of 9.8 eV indicated that the mechanism was reducing Cr(VI) to Cr(III) by ZVI followed by adsorption. The $Fe2p_{3/2}$ and $Fe2p_{1/2}$ peaks from the XPS spectra were found at 711 and 725 eV, respectively. It indicated that iron existed in a +3 state. The XPS spectra have been shown in Figure 6.7. In the study of Cr(VI) remediation by maghemite

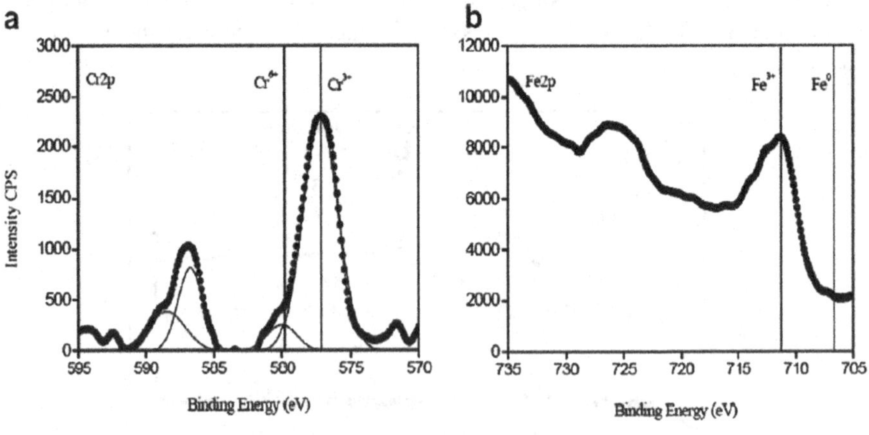

FIGURE 6.7 The (a) high-resolution Cr 2p XPS spectrum and (b) high-resolution Fe 2p XPS spectrum of chitosan-Fe^0. For Cr 2p XPS spectrum, points are experimental data and lines are the results of quantitative Gaussian curve fitting. The two lines indicate the Cr $2p_{3/2}$ peak positions of a Cr(III) species and a Cr(VI) species at 577 and 579 eV, respectively. (Reprinted with permission from (Geng et al., 2009).)

nanoparticles, Hu et al. (2005) reported the occurrence of peaks of $Cr2p_{3/2}$ and $Cr2p_{1/2}$ at 579.8 eV and 587.2 eV, respectively. This indicates that the entire chromium was in the +6 state. It also implied that no chemical redox reaction had taken place. The stability of the nanoparticles was also confirmed.

6.4.5 Other Characterization Techniques

Apart from the characterization techniques already discussed in the previous sections, some other analyses such as TGA analysis, BET analysis, dynamic light scattering (DLS) particle size measurement, etc. are often carried out to get more ideas regarding the heavy metal remediation. TGA analysis is carried out when there is a significant loss in weight with the rise of temperature. Tabesh et al. (2018) showed from the TGA-DTA analysis that within 50°C–120°C there could be a little weight loss due to dehydration. However, after that, a rapid weight loss is observed between 250°C and 400° C. BET analysis was conducted to get an idea about the pore size distribution of the nanoparticles. Chowdhury and Yanful (2013) measured the surface area of the magnetite-maghemite particles by BET analysis and found that as 49.5 m^2/g. The size range covered by the maximum fraction of the nanoparticles can be obtained from the particle size analysis. From the DLS experiment, Recillas et al. found the mean size of the nanoparticles as ~11.7 ± 1.6 nm (Recillas et al., 2010).

6.5 REGENERATION AND REUSABILITY

Regeneration and reusability play an important role in the economics of the heavy metal removal process. A material that can be reused several times without any loss of efficiency is preferred in the long run. For carbon-based nanomaterials, desorption is often achieved by altering the solution pH. In many cases, the metal recovery reaches up to ~100% at pH <2 (Li et al., 2005; Lu et al., 2006). Lu et al. (2007) reused CNTs 11 times for Zn removal without any appreciable loss in adsorption capacity. In many cases, adsorption power is not lost, while in some cases it may get reduced significantly (Deliyanni et al., 2003; Hu et al., 2006).

Hu et al. (2005) found in their investigation that the adsorption of Cr(VI) on maghemite nanoparticles is highly pH-dependent; hence, the desorption of Cr(VI) is possible by controlling the pH. In the study, 0.01 M NaOH was found to be the most effective eluent for the desorption of Cr(VI). It was also noticed that the capacity of maghemite for Cr(VI) adsorption remained almost constant up to six cycles. This indicated that there were no irreversible sites on the surface of the adsorbent.

Huang and Keller (2015) investigated the recovery of Cd^{2+} sorbed onto the regenerable magnetic ligand particle (Mag-Ligand) using a 1% HCl wash to demonstrate the renewability and reusability of the Mag-Ligand. The study of Cd^{2+} removal and Mag-Ligand recovery was conducted during five continuous cycles of regeneration and reuse. Interestingly, the maximum portion of the sorbed Cd^{2+} (>80%, over 8 mg/L) could be recovered, which indicates the easy regeneration of Mag-Ligand. However, some loss (within 10% change) of Cd^{2+} sorption capacity occurred for the regenerated Mag-Ligand after five cycles.

Kataria and Garg (2018) carried out some studies on the desorption of Cd(II) from EDTA-modified iron-oxide-loaded sawdust carbon (EDTA@Fe_3O_4/SC) nanocomposite with five different desorbing solutions. They found that the acidic desorbing solutions HCl, HNO_3, and H_2SO_4 had the highest desorbing efficiencies in the range 99–100%, while $Ca(NO_3)_2$ and $NaNO_3$ showed comparatively poor performance. It was found that HNO_3 (0.05 M) could be a good desorbing solution for Cd(II). In the study, Cd(II) adsorption efficiency was maintained up to 83% after three reuse cycles. On further reuse of the adsorbent, the efficiency decreased to 75% after five cycles. The results indicated that EDTA@Fe_3O_4/SC nanocomposite had a good reusable potential for metal adsorption, and it was economically efficient for water treatment.

Wan et al. (2018) regenerated the exhausted HMO-BC (hydrated manganese oxide-biochar) using a binary solution of 0.2 M HCl and 4 wt% $CaCl_2$. It was found that the preloaded Pb(II) and Cd(II) were almost completely desorbed after only 10-BV(Bed Volume) of regeneration washing (desorption efficiency >97%). During the investigation of the reusability of HMO-BC, batch adsorption-desorption cycles were performed, and the results suggested that HMO-BC could be consecutively used to remove Pb(II) and Cd(II) for at least five runs without any significant loss in adsorption capacity.

Paris et al. (2020) synthesized a novel nanocomposite adsorbent faujasite: cobalt ferrite. They also evaluated the nanocomposite regeneration capacity by magnetic recovery after 24 h suspension, and this presented a high Pb^{2+} ion adsorptive capacity (98.4%) in the first cycle. They found that around 98% of the Pb^{2+} ions were adsorbed in the second cycle. The experimental data suggested that the synthesized faujasite: cobalt ferrite nanocomposite could serve as a promising alternative in adsorption processes, aiming at a synergic effect high adsorptive activity of FAU zeolite and the magnetic activity of cobalt ferrite nanoparticles. The adsorbent recovery from the aqueous medium was made via magnetic force after successive adsorption cycles.

Shin et al. (2011) have found in the recycling study that there was almost no reduction in removal efficiency (>97%) of Cr^{3+} while nitrogen-doped magnetic carbon nanoparticles (N-MCNPs) were reused up to five times. The N-MCNPs in the aqueous solution were readily separated by an external magnetic field. Their study revealed that N-MCNPs could be a potential and highly efficient reusable and recoverable adsorbent for heavy metal ion removal.

Zhang et al. (2013b) found in their study that mercury adsorbed on the sorbents could be desorbed with 1 M HCl containing 3 wt.% of thiourea. Moreover, the sorbents showed good reusability. Tabesh et al. (2018) found in their study that adsorbent regeneration could be possible up to three cycles, maintaining almost constant adsorption capacity. The study suggested that the synthesized Al_2O_3 nanoparticles could be considered a potential adsorbent for Pb(II) and Cd(II) removal.

Godiya et al. (2019) used bilateral hydrogel-containing carboxymethyl cellulose (CMC) and polyacrylamide (PAM) for adsorption of heavy metals. After the adsorption of the metal ions, dried hydrogel (20 mg) was immersed into the EDTA solution (10 mL, 0.1 M) for 30 min and then washed with the DI water three times (3 × 100 mL) to regenerate the sorbent. The regenerated hydrogel was then reused for the next adsorption experiment and found that after three cycles, the adsorption capacities of the hydrogel toward the metal ions were still well-maintained with a slight decrease after each cycle. It was also found that the adsorption efficiencies for the pristine CMC hydrogel decreased from 38.1% to 28.8%, 29.1% to 14.9%, and 35.5% to 30.6% for the Cu(II), Pb(II), and Cd(II) ions, respectively, and the adsorption efficiencies for the CMC/PAM hydrogel decreased from 73.8% to 67.1%, 90.9% to 83.3%, and 70.2% to 63.3% for the Cu(II), Pb(II), and Cd(II) ions, respectively, after the first cycle. The study suggested that the prepared hydrogels had an excellent regeneration property, which might render the potential for practical application on an industrial scale. In their study, Suman et al. (2015) found that the prepared composite could be reused up to five cycles and all the dyes and heavy metals removed could be re-collected just by allowing the different pH solutions to pass through the column.

6.6 FUTURE SCOPE AND CHALLENGES

Nanomaterial-mediated technologies have immense prospects in wastewater treatment. The field of nanoscience and nanotechnology has been continuously offering us newer materials and innovative technologies, which may be very fruitful for the upcoming generation. Since the discovery of graphene in the early 21st century, various graphene-supported nanomaterials are being developed for environmental remediation purposes. Similarly, the discovery of MXenes (a 2D material) in 2011 has opened the door for many more new materials. Ihsanullah (2020) documented in the recent review the progress of MXenes mediated nanoparticles for environmental remediation purposes. Some exceptional properties of MXenes, such as possession of the high surface area, activation of metallic hydroxide sites, ease of functionalization, and biocompatibility, have made the materials unique. Due to all these properties, it has now become one of the most promising emerging candidates for wastewater treatment. Because of its excellent adsorbing and reducing property, it has already proved its efficiency in the field of heavy metal remediation. Ying et al. (2015) applied $Ti_3C_2T_x$ to remove hexavalent chromium from water bodies through reduction followed by adsorption. Other research shows that various other metals such as Hg, Cd, Cu, etc. can also be removed using MXenes. Siddiqui et al. (2019) presented the latest technologies for arsenic remediation using nanomaterials in the recently published review. Various nanoparticles have already been proved to be efficient in the past for arsenic removal. Although many of them can be easily regenerated and reused, yet they suffer from various limitations such as possession of less active sites on the surface, ineffectiveness toward low arsenic concentration. In an effort to overcome these limitations, an attempt has been made to dope the virgin metal nanoparticles with various other metals such as copper, manganese, lanthanide, cerium, etc. Due to such modification, a magnetic property was developed, which was effective toward As(III) remediation. Organic acids made another novel modification of the virgin nanoparticles. This led to the incorporation of various functional groups on the surface. The presence of various functional groups often helped in the oxidation of As(III) to As(V) and also helped in trapping arsenic at various pH. Nata et al. (2011) removed As(V) from water through the application of amine-rich Fe_3O_4/bacterial cellulose nanocomposite.

While the introduction of nanoparticles has opened a new door for several technologies, it is also true that there are various limitations associated with these nanoparticles. There are several risks associated with the use of nanoparticles. Moreover, many biological data indicated the toxicity of nanoparticles toward various organisms. After the use, they are also considered potential waste and need to be treated using proper solid waste management techniques to eradicate the problem. A problem often arises due to polydispersity, size and shape diversity, and reproducibility in the method (Ali et al., 2016). Issues also arise regarding the use of nano ZVI, since it tends to get oxidized under ambient atmospheric conditions. Moreover, due to poor mechanical strength and excess pressure drop during column operation, they are often not considered adsorbent for column studies (Hua et al., 2012). Compared to the traditional bulk materials, their cost of production is often high. But on the other hand, large-scale production can yield an optimal balance between cost and benefit. In most cases, the synthesis of nanoparticles needs a suitable template, and its efficiency is also increased in the presence of support (Wan et al., 2018). However, this support material may incur an extra cost. In some instances, difficulty is felt when separating spent adsorbent from the treated solution. This difficulty has forced scientists to combine the adsorption technique with the filtration technique in the same chamber (Nasir et al., 2019).

Declaration of interest: None

ACKNOWLEDGMENT

Authors express their appreciation to the Indian Institute of Technology (IIT) Kharagpur for providing the requisite facilities and to the Ministry of Human Resource Development, Government of India for financial support.

REFERENCES

Ahmadi, M., M. Foladivanda, N. Jafarzadeh, B. Ramavandi, S. Jorfi and B. Kakavandi. 2017 Synthesis of chitosan zero-valent iron nanoparticles-supported for cadmium removal: characterization, optimization and modeling approach. *J. Water Supply Res. Technol.* 66: 116–130. doi:10.2166/aqua.2017.027

Ali, A., H. Zafar, M. Zia, I. Haq, A. R. Phull, J. S. Ali and A. Hussain. 2016 Synthesis, characterization, applications and challenges of iron oxide nanoparticles. *Nanotechnol. Sci. Appl.* 9: 49–67. doi:10.2147/NSA.S99986

Alidokht, L., A. R. Khataee, A. Reyhanitabar and S. Oustan. 2011 Reductive removal of Cr(VI) by starch stabilized Fe^0 nanoparticles in aqueous solution. *Desalination*, 270: 105–110. doi:10.1016/j.desal.2010.11.028

Al-Qahtani, K. M. 2017 Cadmium removal from aqueous solution by green synthesis zero valent silver nanoparticles with *Benjamina* leaves extract. *Egypt J. Aquat. Res.* 43: 269–274. doi:10.1016/j.ejar.2017.10.003

Badruddoza, A. Z. M., Z. B. Z. Shawon, T. W. J. Daniel, K. Hidajat and M. S. Uddin. 2013 Fe_3O_4/cyclodextrin polymer nanocomposites for selective heavy metals removal from industrial wastewater. *Carbohydr. Polym.* 91: 322–332. doi:10.1016/j.carbpol.2012.08.030

Bai, R. S. and T. E. Abraham. 2002 Studies on enhancement of Cr(VI) biosorption by chemically modified biomass of *Rhizopus nigricans*. *Water Res.* 36: 1224–1236. doi:10.1016/S0043-1354(01)00330-X

Bertagnolli, C., A. Uhart, J. Dupin, M. G. C. Silva, E. Guibal and J. Desbrieres. 2014 Biosorption of chromium by alginate extraction products from *Sargassum filipendula*: Investigation of adsorption mechanisms using X-ray photoelectron spectroscopy analysis. *Bioresour. Technol.* 164: 264–269. doi:10.1016/j.biortech.2014.04.103

Cai, G., G. Zhao, X. Wang and S. Yu. 2010 Synthesis of polyacrylic acid stabilized amorphous calcium carbonate nanoparticles and their application for removal of toxic heavy metal ions in water. *J. Phys. Chem. C* 114: 12948–12954. doi:10.1021/jp103464p

Cao, J. and W. Zhang. 2006 Stabilization of chromium ore processing residue (COPR) with nanoscale iron particles. *J. Hazard. Mater. B* 132: 213–219. doi:10.1016/j.jhazmat.2005.09.008

Chen, G., S. Guan, G. Zeng, X. Li, A. Chen, C. Shang, Y. Zhou, H. Li and J. He. 2013 Cadmium removal and 2,4-dichlorophenol degradation by immobilized *Phanerochaete chrysosporium* loaded with nitrogen-doped TiO_2 nanoparticles. *Appl. Microbiol. Biotechnol.* 97: 3149–3157. doi:10.1007/s00253-012-4121-1

Chen, K., J. He, Y. Li, X. Cai, K. Zhang, T. Liu, Y. Hu, D. Lin, L. Kong and J. Liu. 2017 Removal of cadmium and lead ions from water by sulfonated magnetic nanoparticle adsorbents. *J. Colloid Interf. Sci.* 494: 307–316. doi:10.1016/j.jcis.2017.01.082

Chen, X., K. F. Lam and K. L. Yeung. 2011 Selective removal of chromium from different aqueous systems using magnetic MCM-41 nanosorbents. *Chem. Eng. J.*, 172: 728–734. doi:10.1016/j.cej.2011.06.042

Cho, K., X. Wang, S. Nie, Z. Chen and D. M. Shin. 2008 Therapeutic nanoparticles for drug delivery in cancer. *Clin. Cancer Res.* 14: 1310–1316. doi:10.1158/1078-0432.CCR-07-1441

Chowdhury, S. R. and E. K. Yanful. 2013 Kinetics of cadmium(II) uptake by mixed maghemite-magnetite nanoparticles. *J. Environ. Manage.* 129: 642–651. doi:10.1016/j.jenvman.2013.08.028

Contreras, A. R., A. Garcia, E. Gonzalez, E. Casals, V. Puntes, A. Sanchez, X. Font and S. Recillas. 2012 Potential use of CeO_2, TiO_2 and Fe_3O_4 nanoparticles for the removal of cadmium from water. *Desalin. Water Treat.* 41: 296–300. doi:10.1080/19443994.2012.664743.

Cui, H., Q. Li, S. Gao and J. K. Shang. 2012 Strong adsorption of arsenic species by amorphous zirconium oxide nanoparticles. *J. Ind. Eng. Chem.* 18: 1418–1427. doi:10.1016/j.jiec.2012.01.045

Daraei, P., S. S. Madaeni, N. Ghaemi, E. Salehi, M. A. Khadivi, R. Moradian and B. Astinchap. 2012 Novel polyethersulfone nanocomposite membrane prepared by PANI/Fe_3O_4 nanoparticles with enhanced performance for Cu(II) removal from water. *J. Mem. Sci.* 415–416: 250–259. doi:10.1016/j.memsci.2012.05.007

Deliyanni, E. A., D. N. Bakoyannakis, A. I. Zouboulis and K. A. Matis. 2003 Sorption of As(V) ions by akaganeite-type nanocrystals. *Chemosphere.* 50: 155–163. doi:10.1016/S0045-6535(02)00351-X

Ehrampoush, M. H., M. Miria, M. H. Salmani and A. H. Mahvi. 2015 Cadmium removal from aqueous solution by green synthesis iron oxide nanoparticles with tangerine peel extract. *J. Environ. Health Sci. Eng.* 13: 84. doi:10.1186/s40201-015-0237-4

Fakhri, A. 2015 Investigation of mercury (II) adsorption from aqueous solution onto copper oxide nanoparticles: Optimization using response surface methodology. *Process Saf. Environ. Prot.* 93: 1–8. doi:10.1016/j.psep.2014.06.003

Fan, L., C. Luo, M. Sun, X. Li and H. Qiu. 2013 Highly selective adsorption of lead ions by water-dispersible magnetic chitosan/graphene oxide composites. *Colloids Surf B: Biointerfaces.* 103: 523–529. doi:10.1016/j.colsurfb.2012.11.006

Fato, F. P., D. Li, L. Zhao, K. Qiu and Y. Long. 2019 Simultaneous removal of multiple heavy metal ions from river water using ultrafine mesoporous magnetite nanoparticles. *ACS Omega*, 4: 7543–7549. doi:10.1021/acsomega.9b00731

Feng, L., M. Cao, X. Ma, Y. Zhu and C. Hu. 2012 Superparamagnetic high-surface-area Fe_3O_4 nanoparticles as adsorbents for arsenic removal. *J. Hazard. Mater.* 217–218: 439–446. doi:10.1016/j.jhazmat.2012.03.073

Geng, B., Z. Jin, T. Li and X. Qi. 2009 Kinetics of hexavalent chromium removal from water by chitosan-Fe^0 nanoparticles. *Chemosphere* 75: 825–830. doi:10.1016/j.chemosphere.2009.01.009

Ghaemi, N. 2016 A new approach to copper ion removal from water by polymeric nanocomposite membrane embedded with γ-alumina nanoparticles. *Appl. Surf. Sci.* 364: 221–228. doi:10.1016/j.apsusc.2015.12.109

Ghaemi, N., S. S. Madaeni, P. Daraei, H. Rajabi, S. Zinadini, A. Alizadeh, R. Heydari, M. Beygzadeh and S. Ghouzivand. 2015 Polyethersulfone membrane enhanced with iron oxide nanoparticles for copper removal from water: Application of new functionalized Fe_3O_4 nanoparticles. *Chem. Eng. J.* 263: 101–112. doi:10.1016/j.cej.2014.10.103

Gholami, A., A. R. Moghadassi, S. M. Hosseini, S. Shabani and F. Gholami. 2014 Preparation and characterization of polyvinyl chloride based nanocomposite nanofiltration-membrane modified by iron oxide nanoparticles for lead removal from water. *J. Ind. Eng. Chem.* 20: 1517–1522. doi:10.1016/j.jiec.2013.07.041

Godiya, C. B., X. Cheng, D. Li, Z. Chen and X. Lu. 2019 Carboxymethyl cellulose/polyacrylamide composite hydrogel for cascaded treatment/reuse of heavy metal ions in wastewater. *J. Hazard. Mater.* 364: 28–38. doi:10.1016/j.jhazmat.2018.09.076

Gong, J., L. Chen, G. Zeng, F. Long, J. Deng, Q. Niu and X. He. 2012 Shellac-coated iron oxide nanoparticles for removal of cadmium (II) ions from aqueous solution. *J. Environ. Sci.* 24: 1165–1173. doi:10.1016/S1001-0742(11)60934-0

Goswami, A., P. K. Raul and M. K. Purkait. 2012 Arsenic adsorption using copper(II) oxide nanoparticles. *Chem. Eng. Res. Des.* 90: 1387–1396. doi:10.1016/j.cherd.2011.12.006

Guo, D., G. Xie and J. Luo. 2014 Mechanical properties of nanoparticles: basics and applications. *J. Phys. D: Appl. Phys.* 47: 013001. doi:10.1088/0022-3727/47/1/013001

Hamidi, M., A. Azadi and P. Rafiei. 2008 Hydrogel nanoparticles in drug delivery. *Adv. Drug Deliv. Rev.* 60: 1638–1649. doi:10.1016/j.addr.2008.08.002

Hao, L., H. Song, L. Zhang, X. Wan, Y. Tang and Y. Lv. 2012 SiO_2/graphene composite for highly selective adsorption of Pb(II) ion. *J. Colloid Interf. Sci.* 369: 381–387. doi:10.1016/j.jcis.2011.12.023

Haruta, M., T. Kobayashi, H. Sano and N. Yamada. 1987 Novel gold catalysts for the oxidation of carbon monoxide at a temperature far below 0°C. *Chem. Lett.* 405–408. doi:10.1246/cl.1987.405

He, Y. Q., N. N. Zhang and X. D. Wang. 2011 Adsorption of graphene oxide/chitosan porous materials for metal ions. *Chin. Chem. Lett.* 22: 859–862. doi:10.1016/j.cclet.2010.12.049

Hu, J., G. Chen and I. M. C. Lo. 2005 Removal and recovery of Cr(VI) from wastewater by maghemite nanoparticles. *Water Res.* 39: 4528–4536. doi:10.1016/j.watres.2005.05.051

Hu, J., G. Chen and I. M. C. Lo. 2006 Selective removal of heavy metals from industrial wastewater using maghemite nanoparticle: Performance and mechanisms. *J. Environ. Eng.*, 132: 709–715. doi:10.1061/(ASCE)0733-9372(2006)132:7(709)

Hua, M., S. Zhang, B. Pan, W. Zhang, L. Lv. and Q. Zhang. 2012 Heavy metal removal from water/wastewater by nanosized metal oxides: A review. *J. Hazard. Mater.* 211–212: 317–331. doi:10.1016/j.jhazmat.2011.10.016

Huang, Y. and A. A. Keller. 2015 EDTA functionalized magnetic nanoparticle sorbents for cadmium and lead contaminated water treatment. *Water Res.* 80: 159–168. doi:10.1016/j.watres.2015.05.011

Huang, Z., X. Zheng, W. Lv, M. Wang, Q. Yang and F. Kang. 2011 Adsorption of lead(II) ions from aqueous solution on low-temperature exfoliated graphene nanosheets. *Langmuir.* 27: 7558–7562. doi:10.1021/la200606r

Ihsanullah, I. 2020 MXenes (two-dimensional metal carbides) as emerging nanomaterials for water purification: Progress, challenges and prospects. *Chem. Eng. J.*, 388: 124340. doi:10.1016/j.cej.2020.124340

Jabeen, H., V. Chandra, S. Jung, J. W. Lee, K. S. Kim and S. B. Kim. 2011 Enhanced Cr(VI) removal using iron nanoparticle decorated graphene. *Nanoscale* 3: 3583–3585. doi:10.1039/c1nr10549c

Jabeen, H., K. C. Kemp and V. Chandra. 2013 Synthesis of nano zerovalent iron nanoparticles–Graphene composite for the treatment of lead contaminated water. *J Environ. Manage.* 130: 429–435. doi:10.1016/j.jenvman.2013.08.022

Jawor, A. and E. M. V. Hoek. 2010 Removing cadmium ions from water via nanoparticle-enhanced ultrafiltration. *Environ. Sci. Technol.* 44: 2570–2576. doi:10.1021/es902310e

Jiang, W., M. Pelaez, D. D. Dionysiou, M. H. Entezari, D. Tsoutsou and K. O'Shea. 2013 Chromium(VI) removal by maghemite nanoparticles. *Chem. Eng. J.* 222: 527–533. doi:10.1016/j.cej.2013.02.049

Kataria, N. and V. K. Garg. 2018 Green synthesis of Fe_3O_4 nanoparticles loaded sawdust carbon for cadmium(II) removal from water: Regeneration and mechanism. *Chemosphere* 208: 818–828. doi:10.1016/j.chemosphere.2018.06.022

Khan, I., K. Saeed and I. Khan. 2019 Nanoparticles: Properties, applications and toxicities. *Arab. J. Chem.* 12: 908–931. doi:10.1016/j.arabjc.2017.05.011

Klerk, R. J. D., Y. Jia, R. Daenzer, M. A. Gomez and G. P. Demopoulos. 2012 Continuous circuit coprecipitation of arsenic (V) with ferric ion by lime neutralization: Process parameter effects on arsenic removal and precipitate quality. *Hydrometallurgy* 111–112: 65–72. doi:10.1016/j.hydromet.2011.10.004

Koju, N. K., X. Song, Q. Wang, Z. Hu and C. Colombo. 2018 Cadmium removal from simulated groundwater using alumina nanoparticles: behaviors and mechanisms. *Environ. Pollut.* 240: 255–266. doi:10.1016/j.envpol.2018.04.107

Ku, Y. and I. Jung. 2001 Photocatalytic reduction of Cr(VI) in aqueous solutions by UV irradiation with the presence of titanium dioxide. *Water Res.* 35: 135–142. doi:10.1016/S0043-1354(00)00098-1

Li, Y., Z. Di, J. Ding, D. Wu, Z. Luan and Y. Zhu. 2005 Adsorption thermodynamic, kinetic and desorption studies of Pb^{2+} on carbon nanotubes. *Water Res.* 39: 605–609. doi:10.1016/j.watres.2004.11.004

Li, Y., J. Ding, Z. Luan, Z. Di, Y. Zhu, C. Xu, D. Wu and B. Wei. 2003 Competitive adsorption of Pb^{2+}, Cu^{2+} and Cd^{2+} from aqueous solutions by multiwalled carbon nanotubes. *Carbon*, 41: 2787–2792. doi:10.1016/S0008-6223(03)00392-0

Lin, J., B. Su, M. Sun, B. Chen and Z. Chen. 2018 Biosynthesized iron oxide nanoparticles used for optimized removal of cadmium with response surface methodology. *Sci. Total Environ.* 627: 314–321. doi:10.1016/j.scitotenv.2018.01.170

Liu, F., K. Zhou, Q. Chen, A. Wang and W. Chen. 2019 Application of magnetic ferrite nanoparticles for removal of Cu(II) from copper-ammonia wastewater. *J. Alloys Compd.* 773: 140–149. doi:10.1016/j.jaalcom.2018.09.240

Liu, T., X. Yang, Z. Wang and X. Yan. 2013 Enhanced chitosan beads-supported Fe^0-nanoparticles for removal of heavy metals from electroplating wastewater in permeable reactive barriers. *Water Res.* 47: 6691–6700. doi:10.1016/j.watres.2013.09.006

Lu, C., H. Chiu and H. Bai 2007 Comparisons of adsorbent cost for the removal of zinc(II) from aqueous solution by carbon nanotubes and activated carbon. *J. Nanosci. Nanotechnol.* 7: 1647–1652. doi:10.1166/jnn.2007.349

Lu, C., H. Chiu and C. Liu. 2006 Removal of zinc(II) from aqueous solution by purified carbon nanotubes: kinetics and equilibrium studies. *Ind. Eng. Chem. Res.* 45: 2850–2855. doi:10.1021/ie051206h

Lunge, S., S. Singh and A. Sinha. 2014 Magnetic iron oxide (Fe_3O_4) nanoparticles from tea waste for arsenic removal. *J. Magn. Magn. Mater.* 356: 21–31. doi:10.1016/j.jmmm.2013.12.008

Ma, H., Y. Zhang, Q. Hu, D. Yan, Z. Yu, M. Zhai. 2012 Chemical reduction and removal of Cr(VI) from acidic aqueous solution by ethylenediamine-reduced graphene oxide. *J. Mater. Chem.* 22: 5914–5916. doi:10.1039/C2JM00145D

Madadrang, C. J., H. Y. Kim, G. Gao, N. Wang, J. Zhu, H. Feng, M. Gorring, M. L. Kasner and S. Hou. 2012 Adsorptipn behaviour of EDTA-graphene oxide for Pb(II) removal. *ACS Appl. Mater. Interf.* 4: 1186–1193. doi:10.1021/am201645g

Mahdavi, S., M. Jalali and A. Afkhami. 2013 Heavy metals removal from aqueous solutions using TiO_2, MgO, and Al_2O_3 nanoparticles. *Chem. Eng. Commun.* 200: 448–470. doi:10.1080/00986445.2012.686939

Martinson, C. A. and K. J. Reddy. 2009 Adsorption of arsenic (III) and arsenic (V) by cupric oxide nanoparticles. *J. Colloid Interf. Sci.* 336: 406–411. doi:10.1016/j.jcis.2009.04.075

Meng, X., S. Bang and G. P. Korfiatis. 2000 Effect of silicate, sulfate, and carbonate on arsenic removal by ferric chloride. *Water Res.* 34: 1255–1261. doi:10.1016/S0043-1354(99)00272-9

Mohapatra, P., S. K. Samantaray and K. Parida. 2005 Photocatalytic reduction of hexavalent chromium in aqueous solution over sulphate modified titania. *J. Photochem. Photobiol. A: Chem.*, 170: 189–194. doi:10.1016/j.jphotochem.2004.08.012

Mukherjee, A., P. Bhattacharya, K. Savage, A. Foster and J. Bundschuh. 2008 Distribution of geogenic arsenic in hydrologic systems: Controls and challenges. *J. Cont. Hydrol.* 99: 1–7. doi:10.1016/j.jconhyd.2008.04.002

Mukherjee, R., P. Bhunia and S. De. 2016 Impact of graphene oxide on removal of heavy metals using mixed matrix membrane. *Chem. Eng. J.* 292: 284–297. doi:10.1016/j.cej.2016.02.015

Namasivayam, C. and M. V. Sureshkumar. 2008 Removal of chromium (VI) from water and wastewater using surfactant modified coconut coir pith as a biosorbent. *Bioresour. Technol.* 99: 2218–2225. doi:10.1016/j.biortech.2007.05.023

Nasir, A. M., P. S. Goh, M. S. Abdullah, B. C. Ng and A. F. Ismail. 2019 Adsorptive nanocomposite membranes for heavy metal remediation: Recent progresses and challenges. *Chemosphere*, 232: 96–112. doi:10.1016/j.chemosphere.2019.05.174

Nata, I. F., M. Sureshkumar and C. Lee. 2011 One-pot preparation of amine-rich magnetite/bacterial cellulose nano-composite and its application for arsenate removal. *RSC Adv.*, 1: 625–631. doi:10.1039/c1ra00153a

Nayak, A. K. and A. Pal. 2017 Green and efficient biosorptive removal of methylene blue by *Abelmoschus esculentus* seed: Process optimization and multi-variate modelling. *J. Environ. Manage.*, 200: 145–159. doi:10.1016/j.jenvman.2017.05.045

Neeraj, G., S. Krishnan, P. S. Kumar, K. R. Shriaishvarya and V. V. Kumar. 2016 Performance study on sequestration of copper ions from contaminated water using newly synthesized high effective chitosan coated magnetic nanoparticles. *J. Mol. Liq.* 214: 335–346. doi:10.1016/j.molliq.2015.11.051

Ngwabebhoh, F. A. and U. Yildiz. 2019. Nature-derived fibrous nanomaterial toward biomedicine and environmental remediation: Today's state and future prospects. *J. Appl. Polym. Sci.*, 47878: 1–21. doi:10.1002/APP.47878

Pacheco, S., J. Tapia, M. Medina and R. Rodriguez. 2006 Cadmium ions adsorption in simulated wastewater using structured alumina-silica nanoparticles. *J. Non-Cryst. Solids*, 352: 5475–5481. doi:10.1016/j.noncrysol.2006.09.007

Pal, P. and A. Pal. 2017 Surfactant-modified chitosan beads for cadmium ion adsorption. *Int. J. Biol. Macromol.* 104: 1548–1555. doi:10.1016/j.ijbiomac.2017.02.042

Parham, H., B. Zargar and R. Shiralipour. 2012 Fast and efficient removal of mercury from water samples using magnetic iron oxide nanoparticles modified with 2-mercaptobenzothiazole. *J. Hazard. Mater.* 205–206: 94–100. doi:10.1016/j.jhazmat.2011.12.026

Paris, E. C., J. O. D. Malafatti, H. C. Musetti, A. Manzoli, A. Zenatti and M. T. Escote. 2020. Faujasite zeolite decorated with cobalt ferrite nanoparticles for improving removal and reuse in Pb^{2+} ions adsorption. *Chin. J. Chem. Eng.* 28: 1884–1890. doi:10.1016/j.cjche.2020.04.019

Park, S. and Y. Jang. 2002 Pore structure and surface properties of chemically modified activated carbons for adsorption mechanism and rate of Cr(VI). *J. Colloid Interf. Sci.* 249: 458–463. doi:10.1006/jcis.2002.8269

Ponder, S. M., J. G. Darab and T. E. Mallouk. 2000 Remediation of Cr(VI) and Pb(II) aqueous solutions using supported, nanoscale zero-valent iron. *Environ. Sci. Technol.* 34: 2564–2569. doi:10.1021/es9911420

Rahimi, S., R. M. Moattari, L. Rajabi, A. A. Derakhshan and M. Keyhani. 2015 Iron oxide/hydroxide (α, γ-FeOOH) nanoparticles as high potential adsorbents for lead removal from polluted aquatic media. *J. Ind. Eng. Chem.* 23: 33–43. doi:10.1016/j.jiec.2014.07.039

Rao, G. P., C. Lu and F. Su. 2007 Sorption of divalent metal ions from aqueous solution by *carbon* nanotubes: A review. *Sep. Purif. Technol.*, 58: 224–231. doi:10.1016/j.seppur.2006.12.006

Rashtogi, A., D. K. Tripathi, S. Yadav, D. K. Chauhan, M. Zivcak, M. Ghorbanpour, N. I. El-Sheery and M. Brestic. 2019 Application of silicon nanoparticles in agriculture. *3 Biotech* doi:10.1007/s13205-019-1626-7

Ratnaike, R. N. 2003 Acute and chronic arsenic toxicity. *Postgrad. Med. J.* 79: 391–396. doi:10.1136/pmj.79.933.391

Recillas, S., J. Colon, E. Casals, E. Gonzalez, V. Puntes, A. Sanchez and X. Font. 2010 Chromium VI adsorption on cerium oxide nanoparticles and morphology changes during the process. *J. Hazard. Mater.* 184: 425–431. doi:10.1016/j.jhazmat.2010.08.052

Recillas, S., A. García, E. González, E. Casals, V. Puntes, A. Sánchez and X. Font 2011. Use of CeO_2, TiO_2 and Fe_3O_4 nanoparticles for the removal of lead from water: Toxicity of nanoparticles and derived compounds. *Desalination* 277: 213–220. doi:10.1016/j.desal.2011.04.036

Ruvarac-Bugarcic, I. A., Z. V. Saponjic, S. Zec, T. Rajh and J. M. Nedeljkovic. 2005 Photocatalytic reduction of cadmium on TiO_2 nanoparticles modified with amino acids. *Chem. Phys. Lett.* 407: 110–113. doi:10.1016/j.cplett.2005.03.058

Sabir, S., M. Arshad and S. K. Chaudhari. 2014 Zinc oxide nanoparticles for revolutionizing agriculture: Synthesis and applications. *Sci. World J.*. doi:10.1155/2014/925494

Selvi, K., S. Pattabhi and K. Kadirvelu. 2001 Removal of Cr(VI) from aqueous solution by adsorption onto activated carbon. *Bioresour. Technol.* 80: 87–89. doi:10.1016/S0960-8524(01)00068-2

Sethi, P. K., D. Khandelwal and N. Sethi. 2006 Cadmium exposure: Health hazards of silver cottage industry in developing countries. *J. Med. Toxicol.* 2: 14–15.

Shahzad, K., M. B. Tahir, M. Sagir and M. R. Kabli. 2019 Role of $CuCo_2S_4$ in Z scheme $MoSe_2/BiVO_4$ composite for efficient photocatalytic reduction of heavy metals. *Ceram. Int.*, 45: 23225–23232. doi:10.1016/j.ceramint.2019.08.018

Sharma, Y. C., V. Srivastava, V. K. Singh, S. N. Kaul and C. H. Weng. 2009 Nano-adsorbents for the removal of metallic pollutants from water and wastewater. *Environ. Technol.*, 30: 583–609. doi:10.1080/09593330902838080

Sheshdeh, R. K., M. R. K. Nikou, K. Badii, N. Y. Limaee and G. Golkarnarenji. 2014 Equilibrium and kinetics studies for the adsorption of Basic Red 46 on nickel-oxide nanoparticles-modified diatomite in aqueous solutions. *J. Taiwan Inst. Chem. Eng.* 45: 1792–1802. doi:10.1016/j.jtice.2014.02.020

Shin, K., J. Hong and J. Jang. 2011 Heavy metal ion adsorption behavior in nitrogen-doped magnetic carbon nanoparticles: Isotherms and kinetic study. *J. Hazard. Mater.* 190: 36–44. https://do.org/10.1016/j.jhazmat.2010.12.102

Siddiqui, S. I., M. Naushad and S. A. Chaudhry. 2019 Promising prospects of nanomaterials for arsenic water remediation: A comprehensive review. *Process Saf. Environ.*, 126: 60–97. doi:10.1016/j.psep.2019.03.037

Singh, D., A. Tiwari and R. Gupta. 2012 Phytoremediation of lead from wastewater using aquatic plants. *J. Agr. Technol.* 8: 1–11. http://www.ijat-aatsea.com

Skubal, L. R., N. K. Meshkov, T. Rajh and M. Thurnauer. 2002 Cadmium removal from water using thiolactic acid-modified titanium dioxide nanoparticles. *J. Photochem. Photobiol. A* 148: 393–397. doi:10.1016/S1010-6030(02)00069-2

Stafiej, A. and K. Pyrzynska. 2007 Adsorption of heavy metal ions with carbon nanotubes. *Sep. Purif. Technol.* 58: 49–52. doi:10.1016/j.seppur.2007.07.008

Suganthi, S. H. and K. Ramani. 2017 A novel single step synthesis and surface functionalization of iron oxide magnetic nanoparticles and thereof for the copper removal from pigment industry effluent. *Sep. Purif. Technol.* 188: 458–467. doi:10.1016/j.seppur.2017.07.059

Suman, A. Kardam, M. Gera and V. K. Jain. 2015 A novel reusable nanocomposite for complete removal of dyes, heavy metals and microbial load from water based on nanocellulose and silver nano-embedded pebbles. *Environ. Technol.* 36: 706–714. doi:10.1080/09593330.2014.959066

Sun, Q., Y. Yang, Z. Zhao, Q. Zhang, X. Zhao, G. Nie, T. Jiao and Q. Peng. 2018. Elaborately design of polymeric nanocomposite with Mg(II)-buffering nanochannels for highly efficient and selective heavy metal removal from water: Case study for Cu(II). *Environ. Sci: Nano*, 5: 2440–2451. doi:10.1039/C8EN00611C

Tabesh, S., F. Davar and M. R. Loghman-Estarki. 2018 Preparation of γ-Al_2O_3 nanoparticles using modified sol-gel method and its use for the adsorption of lead and cadmium ions. *J. Alloys Compd.* 730: 441–449. doi:10.1016/j.jallcom.2017.09.246

Tiwari, J. N., R. N. Tiwari and K. S. Kim. 2012 Zero-dimensional, one-dimensional, two-dimensional and three-dimensional nanostructured materials for advanced electrochemical energy devices. *Prog. Mater. Sci.* 57: 724–803. doi.10.1016/j.pmatsci.2011.08.003

Turk, T. and I. Alp. 2014 Arsenic removal from aqueous solutions with Fe-hydrotalcite supported magnetite nanoparticle. *J. Ind. Eng. Chem* 20: 732–738. doi:10.1016/j.jiec.2013.06.002

Vojoudi, H., A. Badiei, S. Bahar, G. M. Ziarani, F. Faridbod and M. R. Ganjali. 2017 A new nano-sorbent for fast and efficient removal of heavy metals from aqueous solutions based on modification of magnetic mesoporous silica nanospheres. *J. Magn. Magn. Mater.* 441: 193–203. doi:10.1016/j.jmmm.2017.05.065

Wan, S., J. Wu, S. Zhou, R. Wang, B. Gao and F. He. 2018 Enhanced lead and cadmium removal using biochar-supported hydrated manganese oxide (HMO) nanoparticles: Behavior and mechanism. *Sci. Total Environ.* 616–617: 1298–1306. doi:10.1016/j.scitotenv.2017.10.188

Wang, H., A. Zhou, F. Peng, H. Yu and J. Yang. 2007b Mechanism study on adsorption of acidified multiwalled carbon nanotubes to Pb(II). *J. Colloid Interf. Sci.* 316: 277–283. doi:10.1016/j.jcis.2007.07.075

Wang, L., N. Wang, L. Zhu, H. Yu and H. Tang. 2008 Photocatalytic reduction of Cr(VI) over different TiO_2 photocatalysts and the effects of dissolved organic species. *J. Hazard. Mater.* 152: 93–99. doi:10.1016/j.jhazmat.2007.06.063

Wang, L., H. Xu, Y. Qiu, X. Liu, W. Huang, N. Yan and Z. Qu. 2020 Utilization of Ag nanoparticles anchored in covalent organic frameworks for mercury removal from acidic wastewater. *J. Hazard. Mater.* 389: 121824. doi:10.1016/j.jhazmat.2019.121824

Wang, S., W. Gong, X. Liu, Y. Yao, B. Gao and Q. Yue. 2007a Removal of lead(II) from aqueous solution by adsorption onto manganese oxide-coated carbon nanotubes. *Sep. Purif. Technol.* 58: 17–23. doi:10.1016/j.seppur.2007.07.006

Wu, Y., H. Luo, H. Wang, C. Wang, J. Zhang and Z. Zhang. 2013 Adsorption of hexavalent chromium from aqueous solutions by graphene modified with cetyltrimethylammonium bromide. *J. Colloid Interf. Sci.* 394: 183–191. doi:10.1016/j.jcis.2012.11.049

Wu, Y., H. Pang, Y. Liu, X. Wang, S. Yu, D. Fu, J. Chen and X. Wang. 2019 Environmental remediation of heavy metal ions by novel-nanomaterials: A review. *Environ. Pollut.*, 246: 608–620. doi:10.1016/j.envpol.2018.12.076

Xiang-Feng, K., Y. Bin, X. Heng, Z. Yang, X. Sheng-Guo, X. Bao-Qiang and W. Shi-Xing. 2014 Selective removal of heavy metal ions from aqueous solutions with surface functionalized silica nanoparticles by different functional groups. *J. Cent. South Univ.*, 21: 3575–3579. doi:10.1007/s11771-014-2338-0

Xu, D., X. Tan, C. Chen and X. Wang. 2008 Removal of Pb(II) from aqueous solution by oxidized multiwalled carbon nanotubes. *J. Hazard. Mater.* 154: 407–416. doi:10.1016/j.jhazmat.2007.10.059

Xu, Y. and D. Zhao. 2007 Reductive immobilization of chromate in water and soil using stabilized iron nanoparticles. *Water Res.* 41: 2101–2108. doi:10.1016/j.watres.2007.02.037

Ying, Y., Y. Liu, X. Wang, Y. Mao, W. Cao, P. Hu, X. Peng. 2015 Two-dimensional titanium carbide for efficiently reductive removal of highly toxic chromium(VI) from water. *ACS Appl. Mater. Interfaces*, 7: 1795–1803. doi:10.1021/am5074722.

Youssef, A. M. and F. M. Malhat. 2014. Selective removal of heavy metals from drinking water using titanium dioxide nanowire. *Macromol. Symp.*, 337: 96–101. doi:10.1002/masy.201450311

Yu, X., S. Tong, M. Ge, J. Zuo, C. Cao and W. Song. 2013 One-step synthesis of magnetic composites of cellulose@iron oxide nanoparticles for arsenic removal. *J. Mater. Chem. A* 1: 959–965. doi:10.1039/C2TA00315E

Zhang, Q., B. Pan, S. Zhang, J. Wang, W. Zhang and L. Lv. 2011. New insights into nanocomposite adsorbents for water treatment: A case study of polystyrene-supported zirconium phosphate nanoparticles for lead removal. *J. Nanopart. Res.* 13: 5355–5364. doi:10.1007/s11051-011-0521-x

Zhang, S., Y. Zhang, J. Liu, Q. Xu, H. Xiao, X. Wang, H. Xu and J. Zhou. 2013b Thiol modified Fe_3O_4@ SiO_2 as a robust, high effective, and recycling magnetic sorbent for mercury removal. *Chem. Eng. J.* 226: 30–38. doi:10.1016/j.cej.2013.04.060

Zhang, Y., H. Ma, J. Peng, M. Zhai and Z. Yu. 2013a Cr(VI) removal from aqueous solution using chemically reduced and functionalized graphene oxide. *J. Mater. Sci.* 48: 1883–1889. doi:10.1007/s10853-012-6951-8

Zhu, J., S. Wei, H. Gu, S. B. Rapole, Q. Wang, Z. Luo, N. Haldolaarachchige, D. P. Young and Z. Guo. 2012 One pot synthesis of magnetic graphene nanocomposites decorated with core@double-shell nanoparticles for fast chromium removal. *Environ. Sci. Technol.* 46: 977–985. doi:10.1021/es2014133

7 Photoactive Polymer for Wastewater Treatment

Ridha Djellabi and Claudia Letizia Bianchi
Università degli Studi di Milano, Dip. Chimica and INSTM-UdR Milano, Milan, Italy

Muhammad Rizwan Haider
Harbin Institute of Technology Shenzhen, Shenzhen, China

Jafar Ali
University of Sialkot, Sialkot, Pakistan

Ermelinda Falletta, Marcela Frias Ordonez, Anna Bruni, and Marta Sartirana
Università degli Studi di Milano, Milan, Italy

Ramadan Geioushy
Central Metallurgical R&D Institute (CMRDI), Helwan, Egypt

CONTENTS

7.1 Introduction 218
7.2 Water Remediation by Photoactive Polymer-Based Photocatalysts 219
7.2.1 Origin of Photoactivity on Polymer-Based Photocatalysts System 219
7.2.2 Removal of Organic Contaminants and Heavy Metals by Polymer Photocatalysts 220
7.2.3 ROSs Photoproduced by Photoactive Polymers 225
7.3 Water Disinfection by Photoactive Polymer-Based Photocatalysts 227
7.4 Combination of Inorganic Semiconductors and Photoactive Polymers 228
7.5 Photoactive Polymer-Based Membranes 231
7.6 Conclusions and Future Aspects 233
References 234

DOI: 10.1201/9781003118749-7

7.1 INTRODUCTION

Nowadays, the purification of wastewaters is a serious issue due because of the dramatically rise in the worldwide people and manufacturing activities (Nesaratnam 2014, Djellabi 2015). Many technologies are being applied at large scale (Riffat 2012, Pooi and Ng 2018), while many novel advanced methods/materials have attracted investments of billions of dollars in scientific research last decades as successful alternative approaches for water and wastewater purification (Bethi, Sonawane et al. 2016, Naushad & Lichtfouse 2019, Taghipour, Hosseini et al. 2019). Purification efficiency versus time, safety, and cost are important parameters in choosing a technology for the types of pollutants in wastewaters.

The use of inexhaustible sunlight as a sustainable and free potential energy for environmental remediation is one of the highly suggested approaches to fight against the huge industrialization and environmental pollution (Chakrabarti 2017). Solar photocatalysis has attracted much attention and huge scientific research because of its potential to transform sunlight into redox species for environmental remediation and energy production (Zhang, Wang et al. 2014, Bora & Mewada 2017). Since the discovery of photocatalysis (Fujishima & Honda 1972), the research community has made great efforts to develop efficient and low-cost photocatalytic materials for use in solar photocatalysis at a large scale. Over more than three decades, a pool of studies has been reported on the synthesis of visible light-responsive photocatalysts using different approaches such as modification of TiO_2 with metal/nonmetal (Daghrir, Drogui et al. 2013, Shayegan, Lee et al. 2018), use of an organic sensitizer dye-sensitizing (Li, Shi et al. 2018, Youssef, Colombeau et al. 2018), construction of heterojunction systems (Low, Yu et al. 2017, Shah, Fiaz et al. 2020) or hybridization of titania with carbon biomass (Djellabi, Yang et al. 2019). In semiconductor photocatalysis, the photoefficiency is due to absorption of light by a semiconductor, resulting in its excitation followed by electron/hole pairs formation. Such a process depends mostly on the optical and physical properties of the semiconductor. One of the most important parameters of a given semiconductor in photocatalysis is its band gap, the key for absorbing enough light to produce redox species (Marcelino & Amorim 2019). However, it is worth noting that the evaluation of a photocatalyst efficiency is not only accounted by its light absorption (excitation step), but also by the account of the separated photoproduced charges (longer-lived polaron pairs) leading to improve the generation redox species yield in the medium (Ohtani 2013, Qian, Zong et al. 2019).

From the real application of solar photocatalysis point of view, it is recommended to use photocatalysts with low band gaps that can absorb a good portion of sunlight. However, extremely fine band gaps may likely lead to accelerating the electron/hole charges recombination, in turn decreasing the photocatalytic efficiency (Dmitriev and Mocker 1995). Therefore, to select a photocatalyst, a trade-off between solar light harvesting and the ability to separate the charge should be obtained.

Recently, much work has been done to develop polymer-based photocatalysts as a new family for photocatalysis applications (McTiernan, Pitre et al. 2014, Zhang, Liu et al. 2016, Zhang, Wang et al. 2017, Farooq, Tahir et al. 2019, Qiang, Chen et al. 2020) including wastewater purification (Vyas, Lau et al. 2016, Yuan, Floresyona et al. 2019, Dai and Liu 2020, Kumar, Travas-Sejdic et al. 2020). Photoactive polymer-based photocatalysts can stand out in the photocatalysis field over inorganic semiconductor photocatalysts due to their several advantages, such as: (i) photoactive polymer-based photocatalysts have tunable energy levels for redox reactions; (ii) easy synthesis and good chemical stability against photobleaching; (iii) excellent visible light response; (iv) excellent electrical conductivity and high carrier mobility along the polymer backbone, which results in better charges separation and transport properties. The study on photoactive conjugated polymers as photocatalysts was reported in the 1980s when scientists in Japan showed the photocatalytic reduction of H^+ to H_2 on the surface of poly(*p*-phenylene) could reduce protons to hydrogen under light irradiation (Matsuoka, Fujii et al. 1991, Wang, Maeda et al. 2009b). Nowadays, such materials have been undergoing fast development because of their potential application in various domains, including energy production, photosynthesis, and environmental remediation (Zhi, Li et al. 2017, Cao, Li et al. 2019, Wang, Vogel et al. 2019, Keshavarzi, Cao et al. 2020). This chapter discusses the fabrication of photoactive polymer-based photocatalysts for water treatment/disinfection, including the mechanistic pathways for the photocatalytic production of reactive oxygen species (ROSs). In addition, the combination of photoactive polymers with inorganic semiconductors for enhanced photoefficiency is addressed. Finally, the development of photoactive polymer-based membranes for water remediation will be discussed.

7.2 WATER REMEDIATION BY PHOTOACTIVE POLYMER-BASED PHOTOCATALYSTS

7.2.1 Origin of Photoactivity on Polymer-Based Photocatalysts System

Designing photoactive polymers for photocatalysis applications gives many advantages, such as high visible light absorption, which is the primary step in the photocatalytic mechanism, and efficient charge carrier mobilities (n- and p-type). The characteristics of photoactive-based polymers, including the physical and optical, can be obtained by choosing appropriate molecules for building the polymer blocks, and adjusting the synthesis conditions and the degree of polymerization (Xu, Jin et al. 2013). Several classes of polymer-based photocatalysts have been fabricated and applied (Figure 7.1).

The origin of the photoactivity in polymer systems is similar to that of inorganic semiconductor systems. The irradiation of the photoactive polymer surface by a light source results in the excitation of polymer (Coulomb-correlated electron-hole pairs) (Wang, Jin et al. 2020a). The photoproduced electron-hole afterward diffuses to the polymer interface in the form of redox separated charges, which in turn can undergo bulk or surface recombination, or react with oxygen species in the medium to form ROSs.

In photoactive polymers, the visible light-harvesting ability is due to the optical band gap (Wang, Maeda et al. 2009b, Ghosh, Kouamé et al. 2015). The sp2 hybridized and the pz orbitals are oriented perpendicular to the chain of the polymer blocks, which leads to π-delocalization and the concomitant energy band gap (Eg) by a HOMO-LUMO. In such a system, polymers with extended conjugation length are expected to enhance the exciton/polaron migration along the polymer backbone (Calik, Auras et al. 2014, Vyas, Lau et al. 2016). The $n \rightarrow \pi^*$ transitions in the polymer backbone can also strengthen the light absorption toward visible wavelengths (Li, Zhang et al. 2010). It is

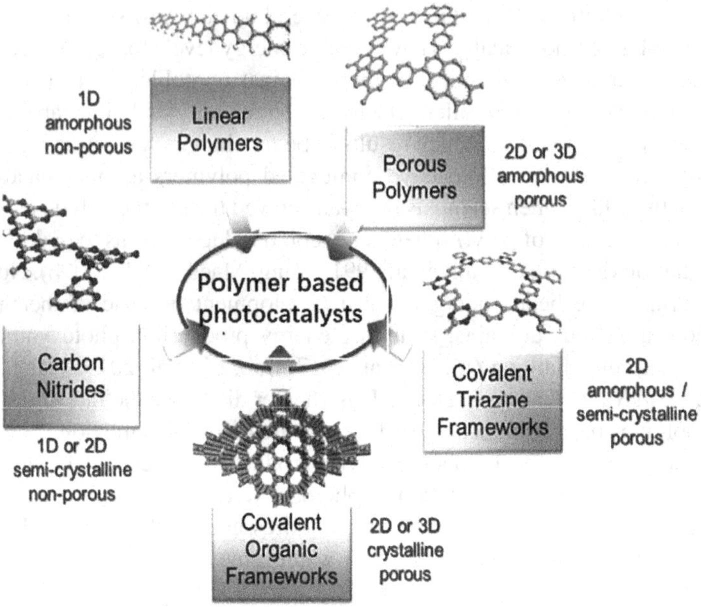

FIGURE 7.1 Classification of various photoactive polymers. Reproduced with permission from (Vyas, Lau et al. 2016).

important to mention that $\pi \rightarrow \pi^*$ transitions can result in better photocatalytic activity compared with $n \rightarrow \pi^*$; however, $n \rightarrow \pi^*$ transitions can improve the absorption of visible light (Chen, Wang et al. 2014, Djellabi, Zhao et al. 2020c). R. Djellabi et al. (2020c) reported that the combination of both $n \rightarrow \pi^*$ and $\pi \rightarrow \pi^*$ transitions in the same polymer platform could help to improve both the absorption of visible light and photoactivity. On top of that, these transitions ensure a longer-lived charge carrier after the electron excitation, resulting in better redox species production.

7.2.2 Removal of Organic Contaminants and Heavy Metals by Polymer Photocatalysts

In recent years, photoactive polymers have shown excellent potential for the photo-oxidation of organic pollutants in water under UV and visible irradiation (Wang, Vogel et al. 2019). Graphitic carbon nitride (g-C_3N_4) has been widely reported as the most fascinating successful and stable conjugated polymer, which has become a novel research hotspot for solar photocatalysis applications (Wen, Xie et al. 2017, Ismael and Wu 2019). The fabrication of g-C_3N_4 can be done *via* the thermal decomposition of nitrogen-containing precursors including melamine (Bellardita, García-López et al. 2018), urea (Kim, Park et al. 2018), cyanamide (Inagaki, Tsumura et al. 2019) and thiourea (Hua, Qu et al. 2019). Many reviews have addressed the synthesis routes, modification and use of g-C_3N_4 based photocatalysts for energy production and water/air purification (Zhao, Sun et al. 2015a, Ong, Tan et al. 2016, Ismael and Wu 2019, Reddy, Reddy et al. 2019, Zhang, Li et al. 2019, Qi, Liu et al. 2020, Rono, Kibet et al. 2020, Wu, Zhang et al. 2020). g-C_3N_4 exhibits a quite low band gap

(~2.7 eV), and its separated polaron pairs could live for around 200 fs (Merschjann, Tschierlei et al. 2015). g-C_3N_4 based photocatalysts were largely used for the photo-oxidation of organic pollutants (Xu, Wang et al. 2017, Ismael and Wu 2019, Su, Zhang et al. 2020, Wang, Nie et al. 2020b), photocatalytic reduction of heavy metals/noble metals (Lu, Chen et al. 2016, Moneim, Gad-Allah et al. 2016, Patnaik, Das et al. 2018, Qiao, Wu et al. 2020) and for disinfection (Huang, Ho et al. 2014, Zhang, Li et al. 2018a, 2019, Liu, Ma et al. 2020a). Apart from g-C_3N_4, several groups have fabricated various photoactive polymers with different optical and physical properties. For example, (Ghosh, Kouamé et al. 2015) have fabricated poly(diphenylbutadiyne) (PDPB) based photoactive polymer *via* the polymerization of 1,4-diphenylbutadiyne under UV irradiation. The band gap of PDPB was evaluated experimentally and found to be 1.95 eV. The photoactivity of bulk PDPB and nano-PDPB was checked against the photooxidation of methyl orange and phenol, and showed good photoactivity under UV and visible light. Interestingly, nano-PDPB showed an enhanced photoactivity under visible light (>450 nm) even better than Ag-modified TiO_2. Djellabi et al. (2020c) have synthesized a novel photoactive polymer based on electron-rich Tris(4-carbazoyl-9-ylphenyl)amine (TCTA) (band gap of 3.4 eV), which was obtained by heat polymerization with PVP as showed in Figure 7.2 The authors reported that the obtained polymer could fully absorb the visible light up to 900 nm. In addition, the increase of TCTA content leads to further red-shifting of the absorption, as well as an improvement of the electrochemical activity and the photoactivity.

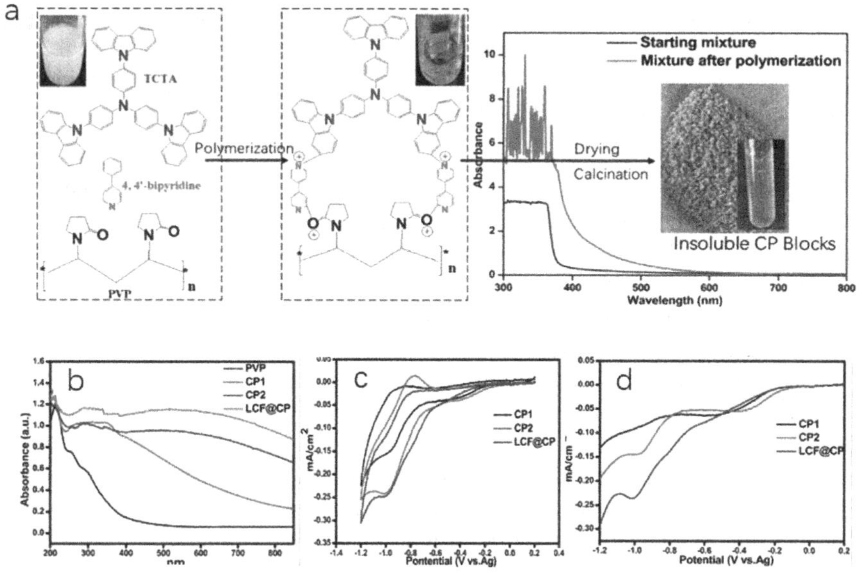

FIGURE 7.2 (a): Synthesis steps of TCTA based photoactive polymer via heat polymerization. (b): UV-DRS of PVP, CP1 (photoactive polymer with low TCTA content), CP2 (photoactive polymer with high TCTA content) and LCF@CP (CP1 supported on olive pits biomass). (c) CV curves of CP1 and LCF@CP. (d): LSV curves of CP and LCF@CP. Figures (a–d) were reproduced with permission from Djellabi et al. (2020c).

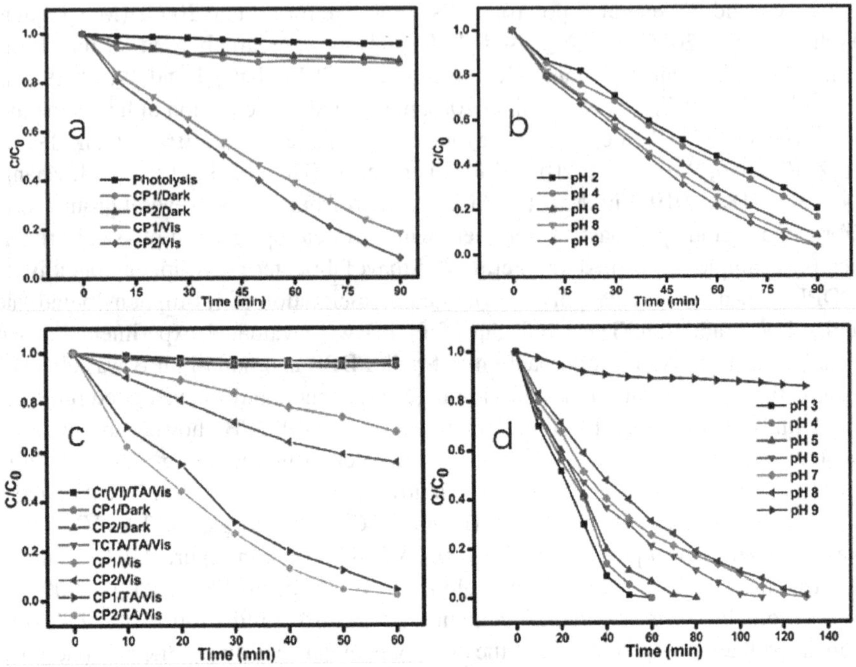

FIGURE 7.3 (a): Photocatalytic oxidation of MB (10 ppm) at pH 6 under visible light using CP1 and CP2, [photocatalyst]: 0.25 g/L; (b): Effect of pH on the photooxidation of MB by CP2 under visible light. (c) Photocatalytic reduction of Cr(VI) (20 ppm) at pH 3 using pure TCTA, CP1 and CP2 under visible light, [Cr(VI)]: 20 ppm, pH: 3, [Photocatalyst]: 0.25 g/L (d): Effect of pH on the photocatalytic reduction of Cr(VI) under visible light using CP2. Figures (a–d) were reproduced with permission from Djellabi et al. (2020c).

The materials showed great ability toward the photooxidation of dyes and the photoreduction of Cr(VI) under UV and visible light (420 nm) (Figure 7.3). The authors mentioned that TCTA photoactive polymer has a great photocatalytic reduction at pH values from 3 to 8, unlike conventional inorganic-based photocatalysts (e.g., TiO_2 (Djellabi and Ghorab 2015)). For enhanced surface area and excellent physical stability, the TCTA based polymer was fixed on lignocellulosic biomass.

Keshavarzi et al. (2020) have synthesized a novel conducting polymer via polymerization of barbituric acid (BA) at different temperatures for the photocatalytic oxidation of RhB under visible light. The authors reported that the bandgap becomes narrower upon increasing the condensation temperature, while above 305°C carbon-like polymers were obtained. On the other hand, it was found that BA photoactive polymer sample condensed at 300°C, which exhibited an optimum conjugated electronic system (e.g., 2.20 eV), has the best performance compared to the BA sample condensed at 350°C (e.g., 1.86 eV).

Xiaojiao Yuan et al. (2019) fabricated photoactive polymer polypyrole nanostructure photocatalysts in three ways: radiolysis without any template (PPy-NS-γ), chemical oxidation of the polypyrene monomer in hexagonal mesophases (PPy-NS-c) and chemical oxidation in the presence of $FeCl_3$ as oxidant (PPy-bulk). PPy-NS-c and PPy-NS-γ photocatalysts have better photoactivity toward the oxidation of phenol compared with PPy-bulk. On top of that, the photoactivity under visible light of PPy nanostructures was similar to that of Ag-modified TiO_2.

Campuzano et al. reported that a non-conjugated metal-free polymer, Poly(1,4-diamine-9,10-dioxoanthracene-alt-[bencen1,4-dioic acid]), COP, has an excellent visible light response (Irigoyen-Campuzano, González-Béjar et al. 2017). COP was obtained via the reaction of teraphtaloyl chloride with 1,4-diaminoanthraquinone in dimethylformamide. The photoactive polymers showed an excellent photocatalytic activity toward 2,5-dichlorophenol in water under air and sunlight irradiation. The tests of the COP-based photocatalysts were carried out under natural solar light, and the efficiency was compared with TiO_2 in a fixed bed reactor, as shown in Figure 7.4a. The results of direct photolysis and the photocatalytic oxidation of 2,5-dichlorophenol by COP and TiO_2 are shown in Figure 7.4b.

Wang et al. (2018a) have prepared different microporous/mesoporous conjugated polymers based on 9,9′-bifluorenylidene-based using Suzuki polycondensation (Figure 7.5). These polymers showed a high specific surface area from 590 to 1306 $m^2\ g^{-1}$. Such photoactive polymers showed great adsorption ability up to 1905 $mg.g^{-1}$ toward RhB dye and also excellent photocatalytic oxidation of organic pollutants in water under visible light (>450 nm). The authors also mentioned that such photoactive polymers have great recyclability.

Among conducting polymers, polyaniline (PANI) is unique for its characteristics, such as redox properties, environmental stability, unique doping/dedoping process that allows us to quickly switch from insulator to conductor and *vice versa* (Della Pina, Falletta et al. 2011).

Even though PANI/inorganic semiconductors photocatalysts have been deeply investigated (see paragraph 7.4) concerning the use of pristine PANI as photoactive material for pollutants degradation, the scientific literature is lacking. Riaz et al. (2014) demonstrated the ability of this polymer to degrade an azo dye (acid orange)

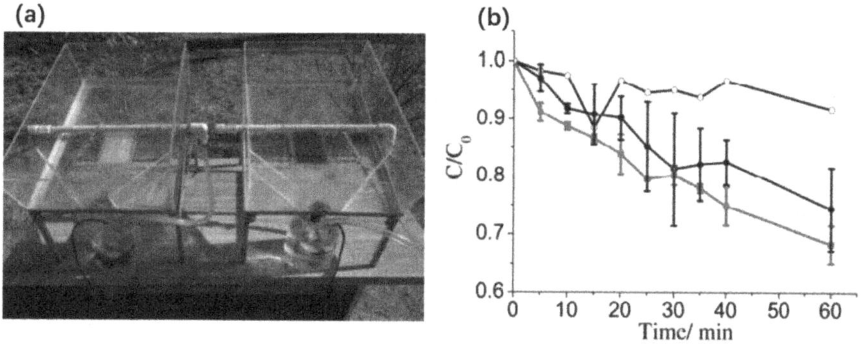

FIGURE 7.4 (a): 1-L tank solar photoreactor with supported TiO_2 (while) or support COP (red), (b): Comparison of the photocatalytic activities of COP (red), TiO_2 (black) and photocatalysis (empty black circles) under sunlight. pH:3, 2,5-dichlorophenol: 20 ppm. Reproduced with permission from (Irigoyen-Campuzano, González-Béjar et al. 2017).

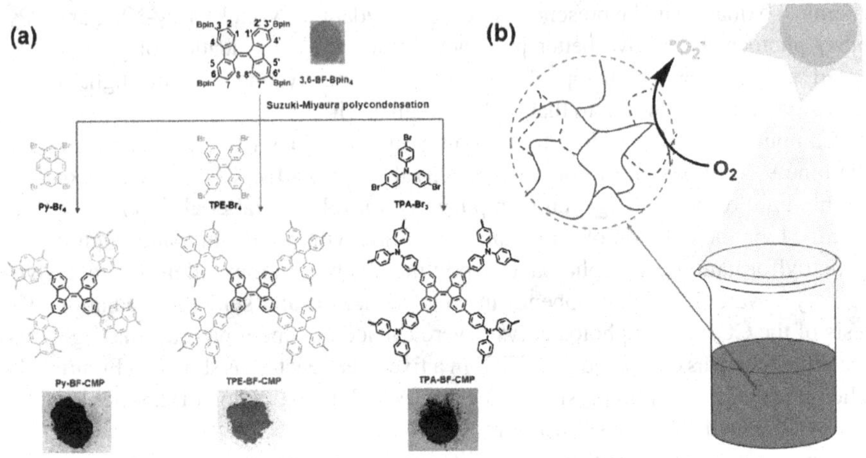

FIGURE 7.5 (a): Synthesis of 9,9′-bifluorenylidene-based based photocatalysts using Suzuki polycondensation, (b): The photocatalytic oxidation of RhB by 9,9′-bifluorenylidene-based photocatalyst under visible light. Reproduced with permission from (Wang, Xie et al. 2018a).

under microwave irradiation, and numerous other researchers reported the PANI ability to work as an absorbing material (Cionti, Della Pina et al. 2018, 2020, Duhan and Kaur 2020).

However, despite the extraordinary results, the investigation of pristine PANI as photocatalyst for water remediation has been incomprehensibly neglected for many years. Only recently, Saha et al. (2020) demonstrated the ability of PANI to photodegrade dyes, reaching for MO 42% degradation while for MB 97% degradation in less than 3 h. Table 7.1 lists several

studies reported on the use of free-metal photoactive polymers for the removal of different types of pollutants in water.

TABLE 7.1
Photocatalytic Removal of Environmental Contaminants over Metal-Free Photoactive Polymers Based Photocatalysts

Photoactive Polymer	Pollutant/Light Irradiation	References
Poly(1,4-diamine-9,10-dioxoanthracene-alt-[bencen1,4-dioic acid])	2,5-DCP, Solar light	Irigoyen-Campuzano, González-Béjar et al. (2017)
Tris(4-carbazoyl-9-ylphenyl) amine@PVP	MB, Cr(VI), visible light (>420 nm)	Djellabi et al. (2020c)
Polyaniline/Graphene	Rose Bengal dye, Xenon arc lamp illumination	Ameen, Seo et al. (2012)
9,9′-bifluorenylidene	RhB, visible light (λ >450 nm)	Wang, Xie et al. (2018a)
Polymeric carbon nitride foam	Tetracycline hydrochloride, visible light	Wang, Wu et al. (2018b)
PAN-carbon dots	Cr(VI), visible light	Xiao, Cheng et al. (2019)

(Continued)

TABLE 7.1 (Continued)
Photocatalytic Removal of Environmental Contaminants over Metal-Free Photoactive Polymers Based Photocatalysts

Poly(diphenylbutadiyne)	Methyl orange, phenol, UV and visible light (>450 nm)	Ghosh, Kouamé et al. (2015)
Poly(3,4-ethylenedioxythiophene)	Methyl orange, phenol, UV and visible light (>450 nm)	Ghosh, Kouamé et al. (2015)
Polypyrrole	Methyl orange, phenol, UV, and visible light (>400 nm)	Yuan, Floresyona et al. (2019)
Poly(3-hexylthiophene)	Phenol, UV and visible light (>400 nm)	Floresyona, Goubard et al. (2017)
Polycation@polyacrylic acid	NADH Cofactor, RhB, visible light (>420 nm)	Byun, Landfester et al. (2019)
Poly(1,3,4-oxadiazole)s	Dyes: MO, RhB and MB, UV light (365 nm)	Ran, Duan et al. (2018)
Zincphthalocyanine	RhB, visible light (>420 nm)	Cai, Li et al. (2018)
Ethynylbenzene-based polymer	RhB, phenol, and tetracycline, visible light (>420 nm)	Wang, Yang et al. (2018c)
Zwitterionic PDAT	Dyes: RB, CR, MO, and AB, visible light (λ >410 nm)	Tang, Gong et al. (2019)
P3HT and MEH-PPV	Dyes: Alizarin S, Alizarin G, Orange G, and Remazol brilliant blue; UV irradiation	Muktha, Madras et al. (2007a)
$g\text{-}C_3N_4$ nanosheets	Oxidation of $Ag(CN)_2^-$, Recovery of Ag, visible light (>420 nm)	Qiao, Wu et al. (2020)
Conjugated microporous polymer@P-PAN	RhB, Visible light (>450 nm)	Lee, Noh et al. (2020)
(SCN)n polymer	RhB, Cr(VI), Visible light (λ = 420, 525, 620 nm)	Gong, Li et al. (2020)
Perylene@ECUT-CO/ECUTSO	UO_2^{2+}, Visible light (>400 nm)	Yu, Zhu et al. (2020)
Heptazine-based (OCN) polymer	Enrofloxacin, Visible light (>400 nm)	Du, Chen et al. (2020)
Porphyrin	RhB, MB, Visible light (>420 nm)	Li, Zhao et al. (2020)
Perylene@triazine@heptazine	Benzylamine, Visible light (>420 nm)	Zhou, Wang et al. (2020)
Boron nitride@thiophene-triazine	Sulfides, Visible light (>400 nm)	Lan, Li et al. (2020)
Phenothiazine@carbazole/benzene		

7.2.3 ROSs Photoproduced by Photoactive Polymers

Photoactive polymers have been successfully applied to efficiently remove a wide range of pollutants in water and wastewater, which is accomplished through oxidation by photoproduced ROSs and/or photoreduction by direct electron transfer. ROSs (•OH, $O_2^{\bullet -}$, H_2O_2 etc.) are produced through interaction between electron/hole pairs and H_2O/O_2 molecules in the system. Illumination of the photoactive polymers with an appropriate source of light generates electrons at valence band (VB) that migrate to the conduction band (CB), leaving the same amount of positively charged holes (hν) in VB. Subsequently, both the electrons and hν escape the recombination and move to the photocatalyst surface to react with the adsorbed species of electron donors and acceptors (Nowakowska and Szczubia, 2017). The electrons reduce molecular oxygen to superoxide radical $O_2^{\bullet-}$ (Eq. 1).

$$O_2 + e^- \rightarrow O_2^{\bullet-}\ Eo = -0.33\,V\,vs.SHE \tag{7.1}$$

As the produced $O_2^{\bullet-}$ is highly reactive, it can react further to produce H_2O_2 and $^{\bullet}OH$ according to the following reactions:

$$O_2^{\bullet-} + 2H^+ \rightarrow H_2O_2 \tag{7.2}$$

$$H_2O_2 + O_2^{\bullet-} \rightarrow {}^{\bullet}OH + O_2 + {}^{-}OH \tag{7.3}$$

$$H_2O_2 + hv \rightarrow 2\,{}^{\bullet}OH \tag{7.4}$$

Meanwhile, the positively charged holes photoproduced in HOMO, depending on the energy level, could also generate $^{\bullet}OH$ by direct reaction with H_2O or ^{-}OH (Eq. 5) (Yuan et al., 2019):

$$H_2O + hv \rightarrow {}^{\bullet}OH + H^+ \quad Eo = -2.73\,V\,vs.SHE \tag{7.5}$$

The photogenerated highly reactive radicals are then responsible for catalytic degradation of pollutants in water. For example, Djellabi et al. (2020c) reported that a TCTA@PVP-based polymer used as a photocatalyst could achieve 98% oxidation of MB in the presence of 0.25 g/L catalyst under visible light. In order to understand the underlying degradation mechanism, authors used tert-Butyl alcohol, *p*-benzoquinone and sodium oxalate as $^{\bullet}OH$, $O_2^{\bullet-}$ and hole scavengers, respectively. The results showed that only 33, 22.5 and 47.8% of MB could be degraded, suggesting the crucial role of $^{\bullet}OH$, $O_2^{\bullet-}$ and h*v* in degradation with $O_2^{\bullet-}$ as the main oxidant. In another study, a photoactive polymer, polydiphenylbutadiyne (PDPB) nanofibers showed 75% oxidation of MO; however, only 1% degradation was observed when the oxygen was purged with argon (Ghosh et al., 2015a). Obviously, the generation of $O_2^{\bullet-}$ radicals was suppressed in the absence of dissolved O_2, which decreased the degradation efficiency of the catalyst. Similarly, Yuan et al. (2019) reported the decrease in phenol degradation performance of polypyrene nanostructure-based photoactive polymer in the absence of O_2 and the presence of Cu^{2+} under UV irradiation. The Cu^{2+} acts as an electron scavenger that inhibits $O_2^{\bullet-}$ generation, resulting in decreased degradation performance.

Keshavarzi et al. (2020) reported that the photocatalytic degradation of RhB on BA condensed based photoactive polymer could not proceed *via* the photocatalytic reduction of dissolved oxygen. However, it takes place mostly through the photoproduced positive holes on the VB of the polymer, along with the participation of photoproduced superoxide anion radicals. Yuan et al. (2019) found out that the photoproduced $O_2^{\bullet-}$ species were the key oxidizing agents for the oxidation of phenol and MO using a polypyrene-nanostructure-based photocatalyst under visible light, while the surface $HO^{\bullet}$ radicals can also be produced. Many reports mentioned that singlet oxygen could be produced on the surface of photoactive polymer-based photocatalysts under irradiation (Urakami, Zhang et al. 2013, Zhang, Kopetzki et al. 2013, Ma, Ghasimi et al. 2015, Wang, Ghasimi et al. 2015).

7.3 WATER DISINFECTION BY PHOTOACTIVE POLYMER-BASED PHOTOCATALYSTS

The presence of bacterial species in water and wastewater is one of the main worries of environmental and public health; therefore, the scientific community has made much effort to develop efficient antibacterial materials and approaches (Li, Mahendra et al. 2008, Feng, Liu et al. 2017, Dutta, Singh et al. 2019). Photoactive polymers have been used for the bacteria inactivation under light irradiation. Such soft materials can produce ROSs under light, which can attack bacterial species. Djellabi et al. (2020a) have used a TCTA-based photoactive polymer, supported into magnetic Fe_3O_4 for the photo-inactivation of *E.Coli* and *Pseudomonas Aeruginosa* (*PA*) under UV-visible light. As shown in Figure 7.6, CFU reduction effectiveness by CP@Fe_3O_4 under UV-visible light was found to be around 85% and 90% for *E. coli* and *PA*, respectively. Plate test showed effective inhibition of *E.coli* and *PA species* by CP@ Fe_3O_4/UV-visible system, compared to photolysis experiments under UV-visible wherein an intense growth can be observed in the plates. The photocatalytic inactivation of bacteria occurs via ROSs including $O_2^{\bullet-}$, h^+, $^{\bullet}OH$ and H_2O_2 photoproduced by the photoactive polymer under a light. Unlike the photooxidation of organic pollutants, ROSs with longer, e.g., lifetime H_2O_2 exhibit higher inactivation performance than ROSs with a short lifetime (Bianchi, Cerrato et al. 2020).

Zerdin et al. (2009) studied the antibacterial activity of anthraquinone-containing photoactive polymer against *Bacillus cereus* spores under UV-A light. The authors reported that the irradiation of *Bacillus cereus* spores in the presence of photoactive polymer leads to a significant inactivation in comparison with the equivalent irradiation of *Bacillus cereus* spores on the surface of inert polymer substrates. The analysis

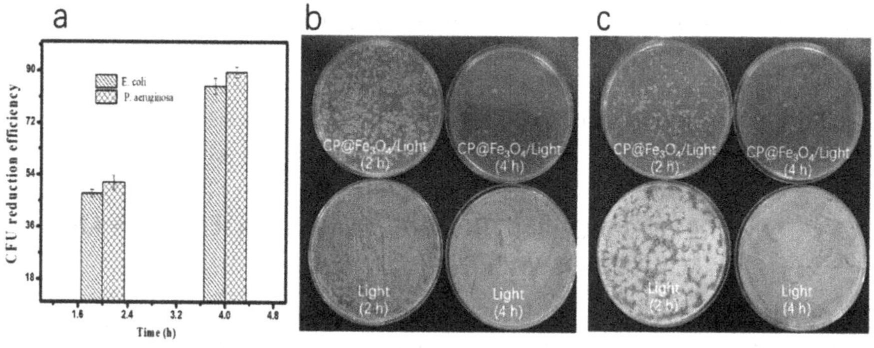

FIGURE 7.6 (a): Photocatalytic CFU reduction efficacy of CP@Fe_3O_4 for *E. coli* and *PA*. (b) and (c): Plate test results after incubation (37°C, 14 h) for *E. coli* and *PA*, respectively. Reproduced with permission from (Djellabi, Ali et al. 2020).

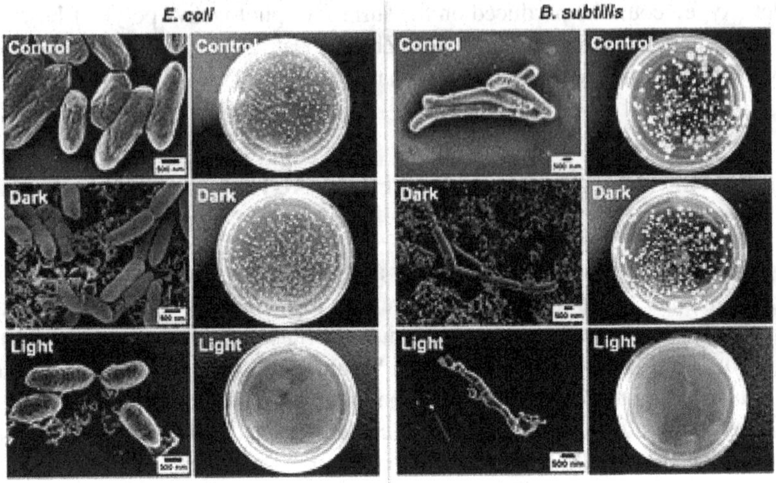

FIGURE 7.7 SEM images and plate results showing the inactivation of bacterial species *Bacillus subtilis* colonies using conjugated microporous photoactive polymer nanoparticles under visible light. Reproduced with permission from (Ma, Ghasimi et al. 2016).

showed that the inactivation was due to the photoproduction of $^1O_2$ and possibly H_2O_2 on the surface of the photoactive polymer under light.

Ma et al. (2016) investigated the antibacterial disinfection abilities of conjugated microporous photoactive polymer nanoparticles against *E. coli* K-12 and *Bacillus subtilis* in water under visible light. Figure 7.7 shows the SEM images and images of plate tests in different conditions, wherein it can be noticed the effective inhibition of both bacterial species growth in cases of a photoactive polymer under light compared with control experiments (control and dark). The authors reported that the addition of scavenger of singlet oxygen (TEMP) causing a decrease in the inhibition efficiency of *E. coli* K-12 to 30%, indicating the predominant role of $^1O_2$ for the photocatalytic degradation of *E. coli* K-12. The scavenging of photoproduced electrons by $NaNO_3$ also significantly limits inactivation efficiency, as it inhibits the reduction of oxygen species to ROSs (e.g., $^1O_2$). In addition, the scavenging of positive holes leads to a decrease in the inactivation up to 60%, which indicates that the degradation of *E. coli* K-12 can take place by direct oxidation with positive holes on the surface of the photoactive polymer.

7.4 COMBINATION OF INORGANIC SEMICONDUCTORS AND PHOTOACTIVE POLYMERS

To better absorption of visible light absorption and photocatalytic synergistic effects, many efforts have been made on the fabrication of organic/inorganic heterojunction based photocatalyst systems (Jana, Bhattacharyya et al. 2015, Sardar, Kar et al. 2015, Zhang, Liu et al. 2016, Ghosh, Remita et al. 2018, Liras, Barawi et al. 2019, El-Naggar and Shoueir 2020). For example, several conjugated photoactive polymers have been used to modify the conventional TiO_2 such as poly(*p*-phenylenevinylene)s and poly(thiophenes) (it was the first report on the photo-induced electron transfer from photoactive conjugated polymer to TiO_2 (Van Hal, Christiaans et al. 1999)), poly(ethylene terephthalate) (Malesic-Eleftheriadou, Evgenidou et al. 2019), polyvinyl alcohol (Song, Zhang et al. 2016), polypropylene (Ramazanov, Hajiyeva et al. 2018), polyethylene (Romero Saez, Jaramillo et al. 2017), polyaniline (Merkulov, Despotović et al. 2018), polypyrrole (Yao, Shi et al. 2016, Liu, Zhou et al. 2020b), polythiophene (Liang and Li 2009), poly(3-hexylthiophene) (Muktha, Mahanta et al. 2007b), poly(3-hexylthiophene) (Wang, Zhang et al. 2009a), polyisoprene (Luo, Bao et al. 2012), polydopamine (Mao, Lin et al. 2016), PVA and PVA_D MO and phenols (Lei, Wang et al. 2014), diketopyrrolopyrrole carbazole (Yang, Yu et al. 2018), poly(benzothiadiazole) (Hou, Zhang et al. 2017), 1,4-diethynylbenzene and 4,7-dibromobenzo[c][1,2,5]thiadiazole (Xiang, Wang et al. 2018), graphitic-C_3N_4 (Sun, Lv et al. 2015, Alenizi, Kumar et al. 2019). Of these photoactive polymers, polyaniline (PANI), due to its efficiency and easy-fabrication (Gottam, Srinivasan et al. 2017), has been widely used to decorate several types of inorganic semiconductors rather that TiO_2 such as Ag_3PO_4 (Bu and Chen 2014), Bi_2SnTiO_7 (Yang and Luan 2012), CdS (Zhang and Zhu 2010), Fe_3O_4 (Yang, Reddy et al. 2016), $Bi_{12}O_{17}Cl_2$ (Xu, Ma et al. 2019), $BiPO_4$ (Yu, Cheng et al. 2018), BiOI (Yan, Zhang et al. 2018), BiOBr (Hao, Gong et al. 2017), BiOCl (Hao, Gong et al. 2017), ZnO (Pei, Ding et al. 2014), MnO_2 (Xu, Zhang et al. 2014), CeO_2 (Samai and Bhattacharya 2018), $Bi2WO_6$ (Wang, Xu et al. 2014), $Bi_2O_2CO_3$ (Zhao, Zhou et al. 2015b), SnS_2 (Zhang, Zhang et al. 2018b), Bi_3NbO_7 (Wu, Liang et al. 2012), $SrTiO_3$ (Shahabuddin, Muhamad Sarih et al. 2016), PbS (Shaban, Rabia et al. 2017), Co_3O_4 (Shahabuddin, Sarih et al. 2015), SiO_2 (Liu, Miao et al. 2014), zirconium (IV) silicophosphate (Pathania, Sharma et al. 2014), NiO (Ahuja, Ujjain et al. 2018) AND Bi2Se3 (Chatterjee, Ahamed et al. 2019).

Recently, the ability of nanostructured metal oxides Fe_3O_4 (Della Pina, Rossi et al. 2012), $CoFe_2O_4$ (Della Pina, Ferretti et al. 2015), MFe_2O_4 nanoferrites (Falletta, Ponti et al. 2015), TiO_2 (Cionti, Della Pina et al. 2018) was demonstrated to catalyze the oxidative polymerization of *N*-(4-aminophenyl) aniline (aniline dimer) to produce PANI/metal oxide NPs composites. This innovative approach is strongly interesting for at least two reasons. Firstly, using *N*-(4-aminophenyl)aniline instead of aniline monomer reduces the production of toxic and carcinogenic by-products (mainly *trans*-azobenzene, benzidine, and heavy metals). Moreover, unlike aniline that is toxic and carcinogenic, *N*-(4-aminophenyl)aniline shows only irritant activity and allows a safer manipulation from researchers. Although so far, these PANI-based composites have not yet been investigated as photocatalysts, they represent an interesting new generation of materials for these kinds of applications.

Figure 7.8a shows the energy edge relative for the famous inorganic semiconductors and widely reported conjugated photoactive polymers. Five possible mechanistic pathways could occur in inorganic semiconductor-photoactive polymers heterojunction systems, as shown in Figure 7.8b–d. (I)

In the sensitization system, the photoactive polymer acts as a photosensitizer, wherein the photoactive polymer absorbs visible light and produces electrons. Such photoelectrons are moved to the CB of the inorganic semiconductor, resulting

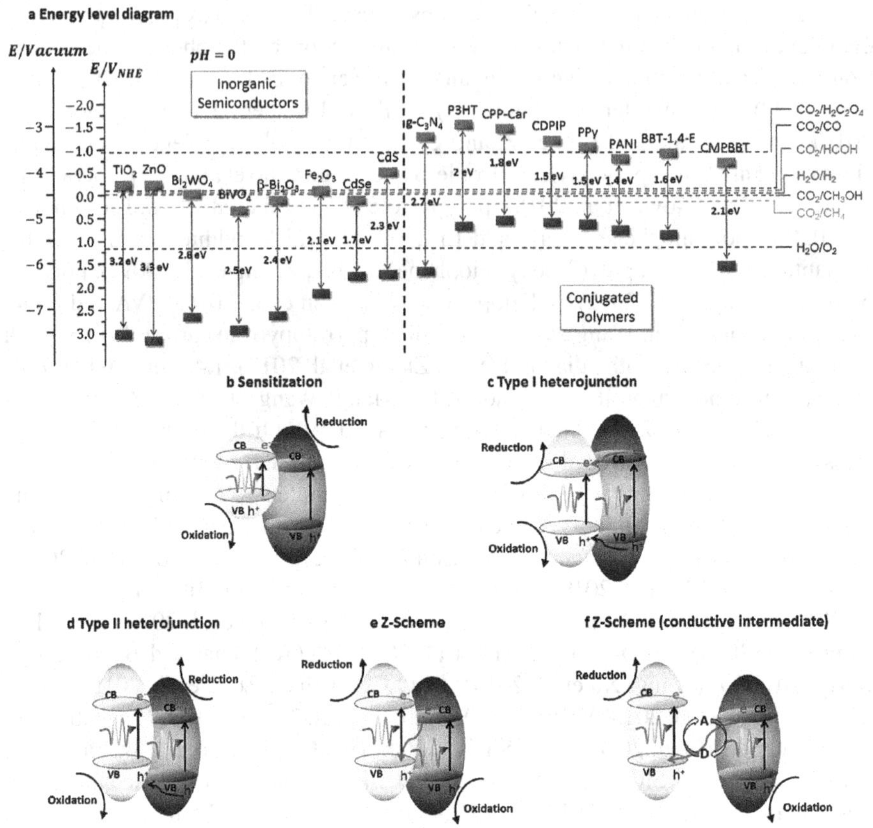

FIGURE 7.8 (a): Energy level diagram of famous inorganic semiconductors vs photoactive polymers and (b–f): types of charge-transfer mechanistic pathways in inorganic-organic hybrid systems. Reproduced with permission from (Liras, Barawi et al. 2019).

in effective charges separation and enhanced photoactivity (Figure 7.8b). (II) In the Type I heterojunction system, both inorganic semiconductor and photoactive polymer can absorb the light; meanwhile, both photoproduced electrons and positive holes move from CB and VB of a broader band gap semiconductor to CB and VB of shorter band gap semiconductor, respectively (Figure 7.8c). (III) In a Type II heterojunction system, the relative band gap edge positions allow the movement of photoproduced electrons from CB with higher to lower energy potential, while the photoproduced positive holes are transferred from lower to higher energy potentials as shown in Figure 7.8d. (IV) Z scheme system has been inspired by the Photosystem I and Photosystem II mechanism of natural photosynthesis. The photoproduced electrons move from the CB of one semiconductor to the VB of the other one (Figure 7.8e). (V) In the Z scheme (conductive intermediate) system, the inorganic semiconductor and the photoactive polymer are associated in sequence with reversible redox pair shuttles (donors of acceptor of electrons/holes) or a conductive medium (Figure 7.8e).

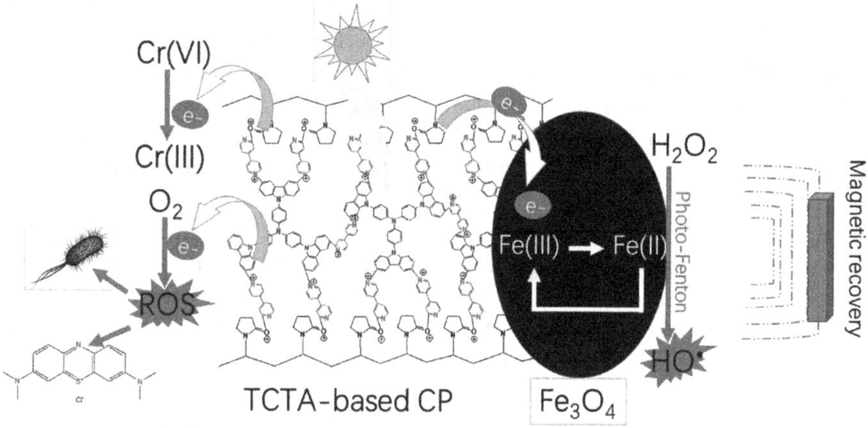

FIGURE 7.9 Mechanistic pathways for photocatalytic oxidation/reduction and bacteria inactivation by CP@Fe_3O_4 system. Reproduced with permission from (Djellabi, Ali et al. 2020b).

Djellabi et al. (2020) synthesized a composite consisting of an TCTA@PVP photoactive lymer (CP) edited with CuO nanoparticles for photocatalytic water treatment/disinfection under visible light. The composite showed a great photocatalytic efficiency under visible light (λ >420 nm) toward the oxidation of MB and the inactivation of *E. coli* species. The visible light irradiation can photoexcite the TCTA photoactive polymer, while the photoproduced electrons can induce redox recycling of Cu(I) and Cu(II), which further enhances the production of more oxidative species. As shown in Figure 7.9, a visible light-responsive magnetic recoverable electron-rich TCTA@Fe_3O_4 was reported by Djellabi et al. (2020) for photocatalytic dye oxidation, Cr(VI) reduction and bacteria (*Escherichia coli* and *Pseudomonas Aeruginosa*) inactivation under visible light. It was mentioned that the transfer of charges from visible light-responsive TCTA based photoactive polymer to Fe_3O_4 may reduce Fe(III) to Fe(II), which favors the Fenton reaction. In general, conducting polymers may help to reduce the release of by-products in water; for example, Sandikly et al. (2021) reported that the treated water by PANI-TiO_2 was less toxic to algae cells, while water treatment with naked TiO_2 was very toxic even more than the started sulfaquinoxaline solution.

7.5 PHOTOACTIVE POLYMER-BASED MEMBRANES

Membrane technology has revolutionized the water and wastewater treatment processes. Polymeric membrane applications have gained considerable attention due to their superior performance and low cost (Zhang, Li et al. 2020). Ultrafiltration/polymeric membranes have been widely used in industrial applications owing to super-selectivity, superior permeability, and low energy consumption. These polymer-based materials with tailored pore sizes are widely preferred in sterile filtration, water treatment, and hemodialysis. However, membrane performance decreases with time due to membrane fouling. Membrane fouling dramatically decreases the system productivity and overall performance. Thus, frequent cleaning of membrane is needed because it reduces the durability and increases the cost. Various approaches have been employed to mitigate membrane fouling (Rana, Nazar et al. 2018, Ali, Wang et al. 2019). Embedding the functional nanomaterials into polymeric membranes is an emerging strategy to enhance antifouling, hydrophilicity, and improve wastewater treatment. Although embedding metal oxide nanomaterials into the casting solution can improve the integrated performance only via physical interaction between nanoparticles and aqueous solution (Wang, Yong et al. 2018d).

Many types of chemical pollutants or bacterial species can be deposited on the surface or within the pores of the membranes, thereby declining the permeate flux (Yang, Xu et al. 2020). Therefore, the fabrication of new nanomaterials is necessary to improve the antifouling properties of the membrane via physical or chemical routes. These functional nanomaterials (photocatalysts) can degrade organic/inorganic pollutants and metal oxides on the surface of photoactive membranes by reacting with strong solid superacid (Zhu, Tian et al. 2012). The use of photoactive nanomaterials has additional advantages which can impart evident improvements in the membrane properties. Photoactive materials are irradiated with a specific wavelength of light to excite into a triplet state. The excess energy can be transferred to O_2 in its resting/ground state, resulting in $^1O_2$. The singlet oxygen is a selective oxidant and very reactive due to its electrophilic nature. In addition, singlet oxygen is cytotoxic in nature and can also be used as a disinfectant. Thus, the use of photoactive polymer-based membranes has great potential in water purification stations as they can eliminate organic substances or bacterial species when their molecular size is too low to be eliminated by micro- or ultrafiltration.

The development of efficient photoactive membranes remains challenging due to the physical loading of photocatalytic nanomaterials on the membrane surface, which can affect the membrane pores. Here we have summarized some recent fabrication strategies for photoactive polymeric membranes. In a recent study, chemical grafting of conjugated polyelectrolytes (CPE) was carried out on the surface of polyvinylidene difluoride (PVDF) membrane to produce visible light-based polymeric membranes with high stability (Jeong, Byun et al. 2021). After chemical tethering of the CPE surface and anion exchange, it was found that the surface hydrophilicity was significantly enhanced without disturbing water permeability and porosity. The As-synthesized photoactive polymeric membrane exhibited excellent photodegradation of organic molecules, photoreduction of hexavalent chromium, and photo-inactivation of mixed culture biofilm. Under visible light irradiation, the antifouling activity recovered the 97% flux even after repetitive filtration cycles.

Another study reported the synthesis of photoactive polymer membrane by immobilizing the two different photosensitizers on the commercial polyethersulfone membrane. In this method, electron-beam-mediated grafting was performed to obtain a highly reactive surface of membranes. Photosensitizers generated singlet oxygen upon activation by visible light, which degraded the pharmaceutical compounds. Immobilization of photosensitizers improves the hydrophilicity and permeance without obviously affecting morphological and porosity properties of the membrane, as

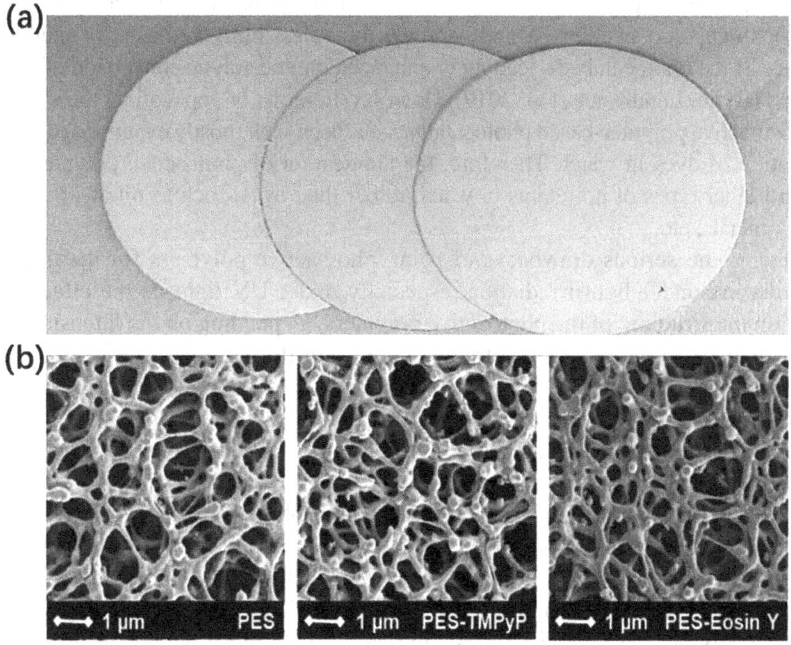

FIGURE 7.10 Images of pure PES membrane (left), and after fixation of TMPyP (middle) and Eosin Y (right). (b) SEM images, respectively. Reproduced with permission from (Becker-Jahn, Griebel et al. 2020).

shown in Figure 7.10 (Becker-Jahn, Griebel et al. 2020). A nanofibrous membrane with enhanced visible light absorption was used for antibacterial wound dressing. The photoactive wound dressing was fabricated in two steps; firstly, the photoactive polymer was developed via mini-emulsion polymerization. Later on, the polymer was embedded in PVA hydrogel nanofibers by colloid-electrospinning, followed by crosslinking by glutaraldehyde/HCl vapor to get a water-insoluble material. The membrane synthesized via this method was mechanically stable with excellent disinfecting characteristics against Gram-positive and Gram-negative bacteria (Jiang, Ma et al. 2018). The underlying mechanism of action for photoactive polymeric membranes is yet to be fully explored. Under the action of photons, photocatalysts generate highly oxidizing free radicals allowing the degradation of compounds adsorbed on its surface.

7.6 CONCLUSIONS AND FUTURE ASPECTS

Recently, the use of photoactive polymers, including conjugated and non-conjugated, showed their great ability for visible light absorption, as well as a good photocatalytic efficiency, in some cases, better than in doped inorganic semiconductors. In addition, the combination of photoactive polymers with inorganic semiconductors has been proved to be very effective for enhancing the visible light response and photoactivity.

Unlike inorganic photocatalysts, photoactive-polymer-based photocatalysts exhibit hydrophilicity and swelling, which improves the accessibility to the active sites on the surface of the photocatalysts, leading to enhanced photocatalytic conversion of the pollutants (Byun, Landfester et al. 2019). However, in terms of application, the efficiency of photoactive polymer-based photocatalysts has been well mostly examined toward the oxidation of dyes in water. Therefore, the application of photoactive polymers looks toward other types of pollutants in water, rather than dyes, such as nitrate, pesticides, heavy metals, etc.

One of the serious drawbacks of using photoactive polymers for the treatment remediation under light irradiation, especially under UV light, is the effect of the light on the structure of the photoactive polymers. Depending on the intensity of UV light, damage to the polymeric structure can occur during long-term photocatalytic treatment, which leads to a decrease in photoactivity/resistibility. It may also lead to contaminate the water from the organic matter released from the polymer mass. Therefore, before choosing a photoactive polymer as a photocatalyst, extensive thermomechanical tests should be considered.

Unlike inorganic nanomaterials, polymer-based photocatalysts are not well stable due to operating conditions such as pH. For example, Irigoyen-Campuzano, González-Béjar et al. (2017) observed that photoactive polymer was released from the glass substrate at pH 10, unlike TiO_2 nanoparticles. In order to transfer the photoactive polymer-based photocatalysts to real-world application, the development of processes for their better fixation on the bed photoreactors is recommended.

REFERENCES

Ahuja, P., S. K. Ujjain, I. Arora and M. Samim (2018). "Hierarchically grown NiO-decorated polyaniline-reduced graphene oxide composite for ultrafast sunlight-driven photocatalysis." *ACS Omega* **3**: 7846–7855.

Alenizi, M., R. Kumar, M. Aslam, F. Alseroury and M. Barakat (2019). "Construction of a ternary gC 3 N 4/TiO_2@ polyaniline nanocomposite for the enhanced photocatalytic activity under solar light." *Scientific Reports* **9**: 1–8.

Ali, J., L. Wang, H. Waseem, H. M. A. Sharif, R. Djellabi, C. Zhang and G. Pan (2019). "Bioelectrochemical recovery of silver from wastewater with sustainable power generation and its reuse for biofouling mitigation." *Journal of Cleaner Production* **235**: 1425–1437.

Ameen, S., H.-K. Seo, M. S. Akhtar and H. S. Shin (2012). "Novel graphene/polyaniline nanocomposites and its photocatalytic activity toward the degradation of rose Bengal dye." *Chemical Engineering Journal* **210**: 220–228.

Becker-Jahn, J., J. Griebel, S. Glaß, P. Langowski, S. Nieß and A. Schulze (2020). "Photoactive polymer membranes for degradation of pharmaceuticals from water." *Catalysis Today* **364**: 256–262.

Bellardita, M., E. I. García-López, G. Marcì, I. Krivtsov, J. R. García and L. Palmisano (2018). "Selective photocatalytic oxidation of aromatic alcohols in water by using P-doped g-C3N4." *Applied Catalysis B: Environmental* **220**: 222–233.

Bethi, B., S. H. Sonawane, B. A. Bhanvase and S. P. Gumfekar (2016). "Nanomaterials-based advanced oxidation processes for wastewater treatment: A review." *Chemical Engineering and Processing-Process Intensification* **109**: 178–189.

Bianchi, C. L., G. Cerrato, B. Bresolin, R. Djellabi and S. Rtimi (2020). "Digitally printed AgNPs doped TiO_2 on commercial porcelain-grès tiles: Synergistic effects and continuous photocatalytic antibacterial activity." *Surfaces* **3**: 11–25.

Bora, L. V. and R. K. Mewada (2017). "Visible/solar light active photocatalysts for organic effluent treatment: Fundamentals, mechanisms and parametric review." *Renewable and Sustainable Energy Reviews* **76**: 1393–1421.

Bu, Y. and Z. Chen (2014). "Role of polyaniline on the photocatalytic degradation and stability performance of the polyaniline/silver/silver phosphate composite under visible light." *ACS Applied Materials & Interfaces* **6**: 17589–17598.

Byun, J., K. Landfester and K. A. Zhang (2019). "Conjugated polymer hydrogel photocatalysts with expandable photoactive sites in water." *Chemistry of Materials* **31**: 3381–3387.

Cai, L., Y. Li, Y. Li, H. Wang, Y. Yu, Y. Liu and Q. Duan (2018). "Synthesis of zincphthalocyanine-based conjugated microporous polymers with rigid-linker as novel and green heterogeneous photocatalysts." *Journal of Hazardous Materials* **348**: 47–55.

Calik, M., F. Auras, L. M. Salonen, K. Bader, I. Grill, M. Handloser, D. D. Medina, M. Dogru, F. Löbermann and D. Trauner (2014). "Extraction of photogenerated electrons and holes from a covalent organic framework integrated heterojunction." *Journal of the American Chemical Society* **136**: 17802–17807.

Cao, X., Y. Li, B. Liu, A. Gao, J. Cao, Y. Yu and X. Hei (2019). "A fluorescent conjugated polymer photocatalyst based on Knoevenagel polycondensation for hydrogen production." *New Journal of Chemistry* **43**: 7093–7098.

Chakrabarti, S. (2017). *Solar Photocatalysis For Environmental Remediation*, The Energy and Resources Institute (TERI), New Delhi, India.

Chatterjee, M. J., S. T. Ahamed, M. Mitra, C. Kulsi, A. Mondal and D. Banerjee (2019). "Visible-light influenced photocatalytic activity of polyaniline-bismuth selenide composites for the degradation of methyl orange, rhodamine B and malachite green dyes." *Applied Surface Science* **470**: 472–483.

Chen, Y., B. Wang, S. Lin, Y. Zhang and X. Wang (2014). "Activation of $n \rightarrow \pi^*$ transitions in two-dimensional conjugated polymers for visible light photocatalysis." *The Journal of Physical Chemistry C* **118**: 29981–29989.

Cionti, C., C. Della Pina, D. Meroni, E. Falletta and S. Ardizzone (2018). "Triply green polyaniline: UV irradiation-induced synthesis of a highly porous PANI/TiO_2 composite and its application in dye removal." *Chemical Communications* **54**: 10702–10705.

Cionti, C., C. Della Pina, D. Meroni, E. Falletta and S. Ardizzone (2020). "Photocatalytic and oxidative synthetic pathways for highly efficient PANI-TiO_2 nanocomposites as organic and inorganic pollutant sorbents." *Nanomaterials* **10**: 441.

Daghrir, R., P. Drogui and D. Robert (2013). "Modified TiO_2 for environmental photocatalytic applications: a review." *Industrial & Engineering Chemistry Research* **52**: 3581–3599.

Dai, C. and B. Liu (2020). "Conjugated polymers for visible-light-driven photocatalysis." *Energy & Environmental Science* **13**: 24–52.

Della Pina, C., E. Falletta and M. Rossi (2011). "Conductive materials by metal catalyzed polymerization." *Catalysis Today* **160**: 11–27.

Della Pina, C., A. M. Ferretti, A. Ponti and E. Falletta (2015). "A green approach to magnetically-hard electrically-conducting polyaniline/$CoFe_2O_4$ nanocomposites." *Composites Science and Technology* **110**: 138–144.

Della Pina, C., M. Rossi, A. M. Ferretti, A. Ponti, M. L. Faro and E. Falletta (2012). "One-pot synthesis of polyaniline/Fe_3O_4 nanocomposites with magnetic and conductive behaviour. Catalytic effect of Fe_3O_4 nanoparticles." *Synthetic Metals* **162**: 2250–2258.

Djellabi, R. (2015). "*Contribution de la photocatalyse à l'élimination des polluants industriels*." Doctorat-LMD en Chimie, Universite Badji, Mokhtar–Annaba.

Djellabi, R., J. Ali, B. Yang, M. R. Haider, P. Su, C. L. Bianchi and X. Zhao (2020a). "Synthesis of magnetic recoverable electron-rich TCTA@ PVP based conjugated polymer for photocatalytic water remediation and disinfection." *Separation and Purification Technology* **250**: 116954.

Djellabi, R., J. Ali, X. Zhao, A. N. Saber and B. Yang (2020b). "CuO NPs incorporated into electron-rich TCTA@ PVP photoactive polymer for the photocatalytic oxidation of dyes and bacteria inactivation." *Journal of Water Process Engineering* **36**: 101238.

Djellabi, R. and M. Ghorab (2015). "Photoreduction of toxic chromium using TiO2-immobilized under natural sunlight: effects of some hole scavengers and process parameters." *Desalination and Water Treatment* **55**: 1900–1907.

Djellabi, R., B. Yang, K. Xiao, Y. Gong, D. Cao, H. M. A. Sharif, X. Zhao, C. Zhu and J. Zhang (2019). "Unravelling the mechanistic role of Ti-OC bonding bridge at Titania/Lignocellulosic biomass interface for Cr (VI) photoreduction under visible light." *Journal of Colloid and Interface Science* **553**: 409–417.

Djellabi, R., X. Zhao, C. L. Bianchi, P. Su, J. Ali and B. Yang (2020c). "Visible light responsive photoactive polymer supported on carbonaceous biomass for photocatalytic water remediation." *Journal of Cleaner Production* **269**: 122286.

Dmitriev, A. V. and M. Mocker (1995). "Recombination and ionization in narrow gap semiconductors." *Physics Reports* **257**: 85–131.

Du, R., P. Chen, Q. Zhang and G. Yu (2020). "The degradation of enrofloxacin by a non-metallic heptazine-based OCN polymer: Kinetics, mechanism and effect of water constituents." *Chemosphere* **273**: 128435.

Duhan, M. and R. Kaur (2020). "Adsorptive removal of methyl orange with polyaniline nanofibers: an unconventional adsorbent for water treatment." *Environmental Technology* **41**: 2977–2990.

Dutta, V., P. Singh, P. Shandilya, S. Sharma, P. Raizada, A. K. Saini, V. K. Gupta, A. Hosseini-Bandegharaei, S. Agarwal and A. Rahmani-Sani (2019). "Review on advances in photocatalytic water disinfection utilizing graphene and graphene derivatives-based nanocomposites." *Journal of Environmental Chemical Engineering* **7**: 103132.

El-Naggar, M. E. and K. Shoueir (2020). "Recent advances in polymer/metal/metal oxide hybrid nanostructures for catalytic applications." *Journal of Environmental Chemical Engineering* **8**: 104175.

Falletta, E., A. Ponti, A. Sironi, A. Ferretti and C. Della Pina (2015). "Nanoferrites as catalysts and fillers for polyaniline/nanoparticle composites preparation." *Journal of Advanced Catalysis Science and Technology* 2 : 8–16.

Farooq, S., A. A. Tahir, U. Krewer and S. Bilal (2019). "Efficient photocatalysis through conductive polymer coated FTO counter electrode in platinum free dye sensitized solar cells." *Electrochimica Acta* **320**: 134544.

Feng, Y., L. Liu, J. Zhang, H. Aslan and M. Dong (2017). "Photoactive antimicrobial nanomaterials." *Journal of Materials Chemistry B* **5**: 8631–8652.

Floresyona, D., F. Goubard, P.-H. Aubert, I. Lampre, J. Mathurin, A. Dazzi, S. Ghosh, P. Beaunier, F. Brisset and S. Remita (2017). "Highly active poly (3-hexylthiophene) nanostructures for photocatalysis under solar light." *Applied Catalysis B: Environmental* **209**: 23–32.

Fujishima, A. and K. Honda (1972). "Electrochemical photolysis of water at a semiconductor electrode." *Nature* **238**: 37.

Ghosh, S., N. A. Kouamé, L. Ramos, S. Remita, A. Dazzi, A. Deniset-Besseau, P. Beaunier, F. Goubard, P.-H. Aubert and H. Remita (2015a). "Conducting polymer nanostructures for photocatalysis under visible light." *Nature Materials* **14**: 505–511.

Ghosh, S., N. A. Kouamé, S. Remita, L. Ramos, F. Goubard, P.-H. Aubert, A. Dazzi, A. Deniset-Besseau and H. Remita (2015b). "Visible-light active conducting polymer nanostructures with superior photocatalytic activity." *Scientific Reports* **5**: 1–9.

Ghosh, S., H. Remita and R. N. Basu (2018). "Visible-light-induced reduction of Cr (VI) by PDPB-ZnO nanohybrids and its photo-electrochemical response." *Applied Catalysis B: Environmental* **239**: 362–372.

Gong, J., Y. Li, Y. Zhao, X. Wu, J. Wang and G. Zhang (2020). "Metal-free polymeric (SCN) n photocatalyst with adjustable bandgap for efficient organic pollutants degradation and Cr (VI) reduction under visible-light irradiation." *Chemical Engineering Journal* **402**: 126147.

Gottam, R., P. Srinivasan, D. D. La and S. V. Bhosale (2017). "Improving the photocatalytic activity of polyaniline and a porphyrin via oxidation to obtain a salt and a charge-transfer complex." *New Journal of Chemistry* **41**: 14595–14601.

Hao, X., J. Gong, L. Ren, D. Zhang, X. Xiao, Y. Jiang, F. Zhang and Z. Tong (2017). "Preparation of polyaniline modified BiOBr with enhanced photocatalytic activities." *Functional Materials Letters* **10**: 1750040.

Hou, H.-J., X.-H. Zhang, D.-K. Huang, X. Ding, S.-Y. Wang, X.-L. Yang, S.-Q. Li, Y.-G. Xiang and H. Chen (2017). "Conjugated microporous poly (benzothiadiazole)/ TiO_2 heterojunction for visible-light-driven H_2 production and pollutant removal." *Applied Catalysis B: Environmental* **203**: 563–571.

Hua, S., D. Qu, L. An, W. Jiang, Y. Wen, X. Wang and Z. Sun (2019). "Highly efficient p-type Cu3P/n-type g-C3N4 photocatalyst through Z-scheme charge transfer route." *Applied Catalysis B: Environmental* **240**: 253–261.

Huang, J., W. Ho and X. Wang (2014). "Metal-free disinfection effects induced by graphitic carbon nitride polymers under visible light illumination." *Chemical Communications* **50**: 4338-4340.

Inagaki, M., T. Tsumura, T. Kinumoto and M. Toyoda (2019). "Graphitic carbon nitrides (g-C3N4) with comparative discussion to carbon materials." *Carbon* **141**: 580-607.

Irigoyen-Campuzano, R., M. González-Béjar, E. Pino, J. B. Proal-Nájera and J. Pérez-Prieto (2017). "A metal-free, nonconjugated polymer for solar photocatalysis." *Chemistry–A European Journal* **23**: 2867–2876.

Ismael, M. and Y. Wu (2019). "A mini-review on the synthesis and structural modification of gC 3 N 4-based materials, and their applications in solar energy conversion and environmental remediation." *Sustainable Energy & Fuels* **3**: 2907–2925.

Jana, B., S. Bhattacharyya and A. Patra (2015). "Conjugated polymer P3HT–Au hybrid nanostructures for enhancing photocatalytic activity." *Physical Chemistry Chemical Physics* **17**: 15392–15399.

Jeong, E., J. Byun, B. Bayarkhuu and S. W. Hong (2021). "Hydrophilic photocatalytic membrane via grafting conjugated polyelectrolyte for visible-light-driven biofouling control." *Applied Catalysis B: Environmental* **282**: 119587.

Jiang, S., B. C. Ma, W. Huang, A. Kaltbeitzel, G. Kizisavas, D. Crespy, K. A. Zhang and K. Landfester (2018). "Visible light active nanofibrous membrane for antibacterial wound dressing." *Nanoscale Horizons* **3**: 439–446.

Keshavarzi, N., S. Cao and M. Antonietti (2020). "A new conducting polymer with exceptional visible-light photocatalytic activity derived from barbituric acid polycondensation." *Advanced Materials* **32**: 1907702.

Kim, J.-G., S.-M. Park, M. E. Lee, E. E. Kwon and K. Baek (2018). "Photocatalytic co-oxidation of As (III) and orange G using urea-derived g-C3N4 and persulfate." *Chemosphere* **212**: 193–199.

Kumar, R., J. Travas-Sejdic and L. P. Padhye (2020). "Conducting polymers-based photocatalysis for treatment of organic contaminants in water." *Chemical Engineering Journal Advances v*: 100047.

Lan, X., Q. Li, Y. Zhang, Q. Li, L. Ricardez-Sandoval and G. Bai (2020). "Engineering donor-acceptor conjugated organic polymers with boron nitride to enhance photocatalytic performance towards visible-light-driven metal-free selective oxidation of sulfides." *Applied Catalysis B: Environmental* **277**: 119274.

Lee, J. J., W. Noh, T.-H. Huh, Y.-J. Kwark and T. S. Lee (2020). "Synthesis of conjugated microporous polymer and its embedding in porous nanofibers for visible-light-driven photocatalysis with reusability." *Polymer* **211**: 123060.

Lei, P., F. Wang, S. Zhang, Y. Ding, J. Zhao and M. Yang (2014). "Conjugation-grafted-TiO_2 nanohybrid for high photocatalytic efficiency under visible light." *ACS Applied Materials & Interfaces* **6**: 2370–2376.

Li, M., H. Zhao and Z.-Y. Lu (2020). "Porphyrin-based porous organic polymer, Py-POP, as a multifunctional platform for efficient selective adsorption and photocatalytic degradation of cationic dyes." *Microporous and Mesoporous Materials* **292**: 109774.

Li, Q., S. Mahendra, D. Y. Lyon, L. Brunet, M. V. Liga, D. Li and P. J. Alvarez (2008). "Antimicrobial nanomaterials for water disinfection and microbial control: potential applications and implications." *Water Research* **42**: 4591–4602.

Li, X., J.-L. Shi, H. Hao and X. Lang (2018). "Visible light-induced selective oxidation of alcohols with air by dye-sensitized TiO_2 photocatalysis." *Applied Catalysis B: Environmental* **232**: 260–267.

Li, Y., J. Zhang, Q. Wang, Y. Jin, D. Huang, Q. Cui and G. Zou (2010). "Nitrogen-rich carbon nitride hollow vessels: synthesis, characterization, and their properties." *The Journal of Physical Chemistry B* **114**: 9429–9434.

Liang, H.-C. and X.-Z. Li (2009). "Visible-induced photocatalytic reactivity of polymer–sensitized titania nanotube films." *Applied Catalysis B: Environmental* **86**: 8–17.

Liras, M., M. Barawi and A. Víctor (2019). "Hybrid materials based on conjugated polymers and inorganic semiconductors as photocatalysts: from environmental to energy applications." *Chemical Society Reviews* **48**: 5454–5487.

Liu, H., S. Ma, L. Shao, H. Liu, Q. Gao, B. Li, H. Fu, S. Fu, H. Ye and F. Zhao (2020a). "Defective engineering in graphitic carbon nitride nanosheet for efficient photocatalytic pathogenic bacteria disinfection." *Applied Catalysis B: Environmental* **261**: 118201.

Liu, J., S. Zhou, P. Gu, T. Zhang, D. Chen, N. Li, Q. Xu and J. Lu (2020b). "Conjugate Polymer-clothed TiO_2@ $V2O_5$ nanobelts and their enhanced visible light photocatalytic performance in water remediation." *Journal of Colloid and Interface Science* **578**: 402–411.

Liu, Z., Y.-E. Miao, M. Liu, Q. Ding, W. W. Tjiu, X. Cui and T. Liu (2014). "Flexible polyaniline-coated TiO2/SiO2 nanofiber membranes with enhanced visible-light photocatalytic degradation performance." *Journal of Colloid And Interface Science* **424**: 49–55.

Low, J., J. Yu, M. Jaroniec, S. Wageh and A. A. Al-Ghamdi (2017). "Heterojunction photocatalysts." *Advanced Materials* **29**: 1601694.

Lu, C., R. Chen, X. Wu, M. Fan, Y. Liu, Z. Le, S. Jiang and S. Song (2016). "Boron doped g-C3N4 with enhanced photocatalytic UO22+ reduction performance." *Applied Surface Science* **360**: 1016–1022.

Luo, Q., L. Bao, D. Wang, X. Li and J. An (2012). "Preparation and strongly enhanced visible light photocatalytic activity of TiO_2 nanoparticles modified by conjugated derivatives of polyisoprene." *The Journal of Physical Chemistry C* **116**: 25806–25815.

Ma, B. C., S. Ghasimi, K. Landfester, F. Vilela and K. A. Zhang (2015). "Conjugated microporous polymer nanoparticles with enhanced dispersibility and water compatibility for photocatalytic applications." *Journal of Materials Chemistry A* **3**: 16064–16071.

Ma, B. C., S. Ghasimi, K. Landfester and K. A. Zhang (2016). "Enhanced visible light promoted antibacterial efficiency of conjugated microporous polymer nanoparticles via molecular doping." *Journal of Materials Chemistry B* **4**: 5112–5118.

Malesic-Eleftheriadou, N., E. N. Evgenidou, G. Z. Kyzas, D. N. Bikiaris and D. A. Lambropoulou (2019). "Removal of antibiotics in aqueous media by using new synthesized bio-based poly (ethylene terephthalate)-TiO_2 photocatalysts." *Chemosphere* **234**: 746–755.

Mao, W.-X., X.-J. Lin, W. Zhang, Z.-X. Chi, R.-W. Lyu, A.-M. Cao and L.-J. Wan (2016). "Core–shell structured TiO_2@ polydopamine for highly active visible-light photocatalysis." *Chemical Communications* **52**: 7122–7125.

Marcelino, R. B. and C. C. Amorim (2019). "Towards visible-light photocatalysis for environmental applications: Band-gap engineering versus photons absorption—a review." *Environmental Science and Pollution Research* **26**: 4155–4170.

Matsuoka, S., H. Fujii, T. Yamada, C. Pac, A. Ishida, S. Takamuku, M. Kusaba, N. Nakashima and S. Yanagida (1991). "Photocatalysis of oligo (p-phenylenes): photoreductive production of hydrogen and ethanol in aqueous triethylamine." *The Journal of Physical Chemistry* **95**: 5802–5808.

McTiernan, C. D., S. P. Pitre and J. C. Scaiano (2014). "Photocatalytic dehalogenation of vicinal dibromo compounds utilizing Sexithiophene and visible-light irradiation." *ACS Catalysis* **4**: 4034–4039.

Merkulov, D. V. Š., V. N. Despotović, N. D. Banić, S. J. Armaković, N. L. Finčur, M. J. Lazarević, D. D. Četojević-Simin, D. Z. Orčić, M. B. Radoičić and Z. V. Šaponjić (2018). "Photocatalytic decomposition of selected biologically active compounds in environmental waters using TiO2/polyaniline nanocomposites: Kinetics, toxicity and intermediates assessment." *Environmental pollution* **239**: 457–465.

Merschjann, C., S. Tschierlei, T. Tyborski, K. Kailasam, S. Orthmann, D. Hollmann, T. Schedel-Niedrig, A. Thomas and S. Lochbrunner (2015). "Complementing graphenes: 1D interplanar charge transport in polymeric graphitic carbon nitrides." *Advanced Materials* **27**: 7993–7999.

Moneim, S. M. A., T. A. Gad-Allah, M. El-Shahat, A. M. Ashmawy and H. S. Ibrahim (2016). "Novel application of metal-free graphitic carbon nitride (g-C3N4) in photocatalytic reduction: Recovery of silver ions." *Journal of Environmental Chemical Engineering* **4**: 4165–4172.

Muktha, B., G. Madras, T. Guru Row, U. Scherf and S. Patil (2007a). "Conjugated polymers for photocatalysis." *The Journal of Physical Chemistry B* **111**: 7994–7998.

Muktha, B., D. Mahanta, S. Patil and G. Madras (2007b). "Synthesis and photocatalytic activity of poly (3-hexylthiophene)/TiO2 composites." *Journal of Solid State Chemistry* **180**: 2986–2989.

Naushad, M. and E. Lichtfouse (2019). *Green Materials for Wastewater Treatment*, Springer, Switzerland.

Nesaratnam, S. T. (2014). *Water Pollution Control*, John Wiley & Sons, Chichester, West Sussex.

Nowakowska, M., and K. Szczubiałka (2017). Photoactive polymeric and hybrid systems for photocatalytic degradation of water pollutants. *Polymer Degradation and Stability*, 145: 120–141.

Ohtani, B. (2013). "Titania photocatalysis beyond recombination: A critical review." *Catalysts* **3**: 942–953.

Ong, W.-J., L.-L. Tan, Y. H. Ng, S.-T. Yong and S.-P. Chai (2016). "Graphitic carbon nitride (g-C3N4)-based photocatalysts for artificial photosynthesis and environmental remediation: Are we a step closer to achieving sustainability?" *Chemical Reviews* **116**: 7159–7329.

Pathania, D., G. Sharma, A. Kumar and N. Kothiyal (2014). "Fabrication of nanocomposite polyaniline zirconium (IV) silicophosphate for photocatalytic and antimicrobial activity." *Journal of Alloys and Compounds* **588**: 668–675.

Patnaik, S., K. K. Das, A. Mohanty and K. Parida (2018). "Enhanced photo catalytic reduction of Cr (VI) over polymer-sensitized g-C3N4/ZnFe2O4 and its synergism with phenol oxidation under visible light irradiation." *Catalysis Today* **315**: 52–66.

Pei, Z., L. Ding, M. Lu, Z. Fan, S. Weng, J. Hu and P. Liu (2014). "Synergistic effect in polyaniline-hybrid defective ZnO with enhanced photocatalytic activity and stability." *The Journal of Physical Chemistry C* **118**: 9570–9577.

Pooi, C. K. and H. Y. Ng (2018). "Review of low-cost point-of-use water treatment systems for developing communities." *NPJ Clean Water* **1**: 1–8.

Qi, K., S.-y. Liu and A. Zada (2020). "Graphitic carbon nitride, a polymer photocatalyst." *Journal of the Taiwan Institute of Chemical Engineers*.

Qian, R., H. Zong, J. Schneider, G. Zhou, T. Zhao, Y. Li, J. Yang, D. W. Bahnemann and J. H. Pan (2019). "Charge carrier trapping, recombination and transfer during TiO_2 photocatalysis: An overview." *Catalysis Today* **335**: 78–90.

Qiang, H., T. Chen, Z. Wang, W. Li, Y. Guo, J. Yang, X. Jia, H. Yang, W. Hu and K. Wen (2020). "Pillar [5] arene based conjugated macrocycle polymers with unique photocatalytic selectivity." *Chinese Chemical Letters*.

Qiao, M., X. Wu, S. Zhao, R. Djellabi and X. Zhao (2020). "Peroxymonosulfate enhanced photocatalytic decomposition of silver-cyanide complexes using g-C3N4 nanosheets with simultaneous recovery of silver." *Applied Catalysis B: Environmental* **265**: 118587.

Ramazanov, M., F. Hajiyeva and A. Maharramov (2018). "Structure and properties of PP/TiO2 based polymer nanocomposites." *Integrated Ferroelectrics* **192**: 103–112.

Ran, X., L. Duan, X. Chen and X. Yang (2018). "Photocatalytic degradation of organic dyes by the conjugated polymer poly (1, 3, 4-oxadiazole) s and its photocatalytic mechanism." *Journal of materials science* **53**: 7048–7059.

Rana, S., U. Nazar, J. Ali, Q. u. A. Ali, N. M. Ahmad, F. Sarwar, H. Waseem and S. U. U. Jamil (2018). "Improved antifouling potential of polyether sulfone polymeric membrane containing silver nanoparticles: self-cleaning membranes." *Environmental Technology* **39**: 1413–1421.

Reddy, K. R., C. V. Reddy, M. N. Nadagouda, N. P. Shetti, S. Jaesool and T. M. Aminabhavi (2019). "Polymeric graphitic carbon nitride (g-C3N4)-based semiconducting nanostructured materials: synthesis methods, properties and photocatalytic applications." *Journal of Environmental Management* **238**: 25–40.

Riaz, U., S. Ashraf and M. Aqib (2014). "Microwave-assisted degradation of acid orange using a conjugated polymer, polyaniline, as catalyst." *Arabian Journal of Chemistry* **7**: 79–86.

Riffat, R. (2012). *Fundamentals of Wastewater Treatment and Engineering*, CRC Press, London.

Romero Saez, M., L. Jaramillo, R. Saravanan, N. Benito, E. Pabón, E. Mosquera and F. Gracia Caroca (2017). "Notable photocatalytic activity of TiO_2-polyethylene nanocomposites for visible light degradation of organic pollutants." *eXPRESS Polymer Letters* **11**: 899–909.

Rono, N., J. K. Kibet, B. S. Martincigh and V. O. Nyamori (2020). "A review of the current status of graphitic carbon nitride." *Critical Reviews in Solid State and Materials Sciences* **46**: 189–217.

Saha, S., N. Chaudhary, A. Kumar and M. Khanuja (2020). "Polymeric nanostructures for photocatalytic dye degradation: Polyaniline for photocatalysis." *SN Applied Sciences* **2**: 1115.

Samai, B. and S. C. Bhattacharya (2018). "Conducting polymer supported cerium oxide nanoparticle: enhanced photocatalytic activity for waste water treatment." *Materials Chemistry and Physics* **220**: 171–181.

Sandikly, N., M. Kassir, M. El Jamal, H. Takache, P. Arnoux, S. Mokh, M. Al-Iskandarani and T. Roques-Carmes (2021). "Comparison of the toxicity of waters containing initially sulfaquinoxaline after photocatalytic treatment by TiO_2 and polyaniline/TiO_2." *Environmental Technology* **42**: 419–428.

Sardar, S., P. Kar, H. Remita, B. Liu, P. Lemmens, S. K. Pal and S. Ghosh (2015). "Enhanced charge separation and FRET at heterojunctions between semiconductor nanoparticles and conducting polymer nanofibers for efficient solar light harvesting." *Scientific Reports* **5**: 1–14.

Shaban, M., M. Rabia, A. M. Abd El-Sayed, A. Ahmed and S. Sayed (2017). "Photocatalytic properties of PbS/graphene oxide/polyaniline electrode for hydrogen generation." *Scientific Reports* **7**: 1–13.

Shah, J. H., M. Fiaz, M. Athar, J. Ali, M. Rubab, R. Mehmood, S. U. U. Jamil and R. Djellabi (2020). "Facile synthesis of N/B-double-doped Mn_2O_3 and WO_3 nanoparticles for dye degradation under visible light." *Environmental Technology* **41**: 2372–2381.

Shahabuddin, S., N. Muhamad Sarih, S. Mohamad and J. Joon Ching (2016). "$SrtiO_3$ nanocube-doped polyaniline nanocomposites with enhanced photocatalytic degradation of methylene blue under visible light." *Polymers* **8**: 27.

Shahabuddin, S., N. M. Sarih, F. H. Ismail, M. M. Shahid and N. M. Huang (2015). "Synthesis of chitosan grafted-polyaniline/Co_3O_4 nanocube nanocomposites and their photocatalytic activity toward methylene blue dye degradation." *RSC Advances* **5**: 83857–83867.

Shayegan, Z., C.-S. Lee and F. Haghighat (2018). "TiO_2 photocatalyst for removal of volatile organic compounds in gas phase–A review." *Chemical Engineering Journal* **334**: 2408–2439.

Song, Y., J. Zhang, L. Yang, S. Cao, H. Yang, L. Jiang, Y. Dan, P. Le Rendu and T. Nguyen (2016). "Photocatalytic activity of TiO_2 based composite films by porous conjugated polymer coating of nanoparticles." *Materials Science in Semiconductor Processing* **42**: 54–57.

Su, P., J. Zhang, K. Xiao, S. Zhao, R. Djellabi, X. Li, B. Yang and X. Zhao (2020). "C3N4 modified with single layer ZIF67 nanoparticles for efficient photocatalytic degradation of organic pollutants under visible light." *Chinese Journal of Catalysis* **41**: 1894–1905.

Sun, Q., K. Lv, Z. Zhang, M. Li and B. Li (2015). "Effect of contact interface between TiO_2 and g-C_3N_4 on the photoreactivity of g-C_3N_4/TiO_2 photocatalyst:(0 0 1) vs (1 0 1) facets of TiO_2." *Applied Catalysis B: Environmental* **164**: 420–427.

Taghipour, S., S. M. Hosseini and B. Ataie-Ashtiani (2019). "Engineering nanomaterials for water and wastewater treatment: Review of classifications, properties and applications." *New Journal of Chemistry* **43**: 7902–7927.

Tang, Q., J. Gong and Q. Zhao (2019). "Efficient organic pollutant degradation under visible-light using functional polymers of intrinsic microporosity." *Catalysis Science & Technology* **9**: 5383–5393.

Urakami, H., K. Zhang and F. Vilela (2013). "Modification of conjugated microporous polybenzothiadiazole for photosensitized singlet oxygen generation in water." *Chemical Communications* **49**: 2353–2355.

Van Hal, P. A., M. P. Christiaans, M. M. Wienk, J. M. Kroon and R. A. Janssen (1999). "Photoinduced electron transfer from conjugated polymers to TiO_2." *The Journal of Physical Chemistry B* **103**: 4352–4359.

Vyas, V. S., V. W.-h. Lau and B. V. Lotsch (2016). "Soft photocatalysis: Organic polymers for solar fuel production." *Chemistry of Materials* **28**: 5191–5204.

Wang, B., Z. Xie, Y. Li, Z. Yang and L. Chen (2018a). "Dual-functional conjugated nanoporous polymers for efficient organic pollutants treatment in water: a synergistic strategy of adsorption and photocatalysis." *Macromolecules* **51**: 3443–3449.

Wang, D., J. Zhang, Q. Luo, X. Li, Y. Duan and J. An (2009a). "Characterization and photocatalytic activity of poly (3-hexylthiophene)-modified TiO2 for degradation of methyl orange under visible light." *Journal of Hazardous Materials* **169**: 546–550.

Wang, H., S. Jin, X. Zhang and Y. Xie (2020a). "Excitonic effects in polymeric photocatalysts." *Angewandte Chemie* 132 : 23024–23035.

Wang, H., Y. Wu, M. Feng, W. Tu, T. Xiao, T. Xiong, H. Ang, X. Yuan and J. W. Chew (2018b). "Visible-light-driven removal of tetracycline antibiotics and reclamation of hydrogen energy from natural water matrices and wastewater by polymeric carbon nitride foam." *Water Research* **144**: 215–225.

Wang, J., H. Yang, L. Jiang, S. Liu, Z. Hao, J. Cheng and G. Ouyang (2018c). "Highly efficient removal of organic pollutants by ultrahigh-surface-area-ethynylbenzene-based conjugated microporous polymers via adsorption–photocatalysis synergy." *Catalysis Science & Technology* **8**: 5024–5033.

Wang, T., C. Nie, Z. Ao, S. Wang and T. An (2020b). "Recent progress in gC_3N_4 quantum dots: Synthesis, properties and applications in photocatalytic degradation of organic pollutants." *Journal of Materials Chemistry A* **8**: 485–502.

Wang, W., J. Xu, L. Zhang and S. Sun (2014). "Bi2WO6/PANI: An efficient visible-light-induced photocatalytic composite." *Catalysis Today* **224**: 147–153.

Wang, X., K. Maeda, A. Thomas, K. Takanabe, G. Xin, J. M. Carlsson, K. Domen and M. Antonietti (2009b). "A metal-free polymeric photocatalyst for hydrogen production from water under visible light." *Nature Materials* **8**: 76–80.

Wang, Y., A. Vogel, M. Sachs, R. S. Sprick, L. Wilbraham, S. J. Moniz, R. Godin, M. A. Zwijnenburg, J. R. Durrant and A. I. Cooper (2019). "Current understanding and challenges of solar-driven hydrogen generation using polymeric photocatalysts." *Nature Energy* **4**: 746–760.

Wang, Y., M. Yong, S. Wei, Y. Zhang, W. Liu and Z. Xu (2018d). "Performance improvement of hybrid polymer membranes for wastewater treatment by introduction of micro reaction locations." *Progress in Natural Science: Materials International* **28**: 148–159.

Wang, Z. J., S. Ghasimi, K. Landfester and K. A. Zhang (2015). "Molecular structural design of conjugated microporous poly (benzooxadiazole) networks for enhanced photocatalytic activity with visible light." *Advanced Materials* **27**: 6265–6270.

Wen, J., J. Xie, X. Chen and X. Li (2017). "A review on g-C3N4-based photocatalysts." *Applied Surface Science* **391**: 72–123.

Wu, W., S. Liang, L. Shen, Z. Ding, H. Zheng, W. Su and L. Wu (2012). "Preparation, characterization and enhanced visible light photocatalytic activities of polyaniline/Bi_3NbO_7 nanocomposites." *Journal of Alloys and Compounds* **520**: 213–219.

Wu, X., X. Zhang, S. Zhao, Y. Gong, R. Djellabi, S. Lin and X. Zhao (2020). "Highly-efficient photocatalytic hydrogen peroxide production over polyoxometalates covalently immobilized onto titanium dioxide." *Applied Catalysis A: General* **591**: 117271.

Xiang, Y., X. Wang, X. Zhang, H. Hou, K. Dai, Q. Huang and H. Chen (2018). "Enhanced visible light photocatalytic activity of TiO_2 assisted by organic semiconductors: A structure optimization strategy of conjugated polymers." *Journal of Materials Chemistry A* **6**: 153–159.

Xiao, J., Y. Cheng, C. Guo, X. Liu, B. Zhang, S. Yuan and J. Huang (2019). "Novel functional fiber loaded with carbon dots for the deep removal of Cr (VI) by adsorption and photocatalytic reduction." *Journal of Environmental Sciences* **83**: 195–204.

Xu, H., J. Zhang, Y. Chen, H. Lu, J. Zhuang and J. Li (2014). "Synthesis of polyaniline-modified MnO_2 composite nanorods and their photocatalytic application." *Materials Letters* **117**: 21–23.

Xu, J., Z. Wang and Y. Zhu (2017). "Enhanced visible-light-driven photocatalytic disinfection performance and organic pollutant degradation activity of porous g-C3N4 nanosheets." *ACS Applied Materials & Interfaces* **9**: 27727–27735.

Xu, Y., S. Jin, H. Xu, A. Nagai and D. Jiang (2013). "Conjugated microporous polymers: design, synthesis and application." *Chemical Society Reviews* **42**: 8012–8031.

Xu, Y., Y. Ma, X. Ji, S. Huang, J. Xia, M. Xie, J. Yan, H. Xu and H. Li (2019). "Conjugated conducting polymers PANI decorated Bi12O17Cl2 photocatalyst with extended light response range and enhanced photoactivity." *Applied Surface Science* **464**: 552–561.

Yan, C., Z. Zhang, W. Wang, T. Ju, H. She and Q. Wang (2018). "Synthesis and characterization of polyaniline-modified BiOI: A visible-light-response photocatalyst." *Journal of Materials Science: Materials in Electronics* **29**: 18343–18351.

Yang, L., Y. Yu, J. Zhang, F. Chen, X. Meng, Y. Qiu, Y. Dan and L. Jiang (2018). "In-situ fabrication of diketopyrrolopyrrole-carbazole-based conjugated polymer/TiO_2 heterojunction for enhanced visible light photocatalysis." *Applied Surface Science* **434**: 796–805.

Yang, R.-B., P. M. Reddy, C.-J. Chang, P.-A. Chen, J.-K. Chen and C.-C. Chang (2016). "Synthesis and characterization of Fe3O4/polypyrrole/carbon nanotube composites with tunable microwave absorption properties: Role of carbon nanotube and polypyrrole content." *Chemical Engineering Journal* **285**: 497–507.

Yang, W., H. Xu, W. Chen, Z. Shen, M. Ding, T. Lin, H. Tao, Q. Kong, G. Yang and Z. Xie (2020). "A polyamide membrane with tubular crumples incorporating carboxylated single-walled carbon nanotubes for high water flux." *Desalination* **479**: 114330.

Yang, Y. and J. Luan (2012). "Synthesis, property characterization and photocatalytic activity of the novel composite polymer polyaniline/Bi_2SnTiO_7." *Molecules* **17**: 2752–2772.

Yao, T., L. Shi, H. Wang, F. Wang, J. Wu, X. Zhang, J. Sun and T. Cui (2016). "A simple method for the preparation of TiO_2/Ag-AgCl@ polypyrrole composite and its enhanced visible-light photocatalytic activity." *Chemistry–An Asian Journal* **11**: 141–147.

Youssef, Z., L. Colombeau, N. Yesmurzayeva, F. Baros, R. Vanderesse, T. Hamieh, J. Toufaily, C. Frochot, T. Roques-Carmes and S. Acherar (2018). "Dye-sensitized nanoparticles for heterogeneous photocatalysis: Cases studies with TiO_2, ZnO, fullerene and graphene for water purification." *Dyes and Pigments* **159**: 49–71.

Yu, F., Z. Zhu, S. Wang, Y. Peng, Z. Xu, Y. Tao, J. Xiong, Q. Fan and F. Luo (2020). "Tunable perylene-based donor–acceptor conjugated microporous polymer to significantly enhance photocatalytic uranium extraction from seawater." *Chemical Engineering Journal* **412**: 127558.

Yu, W. J., Y. Cheng, T. Zou, Y. Liu, K. Wu and N. Peng (2018). "Preparation of $BiPO_4$-polyaniline hybrid and its enhanced photocatalytic performance." *Nano* **13**: 1850009.

Yuan, X., D. Floresyona, P.-H. Aubert, T.-T. Bui, S. Remita, S. Ghosh, F. Brisset, F. Goubard and H. Remita (2019). "Photocatalytic degradation of organic pollutant with polypyrrole nanostructures under UV and visible light." *Applied Catalysis B: Environmental* **242**: 284–292.

Zerdin, K., M. A. Horsham, R. Durham, P. Wormell and A. D. Scully (2009). "Photodynamic inactivation of bacterial spores on the surface of a photoactive polymer." *Reactive and Functional Polymers* **69**: 821–827.

Zhang, C., Y. Li, D. Shuai, Y. Shen, W. Xiong and L. Wang (2019). "Graphitic carbon nitride (g-C3N4)-based photocatalysts for water disinfection and microbial control: A review." *Chemosphere* **214**: 462–479.

Zhang, C., Y. Li, D. Shuai, W. Zhang, L. Niu, L. Wang and H. Zhang (2018a). "Visible-light-driven, water-surface-floating antimicrobials developed from graphitic carbon nitride and expanded perlite for water disinfection." *Chemosphere* **208**: 84–92.

Zhang, F., Y. Zhang, G. Zhang, Z. Yang, D. D. Dionysiou and A. Zhu (2018b). "Exceptional synergistic enhancement of the photocatalytic activity of SnS_2 by coupling with polyaniline and N-doped reduced graphene oxide." *Applied Catalysis B: Environmental* **236**: 53–63.

Zhang, H., G. Liu, L. Shi, H. Liu, T. Wang and J. Ye (2016). "Engineering coordination polymers for photocatalysis." *Nano Energy* **22**: 149–168.

Zhang, H. and Y. Zhu (2010). "Significant visible photoactivity and antiphotocorrosion performance of CdS photocatalysts after monolayer polyaniline hybridization." *The Journal of Physical Chemistry C* **114**: 5822–5826.

Zhang, K., D. Kopetzki, P. H. Seeberger, M. Antonietti and F. Vilela (2013). "Surface area control and photocatalytic activity of conjugated microporous poly (benzothiadiazole) networks." *Angewandte Chemie* **125**: 1472–1476.

Zhang, T., X. Wang and X. Zhang (2014). "Recent progress in TiO_2-mediated solar photocatalysis for industrial wastewater treatment." *International Journal of Photoenergy* **2014**: 12.

Zhang, X.-H., X.-P. Wang, J. Xiao, S.-Y. Wang, D.-K. Huang, X. Ding, Y.-G. Xiang and H. Chen (2017). "Synthesis of 1, 4-diethynylbenzene-based conjugated polymer photocatalysts and their enhanced visible/near-infrared-light-driven hydrogen production activity." *Journal of Catalysis* **350**: 64–71.

Zhang, Y., Q. Li, Q. Gao, S. Wan, P. Yao and X. Zhu (2020). "Preparation of Ag/β-cyclodextrin co-doped TiO_2 floating photocatalytic membrane for dynamic adsorption and photoactivity under visible light." *Applied Catalysis B: Environmental* **267**: 118715.

Zhao, Z., Y. Sun and F. Dong (2015a). "Graphitic carbon nitride based nanocomposites: A review." *Nanoscale* **7**: 15–37.

Zhao, Z., Y. Zhou, F. Wang, K. Zhang, S. Yu and K. Cao (2015b). "Polyaniline-decorated {001} facets of Bi2O2CO3 nanosheets: in situ oxygen vacancy formation and enhanced visible light photocatalytic activity." *ACS Applied Materials & Interfaces* **7**: 730–737.

Zhi, Y., K. Li, H. Xia, M. Xue, Y. Mu and X. Liu (2017). "Robust porous organic polymers as efficient heterogeneous organo-photocatalysts for aerobic oxidation reactions." *Journal of Materials Chemistry A* **5**: 8697–8704.

Zhou, J., Y. Wang, Z. Cui, Y. Hu, X. Hao, Y. Wang and Z. Zou (2020). "Ultrathin conjugated polymer nanosheets as highly efficient photocatalyst for visible light driven oxygen activation." *Applied Catalysis B: Environmental* **277**: 119228.

Zhu, X., C. Tian, S. M. Mahurin, S.-H. Chai, C. Wang, S. Brown, G. M. Veith, H. Luo, H. Liu and S. Dai (2012). "A superacid-catalyzed synthesis of porous membranes based on triazine frameworks for CO_2 separation." *Journal of the American Chemical Society* **134**: 10478–10484.

8 Plasmonic Nanomaterials for Remediation of Water and Wastewater

Ewa Kowalska and Kenta Yoshiiri
Institute for Catalysis (ICAT) and Graduate School of Environmental Science, Hokkaido University, Sapporo, Japan

Maya Endo-Kimura
Institute for Catalysis (ICAT), Hokkaido University, Sapporo, Japan

Tharishinny Raja-Mogan
Institute for Catalysis (ICAT) and Graduate School of Environmental Science, Hokkaido University, Sapporo, Japan

Oliwia Paszkiewicz
West Pomeranian University of Technology in Szczecin, Szczecin, Poland

Zhishun Wei
Hubei University of Technology, Wuhan, China

Kunlei Wang
Institute for Catalysis (ICAT), Hokkaido University, Sapporo, Japan
Northwest Research Institute, Lanzhou, P.R. China

Marcin Janczarek
Poznan University of Technology, Poznan, Poland

DOI: 10.1201/9781003118749-8

Agata Markowska-Szczupak
West Pomeranian University of Technology in Szczecin,
Szczecin, Poland

CONTENTS

8.1 Introduction 246
8.2 Decomposition of Chemical Compounds 249
8.2.1 Degradation of Organic Compounds 249
8.2.2 Application of Plasmonic Nanomaterials for Inorganic Pollutants.....258
8.3 Decomposition of Microorganisms 259
8.4 Summary and Conclusions 266
References 266

8.1 INTRODUCTION

Photocatalysis under solar radiation ("solar photocatalysis") has been proposed as a possible and reasonable solution for most serious problems facing humanity, i.e., environment, water and energy (Smalley, 2003), since photocatalyst might be activated under irradiation with natural solar light causing various redox reaction without the participation of any expensive or hazardous chemical compounds. For example, abundant titanium (IV) oxide (titania; TiO_2) under UV irradiation and in aerobic conditions might decompose completely various pollutants, such as phenols (Ilisz et al., 1999; Loddo et al., 1999), dyes (Kowalska et al., 2008), pesticides (Zaleska et al., 2000), herbicides (Ollis et al., 1991; Pathirana and Maithreepala, 1997), cyanoginosins (Khedr et al., 2019a, 2019b), and even microorganisms and their toxins (Markowska-Szczupak et al., 2011; Markowska-Szczupak et al., 2015; Markowska-Szczupak et al., 2018), resulting in efficient wastewater treatment and water purification (Minero et al., 1996). Moreover, excited titania might convert solar energy into "green" fuels, such as hydrogen (Fujishima and Honda, 1972) and methane (Kraeutler and Bard, 1978), and photocurrent (Bach et al., 1998; Nischk et al., 2016). However, two shortcomings of photocatalysts, i.e., charge carriers' recombination (causing lower quantum yields (QY) of photocatalytic reaction than 100%) and bandgap limitations, should be overcome for broad commercialization. Considering the bandgap, its wide value, e.g., ca. 3.0 eV (UV range of excitation; TiO_2 and ZnO), is perfect for redox properties (to drive both oxidation and reduction reactions efficiently), but at the same time, wide-bandgap means inactivity under visible light (vis) range of solar radiation. Accordingly, wide-bandgap semiconductors have been modified or doped with various components to increase QY and extend their photoabsorption properties toward visible light (Zaleska, 2008). Although doping might result in the effective narrowing of bandgap, and thus the appearance of vis-activity, it might also decrease UV activity since dopants might work as recombination centers. Therefore, surface modification with noble metals (NMs) seems to be highly attractive, as NMs have been known for UV-activity enhancement (electron scavengers (Kraeutler and Bard, 1978; Ohtani et al., 1993; Pichat et al., 1981)) and activation of wide-bandgap semiconductors toward vis range of solar spectrum due to plasmonic properties (Kowalska et al., 2009; Tian et al., 2005), i.e., localized surface plasmon resonance (LSPR) for deposits of NMs.

In brief, the mechanism of heterogeneous photocatalysis is composed of three main steps, i.e., (i) excitation of semiconductor (charge carriers' generation), (ii) migration/transfer of charge carriers, and (iii) resultant redox reactions on the surface of the photocatalyst, as shown in Figure 8.1 (left). Semiconductor must be excited by irradiation with light energy larger or equal to its bandgap, e.g., ca. 390–400 nm for titania, resulting in the transfer of electrons from the valence band (VB) to the conduction band (CB), and thus electrons (e^-) and holes (h^+) are formed in CB and VB, respectively. Charge carriers' migration means either their transfer to the particle surface or unwanted bulk/surface recombination. The redox reactions consider the direct reactions of photogenerated charge carriers (e^-/h^+) with adsorbed pollutants or molecular O_2 and H_2O, leading to the formation of reactive oxygen species (ROS). The most famous ROS is hydroxyl radical, a highly powerful oxidant able to decompose completely all organic compounds. As mentioned above, the e^-/h^+ recombination results in much lower efficiency than expected. Therefore, modification with NMs increases activity since NMs work as electron scavengers, as shown in Figure 8.1 (right). Moreover, NMs might also play another function, being co-catalysts. For example, molecular hydrogen formation occurs on their surface during alcohol dehydrogenation (Wang et al., 2018).

In the case of vis excitation, the mechanism of plasmonic photocatalysis has not been fully clarified and understood yet. There are three main hypotheses: (i) an electron transfer ("hot" electrons' migration; Figure 8.2 (left)), (ii) energy transfer (Figure 8.2 (right)), and (iii) plasmonic heating ("thermal effect"). Additionally, enhanced scattering on NMs' deposits has been considered to improve the light-harvesting efficiency.

For the mechanism of "hot" electron transfer, NMs work as "plasmonic sensitizers," and thus plasmonic electrons are transferred from NMs (above Schottky barrier)

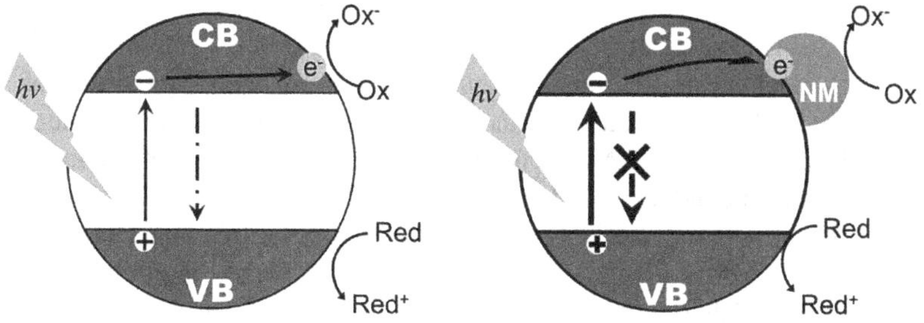

FIGURE 8.1 The schematic mechanisms of heterogeneous photocatalysis in the absence (left) and presence (right) of adsorbed NMs; Ox and Red – reagents adsorbed on the semiconductor surface being reduced and oxidized, respectively; - electron; + hole.

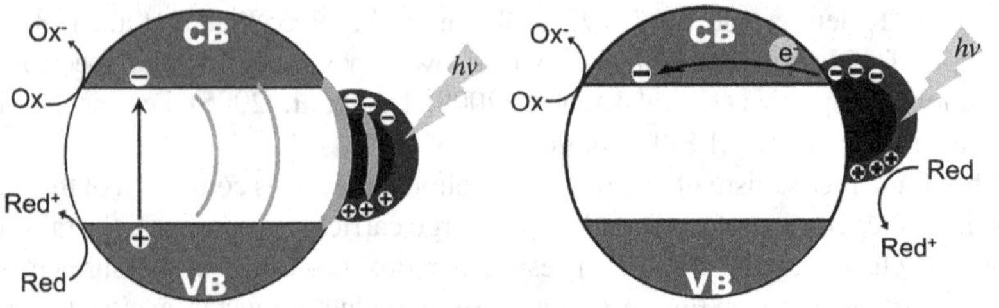

FIGURE 8.2 The schematic mechanisms of plasmonic photocatalysis according to energy transfer (left) and electron transfer (right); Ox and Red – reagents adsorbed on the semiconductor surface being reduced and oxidized, respectively; - electron; + hole.

to CB of semiconductor (Tian et al., 2005). Then, the "hot" electrons might reduce the adsorbed oxygen causing ROS formation. It should be pointed out that this might result in unwanted oxidation of NMs, and thus oxidation of some compounds (e.g., organic pollutants) on the surface of electron-deficient NMs must happen (Kowalska et al., 2009). Many reports confirm directly or indirectly the mechanism of electron transfer by various techniques and methods, e.g., femtosecond absorption spectroscopy (Du et al., 2009a, 2009b), electron paramagnetic resonance (Caretti et al., 2015; Priebe et al., 2015), time-resolved microwave conductivity (TRMC) method (Mendez-Medrano et al., 2016; Wei et al., 2018a), electrochemical study (Sakai et al., 2009a, b).

In contrast to the electron transfer mechanism, the energy transfer mechanism does not involve unwanted oxidation of NMs, but the respective energy levels (NM and semiconductor) should overlap. Accordingly, energy transfer is more probably for silver-modified titania photocatalysts with similar energies of about 3.0 eV for AgNPs (depending on the properties of Ag deposits, i.e., longer wavelengths of LSPR for larger and non-spherical nanoparticles (NPs)) and 3.0–3.2 eV for TiO_2, than for copper- and gold-modified titania photocatalysts with much different energy levels (LSPR at ca. 2.0–2.4 eV for spherical NPs of copper and gold). Accordingly, only semiconductors with vis-activity (e.g., with narrower bandgap or modified/doped) might be considered for energy transfer from LSPR of Au/Cu, e.g., N-modified TiO_2 (Bouhadoun et al., 2015) and defective titania, that is, with crystalline defects (Hou et al., 2011b; Bielan et al., 2020b) and disorders (Seh et al., 2012).

The third mechanism, i.e., thermal activation, has also been suggested for some reactions, that is, the local heating around NMs causes the cleavage of chemical bonds (Chen et al., 2008a, 2008b). Additionally, it has been proposed that the mechanism depends on the semiconductor kind, e.g., both gold-modified ZnO and TiO_2 are active at visible-light irradiation, but only Au-modified ZnO is active under heating in the dark (Hou et al., 2011a; Wang et al., 2013b). Although some reports have proposed plasmonic heating (Chen et al., 2008a, 2008b; Mohamed and Aazam, 2011; Trammell et al., 2012; Wang et al., 2013a), various studies have rejected plasmonic heating as the main mechanism of plasmonic photocatalysis because of the lack of activity for unsupported NMs and NM-modified insulators as compared with highly active NM-modified semiconductors (Son et al., 2010; Silva et al., 2011; Hou et al., 2011b; Liu et al., 2011; Cushing et al., 2012; Mukherjee et al., 2013). Additionally, activation-energy studies have rejected the plasmonic heating for decomposition of organic compounds (Kominami et al., 2011) and photocurrent generation (Nishijima et al., 2010).

Despite the unclear mechanism, which can also depend on the morphology and properties, it is clear that plasmonic photocatalysts are active in many reactions, including conversion of solar energy (Sakai et al., 2009a, b; Ilie et al., 2011), degradation of organic compounds (Kowalska et al., 2010) and microorganisms (Endo et al., 2018), organic synthesis (Mkhalid, 2015), and even cancer treatment (Seo et al., 2011). Accordingly, this chapter presents plasmonic photocatalysts for possible water and wastewater treatment.

8.2 DECOMPOSITION OF CHEMICAL COMPOUNDS

8.2.1 Degradation of Organic Compounds

At first, NM-modified semiconductors were tested for UV-activity enhancement (inhibition of e^-/h^+ recombination), e.g., during oxidative purification of liquid phase containing model pollutants, such as oxalic acid (Iliev et al., 2007), formic acid (Bouhadoun et al., 2015), stearic acid (Kafizas et al., 2009), salicylic acid (Pap et al., 2013; Grčić et al., 2020), various antibiotics (tetracycline hydrochloride) (Smirnova et al., 2010), trimethoprim (Oros-Ruiz et al., 2013)), dyes (Plasmocorinth B, methylene blue (MB) (Li and Li, 2002; Wang et al., 2013b) congo red (Pearson et al., 2012), rhodamine 6G (R6G) (Yi et al., 2015) and methyl orange (MO) (Sun et al., 2014)). It should be pointed out that not only during oxidation pathways but also during reductive decomposition, activity enhancement was observed after NM deposition, e.g., for degradation of 2,2′-dipyridyl disulfide (RSSR), nitrobenzene and MB (Costi et al., 2008). It was proven that enhanced activity was caused by an inhibition of e^-/h^+ recombination, e.g., by photoluminescence spectroscopy (Kiyonaga et al., 2008) and TRMC study (Kowalska et al., 2008; Kowalska et al., 2015a, b). Typically, high improvement of photocatalytic activity is obtained after modification of the semiconductor surface with NMs. For example, activity was enhanced ca. two times for degradation of: (i) oxalic acid after deposition of 5-nm sized gold on P25 titania (P25 – famous titania (anatase/rutile/amorphous) photocatalyst with one of the best photocatalytic performances for different reaction systems, and thus commonly used as a standard (Iliev et al., 2007; Ohtani et al., 2010); (ii) congo red after gold was deposited on titania nanotubes (Pearson et al., 2012); and (iii) MB after Au (0.5 mol%) was deposited on anatase (one of titania polymorphs) (Li and Li, 2002). Additionally, four times higher activity was obtained during: (i) reduction of RSSR when13-nm gold NPs were deposited on titania (Kiyonaga et al., 2008); (ii) oxidation of MB when titania was modified with either 2–5-nm gold NPs or gold NRs of 7–15-nm diameter and 40–110-nm length (Okuno et al., 2015); and (iii) oxidation of R6G when ZnO nanorods were decorated with 15-nm Au NPs (Yi et al., 2015). The highest activity enhancement, i.e., more than 20 times, was obtained during oxidation of benzene after titania nanotubes (TNTs) modification with 1 wt% of gold (Awate et al., 2009). However, in some cases, only a slight increase of activity was observed after gold deposition, e.g., during degradation of Plasmocorinth B (Armelao et al., 2007) and MB (Wang et al., 2013b). In some cases, even inhibition of photocatalytic activity was obtained, e.g., for inactivation of *Bacillus subtilis* (Armelao et al., 2007). The impact of NM content has also been investigated with different optima. For example, for 0.25, 0.5, 1.1, 2.2, and 3.1 mol% of gold, the highest activity of MB degradation was found for samples containing 0.5 mol% gold. It has been suggested that too large content of NMs might enhance charge carriers' recombination instead of hindering it (Li and Li, 2002). Therefore, it has been postulated that NMs might also work as recombination centers causing diminished activity.

The first study of solar radiation was probably performed in 2006, showing twice higher activity of Au/TiO_2 than that by bare titania (Sonawane and Dongare, 2006). It was proposed that activity could be enhanced by, i.e., (1) Fermi level shift toward negative value (electrons' accumulation), (2) enhanced interface charge transfer, (3) light harvesting. Experiments under simulated solar irradiation also showed similar enhancement of activity for congo red degradation on TNTs after their dual modification with 12-phosphotungstic acid (PTA) and with gold NPs (Pearson et al., 2012). Similarly, experiments performed in Egypt under solar radiation showed twice higher activity for bleaching of R6G after deposition of gold on titania/polyethene beads (Elfeky and Al-Sherbini, 2011). It should be underlined that under natural solar radiation, the difference between the activity of titania and plasmon-assisted catalysis by NMs could not be observed. Accordingly, though these studies are very important for possible photocatalyst commercialization, but could not be applied to the discussion on the mechanism of vis-responsive photocatalysts.

Interestingly, under vis irradiation the first study showed inactivity of gold/titania photocatalysts (Awate et al., 2008). It is thought that this inactivity could be caused by the properties of those photocatalysts (too small gold NPs of nano-size, and thus without LSPR), or too slow reaction of acetone oxidation (Awate et al., 2008). It should be pointed out that under vis irradiation the reaction is even two to three orders of magnitude slower than the same reaction under UV (titania photocatalysis), probably due to fast back electron transfer (Kowalska et al., 2010). Similarly, lack of activity improvement after titania modification with gold during degradation of formic acid under visible-light irradiation was reported (Bouhadoun et al., 2015), where co-modification of titania with nitrogen was necessary for activity improvement. Then, successfully vis-activity of plasmonic photocatalyst (Au/TiO_2) was confirmed during decomposition of methyl tert-butyl ether (MTBE) under 495-nm irradiation (Rodriguez-Gonzalez et al., 2008).

Usually, the activity of plasmonic photocatalysts are tested for dyes' degradation, but it should be stressed that vis response of modified wide-bandgap semiconductors must not be examined for dyes as model molecules (Yan et al., 2006; Ohtani, 2010), due to the sensitization mechanism, i.e., activation of titania by dye, e.g., well-known dye-sensitized solar cells (DSSCs). The participation of both mechanisms, that is, titania sensitization (at 600–700 nm) and plasmonic excitation (400–1000 nm), was clearly shown by action spectrum experiments for MB oxidation on gold NPs and NRs deposited on TiO_2/SiO_2 composite (Okuno et al., 2015). Degradation of dyes (congo red (Pearson et al., 2012), MB (Bian et al., 2014; Yang et al., 2020), Rhodamine B (RhB) (Bian et al., 2014), MO (Cushing et al., 2012; Sun et al., 2014), Acid Orange 7 (AO7) (Wang et al., 2012), amaranth and Reactive Red 120 (Ryu et al., 2020)) under visible light on NM-modified titania was often reported, as exemplarily

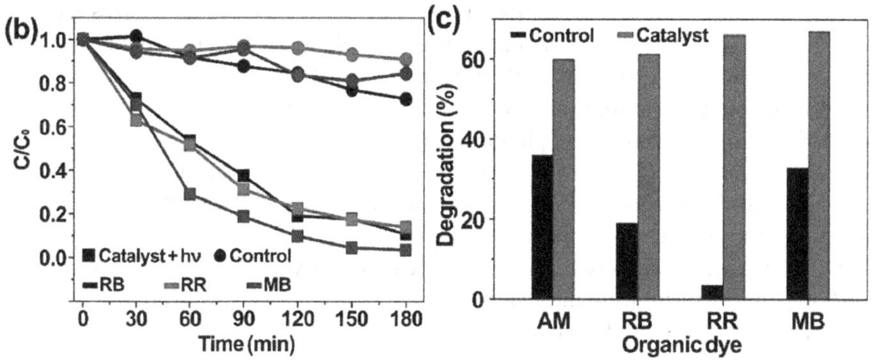

FIGURE 8.3 Efficiency of bimetallic photocatalysts (AgCl@AgAu) during dyes' decomposition (amaranth (AM), rhodamine B (RhB), reactive red 120 (RR), and methylene blue (MB)) under visible-light irradiation: (left) time-dependent activity with (square) and without (circle) photocatalysts, and (right) degradation in the natural environment (Han River, Seoul) after 3 h with (red) and without (black) photocatalysts. Adapted from (Ryu et al., 2020). Copyrights 2020 Creative Commons Attribution.

shown in Figure 8.3. Of course, these studies are very important and crucial for the environmental application of plasmonic photocatalysts. However, they should not be used for mechanism clarification, i.e., as proof of visible-light response. Okuno et al. have proposed three mechanisms of MB oxidation on Au/TiO_2-SiO_2 under broad solar spectrum, i.e., (i) separation of charge carriers under UV (Au as an electron trap), (ii) titania sensitization by dye at 600–700 nm, and (iii) plasmonic excitation (transfer of electrons from Au to titania CB) at 400–1000 nm (Okuno et al., 2015).

The visible-light activity of plasmonic photocatalysts have been proven for colorless compounds, i.e., during oxidative degradation of 2-propanol to acetone (Kowalska et al., 2012; Liu et al., 2013; Liu et al., 2014a, b) and carbon dioxide (Liu et al., 2013), cinnamyl alcohol to a, bcinnamaldehyde (Naya et al., 2015), methyl parathion to 4-nitrophenolate (Trammell et al., 2012), phenol (Wang et al., 2012) to hydroquinone, benzoquinone, muconic aldehyde, and acids (muconic and oxalic) (Zielińska-Jurek et al., 2011), and sulfamethazine (Yu et al., 2019). It has been proposed that mineralization (complete oxidative decomposition) of some compounds does not happen because of the dipole effect stabilizing the by-products (Rodriguez-Gonzalez et al., 2008).

Dependence of activity on the amount of NM has also been extensively examined. For example, an optimal gold amount of 1 wt% has been reported, and an increase in its content (2.5 and 5 wt%) has resulted in activity decrease, possibly due to enhanced recombination (gold as recombination center) (Wang et al., 2012a, b). Similarly, 1 wt% of gold has been found an optimal value, resulting in an almost twofold increase in the degradation rate of RhB and 2,4-dichlorphenol compared with bare Bi_2O_3 samples under vis irradiation. It has been proposed that the best content of Au (optimum at 1 wt%) results from the balance between various properties, i.e., NP size, morphology and surface coverage with hydroxyl groups (Jiang et al., 2012). Kowalska et al. has shown that the optimal content of gold also depends on the titania properties

(Kowalska et al., 2012), i.e., for titania with fine particles (crystallite sizes smaller than 15 nm) an increase in gold content results in the preparation of photocatalysts with very dark color ("shielding effect"), and thus with low activity. In contrast, an increase in gold content for titania with large particles (crystallite sizes of ca. 60-nm) results in a three-step growth of photocatalytic activity. It has been suggested that the first activity increase (0.05–0.5 wt%) might be caused by enhanced light harvesting, i.e., more gold deposits, similar to the well-known dependence of photocatalytic activity on the amount of photocatalyst, reaching a plateau for optimum light absorption (Herrmann, 1999; Kisch, 2010), whereas the next steps at 2.25–4 wt% and 6–10 wt% might result from the change in the gold properties, i.e., from fine NPs into larger and non-spherical deposits.

Nowadays, various studies have focused on activity enhancement by morphology design. For example, faceted titania photocatalysts show excellent activity because of hindered recombination of charge carriers. Two types of faceted anatase have been commonly investigated, that is, octahedral anatase particles (OAPs) possessing only {101} facets and decahedral anatase particles (DAPs) with both {101} and {001} facets (at the top/bottom of OAP crystals). It has been shown that high activity of OAPs results from large content of shallow electron traps (ETs), and thus the longer lifetime of photogenerated electrons (Wei et al., 2014, 2015). Whereas high activity of DAP results from an intrinsic property of these nanocrystals, i.e., spatial separation of e^-/h^+, since holes migrate to {001} facets, whereas electrons to {101} ones (Tachikawa et al., 2011). Probably the first report on faceted-based plasmonic photocatalysts was published by Janczarek et al. for Ag- and Cu-modified DAP (mono- and bimetallic photocatalysts (Figure 8.4) by the photodeposition method (Janczarek et al., 2017). The photodeposition method is very convenient and efficient since all metal cations are usually completely deposited on the semiconductor surface (Kowalska et al., 2012), as metal cations are directly reduced by photogenerated electrons. Interestingly, the order of NM deposition has governed the properties of formed deposits, and thus the overall activity (Janczarek et al., 2017), e.g., the sequential deposition causes that the secondly deposited metal has also been formed on the surface of the first metal, which is reasonable considering that electrons photogenerated under UV irradiation are first scavenged by already deposited metal; hence adsorption of cations occurs on the electron-rich metal (Figure 8.4c–d).

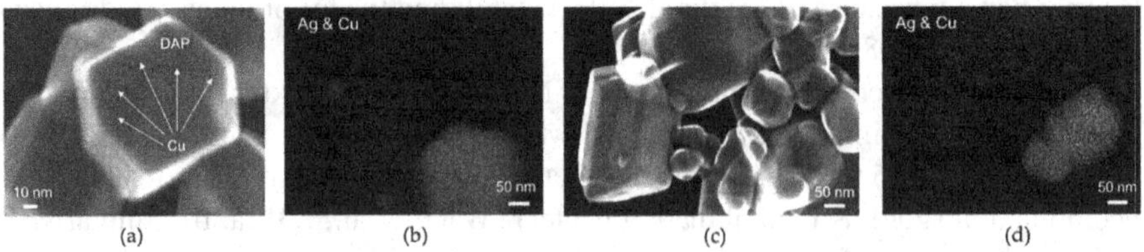

FIGURE 8.4 SEM images (a, c) and EDS mapping (b, d) of decahedral anatase particles (DAP) modified with: (a) fine copper NPs, (b) large silver and fine copper NPs (co-deposition method), and (c–d) bimetallic NPs prepared by copper deposition on the Ag/DAP; where silver and copper are shown in red and yellow, respectively. Adapted from (Janczarek et al., 2017). Copyright 2017 Creative Commons Attribution.

Therefore, both monometallic and bimetallic deposits have been deposited on DAP samples. On the contrary, the competition between metals for photogenerated electrons occurs during co-deposition, causing the formation of large Ag NPs and fine Cu NPs (Figure 8.4b). Interestingly, though under UV irradiation, NM/DAP samples exhibit high efficiency enhancement (more than threefold), only a monometallic sample containing silver shows high activity under vis (> 455 nm) irradiation. It has been proposed that it could be caused by easy oxidation of copper, and thus mainly Cu_xO is present in those samples rather than plasmonic zero-valent cupper.

Indeed, it has also been confirmed for another faceted titania (OAP) that visible-light activity increases in the following order: Cu < Ag < Au, which correlates with the content of zero-valent metal in those samples (Wei et al., 2017a), as shown in Figure 8.5. It has been observed that activity under vis irradiation increases in the following order: Cu/OAP < Ag/OAP <Au/OAP, which correlates well with the surface oxidation state of metals, i.e., zero-valent state governs vis-activity. Additionally, the properties of NM-deposits have been investigated by changing the conditions during photodeposition, i.e., from anaerobic to aerobic, confirming that zero-valent metal is crucial for vis-activity (higher content under anaerobic conditions), as shown in Figure 8.5. Moreover, the influence of NM size has been investigated in other studies, showing that activity for 2-propal oxidation increases when the size of NPs of NMs increases, possibly because of strong plasmonic field enhancement, as shown in Figure 8.6 (Wei et al., 2017b; 2018a). Additionally, it has been confirmed by TRMC that "hot" electrons should participate in the mechanism of vis degradation of organic compounds as the decay of TRMC signal correlates with the activity, as shown in Figure 8.7.

Finally, two types of faceted (OAP vs. DAP) plasmonic photocatalysts have been compared (Wei et al., 2018b). It must be mentioned that DAP samples usually show the best photocatalytic performance among various titania samples (both commercial and self-prepared), even though they do not have the best surface properties (5–20 $m^2g^{-1}$of specific surface area (Janczarek et al., 2016), because of spatial separation of charge carriers, as already mentioned. This has also been confirmed for NM-modified

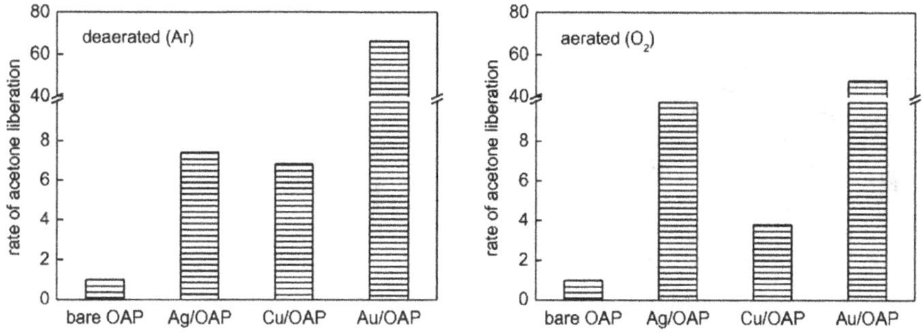

FIGURE 8.5 Rate of acetone generation during oxidative decomposition of 2-propanol on NM-modified OAP samples prepared in the absence (left) and the presence (right) of oxygen during NM photodeposition. Adapted from (Wei et al., 2017a). Copyright 2017 Creative Commons Attribution.

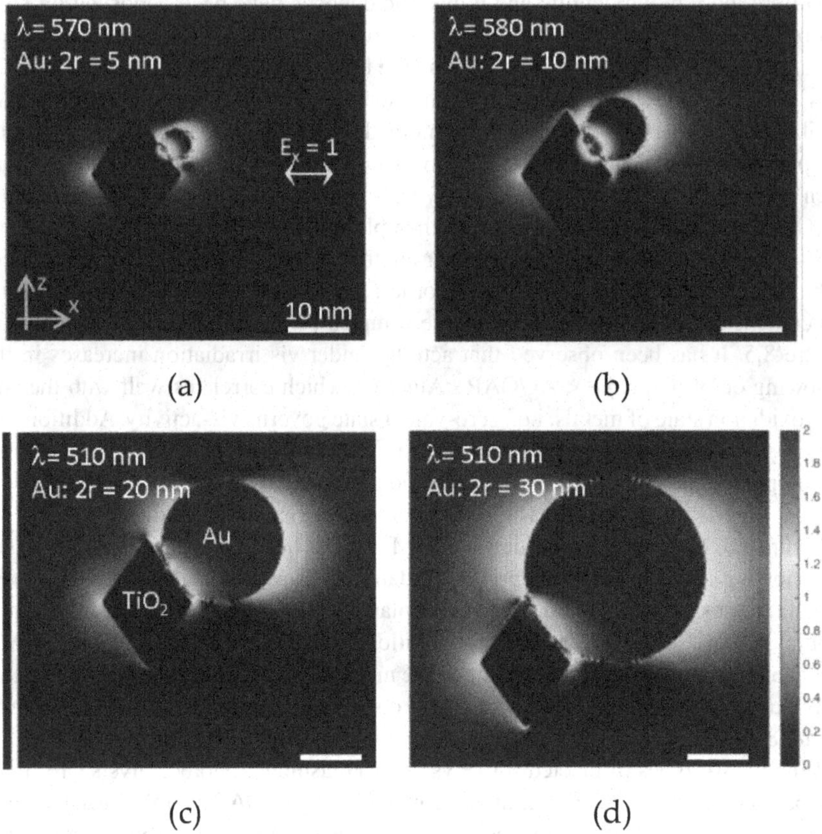

FIGURE 8.6 Finite-difference time-domain (FDTD) data showing enhancement of light intensity for OAP modified with various sizes of gold NPs: (a) 5 nm, (b) 10 nm, (c) 20 nm, and (d) 30 nm. Adapted from (Wei et al., 2017b). Copyright 2017 Creative Commons Attribution.

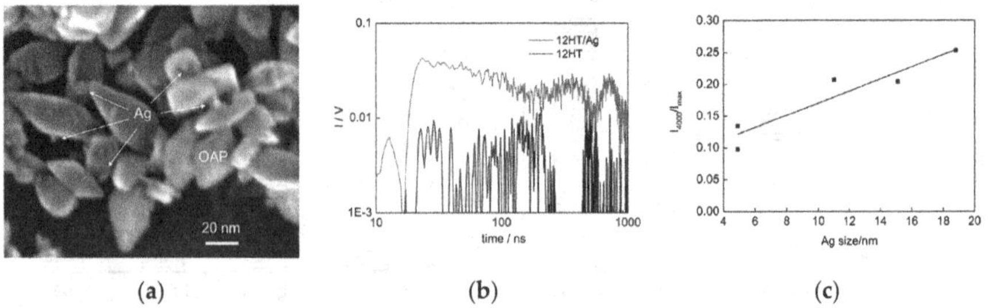

FIGURE 8.7 Properties of silver-modified OAP samples: (a) SEM image, (b) TRMC signal (bare OAP-12HT; silver-modified OAP-12HT/Ag), (c) TRMC signal decay depending on the size of silver NPs. Adapted from (Wei et al., 2018a). Copyright 2017 Creative Commons Attribution.

samples under UV irradiation since DAP samples (pristine and silver-, gold- and copper-modified) are more active than respective OAP samples. Interestingly, under visible-light irradiation, OAP samples show better activity than respective DAP ones. Moreover, it has been found that the activity of Au/OAP sample is even much higher (ca. ten times) than the activity of gold-modified commercial titania samples with similar surface properties; Figure 8.8a), confirming the fast transfer of mobile "hot" electrons (Figure 8.8b). Then, why is the performance of highly active DAP the worst among 16 tested samples? It has been proposed that due to spatial separation of charge carriers in DAP (as presented above), i.e., transfer of electrons to {101} and transfer of holes to {001}, the back transfer of "hot" electrons to {101} facets (Au → TiO_2{101} → Au), results in negligible activity under plasmonic excitation, as shown in Figure 8.8c.

Song et al. (2016) have presented interesting comparison between 2-D (rods: Au/RR) and 3-D (hierarchical structure: Au/3DR) plasmonic photocatalysts. It has been found that during degradation of p-nitrosodimethylaniline under visible-light irradiation ($\lambda > 455$ nm), Au/3DR is more efficient than Au/RR. Interestingly, the bathochromic shift of LSPR and enhanced plasmonic coupling under irradiation has been observed for in Au/3DR sample. The study on irradiation-angle dependent activity has been performed, and it has been found that the transverse resonance of LSPR is more dominant than the longitudinal resonance, and the hypsochromic shift of LSPR occurs when the tilted angle is lower than 45°. In contrast, with an increase in the tilted angle above 45°, the longitudinal resonance causes the bathochromic shift of LSPR. Interestingly, random orientation of rods in Au/RR sample results in negatively averaged plasmonic coupling effects, whereas rods are uniformly oriented in Au/3DR sample, hence most gold NPs might be vertically irradiated, resulting in highly intensive and red-shifted LSPR, and thus with high activity.

The (nano)structure of photonic crystals (PCs; periodic nanostructures) is probably one of the most interesting materials, broadly investigated for various applications, such as optoelectronic devices (Khaleque & Hattori, 2016; Chen et al., 2019), sensors (Dash & Jha, 2014) photocatalysis (Sordello et al., 2011; Rahul & Sandhyarani, 2015), surface-enhanced Raman spectroscopy (SERS) (Li et al., 2018) and heterogeneous photocatalysis (Raja-Mogan et al., 2020). The uniqueness of such

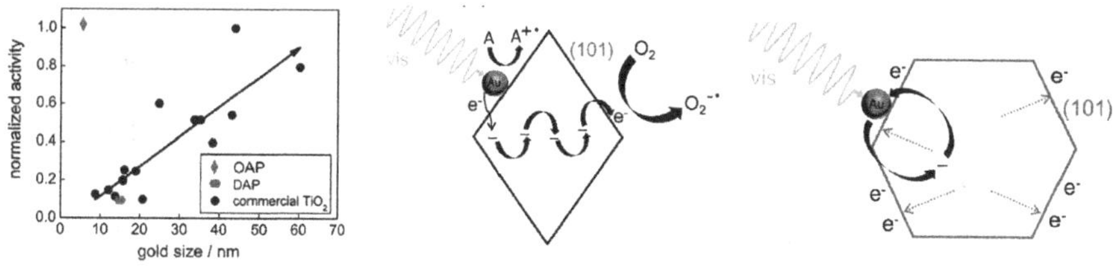

(a) (b) (c)

FIGURE 8.8 (a) The comparison of vis-activity for gold-modified commercial and faceted titania samples, and (b-c) the schematic mechanisms of "hot" electron migration under vis irradiation for Au/OAP (b) and Au/DAP (c). Adapted from (Wei et al., 2018b). Copyright 2018 Creative Commons Attribution.

materials originates from the appearance of photonic band gap (PBG) due to its periodicity in refractive index (high and low dielectric media) of the structure, resulting in 'slow photons' propagation at the edges of the stop bands, which could be utilized efficiently for enhanced photoabsorption (Yablonovitch, 1993).

PCs-based nanomaterials are well explored and considered promising alternatives for reducing environmental pollutions (Wu et al., 2014; Zhang et al., 2018). Accordingly, PCs coupled with NMs are routed to be excellent photocatalysts to treat and purify water/wastewater containing pollutants due to their synergistic effects of photonic and plasmonic effects, as summarized in Table 8.1. Multiscattering of photons within the PCs structure enables the extension of light pathway duration, which might be utilized efficiently by NM NPs, and intensify

TABLE 8.1
Photodegradation and Detection of Pollutants by Noble Metals Modified PCs

NM-Modified PCs	Application	Tested Pollutants	Efficiency	References
TiO_2 PC/Au NPs film	PD, $\lambda > 420$ nm	2,4-dichlorophenol	2.3-fold higher[a]	Lu et al. (2012)
TiO_2 IO /Ag NPs film	PD, $\lambda = 254$ nm and $\lambda =$ 400–760 nm	MB	enh. DR	Zhao et al. (2012)
Au-TiO_2-NAA-DBRs	PD, stimulated solar light	MB	56% enh.[a]	Lim et al. (2019)
TiO_2-IO Au/AgNPs film	PD, $\lambda > 365$ nm	Acetylene	62% enh.[a]	Temerov et al. (2020)
i-Pt-TiO_2-o film	PD, $\lambda > 400$ nm	AO 7	4-fold enh.[a]	Chen et al., 2008a, 2008b
Au NP/ ZnO-PCs film	PD, $\lambda > 420$ nm	RhB, 4-chlorophenol	24.8x that commercial ZnO	Meng et al. (2015)
rGO/Pt/3DOM TiO_2 powder	PD, $\lambda > 420$ nm	MO	enh.DR	Huo et al. (2019)
Ag/io-$BiVO_4$	PD, $\lambda > 420$ nm	MB	3-fold enh.[a]	Fang et al. (2016)
Au/ TiO_2-BMPC	PD, $\lambda > 420$ nm	MO	7-fold higher[a]	Wang et al. (2016a, 2016b)
Ag@AAO DHPCs	SERS, $\lambda = 665$ nm	RhB	enh. detection	Wang et al. 2021
SiO_2 OPCs/ AuNS	SERS, $\lambda = 532$ nm	Phenoxybenzoic acid	enh. low detection limit	Wang et al., (2019)
Au NPs@IOT	PD, $\lambda = 450$ nm, $\lambda = 530$ nm	Acetic acid	enh. activity[a]	Raja-Mogan et al. (2021)

[a] enhanced activity as compared to the respective pristine sample; AO – acid orange; Au NPs@IOT – TiO_2 IO containing Au NP per void; BMPC – biomorphic PC; DR – degradation rate; enh. – enhanced; i-Pt-TiO_2-o film – TiO_2-IO loaded with Pt nanoclusters; MB – methylene blue; MO – methyl orange; NAA-DBRs – functionalized nanoporous anodic alumina distributed Bragg reflectors; OPCs/AuNS – opal PC/ gold nanostars; PD – photodegradation; rGO/Pt/3DOM TiO_2 – three-dimensionally ordered macroporous TiO_2 with Pt and reduced graphene oxide; SERS – surface-enhanced Raman spectroscopy; RhB – Rhodamine B.

their plasmonic effect, causing an enhanced photocatalytic performance (Zhao et al., 2012, Wang et al., 2019).

Significant enhancement in the photodecomposition of organic pollutants could be realized when PBG edges and LSPR wavelengths overlap via the tuning of the nanostructures' fabrication. For example, Lu et al. (2012) have shown about 2.3-fold enhanced photodegradation of 2–4 dichlorophenol in comparison to pristine Au NP/ TiO_2 photocatalyst when precise tuning of the plasmonic-based TiO_2 inverse opal (IO) film has been achieved to allow the overlapping between blue-edge of PBG and LSPR wavelengths region. On the other hand, the optimum amount of NM deposition on PCs is the prerequisite condition for efficient pollutants degradation. For instance, 1.3 wt% of platinum loading on TiO_2 IO does not exhibit any improvement of the dye degradation rate as compared to that for pristine IO, whereas 2 wt% has increased the efficiency by 3.5–4 times (Chen et al., 2008a, 2008b). Additionally, the excess amount of NM (e.g., silver) might result in NM aggregation, blocking the propagation of the slow photons and charge carriers within the PC, and thus the loss of both photonic and LSPR effect, as shown by a decrease in degradation of MB (Chen et al., 2014).

For easy application, the magnetic separation of plasmonic photocatalysts has been proposed by Bielan et al. using a magnetic core covered with silica interlayer and a titania shell modified with NPs of platinum and copper, as shown in Figure 8.9 (Bielan et al., 2020a; Bielan et al., 2020b). It has been proven that magnetic-based materials show perfect reusability during penol degradation. Interestingly, it has been found that the activity of Pt/TiO_2 has decreased when prepared on a magnetic core, possibly due to trapping of "plasmonic" electrons by ferrous ferric oxide ($Pt \rightarrow TiO_2 \rightarrow SiO_2@Fe_3O_4$; (Bielan et al., 2020a)), similarly as suggested for bimetallic (Kowalska et al., 2014a) and hybrid (co-modified with ruthenium dye (Kowalska et al., 2015a, b; Zheng et al., 2016a, 2016b)) plasmonic photocatalysts.

It should be mentioned that plasmonic nanoparticles (those with LSPR resonance at visible-light region) have also been used as "dark" catalysts (not photocatalyst). For example, Ag NPs of ca. 15-nm diameter (Figure 8.10 (left)),

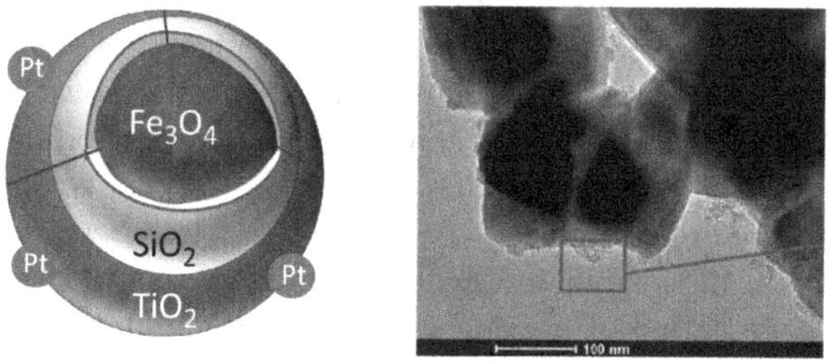

(a) (b)

FIGURE 8.9 Platinum-modified core/interlayer/shell (i.e., Fe_3O_4)/SiO_2/TiO_2) photocatalysts: (**a**) the schematic structure; adapted from (Wei et al., 2020). Copyrights 2020 Creative Commons Attribution; and (**b**) TEM image (platinum shown in red rectangle). Adapted from (Bielan et al., 2020b). Copyright 2020 Creative Commons Attribution.

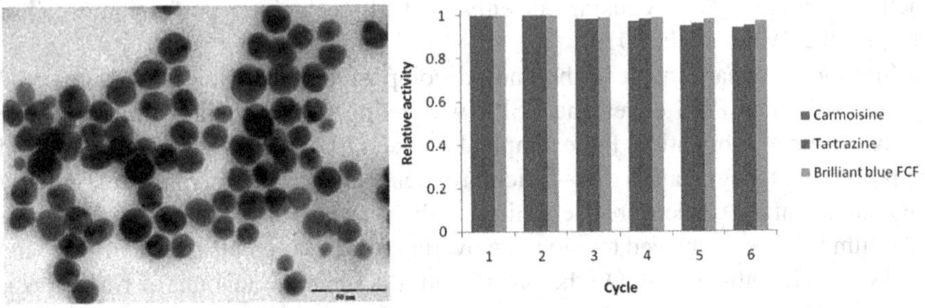

FIGURE 8.10 (left) TEM image of obtained Ag NPs and (right) reusability of Ag NPs for reduction of organic dyes with $NaBH_4$. Adapted from (David and Moldovan, 2020). Copyright 2020 Creative Commons Attribution.

synthesized with the help of *Viburnum opulus* L. fruit extract, which can work both as a reducing agent of Ag^+ and a stabilizing agent for formed Ag NPs, have shown high activity and stability (Figure 8.10 (right)) for catalytic reduction of dyes (tartrazine, carmoisine, and brilliant blue FCF) with $NaBH_4$ (David and Moldovan, 2020).

8.2.2 Application of Plasmonic Nanomaterials for Inorganic Pollutants

Plasmonic photocatalysts have also been proposed for the removal of inorganic pollutants, e.g., heavy metals and radioactive anions (iodine). Wang et al. (2020) have proposed titania nanotubes (TNT) modified with silver, prepared on ITO glass by magnetron sputtering and anodization, to remove chromium. Ag/TNT shows high activity, causing almost complete removal of Cr(VI) (99.1%) during 90-min irradiation with visible light. Liu et al. have prepared 3D Ag_2O–Ag/TiO_2 composites to remove I^- from salt lake water, deionized water and seawater (Liu et al., 2015). It has been proposed that the composite exhibits some synergy between (i) enhanced adsorption abilities of silver oxide for I^-, and (ii) the photocatalytic activity of silver-modified titania for oxidation of iodine anion under vis irradiation. It should be underlined that the composite shows high adsorption ability, selectivity, reusability, and removal efficiency for iodine anions under vis, which suggests that Ag_2O–Ag/TiO_2 might be a prospective photocatalyst for radioactive wastewater treatment.

Although plasmonic nanomaterials for water and wastewater are mainly suggested for the degradation of organic and inorganic pollutants, and inactivation of microorganisms (as discussed in the next section), they have also been proposed as sensors for pollutants in water/wastewater. For example, it has been shown that bi-functionalized Ag NPs might be used to sense Hg^{2+} (Prosposito et al., 2019). It has been found that Ag NPs work as a selective sensor for mercury cations among 16 different metal ions tested in the range of 1–10 ppm, as shown in Figure 8.11.

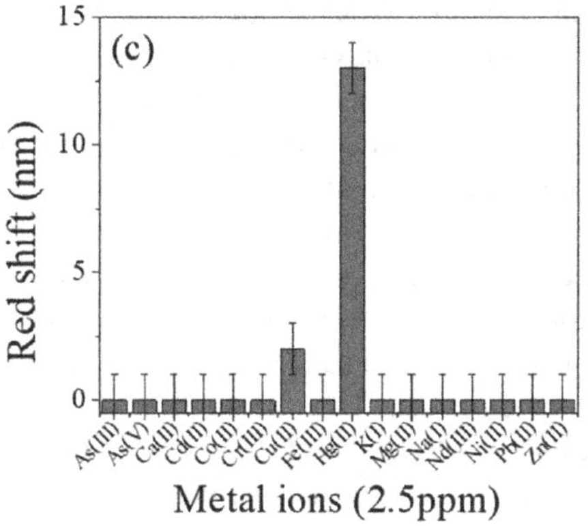

FIGURE 8.11 Bathochromic shift of the Ag LSPR peak in the presence of mercury cations (Hg^{2+}) in comparison to other metal ions tested at a concentration of 2.5 ppm. Adapted from (Prosposito et al., 2019). Copyright 2019 Creative Commons Attribution.

8.3 DECOMPOSITION OF MICROORGANISMS

Advanced oxidation processes (AOPs) have been used in water treatment and disinfection, since the complete oxidation of microorganisms, the toxins they release, and their nucleic acid is possible by highly active ROS, particularly the most active hydroxyl radicals ($^{\bullet}OH$). The ideal disinfectant should fulfill the following criteria: (i) high antibacterial efficiency in a short time; (ii) wide spectrum of inactivated microorganisms; (iii) safety and durability of process (including protection against re-infection in water supply network; (iv) low-content of disinfection by-products; (v) absence of toxicity (or very low toxicity) in humans, animals, and environment; (vi) lack of impact on water organoleptic characteristic, e.g., taste and color. In the last decades, the crisis of environmentally resistant antibiotics, resistant pathogenic bacterial strains (ARB) and new zoonotic viruses (including SARS-CoV-2) is one of the biggest public health challenges. Hence, more and more attention is paid to disinfection methods leading to the termination of bacterial antibiotic-resistant *genes* (*ARGs*) or naked DNA horizontal transfer (Zhang et al., 2019; Jin et al., 2020).

Since the first report by Matsunaga et al. (1985), it has been known that titania and platinum (Pt)-loaded titania might inactivate yeast *Saccharomyces cerevisiae.* After that, other studies have confirmed the disruption of the cell wall, cell membranes, and leakage of K^+ (not the direct cause of cell death), considerable loss (even 97%) and oxidation of intracellular coenzyme A (CoA) under irradiation (Matsunaga et al., 1988; Saito et al., 1992; Huang et al., 2000). As CoA is important for metabolism (cell respiration), its inactivation leads to cell death. Other mechanisms of bacterial inactivation by titania have also been proposed, i.e., peroxidation of cell membrane phospholipids (Kiwi & Nadtochenko, 2004, 2005), by inducing oxidative stress (Foster et al., 2011; Lin et al., 2014) and direct DNA damage (Huang et al., 2000) by ROS ($^{\bullet}OH$, $O_2^{\bullet-}$ and H_2O_2), formed from water and oxygen by photogenerated charge carriers on titania (Cho et al., 2004; Van Grinsven et al., 2012). Moreover, it is proposed that surface charge carriers (e^-/h^+) also correlate with the bacterial inactivation

(Guo et al., 2015; Wang et al., 2012) as the direct contact of titania with bacteria has increased the oxidative damage. Analysis of scanning electron microscopy (SEM) images indicates that ROS rich plasma membrane (PM) until 30 min of photocatalytic reaction, cell wall located outside PM (vary in composition between species) is partially disrupted within 1 to 3 min, whereas further decomposition occurs after 60 to 120 min (Saito et al., 1992). Accompanying leakage of cellular material has a significant effect on cell division (Foster et al., 2011), but also might help in the release of DNA and mobile genetic elements (MGEs), e.g., plasmids (carry genes for managing their lifecycles and resistance to various antibiotics) and insertion sequences from "dead bacteria" cell to the surroundings (Jin et al., 2020). Kim (2013) has demonstrated that bacteria treated with 60-s UVC irradiation in the presence of titania have ceased to reproduce. Further increase in treatment time (6 min) has caused that all *Escherichia coli* plasmid DNA to become linear. Similarly, Hwangbo et al. (2019) have shown the structural changes in the plasmid DNA and ARGs-containing plasmids after 6 h of the photocatalytic process. Given that, total bacteria cell mineralization in water is required. According to Jacoby et al. (1998), the time necessary for total mineralization of *E. coli* on TiO_2-coated glass is about 75 hours. Although it is well known that AOPs methods result in efficient mineralization (i.e., producing only simple molecules, such as CO_2 and H_2O), which indicates that harmful by-products are not formed, there is a considerable lack of knowledge on how much time is needed for total mineralization/oxidation of bacterial compounds, such as peptidoglycan or murein consist of NAM and NAG and their copolymers or DNA.

As described in the previous section, plasmonic photocatalysts possess some advantages compared with bare photocatalysts under solar light irradiation, i.e., the intrinsic antimicrobial property of noble metals and visible-light absorption. Although the intrinsic activity of noble metals is not a plasmonic photocatalysis, they highly participate in the overall activity. Thus the antimicrobial effects and mechanisms of typical NMs are shortly described nex.

Although various metals have been studied for their antimicrobial activities, e.g., silver, copper, zinc, bismuth, gold, etc., silver and copper are considered the most effective ones. The following mechanism of antibacterial activity of silver (ion) has been proposed: (i) adsorption of Ag cations at cell walls of bacteria (negatively charged), causing the disintegration of the cell walls and the PM (collapse of membrane potential (Lok et al., 2006)), and then lethal leakage of substances (including protons) outside the cell (Lok et al., 2006; Varun Sambhy et al., 2006; Chwalibog et al., 2010), and (ii) interaction with thiol groups of essential transport or/and respiratory enzymes, causing the uncoupling of respiration from ATP synthesis (Holt & Bard, 2005). It should be noted that some studies suggest that the silver release from the carrier is crucial for bactericidal activity (Akhavan, 2009; van Grieken et al., 2009; Li et al., 2011; Lin et al., 2010; Liga et al., 2011). Additionally, Ag NPs possess the microbial activity for other microorganisms, such as algae, fungi, and viruses. For instance, it is proposed that Ag NPs might interact with viral surface glycoproteins (Figure 8.12), resulting in the hindering of: (i) virus binding to target cells; and (ii) virus replication when silver penetrates its cells (Speshock et al., 2010; Galdiero et al., 2011). The important point of the bactericidal mechanism of copper (cation) is also the adsorption on the bacteria surface, followed by (i) the degeneration of the

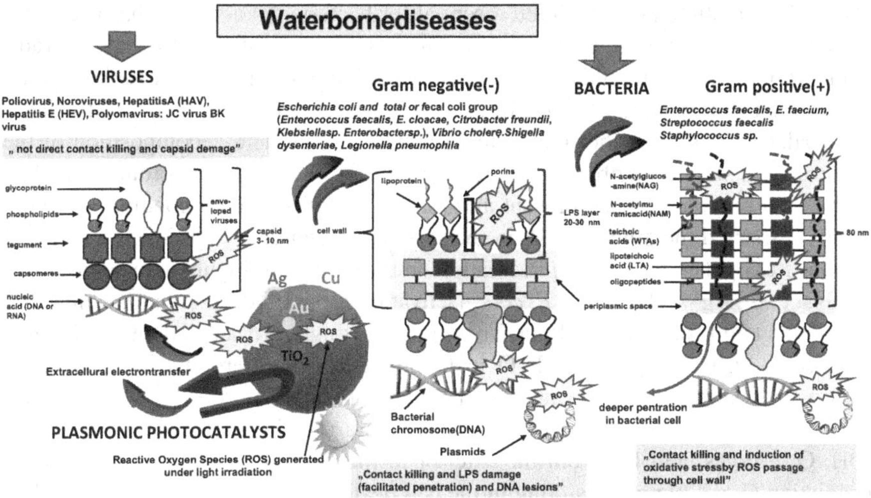

FIGURE 8.12 Killing mechanism of plasmonic photocatalyst against some waterborne pathogens.

proteins (Sunada et al., 2012), and/or (2) induced oxidative stress (Deng et al., 2017). Moreover, the cumulation of copper cations inside bacterial cells has been suggested as one of the main mechanisms of inactivation of bacteria (Dankovich & Gray, 2011). It should be pointed out that cupric oxide is more active than cuprous oxide and silver (Sunada et al., 2012). When compared with silver and copper, other metals seem to be not as highly active against microorganisms. The contradictory results have been obtained for the antimicrobial activity of gold, i.e., gold does not affect the viability of bacteria (Amin et al., 2009; Zhang et al., 2013; Zheng et al., 2017), probably because of the inability to surround the bacterial cells (Ratte, 1999), or shows high activity (Rai et al., 2010; Badwaik et al., 2012; Cui et al., 2012; Zheng et al., 2017). The possible antimicrobial mechanisms of gold have been proposed, including (i) the change in the membrane potential, (ii) activity inhibition of ATP synthase to decrease the ATP level, (iii) inhibition of the ribosome subunit for tRNA binding Cui et al. (2012), and (iv) inhibition of H^+-ATPase-mediated proton pumping (Ahmad et al., 2013). Moreover, it has been found that bismuth NPs show bactericidal activity, possibly due to the replacement of other metals, e.g., Zn(II) in the enzyme, by Bi(III) (Manhart, 1990), resulting in the disruption of bacterial cells.

Therefore, due to the potential antimicrobial activity of metal NPs, plasmonic photocatalysts with silver (Kubacka et al., 2008; van Grieken et al., 2009; Mai et al., 2010; Liga et al., 2011; Castro et al., 2012; Padervand et al., 2012; X. Wang et al., 2012; Kowalska et al., 2015a, b), copper (copper oxide) (Baghriche et al., 2012; Qiu et al., 2012; Liu et al., 2014a, b; Rtimi et al., 2016; Endo-Kimura et al., 2020) and gold (Endo et al., 2018, 2019; J. Li et al., 2014) have been intensively studied. Among them, Ag-modified titania has been intensively investigated, focusing on the intrinsic property of silver under dark conditions, and improved activity of TiO_2 under UV/vis irradiation. The concentration of silver cations, their adsorption at bacteria surface, and ROS formation are considered essential for bactericidal action. It has been suggested that an increase in Ag^+ concentration on the surface of TiO_2 by the pre-irradiation with visible light enhances the bactericidal activity in the dark (Castro et al., 2012), indicating the significance of electron transfer between silver and titania by vis irradiation. On the other hand, importantly, the complete decomposition of the bacterial cell should be obtained (Stentzel et al., 2017). It has been found that Ag–AgBr/TiO_2 composites also present antibacterial activity in dark conditions, due to evenly dispersed metallic Ag NPs in the systems (Wang et al., 2012). As a result of various ROS generation during irradiation on the surface of highly active plasmonic catalyst (e.g., Ag–AgBr/TiO_2), *E. coli* inactivation is possible alike under white LED irradiation and visible indoor light. These radicals might break covalent bonds, such as C-N, C-O, C-C, C-H, and H-O in biomolecules like carbohydrates, proteins, amino acids, and DNA. The function of ROS for the decomposition of bacteria has also been studied, e.g., Ag/AgBr/TiO_2 nanotube arrays induce oxidative attack by $^{\bullet}OH$, $O_2^{\bullet -}$, holes, and Br^0 from the exterior to the interior of the *E. coli* under vis irradiation ($\lambda > 420$ nm), resulting in the cell death, as shown in (Hou et al., 2012). The bactericidal efficacy depends on the degree of adhesion (photocatalysts/bacteria), resulting from a charge on their surface. As shown by Hu et al. (2007) and Tallósy et al. (2016), the electrostatic attraction existing between negatively charged bacteria and photocatalysts leads to their tightly bounding what causes an increase in antimicrobial action since the lifespan of ROS is very short of ca. 10^{-9} s (Cohn et al., 2006). Likewise, the pH value of the reaction environment should also be considered since the Zeta potential of a bacterial surface depends on it (Hu et al., 2007). On the other hand, the cell wall structure is also crucial in influencing the bacterial sensitivity to plasmonic photocatalysts. As shown in Figure 8.12, the cell wall of Gram-positive bacteria is composed of only thick peptidoglycan (mesh-like layer of N-acetylmuramic acid and N-acetylglucosamine polysaccharide chains) and periplasmic space (concentrated gel-like matrix). Therefore, it is much more vulnerable for ROS passage than that in the cell wall of Gram-negative bacteria. In mentioned organisms, a thin layer of peptidoglycan is enveloped by an outer semi-permeable membrane, consisting of a complex layer of lipids, LPS, and proteins. The radicals generated during the photocatalytic process react with the lipid constituents of the membrane rather than pass through it (Page et al., 2007). This mechanism has been confirmed by a recent study on Ag/TiO_2plasmonic photocatalyst against Gram-positive bacteria methicillin-resistant *Staphylococcus aureus* MRSA (Tallósy et al., 2016).

The mineralization of bacterial cells by Ag/TiO_2 has been elucidated by the estimation of generated CO_2, which is much higher under irradiation with visible light than under dark conditions (as well as higher inactivation of bacteria) (Endo et al., 2019). Interestingly, it has been suggested that the bactericidal activity of Ag-modified TiO_2 is higher in vis than that in UV (Wei et al., 2017), probably because of the surface oxidation of silver (electron transfer mechanism under plasmonic excitation). In contrast, under UV, electron-rich silver (after electron transfer from TiO_2 to Ag) might result in repulsion between Ag surface and negatively charged bacterial surface (Wei et al., 2017). Moreover, it has been found that UV irradiation might stabilize Ag deposits, though dark leaching of silver cations has increased the bactericidal activity (van Grieken et al., 2009). The physical properties of Ag/TiO_2 are highly associated with the activity. It has been found that Ag-modified TiO_2 with a flower-like structure is more active than spherical composites (*E.coli* and *Staphylococcus aureus*) because of its efficient generation of ROS and silver release (Ye et al., 2017). Unfortunately, some studies indicate that plasmonic photocatalysts (Ag/AgX-CNTs and AgCl/Ag) exhibit significant loss of antimicrobial activity after 3–5 times of use in water disinfection (Shi et al., 2014; Cheikhrouhou et al., 2019). Most likely, a decrease in photocatalytic activity is caused by adsorption of decomposed bacterial structures and organic debris onto the photocatalysts' surface, thus blocking the active sites (Shi et al., 2014).

Silver-modified titania has also been applied as the antifungal and antiviral agent (Zielińska et al., 2010; Kowal et al., 2011; Liga et al., 2011; Wei et al., 2017; Wysocka et al., 2019). For instance, faceted titania (OAP) modified with NPs of silver exhibits higher activity against model yeast *Candida albicans* than OAP modified with NPs of copper, gold, and platinum under both vis and in the dark (Wei et al., 2017). It is thought that the fungicidal activity of Ag-modified titania depends on Ag content (Zielińska et al., 2010), and the adjustment of Ag/TiO_2 dose or/and irradiation duration are necessary for the efficient fungal inactivation (Kowal et al., 2011). Interestingly, titania modified with two NMs (Pt/Ag and Ag/Cu) shows higher activity than monometallic ones, because of the efficient formation of superoxide anions (Wysocka et al., 2019). As for antiviral activity, Liga et al. (2011) have proposed that the viral adsorption at the silver and silver leaching participate in the inactivation of bacteriophage MS2. It is well known that viruses are more sensitive to disinfection factors than bacteria because of their simple structure. Most of them have one type of nucleic acid: RNA or DNA (Figure 8.12). Despite having their genetics enclosed within a protein coat or capsid (an enveloped virus has an additional lipid membrane), composed of different proteins or multiple copies of the same protein, they cannot reproduce by themselves. It is thought that the photo-killing mechanism by plasmonic catalyst is entirely connected with hydroxyl radicals that induced decomposition of protein capsids (Gerrity et al., 2008). However, a different view has been presented by Nazari et al. (2017), i.e., plasmon excitation significantly reduces viral fusion with the host cells. It means that plasmonic photocatalyst might significantly influence the early stages of the viral infection rather than the viral genome or enzymes. Additionally, virus inactivation does not require direct contact between capsid and ROS, occurring even if ROS formation is limited (Nazari et al., 2017). The outbreak of coronavirus disease in 2019 has caused that relevant discoveries are *rapidly* implementable. Hence, silver nanocluster/silica composite has been incorporated on facial masks, protected against SARS-CoV-2 (Talebian et al. 2020).

Copper in Cu-modified titania exists in mixed oxidation states, i.e., Cu(0), Cu^+, and Cu^{2+}, since copper can be easily oxidized in the ambient condition. Therefore, a broad absorption spectrum is obtained in the vis range (Qiu et al., 2012; Endo-Kimura et al., 2020). It has been proposed that the bactericidal activity of Cu_2O/TiO_2 is higher than both CuO/TiO_2 and $CuNPs/TiO_2$. The optimal ratio of Cu_2O and CuO in Cu_xO/TiO_2 photocatalyst ($Cu^I/Cu^{II} = 1.3$) is important to achieve high antibacterial performance under vis irradiation and dark conditions (Qiu et al., 2012). The proposed mechanisms of Cu-modified titania are similar to a silver-modified one, i.e.,

(i) degeneration of the surface proteins (Sunada et al., 2012) and (ii) oxidative stress by the adsorbed copper ions (Deng et al., 2017) and/or accumulated copper ions inside bacteria (Dankovich & Smith, 2014). Moreover, Qiu et al. (2012) have proposed that vis-initiated antimicrobial activity arises from multielectron reduction by Cu^{II}, activated by interfacial charge transfer from the VB of TiO_2 to the Cu_xO clusters, whereas dark activity corresponds to Cu^{I}. It has been revealed by the scavenger test (dimethyl sulfoxide and superoxide dismutase) under actinic light and anaerobic conditions that the photogenerated holes on titania and copper cations are responsible for the inactivation of *E. coli* (Rtimi et al., 2016).

It should be pointed out that fouling is a major problem of filtration processes during water/wastewater treatment and thus should be prevented (Nguyen et al. 2012). Filamentous (molds) fungi, such as *Aspergillus* sp., *Cladosporium* sp., *Botrytis* sp., *Alternaria* sp., and *Penicillium* sp., are common fouling organisms (Chaves et al., 2020). Accordingly, the fungistatic activity of plasmonic photocatalyst Cu/TiO_2 against *Aspergillus niger* has been observed. Moreover, inhibition of mold sporulation (as a model of action) has been confirmed (Figure 8.13) by Endo-Kimura et al. (2020). It is known from other reports that copper NPs are less toxic than silver NPs against some species of fungi (Jafari et al., 2015).

It should be pointed out that though gold NPs do not show as high antimicrobial activity as silver, gold-modified titania photocatalyst induces ROS generation (by the plasmonic photocatalysis) under irradiation (He et al., 2013; Noimark et al., 2014; Guo et al., 2015; Endo et al., 2018) and also the transfer of electrons between gold and bacteria on the interface (extracellular electron transfer) (Li et al., 2014; Wang et al., 2016), resulting in a lethal collapse of bacterial cells. Additionally, it has been found that Au/TiO_2 suppresses the sporulation of mold fungi efficiently under indoor light, indicating the inhibition of fungi proliferation (Kowalska et al. 2014b; Endo et al. 2018).

The change in the surface properties (oxidation state) of silver and copper, attributed to the electron transfer under light irradiation, might enhance the adsorption of the photocatalyst on the cell surface, causing the enhanced inactivation of

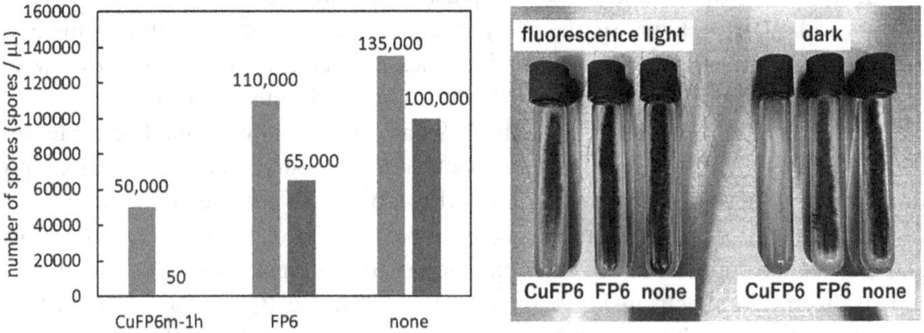

FIGURE 8.13 (left) Number of *Aspergillus niger* spores in the dark (gray) and under fluorescence light (green), and (right) exemplary photographs of fungal growth for Cu/TiO_2, bare TiO_2, and in the absence of photocatalyst (none). Adapted with permission from (Endo-Kimura et al., 2020). Copyright 2020, Creative Commons Attribution.

microorganisms. Plasmonic photocatalysts are generally non-selective agents; therefore, it should be expected that they should work against the wild spectrum of viruses, Gram-positive and Gram-negative bacteria (including antibiotic-resistant pathogenic species ARB), filamentous fungi, and protozoa. It should be mentioned that the efficacy and duration of photocatalytic processes are difficult for comparison because of the diversity in methods used (type of photocatalyst and light, used strains of microorganisms etc.), and other precise details of the experiments performed. However, raw water from rivers, groundwater etc. should be used for practical application since the presence of various components and variety of microorganism could influence the disinfection performance. One of the latest works has demonstrated 96% removal of *E. coli* from the Songhua River (China) on Ag/g-C_3N_4(10 mg/L) photocatalyst after vis irradiation for 90 min, whereas a decrease in photocatalyst content to 1 mg/L results in 83% removal (Wei et al., 2020).

Titania modified with Au and Pt to generate large content of diffusible ROS of long lifetime breaks new ground (Leong et al., 2018). The potential applications of plasmonic photocatalyst are shown in Figure 8.14. For example, photocatalyst might offer great possibilities in various configurations of water photo-disinfection. However, the main limitation of the common use of this process is the final separation of photocatalyst from the treated stream. For that reason, immobilization of photocatalyst is increasingly studied. One of the possible solutions is *cement/mortar*-based systems for the construction of water storage tanks. As described by Janus et al. (2019), both bare titania and N,C co-doped photocatalysts have increased the antibacterial properties of concretes and caused *E. coli* removal from water after 60 min. of artificial solar irradiation. Moreover, this system could be easily anticipated to eliminate chlorine-resistant protozoan parasites, such as *Giardia lamblia*, *Naegleria fowleri*, *Acanthamoeba* spp., *Entamoeba histolytica*, *Cryptosporidium parvum*, *Cyclospora cayetanesis*, and *Isospora belli* (Marshal, 1997). It has been

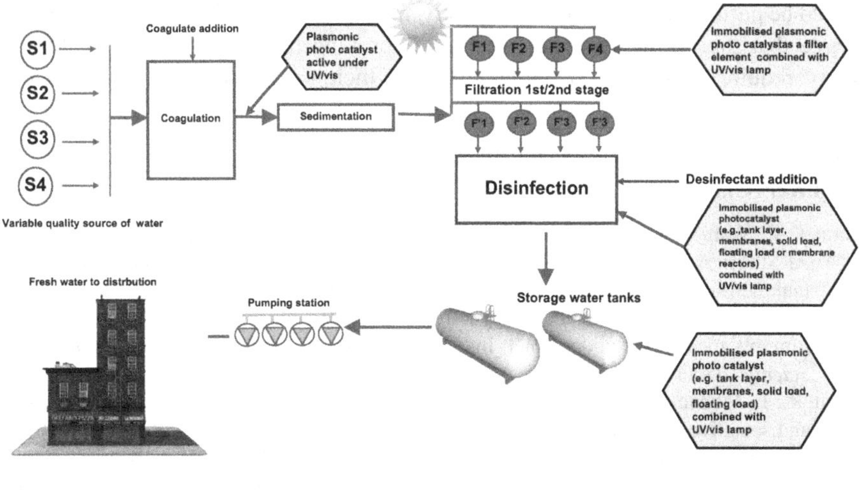

FIGURE 8.14 Possible applications of plasmonic photocatalysts for water remediation technology.

demonstrated that photocatalytic ceramic TiO_2 fibers significantly reduce *C. parvum* and *G. lamblia* even at a very short exposure time (9 min) under UV irradiation (Navalon et al., 2009). An alternative would be to design a photocatalyst separating system by employing commercial membrane systems to separate the photocatalysts from the water. There is no need for photocatalyst immobilization in such membrane reactors, since the powdered plasmonic photocatalysts could be suspended, and easily recovered and reused (Wang et al. 2012; Dong et al., 2015). Further pilot-scale studies of various configurations are needed, but it is reasonable to expect that a photocatalytic treatment process with high efficacy plasmonic photocatalyst active under solar radiation might be applied in the near future.

8.4 SUMMARY AND CONCLUSIONS

The chapter highlights the current role and perspectives of plasmonic nanomaterials designed for water and wastewater treatment. Many efforts have been devoted to the remediation of water contaminants with plasmonic photocatalysts due to their beneficial properties, such as high photocatalytic activity manifested in the decomposition of both toxic compounds (organic/inorganic) and microorganisms, and also tunable optical properties that guarantee visible light activity. Most recent publications indicate that plasmonic photocatalysts with advanced structures, such as faceted oxides and PCs, might be a versatile material with wide structural and functional variations. Another perspective functionality, extending the range of application, which might be incorporated to plasmonic nanomaterials, is a magnetic response of photocatalyst's particles enabling an efficient separation of material and then its reuse. The mechanical issues describing the photocatalytic activity of various plasmonic materials still require better clarification at some points, e.g., the role of selective deposition of NMs NPs on faceted titania or killing mechanism against waterborne pathogens. From an application perspective, it is important to emphasize that the cost of plasmonic photocatalysts and simultaneously associated remediation technologies should be particularly considered. Recently, plasmonic materials have matured to the degree that the following issues, such as improving their stability and ecological safety, require more detailed studies to ensure their wide application in the nearest future.

REFERENCES

Ahmad, T., Wani, I.A., Lone, I.H., Ganguly, A., Manzoor, N., Ahmad, A., Ahmed, J., & Al-Shihri, A.S. (2013). Antifungal activity of gold nanoparticles prepared by solvothermal method. *Materials Research Bulletin*, *48*(1), 12–20.

Akhavan, O. (2009). Lasting antibacterial activities of Ag-TiO_2/Ag/a-TiO_2 nanocomposite thin film photocatalysts under solar light irradiation. *Journal of Colloid and Interface Science*, *336*(1), 117–124.

Amin, R.M., Mohamed, M.B., Ramadan, M.A., Verwanger, T., & Krammer, B. (2009). Rapid and sensitive microplate assay for screening the effect of silver and gold nanoparticles on bacteria. *Nanomedicine*, *4*(6), 637–643.

Armelao, L., et al. (2007). Photocatalytic and antibacterial activity of TiO_2 and Au/TiO_2 nanosystems. *Nanotechnology*, *18*(37), 375709.

Awate, S. V., et al. (2008). Effect of gold dispersion on the photocatalytic activity of mesoporous titaniafor the vapor-phase oxidation of acetone. *International Journal of Photoenergy*, Article ID 789149.

Awate, S.V., et al. (2009). Photocatalytic mineralization of benzene over gold containing titania nanotubes: Role of adsorbed water and nanosize gold crystallites. *Catalysis Today*, *141*(1–2), 144–151.

Bach, U., et al. (1998). Solid-state dye-sensitized mesoporous TiO_2 solar cells with high proton-to-electron conversion efficiencies. *Nature*, *395*, 583–585.

Badwaik, V.D., Vangala, L.M., Pender, D.S., Willis, C.B., Aguilar, Z.P., Gonzalez, M.S., Paripelly, R., & Dakshinamurthy, R. (2012). Size-dependent antimicrobial properties of sugarencapsulated gold nanoparticles synthesized by a green method. *Nanoscale Research Letters*, *7*, 1–11.

Baghriche, O., Rtimi, S., Pulgarin, C., Sanjines, R., & Kiwi, J. (2012). Innovative TiO_2/Cu nanosurfaces inactivating bacteria in the minute range under low-intensity actinic light. *ACS Applied Materials and Interfaces*, *4*(10), 5234–5240.

Bian, Z.F., et al. (2014). Au/TiO_2 superstructure-based plasmonic photocatalysts exhibiting efficient charge separation and unprecedented activity. *Journal of the American Chemical Society*, *136*(1), 458-465.

Bielan, Z., et al. (2020a). Mono- and bimetallic (Pt/Cu) titanium(IV) oxide core–shell photocatalysts with UV/Vis light activity and magnetic separability. *Catalysis Today*, *31*, 105814.

Bielan, Z., et al. (2020b.) Defective TiO_2 core-shell magnetic photocatalyst modified with plasmonic nanoparticles for visible light-induced photocatalytic activity. *Catalysts*, *10*(6), 672.

Bledowski, M., Wang, L., Ramakrishnan, A., Khavryuchenko, O.V., Khavryuchenko, V.D., Ricci, P.C., Strunk, J., Cremer, T., Kolbeck, C., & Beranek, R. (2011). Visible-light photocurrent response of TiO_2-polyheptazine hybrids: Evidence for interfacial charge-transfer absorption. *Physical Chemistry Chemical Physics*, *13*(48), 21511–21519.

Bouhadoun, S., Guillard, C., Dapozze, F., Singh, S., Amans, D., Bouclé, J., & Herlin-Boime, N. (2015). One step synthesis of N-doped and Au-loaded TiO_2 nanoparticles by laser pyrolysis: Application in photocatalysis. *Applied Catalysis B: Environmental*, *174–175*, 367–375.

Caretti, I., et al. (2015). Light-induced processes in plasmonic gold/TiO_2 photocatalysts studied by electron paramagnetic resonance. *Topics in Catalysis*, *58*(12–13), 776–782.

Castro, C.A., Osorio, P., Sienkiewicz, A., Pulgarin, C., Centeno, A., & Giraldo, S.A. (2012). Photocatalytic production of $^1O_2$ and OH mediated by silver oxidation during the photoinactivation of *Escherichia coli* with TiO_2. *Journal of Hazardous Materials*, *211–212*, 172–181.

Chaves, A.F.A. et al. (2020). Chapter 5 – The role of filamentous fungi in drinking water biofilm formation, The role of filamentous fungi in drinking water biofilm formation, Editor(s): Manuel Simoes, Anabel Borges, Lucia Chaves Simoes, *Recent Trends in Biofilm Science and Technology*, Academic Press, 101–125.

Cheikhrouhou, W., et al. (2019). AgCl/Ag functionalized cotton fabric: An effective plasmonic hybrid material for water disinfection under sunlight, *Solar Energy*, *183*, 653–664.

Chen, X., Zhu, H.Y., Zhao, J.C., Zheng, Z.F., & Gao, X.P. (2008a). Visible-light-driven oxidation of organic contaminants in air with gold nanoparticle catalysts on oxide supports. *Angewandte Chemie – International Edition*, *47*(29), 5353–5356.

Chen, J.I.L., Loso, E., Ebrahim, N., & Ozin, G.A. (2008b). Synergy of slow photon and chemically amplified photochemistry in platinum nanocluster-loaded inverse titania opals. *Journal of the American Chemical Society*, *130*, 5420–5421.

Chen, Z., Fang, L., Dong, W., Zheng, F., Shen, M., & Wang, J. (2014). Inverse opal structured Ag/TiO_2 plasmonic photocatalyst prepared by pulsed current deposition and its enhanced visible light photocatalytic activity. *Journal of Materials Chemistry A*, *2*, 824–832.

Chen, K., Zhou, X., Cheng, X., Qiao, R., Cheng, Y., Liu, C., Xie, Y.,Yu, W., Yao, F., Sun, Z., Wang, F., Liu, K., & Liu, Z. (2019). Graphene photonic crystal fibre with strong and tunable light–matter interaction. *Nature Photonics, 13*, 754–759.

Cho, M., Chung, H., Choi, W., & Yoon, J. (2004). Linear correlation between inactivation of *E. coli* and OH radical concentration in TiO_2 photocatalytic disinfection. *Water Research, 38*(4), 1069–1077.

Chwalibog, A., Sawosz, E., Hotowy, A., Szeliga, J., Mitura, S., Mitura, K., Grodzik, M., Orlowski, P., & Sokolowska, A. (2010). Visualization of interaction between inorganic nanoparticles and bacteria or fungi. *International Journal of Nanomedicine, 5*(1), 1085–1094.

Cohn, C. et al., (2006). Pyrite-induced hydroxyl radical formation and its effect on nucleic acids. *Geochem. Trans. Geochemucal Transistion Journal*, 4(7), 3.

Costi, R., et al. (2008). Visible light-induced charge retention and photocatalysis with hybrid CdSe-Au nanodumbbells. *Nano Letters*, 8(2), 637–641.

Cui, Y., Zhao, Y., Tian, Y., Zhang, W., Lü, X., & Jiang, X. (2012). The molecular mechanism of action of bactericidal gold nanoparticles on *Escherichia coli. Biomaterials, 33*(7), 2327–2333.

Cushing, S.K., Li, J., Meng, F., Senty, T.R., Suri, S., Zhi, M., Li, M., Bristow, A.D., & Wu, N. (2012). Photocatalytic activity enhanced by plasmonic resonant energy transfer from metal to semiconductor. *Journal of the American Chemical Society, 134*(36), 15033–15041.

Dankovich, T.A., & Gray, D.G. (2011). Bactericidal paper impregnated with silver nanoparticles for emergency water disinfection. *SI, 45*(c), COLLSYMP-513.

Dankovich, T.A., & Smith, J.A. (2014). Incorporation of copper nanoparticles into paper for point-of-use water purification. *Water Research, 63*, 245–251.

Dash, J.N. & Jha, R. Graphene-based birefringent photonic crystal fiber sensor using surface plasmon resonance (2014). *IEEE Photonics Technology Letters, 26,* 1092–1094.

David, L., & Moldovan, B. (2020). Green synthesis of biogenic silver nanoparticles for efficient catalytic removal of harmful organic dyes. *Nanomaterials, 10*(2), 202.

Deng, C.H., Gong, J.L., Zeng, G.M., Zhang, P., Song, B., Zhang, X.G., Liu, H.Y., & Huan, S.Y. (2017). Graphene sponge decorated with copper nanoparticles as a novel bactericidal filter for inactivation of *Escherichia coli. Chemosphere, 184*, 347–357.

Dong, S., et al. (2015). Recent developments in heterogeneous photocatalytic water treatment using visible lightresponsive photocatalysts: A review. *RSC Advances, 5*, 14610.

Du, L.C., et al. (2009a). Plasmon-induced charge separation and recombination dynamics in gold-tio_2 nanoparticle systems: Dependence on TiO_2 particle size. *Journal of Physical Chemistry C, 113*(16), 6454–6462.

Du, L., et al. (2009b). Plasmon induced electron transfer at gold-TiO_2 interface under femtosecond near-IR two photon excitation. *Thin Solid Films, 158*(2), 861–864.

Elfeky, S.A., and Al-Sherbini, A.S.A. (2011). Photo-oxidation of rhodamine-6-G via TiO_2 and Au/TiO_2-bound polythene beads. *Journal of Nanomater, 2011,* 570438.

Endo-Kimura, M., Wang, K., Wei, Z., Ohtani, B., Markowska-Szczupak, A., & Kowalska, E. (2020). Vis-responsive copper-modified titania for decomposition of organic compounds and microorganisms. *Catalysts, 10*, 1194.

Endo, M., Janczarek, M., Wei, Z., Wang, K., Markowska-Szczupak, A., Ohtani, B., & Kowalska, E. (2019). Bactericidal properties of plasmonic photocatalysts composed of noble metal nanoparticles on faceted anatase titania. *Journal of Nanoscience and Nanotechnology, 19*(1), 442–452.

Endo, M., Wei, Z., Wang, K., Karabiyik, B., Yoshiiri, K., Rokicka, P., Ohtani, B., Markowska-Szczupak, A., & Kowalska, E. (2018). Noble metal-modified titania with visible-light activity for the decomposition of microorganisms. *Beilstein Journal of Nanotechnology, 9*(1), 829–841.

Fang, L., Nan, F., Yang, Y., & Cao, D.W. (2016). Enhanced photoelectrochemical and photocatalytic activity in visible-light-driven Ag/$BiVO_4$ inverse opals. *Applied Physics Letters*, *108*, 093902, 1–4.

Foster, H.A. et al. (2011). Photocatalytic disinfection using titanium dioxide: Spectrum and mechanism of antimicrobial activity. *Appl Microbiol Biotechnol*, 90, 1847–1868.

Fujishima, A., & Honda, K. (1972). Electrochemical photolysis of water at a semiconductor electrode. *Nature (London United Kingdom)*,*238*(5358), 37–38.

Galdiero, S., Falanga, A., Vitiello, M., Cantisani, M., Marra, V., & Galdiero, M. (2011). Silver nanoparticles as potential antiviral agents. *Molecules*, *16*, 8894–8918.

Gerrity, D. et al. (2008). Photocatalytic inactivation of viruses using titanium dioxide nanoparticles and low-pressure UV light. *Journal of Environtal Science and Health, Part A*,*43*(11), 1261–1270.

Grčić, I., et al. (2020). Enhanced visible-light driven photocatalytic activity of Ag@TiO_2 photocatalyst prepared in chitosan matrix *Catalysts*, *10*, 763.

Guo, L., Shan, C., Liang, J., Ni, J., & Tong, M. (2015). Bactericidal mechanisms of Au@TNBs under visible light irradiation. *Colloids and Surfaces B: Biointerfaces*, *128*, 211–218.

He, W., Huang, H., Yan, J., & Zhu, J. (2013). Photocatalytic and antibacterial properties of Au-TiO_2 nanocomposite on monolayer graphene: From experiment to theory. *Journal of Applied Physics*, *114*(20), 204701.

Herrmann, J.M. (1999). Heterogeneous photocatalysis: Fundamentals and applications to the removal of various types of aqueous pollutants. *Catalysis Today*, *53*(1), 115–129.

Holt, K.B., & Bard, A.J. (2005). Interaction of silver(I) ions with the respiratory chain of *Escherichia coli*: An electrochemical and scanning electrochemical microscopy study of the antimicrobial mechanism of micromolar Ag. *Biochemistry*, *44*(39), 13214–13223.

Hou, W.B., et al. (2011a). Photocatalytic conversion of CO_2 to hydrocarbon fuels via plasmon-enhanced absorption and metallic interband transitions. *ACS Catalysis*, 1(8):929–936.

Hou, W., et al. (2011b). Plasmonic enhancement of photocatalytic decomposition of methyl orange under visible light. *Journal of Catalysis*, *277*, 149–153.

Hou, Y., Li, X., Zhao, Q., Chen, G., & Raston, C.L. (2012). Role of hydroxyl radicals and mechanism of *Escherichia coli* inactivation on Ag/AgBr/TiO_2 nanotube array electrode under visible light irradiation. *Environmental Science and Technology*, *46*(7), 4042–4050.

Huo, J.W., Yuan, C., & Wang, Y. (2019). Nanocomposites of three-dimensionally ordered porous TiO_2 decorated with Pt and reduced graphene oxide for the visible-light photocatalytic degradation of waterborne pollutants. *ACS Applied Nano Materials, 2*, 2713–2724.

Hu, Ch., et al., (2007). Photocatalytic degradation of pathogenic bacteria with AgI/TiO_2 under visible light Irradiation, *Langmuir*, 23(9), 4982–4987.

Huang, Z., Maness, P.-C., Blake, D. M., Wolfrum, E. J., Smolinski, S. L., & Jacoby, W. A. (2000). Bactericidal mode of titanium dioxide photocatalysis. *Journal of Photochemistry and Photobiology A: Chemistry*, *130*, 163–170.

Hwangbo, M., et al. (2019). Effectiveness of zinc oxide-assisted photocatalysis for concerned constituents in reclaimed wastewater: 1, 4-Dioxane, trihalomethanes, antibiotics, antibiotic resistant bacteria (ARB), and antibiotic resistance genes (ARGs). *Science of The Total Environment*, *649*, 1189–1197,

Ilie, M., et al. (2011). Improving TiO_2 activity in photo-production of hydrogen from sugar industry wastewaters. *International Journal of Hydrogen Energy*, *36*(24), 15509–15518.

Iliev, V., et al. (2007). Influence of the size of gold nanoparticles deposited on TiO_2 upon the photocatalytic destruction of oxalic acid. *Journal of Molecular Catalysis A: Chemical, 263*(1–2), 32–38.

Ilisz, I., Laszlo, Z., and Dombi, A. (1999). Investigation of the photodecomposition of phenol in near-UV-irradiated aqueous TiO_2 suspensions. I. Effect of charge-trapping species on the degradation kinetics. *Applied Catalysis, A: General, 180*(1–2), 25–33.

Jacoby, W.A., et al. (1998). Mineralization of bacterial cell mass on a photocatalytic surface in air. *Environmental Science & Technolology, 32*(17), 2650–2653.

Jafari, A., Pourakbar, L., Farhadi, K., Mohamadgolizad, L., & Goosta, Y. (2015). Biological synthesis of silver nanoparticles and evaluation of antibacterial and antifungal properties of silver and copper nanoparticles. *Turkish Journal of Biology, 39*(4), 556–561.

Janczarek, M., Kowalska, E., & Ohtani, B. (2016). Decahedral-shaped anatase titania photocatalyst particles: Synthesis in a newly developed coaxial-flow gas-phase reactor. *Chemical Engineering Journal, 289*, 502–512.

Janczarek, M., et al. (2017). Silver- and copper-modified decahedral anatase tiania particles as visible light-responsive plasmonic photocatalyst. *Journal of Photonics for Energy*, 7(1), 1–16.

Janus, M. et al., (2019). Bacterial inactivation on concrete plates loaded with modified TiO_2 photocatalysts under visible light irradiation. *Molecules, 24*(17), 3026.

Jiang, H.Y., Cheng, K., and Lin, J. (2012). Crystalline metallic Au nanoparticle-loaded alpha-Bi_2O_3 microrods for improved photocatalysis. *Physical Chemistry Chemical Physics, 14*(35), 12114–12121.

Jin, M., et al. (2020). Chlorine disinfection promotes the exchange of antibiotic resistance genes across bacterial genera by natural transformation. *ISME Journal, 14*, 1847–1856.

Kafizas, A., et al. (2009). Titanium dioxide and composite metal/metal oxide titania thin films on glass: A comparative study of photocatalytic activity. *Journal of Photochemistry and Photobiology A: Chemistry,204*(2–3), 183–190.

Khaleque, A., & Hattori, H.T. (2016). Absorption enhancement in graphene photonic crystal structures. *Applied Optics, 55*, 2936–2942.

Khedr, T.M., et al. (2019a). A comparative study of microcystin-LR degradation by UV-A, solar and visible light irradiation using bare and C/N/S-modified titania. *Catalysts, 9*, 877.

Khedr, T.M., et al. (2019b). Photodegradation of microcystin-LR using visible light-activated C/N-co-modified mesoporous TiO_2 photocatalyst. *Materials, 12*(7), 1027.

Kim, S. (2013). Bacterial inactivation in water, DNA strand breaking, and membrane damage induced by ultraviolet-assisted titanium dioxide photocatalysis. *Water Research, 47*(13), 4403–4411.

Kisch, H. (2010). On the problem of comparing rates or apparent quantum yields in heterogeneous photocatalysis. *Angewandte Chemie International Edition, 49*, 9588-9589.

Kiwi, J., & Nadtochenko, V. (2004). New evidence for TiO_2 photocatalysis during bilayer lipid peroxidation. *Journal of Physical Chemistry B, 108*(45), 17675–17684.

Kiwi, J., & Nadtochenko, V. (2005). Evidence for the mechanism of photocatalytic degradation of the bacterial wall membrane at the TiO_2 interface by ATR-FTIR and laser kinetic spectroscopy. *Langmuir, 21*(10), 4631–4641.

Kiyonaga, T., et al. (2008). Size-dependence of Fermi energy of gold nanoparticles loaded on titanium(IV) dioxide at photostationary state. *Physical Chemistry Chemical Physics, 10*(43), 6553–6561.

Kominami, H., Tanaka, A., & Hashimoto, K. (2011). Gold nanoparticles supported on cerium(IV) oxide powder for mineralization of organic acids in aqueous suspensions under irradiation of visible light of $\lambda = 530$ nm *Applied Catalysis A: General*, 397(1–2), 121–126.

Kowal, K., Wysocka-Król, K., Kopaczyńska, M., Dworniczek, E., Franiczek, R., Wawrzyńska, M., Vargová, M., Zahoran, M., Rakovský, E., Kuš, P., Plesch, G., Plecenik, A., Laffir, F., Tofail, S. A. M., & Podbielska, H. (2011). In situ photoexcitation of silver-doped titania nanopowders for activity against bacteria and yeasts. *Journal of Colloid and Interface Science, 362*(1), 50–57.

Kowalska, E., Abe, R., & Ohtani, B. (2009). Visible light-induced photocatalytic reaction of gold-modified titanium(IV) oxide particles: Action spectrum analysis. *Chemical Communications*, (2), 241–243.

Kowalska, E., Remita, H., Colbeau-Justin, C., Hupka, J., & Belloni, J. (2008). Modification of titanium dioxide with platinum ions and clusters: Application in photocatalysis.

Kowalska, E., et al. (2010). Visible-light-induced photocatalysis through surface plasmon excitation of gold on titania surfaces. *Physical Chemistry Chemical Physics,12*(10), 2344–2355.

Kowalska, E., Rau, S., Ohtani, B. (2012). Plasmonic titania photocatalysts active under UV and visible-light irradiation: Influence of gold amount, size, and shape. *Journal of Nanotechnology*, *2012*, 1–11.

Kraeutler, B., & Bard, A.J. (1978). Heterogeneous photocatalytic preparation of supported catalysts. photodeposition of platinum on TiO_2 powder and other substrates. *Journal of the American Chemical Society*, *100*(13), 4317–4318.

Kubacka, A., Ferrer, M., Martínez-Arias, A., & Fernández-García, M. (2008). Ag promotion of TiO_2-anatase disinfection capability: Study of *Escherichia coli* inactivation. *Applied Catalysis B: Environmental*, *84*(1–2), 87–93.

Leong, K.H., et al. (2018). Mechanistic insights into plasmonic photocatalysts in utilizing visible light. *Beilstein Journal of Nanotechnology*, *9*, 628–648.

Li, F.B., & Li, X.Z. (2002). Photocatalytic properties of gold/gold ion-modified titanium dioxide for wastewater treatment. *Applied Catalysis A.-General, 228*(1–2), 15–27.

Li, G., Nie, X., Gao, Y., & An, T. (2016). Can environmental pharmaceuticals be photocatalytically degraded and completely mineralized in water using g-C_3N_4/TiO_2 under visible light irradiation? -Implications of persistent toxic intermediates. *Applied Catalysis B: Environmental*, *180*, 726–732.

Li, J., Zhou, H., Qian, S., Liu, Z., Feng, J., Jin, P., & Liu, X. (2014). Plasmonic gold nanoparticles modified titania nanotubes for antibacterial application. *Applied Physics Letters*, *104*(26), 261110.

Li, M., Noriega-Trevino, M.E., Nino-Martinez, N., Marambio-Jones, C., Wang, J., Damoiseaux, R., Ruiz, F., & Hoek, E.M.V. (2011). Synergistic bactericidal activity of Ag-TiO_2 nanoparticles in both light and dark conditions. *Environmental Science and Technology*, *45*(20), 8989–8995.

Li, X.H., Wu, Y., Shen, Y.H., Sun, Y., Yang, Y., & Xie, A.J. (2018). A novel bifunctional Ni-doped TiO_2 inverse opal with enhanced SERS performance and excellent photocatalytic activity. *Applied Surface Science, 427,* 739–744.

Liga, M.V., Bryant, E.L., Colvin, V.L., & Li, Q. (2011). Virus inactivation by silver doped titanium dioxide nanoparticles for drinking water treatment. *Water Research*, *45*(2), 535–544.

Lim, S.Y., Law, C.S., Liu, L., Markovis, M., Abell, A.D., & Santos, A. (2019). Integrating surface plasmon resonance and slow photon effects in nanoporous anodis alumina photonic crystals for photocatalysis. *Catalysis Science & Technology, 9*, 3158–3176.

Lin, W.C., Chen, C.N., Tseng, T.T., Wei, M.H., Hsieh, J.H., & Tseng, W.J. (2010). Micellar layer-by-layer synthesis of TiO_2/Ag hybrid particles for bactericidal and photocatalytic activities. *Journal of the European Ceramic Society*, *30*(14), 2849–2857.

Lin, W.C., et al., (2014). Toxicity of TiO_2 nanoparticles to *Escherichia coli*: Effects of particle size, crystal phase and water chemistry, *PLoS One*, 9(10), e110247.

Liu, L.Q., Ouyang, S.X., and Ye, J.H. (2013). Gold-nanorod-photosensitized titanium dioxide with wide-range visible-light harvesting based on localized surface plasmon resonance. *Angewandte Chemie*, 52(26),6689–6693.

Liu, S., et al. (2015). Efficient removal of radioactive iodide ions from water by three-dimensional Ag_2O–Ag/TiO_2 composites undervisible light irradiation. *Journal of Hazardous Materials* 284 171–181.

Liu, Z., et al. (2011). Plasmon resonance enhancement of photocatalytic water splitting under visible illumination. *Nano Letters*, 11, 1111–1116.

Loddo, V., et al. (1999). Preparation and characterisation of TiO_2 (anatase) supported on TiO_2 (rutile) catalysts employed for 4-nitrophenol photodegradation in aqueous medium and comparison with TiO_2 (anatase) supported on Al_2O_3. *Applied Catalysis B: Environmental.* 20, 29–45.

Lok, C.-N., Ho, C.-M., Chen, R., He, Q.-Y., Yu, W.-Y., Sun, H., Tam, P. K.-H., Chiu, J.-F., & Che, C.-M. (2006). Proteomic analysis of the mode of antibacterial action of silver nanoparticles. *Journal of Proteome Research*, *5*, 916–924.

Lu, Y., Yu, H.T., Chen, S., Quan, X., & Zhao, H.M. (2012). Integrating plasmonic nanoparticles with TiO_2 photonic crystal for enhancement of visible-light-driven photocatalysis. *Environmental Science & Technology*, *46*, 1724–1730.

Mai, L., Wang, D., Zhang, S., Xie, Y., Huang, C., & Zhang, Z. (2010). Synthesis and bactericidal ability of Ag/TiO_2 composite films deposited on titanium plate. *Applied Surface Science*, *257*(3), 974–978.

Manhart, M. D. (1990). In vitro antimicrobial activity of bismuth subsalicylate and other bismuth salts. *Reviews of Infectious Diseases*, *12*(February), S11–S15.

Markowska-Szczupak, A., et al. (2015). Effect of water activity and Titania P25 photocatalyst on inactivation of pathogenic fungi – contribution to the protection of public health. *Central European Journal of Public Health*, 23(3),267–271.

Markowska-Szczupak, A., et al. (2018). Photocatalytic water disinfection under solar irradiation by D-glucose-modified titania. *Catalysts*, 8(8), 316.

Markowska-Szczupak, A., Ulfig, K., & Morawski, W.A. (2011). The application of titanium dioxide for deactivation of bioparticulates: An overview. *Catalysis Today*, 161(9),249–257.

Marshall, M.M. (1997). Waterborne protozoan pathogens, *Clinical Microbiological Review*, 10(1), 67–85.

Matsunaga, T., Tomoda, R., Nakajima, T., Nakamura, N., & Komine, T. (1988). Continuous-sterilization system that uses photosemiconductor powders. *Applied and Environmental Microbiology*, *54*(6), 1330–1333.

Matsunaga, T., Tomoda, R., Nakajima, T., & Wake, H. (1985). Photoelectrochemical sterilization of microbial cells by semiconductor powders. *FEMS Microbiology Letters*, *29*(1–2), 211–214.

Mendez-Medrano, M.G., et al. (2016). Surface modification of TiO_2 with Au nanoclusters for efficient water treatment and hydrogen generation under visible light. *Journal of Physical Chemistry C*, 120(43), 25010–25022.

Meng, S.G., Li, D.Z., Fu, X.L., & Fu, X.Z. (2015). Integrating photonic bandgaps with surface plasmon resonance for the enhancement of visible-light photocatalytic performance. *Journal of Materials Chemistry A*, *3*, 23501–23511.

Minero, C., et al. (1996). Large solar plant photocatalytic water decontamination: Effect of operational parameters. *Solar Energy* 56(5):421–428.

Mkhalid, I.A. (2015). Visible light photocatalytic synthesis of aniline with an Au/$LaTiO_3$ nanocomposites. *Journal of Alloys and Compounds*, 631, 298–302.

Mohamed, R.M., and Aazam, E.S. (2011). Characterization and catalytic properties of nano-sized Au metal catalyst on titanium containing high mesoporous silica (Ti-HMS) synthesized by photo-assisted deposition and impregnation methods. *International Journal of Photoenergy*, 2011, 137328

Mukherjee, S., et al. (2013). Hot electrons do the impossible: Plasmon-induced dissociation of H_2 on Au. *Nano Letters*, 13(1), 240–247.

Noimark, S., Page, K., Bear, J.C., Sotelo-Vazquez, C., Quesada-Cabrera, R., Lu, Y., Allan, E., Darr, J.A., & Parkin, I.P. (2014). Functionalised gold and titania nanoparticles and surfaces for use as antimicrobial coatings. *Faraday Discussions*, *175*, 273–287.

Navalon, S. et al. (2009). Photocatalytic water disinfection of Cryptosporidium parvum and Giardia lamblia using a fibrous ceramic TiO_2 photocatalyst. *Water Science & Technology*, 59(4), 639–645.

Naya, S.-I., et al. (2015). A bi-overlayer type plasmonic photocatalyst consisting of mesoporous Au/TiO_2 and CuO/SnO_2 films separately coated on FTP. *Physical Chemistry Chemical Physics*, 17, 18004–18010.

Nazari, M. et al. (2017). Plasmonic enhancement of selective photonic virus inactivation. *Scientific Reports*, 7(1), 11951.

Nguyen, T. et al. (2012). Biofouling of water treatment membranes: A review of the underlying causes, monitoring techniques and control measures. *Membranes*, 2(4), 804–840.

Nischk, M., et al. (2016). Enhanced photocatalytic, electrochemical and photoelectrochemical properties of TiO2 nanotubes arrays modified with Cu, AgCu and Bi nanoparticles obtained via radiolytic reduction. *Applied Surface Science*, 387, 89–102.

Nishijima, Y., et al. (2010). Plasmon-assisted photocurrent generation from visible to near-infrared wavelength using a Au-nanorods/TiO_2 electrode. *The Journal of Physical Chemistry Letters*, 1, 2031–2036.

Ohtani, B. (2010). Photocatalysis A to Z – What we know and what we do not know in a scientific sense. *Journal of Photochemistry and Photobiology C: Photochemistry Reviews*, 11(4), 157–178.

Ohtani, B., et al. (1993). Photocatalytic reaction of neat alcohols by metal-loaded titanium(IV) oxide particles. *Journal of Photochemistry and Photobiology, A: Chemistry*, 70(3), 265–272.

Ohtani, B., et al. (2010). What is degussa (Evonik) P25? Crystalline composition analysis, reconstruction from isolated pure particles and photocatalytic activity test. *Journal of Photochemistry and Photobiology a-Chemistry*, 216(2–3), 179–182.

Okuno, T., et al. (2015). Three modes of high-efficient photocatalysis using composites of TiO_2-nanocrystallite-containing mesoporous SiO_2 and Au nanoparticles. *Journal of Sol-Gel Science and Technology*, 74(3), 748–755.

Ollis, D.F., Pelizzetti, E., & Serpone, N. (1991). Photocatalyzed destruction of water contaminants. *Environmental Science and Technology*, 25(9), 1522–1529.

Oros-Ruiz, S., Zanella, R., & Prado, B. (2013). Photocatalytic degradation of trimethoprim by metallic nanoparticles supported on TiO_2-P25. *Journal of Hazardous Materials*, 263, 28–35.

Padervand, M., Reza Elahifard, M., Vatan Meidanshahi, R., Ghasemi, S., Haghighi, S., & Reza Gholami, M. (2012). Investigation of the antibacterial and photocatalytic properties of the zeolitic nanosized AgBr/TiO_2 composites. *Materials Science in Semiconductor Processing*, *15*(1), 73–79.

Page, K. et al. (2007). Titania and silver – titania composite films on glass-potent antimicrobial coatings. *Journal of Material Chemistry*, 17, 95–104.

Pap, Z., et al. (2013). Behavior of gold nanoparticles in a titania aerogel matrix: Photocatalytic activity assessment and structure investigations. *Chinese J. Catal.*, 34(4), 734–740.

Pathirana, H.M.K.K., & Maithreepala, R.A. (1997). Photodegradation of 3,4-dichloropropionamide in aqueous TiO_2 suspensions. *Journal of Photochemistry and Photobiology, A: Chemistry*, 102(2–3), 273–277.

Pearson, A., et al. (2012). Decoration of TiO_2 nanotubes with metal nanoparticles using polyoxometalate as a UV-switchable reducing agent for enhanced visible and solar light photocatalysis. *Langmuir*, 28(40), 14470–14475.

Pichat, P., et al. (1981). Photocatalytic hydrogen production from aliphatic alcohols over a bifunctional platinum on titanium dioxide catalyst. *Nouveau Journal de Chimie*, 5(12), 627–636.

Priebe, J.B., et al. (2015). Solar hydrogen production by plasmonic Au-TiO_2 catalysts: Impact of synthesis protocol and TiO_2 phase on charge transfer efficiency and H_2 evolution rates. *ACS Catalysis*, 5(4), 2137–2148.

Prosposito, P., et al. (2019). Bifunctionalized silver nanoparticles as Hg^{2+} plasmonic sensor in water: Synthesis, characterizations, and ecosafety. *Nanomaterials*, 9, 1353.

Qiu, X., Miyauchi, M., Sunada, K., Minoshima, M., Liu, M., Lu, Y., Li, D., Shimodaira, Y., Hosogi, Y., Kuroda, Y., & Hashimoto, K. (2012). Hybrid Cu_xO/TiO_2 nanocomposites as risk-reduction materials in indoor environments. *ACS Nano*, *6*(2), 1609–1618.

Rai, A., Prabhune, A., & Perry, C.C. (2010). Antibiotic mediated synthesis of gold nanoparticles with potent antimicrobial activity and their application in antimicrobial coatings. *Journal of Materials Chemistry*, *20*(32), 6789–6798.

Rahul, T.K., & Sandhyarani. N. (2015).Nitrogen-Fluorine co-doped titania inverse opals for enhanced solar light driven photocatalysis. *Nanoscale, 7,* 18259–18270.

Raja-Mogan, T., Ohtani, B., & Kowalska, E. (2020). Photonic crystals for plasmonic photocatalysis. *Catalysts*, 10(8), 827.

Raja-Mogan, T., Lehoux, A., Takashima, M., Kowalska, E., & Ohtani, B. (2021). Slow photon-induced enhancement of photocatalytic activity of gold nanoparticle-incorporated titania in-verse opal. *Chem. Letter. Advance online publication.* https://doi.org/10.1246/cl.200804.

Ratte, H.T. (1999). Bioaccumulation and toxicity of silver compounds: A review. *Environmental Toxicology and Chemistry*, *18*(1), 89–108.

Rodriguez-Gonzalez, V., et al. (2008). MTBE visible-light photocatalytci decomposition over Au/TiO_2 and Au/TiO_2-Al_2O_3 sol-gel prepared catalysts. *Journal of Molecular Catalysis A: Chemical*, 281, 93–98.

Rtimi, S., Giannakis, S., Sanjines, R., Pulgarin, C., Bensimon, M., & Kiwi, J. (2016). Insight on the photocatalytic bacterial inactivation by co-sputtered TiO_2-Cu in aerobic and anaerobic conditions. *Applied Catalysis B: Environmental*, *182*, 277–285.

Ryu, H.-J., et al. (2020). Structurally and compositionally tunable absorption properties of AgCl@AgAu nanocatalysts for plasmonic photocatalytic degradation of environmental pollutants. *Catalysts*, *10*, 405.

Saito, T. et al., (1992). Mode of photocatalytic bactericidal action of powdered semiconductor TiO_2 on mutans streptococci, *Journal of Photochemistry and Photobiology B: Biology*, *14*, 369–379,

Seh, Z.W., et al. (2012). Janus Au-TiO_2 photocatalysts with strong localization of plasmonic near fields for efficient visible-light hydrogen generation. *Advanced Materials*, *24*, 2310–2314.

Seo, J.H., et al. (2011). Cytotoxicity of serum protein-adsorbed visible-light photocatalytic Ag/AgBr/TiO_2 nanoparticles. *Journal of Hazardous Materials*, *198*, 347–355.

Shi, H., et al. (2014). Visible-light-driven photocatalytic inactivation of E. coli by Ag/AgX-CNTs (X=Cl, Br, I). plasmonic photocatalysts: Bacterial performance and deactivation mechanism, *Applied Catalysis B: Environmental*, *158–159*, 301–307.

Silva, C.G., et al. (2011). Influence of excitation wavelength (UV or visible light) on the photocatalytic activity of titania containing gold nanoparticles for the generation of hydrogen or oxygen from water. *Journal of the American Chemical Society*, *133*, 595–602.

Smalley, R.E. (2003). *Nanotechnology, Energy and People, in MIT Forum*. River Oaks, USA.

Smirnova, N., et al. (2010). Photoelectrochemical and photocatalytic properties of mesoporous TiO_2 films modified with silver and gold nanoparticles. *Surface and Interface Analysis*, *42*(6–7), 1205–1208.

Son, M.S., et al. (2010). Surface plasmon enhanced photoconductance and single electron effects in mesoporous titania nanofibers loaded with gold nanoparticles. *Applied Physics Letters* 96:023115.

Sonawane, R.S., & Dongare, M.K. (2006). Sol-gel synthesis of Au/TiO_2 thin films for photocatalytic degradation of phenol in sunlight. *Journal of Molecular Catalysis A: Chemical 243*(1), 68–76.

Song, C.K., et al. (2016). Exploring crystal phase and morphology in the TiO_2 supporting materials used for visible-light driven plasmonic photocatalyst. *Applied Catalysis B-Environmental, 198*, 91–99.

Sordello, F., Duca, C., Maurino, V., & Minero, C. (2011). Photocatalytic metamaterials: TiO_2 inverse opals. *Chemical Communications, 47,* 6147–6149.

Speshock, J.L., Murdock, R.C., Braydich-Stolle, L.K., Schrand, A.M., & Hussain, S.M. (2010). Interaction of silver nanoparticles with HIV-1. *Journal of Nanobiotechnology, 8*, 19–27.

Stentzel, S., Teufelberger, A., Nordengrün, M., Kolata, J., Schmidt, F., van Crombruggen, K., Michalik, S., Kumpfmüller, J., Tischer, S., Schweder, T., Hecker, M., Engelmann, S., Völker, U., Krysko, O., Bachert, C., & Bröker, B.M. (2017). Staphylococcal serine protease–like proteins are pacemakers of allergic airway reactions to *Staphylococcus aureus*. *Journal of Allergy and Clinical Immunology, 139*(2), 492–500.

Sun, J.W., et al. (2014). Gold-titania/protonated zeolite nanocomposite photocatalysts for methyl orange degradation under ultraviolet and visible irradiation. *Materials Science in Semiconductor Processing, 25*, 286–293.

Sunada, K., Minoshima, M., & Hashimoto, K. (2012). Highly efficient antiviral and antibacterial activities of solid-state cuprous compounds. *Journal of Hazardous Materials, 235–236*, 265–270.

Tallósy, S.P., et al., (2016). Adhesion and inactivation of Gram-negative and Gram-positive bacteria on photoreactive TiO_2/polymer and Ag–TiO_2/polymer nanohybrid films. *Applied Surface Science, 371*, 139–150.

Talebian, S., et al. (2020). Nanotechnology-based disinfectants and sensors for SARS-CoV-2. *Nature Nanotechnology, 15*, 618–621.

Temerov, F., Ankudze, B., & Saarinen, J.J. (2020). TiO_2 inverse opal structures with facile decoration of precious metal nanoparticles for enhanced photocatalytic activity. *Materials Chemistry and Physics, 242,* 122471, 1–6.

Tian, Y., Notsu, H., & Tatsuma, T. (2005). Visible-light-induced patterning of Au- and Ag-TiO_2 nanocomposite film surfaces on the basis of plasmon photoelectrochemistry. *Photochemical and Photobiological Sciences, 4*(8), 598–601.

Trammell, S.A., et al. (2012). Accelerating the initial rate of hydrolysis of methyl parathion with laser excitation using monolayer protected 10 nm Au nanoparticles capped with a Cu(bpy) catalyst. *Chemical Communications, 48*(34), 4121–4123.

van Grieken, R., Marugán, J., Sordo, C., Martínez, P., & Pablos, C. (2009). Photocatalytic inactivation of bacteria in water using suspended and immobilized silver-TiO_2. *Applied Catalysis B: Environmental, 93*(1–2), 112–118.

Van Grinsven, H.J.M., Ten Berge, H.F.M., Dalgaard, T., Fraters, B., Durand, P., Hart, A., Hofman, G., Jacobsen, B.H., Lalor, S.T.J., Lesschen, J. P., Osterburg, B., Richards, K.G., Techen, A. K., Vertès, F., Webb, J., & Willems, W.J. (2012). Management, regulation and environmental impacts of nitrogen fertilization in northwestern Europe under the Nitrates Directive; a benchmark study. *Biogeosciences, 9*(12), 5143–5160.

Sambhy, V., MacBride, M.M., Peterson, B.R., & Sen, A. (2006). Silver bromide nanoparticle/polymer composites: Dual action tunable antimicrobial materials. *Journal of American Chemical Society, 128*, 9798–9808.

Wang, K.L., et al. (2018). Interparticle electron transfer in methanol dehydrogenation on platinum-loaded titania particles prepared from P25. *Catalysis Today, 303*, 327–333.

Wang, S., et al. (2020). Preferentially oriented Ag-TiO_2 nanotube array film: An efficient visible-light-driven photocatalyst. *Journal of Hazardous Materials, 399*, 123016.

Wang, X.D., et al. (2013a). Enhanced photocatalytic activity: Macroporous electrospun mats of mesoporous Au/TiO_2 nanofibers. *Chem Catchem*,*5*(9), 2646–2654.

Wang, C., Ranasingha, O., Natesakhawat, S., Ohodnicki, P.R., Andio, M., Lewis, J.P., & Matranga, C. (2013b). Visible light plasmonic heating of Au–ZnO for the catalytic reduction of CO_2. *Nanoscale*, *5*(15), 6968–6974.

Wang, G., Feng, H., Gao, A., Hao, Q., Jin, W., Peng, X., Li, W., Wu, G., & Chu, P.K. (2016a). Extracellular electron transfer from aerobic bacteria to Au-loaded TiO_2 semiconductor without light: A new bacteria-killing mechanism other than localized surface plasmon resonance or microbial fuel cells. *ACS Applied Materials and Interfaces*, *8*(37), 24509–24516.

Wang, X., Tang, Y., Chen, Z., & Lim, T.T. (2012). Highly stable heterostructured Ag-AgBr/TiO_2 composite: A bifunctional visible-light active photocatalyst for destruction of ibuprofen and bacteria. *Journal of Materials Chemistry*, *22*(43), 23149–23158.

Wang, Y.F., Xiong, D.B., Zhang, W., Su, H.L., Liu, Q.L., Gu, J.J., Zhu, S.M., & Zhang, D. (2016b). Surface plasmon resonance of gold nanocrystals coupled with slow-photon-effect of biomorphic TiO_2 photonic crystals for enhanced photocatalysis under visible-light. *Catalysis Today, 274,* 15–21.

Wang, H., Wu, Y., & Song, H. (2019). Synergistic effects of photonic crystal and gold nanostars for quantitative SERS detection of 3-Phenoxybenzoic acid. *Applied Surface Science, 476,* 587–593.

Wang, X-G., Wang, J., Jiang, Z.-J., Tao, D.-W., Zhang, X.-Q., & Wang, C.-W. (2021). Silver loaded anodic aluminum oxide dual-bandgap heterostructure photonic crystals and their application for surface enhanced Raman scattering. *Applied Surface Science, 544,* 148881, 1–8.

Wei, Z., et al. (2015). Morphology-dependent photocatalytic activity of octahedral anatase particles prepared by ultrasonication-hydrothermal reaction of titanates. *Nanoscale*, *7*(29), 12392–12404.

Wei, Z., Endo, M., Wang, K., Charbit, E., Markowska-Szczupak, A., Ohtani, B., & Kowalska, E. (2017a). Noble metal-modified octahedral anatase titania particles with enhanced activity for decomposition of chemical and microbiological pollutants. *Chemical Engineering Journal, 318*, 121–134.

Wei, Z., et al. (2018a). Silver-modified octahedral anatase particles as plasmonic photocatalyst. *Catalysis Today*, *310*, 19–25.

Wei, Z., et al. (2018b). Noble metal-modified faceted anatase titania photocatalysts: Octahedron versus decahedron. *Applied Catalysis B–Environmental*, 237, 574–587.

Wei, Z., et al. (2020). Morphology-governed performance of plasmonic photocatalysts. *Catalysts*, *10*(9), 1070.

Wei, Z., et al. (2017b). Size-controlled gold nanoparticles on octahedral anatase particles as efficient plasmonic photocatalyst. *Appl. Catal. B: Environ.*, *206*, 393–405.

Wei, Z. S., Kowalska, E., Ohtani, B. (2014). Enhanced photocatalytic activity by particle morphology: Preparation, characterization, and photocatalytic activities of octahedral anatase titania particles. *Chemistry Letters,* 43(3), 346–348.

Wu, M., Liu, J., Jin, J.,Wang, C., Huang, S.Z., Deng, Z., Li, Y., & Su, B.L. (2014). Probing significant light absorption enhancement of titania inverse opal films for highly exalted photocatalytic degradation of dye pollutants. *Applied Catalysis B: Environmental, 150,* 411–420.

Wysocka, I., Markowska-Szczupak, A., Szweda, P., Ryl, J., Endo-Kimura, M., Kowalska, E., Nowaczyk, G., & Zielińska-Jurek, A. (2019). Gas-phase removal of indoor volatile organic compounds and airborne microorganisms over mono- and bimetal-modified (Pt, Cu, Ag) titanium(IV) oxide nanocomposites. *Indoor Air*, *29*(6), 979–992.

Yablonovitch, E. (1993). Photonic band-gap structures. *Journal of the Optical Society of America B, 10,* 283–295.

Yan, X., et al. (2006). Is Methylene blue an appropriate substrate for a photocatalytic activity test? A study with visible-light responsive titania. *Chem. Phys. Lett.*,*429*(4–6), 606–610.

Yang, X., et al. (2020). The use of tunable optical absorption plasmonic Au and Ag decorated TiO_2 structures as efficient visible light photocatalysts *Catalysts, 10,* 139.

Ye, J., Cheng, H., Li, H., Yang, Y., Zhang, S., Rauf, A., Zhao, Q., & Ning, G. (2017). Highly synergistic antimicrobial activity of spherical and flower-like hierarchical titanium dioxide/silver composites. *Journal of Colloid and Interface Science, 504,* 448–456.

Yi, Z., et al. (2015). Surface-plasmon-enhanced band emission and enhanced photocatalytic activity of Au nanoparticles-decorated ZnO nanorods. *Plasmonics, 10,* 1373–1380

Yu, J., et al. (2019). Duality in the mechanism of hexagonal ZnO/Cu_xO nanowires inducing sulfamethazine degradation under solar or visible light *Catalysts, 9,* 916.

Zaleska, A. (2008). Doped-TiO_2: A review. *Recent Patents on Engineering, 2*(3), 157–164.

Zaleska, A., et al. (2000). Photocatalytic degradation of lindane, p, p'-DDT and methoxychlor in an aqueous environment. *Journal of Photochemistry and Photobiology, A: Chemistry, 135*(2–3), 213–220.

Zheng, S., et al. (2016a). Titania modification with ruthenium(II) complex and gold nanoparticles for photocatalytic degradation of organic compounds. *Photochemical & Photobiological Sciences, 15,* 69–79.

Zheng, S. Z., et al. (2016b). Mono- and dual-modified titania with a Ruthenium(II) complex and silver nanoparticles for photocatalytic degradation of organic compounds. *Journal of Advanced Oxidation Technologies,* 19(2), 208–217.

Zhang C., et al. (2019). Higher functionality of bacterial plasmid DNA in water after peracetic acid disinfection compared with chlorination, *Science of The Total Environment, 685,* 419–427.

Zhang, W., Li, Y., Niu, J., & Chen, Y. (2013). Photogeneration of reactive oxygen species on uncoated silver, gold, nickel, and silicon nanoparticles and their antibacterial effects. *Langmuir, 29*(15), 4647–4651.

Zheng, K., Setyawati, M.I., Leong, D.T., & Xie, J. (2017). Antimicrobial gold nanoclusters. *ACS Nano, 11*(7), 6904–6910.

Zhang, Y., Wang, L., Liu, D., Gao, Y., Song, C., Shi, Y., Qu, D., & Shi, J. (2018). Morphology effect of honeycomb-like inverse opal for efficient photocatalytic water disinfection and photodegradation of organic pollutant. *Molecular Catalysis, 444,* 42–52.

Zhao, Y.X., Yang, B.F., Xu, J., Fu, Z.P., Wu, M., & Li, F. (2012). Facile synthesis of Ag nanoparticles supported on TiO_2 inverse opal with enhanced visible-light photocatalytic activity. *Thin Solid Films, 520,* 3515–3522.

Zielińska-Jurek, A., et al. (2011). Preparation and characterization of monometallic (Au) and bimetallic (Ag/Au) modified-titania photocatalysts activated by visible light. *Applied Catalysis B: Environmental, 101*(3–4), 504–514

Zielińska, A., Kowalska, E., Sobczak, J. W., Łacka, I., Gazda, M., Ohtani, B., Hupka, J., & Zaleska, A. (2010). Silver-doped TiO_2 prepared by microemulsion method: Surface properties, bio- and photoactivity. *Separation and Purification Technology, 72*(3), 309–318.

9 Magnetic Nanomaterials for Wastewater Remediation

Ramalingam Suhasini and Viruthachalam Thiagarajan
Bharathidasan University, Tiruchirappalli, India

CONTENTS

9.1 Introduction 279
9.2 Source of Water Pollution 280
9.3 Available Techniques for Water Remediation 281
9.4 Iron Oxide Nanoparticles for Wastewater Treatment 282
9.4.1 Zero-Valent Iron Nanoparticles (ZVI NPs) 282
9.4.2 Hematite 285
9.4.3 Magnetite 286
9.4.4 Maghemite 288
9.5 Nano-Adsorbents 289
9.6 Adsorption Method Protocol 290
9.7 Factors Affecting Adsorption Processes 291
9.8 Various Mechanisms Involved in the Removal of Pollutants 292
9.9 Magnetic Graphene/GO-Based Composites 293
9.10 Nickel Based Nanoparticles 294
9.11 Cobalt-Based Nanoparticles 295
9.12 Magnetic Polymer Nanocomposites 295
9.13 Magnetic Nanomaterials for Removal of Toxic Organic, Pharmaceutical, and Pesticides Compounds 297
9.14 Drawbacks of Using Nanomaterials in Water Treatment 297
9.15 Conclusion and Future Perspectives 298
Acknowledgment 298
References 298

9.1 INTRODUCTION

For the very survival of living entities since the early days of evolution of life on our planet, water is considered one of the most indispensable requirements to fulfill diverse demands in every part of the life cycle. Out of the total available water on

DOI: 10.1201/9781003118749-9

the Earth's crust, 99.7 % is in the forms of oceans, soil-associated, icebergs, vaporized in atmosphere or aerosol, thereby proving unfit for drinking purposes. Only 0.3 % becomes directly useful for the utilities of humankind. Despite its less availability, the release of huge loads of industrial pollutants (mainly spanning heavy metals and dyes), (Schwarzenbach et al., 2010) either intentionally or accidently to the environment, domestic, chemical, and other wastes lead to severe water contamination, all of which solely stem from human activities. Groundwater containing dissolved contaminants moves along with the water to nearby wells (source of drinking water). Such continuous source of contamination moving to the groundwater results in the plume therein (an area of contaminated groundwater). Globally, there is a high demand for water purification to meet the various needs of humanity due to the enhanced depletion of groundwater resources (Famiglietti, 2014; Prathna et al., 2018; Bhateria & Singh, 2019). Among the various materials available for water purification, magnetic nanomaterials play a key role in water remediation due to their unique physical, chemical, and magnetic properties (Natarajan et al., 2019). Specifically, the area of magnetite (Fe_3O_4) nanomaterials is attracting huge interest, especially toward wastewater treatment issues, thanks to their superparamagnetic property. After appropriate applications directly using an external magnetic field, the separation of magnetic adsorbents from their source becomes practically feasible, even without any other additional methods such as filtration and centrifugation (Gutierrez et al., 2017; Natarajan et al., 2019; Biftu et al., 2020; Fadillah et al., 2020; Wang et al., 2020a; Yadav et al., 2020; Natarajan et al., 2020; Luo et al., 2021; Leonel et al., 2021). Addressing the subcritical areas, in the current chapter, a detailed note on applications of magnetic nanoparticles and nanocomposites used for wastewater remediation are discussed.

9.2 SOURCE OF WATER POLLUTION

Majorly classified two different sources that lead to water pollution are point and non-point sources. The former is named as point sources as they hold directly identifiable starting points like effluents coming out from various industrial operations, an oil spill from a tanker, etc. Another source is called a non-point source, where the pollutants are released from different sources of origin, thereby becoming difficult to attribute the cause to a single point. One of the setback causative reasons of non-point source pollution is urban runoff which includes suspended and dissolved solids, pesticides from landscapes, toxic metals, oil, and grease, synthetic organic waste, mostly present in high concentrations (Speight, 2019).

Significant sources of water pollution are listed below:

- Acid rain
- Alien species
- Climate changes
- Disruption of sediments
- Radioactive wastes
- Industrial and agro-chemical wastes
- Sewage wastes

- Thermal wastes
- Urbanization

9.3 AVAILABLE TECHNIQUES FOR WATER REMEDIATION

Based on the source of water pollution, traditional as well as modern approaches have been used to remove different types of contaminants. In conventional techniques, columns of bed and bank materials were used, in which the contaminated water is filtered by passing through it. In order to protect water from organic matter, riverbank filtration systems were used. Leaching, hydrolysis, precipitation, oxidation, and reduction occurred naturally on the sub-surface processes on the Earth's crust that help in the naturalized treatment of water. Even though the above methods were reasonably good self-purification techniques, they suffer from serious environmental challenges pertaining to the compromised potable water quality, resulting from the emerging organic contaminants (EOCs) detected therein. These methods are also not typically designed to remove contaminants from the wastewater, thereby prompting the creation of new techniques. Therefore, to eliminate such heavily mixed contaminants, various effective and rapid methods such as membrane filtration (Geise et al., 2010; Deng et al., 2019), ion exchange (Subban & Gadgil, 2019), coagulation and flocculation (Wei et al., 2018; Kim et al., 2019), different chemical processes (Ruiz-Bevia & Fernandez-Torres, 2019), photocatalyst, and adsorption technologies have been used (Bonilla-Petriciolet et al., 2019; Yu et al., 2019a). Due to their extraordinary surface charge, and redox activity characteristics, magnetic nanomaterials make a potential fit as effective adsorbents, mainly finding applications in the removal of composite pollutants from aqueous systems (Martinez-Boubeta & Simeonidis, 2019). Schematic representations of various water pollutants and different nanomaterials used for water remediation are presented in Figure 9.1.

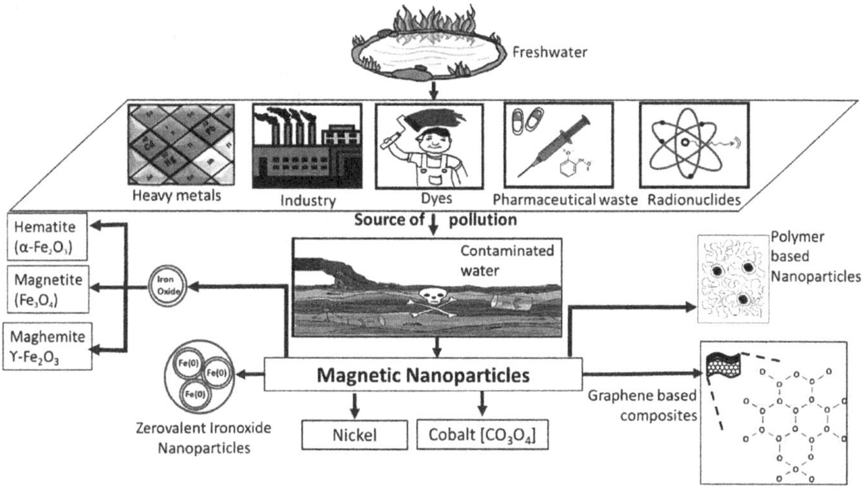

FIGURE 9.1 Schematic representation of various water pollutants and different nanomaterials used for water remediation.

9.4 IRON OXIDE NANOPARTICLES FOR WASTEWATER TREATMENT

Naturally, iron oxide NPs exist in various structures; among them, few possess temperature-induced phase transition polymorphism known as hematite (α-Fe_2O_3), magnetite (Fe_3O_4), and maghemite (γ-Fe_2O_3) (Cornell & Schwertmann, 2003). Magnetic iron oxide nanoparticles (MIONPs), which include the aforementioned three polymorphisms, are synthesized using environmentally friendly methods with efficient surface modification to minimize the toxicity and maximize the separation efficiency by enhanced magnetic moments for each carrier molecule (Natarajan et al., 2019). Schematic representation of various synthetic methods available for MIONPs synthesis and surface modification materials removing pollutants from wastewater is presented in Figure 9.2.

9.4.1 Zero-Valent Iron Nanoparticles (ZVI NPs)

Zero-valent iron nanoparticles (nZVI) are well-suited for applications in water pollution treatment, especially owing to their reduction potential and subsequent ability as reducing agents. Their advantages contributing to such applications include their non-toxic nature, abundance, low cost, ease in synthesis and bioavailability (Zhu & Chen, 2019). They can be used to eliminate a wide selection of organic molecules, given the availability of dissolved oxygen in the test system, achieved by degradation and oxidation to yield hydrogen peroxide. By two-electron transportation and the Fenton reaction, H_2O_2 is reduced to water by the combination of H_2O_2 and Fe^{2+} along with the generation of hydroxyl radicals (OH°) which possess influential oxidizing capability for organic compounds (Fu et al., 2014a). Contaminants, like halogenated organic compounds (Liang et al., 2014), nitroaromatic compounds (Xiong et al., 2015), organic dyes (Hoag et al., 2009), phenols (Wang et al., 2013a), heavy metals (Arancibia et al., 2016), phosphates (Markova et al., 2013), nitrates (Muradova et al., 2016), metalloids (Ling et al., 2015), and radio elements (Ling & Zhang, 2015), were successfully removed from the water by nZVI. ZVI/titanium dioxide evolved from activated carbon via the sol-gel method has been shown effective for 2,4-Dichlorophenoxyacetic acid degradation. Following a pseudo-second-order

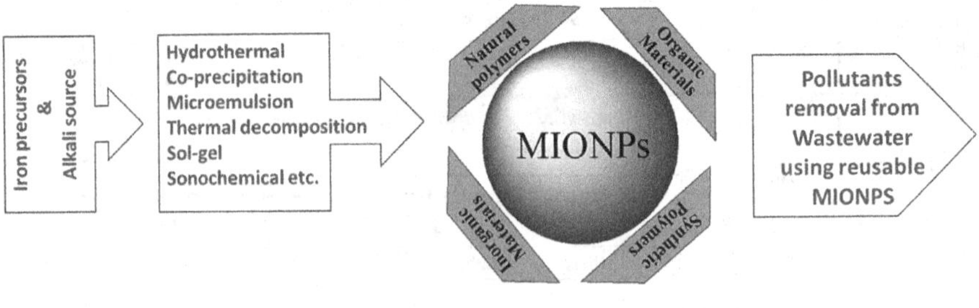

FIGURE 9.2 Various methods used for the synthesis MIONPs and their surface modification for wastewater treatment.

kinetic method, 46.44% adsorption and 86.37% degradation efficiency have been obtained (JokarBaloochi et al., 2018). Solid powder state and small particle size of ZVI NPs lead to agglomerates with surrounding media and thereby to their low mobility into the pollutants. Traces of Chromium (VI) could be effectively handled using ZVI nanoparticles functionally modified with amino groups and supported by vermiculite (Zhao et al., 2019a). For the quality removal of uranium (VI) in drainage systems, similar ZVI nanoparticles, but with DLH (double-layered hydroxide) coatings resulted in a remarkable adsorption capacity of 176 mg. g^{-1} (Yu et al., 2019b). Despite the many advantages of ZVI NPs, they are also not devoid of certain limitations like aggregation, oxidation, and difficulty in separation from degraded systems. To enhance the reactivity of ZVI, doping with other metals (Liou et al., 2005) is required, with an overall goal to prevent aggregation, achieve the desired surface coating and conjugation (Singh & Mirsa, 2015) and encapsulation in a matrix, facilitating the separation of ZVI from degraded systems (Li et al., 2016). The merits and demerits of some important nanoparticles obtained from diverse synthetic methods are presented in Table 9.1.

TABLE 9.1
Merits and Demerits of Some Important Nanoparticles Obtained from Diverse Synthesis Methods

Nanoparticles Obtained from Different Synthetic Methods	Advantage	Disadvantage	Application	References
Zero-valent iron/ dioxide titanium Nanocomposites from Sol-gel method	Maximum degradation performance at minimum consumption number of nanocomposites, hydrogen peroxide, neutral pH	Degradation decreases after the fourth cycle	2,4 Dichlorophenoxy acetic acid Herbicide degradation	(Hoag et al., 2009)
Amino-functionalized vermiculite-supported nanoscale zero-valent iron from Chemical reduction method	100% removal efficiency	Surface is covered by corrosion products hindered the reusability	Removal of Cr(VI)	(Wang et al., 2013a)
LDH@nZVI composites from in-situ growth method	Super resistibility to hostile environment	After six cycles, slight decrease in removal efficiency	Removal of U (VI)	(Arancibia et al., 2016)

(Continued)

TABLE 9.1 (Continued)

Merits and Demerits of Some Important Nanoparticles Obtained from Diverse Synthesis Methods

Nanoparticles Obtained from Different Synthetic Methods	Advantage	Disadvantage	Application	References
Fe_2O_3/3DOM $BiVO_4$ from wetness impregnation method	Excellent photocatalytic stability, 98% degradation, unique porous architecture, high surface area, good-light harvesting ability, high adsorbed oxygen species concentration	0.15 mL H_2O_2 insufficient for total degradation, if 1.20 mL H_2O_2 used photocatalytic activity decreased. low efficiency of separation	Degradation of 4-nitrophenol	(Li et al., 2016)
Fe_2O_3/C-g-C_3N_4 Z-scheme heterojunction by facile one-step carbonizing process	Double the time efficiency increased compared to pristine hematite	Fg hetero junction photocatalyst shows less degradation compared to FCg heterojunction photocatalyst	Rhodamine B (RhB) degradation	(Wu et al., 2015)
AgBr/Ag_3PO_4@ natural hematite by in-situ deposition method	Deposition of silver species greatly reduce the band gap which enhances the photocatalytic activity	At acidic conditions – degradation accelerated At basic conditions – degradation depressed	Degraded four antibiotics	(Krehula et al., 2017)
Fe_2O_3/GO/WO_3 nanocomposites prepared by ex-situ method	Efficient electron separation and transformation which increased photocatalytic activity	Photocatalytic degradation is inhibited for phenol in the presence of benzoquinone	Degradation of phenol	(Teja et al., 2019)
Fe_2O_3/ZnO/$ZnFe_2O_4$ composites by simple hydrothermal reaction with thermal treatment	Best photocatalytic activity –95.7% within 60 min	Only three cycles performed, not to selective	Degradation of RhB	(Tadic et al., 2019)
3D flower-like α-Fe_2O_3 by urea-assisted hydrothermal route	Prevents aggregation, enhanced surface area alone with multiple spaces and pores	Adsorption process is slow after 30 min	To remove heavy metal like As(V) and Cr(VI)	(Santhosh et al., 2019)

9.4.2 Hematite

Hematite is considered highly stable in the surrounding environmental conditions, naturally existing in rocks and soils (Wu et al., 2015). In the crystal structure of α-Fe_2O_3, ferric ions hold the two-thirds of B-site (octahedral), and oxygen ions exhibit close packing in a hexagonal fashion (Krehula et al., 2017). At room temperature, hematite behaves as weakly ferromagnetic; in addition, it transforms into an anti-ferromagnetic phase in bulk form at Morin temperature (T_M = −13.15°C); it acquires paramagnetic properties while crossing its Curie temperature, (Tc = 682.85°C) (Tadic et al., 2019; Teja & Koh, 2019). In the visible spectrum, hematite absorbs around 43% of total light, thereby finding immense applications in water treatment (Santhosh et al., 2019). For the removal of heavy metal ions from pointed tap water, nanohematite is considered an effective adsorbent (Shipley et al., 2013). Hematite is incorporated with other semiconducting materials to enhance photocatalytic activity; associated parameters such as low conductivity and efficiency of separation inhibit this process (Zhang et al., 2017).

Synthesis of hematite by Kang et al. (2019) includes Z-scheme heterojunction photocatalyst based on graphitic carbon nitride, while pristine hematite synthesis showed lower yields using graphitic carbon nitride heterojunction photocatalyst; the efficiencies improved with double the time for rhodamine B photocatalytic degradation. Chen et al. (2019) synthesized novel $AgBr/Ag_3PO_4$ with natural hematite heterojunction through facile approach, consisting of molar ratio of [Ag]: [Hematite] 1.5: 1. Silver deposition leads to a decrease in the band gap, with the photocatalytic degradation rate constants of0.16, 0.19, 0.34 and 0.10 min^{-1}respectively, for four antibiotics [ciprofloxacin (CIP), norfloxacin (NOR), sulfadiazine (SDZ), and tetracycline (TTC)], at pH = 3. Iron oxide NPs were combined with tungsten trioxide (WO_3), making them exhibit poor photocatalytic activity upon use. This property has been exploited for achieving improved photocatalytic activity, in turn for diverse applications. Mohamed (2019) has synthesized ($Fe_2O_3/GO/WO_3$) hematite, graphene oxide with tungsten trioxide, all of which use Z-scheme photocatalyst for demonstrating applications in the degradation of inorganic dyes. To increase the electron-hole pair separation, transport of electrons at the graphene oxide interface containing holes of hematite's valence band occurred in the conduction band of tungsten trioxide, with relatively better recombination for both (hematite and tungsten trioxide) systems. Li et al. (2018) reported rhodamine B's improved photocatalytic degradation efficiency using a hydrothermal method to synthesize ternary hematite/zinc oxide/zinc ferrite (ZFO). This is achieved within an hour, reaching 95.7% and exhibiting stability for three cycles.

Synthesis of α-Fe_2O_3 microstructures in the form of three-dimensional floral patterns has been achieved, which effectively prevents aggregation, possessing enhanced surface area along with multiple spaces and pores that help to interact with contaminants and exhibiting maximum adsorption capacities for As (V) and Cr (VI) (Liang et al., 2013). Hematite is also an economical and abundant option for the elimination of organic dyes and heavy metal ions. Carbon-carbon hematite (α-Fe_2O_3@AC) (Li et al., 2019a) for the removal of Cr (VI) from water and pinewood biomass (PB) with natural hematite admixture (Wang et al., 2019) at variable temperature are used for maximum sorption capacities, 173 and 359 mmol.kg^{-1} for Cd^{2+} and Cu^{2+}, respectively, with an increase in pH up to 5. Hematite/aluminum oxide composites (Ravindranath et al., 2017) with high surface area, porous structure, and adsorption properties showed high coherence toward removing Hg^{2+}, Cd^{2+}, Pb^{2+} cations from actual samples. Hematite NPs synthesized in the presence of oleic acid (Hashemzadeh et al., 2019) have been used to remove cobalt-60 radio-cations. Hematite is also used for the removal of U (VI) (Shuibo et al., 2019) from aqueous solution; its maximum adsorption capabilities were influenced by temperature. Pb(II) ions can be removed by 3D graphene nanocomposite with $NiFe_2O_4$ nanoparticles (Nasiri et al., 2018), by varying pH and magnetic chistosan-functionalized materials. Fe_3O_4/MgAl-LDH (Shou et al., 2015) shows different adsorption capabilities for Co^{2+} in different pH values. At low pH, the surface of Fe_3O_4/MgAl-LDH shows a positive charge and hence, there will be low adsorption of Co^{2+} (positively charged) due to electrostatic repulsion. At high pH, the surface of Fe_3O_4/MgAl-LDH shows a negative charge, enhancing the surface complexes with Co^{2+}. High adsorption due to the occurrence of precipitates on the surface has been reported with $Co(OH)_2$ and $CoCO_3$ at pH = 8.0.

9.4.3 Magnetite

Magnetite contains both ferrous (Fe^{2+}) and ferric (Fe^{3+}) ions (Bhateria & Singh, 2019), displaying a cubic inverse spinel phase structure (Okube et al., 2012) wherein antiparallel spin moments are noticed, with tetrahedral (A-site) and octahedral (B-site) producing a couple of magnetic sublattices (Noh et al., 2014). With a small band gap energy (0.1eV) reflecting on the least resistivity for any mineral oxide (5×10^{-5} Ωm), magnetite exhibits both negative type (n-type) and positive type (p-type) semiconductor behavior (Wu et al., 2015). As against the chemical as well as structural features contributing to their stability, the magnetization values for magnetite are even lower than the half of ZVI.

The adsorbent properties of magnetite greatly rely on effective surfaces, which does not change any magnetism effect. Recently, biocompatible shells or surfactants were used for magnetite NPs used as an adsorbent (Farahbakhsh et al., 2019; Mahalingam & Ahn, 2018) synthesized reduced graphene oxide/magnetite/nickel oxide (rGO-Fe_3O_4-NiO) in the form of hybrid nanocomposite photocatalyst, which exhibited improved efficiency for degrading inorganic dyes. Diverse magnetite and zinc oxide (Fe_3O_4-ZnO) core-shell based adsorbents, zinc oxide-coated Fe_3O_4 NPs, SiO_2 and ZnO-coated magnetite core, and SiO_2 (3-aminopropyl) triethoxysilane (APTS) and ZnO-coated magnetite core have all been robustly prepared and reported by Atla et al. (2018) Among the materials, coating with silica enhanced the transport efficiency of MB dye, as well as the synergistic effect of adsorption.

Banic et al. (2019), synthesized tungsten trioxide and magnetite (WO_3/Fe_3O_4) catalyst with the help of UV light for degrading thiacloprid with an efficiency that improves over the catalyst dosage used, thereby improving the absorption of photons with an increase of the surface area of the catalyst. In addition to that, increase in the tungsten content results in easy magnetic separation.

The effective separation of nanoparticles from contaminated water sources poses to be a challenge mainly due to their smaller size range. But magnetic hematite is easily separated and recovered from the waste water by applying external magnetic field, and therefore utilized to eliminate heavy metals from water (Ngomsik et al., 2012; Lei et al., 2014). It shows good affinity to various heavy metals empowering direct adsorption of their oxy-ionic forms. Increased capacity of adsorption of lead ions have been noticed with magnetite NPs immobilised using carboxymethyl cellulose-immobilized magnetite NPs (CMC- Fe_3O_4) (Fan et al., 2019), whose maximum adsorption capacity was 152 $mg.g^{-1}$ compared to pure magnetite NPs. Cr (VI) elimination with the highest capacity of adsorption (when compared to other metal ions) has been achieved using magnetite-based composites of chitosan@bentonite (Fe_3O_4-CS@BT) (Feng et al., 2019).

Radionuclide wastes such as uranium are effectively removed during wastewater treatment by using hydrothermally synthesized Fe_3O_4/porous carbonaceous materials (Fe_3O_4/PCMs) (Guo et al., 2019), with high stability and recoverability. The maximum adsorption capacity of U(VI) by using Fe_3O_4/PCMs NPs is 123 $mg.g^{-1}$ at 55°C. Adsorbent complexes such as the silicon Schiff base complex (M/SiO_2-Si-SBC) (Khan et al., 2018) decorated with magnetite nanorod has been used for wastewater treatment, specifically to remove U(VI) and Pb(II) contaminants. The maximum adsorption capacity of this complex was 6.5×10^{-4} $mol.g^{-1}$ for lead ions and 4.82×10^{-4} $mol.g^{-1}$ for U (VI). Ding et al. (2015) reported fungus-magnetite as a suitable agent to remove radionuclides such as Th(IV) (with adsorbent removal capacity of 280.8 $mg.g^{-1}$ @ pH 3), Sr(II) (100.9 $mg.g^{-1}$ @ pH 5), and U(VI) (223.9 $mg.g^{-1}$ @ pH 5). Fe_3O_4 nanoparticles with 30 nm size perform excellently to handle Cr(VI) from water (by associated reduction and precipitation), with a capacity of uptake extending 4.5 mg/g at a residual concentration of Cr(VI), 50 μg/L (Simeonidis et al., 2015). The effect of particle size and contact time play a crucial role for estimating the efficiency; hence, increasing the particle size would have a clear impact on the yield during Cr(VI) removal (Simeonidis et al., 2014). The use of a column bed for holding the nanoparticles, in place of a dispersion for shorter periods in a contact tank, will increase the contact period between particles and polluted water which, in turn, almost doubles the uptake capacity (Karprara et al., 2016). Uptake of Cr(VI) on Fe_3O_4 nanoparticles is done by capture mechanism combining diverse steps including Cr(VI) reduction by Fe^{2+} sites on its tetrahedral sites, and subsequent diffusion, adsorption and complexation upon particle surfaces (Gorski et al., 2012; Pinakidou et al., 2016).

Removal of Hg^{2+} and U(VI) was achieved by reducing the potential of Fe_3O_4 nanoparticles (Wiatrowski et al., 2009; Singer et al., 2012; Pasakarnis et al., 2013). The performance of Fe_3O_4 nanoparticles when encapsulated on graphene oxide showed uptake capacity for Se(IV) (residual concentration of 10 μg/L) @1.9 mg/g (Fu et al., 2014b). Fe_3O_4 nanoparticles deserves broader utilities in the treatment of water wastes for the adsorption of arsenic species. Oleic-acid-coated nanoparticles with a size of 12 nm (Yavuz et al., 2006) is reported to be the best for the removal of As (III) and As (V) (@2 mg/g) with a residual concentration of 10μg/L; non-surfactant based alternative approaches could be tried for further optimization. In hollow nanoparticles, the concentration of cations was high, which adds to increased surface area simultaneously increasing the adsorption capacity (Balcells et al., 2016). Electrochemically synthesized Fe_3O_4 nanorods show potential elimination of Pb^{2+}, Cd^{2+}, and Ni^{2+}, however, with an estimated effective capacity at lower concentration levels (Karami, 2013).

Inactivation of bacterial species such as *Staphylococcus aureus* could be attributed to the adsorption by magnetite nanoparticles rather than any toxic interaction (Tran et al., 2010). Effective removal of various organic contaminants including pesticides and dyes, as well as organochlorine or organophosphorus pesticides (Zheng et al., 2014), triazine herbicides (He et al., 2014), and bromelain (Chen & Huang, 2004) by Fe_3O_4 nanoparticles have been recently reported. Huang and Keller (2015) used purchased maghemite nanoparticles with 30 nm size, subjected to toluene and 3-aminopropyl-triethoxysilane (APTES) treatment to include an amino group, followed by the final addition of pyridine and EDTA. The particles exhibited polydispersity in their porosity and size range, from 1 to 10 µm, exploited for rapid removal of Pb^{2+} with high sorption capacity of 100.2 mg.g^{-1} within a timeframe of less than 15 min.

9.4.4 Maghemite

Maghemite (γ-Fe_2O_3) possesses a cubic structure where Fe^{3+} ions are split between A-sites (8 Fe^{3+} ions per unit cell) and B-sites, reflecting on its high chemical stability in the absence of any reducing activity (Wu et al., 2015). Compared to other materials, maghemite renders high efficiency of separation from wastewater using a magnetic field, low-cost synthesis, and non-toxicity leads to various applications in wastewater treatment (Leone et al., 2018).

Methylene blue is eliminated with a greater adsorption capacity of 905.5 mg.g^{-1} by coating the maghemite with carbon beads of multiwalled nature functionalized with magnetic alginate (Boukhalfa et al., 2019). The magnetite@maghemite with oxalate improved the degradation of orange II. Iron ions get removed from the catalyst in order to enhance Fenton reaction with an increase in the pH of the solution (Dai et al., 2018). Conducting polymers used as functionalization elements for magnetic materials exhibit properties of chelating agents, which significantly contribute to the effective removal of several pollutants by electrostatic interactions. For the removal of Cr (VI) and Cu (II) ions from water, Polypyrrole/maghemite (PPY/γ-Fe_2O_3) and polyaniline/maghemite (PANI/γ-Fe_2O_3) are in use, with a maximum adsorption capacity of 209 mg.g^{-1} and 171 mg.g^{-1}, respectively (Chavez-Guajardo et al., 2015). Maghemite NPs have been used to remove Pb^{2+} and Cu^{2+} from water whose specific surface area is 79 m (Bhateria & Singh, 2019).g^{-1} and saturation magnetization is 45 emu.g^{-1} at 27°C, with a maximum adsorption capacity of 68.9 and 34.0 mg.g^{-1}, respectively (Rajput et al., 2017). Maghemite nanoparticles bind with As(V) oxy-ions, which act as better adsorbents than magnetite owing to their adsorption on the γ-Fe_2O_3 structure or post-hydrolysis surface generated upon contact with water (Tuutijarvi et al., 2009). Thin γ-Fe_2O_3 layer on Fe_2O_3 or ZVI nanoparticles have increased specific surface area and/or secondary porosity, indicating their performance against arsenic (Martinez-Boubeta et al., 2006). Maghemite nanoparticles (23 nm) prepared by electrochemical route to remove As(V) seem to hold greater potential for use in drinking water treatment (Park et al., 2009).

For the degradation of orange G dye, sonocatalytic activities of maghemite along with TiO_2 magnetic catalysts were made (Pang et al., 2016). The treatment efficiency was improved by the inclusion of titanium dioxide nanotubes, where higher specific surface area allowed higher reactant dye to get adsorbed on the catalyst surface with higher pore volume, thereby resulting in increased rapid diffusion during the entire sonocatalytic reaction. γ-Fe_2O_3 shows fast kinetics, involving a two stage mechanism that includes surface diffusion and subsequent intraparticle diffusion, where Cr(VI) is adsorbed in relatively higher concentration (Jiang et al., 2013) It is also feasible for demonstrating applications of maghemite nanoparticles in the form of selenium or molybdenum adsorbents, with no compromise on their performances (Lounsbury et al., 2016). Readily purchased maghemite nanoparticles with an average size of 30 nm were mixed with 3-aminopropyltriethoxysilane (APTES) which shows a very high porosity in the nanoscale. The prepared composites displayed negative surface charges, used for fast removal of Pb^{2+} rapidly within a 15 min timeframe, with high sorption capacity of 100.2mg. g^{-1} (Huang & Keller, 2015). Through aging a mixture of magnetite NPs, sand and Portland cement, concrete/maghemite nanocomposites (CM nano) are obtained, finding use as adsorbent in the remediation of As(V). This is reported to decrease the concentration of As (V) in water from 10 ppm to 10 ppb (Park et al., 2009; Hernandez-Flores et al., 2018) synthesized maghemite NPs for separation of As (V) from aqueous solution, where the average size and surface area of the maghemite NPs were in the range of 11–23 nm and 41–49 m (Bhateria & Singh, 2019).g^{-1}. The maximum adsorption capacity was 4.643 mg.g^{-1} at pH = 7.

9.5 NANO-ADSORBENTS

Iron oxide nanoparticle is more economical in terms of reusability in the water treatment through simple desorption process (Hao et al., 2010). They can be reused and at the same time utilized for prolonged timeframes during the treatment of pollutants, which make them a cost-effective solution. They possess special features like improved BET surface area, microporous nature, high dispersion, and most importantly, being environment friendly (Gupta et al., 2015). Water treatment reached the next level when iron oxide nanoparticles were used as nano-adsorbents combined with other suitable composite materials, in which performances were noticed to be very high (Gutierrez et al., 2017). Alumina-based nanocomposites show high and efficient adsorption for various heavy metals. By creating a layer on the surface of the nanocomposites, naphthol green B is adsorbed in high order when enclosing iron nanoparticles on the surface of halloysite nanotubes (Gupta et al., 2017). However, economic reuse and regeneration are areas of dispute, which can be modified by hydrous manganese oxide (HMO) and multi-wall carbon nanotubes (MWCNT) (Tarigh & Shemirani, 2013). Functionalized μm-size magnetic particles give more selective separation for Zinc and Nickel from aqueous solutions. Cyclodextrin iron nanomaterial/reduced graphene nanocomposite was vortexed with high viscous honey fluids having higher molecular recognition capacity as well as adsorption, which is directly detected in gas chromatography-mass spectrometry (GC-MS) (Mahpishanian & Sereshti, 2017). In order to utilize adsorbent in remediation of water, it should hold the following four important features:

i. capability to remove greater quantities of contaminants,
ii. high porosity,
iii. greater reactivity, and
iv. a non-toxic nature (Crane & Scott, 2012)

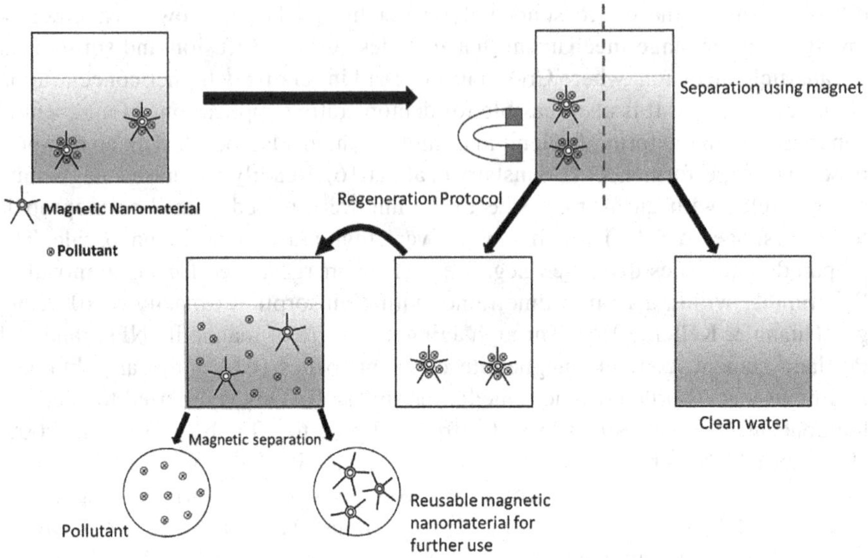

FIGURE 9.3 Schematic representation of separation of water pollutants from wastewater using magnetic materials and their regeneration.

Nanocomposite Fe_2O_3/biochar from pyrolysis of cottonwood shows outstanding efficiency for removing aqueous As(V) (Zhang et al., 2013). Nano-adsorbents are classified into diverse heads depending on their adsorption characteristics. This includes metallic nanoparticles, nanostructured mixed oxides, magnetic as well as metallic oxide NPs, carbonaceous nanomaterials, etc. Numerous factors like size, surface adsorption, chemical composition, crystal structure, agglomeration behavior, arbitrary dimension, solubility, etc. will help in effective control over the nano-adsorbent properties (Rasmussen et al., 2016). The advantages of using magnetic nanoparticles over other nanoparticles for water remediation are its easy separation using an external magnetic field, recycling, cost-effectiveness, and in preparing highly dispersed solutions without any sedimentation in the absence of a magnetic field. Schematic representation of the separation of water pollutants from wastewater using magnetic materials and their regeneration is presented in Figure 9.3.

9.6 ADSORPTION METHOD PROTOCOL

Various concentrations of adsorbate solutions with a desirable quantity of magnetic adsorbents were taken in a flask, and shaken at fixed rpm up to a predefined contact time at a required temperature. After the adsorption, the adsorbents were removed from the solution using an external magnetic field. In the case of adsorption of dyes, the unabsorbed dye concentrations were measured from the absorption maximum of the dye at different time intervals. For the metal ion adsorption, the initial and final metal ion concentrations were measured with an atomic absorption/emission spectrometer.

The adsorption percentage and the adsorption capacity were calculated using the following formula,

$$\text{Percentage adsorption} = \frac{(\mathbf{C_o} - \mathbf{C_e}) \times \mathbf{100}}{\mathbf{C_o}}$$

$$\text{Adsorption capacity}, q_e\,(\text{mg/g}) = \frac{(\mathbf{C_o} - \mathbf{C_e}) \times \mathbf{v}}{\mathbf{w} \times \mathbf{1000}}$$

where

C_0 = initial dye concentration
C_e = equilibrium dye concentration (mg/L)
q_e = amount of dye adsorbed per unit gram of the adsorbent
v = volume of dye solution
w = weight of the adsorbent (g).

Generally, the concentration of the adsorbate was selected based on the concentration range of the contaminated water. The impact of important adsorption parameters such as adsorption dose, initial adsorbate concentration, adsorption rate, temperature, pH, and reusability should be checked before going for a real-time application. For reusability applications, the adsorbent was regenerated by washing three times with DD water and ethanol after each adsorption cycle. In special cases, the adsorbent was washed with an acid or alkali before going for a water and ethanol wash (Natarajan et al., 2020).

9.7 FACTORS AFFECTING ADSORPTION PROCESSES

Temperature, pH, adsorbent dose, incubation/contact time are some of the critical factors affecting the adsorption of heavy metals from wastewater while using the nano-adsorbents. Out of all these, pH plays a vital role in the adsorption of heavy metals (Srivastava et al., 2015). At pH 5.5, adsorption for Zn (II) on magnetic nano-adsorbent is considered as the maximum adsorption, but this decreases with an increase in pH. Contact time plays a vital role in the adsorption process, especially to remove heavy metals from an aqueous solution. While increasing the contact time, the adsorption also increases. But in some cases, blockage of active sites will decrease the adsorption of heavy metals as time increases (Shirsath & Shirivastava, 2015). The equilibrium time considered for the maximum adsorption of Pb(II) and Cr(III) from wastewater is reported to be 120 min (Lingamdinne et al., 2016).

In the context of separation of dyes, a broad group of nano-adsorbents in various forms and dimensionalities (D), such as nanoparticles (0D), nanofibers and nanotubes (1D), nanosheets (2D), and nanoflowers (3D) are under broad examination. The 0D nano-adsorbents (within 1 and 100 nm dimensions), based on their large porosity, small size, high surface area, expand the scope of their interaction with the dye molecules increasing their adsorption efficiency with pollutant binding sites. Once attaining saturation, they are pre-eminently regenerated via different chemical processes (Yang & Xing, 2007). Amorphized transition metal oxide nanoparticles of

Fe_2O_3, CoO, and NiO have been used for MB adsorption, from laser irradiation in liquid, with a maximum adsorption capacity of as much as 10584.6 $mg.g^{-1}$ (Li et al., 2015). A notable aspect of the recovery of the adsorbent nanoparticles relates to their potential applications attributed to their small size. Filtration and sedimentation techniques were traditionally exploited for the separation of the nanoparticles from treated wastewater where small dimensions of such adsorbents developing a blockage of the filters lead to creating a release of nanoparticles, in turn causing problems such as secondary pollution (Reddy & Lee, 2013)

9.8 VARIOUS MECHANISMS INVOLVED IN THE REMOVAL OF POLLUTANTS

Two main mechanisms are involved in the removal of pollutants from the wastewater (Gutierrez et al., 2017; Natarajan et al., 2019; Biftu et al., 2020; Fadillah et al., 2020; Natarajan et al., 2020; Yadav et al., 2020; Wang et al., 2020a; Leonel et al., 2021; Luo et al., 2021). An illustration of methods and mechanisms involved in removing various pollutants from wastewater is presented in Figure 9.4. The first one is the adsorption of pollutants on the surface of magnetic nanocomposites through complexation or electrostatic interaction. The adsorption mechanisms occur through the following pathways:

i. Electrostatic interaction or charge neutralization
   Ionic dyes, as well as ions, could be removed through this mechanism using surface-functionalized materials on the iron oxide nanoparticles. The charged dyes/ions might be adsorbed between the interlayers of surface-functionalized materials by replacing its counterions. The adsorption process occurs via electrostatic or polarization interaction.
ii. Surface complexation and group replacement
   In this mechanism, the surface-functionalized MIONPs hydroxyl groups are replaced by pollutants such as arsenic, thallium, antimony, and phosphate

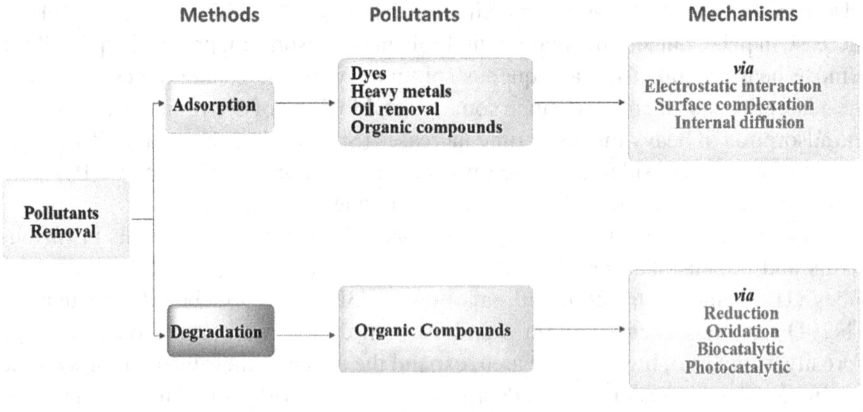

FIGURE 9.4 Illustration of methods and mechanisms involved in the removal of various pollutants from wastewater.

groups to form outer sphere or inner sphere complexes on the surface (like -O-As, -O-Sb).

iii. Internal diffusion

It occurs through two consecutive steps by transferring contaminants from the solution to the material surface and then diffusing into the surface pores. This mechanism depends on the surface properties as well as the pore size of the materials.

In addition to the adsorption mechanisms just discussed, the pollutants are removed via oxidation, reduction, or degradation by surface-functionalized materials. Among the mechanisms, adsorption would be preferred due to the complete removal of pollutants from the wastewater without leaching out any surface-functionalized materials into the water.

9.9 MAGNETIC GRAPHENE/GO-BASED COMPOSITES

Only very little energy is required for magnetic separation over other conventional adsorbent separation methods applied for environmental remediation, contributing to greater separation resulting from very small particle size. The combinatory use of graphene/GO together with magnetic particles (MRGO), which reduces the chance of relentless agglomeration, restacking nanosheets, provides higher adsorption through a preponderant attainable surface area of graphene/GO materials (Sun et al., 2011). Synthesis of magnetic graphene/GO-based nanoparticles includes methods such as solvothermal (of immense utilitarian value through one-pot non-hazardous approach) (Sun et al., 2011), crosslinking (Fan et al., 2012), hydrothermal synthesis (Bharath et al., 2017), green synthesis employing extracts from plant sources, either as basic molecules/reducing agents (Thakur & Karak, 2014), and in-situ chemical co-precipitation (Wang et al., 2011; Wang et al., 2013b; Liu et al., 2014).

G/Fe_3O_4 was automated for adsorbent use in the removal of dye fuchsine in two different ways: first exploiting the van der Waals' interactions between atoms of carbon seen in the graphene sheet, with the hexagonal pattern of the G/Fe_3O_4 and then with the dye's aromatic rings. An increase in the adsorption was noticed with the increase in adsorbent dosage from 0.2 to 0.4 g L^{-1}, where at pH 5.5, the best adsorption performance was obtained. This corresponds to the amido groups in the dye with weak basic nature (Wang et al., 2011).

Reduced graphene/Fe_3O_4 nanoparticle hybrids are utilized to eliminate dye from water where the efficiency can be tuned by the hybrid's BET surface area, with an increase in the composition of rGO, for example. The adsorption capacity of the RGF-1 for RhB was 36.6 mg/g (Geng et al., 2012). Magnetic xanthate GO has been utilized, exploiting its enhanced adsorption capacity of 526.32 mg/g, for the separation of MB from water samples (Cui et al., 2015). GO was combined with magnetic chitosan to form magnetic bioadsorbent (MCGO), showing an improved MB dye adsorption, mainly because of the strong interactions between the positively charged dye molecules and negatively charged MCGO. The reason behind the high adsorption is the establishment of chemical bonds of high binding strength because of the reactions occurring in the carboxyl groups of GO amine group of magnetic chitosan, resulting in rougher surface occupied by an increased number of active adsorption sites (Fan et al., 2012). Without much effort, porous graphene shows diffusion into 2D structures, possesses greater surface area, and is useful in effective separation of pollutants of organic nature. Graphene with magnetic nanoparticles constitute the major part of the composites enabling easier separation and high utility. The separation of crystal violet dye from aqueous solution with rapid adsorption (5 min, with 460 mg/g capacity at an optimum pH of 7) is a relevant example (Bharath et al., 2017).

When sulfonation of the graphene increases, surface zeta potential increases, subsequently leading to dispersion in aqueous samples. Cationic dyes like safranine T, NR, Victoria blue, and anionic dyes like MO, brilliant yellow, and alizarin red were tested for sorption using magnetic sulfonic graphene ($G\text{-}SO_3H/Fe_3O_4$), whose adsorption capacity improved with pH change from 2 to 6. The mechanism behind this will be ionic interactions between the dyes (positively charged) and the adsorbent (negatively charged), with the effective separation of amino groups, at pH lower than 6, there will be a decrease in the electron-donating strength of the composites (Wang et al., 2015).

Some drawbacks of magnetic graphene include poor dependability, in a comparative manner having poor performance in adsorption characteristics, hazardous nature (toxicity), and expensive nature. Magnetic graphene hybrids that are functionalized are considered to be an effective replacement for better utility. Cyclodextrin (CDs) natural molecules, from the adsorption perspective, can form complexes resulting from host–guest inclusion with contaminants of aromatic nature through weaker intermolecular forces which bestow enhanced capacity of adsorbents. β-cyclodextrin magnetic GO (mGO-β-CD) developed for separation of rhodamine 6G (R6G) and acid fuchsin (AF) from water achieved treatment within 2 min, with a maximum adsorption capacity after 8 min. β-CD enormously improves the adsorption capacity with about 330% increase for R6G and 170% for AF compared to mGO (Liu et al., 2014).

9.10 NICKEL BASED NANOPARTICLES

Nickel oxide is considered a good adsorbent based on its chemical and magnetic properties (Nateghi et al., 2012). The sorption properties of nano-NiO are mainly established through reduction (redox potential is −0.24 V), and nano-NiO establishes the same through complex formation. Using the co-precipitation method, nano-NiO particles were synthesized and utilized to remove sulfate and nitrate, which showed effective removal at pH 7 from water (Rafique et al., 2012). Nickel acetate is used as a precursor for the synthesis of nano-NiO used as an effective adsorbent for chromium ion considering different physicochemical factors like pH, sorbent dosage, agitation time, etc. (Srivastava et al., 2014). Nickel nitrate and sodium hydroxide-aided hydrothermal synthesis of $\beta\text{-}Ni(OH)_2$ has been effectively used for eliminating rhodamine B from polluted water sources (Motahari et al., 2015). Oxalic acid is used with nickel nitrate to synthesize nano Ni which is used as an adsorbent for effective removal of Pb^{2+} pollutant from aqueous systems (Mahmoud et al., 2015). Catalysts like poly(acrylamide-co-acryl acid) loaded with nano-Ni particles have been used for the removal of mono azo orange II dye from wastewater (Tang et al., 2014). The modification of adsorbent diatomite surface using nano-NiO oxide was exploited for the effective extraction of basic red 46 (Khalighi et al., 2012). By using $Ni(NO_3)_2$ as a starting material, Ni nanoparticles were synthesized for the removal of dyes including eosin, crystal violet, orange II, and anions such as NO_3^- and SO_4^{2-} from contaminated water in an effective manner (Pandian et al., 2015).

9.11 COBALT-BASED NANOPARTICLES

The use of transition metals containing nanoparticles attracted recognition in the field of photocatalyst, with suitable band gap and flat band potential. The use of cobalt and its oxide as a catalyst in the degradation of organic pollutants like dyes have also been reported. Few examples include the simple, cost-effective sol-gel method, where Co_3O_4 nanoparticle was prepared for rhodamine B and direct red 80's degradation in visible light. This achieved 78% degradation for 0.3 gL^{-1} of catalyst used in the process (Dhas et al., 2015). Iron and nickel-doped cobalt oxide were used to degrade 95% eosine blue (EB) in aqueous solution (Koli et al., 2018). From Ni(II) and Co(II) Schiff base complexes, Co_3O_4 and NiO nanoparticles were prepared which reveal satisfying photocatalytic activity, and in the presence of hydrogen peroxide degrades methylene blue under UV illumination with 55.71% in 420 min (Nassar et al., 2017). Nanofiber-based Co_3O_4/Fe_2O_3 composite with 20 nm diameter shows catalytic degradation of acridine orange (AO) and brilliant cresyl blue (BCB) under solar light (Fei et al., 2012). By using chemical and microwave methods, Co and Co_3O_4 nanoparticles were synthesized, showing degradation of murexide and EBT dyes with increase in sunlight radiation time and catalyst load. Degradation efficiencies toward murexide by chemical and microwave synthesized cobalt nanoparticle has been reported to be 43.6% and 34% at 0.25^{-1} and 0.2 gL^{-1} catalyst load with 40 min sunlight radiation (Adekunle et al., 2020). Uddin and Baig reported that cobalt-oxide nanoparticles could be used to remove methyl orange dyes effectively and efficiently (Uddin & Baig, 2019). A new $CoFe_2O_4$@WTRs (drinking water treatment residuals) was synthesized through altered combinatorial method co-precipitation and calcination, showing superior catalytic reactivity toward peroxymonosulfate, (atrazine) (10 μM) with 0.03 gL^{-1} $CoFe_2O_4$@WTRs, 0.20 mM PMS, with initial pH 4.01 (Li et al., 2019b).

9.12 MAGNETIC POLYMER NANOCOMPOSITES

Magnetic polymer nanocomposites encompass a state of materials consisting of an inorganic magnetic component with any case of one dimension in the nanometer range, falling under the category of particles, fibers or lamellae encapsulated in an organic polymer (Mai & Yu, 2006). Organic and inorganic synergies will add up new properties to magnetic polymer nanocomposites, which cannot be attained in either single organic or individual inorganic components. Generally, magnetic nanocomposites were synthesized based on *in situ* (right in place, with other phase materials) and *ex situ* techniques (mixing of pre-synthesized nanostructures and polymers) (Nathani et al., 2004; Wilson et al., 2004). Diverse types of nanofillers and polymers exist, for the preparation of polymer nanocomposites differing in the range of cross-linking process (Zhou et al., 2012), grafting (Tirtom et al., 2012), simple impregnation (Auta & Hameed, 2014), intercalation, and template synthesis (Azeez et al., 2013). Water soluble and insoluble polymers nanocomposites are used to improve the efficiency and selectivity of the adsorption by facilitating new surface interaction, which enhances the adsorption performance (Rivas et al., 2018).

Thiol-containing polymer-encapsuled magnetic NPs show appreciable potential to be utilized as promising adsorbents (Shin & Jang, 2007). In some reports, magnetite with polypropylene as the polymer matrix synthesized with an increased concentration of magnetite results in the increase of Young's modulus, exhibiting a superparamagnetic behavior that can be put upon in environmental applications (Shirinova et al., 2016). Nanoparticles with Davankov-type hyper-crosslinked-polymer (HCP), which is based on poly(DVB-co-VBC) precursor, have been reported in studies pertaining to adsorption of benzene-ring-like containing dyes such as MB, nigrosine, AO, etc., with an escalated adsorption efficiencies of 96%, 97%, and 94%, respectively (Liu et al., 2015). There is a demand for adsorbent materials with larger surface area, open porosity, an abundance of functional groups, altogether targeting toward maximizing the adsorption efficiency. The nanoparticles made of polymer, which is obtained through the Schiff base reaction and rich in amine porous organic polymer (RAPOP), exhibits numerous characteristics like increased specific surface area (368.05 m (Bhateria & Singh, 2019).g^{-1}), enormous porosity (0.651 cm .g^{-1} (Prathna et al., 2018), and negligible density, effectively helping in the removal of 454.545 mg.g^{-1} MO from aqueous solution conveniently in an acidic condition (Ou et al., 2016).

Amino-functionalized magnetic cellulose composites were synthesized for use in the adsorption of Cr^{6+}. Different forms of Cr^{6+}: $Cr_2O_7^{2-}$, $HCr_2O_7^-$, CrO_4^{2-}, $HCrO_4^-$, H_2CrO_4 exist depending upon the solution pH. In lower pH below 2, neutral form (H_2CrO_4) was formed. The maximum adsorption uptake of Cr^{6+} by amino-functionalized magnetic cellulose composites was 171.5 mg/g, achieved at pH 2.0 (Sun et al., 2014). Another report showed nanocomposites with polypyrrole-organically modified montmorillonite clay which showed significant adsorption efficiency of 55.55% during the adsorption of Cr^{6+} from aqueous solution (Setshedi et al., 2013). In addition, the role of pH in effluent solution is critical in determining the mechanism of removal of metal cations and their preference for specific polymers for defined action (Beaugeard et al., 2020; Gao et al., 2018). Through electrostatic interaction or hydrogen bonding, amino groups showed interaction with the silanol group, subsequently blocking the feasibility of surface silanol groups and proving constructive for the anionic contaminant's removal (Iftekhar et al., 2020).

EDTA and its derivatives were used to stabilize the superparamagnetic nanoparticles in their core-shell content to obtain nanoparticles with improved chemical stability, especially for the removal of Pb, Cu, Zn, and Cd in an acidic medium in an efficient manner (Kobylinska et al., 2020). In the midst of many examples, certain polymers have been well established concerning the removal of metal cations: polyamine in the presence of HDPE (Zhao et al., 2019b), diphosphonic (Fila et al., 2019), carboxylic-amine-phosphonic acid (Wang et al., 2020b), bifunctional polymer-supported aminophosphonic acid, to name a few (Zidan et al., 2020).

9.13 MAGNETIC NANOMATERIALS FOR REMOVAL OF TOXIC ORGANIC, PHARMACEUTICAL, AND PESTICIDES COMPOUNDS

Like dyes and metal ions, organic molecules such as pesticides and pharmaceuticals compounds are also dangerous to the environment and human beings once disposed of in water. These toxic materials should be removed from the environment for balanced growth even though the removal procedures are complicated. Among the methods available for removing these pollutants, the adsorption method would be effective, fast, simple, and cost-effective. $MnFe_2O_4$ infused chitosan microspheres (CMMS) are used for the removal of methylene blue from pharmaceutical wastewater with an adsorption capacity of 371.7 mg/g and 96% removal percentage at pH 6 (Jyothi et al., 2019). The drug oxcarbazepine can be removed from wastewater using iron oxide nanoparticles, and the removal efficiency depends on the pH, with a maximum adsorption capacity of 99.5% at pH 2.5 (Parashar et al., 2020). Removal of pesticides from the wastewater also gained important attention to safeguard human health and ecosystems. Magnetic biochar materials are prepared using $Fe(NO_3)_3$ deposited on corn husks and followed by carbonization, and it efficiently adsorbs paraquat pesticide with an adsorption capacity of 34.43 mg/g and 90% percentage removal (Damdib et al., 2019). Special kinds of sodium dodecylsulfate-coated magnetic chitosan nanoparticles (MHMS-MCNPs) removes diazinon, phosalone, and chlorpyrifos from wastewater (Bandforuzi & Hadjmohammadi, 2019). Superparamagnetic magnetite nanoparticles are used to remove indigo carmine (IC) dye from wastewater and exhibit a maximum adsorption capacity of 17.45 mg g^{-1} at pH 4.0 and 150 min contact time (Damascenoa et al., 2020). SiO_2-stabilized Fe_3O_4NPs efficiently degrade fuchsin up to five cycles with 98% efficiency at room temperature (Ning et al., 2018). Magnetic core-modified silver nanoparticles (Fe_3O_w@AgNPs) adsorb ibuprofen (an organic micropollutant) from an aqueous solution with 93% removal efficiency (Vicente-Martinez et al., 2020). Amine-rich functionalized on rice huck cellulose-based magnetic nanoparticles (RHB-MH_2) adsorb Pb (II) ion up to 680.19 mg/g adsorption capacity at pH = 5. The higher adsorption capacity for RHB-MH_2 is due to amine functionalization on the surface of magnetic nanoparticles (Nata et al., 2020). Hyperbranched polyglycerol grafted magnetic iron oxide nanoparticles (HPG-MNPs) acts as a promising adsorption material to remove heavy metal ions such as Cu, Ni, and Al with a maximum adsorption capacity of 0.700, 0.451, and 0.790 mg.mg^{-1}, respectively (Almomani et al., 2020).

9.14 DRAWBACKS OF USING NANOMATERIALS IN WATER TREATMENT

Different nanomaterials have been under trial with several possibilities to treat wastewater based on the particles' morphology. The primary challenge is the lack of properly validated information about their specific use in wastewater treatment concerning their release into the environment, residual traces, post-treatment, potential possibilities for accumulation, etc. (Bushra et al., 2014). To date, only a small number of nanoparticles are commercially available for direct use in wastewater treatment.

However, another challenge is based on their adverse effects on human health caused by newly synthesized nanoparticles due to their toxicology, thereby also resulting in adverse effects on the environment (Chen et al., 2015). While considering nanosorbents, reusability plays a big challenge, underlining the need to concentrate more on recovery processes in an easily doable manner. Fundamental challenges were found to remove nanoparticles using a wastewater treatment plant by the United States Environmental Protection Agency (USEPA) (Schlosser, 2020; Lu & Astruc, 2020).

9.15 CONCLUSION AND FUTURE PERSPECTIVES

In this chapter, various types of magnetic nanomaterials such as zero-valent metal nanoparticles, iron oxide nanoparticles, nanocomposites, polymer-based nanomaterials, etc. were discussed in detail, with a specific focus on their synthesis, properties, and potential applications toward water remediation. Nanomaterials surpassingly act as promising tools for water and wastewater treatment based upon their synthesis, surface modifications, and application mechanisms, thus helping the adverse conditions to handle water treatment and global freshwater requirements. Nano-engineered water approach is modest with conservative treatment processes. The easy separation and reusability of magnetic nanomaterials can possibly serve as an apt solution toward the supply of drinking water. Due to its superparamagnetic nature and stability, magnetic iron oxide nanoparticles can be easily separated from the wastewater and reused for several cycles. In addition to that, desorbed pollutants from MIONPs can also be recycled for other applications. However, an optimized synthetic procedure along with desired surface functionalization are needed to have MIONPs that can be applied to remove multiple pollutants simultaneously from wastewater. Further, more research is needed to understand the cost effectiveness of MIONPS in terms of both making and expenditures.

ACKNOWLEDGMENT

This work was supported by Department of Science and Technology Nanomission, Government of India [Grant No. DST/NM/NB-2018/10(G)], Science and Engineering Research Board. Ms. RS acknowledge Department of Science and Technology, India [Grant No. YSS/2014/00026] for the fellowship. We thank Dr. S. Narayanan for his help with writing this chapter.

REFERENCES

Adekunle, A.S., Oyekunle, J.A.O., Durosinmi, L.M., Oluwafemi, O.S., Olayanju, D.S., Akinola, A.S., Obisesan, O.R., Akinyele, O.F., and Ajayeoba, T.A. 2020. Potential of cobalt and cobalt oxide nanoparticles as nanocatalyst towards dyes degradation in wastewater. *Nano-Struc Nano-Obj* 21:100405.

Almomani, F., Bhosale, R., Khraisheh, M., kumar, A., and Almomani, T. 2020. Heavy metal ions removal from industrial wastewater using magnetic nanoparticles (MNP). *Appl Surf Sci* 506:144924.

Arancibia, M.N., Baltazar, S.E., and Garcia, A. 2016. Nanoscale zero valent supported by zeolite and montmorillonite: Template effect of the removal of lead ion from an aqueous solution. *J. Hazard, Mater* 301:371–380.

Atla, S., Lin, W.-R., Chien, T.-C., Tseng, M.-J., Shu, J.-C., Chen, C.-C., and Chen, C.-Y. 2018. Fabrication of Fe_3O_4/ZnO magnetite core shell and its application in photocatalysis using sunlight. *Mater Chem Phys* 216:380–386.

Auta, M., and Hameed B.H. 2014. Chitosan-clay composite as highly effective and low-cost adsorbent for batch and fixed-bed adsorption of methylene blue. *Chem. Eng. J.* 237:352–361.

Azeez, A.A., Rhee, K.Y., Park, S.J., and Hui, D. 2013. Epoxy clay nanocomposites-processing, properties and applications: A review. *Compos. Part B.* 45:308–320.

Balcells, L., Martinez-Boubeta, C., Cisneros-Fernandez, J., Simeonidis, K., Bozzo, B., Oro-Sle, J., Bagues, N., Arbiol, J., Mestres, N., and Martinez, B. 2016. One-step route to iron oxide hollow nanocuboids by cluster condensation: Implementation in water remediation technology. *ACS Appl Mater Interfaces* 8:28599–28606.

Bandforuzi, S. R., and Hadjmohammadi, M. R. 2019. Modified magnetic chitosan nanoparticles based on mixed hemimicelle of sodium dodecyl sulfate for enhanced removal and trace determination of three organophosphorus pesticides fromnatural waters. *Anal ChimActa* 1078: 90–100.

Banic, N.D., Abramovic, B.F., Krstic, J.B., SojicMerkulov, D.V., Fincur, N.L., and Mitric, M. N. 2019. Novel WO_3/Fe_3O_4 magnetic photocatalysts: Preparation, characterization and thiacloprid photodegradation. *J IndEngChem* 70:264–275.

Beaugeard, V., Muller, J., Graillot, A., Ding, X., Robin, J.J., and Monge, S. 2020. Acidic polymeric sorbents for the removal of metallic pollution in water: A review. *React.Funct. Polym* 152:104599.

Bharath, G., Alhseinat, E., Ponpandian, N., Khan, M.A., Siddiqui, M.R., Ahmed, F., and Alsharaeh, E.H. 2017. Development of adsorption and electrosorption techniques for removal of organic and inorganic pollutants from wastewater using novel magnetite/porous graphene-based nanocomposites. *Separ. Purif. Technol.* 188:206–218.

Bhateria, R., and Singh, R. 2019. A review on nanotechnological application of magnetic iron oxides for heavy metal removal. *J. Water Process Eng* 31:100845.

Biftu, W. K., Ravindhranath, K., and Ramamoorty, M, 2020. New research trends in the processing and applications of iron-based nanoparticles as adsorbents in water remediation methods. *Nanotechnol Environ Eng.* 5:12.

Bonilla-Petriciolet, A., Mendoza-Castillo, D.I., Dotto, G.L., and Duran-Valle, C.J. 2019. *Adsorption in Water Treatment, Reference Module in Chemistry, Molecular Sciences and Chemical Engineering*. Elsevier

Boukhalfa, N., Boutahala, M., Djebri, N., and Idris, A. 2019. Maghemite/alginate/functionalized multiwalled carbon nanotubes beads for methylene blue removal: Adsorption and desorption studies. *J MolLiq* 275:431–440.

Bushra, R., Shahadat, M., Ahmad, A., Nabi, S.A., Umar, K., Muneer M., Raeissia, A.S., and Owais, M. 2014. Synthesis, characterization, antimicrobial activity and applications of composite adsorbent for the analysis of organic and inorganic pollutants. *J. Hazard. Mater.* 264: 481–489.

Chavez-Guajardo, A.E., Medina-Llamas, J.C., Maqueria, L., Andrade, C.A.S., Alves, K.G.B., de Melo, C.P. 2015. Efficient removal of Cr (VI) and Cu (II) ions from aqueous media byuse of polypyrrole/maghemite and polyaniline/maghemite magnetic nanocomposites. *ChemEng J* 281: 826–839.

Chen D.H., and Huang S.H. 2004. Fast separation of bromelain by polyacrylic acid-bound iron oxide magnetic nanoparticles. *Process Biochem* 39:2207–2211.

Chen, L., Yang, S., Huang, Y., Zhang, B., Kang, F., Ding, D., and Cai, T. 2019. Degradation of antibiotics in multi-component systems with novel ternary AgBr/Ag_3PO_4@natural hematite heterojunctionphotocatalyst under simulated solar light. *J. Hazard. Mater* 371:566–575.

Chen, X.L., O'Halloran, J., Jansen, M.A.K.. 2015. The toxicity of zinc oxide nanoparticles to Lemma minor (L.) is predominantly caused by dissolved Zn, Aquat. *Toxicol* 158:1–13.

Cornell, R.M., and Schwertmann, U. 2003. *The Iron Oxides: Structure, Properties, Reactions, Occurrences and Uses*. John Wiley & Sons

Crane, R., and Scott, T. 2012. Nanoscale zero-valent iron: Prospects for an emerging water treatment technology. *J. Hazard. Mater* 211:112–125.

Cui, L., Guo, X., Wei, Q., Wang, Y., Gao, L., Yan, L., Yan, L., Yan, T., and Du, B. 2015. Removal of mercury and methylene blue from aqueous solution by xanthate functionalized magnetic graphene oxide: sorption kinetic and uptake mechanism. *J. Colloid Interface Sci* 439:112–120.

Dai, H., Xu, S., Chen, J., Miao, X., and Zhu, J. 2018. Oxalate enhanced degradation of Orange II in heterogeneous UV-Fenton system catalyzed by $Fe_3O_4@\gamma\text{-}Fe_2O_3$ composite. *Chemosphere* 199:147–153.

Damascenoa, B.S., Viana da Silvab, A.F., and Vaz de Araujoa, A.C. 2020. Dye adsorption onto magnetic and superparamagnetic Fe_3O_4 nanoparticles: A detailed comparative study. *J Environ ChemEng* 8: 103994.

Damdib, S., Phamornpiboon, P., Siyasukh. A., Thanachayanont, C., Punyapalakul, P., and Tonanon N. 2019. *Paraquat pesticide removal by magnetic biochar derived from corn husk. Pure and Applied Chemistry International Conference (PACCON 2019).* https://www.academia.edu/download/61160304/ee-fp-0120191107-53320-1opnwqv.pdf

Deng, L., Ngo, H.-H., Guo, W., and Zhang, H. 2019. Pre-coagulation coupled with sponge-membrane filtration for organic matter removal and membrane fouling control during drinking water treatment. *Water Res* 157:155–166.

Dhas, C.R., Venkatesh, R., Jothivenkatachalam, K., Nithya, A., Benjamin, B.S., Raj, A.M.E., Jeyadheepan, K., and Sanjeeviraja, C. 2015. Visible light driven photocatalytic degradation of Rhodamine B and direct red using cobalt oxide nanoparticles, *Ceram. Int.* 41: 9301–9313.

Ding, C., Cheng, W., Sun, Y., and Wang, X., 2015. Novel fungus-Fe_3O_4 bio-nanocomposites as high performance adsorbents for the removal of radionuclides. *J Hazard Mater* 295:127–137.

Fadillah, G., Yudha, S.P., Sagadevan, S., Fatimah, I., and Muraza, O. 2020. Magnetic iron oxide/clay nanocomposites for adsorption and catalytic oxidation in water treatment applications. *Open Chem.* 18:1148–1166.

Famiglietti, J.S. 2014. The global groundwater crisis. *Nat Clim Change* 4:945–948.

Fan, H., Ma, X., Zhou, S., Huang, J., Liu, Y., and Liu, Y., 2019. Highly efficient removal of heavy metal ions by carboxymethyl cellulose-immobilized Fe_3O_4 nanoparticles prepared via high-gravity technology. *CarbohydrPolym* 213:39–49.

Fan, L., Luo, C., Sun, M., Li, X., Lu, F., and Qiu, H. 2012. Preparation of novel magnetic chitosan/graphene oxide composite as effective adsorbents toward methylene blue. *Bioresour. Technol* 114:703–706.

Farahbakhsh, F., Ahmadi, M., Hekmatara, S.H., Sabet, M., and Heydari-Bafrooei, E., 2019. Improvement of photocatalyst properties of magnetic NPs by new anionic surfactant. *Mater ChemPhy* 224:279–285.

Fei, Z., He, S., Li, L., Ji, W., and Au, C.T. 2012. Morphology directed synthesis of Co_3O_4 nanotubes based on modified kirkendall effect and its application in CH_4 combustion. *Chem. Commun* 48:853–855.

Feng, G., Ma, J., Zhang, X., Zhang, Q., Xiao, Y., Ma, Q., and Wang, S., 2019. Magnetic natural composite Fe_3O_4-chitosan@bentonite for removal of heavy metals from acid mine drainage. *J Colloid Interface Sci* 538:132–141.

Fila, D., Hubicki, Z., and Kolodynska, D. 2019. Recovery of metals from waste nickel-metal hybride batteries using multifunctional Diphonix resin, *Adsorption* 25: 367–382.

Fu, F., Dionysiou, D.D., and Liu, H. 2014a. The use of zero-valent iron for groundwater remediation and wastewater treatment: A review. *J. Hazard, Mater* 267:194–205.

Fu, Y., Wang, J., Liu, Q., and Zeng, H., 2014b. Water-dispersible magnetic nanoparticles-graphene oxide composites for selenium removal. *Carbon* 77:710–721.

Gao, B., Iftekhar, S., Srivastava, V., Doshi, B., and Sillanpaa, M. 2018. Insights into the generation of reactive oxygen species (ROS) over polythiophene/$ZnIn_2S_4$ based on different modification processing. *Catal. Sci. Techno* 8:2186–2194.

Geise, G.M., Lee, H.-S., Miller, D.J., Freeman, B.D., McGrath, J.E., and Paul, D.R. 2010. Water purification by membranes: The role of polymer science, *J. Polym. Sci. Part B: Polym. Phys* 48:1685–1718.

Geng, Z., Lin, Y., Yu, X., Shen, Q., Ma, L., Li, Z., Pan, N., and Wang, X. 2012. Highly efficient dye adsorption and removal: a functional hybrid of reduced graphene oxide-Fe_2O_3 nanoparticles as an easily regenerative adsorbent. *J. Mater. Chem.* 2 22: 3527–3535.

Gorski, C.A., Handler, R.M., Beard, B.L., Pasakanis, T., Johnson, C.M., and Scherer, M.M. 2012. Fe atom exchange between aqueous Fe^{2+} and magnetite. *Environ SciTechnol* 46:12399–12407.

Guo, H., Wang, H., Zhang, N., Li, J., Liu, J., Alsaedi, A., Hayat, T., Li, Y., and Sun, Y. 2019. Modeling and EXAFS investigation of U(VI) sequestration on Fe_3O_4/PCMs composites. *ChemEng J* 369:736–744.

Gupta, A.K., Ghosal, P.S., and Dubey, B.K., 2017. Hybrid nanoadsorbents for drinking water treatment: A critical review, in: *Hybrid Nanomaterials. Advance in Energy, Environment, And Polymer Nanocomposites*, Wiley, 199.

Gupta, V.K., Tyagi, I., Sadegh, H., Shahryari-Ghoshekand, R., Makhlouf, A.S.H., and Maazinejad, B., 2015. Nanoparticles as adsorbent; a positive approach for removal of noxious metal ions: A review. *SciTechnolDev* 34:195–214.

Gutierrez, A.M., Dziubla, T.D., and Hilt, J.Z. 2017. Recent advances on iron oxide magnetic nanoparticles as sorbents of organic pollutants in water and wastewater treatment, *Rev Environ Health.* 32(1–2):111–117.

Hao, Y.-M., Man, C., and Hu, Z.-B. 2010. Effective removal of Cu (II) ions from aqueous solution by amino-functionalized magnetic nanoparticles. *J Hazard Mater* 184:392–399.

Hashemzadeh, M., Nilchi, A., Hassani, A.H., and Saberi, R. 2019.Synthesis of novel surface-modified hematite nanoparticles for the removal of cobalt-60 radiocations from aqueous solution. *Int. J. Environ. Sci, Technol* 16:775–792.

He, Z., Wang, P., Liu, D., and Zhou, Z., 2014. Hydrohilic-lipohilic balanced magnetic nanoparticles: preparation and application in magnetic solid-phase extraction of organochlorine pesticides and triazine herbicides in environmental water samples. *Talanta* 127:1–8.

Hernandez-Flores, H., Pariona, N., Herrera-Trejo, M., Hdz-Garcia, H.M., and Mtz-Enriquez, A.I., 2018. Concrete/maghemite nanocomposites as novel adsorbents for arsenic removal. *J Mol Struct* 1171:9–16.

Hoag, G.E., Collins, J.B., Holcomb, J.L., Hoag, J.R., Nadagouda, M.N., and Varma, R.S.2009. Degradation of bromothymolblue by greener nano-scale zero-valent iron synthesized using tea polyphenols. *J. Mater. Chem* 19:8671–8677.

Huang, Y., and Keller A.A., 2015. EDTA functionalized magnetic nanoparticle sorbents for cadmium and lead contaminated water treatment. *Water Res* 80:159–168.

Iftekhar, S., Srivastava, V., Sillanpaa, M. Synthesis of hybrid bionanocomposites and their application for the removal of rare-earth elements from synthetic wastewater, in: Sillanpaa, M. (ed) *Advanced Water Treatment: Adsorption.* Elsevier, New York. 2020.

Jiang, W., Pelaez, M., Dionysiou, D.D., Entezari, M.H., Tsoutsou, D., and O'Shea, D., 2013. Chromium (VI) removal by maghemite nanoparticles. *ChemEng J* 222:527–533.

JokarBaloochi, S., SolaimanyNazar, A.Z., and Farhadian, M. 2018. 2,4-Dichlorophenoxyacetic acid herbicide photocatalytic degradation by zero-valent iron/titanium dioxide based on activated carbon. *Environ. Nanotechnol. Monit. Manage* 10:212–222.

Jyothi, M. S., Angadi V. J., Kanakalakshmi T. V., Padaki M., Balakrishna R. G., Soontarapa K. 2019. Magnetic nanoparticles impregnated, cross-linked, porous chitosan microspheres for efficient adsorption of methylene blue from pharmaceutical waste water. *J Polym Environ.* 27:2408–2418.

Kang, M.J., Yu, H., Lee, W., and Cha, H.G. 2019. Efficient Fe_2O_3/C-g-C_3N_4 Z-scheme heterojunction photocatalyst prepared by facile one-step carbonizing process. *J. Phys. Chem. Solids* 130:93–99.

Karami, H. 2013. Heavy metal removal from water by magnetite nanorods. *ChemEng J* 219:209–216.

Karprara, E., Simeonidis, K., Zouboulis, A., and Mitrakas, M. 2016. Rapid small-scale column tests for Cr(VI) removal by granular magnetite. *Water SciTechnol Water Supply* 16: 525–532.

Khalighi, S.R., Khosravi, M.R., Badii, K.H., and Yousefi, L. 2012. Adsorption of acid blue 92 dye on modified diatomite by nickel oxide nanoparticles in aqueous solution. *Prog. Color Colorants Coat* 5:101–116.

Khan, A., Xing, J., Elseman, A.M., Gu, P., Gul, K., Ai, Y., Jehan, R., Alsaedi, A., Hayat, T., and Wang, X., 2018. A novel magnetite nanorod-decorated Si-Schiff base complex for efficient immobilization of U(vi) and Pb(ii) from water solutions. *Dalton Trans* 47:11327–11336.

Kim, K.-W., Shon, W.-J., Oh, M.-K., Yang, D., Foster, R. I., and Lee, K.-Y. 2019. Evaluation of dynamic behavior of coagulation-flocculation using hydrous ferric oxide for removal of radioactive nuclides in wastewater. *Nucl Eng. Technol* 51:738–745.

Kobylinska, N., Kostenko, L., Khainakov, S., and Garcia-Granda, S. 2020. Advanced core-shell EDTA-functionalized magnetite nanoparticles for rapid and efficient magnetic solid phase extraction of heavy metals from water samples prior to the multi-element determination by ICP-OES. *MicrochimActa* 187:289.

Koli, P.B., Kapadnis, K.H., Deshpande, U.G., and Patil, M.R. 2018. Fabrication and characterization of pure and modified Co_3O_4nanocatalyst and their application for photocatalytic degradation of eosine blue dye: A comparative study. *J. Nanostruct. Chem.* 8:453–463.

Krehula, S., Ristic, M., Reissner, M., Kubuki, S., and Music, S. 2017. Synthesis and properties of indium-doped hematite. *J. Alloy. Compd* 695:1900–1907.

Lei, Y., Chen, F., Luo, Y., and Zhang, L. 2014. Three-dimensional magnetic graphene oxide foam/Fe_3O_4 nano-composites as an efficient absorbent for Cr(VI) removal. *J Mater Sci* 49:4236–4245.

Leone, V.O., Pereira, M.C., Aquino, S.F., Oliveira, L.C.A., Correa, S., Ramalho, T.C., Gurgel, L.V.A., and Silva, A.C. 2018. Adsorption of diclofenac on a magnetic adsorbent based on maghemite: Experimental and theoretical studies. *New J Chem* 42:437–449.

Leonel, A.G., Mansur, A.A.P., and Mansur, H.S. 2021. Advanced functional nanostructures based on magnetic iron oxide nanomaterials for water remediation: A review. *Water Res.* 190:116693.

Li, B., Yin, W., Xu, M., Tan, X., Li, P., Gu, J., Chiang, P., and Wu, J. 2019a. Facile modification of activated carbon with highly dispersed nano-sized α-Fe_2O_3 for enhanced removal of hexavalent chromium from aqueous solutions. *Chemosphere* 224: 220–227.

Li, L.H., Xiao, J., Liu, P., and Yang, G.W. 2015. Super adsorption capability from amorphousization of metal oxide nanoparticles for dye removal. *Sci. Rep.* 5: 9028.

Li, X., Jin, B., Huang, J., Zhang, Q., Peng, R., and Chu, S. 2018. Fe_2O_3/ZnO/$ZnFe_2O_4$ composites for the efficient photocatalytic degradation of organic dyes under visible light. *Solid state Sci* 80:6–14.

Li, X., Liu, X., Lin, C., Zhang, H., Zhou, Z., Fan, G., and Ma, J. 2019b. Cobalt ferrite nanoparticles supported on drinking water treatment residuals: An efficient magnetic heterogeneous catalyst to activate peroxymonosulfate for the degradation of atrazine. *Chem. Eng. J* 367:208–218.

Li, X.Y., Ai, L.H., and Jiang, J. 2016. Nanoscale zerovalent iron decorated on grapheme nanosheets for Cr(VI) removal from aqueous solution: surface corrosion retard induced the enhanced performance. *Chem. Eng. J.* 288:789–797.

Liang, D.-W., Yang, Y.-H., Xu, W.-W., Peng, S.-K., Lu, S.-F., and Xiang, Y. 2014. Nonionic surfactants greatly enhances the reductive debromination of polybrominated diphenyl ethers by nanoscale zero-valent iron: Mechanism and kinetics. *J. Hazard, Mater* 278:592–596.

Liang, H., Xu, B., Wang, Z. 2013. Self-assembled 3D flower-like α-Fe_2O_3 microstructures and their superior capability for heavy metal ion removal. *Mater. Chem. Phys* 141:727–734.

Ling, L., Pan, B., and Zhang, W.-X. 2015. Removal of selenium from water with nanoscale zero-valent iron: Mechanisms of intraparticle reduction of Se(IV). *Water Res* 71:274–281.

Ling, L., and Zhang, W.-X. 2015. Enrichment and encapsulation of uranium with iron nanoparticles. *J. Am. Chem. Soc* 137: 2788–2791.

Lingamdinne, L.P., Koduru, J.R., Choi, Y.L., Chang, Y.Y., and Yang, J.K. 2016. Studies on removal of Pb (II) and Cr (III) using graphene oxide based inverse spinel nickel ferrite nanocomposite as sorbent. *Hydrometallurgy* 165:64–72.

Liou, Y.H., Lo, S.-L., Lin, C.-J., Kuan, W.H., and Weng, S.C. 2005. Chemical reduction of an unbuffered nitrate solution using catalyzed and uncatalysednanoscale iron particles. *J. Hazard, Mater* 127:102–110.

Liu, S., Chen, D., Zheng, J., Zeng, L., Jiang, J., Jiang, R., Zhu, F., Shen, Y., Wu, D., and Ouyang, G. 2015. The sensitive and selective adsorption of aromatic compounds with highly crosslinked polymer nanoparticles. *Nanoscale* 7:16943–16951.

Liu, X., Yan, L., Yin, W., Zhou, L., Tian, G., Shi, J., Yang, Z., Xiao, D., Gu, Z., and Zhao, Y. 2014. A magnetic graphene hybrid functionalized with beta-cyclodextrins for fast and efficient removal of organic dyes. *J. Mater. Chem.* 2,12296.

Lounsbury, A.W., Yamani, J.S., Johnston, C.P., Larese-Casanova, P., and Zimmerman, J.B. 2016. The role of counter ions in nano-hematite synthesis: Implications for surface area and selenium adsorption capacity. *J Hazard Mater* 310:117–124.

Lu, F., and Astruc, D. 2020. Nanocatalysts and other nanomaterials for water remediation from organic pollutants. *Coord. Chem. Rev* 408:213180.

Luo, J., Fu, K., Yu, D., Hristovski, K.D., Westerhoff, P., and Crittenden, J.C. 2021. Review of advances in engineering nanomaterial adsorbents for metal removal and recovery from water: Synthesis and microstructure impacts. *ACS EST Engg.* https://doi.org/10.1021/acsestengg.0c00174.

Mahalingam, S., and Ahn, Y.-H.Y.-H. 2018. Improved visible light photocatalytic activity of rGO-Fe_3O_4.NiO hybrid nanocomposites synthesized by in situ facile method for industrial wastewater treatment applications. *New J Chem* 42:4372–4383.

Mahmoud, A.M., Ibrahim, F.A., Shaban, S.A., and Youssef, N.A. 2015. Adsorption of heavy metal ion from aqueous solution by nickel oxide nano catalyst prepared by different methods. *Egypt. J. Pet* 24:27–35.

Mahpishanian, S., and Sereshti, H. 2017. One-step green synthesis of β-cyclodextrin/iron oxide-reduced graphene oxide nanocomposites with high supramolecular recognition capability: Application for vortex-assisted magnetic solid phase extraction of organochlorine pesticides residue from honey samples. *J. Chromatogr. A.*1485:32–43.

Mai, Y.-W., and Yu, Z.-Z. (Eds.), *Polymer Nanocomposites*. CRC Press LLC, Boca Raton, FL, USA, 2006.

Markova, Z., Siskova, K.M., and Filip, J. 2013. Air stable magnetic bimetallic Fe-Ag nanoparticles for advanced antimicrobial treatment and phosphorus removal. *Environ. Sci. Technol* 47:5285–5293.

Martinez-Boubeta, C., and Simeonidis, K. 2019. Chapter 20 – Magnetic nanoparticles for water purification, in: Thomas, S., Pasquini, D., Leu, S.-Y., and Gopakumar, D.A. (Eds.). *Nanoscale Materials in Water Purification*, Elsevier 521–552.

Martinez-Boubeta, C., Simeonidis, K., Angelakeris, M., Pazos-Perez, N., Giersig, M., Delimitis, A., Nalbandian, L., Alexandrakis, V., and Niarchos, D. 2006. Critical radius for exchange bias in naturally oxidized Fe nanoparticles. *Phys Rev B Condens Matter* 74:054430.

Mohamed, H.H. 2019. Rationally designed Fe_2O_3/GO/WO_3 Z-Scheme photocatalyst for enhanced solar light photocatalytic water remediation. *J. Photochem. Photobio. A: Chem* 378:74–84.

Motahari, F., Mozdianfard, M.R., and Salavati-Niasari, M. 2015. Synthesis and adsorption studies of NiO nanoparticles in the presence of H_2acacen ligand, for removing Rhodamine B in wastewater treatment. *Process Saf. Environ* 93:282–292.

Muradova, G.G., Gadjieva, S.R., Di Palma, L., and Vilardi, G. 2016. Nitrates removal by bimetallic nanoparticles in water. *Chem. Eng. Trans* 47:205–210.

Nasiri, R., Arsalani, N., and Panahian, Y., 2018. One-pot synthesis of novel magnetic three-dimensional graphene/chitosan/nickel ferrite nanocomposite for lead ions removal from aqueous solution: RSM modelling design. *J Cleaner Prod* 201:507–515.

Nassar, M.Y., Aly, H.M., Abdelrahman, E.A., and Moustafa, M.E. 2017.Synthesis characterization, and biological activity of some novel Schiff bases and their Co(II) and Ni(II) complexes: A new route for Co_3O_4 and NiO nanoparticles for photocatalytic degradation of methylene blue dye. *J. Mol. Struct* 1143:462–471.

Nata, I.F., Wicakso, D.R., Miwan, A., Irwan, C., Ramadhani, D., and Ursulla. 2020. Selective adsorption of Pb(II) ion on amine-rich functionalized rice husk magnetic nanoparticle biocomposites in aqueous solution. *J Environ Chem Eng* 8: 104339.

Natarajan, S., Anitha, V., Gajula, G.P., and Thiagarajan, V.2020. Synthesis and characterization of magnetic superadsorbent Fe_3O_4-PEG-Mg-Al-LDH nano-composites for ultra-high removal of organic dyes. *ACS Omega* 5:3181–3193.

Natarajan, S., Harini, K., Gajula, G.P., Sarmento, B., Neves-Peterson, M.T., and Thiagarajan, V. 2019. Multifunctional magnetic iron oxide nanoparticles: diverse synthetic approaches, surface modifications, cytotoxicity towards biomedical and industrial applications. *BMC Mat* 1, 2.

Nateghi, N., Bonyadinejad, G.R., Amin, M.M., and Mohammadi, H. 2012. Decolorization of synthetic wastewaters by nickel oxide nanoparticle, *Int. J. Env. Health Eng* 1: 25.

Nathani, H., Gubbala, S., and Misra, R.D.K. 2004. Magnetic behavior of nickel ferrite-polyethylene nanocomposites synthesized by mechanical milling process. *Mater. Sci. Eng. B* 111:95–100.

Ngomsik A.-F., Bee, A., Talbot, D., and Cote, G. 2012. Magnetic solid-liquid extraction of Eu(III), La(III), Ni(II) and Co(II) with maghemite nanoparticles. *Sep Purif Technol* 86:1–8.

Ning, J., Wang, M., Luo, X., Hu, Q., Hou, R., Chen. W., Chen, D., Wang, J., and Liu, J. 2018. SiO_2 stabilized magnetic nanoparticles as a highly effective catalyst for the degradation of basic fuchsin in industrial dye wastewaters. *Molecules* 23: 2573.

Noh, J., Osman, O.I., Aziz, S.G., Winget, P., and Bredas, J. 2014. A density functional theory investigation of the electronic structure and spin moments of magnetite. *Sci Teechnol Adv Mater* 15:044202.

Okube, M., Yasue, T., and Sasaki, S. 2012. Residual-density mapping and site-selective determination of anomalous scattering factors to examine the origin of the Fe K pre-edge peak of magnetite. *J SynchrRadiat* 19:759–767.

Ou, H., You, Q., Li, J., Liao, G., Xia, H., and Wang, D. 2016. A rich-amine porous organic polymer: An efficient and recyclable adsorbent for removal of azo dye and chlorophenol. *RSC Adv.* 6:98487–98497.

Pandian, C.J., Palanivel, R., and Dhananasekaran, S. 2015. Green synthesis of nickel nanoparticles using ocimum sanctum and their application in dye and pollutant adsorption, *Chin. J. Chem. Eng.* 23:1307–1315.

Pang, Y.L., Lim, S., Ong, H.C., and Chong, W.T. 2016. Synthesis, characteristics and sono-catalytic activities of calcined γ-Fe_2O_3 and TiO2 nanotubes/γ-Fe_2O_3 magnetic catalysts in the degradation of Orange G. *UltrasonSonochem* 29:317–327.

Parashar, A., Sikarwar, S., and Jain, R. 2020. Removal of pharmaceuticals from wastewater using magnetic iron oxide nanoparticles (IOPs). *Int. J. Environ. Anal.Chem.* doi: 10.1080/03067319.2020.1716977

Park, H., Myung, N.V., Jung, V., and Choi, H. 2009. As(V) remediation using electrochemically synthesized maghemite nanoparticles. *J Nanopart Res* 11:1981–1989.

Pasakarnis, T.S., Boyanov, M.I., Kemner, K.M., Mishra, B., O'Loughlin, E. J., Parkin, G., and Scherer, M.M. 2013. Influence of chloride and Fe(II) content on the reduction of Hg(II) by magnetite. *Environ SciTechnol* 47:6987–6994.

Pinakidou, F., Katsikini, M., Simeonidis, K., Karprara, E., Paloura, E.C., and Mitrakas, M. 2016. On the passivation mechanism of Fe_3O_4 nanoparticles during Cr(VI) removal from water: A XAFS study. *Appl Surf Sci* 360:1080–1086.

Prathna, T.C., Sharma, S.K., and Kennedy, M. 2018. Nanoparticles in household level water treatment. An overview. *Sep. Purif. Technol* 199:260–270.

Rafique, U., Imtiaz, A., and Khan, A.K.. 2012. Synthesis, characterization and applications of nanomaterials for the removal of emerging pollutants from industrial waste water, Kinetics and Equilibrium model. *J. Water Sust.* 2:233–244.

Rajput, S., Singh, L.P., Pittman, C.U., and Mohan, D. 2017. Lead(Pb^{2+}) and copper(Cu^{2+}) remediation from water using superparamagnetic maghemite (γ-Fe_2O_3) nanoparticles synthesized by Flame Spray Pyrolysis (FSP). *J Colloid Interface Sci* 492:176–190.

Rasmussen, K., González, M., Kearns, P., Sintesa, J.R., Rossia, F., and Sayre, P. 2016. Review of achievements of the OECD working party on manufactured nanomaterials' testing and assessment programme: From exploratory testing to test guidelines. *Regul. Toxicol. Pharmacol.* 74: 147–160.

Ravindranath, R., Roy, P., Periasamy, A.P., Chen, Y.-W., Liang, C.-T., Chang, H.-T. 2017. Fe_2O_3/Al_2O_3microboxes for efficient removal of heavy metal ions. *New. J. Chem* 41:7751–7757.

Reddy, D.H.K., and Lee, S.-M. 2013. Application of magnetic chitosan composites for the removal of toxic metal and dyes from aqueous solutions. *Adv. Colloid Interface Sci.* 201:68–93.

Rivas, B.L., Urbano, B.F., and Sanchez, J. 2018. Water-soluble and insoluble polymers, nanoparticles, nanocomposites and hybrids with ability to remove Hazardous inorganic pollutants in water. *Front Chem* 31:320.

Ruiz-Bevia, F., and Fernandez-Torres, M.J. 2019. Effective catalytic removal of nitrates from drinking water: An unresolved problem?. *J. Cleaner Prod* 217:398–408.

Santhosh, C., Malathi, A., Dhaneshvar, E., Bhatnagar, A., Grace, A.N., and Madhavan, J. 2019. Chapter 16 – Iron oxide nanomaterials for water purification, in: Thomas S, Pasquini, D., Leu, S.-Y., and Gopakumar, D.A. (Eds.), *Nanoscale Materials in Waer Purification*. Elsevier 431–446.

Schlosser, D. 2020. Biotechnologies for water treatment, in *Advanced Nano-Bio Technologies for Water and Soil Treatment.* Springer, Cham, Berlin/Heidelberg, Germany 335–343.

Schwarzenbach, R.P., Egli, T., Hofstetter, T.B., Gunten, U.V., and Wehrli, B. 2010. Global Water Pollution and Human Health. *Annu Rev Environ Resour* 35:109–136.

Setshedi, K.Z., Bhaumik, M., Songwane, S., Onyango, M.S., and Maity, A. 2013. Exfoliated polypyrrole-organically modified montmorillonite clay nanocomposite as a potential adsorbent for Cr(VI) removal. *Chem. Eng. J.* 222: 186–197.

Shin, S., and Jang, J. 2007. Thiol containing polymer encapsulated magnetic nanoparticles as reusable and efficiently separable adsorbents for heavy metal ions. *Chem. Commun.* 4230–4232.

Shipley, H.J., Engates, K.E., and Grover, V.A. 2013. Removal of Pb(II), Cd(II), Cu(II), and Zn(II) by hematite nanoparticles: Effect of sorbent concentration, pH, temperature, and exhaustion. *Environ. Sci. Pollut. Res* 20:1727–1736.

Shirinova, H., Palma, L.D., Sarasini, F., Tirillo, J., Ramazanov, M.A., Hajiyeva, F., Sannino, D., Polichetti, M., and Galluzzi, A. 2016. Synthesis and characterization of magnetic nanocomposites for environmental remediation. *Chem. Eng. Trans.*. 47:103–108.

Shirsath, D.S., and Shirivastava, V.S. 2015. Adsorptive removal of heavy metals by magnetic nanoadsorbent: An equilibrium and thermodynamic, study. *Appl. Nanosci.* 5: 927–935.

Shou, J., Jiang, C., Wang, F., Qiu, M., and Xu, Q. 2015. Fabrication of Fe_3O_4/MgAl-layered double hydroxide magnetic composites for the effective decontamination of Co(II) from synthetic wastewater. *J Mol Liq* 207:216–223.

Shuibo, X., Chun, Z., Xinghuo, Z., Jing, Y., Xiaojian, Z., and Jingsong, W. 2019. Removal of uranium (VI) from aqueous solution by adsorption of hematite. *J Environ Radioact* 100: 162–166.

Simeonidis, K., Kaprara, E., Samaras, T., Angelakeris, M., Pliatsikas, N., Vourlias, G., Mitrakas, M., and Andritsos, N. 2015. Optimizing magnetic nanoparticles for drinking water technology: the case of Cr (VI). *Sci Total Environ* 535:61–68.

Simeonidis, K., Tziomaki, M., Angelakeris, M., Martinez-Boubeta, C., Balcells, L., Monty, C., Mitrakas, M., Vourlias, G., and Andritsos, N. 2014. Development of iron-based nanoparticles for Cr (VI) removal from drinking water. *EPJ Web of Conf.* 40:8007.

Singer, D.M., Chatman, S.M., Ilton, E.S., Rosso, K.M., Banfield, J.F., and Waychunas, G.A. 2012. U(VI) sorption and reduction kinetics on the magnetite (111) surface. *Environ SciTechnol* 46:3821–3830.

Singh, R., and Mirsa, V. 2015. Stabilization of zero-valent iron nanoparticles: Role of polymers and surfactants, in Aliofkhazraei, M. (Ed.), *Handbook of Nanoparticles*, Springer, New York, NY, USA, 1–19.

Speight, J.G. 2019. *Natural Water Remediation: Chemistry and Technology*. Elsevier, Butterworth-Heinemann, Oxford, United Kingdom.

Srivastava, N.K., Jha, M.K., and Sreekrishnan, T.R. 2014. Removal of Cr (Vi) from waste water using NiO nanoparticles. *Int. J. Environ. Sci. Technol* 3:395–402.

Srivastava, R.R., Kim, M.S., Lee, J.C., and Ilyas, S. 2015. Liquid-liquid extraction of rhenium (VII) from an acidic chloride solution using cyanex 923. *Hydrometallurgy* 157:33–38.

Subban, C.V., and Gadgil, A.J. 2019. Electrically regenerated ion-exchange technology for desalination of low-salinity water sources. *Desalination* 465:38–43.

Sun, H., Cao, L., and Lu, L. 2011. Magnetite/reduced graphene oxide nanocomposites: one step solvothermal synthesis and use as a novel platform for removal of dye pollutants. *Nano Res.* 4:550–562.

Sun, X., Yang, L., Li, Q., Zhao, J., Li, X., Wang, X., and Liu, H. 2014. Amino-functionalized magnetic cellulose nanocomposite as adsorbent for removal of Cr(VI): Synthesis and adsorption studies. *Chem. Eng. J.* 241:175–183.

Tadic, M., Trpkov, D., Kopanja, L., Vojnovic, S., and Panjan, M. 2019. Hydrothermal synthesis of hematite (α-Fe_2O_3) nanoparticle forms: Ssynthesis conditions, structure, particle shape analysis, cytotoxicity and magnetic properties. *J. Alloy. Compd* 792:599–60s9.

Tang, M., Huang, G., Zhang, S., Liu, Y., Li, X., Wang, X., and Qiu, X.P.H. 2014. Low-cost removal of organic pollutants with nickel nanoparticle loaded ordered microporous hydrogel as high-performance catalyst, Mater. *Chem. Phys* 145:418–424.

Tarigh, G.D., and Shemirani, F. 2013. Magnetic multi-wall carbon nanotube nanocomposite as an adsorbent for preconcentration and determination of lead (II) and manganese (II) in various matrices. *Talanta* 115:744–750.

Teja, A.S., and Koh, P.-Y.. 2019. Synthesis, properties, and applications of magnetic iron oxide nanoparticles. *Prog, Cryst. Growth Charact. Mater* 55:22–45.

Thakur, S., and Karak, N. 2014. One-step approach to prepare magnetic iron oxide/reduced graphene oxide nanohybrid for efficient organic and inorganic pollutants removal, *Mater. Chem. Phys.* 144: 425–432.

Tirtom, V.N., Dincer, A., Becerik, S., Aydemir, T., and Celik, A. 2012. Comparative adsorption of Ni(II) and Cd(II) ions on epichlorohydrin crosslinked chitosan-clay composite beads in aqueous solution. *Chem. Eng. J.* 197:379–386.

Tran, N., Mir, A., Mallik, D., Sinha, A., Nayar, S., and Webster T.J. 2010. Bactericidal effect of iron oxide nanoparticles on Staphylococcus aureus. *Int J Nanomed* 5:277–283.

Tuutijarvi, T., Lu, J., Sillanpaa, M., and Chen, G. 2009. As(V) adsorption on maghemite nanoparticles. *J Hazard Mater* 166:1415–1420.

Uddin, M.K., and Baig, U. 2019. Synthesis of Co_3O_4 nanoparticles and their performance towards methyl orange dye removal: characterization, adsorption and response surface methodology. *J. clean, Prod* 211:1141–1153.

Vicente-Martinez, Y., Caravaca, M., Soto-Meca, A., and Solana-Gonzalez, R. 2020. Magnetic core-modified silver nanoparticles for ibuprofen removal: An emerging pollutant in waters. *Sci. Rep.* 10:18288.

Wang, C., Feng, C., Gao, Y., Ma, X., Wu, Q., and Wang, Z. 2011. Preparation of a graphene-based magnetic nanocomposite for the removal of an organic dye from aqueous solution, *Chem. Eng. J.* 173: 92–97.

Wang, H., Li, S., Yuan, Y., Liu, X., Sun, T., and Wu, Z. 2020b. Study of the epoxy/amine equivalent ratio on thermal properties, cryogenic mechanical properties, and liquid oxygen compatibility of the bisphenola epoxy resin containing phosphorus. *High perform. Polym* 32:429–443.

Wang, J., Chen, Z., and Chen, B. 2015. Adsorption of polycyclic aromatic hydrocarbons by graphene and graphene oxide nanosheets, *Environ. Sci. Technol* 48:4817–4825.

Wang, L., Shi, C., Wang, L., Pan, L,. Zhang, X., and Zou, J.-J. 2020a. Rational design, synthesis, adsorption principles and applications of metal oxide adsorbents: a review. *Nanoscale* 12: 4790–4815.

Wang, S., Wei, J., Lv, S., Guo, Z., and Jiang, F. 2013b. Removal of organic dyes in environmental water onto magnetic-sulfonic graphene nanocomposite. *Clean. -Soil, Air, Water.* 41:992–1001.

Wang, S., Zhao, M., Zhou, M., Zhao, Y., Li, Y.C., Gao, B., Feng, K., Yin, W., Ok, Y.S., and Wang, X. 2019. Biomass facilitated phase transformation of natural hematite at high temperatures and sorption of Cd^{2+} and Cu^{2+}, *Environ. Int* 124:473–481.

Wang, X.Y., Zhu, M.P., Liu, H.L., Ma, J., and Li, F. 2013a. Modification of Pd-Fe nanoparticles for catalytic dechlorination of 2,4-dichlorophenol. *Sci. Total Environ* 449:157–167.

Wei, H., Gao, B., Ren, J., Li, A., and Yang, H. 2018. Coagulation/flocculation in dewatering of sludge: A review. *Water Res* 143:608–631.

Wiatrowski, H.A., Das, S., Kukkadapu, R., Ilton, E.S., Barkay, T., and Yee, N. 2009. Reduction of Hg(II) to Hg(0) by magnetite. *Environ Sci Technol* 43:5307–5313.

Wilson, J.L., Poddar, P., Frey, N.A., Srikanth, H., Mohomed, K., Harmon, J.P., Kotha, S., and Wachsmuth, J. 2004. Synthesis and magnetic properties of polymer nanocomposites with embedded iron nanoparticles. *J. Appl. Phys* 95:1439.

Wu, W., Wu, Z., Yu, T., Jiang, C., and Kim, W.-S.2015. Recent progress on magnetic iron oxide nanoparticles: synthesis, surface functional strategies and biomedical applications. *Sci. Technol. Adv. Mater.* 16:023501.

Xiong, Z., Lai, B., Yang, P., Zhou, Y., Wang, J., and Fang, S. 2015. Comparative study on the reactivity of Fe/Cu bimetallic particles and zero valent iron (ZVI) under different conditions of N_2, air or without aeration. *J. Hazard, Mater* 297:261–268.

Yadav, V. K., Ali, D., Khan, S. H., Gnanamoorthy, G., Choudhary, N., Yadav, K. K., Thai, V. N., Hussain, S. A., and Manhrdas, S. 2020. Synthesis and characterization of amorphous iron oxide nanoparticles by the sonochemical method and their application for the remediation of heavy metals from wastewater, *Nanomaterials* 10:1551.

Yang, K., and Xing, B. 2007. Desorption of polycyclic aromatic hydrocarbons from carbon nanomaterials in water. *Environ. Pollut.* 145:529–537.

Yavuz, C.T., Mayo, J.T., Yu, W.W., Prakesh, A., Falkner, J.C., Yean, S., Cong, L., Shipley, H. J., Kan, A., Tomson, M., Natelson, D., and Colvin, V. L. 2006. Low-field magnetic separation of monodisperse Fe_2O_3 nanocrystals. *Science* 314:964–967.

Yu, S., Wang, X., Liu, Y., Chen, Z., Wu, Y., Liu Y., Pang, H., Song, G., Chen, J., and Wang, X. 2019b. Efficient removal of uranium(VI) by layered double hydroxides supported nanoscale zero-valent iron: A combined experimental and spectroscopic studies. *Chem. Eng. J.* 365:51–59.

Yu, X., Cui, W., Zhang, F., Guo, Y., and Deng, T. 2019a. Removal of iodine from the salt water used for caustic soda production by ion-exchange resin adsorption. *Desalination* 458:76–83.

Zhang, K., Liu, Y., Deng, J., Xie, S., Lin, H., Zhao, X., Yang, J., Han, Z., and Dai, H. 2017. Fe_2O_3/3DOM $BiVO_4$: High-performance photocatalysts for the visible light-driven degradation of 4-nitrophenol. *Appl. Catal. B* 202:569–579.

Zhang, M., Gao, B., Varnoosfaderani, S., Hebard, A., Yao, Y., and Inyang, M. 2013. Preparation and characterization of a novel magnetic biochar for arsenic removal, *Bioresour. Technol* 130:457–462.

Zhao, R., Zhou, Z., Zhao, X., and Jing, G. 2019a. Enhanced Cr(VI) removal from simulated electroplating rinse wastewater by amino-functionalized vermiculite-supported nanoscale zero-valent iron. *Chemosphere* 218: 458–467.

Zhao, W., Liu, Z., Yuan, Y., Liu, F., Zhu, C., Ling, C., and Li, A. 2019b. Insight into Cu(II) adsorption on polyamine resin in the presence of HEDP by tracking the evolution of amino groups and Cu(II)-HEDP complexes. *ACS Sustainable Chem. Eng* 7: 5256–5263.

Zheng, X., He, L., Duan, Y., Jiang, X., Xiang, G., Zhao, W., and Zhang, S. 2014. Poly(ionic liqid) immobilized magnetic nanoparticles as new adsorbent for extraction and enrichment of organophosphorus pesticides from tea drinks. *J Chromatogr A* 1358:39–45.

Zhou, C.H., Zhang, D., Tong, D.S., Wu, L.M., Yu, W.H., and Ismadji, S. 2012. Paper-like composites of cellulose acetate-organo-montmorillonite for removal of hazardous anionic dye in water. *Chem. Eng. J.* 209:223–234.

Zhu, K., and Chen, C. 2019. Chapter 6 – Application of nZVI and its composites into the treatment of toxic/radioactive metal ions, in: Chen, C.(Ed.), *Interface Science and Technology*. Elsevier 281–330.

Zidan, I.H., Cheira, M.F., Bakry, A.R., and Atia, B.M. 2020 Potentiality of uranium recovery from G. Gatter leach liquor using Duolite ES-467 chelatineg resin: Kinetic, thermodynamic and isotherm features. *Int J Environ Anal Chem* https://doi.org/10.1080/03067319.2020.1748613.

10 Nanofiber Membranes for Wastewater Treatments

N.S. Jamaluddin, N.H. Alias, N.H. Othman, M.S.M. Shayuti, and F. Marpani
Universiti Teknologi MARA, Shah Alam, Selangor, Malaysia

M.H.A. Aziz
Universiti Teknologi Malaysia, Skudai, Malaysia

CONTENTS

10.1 Introduction 309
10.2 Fabrication of Nanofiber Membranes 311
10.3 Recent Applications of Nanofiber Membranes in Treating Various Wastewater 314
10.3.1 Heavy Metal Adsorption 314
10.3.2 Oily Wastewater 317
10.3.3 Dye Degradation 320
10.3.4 Pharmaceutical Wastewater 323
10.4 Conclusion 325
References 326

10.1 INTRODUCTION

In recent years, pollutions are seriously threatening the ecological environment, economic development, as well as human health. In fact, many health problems such as infant mortality, cardiovascular disorder, and allergy are caused by the contamination of water, air, and soil. According to a recent study published in 2017 about health and pollution, nine million premature deaths in the entire world were highly related to environmental pollutions in 2015, and remarkably about 1.8 million deaths were associated with water pollution issues (Landrigan et al., 2017). Up to now, oil spills and industrial wastewater discharges have caused severe effects on water resources. Therefore, researchers urgently need to invent a suitable and effective treatment to solve these significant issues. Over the last decades, conventional strategies like adsorption, biological treatment, filtration, catalysis, and electro-coalescence have

DOI: 10.1201/9781003118749-10

been extensively used for water purification treatment. However, secondary pollution, unrepeatable usage, and high energy consumption still seem to be unavoidable.

Through to the 21st century, massive development in nanoscience and nanotechnology has demonstrated a rapid development, not only in the characterization and formation of nanomaterials but also in their significant functions in countless utilizations. The immense interest in one-dimensional materials has revealed that nanofibers can meet the needs of wastewater purification technology due to their amazing highly porous properties, with excellent pore interconnectivity and enormous specific surface to volume ratio due to small fiber diameter (Su et al., 2017; Alias et al., 2019; Jamil et al., 2019). Figure 10.1 represents the number of published articles on the recent evolution of electrospun nanofibers in wastewater treatment applications.

Nanofiber structure possesses unique features in that it is flexible and can be easily shaped, thus making it versatile to be attached with a rigid support structure if needed. They are manufactured using various developed methods from low to high-volume production such as template (Choi et al., 2005), drawing (Bajakova et al., 2011), self-assembly (Liao et al., 2016), flash freezing (Samitsu et al., 2013), melt-blowing (Han et al., 2016), phase separation (Huang et al., 2014a), melt fibrillation (Ramakrishna et al. 2006), gas jet (Benavides et al., 2012), island-in-sea (Kamiyama et al., 2012), as well as electrospinning (Mirjalili and Zohoori, 2016; Nor et al., 2016). Among these, electrospinning has been regarded as the most reproducible, simple, versatile, scalable and continuous technique to tailor polymeric nanofibers (Kumbar et al., 2008).

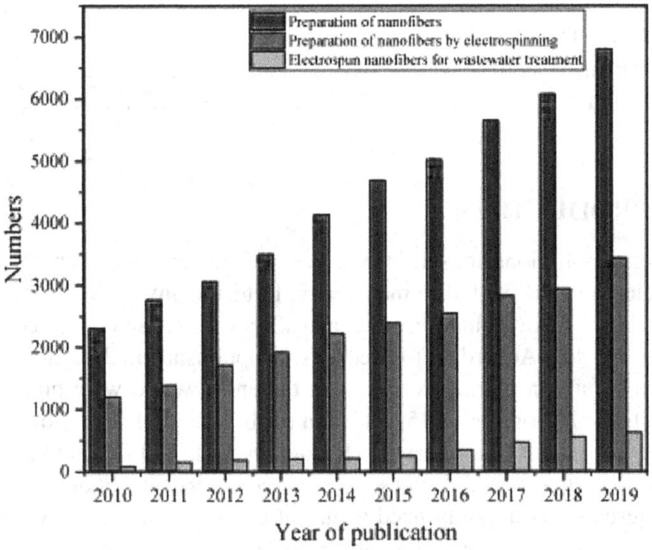

FIGURE 10.1 Illustration of trends for the annual number of publications on nanofibers, electrospun nanofibers, electrospun nanofibers for wastewater treatment over the last 10 years, according to the Web of Science. (adapted from Sofi et al., 2020.)

10.2 FABRICATION OF NANOFIBER MEMBRANES

Electrostatic spinning, or commercially known as "electrospinning," has been known as one of the most versatile and outstanding techniques to develop functional and nonwoven nanofibers with a diameter ranging from nanometers to micrometers (Yang et al., 2015). It was previously stated that the electrospinning technique could manufacture fibers in the diameter range of 3 nm to 6 μm and several meters in length (Kumbar et al., 2008). Currently, the technology for electrospinning is getting well-known in material sciences. The patents on electrospinning working principle were published to Formhal since 1934. It started with an experimental method to produce polymer filaments. However, electrospinning was first patented in the United States, and it was forgotten until the 1990s (Ramakrishna et al., 2006). Interestingly, the spinning of nanofibers is not only based on the prerogative of smart researchers, but it also influenced by the technique found from nature which frequently used by animals such as bees, caterpillars, and spiders, which develop a good territorial environment to grow, reproduce, as well as to catch prey (Stevens, 2014).

Technically, electrospinning is a process that uses three major integral parts: a collector, a high voltage electric current and a syringe pump with a syringe that applies a high electrical field to tailor a polymer solution into an ultrafine fiber. The schematic of the basic electrospinning method is illustrated in Figure 10.2.

Based on the schematic diagram, a metallic needle connected to a syringe is used as a spinneret. When a certain flowrate is applied to the syringe pump, the polymer solution in the syringe can be fed via the spinneret. Subsequently, a uniform electric field distribution is created, and the polymer liquid surface at the end of the needle tip is continuously charged as a high voltage is applied. As the applied voltage increases,

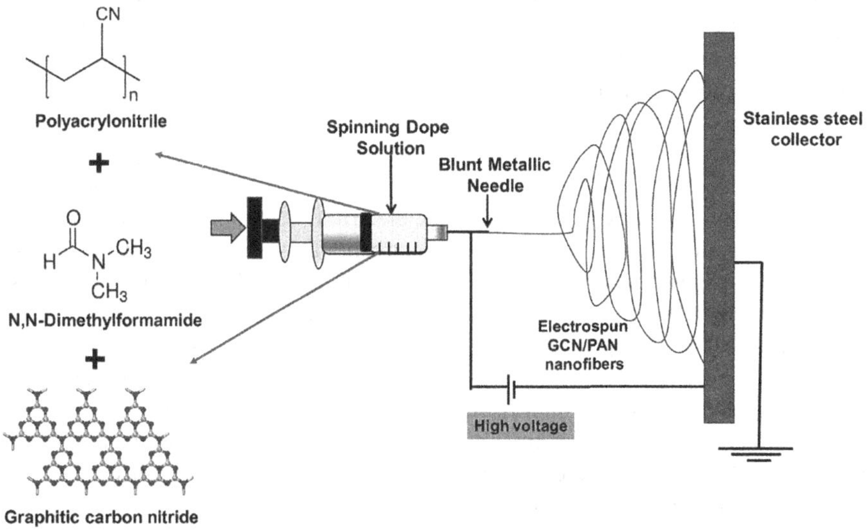

FIGURE 10.2 Schematic illustration of basic electrospinning setup. (adapted from Alias et al., 2020b.)

the mutual electrostatic repulsion between polymer chains increases too. This leads to a counterbalance in surface tension, which can distort the polymer solution droplet shape from a hemisphere to a Taylor cone shape. Once the electric field reaches a critical voltage, the balance between surface tension and electrostatic force will be broken and the jet flow of charged polymer at the tip of the needle will be continuously ejected. Therefore, the solvent in the polymer solution will evaporate and jet diameter will be reduced from hundreds of nanometer ranges. Lastly, nanofibers will be deposited and solidified on the ground collector.

Generally, electrospinning is a straightforward process, but serious attention needs to be given to several parameters that may affect the nanofiber quality, such as polymer spinning solution properties, elasticity, molecular weight, conductivity, viscosity, as well as surface tension (Huang et al., 2014b). In order to conserve the polymer entanglement, a high concentration of polymer is needed during the electrospinning process (Tarus et al., 2016; Xue et al., 2019), but in a controlled manner because if the concentration of polymer solution used is too high, it will lead to polymer deformation (Tijing et al., 2019). Besides that, process parameters such as applied voltage, the distance between needle tip and collector, feed flow rate, and needle inner diameter are important for smooth and good fiber morphology. Moreover, environmental factors like relative humidity, temperature, and air velocity in the electrospinning chamber are important to consider (Mirjalili and Zohoori, 2016). It must be noted that changing the electrospinning process parameters by even a minute can affect the fiber morphology (Bera, 2016). For instance, the incorrect placement of needle and collector during the electrospinning process may result in beads formation on the nanofiber membrane surface, which can severely decrease the nanofiber porosity and the morphology of some fibers may change shape from circular to flat shape.

Recent studies on the electrospinning process have contributed to a new approach in producing mass production of nanofibers (Wang et al., 2019). Hence, continuous studies are being explored in order to find an ideal electrospinning system for nanofibers modification such as process orientation, controlled alignment, and tethering nanofiber with specific functionalities. As an example, Varesano et al. (2010) developed a multiple-jet nozzle suitable for large-scale nanofiber production. This can be achieved by increasing the productivity as well as recovering area with appropriate coordination of jet nozzle to ensure that the electric field current distribution is uniform. Besides multiple-jet nozzle, other electrospinning techniques have been pioneered to achieve different desired nanofiber morphologies like needleless electrospinning (Partheniadis et al., 2020), core–shell electrospinning (Fibrous and Containing, 2019), rotating disk (Alias et al., 2020a), auxiliary electric fields (Carnell et al., 2009) and rotating tube (Alias, et al., 2019). In recent years, several researchers have made an abundance of improvements on the fabrication of core-shell and hollow nanofibers by adopting coaxial electrospinning technique (Pakravan et al., 2012; Fang et al., 2019). A vast development of coaxial electrospinning has been explored and regarded as the potential application in nanocapsule, small encapsulating devices and nanochannel (Li et al., 2010).

Recently, several studies have reported on the applications of nanofibers as a coating layer for membrane support to improve membrane performance. The latest development by Alias, et al. (2019) reported on the newly designed g-C_3N_4/PAN nanofiber coated Al_2O_3 hollow fiber membranes to treat oilfield produced water (OPW). They reported that the nanofiber coating improved the membrane performance with the highest permeate flux of 640 $L{\cdot}m^{-2}{\cdot}h^{-1}$ and oil separation rejection of 99% within 180 min under crossflow filtration of OPW at 2 bar. On the other study by Su and co-workers (2018), they previously reported on nanofiber-coated hollow fiber membrane (N-HFM) via electrospinning with static/flat collectors. In addition, they used a commercial supporting woven tube as a supporting tube to be coated with nanofiber. The resultant membrane exhibited a high flux of 13.2 $L.m^{-2}$ h^{-1} and stable salt rejection of 99.9% within 5 hours of the test. Thus, the fabricated membrane is very recommended for membrane distillation (MD) application.

Dobosz et al. (2017) previously coated cellulose and PS nanofibers on flat sheet PES membrane using electrospinning and no adhesive was used during the coating process. The results obtained demonstrated that 90% of fouling resistance was achieved using polysulfone nanofiber membranes and cellulose nanofiber membranes. The same method was used in a different work by Efome et al. (2016), where they electrospun PVDF nanofibers on PVDF flat sheet membranes. They reported that the flux of nanofiber-coated membrane was tremendously improved and greater than uncoated membrane at 99.98%. On the other hand, Nor et al. (2016) fabricated a nanofiber-coated membrane using TiO_2 nanofiber hot-pressed on flat sheet PVDF membranes (PVDF/e-TiO_2). The results showed that the degradation efficiency of bisphenol A (BPA) increased rapidly from 63% to 85% with the help of PVDF/e-TiO_2 nanofiber membrane under UV light irradiation. Other research works on nanofiber-coating methods on membrane supports were tabulated and described in Table 10.1.

TABLE 10.1
Research Works on Nanofiber Coating on Membrane Supports

Nanofiber Membranes	Coating Method	References
GCN/PAN nanofibers coating Al_2O_3 hollow fiber membrane	The fabricated Al_2O_3 hollow fiber membrane was placed on a rotating wood holder and photocatalytic GCN/PAN nanofibers were directly electrospun on the hollow fiber membranes.	Alias et al. (2019)
PVDF-HFP nanofibers-covered hollow fiber membrane (N-HFM)	Direct electrospinning on commercial supporting woven tube.	Su et al. (2018)
	The supporting tube covered on the negative electrode was used as the inner layer and collector for N-HFM composite membrane.	
Cellulose nanofiber on PES membrane	A layering method was used where a cellulose nanofiber layer was placed on the top surface of PES membrane.	Dobosz et al. (2017)
PS nanofiber on PES membrane	No additional adherent was used for coating.	

(Continued)

TABLE 10.1 (Continued)
Research Works on Nanofiber Coating on Membrane Supports

Nanofiber Membranes	Coating Method	References
	Only the internal O-rings in the crossflow and dead-end testing cells adhered to the nanofiber layers on the membrane.	
PVDF nanofibers on PVDF membrane	Direct electrospinning of PVDF nanofibers on PVDF membrane support.	Efome et al. (2016)
TiO_2 nanofibers coated on PVDF membrane	Hot-press technique was applied and observed at different operating temperatures: 100°C, 160°C, and 180°C for 30 minutes.	Nor et al. (2016)
PVDF nanofibers coated on polypropylene (PP) membrane	Direct electrospinning of PVDF nanofibers on PP membrane surface using nozzle-less electrospinning. This technique can create stronger adhesion to the membrane substrate and generate nanofibers with smaller diameter compared to a single nozzle electrospinning.	Alcoutlabi et al. (2013)
Polyvinylidene fluoride (PVDF) nanofiber-coated fabric substrates	Direct electrospinning of PVDF nanofibers on fabric substrate held on rotating collector.	Bui et al. (2011)
	The nanofiber membrane that has been coated were dried in air at room temperature to remove the residual solvents from the composite membrane.	

10.3 RECENT APPLICATIONS OF NANOFIBER MEMBRANES IN TREATING VARIOUS WASTEWATER

Research on nanofiber membranes in wastewater treatment has increased enormously in recent years due to the excellent and unique properties of nanofiber, such as large specific surface area, diverse morphology, and adjustable wettability. Practically, the nanofiber membrane was applied as a self-standing membrane or as a surface modification layer/template for additives or functional groups. As seen in Figure 10.3, the nanofiber membrane has been extensively used to remove pollutants in contaminated wastewater like heavy metals, oils, dyes, and pharmaceutical contaminants. Therefore, continuous research has been carried out until now purposely to improve the quality of nanofiber membranes. Many researchers optimize the nanofiber performance by varying the process parameters during nanofiber synthesis, surface modifications, and incorporating nanomaterials in the membrane matrix. This section accentuates the recent developments of nanofiber membrane in wastewater treatment, particularly in wastewater containing heavy metal, oil, dyes, and pharmaceutical pollutant.

10.3.1 Heavy Metal Adsorption

Industrial activities such as metallurgical, mining, fossil fuel combustion, and process manufacturing are the major sources of heavy metal discharged in wastewater (Ihsanullah et al., 2016). The discharge of wastewater containing hazardous heavy metals can risk human health and the environment (Azimi et al., 2017). This is mainly

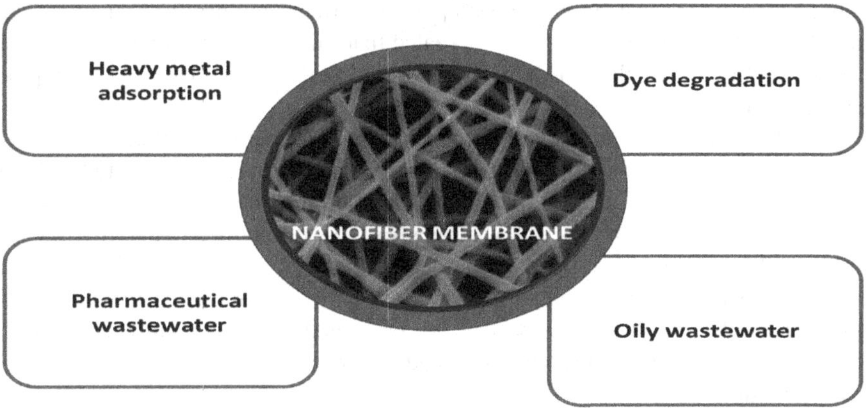

FIGURE 10.3 Widespread use of nanofiber membrane in wastewater treatment.

due to the toxicological characteristics of heavy metals which make them difficult to degrade and which tend to accumulate in the environment (Hashim et al., 2011; Foong et al., 2020). Various effective strategies to remove the toxic heavy metals such as arsenic (As), chromium (Cr), copper (Cu), cobalt (Co), cadmium (Cd), plumbum (Pb), and mercury (Hg) have been discovered by researchers. The treatment technologies include ion exchange (Pepe et al., 2013), membrane separation or adsorption (Khulbe and Matsuura, 2018), chemical precipitation (Zhang and Duan, 2020), electrochemical treatment (Ya et al., 2018), and coagulation flocculation (Carolin et al., 2017). Among them, the membrane adsorption technique is the most promising alternative to remove persistent heavy metals (Huang and Cheng, 2020).

The advantages of the membrane as an adsorbent are: i) low operating pressure; ii) high percentage of pollutant removal; iii) simple process; iv) excellent permeability flux and less land required (Huang and Cheng, 2020). Recently, electrospun nanofiber membrane with abundant active functional sites has superior potential for creating adsorption affinity of the membrane (Wang and Hsiao, 2016). Electrospun nanofiber membrane is regarded as a promising adsorbent due to its unique characteristics toward heavy metal adsorption such as high porosity, controllable pore size, good interconnecting pores, and high potential to incorporate with nanoscale-size materials or functionality (Wang and Hsiao, 2016; Khulbe and Matsuura, 2018; Lee and Kim, 2018; Tijing et al. 2019).

The membrane adsorption process mechanism can be either physical or chemical adsorption. Generally, the membrane adsorption mechanism of heavy metal ions, dyes, or other inorganic/organic pollutants using electrospun nanofiber membrane is similar to the typical adsorption method (Abdullah et al., 2019). The adsorption process is highly dependent on a solid substance's capability to attract the surface of molecules. Physical adsorption originates from the electrostatic force between heavy metal ions and the adsorbent. In contrast, chemical adsorption relies on the construction of a chemical bond between heavy metal ions and functional groups (Cui et al., 2020). Heavy metal ions with positive charge will be absorbed via ionic interactions between the negatively charged nanofiber membrane. On the other hand, coordination bonds can be well developed by surface modification on nanofiber membranes (Chen et al., 2020). This revealed that the active functional sites on the membrane surface are the most crucial factor during the membrane adsorption process (Min et al., 2012).

A wide variety of functional polymers can be used to fabricate different morphology of electrospun nanofiber membrane to adsorb heavy metal ions. Li et al. (2017) studied the potential of the spiral wound module of electrospun chitosan nanofiber membrane for chromium ions removal by adsorptive filtration. At only 10% breakthrough, 2 g/m^2 nanofiber membrane could achieve up to 20.5 mg/g adsorption capacity. It was also reported that the loading capacity of the module greatly depends on the nanofiber density. As expected, the fabricated chitosan membrane demonstrated higher removal of chromium ions and permeate flux compared to the nanofiltration membrane.

Besides chitosan, poly (acrylic acid) (PAA) is another type of functional polymer that provides excellent affinity toward various heavy metal ions. This is due to the molecular chain of PAA contains a high number of carboxyl groups (Sezgin and Balkaya, 2016). Nevertheless, PAA has low stability in aqueous liquid (Zhang et al., 2019). Xiao et al. (2010) fabricated PAA/PVA electrospun nanofiber composite membrane to improve the stability of PAA for efficient copper metal removal. The nanofiber composite membrane has successfully removed 91% of Cu (II) ions in aqueous liquid within 3 hours. Recently, Zhang et al. (2019) also developed PAA/PVA nanofiber membrane to eliminate Cd (II) and Pb (II) ions. Based on the result, the PAA/PVA nanofiber membrane composite obtained 102 mg/g adsorption capacity of Cd (II) ions and 159 mg/g Pb (II) ions. Moreover, it was also reported that the resultant membrane could be regenerated in acidic solution for up to five cycles.

Recent progress in enhancing the maximum adsorption capacity for heavy metal ions is by modifying the nanofiber surface with various functional groups. The purpose of surface modification on electrospun nanofiber membrane is to increase the adsorption capability of the composite membrane. In addition, surface modification also can ensure that the immobilized functional groups are placed on the nanofiber surface (Wang and Hsiao, 2016). The most common functional groups that have been introduced are amino groups, carboxyl groups, and sulfhydryl/thiols groups. Besides, carbonyl and phosphate groups were also reported as excellent functional groups (Pereao et al., 2019; Choi et al., 2020). All these functional groups can provide powerful interactions with heavy metal ions for effective adsorptive separation (Khulbe and Matsuura, 2018).

For instance, Yang et al. (2020) fabricated hydrophobic PAN and hydrophilic polydopamine nanofibrous membrane for chromium (VI) ions removal in aqueous liquid. Membrane synthesis has been successfully done via coelectrospinning of PAN/dopamine and self-polymerization of the polydopamine onto the PAN membrane surface. A homogeneous coating of polydopamine, rich with amine functional groups, has improved the hydrophilicity and water stability of PAN/dopamine nanofiber membrane. Therefore, the resultant nanofibrous composite membrane exhibited higher adsorption performance of Cr (IV) ions (61.65 mg/g) compared with other typical adsorbents (Yang et al., 2020).

TABLE 10.2
Electrospun Nanofiber Membrane for Removal of Heavy Metal Ions in Wastewater

Nanofiber Membranes	Heavy Metals	Removal Performance	References
Chitosan	Cr (IV)	20.5 mg/g	Li et al. (2017)
Chitosan/PVA	Pb (II)	226.12 mg/g	Karim et al. (2019)
	Cd (II)	148.79 mg/g	
PAA/PVA	Cu (II)	91% removal	Xiao et al. (2010)
	Pb (II)	159 mg/g	Zhang et al. (2019)
	Cd (II)	102 mg/g	
PAN/Dopamine / Polydopamine	Cr (IV)	61.65 mg/g	Yang et al. (2020)
MWCNT-PEI/PAN	Cu (II)	112.5 mg/g	Deng et al. (2019)
	Pb(II)	232.7 mg/g	

Besides surface modification of nanofiber membrane, the application of nanotechnology into membrane systems for removal of heavy metals offers a great advantage to improve the composite of nanofiber membrane. The benefits of nanomaterials incorporation in membrane matrix are: i) improve membrane permeability; ii) increase mechanical and thermal stability; iii) reduce fouling issues; and iv) self-cleaning (Qu et al., 2013). Deng et al. (2019) had proved that multiwalled carbon nanotubes (MWCNTs) nanomaterial can enhance the adsorption capacities of PEI/PAN nanofiber. It was reported that the nanofibrous composite demonstrated higher mechanical strength and hydrophilicity compared to pristine PAN nanofiber. The good structure and characteristics of the MWCNT-PEI/PAN composite membrane resulted in high filtration, adsorption, and permeation performance. The reported maximum adsorption capacities for Pb(II) and Cu(II) ions were 232.7 and 112.5 mg/g, respectively. Therefore, the addition of nanomaterials/nanoparticles in electrospun nanofiber membrane is regarded as new promising development in removing heavy metals in wastewater. Table 10.2 shows the removal performance of electrospun nanofiber membrane for heavy metal in aqueous solution.

In summary, developing a functionalized nanofiber membrane as an affinity membrane is a great alternative for heavy metal ions removal in wastewater. A recent approach that can enhance the maximum adsorption capacity for toxic heavy metals ions in water and wastewater is by modifying the nanofiber surface and incorporating nanomaterials into the composite membrane. Nevertheless, improvement and preparation of functionalized nanofiber membrane as an adsorbent still need to be explored, especially on their regeneration capabilities.

10.3.2 Oily Wastewater

There is an urgent need for adequate treatment of oily wastewater, which is mainly discharged from the textile, food, petrochemical, and metallurgical industries. This is because oil-contaminated water can continuously affect groundwater, drinking water resources, and disturb the ecosystem (Junaidi et al., 2020). However, conventional treatments such as skimming, flotation gravity separation, and biochemical treatment could not treat oil-contaminated water efficiently, especially for small size (less than 20 μm) oil droplets (Su et al., 2021). Thus, many researchers are searching for effective, green, and economic treatment of oily wastewater, and recently, nanofiber membranes have come into the picture as the competitive option for the treatment (Li et al., 2019; Cui et al., 2020; Su et al., 2021; Zhang et al., 2021).

Ideally, high-performance nanofiber membranes for oily and water separation must have a high selective affinity towards oil and water as it acts as a driving force (Wang et al., 2016). Besides that, several factors influence the performance of oily wastewater treatment during the separation process, including electrostatic interactions, surface charges, and wettability criteria. In addition, size exclusion also affects the treatment efficiency, where the passage of pressurized water via membrane is allowed, while the oil droplets larger than the membrane pores are prohibited (Jafari et al., 2020). Therefore, wettability selection is important to ensure that the oil droplets always dry and permeate the membrane due to their hydrophobicity (Huang et al., 2018). As a result, there is a rapid expansion of studies on the development of nanofiber membranes with superwettability properties for oily wastewater purification. The properties of the nanofiber membrane for effective water and oil separation can be superhydrophilic/superhydrophobic and superoleophilic/superoleophobic.

Many studies reported that a functional membrane with a superhydrophobic surface or superhydrophobic lipophilic membrane showed a very good performance on oil/water separation. For instance, Ma et al. (2019a) have developed a high-performance superhydrophobic lipophilic membrane based on a fluorine-free system. In particular, researchers combine the fabricated electrospun polyimide (PI) nanofibers with in-situ polymerized polybenzoxazine (PBZ) that contains silica nanoparticles (SNP) via dip coating and electrospinning technique. The composite nanofibrous membrane displayed excellent chemical and thermal stability with high separation efficiency even after 20 separation cycles. The water and oil contact angle for the resultant membrane were above 154° and approached 0°, respectively, which represented their superhydrophobic/superoleophilic properties. The reported separation efficiency was approximately more than 99.0% with 4798 $L.m^{-2}.h^{-1}$ of flux.

However, superhydrophobic lipophilic membrane has several limitations, e.g., its pores can easily be polluted/blocked by oil since lipophilic properties tend to combine with fats or lipids, which may affect the separation and reusability performance of nanofiber membrane. Other than that, water may deposit on the nanofiber membrane surface since water is less dense compared to the oily substance. The deposited water may block or create a barrier between oily substances and the membrane surface (Wang et al., 2020b). Thus, additional pressure is required to overcome this problem but results in high energy consumption. For a better alternative, the oleophobic hydrophilic nanofiber membrane has gained wide attention among current researchers. The membrane surface properties are hydrophilic in air and oleophobic underwater. This type of membrane offers great advantages since water can pass through while repelling oily substances. Therefore, membrane fouling can be reduced and excellent purification of oily wastewater is possible (Chen et al., 2020).

A study conducted by Lv et al. (2020) has proven that a superhydrophilic, underwater superoleophobic nanofiber membrane has a satisfactory repulsion effect and antifouling ability toward anionic surfactants for oily wastewater separation. Briefly, acrylic acid was added to a PAN nanofiber surface to improve the wettability. Hence, the developed membrane exhibited good anti-adhesion properties. In addition, the membrane displayed excellent stability even after 10 cycles with permeate flux of 41,020 ± 805 $Lm^{-2}h^{-1}bar^{-1}$. The separation efficiency for observed oil/water emulsion can be achieved up to 99.9%.

Another work by Zhuang et al. (2020) successfully developed a good hydrophilic-oleophobic surface from bio-cellulose nanofiber membrane. It was reported that low sodium hydroxide concentration in alkali treatment could increase hydroxyl groups on the cellulose molecular chains as well as the hydrophilicity properties. Thus, the alkali-treated bio-cellulose demonstrated more than 99% oil and water separation efficiency with excellent anti-oil fouling. Moreover, even after 10 hours, the nanofiber membrane was still in good shape and recovered approximately 98% of permeate flux. This is mainly due to the low adhesion properties in the oil.

However, the underwater oleophobic surface of the nanofiber membrane is not applicable in MD since hydrophobicity is necessary in MD process conditions to restrict liquid permeation through the membrane pores. One effective method to improve the wettability of MD is by depositing a thin hydrophilic layer on a hydrophobic substrate, thereby resulting in a composite membrane with both hydrophilic and hydrophobic properties. In order to ensure the successful incorporated hydrophilic layer on the substrate, the researcher should ensure that there is no loss on the hydrophobicity of the substrate surface and no changes in mass transfer (Tang et al., 2021).

Hou et al. (2018) have successfully fabricated a novel hydrophilic polyvinyl alcohol/silica particle (PVA-Si) nanofiber on a hydrophobic polytetrafluoroethylene (PTFE) membrane via electrospinning. Then, the nanofiber composite membrane was crosslinked with glutaraldehyde (GA) in order to enhance the in-air hydrophilicity, underwater superoleophobicity, and restrict the PVA-Si hydrogel formation. The developed nanofibrous composite membrane exhibited a stable flux and proven to create anti-oil fouling, especially for MD.

A dual-layered nanofiber membrane composite for MD of oily wastewater was successfully developed by Makanjuola et al. (2019). The bottom layer of the stacked membrane was made from hydrophobic poly (vinylidene fluoride-co-hexafluoropropylene, PVDF-HFP) nanofiber. Meanwhile, the top layer was fabricated from PVDF-HFP/regenerated cellulose in order to enhance the hydrophilicity properties. The fabricated membrane successfully maintained permeate flux at 12.8 $kg.m^2/hr$ with 100% salt rejection from 1000 ppm saline oil feed.

Moreover, very recently, a novel multifunctional deacetylation-cellulose acetate (d-CA) nanofiber membrane with advanced wettability properties for oily wastewater purification was fabricated by W. Wang et al. (2020b). Many ester groups and hydroxyl groups in the d-CA molecule posed various properties: superamphiphilic in air, superhydrophilic under oil, and oleophobic underwater. The composite membrane can not only separate oil and water mixture in harsh environmental conditions, but they also can separate emulsified oil. The reported average separation efficiency

TABLE 10.3
Electrospun Nanofiber Membrane for Oil/Water Separation

Nanofiber Membranes	Type of Membrane	Separation Performance	References
PAN/ carboxylate carbonaceous microspheres	Superhydrophobicity/ superoleophilicity	More than 99% oil and water separation efficiency with 53265 ± 3823 L m^{-2} $h^{-1}bar^{-1}$ maximum membrane permeability.	Lv et al. (2020)
Polyimide/polymerized polybenzoxazine/ silica nanoparticles	Superhydrophobicity / superoleophilicity	More than 99% oil and water separation efficiency with 4798 $L{\cdot}m^{-2}h^{-1}$ flux.	Ma et al. (2019a)
Biocellulose (Gluconacetobacter xylinus)	Superhydrophilicity / superoleophobic	More than 99% oil and water separation efficiency with 94% permeate flux recovered.	Zhuang et al. (2020)
Top layer: PVDF-HEP@ regenerated cellulose Bottom layer: PVDF-HFP	Hydrophilic layer on the substrate on hydrophobic substrate	PH stacked with PH-5RC (67 μm thick) demonstrated the best flux of 12.8 ± 1.1	Makanjuola et al. (2019)
Top layer: Polyvinyl alcohol@silica particle-glutaraldehyde Bottom layer: polytetrafluoroethylene (PTFE)	Hydrophilic layer on the substrate on hydrophobic substrate	Demonstrated 17.53 kg/ m^2h permeate flux.	Hou et al. (2018)
Deacetylation of cellulose acetate	Superamphiphilic in air, superhydrophilic in oil, and oleophobic in water	90% oil and water separation average efficiency.	Wang et al. (2020b)

was above 90%. Table 10.3 shows the performance of electrospun nanofiber membrane for oil and water separation.

The studies have proved that nanofiber membrane has superior performance for oily wastewater treatment. Most of the recent studies reported more than 99% oil and water separation efficiency. Nanofiber membranes like oleophobic, hydrophilic hydrophobic, and lipophilic membranes can be well developed in different scales of application. Some studies have incorporated a hydrophilic layer on hydrophobic substrate in order to improve the vapor permeation and reduce fouling issues in MD. To date, dual functional membranes were discovered and have become a great alternative in the oil/water separation field. Although nanofiber can separate oil and water very effectively, it is necessary to improve the membrane stability and antifouling property.

10.3.3 Dye Degradation

Dye effluents have been greatly found in wastewater, especially from textile industries. The low biodegradability of dye poses great risks to the ecosystem and human health even at low concentrations (Cai et al., 2019). The dye in aqueous liquid is present in the form of ions and it can be classified into two groups: ionic and nonionic dyes. Chemical adsorption has been identified as an efficient method for nonionic dyes adsorption (Cui et al., 2020). Similar to the membrane adsorption mechanism of heavy metal ions, dye adsorption or removal mainly relies on the available functional groups in the nanofiber membrane to create strong chemical bonds or surface complexion with the targeted dyes.

Therefore, many nanofiber membranes have been functionalized with various functional groups to enhance the nanofiber performance for removing dye in wastewater effluents. For instance, Ma et al. (2019b) developed gelatin and calcium alginate (GA) electrospun nanofiber membrane to degrade methylene blue. It was reported that the composite membrane has high reusability and adsorption capacity since gelatin is rich with amino functional groups; thereby, it can create a high electrostatic force to degrade MB. Moreover, the fabricated nanofiber membrane has high porosity, mechanical strength, and large surface area due to the excellent properties of nanofiber and strong interaction between alginate and gelatin chains. The maximum adsorption capacity of GA nanofiber membrane was 1937 mg/g, which is much higher compared to other typical adsorbents such as activated carbon/chitosan (250 mg/g) (Karaer and Kaya, 2016) and carbon (435 mg/g) (Legrouri et al., 2005).

Moreover, Huong et al. (2020) had prepared a modified polyacrylonitrile bovine serum albumin (PAN-BSA) nanofiber membrane via mild hydrolysis and electrospinning to remove toluidine blue O (TDO) dye. The modified PAN with BSA formed P-OOH-BSA protein nanofiber membrane. Based on this research, the maximum adsorption capacity of the nanofiber composite membrane was constant at 434.78 mg/g. In addition, the P-OOH-BSA composite nanofiber membrane demonstrated a good desorption performance in 1 M of sodium hydroxide and 50% glycerol. It was reported that the degradation efficiency of TDO could be recovered up to 97% even after five consecutive cycles.

On the other hand, immense interest has been focused on integrating photocatalyst into the electrospun membranes. Previously, Xu et al. (2015) have fabricated a visible-light-responsive electrospun nanofiber by dispersing PAN with photocatalyst graphitic carbon nitride (g-C_3N_4). The results showed that RhB degradation under sunlight irradiation was more than 90% within 40 min at pH 3, with more than 41% of RhB was initially adsorbed by nanofiber in the first 20 min. The report also concluded that the g-C_3N_4/PAN nanofibers could be reused without decreasing the photocatalytic degradation even after several times used.

Similarly, Alias et al. (2020b) performed a study on photocatalytic degradation of MB using g-C_3N_4/PAN nanofibers under UV light irradiation. The photodegradation performance of the g-C_3N_4/PAN nanofiber was observed at 97.3% oxidation of MB. The findings highlighted that the photodegradation of MB in aqueous solution was due to the synergetic effects between adsorption of MB by PAN nanofibers and photocatalytic degradation of MB by g-C_3N_4 photocatalyst, thus paved new opportunities as a great potential nanofiber membrane for dye removal in aqueous solution.

Recently, Zhang et al. (2021) have successfully immobilized TiO_2 and graphene oxide (GO) into electrospun PAN/β-cyclodextrin (β-CD) nanofibrous membrane. It was found that PAN/β-CD/TiO_2/GO composite membrane with an 8:2 mass ratio of

TiO_2-to-GO exhibited high degradation percentages for MO and MB dyes which were approximately at 90.92% and 93.52%, respectively. Apart from that, this nanofibrous membrane displayed excellent antibacterial properties. In addition, the catalytic degradation efficiency of both types of dyes can remain up to 80% after three cycles. Thus, it has a great potential for the removal of dyes from industrial wastewater.

In addition, the application of nanofiber membrane toward dye degradation under visible light in MD also has been discovered by Guo et al. (2019). Based on the research outcome, the electrospun BiOBr/Ag membrane as photocatalyst achieved a high dye rejection and good flux recovery after the fifth cycle in the MD process.

Additional nanomaterials can be embedded into nanofiber membranes to enhance membrane adsorption capacity performance. As an example, Phan et al. (2020) fabricated PAN nanofibers with zinc oxide nanoparticles and hinokitiol (PA-ZnO-HT) for dye removal. The nanofibrous membrane exhibited high adsorption capacity for reactive red 195 (RR) and reactive blue 19 (RB) for 26 hours without any leaching issue. The reported adsorption capacities for RR and RB were 245.75 and 267.37 mg/g, respectively.

Another study reported on the performance of nanoparticles embedded in membrane matrix for Acid Red 1 catalytic oxidation. According to the author Cai et al. (2019), the large specific surface area of nanofiber membrane offers an effective dispersion of nanoparticles. Thus, cobalt was introduced into the nanofiber matrix via carbothermic reduction of polyacrylonitrile (PAN) and polyvinylpyrrolidone (PVP). In particular, the nanofibrous membrane of PAN/PVP/Co achieved 100 % catalytic oxidation of AR1 within only 6 min. The resultant membrane's reusability performance also was considered high, since AR1 degradation can be retained at 98.1% even after the fourth cycle. This research proved that incorporating photocatalyst embedded in nanofiber membrane can enhance the degradation of pollutants and alleviate secondary pollution. Table 10.4 shows the performance of electrospun nanofiber membrane for dye degradation

TABLE 10.4
Electrospun Nanofiber Membrane for Removal of Dye in Wastewater

Nanofiber Membranes	Type of Dye	Removal Performance	References
Gelatin/calcium alginate	Methylene blue	1937 mg/g	Ma et al. (2019b)
Polyacrylonitrile/ Bovine serum albumin	Toluidine blue O	434.78 mg/g	Huong et al. (2020)
g-C_3N_4/PAN	Methylene blue	Achieved 90% degradation of MB within 40 min	Xu et al. (2015)
g-C_3N_4/PAN	Methylene blue	Achieved 97.3% degradation of MB	Alias et al. (2020a)
PAN/β-cyclodextrin (β-CD)/TiO_2 /graphene oxide (GO)	Methylene blue, methyl oxide	Achieved 93.52 ± 1.83% degradation of MB and 90.92 ± 1.52% degradation of MO	Zhang et al. (2021)

(Continued)

TABLE 10.4 (Continued)
Electrospun Nanofiber Membrane for Removal of Dye in Wastewater

BiOBr/Ag	Mixed dyes (acid red 18, acid yellow 36, methylene blue, and crystal violet)	Achieved 99.9 % dye rejection with 38.4 LMH water flux in membrane distillation	Guo et al. (2019)
PA-ZnO-HT	Reactive blue 19 and reactive red 195	RR 19: 267.37 mg/g RR 195: 245.76 mg/g	Phan et al. (2020)
PAN/PVP/Co	Acid Red 1 (AR1)	Achieved 100 % catalytic oxidation of AR1 within 6 min	Cai et al. (2019)

In recent years, electrospun nanofiber membrane has shown good performance on dye oxidation due to their high porosity and large specific surface area to volume. The main factors affecting the dye degradation performance are the functional groups and material embedded in the nanofiber membrane. The reported percentage reusability of nanofiber membrane was satisfactory, especially when photocatalyst or nanoparticle is embedded in nanofiber. This is mainly due to the high ability of nanofiber to combine with various types of photocatalyst/nanoparticle. However, it should be pointed out that most polymers' physical and chemical structures can be affected under UV light irradiation. Hence, there is an urgent need to develop a suitable adsorbent or photocatalyst-based membrane to efficiently remove dye in wastewater.

10.3.4 Pharmaceutical Wastewater

Wastewater containing pharmaceutical waste, like ibuprofen, sulfonamide, antibiotics, sulfamethoxazole needs to be seriously considered because they cannot be degraded by common biological process treatment (Kumar et al., 2019). Therefore, many researchers are questing for a new efficient treatment. Over the last decades, common technology to treat pharmaceutical wastewater are ozone, liquid chlorine, chlorine dioxide, sodium hypochlorite disinfections, adsorption, flocculation, ion exchange, photodegradation, and electrolyte decomposition (Khalil et al., 2019; Wang et al., 2020a). Among these, the adsorption process is the simplest and effective way to degrade pollutants.

Recently, nanofiber for adsorptive removal of pollutants has drawn interest from the pharmaceutical wastewater field due to its excellent properties. An active functional site is an important factor for an efficient adsorption process. Hence, Kebede, Dube, and Nindi (2019) developed a PVA/moringa seed protein nanofiber composite membrane to remove several drugs (fenoprofen, ibuprofen, ketoprofen, diclofenac) and carbamazepine. The blending of moringa seed protein and PVA has successfully created numerous functional groups, such as hydroxyl, amide, and amine, which can remove the pharmaceutical waste via hydrogen bonding and peptide bond linkage. It was reported that the percentage removal of each pharmaceutical waste varies under different operational conditions (pH, initial pollutant concentration, and adsorbent dosage). Interestingly, the removal percentage range for standard mixture solution under optimum condition was from 84.5% to 97.4%, whereas 77.8% to 96.1%

removal of pharmaceutical waste was obtained in the real wastewater. These results proved the effectiveness of this treatment.

In addition, ultraviolet (UV) light irradiation assistance is considered one of the best techniques for pharmaceutical waste adsorption or removal. However, the limitation of this process is that some of the photocatalysts need to be treated after the photodegradation process in order to gain high reusability performance. Due to this disadvantage, the nanofiber membrane has become the center of attention for photocatalysis improvement in removing pharmaceutical waste as it has a high specific surface area, high aspect ratio, and small fiber diameter (Khalil et al., 2019). Like other pollutant treatments, many recent studies have incorporated nanomaterials into the nanofiber membrane to enhance the performance of nanofiber membrane as an adsorbent for pharmaceutical wastewater purification.

Yilmaz et al. (2020) researched TiO_2 nanoparticles and C-nanofibers with magnetic Fe_3O_4 nanospheres (TiO_2@Fe_3O_4@C-NFs) to remove drug molecules effectively. The nanofiber composite membrane was applied in both photocatalytic system and magnetic solid-phase extraction to remove undesired ibuprofen in pharmaceutical wastewater. It was deduced that in the magnetic solid-phase extraction of ibuprofen, the fabricated TiO_2@Fe_3O_4@C- NFs can be utilized at least 15 times while in photocatalytic degradation, it can be used for at least seven successive cycles without losing its performance. In particular, the TiO_2@Fe_3O_4@C–NF nanofiber membrane can degrade 100% of ibuprofen under UV light irradiation within 125 min. Moreover, Qiu et al. (2019) synthesized a nanosheet-shaped Mo/N-doped TiO_2 nanorods on carbon nanofiber, and this modification had enhanced the photocatalytic activity of TiO_2. The fabricated nanosheet had been used in the degradation of SMX and bisphenol A (BPA) (Qiu et al., 2019)

Bao et al. (2020) incorporated a novel cobalt crystal (Co) into the carbon nanofiber (CNF) membrane using electrospinning and thermal treatment techniques. The nanofibrous membrane was used with a peroxymonosulfate (PMS) activator to remove sulfamethoxazole (SMX) in aqueous liquid. The carbon nanofiber layer was mainly designed to increase the active adsorption sites, whereas the presence of Co can improve the interfacial charge transfer during the catalytic degradation of SMX. It was reported that an increasing the Co can enhance the SMX removal, but as the increment is higher than 8%, no changes were observed. This is mainly due to the limitation of PMS in the solution. On the other hand, Mehrabi et al. (2020) synthesized electrospun PVA–TEOS/grafted Fe_3O_4@SiO_2 magnetic nanofiber as magnetic nanofiber sorbent to recover SMX and trimethoprim (TMP) from aqueous liquid via solid-phase extraction method (SPE). The composite nanofiber membrane as sorbent was able to recover 95.1% SMX and 98.2% TMP. This work has displayed an efficient extraction method to separate pharmaceutical waste and clean water with low-cost nanofiber fabrication tools.

As shown in Table 10.5, it can be observed that composite nanofiber membrane with various modifications has good prospects in eliminating or recovering different types of pharmaceutical pollutants (drugs and carbamazepine) from the wastewater effluent. Several operational condition parameters like pH, initial pollutant concentration, and adsorbent dosage have been determined in order to obtain the best pollutant degradation. Some researchers even can achieve complete removal of drugs.

TABLE 10.5
Electrospun Nanofiber Membrane for Pharmaceutical Wastewater

Nanofiber Membranes	Pharmaceutical Waste	Removal Performance	References
PVA/Moringa seeds protein	Fenoprofen, ibuprofen, ketoprofen, diclofenac, and carbamazepine	77.8–96.1% removal from pharmaceutical wastewater effluent	Kebede et al. (2019)
TiO_2@Fe_3O_4@C-NFs	Ibuprofen	100% removal	Yilmaz et al. (2020)
Mo/N-doped TiO_2 nanorods/CNF	SMX and BPA	63.12% removal	Qiu et al. (2019)
CNF/Co	SMX	100% removal	Bao et al. (2020)
PVA –TEOS / grafted Fe_3O_4@SiO_2	SMX and TNP	95.1% SMX and 98.2% TMP recoveries	Mehrabi et al. (2020)
Polysulfone nanofiber	Antibiotics	98.1–99.9% removal	Zhang et al. (2020)
PAN-based graphene/ SnO_2 carbon nanofibers	SMX	85% removal	Yu et al. (2020)

However, less research reported on the nanofiber membrane/adsorbent reusability. Therefore, the reusability performance of fabricated membrane still needs to be discovered.

10.4 CONCLUSION

Electrospinning has been observed as an extraordinarily favorable method for preparing nanofiber membranes for the efficient treatment of contaminated wastewater. This brilliant technique offers potential advantages in designing and fabricating advanced nanofiber membranes of tunable wettability, high adsorption rate, high permeability, and separation efficiency, which are significant for practical applications, especially in real-lfe industry. Significant performance has been achieved in removing and separating persistent contaminants in wastewater. In this chapter, recent research progress starting from the basic principles of electrospinning technique and the fabrication of nanofibrous materials for effective treatment applications on heavy metal adsorption, dyes degradation, oily wastewater, and pharmaceutical wastewater are summarized. Despite the recent advances toward the development of nanofibrous membranes for wastewater remediation, the remaining challenges must be addressed. First, developing stable and durable surface structures of nanofibers with great wettability properties is still challenging, as most of the fine surface structures can be easily interrupted by external impacts. Besides that, synthesis methods like in situ polymerization are not applicable for large-scale production. Even though electrospun nanofibers can be produced on an industrial scale, the poor mechanical stability of the nanofibers still becomes a substantial challenge. In addition, less attention is given to the fundamental mechanisms and theories during wastewater treatment since many efforts only address the design and selection of various potential separation materials for nanofiber membranes. Although nanofiber membranes will face more challenges in the future, their versatile nanofibrous structure with controllable superwettability still gives them great potential as sustainable materials for wastewater treatment. Therefore, a continuous exploration of nanofiber membranes as a separation medium in the coming decades is significantly important for the great leap-forward development of industrial wastewater treatment with superior performance.

REFERENCES

Abdullah, N., Yusof, N., Lau, W. J., Jaafar, J. and Ismail, A. F. (2019) "Recent trends of heavy metal removal from water/wastewater by membrane technologies", *Journal of Industrial and Engineering Chemistry*, 76, pp. 17–38.

Alcoutlabi, M., Lee, H., Watson, J. V. and Zhang, X. (2013) "Preparation and properties of nanofiber-coated composite membranes as battery separators via electrospinning", *Journal of Materials Science*, 48(6), pp. 2690–2700.

Alias, N. H., Jaafar, J., Samitsu, S., Ismail, A. F., Nor, N. A. M., Yusof, N. and Aziz, F. (2020a) "Mechanistic insight of the formation of visible-light responsive nanosheet graphitic carbon nitride embedded polyacrylonitrile nanofibres for wastewater treatment", *Journal of Water Process Engineering*, 33, p. 101015.

Alias, N. H., Jaafar, J., Samitsu, S., Ismail, A. F., Othman, M. H. D., Rahman, M. A., Othman, N. H., Yusof, N., Aziz, F. and Mohd, T. A. T. (2020b) "Efficient removal of partially hydrolysed polyacrylamide in polymer-flooding produced water using photocatalytic graphitic carbon nitride nanofibres", *Arabian Journal of Chemistry*, 13, pp. 4341–4349.

Alias, N. H., Jaafar, J., Samitsu, S., Matsuura, T., Ismail, A. F., Huda, S., Yusof, N. and Aziz, F. (2019) "Photocatalytic nanofiber-coated alumina hollow fiber membranes for highly efficient oilfield produced water treatment", *Chemical Engineering Journal*, 360, pp. 1437–1446.

Azimi, A., Azari, A., Rezakazemi, M. and Ansarpour, M. (2017) "Removal of heavy metals from industrial wastewaters: A review", *Che BioEng Reviews*, 4(1), pp. 37–59.

Bajakova, J., Chaloupek, J., Lacarin, M. and Lukáš, D. (2011) "'Drawing' – the production of individual nanofibers by experimental method", *Nanocon*, 9, pp. 21–23.

Bao, Y., Tian, M., Lua, S. K., Lim, T. T., Wang, R. and Hu, X. (2020) "Spatial confinement of cobalt crystals in carbon nanofibers with oxygen vacancies as a high-efficiency catalyst for organics degradation", *Chemosphere*, 245, 125407.

Benavides, R. E., Jana, S. C. and Reneker, D. H. (2012) "Nano fibers from scalable gas jet process", *ACS Macro Letters*, 1, pp. 1032–1036.

Bera, B. (2016) "Literature review on electrospinning process (A Fascinating Fiber Fabrication Technique)", *Imperial Journal of Interdisciplinary Research (IJIR)*, 2(8), :972–984.

Bui, N., Laura, M., Hoek, E. M. V. and McCutcheon, J. R. (2011) "Electrospun nanofiber supported thin film composite membranes for engineered osmosis", *Journal of Membrane Science*, 385–386, pp. 10–19.

Cai, N., Chen, M., Liu, M., Wang, J., Shen, L., Wang, J., Feng, X. and Yu, F. (2019) "Meso-microporous carbon nanofibers with in-situ embedded Co nanoparticles for catalytic oxidization of azo dyes", *Journal of Molecular Liquids*, 289, p. 111060.

Carnell, L. S., Siochi, E. J., Wincheski, R. A., Holloway, N. M. and Clark, R. L. (2009) "Electric field effects on fiber alignment using an auxiliary electrode during electrospinning", *Scripta Materialia*, 60(6), pp. 359–361.

Carolin, C. F., Kumar, P. S., Saravanan, A., Joshiba, G. J. and Naushad, M. (2017) "Efficient techniques for the removal of toxic heavy metals from aquatic environment: A review", *Journal of Environmental Chemical Engineering*, 5(3), pp. 2782–2799.

Chen, H., Huang, M., Liu, Y., Meng, L. and Ma, M. (2020) "Functionalized electrospun nanofiber membranes for water treatment: A review", *Science of the Total Environment*, 739, p. 139944.

Choi, H. Y., Bae, J. H., Hasegawa, Y., An, S., Kim, I. S., Lee, H. and Kim, M. (2020) "Thiol-functionalized cellulose nanofiber membranes for the effective adsorption of heavy metal ions in water", *Carbohydrate Polymers*, 234, p. 115881.

Choi, S. S., Chu, B. Y., Hwang, D. S., Lee, S. G., Park, W. H. and Park, J. K. (2005) "Preparation and characterization of polyaniline nanofiber webs by template reaction with electrospun silica nanofibers", *Thin Solid Films*, 477, pp. 233–239.

Cui, J., Li, F., Wang, Y., Zhang, Q., Ma, W. and Huang, C. (2020) "Electrospun nanofiber membranes for wastewater treatment applications", *Separation and Purification Technology*, 250(December 2019), p. 117116.

Deng, S., Liu, X., Liao, J., Lin, H. and Liu, F. (2019) "PEI modified multiwalled carbon nanotube as a novel additive in PAN nanofiber membrane for enhanced removal of heavy metal ions", *Chemical Engineering Journal*, 375, p. 122086.

Dobosz, K. M., Kuo-Leblanc, C. A., Martin, T. J. and Schiffman, J. D. (2017) "Ultrafiltration membranes enhanced with electrospun nanofibers exhibit improved flux and fouling resistance", *Industrial and Engineering Chemistry Research*, 56(19), pp. 5724–5733.

Efome, J. E., Rana, D., Matsuura, T. and Lan, C. Q. (2016) "Enhanced performance of PVDF nanocomposite membrane by nanofiber coating: A membrane for sustainable desalination through MD", *Water Research*, 89, pp. 39–49.

Fang, Y., Zhu, X., Wang, N., Zhang, X., Yang, D. and Nie, J. (2019) "Biodegradable core-shell electrospun nano fibers based on PLA and γ -PGA for wound healing", *European Polymer Journal*, 116(January), pp. 30–37.

Fibrous, E. and Containing, M. (2019) "Fabrication and characterization of core-shell electrospun fibrous mats containing medicinal herbs for wound healing and skin", *Marine Drugs*, 17, p. 27.

Foong, C. Y., Wirzal, M. D. H. and Bustam, M. A. (2020) "A review on nanofibers membrane with amino-based ionic liquid for heavy metal removal", *Journal of Molecular Liquids*, 297, p. 111793.

Guo, J., Yan, D. Y. S., Lam, F. L. Y., Deka, B. J., Lv, X., Ng, Y. H. and An, A. K. (2019) "Self-cleaning BiOBr/Ag photocatalytic membrane for membrane regeneration under visible light in membrane distillation", *Chemical Engineering Journal*, 378, p. 122137.

Han, W., Bhat, G. S. and Wang, X. (2016) "Investigation of nanofiber breakup in the melt-blowing process", *Industrial & Engineering Chemistry Research*, 55, pp. 3150–3156.

Hashim, M. A., Mukhopadhyay, S., Sahu, J. N. and Sengupta, B. (2011) "Remediation technologies for heavy metal contaminated groundwater", *Journal of Environmental Management*, 92(10), pp. 2355–2388.

Hou, D., Ding, C., Li, K., Lin, D., Wang, D. and Wang, J. (2018) "A novel dual-layer composite membrane with underwater-superoleophobic/hydrophobic asymmetric wettability for robust oil-fouling resistance in membrane distillation desalination", *Desalination*, 428, pp. 240–249.

Huang, S., Ras, R. H. A. and Tian, X. (2018) "Antifouling membranes for oily wastewater treatment: Interplay between wetting and membrane fouling", *Current Opinion in Colloid & Interface Science*, 36, pp. 90–109.

Huang, W., Wang, M.-J., Liu, C.-L., You, J., Chen, S.-C., Wang, Y.-Z. and Liu, Y. (2014a) "Phase separation in electrospun nanofibers controlled by crystallization induced self-assembly", *Journal of Materials Chemistry A*, 2, p. 8416.

Huang, Y., Miao, Y. E. and Liu, T. (2014b) "Electrospun fibrous membranes for efficient heavy metal removal", *Journal of Applied Polymer Science*, 131, pp. 1–12.

Huang, Z. and Cheng, Z. (2020) "Recent advances in adsorptive membranes for removal of harmful cations", *Journal of Applied Polymer Science*, 137(13), p. 48579.

Huong, D. T. M., Chai, W. S., Show, P. L., Lin, Y. L., Chiu, C. Y., Tsai, S. L. and Chang, Y. K. (2020) "Removal of cationic dye waste by nanofiber membrane immobilized with waste proteins", *International Journal of Biological Macromolecules*, 164, pp. 3873–3884.

Ihsanullah, A. A., Al-Amer, A. M., Laoui, T., Al-Marri, M. J., Nasser, M. S., Khraisheh, M. and Atieh, M. A. (2016) "Heavy metal removal from aqueous solution by advanced carbon nanotubes: Critical review of adsorption applications", *Separation and Purification Technology*157, pp. 141–161.

Jafari, B., Abbasi, M. and Hashemifard, S. A. (2020) "Development of new tubular ceramic microfiltration membranes by employing activated carbon in the structure of membranes for treatment of oily wastewater", *Journal of Cleaner Production*, 244, p. 118720.

Jamil, N., Alias, N. H., Shahruddin, M. Z. and Othman, N. H. (2019) "A green in situ synthesis of hybrid graphene-based zeolitic imidazolate framework-8 nanofillers using recycling mother liquor", *Key Engineering Materials*, 797, pp. 48–54.

Junaidi, N. F. D., Othman, N. H., Shahruddin, M. Z., Alias, N. H., Marpani, F., Lau, W. and Ismail, A. F. (2020) "Fabrication and characterization of grahene-polyethersulfne (GO-PES) composite flat sheet and hollow fiber membraens for oil-water separation", *Journal of Chemical Technology & Biotechnology*, 95, pp. 1308–1320.

Kamiyama, M., Soeda, T., Nagajima, S. and Tanaka, K. (2012) "Development and application of high-strength polyester nanofibers", *Polymer Journal*, 44(10), pp. 987–994.

Karaer, H. and Kaya, I. (2016) "Synthesis, characterization of magnetic chitosan/active charcoal composite and using at the adsorption of methylene blue and reactive blue4", *Microporous and Mesoporous Materials*, 232, pp. 26–38.

Karim, M. R., Aijaz, M. O., Alharth, N. H., Alharbi, H. F., Al-Mubaddel, F. S. and Awual, M. R. (2019) "Composite nanofibers membranes of poly(vinyl alcohol)/chitosan for selective lead(II) and cadmium(II) ions removal from wastewater", 169, pp. 479–486.

Kebede, T. G., Dube, S. and Nindi, M. M. (2019) "Biopolymer electrospun nanofibres for the adsorption of pharmaceuticals from water systems", *Journal of Environmental Chemical Engineering*, 7(5), p. 103330.

Khalil, A., Nasser, W. S., Osman, T. A., Toprak, M. S. and Muhammed, M. (2019) "Surface modified of polyacrylonitrile nanofibers by TiO 2/MWCNT for photodegradation of organic dyes and pharmaceutical drugs under visible light irradiation", *Environmental Research*, 179(July), p. 108788.

Khulbe, K. C. and Matsuura, T. (2018) "Removal of heavy metals and pollutants by membrane adsorption techniques", *Applied Water Science*, 8(1), p. 19.

Kumar, A., Rana, A., Sharma, G., Naushad, M., Dhiman, P., Kumari, A. and Stadler, F. J. (2019) "Recent advances in nano-Fenton catalytic degradation of emerging pharmaceutical contaminants", *Journal of Molecular Liquids*, 290, p. 111177.

Kumbar, S. G., James, R., Nukavarapu, S. P. and Laurencin, C. T. (2008) "Electrospun nanofiber scaffolds: Engineering soft tissues", *Biomedical Materials*, 3, pp. 1–15.

Landrigan, P. J., Fuller, R., Acosta, N. J. R., Adeyi, O., Arnold, R., Basu, N. N., Baldé, A. B., Bertollini, R., Fuster, V., Greenstone, M., Haines, A., Hanrahan, D., Hunter, D., Khare, M., Krupnick, A., Lanphear, B., Lohani, B., Martin, K., Mathiasen, K. V., McTeer, M. A., Murray, C. J. L., Ndahimananjara, J. D., Perera, F., Potočnik, J., Preker, A. S., Ramesh, J., Rockström, J., Salinas, C., Samson, L. D., Sandilya, K., Sly, P. D., Smith, K. R. and Steiner, A. (2017) "The *Lancet* Commission on pollution and health", *The Lancet*, 391(10119), pp. 462–512.

Lee, H. and Kim, I. S. (2018) "Nanofibers: Emerging progress on fabrication using mechanical force and recent applications", *Polymer Reviews*58(4), pp. 688–716.

Legrouri, K., Khouya, E., Ezzine, M., Hannache, H., Denoyel, R., Pallier, R. and Naslain, R. (2005) "Production of activated carbon from a new precursor molasses by activation with sulphuric acid", *Journal of Hazardous Materials*, 118(1–3), pp. 259–263.

Li, F., Song, Y. and Zhao, Y. (2010) "Core-shell nanofibers: Nano channel and capsule by coaxial electrospinning", in A. Kumar (ed), *Nanofibers*. In Tech, pp. 419–438.

Li, H., Zhu, L., Zhang, J., Guo, T., Li, X., Xing, W. and Xue, Q. (2019) "Applied surface science high-efficiency separation performance of oil-water emulsions of polyacrylonitrile nano fibrous membrane decorated with metal-organic frameworks", *Applied Surface Science*, 476(October 2018), pp. 61–69.

Li, L., Zhang, J., Li, Y. and Yang, C. (2017) "Removal of Cr (VI) with a spiral wound chitosan nanofiber membrane module via dead-end filtration", *Journal of Membrane Science*, 544, pp. 333–341.

Liao, H.-S., Lin, J., Liu, Y., Huang, P., Jin, A. and Chen, X. S. (2016) "Self-assembly mechanism of peptide nanofibers in solution and on substrate surface", *Nanoscale*, 8(31), pp. 14814–14820.

Lv, Y., Ding, Y., Wang, J., He, B., Yang, S., Pan, K. and Liu, F. (2020) "Carbonaceous microsphere/nanofiber composite superhydrophilic membrane with enhanced anti-adhesion property towards oil and anionic surfactant: Membrane fabrication and applications", *Separation and Purification Technology*, 235, 116189.

Ma, W., Zhang, M., Liu, Z., Kang, M., Huang, C. and Fu, G. (2019a) "Fabrication of highly durable and robust superhydrophobic-superoleophilic nanofibrous membranes based on a fluorine-free system for efficient oil/water separation", *Journal of Membrane Science*, 570–571, pp. 303–313.

Ma, Y., Qi, P., Ju, J., Wang, Q., Hao, L., Wang, R., Sui, K. and Tan, Y. (2019b) "Gelatin/alginate composite nanofiber membranes for effective and even adsorption of cationic dyes", *Composites Part B: Engineering*, 162, pp. 671–677.

Makanjuola, O., Ahmed, F., Janajreh, I. and Hashaikeh, R. (2019) "Development of a dual-layered PVDF-HFP/cellulose membrane with dual wettability for desalination of oily wastewater", *Journal of Membrane Science*, 570–571, pp. 418–426.

Mehrabi, F., Mohamadi, M., Mostafavi, A., Hakimi, H. and Shamspur, T. (2020) "Magnetic solid phase extraction based on PVA-TEOS/grafted Fe_3O_4@SiO_2 magnetic nanofibers for analysis of sulfamethoxazole and trimethoprim in water samples", *Journal of Solid State Chemistry*, 292(August), p. 121716.

Min, M., Shen, L., Hong, G., Zhu, M., Zhang, Y., Wang, X., Chen, Y. and Hsiao, B. S. (2012) "Micro-nano structure poly(ether sulfones)/poly(ethyleneimine) nanofibrous affinity membranes for adsorption of anionic dyes and heavy metal ions in aqueous solution", *Chemical Engineering Journal*, 197, pp. 88–100.

Mirjalili, M. and Zohoori, S. (2016) "Review for application of electrospinning and electrospun nanofibers technology in textile industry", *Journal of Nanostructure in Chemistry*, pp. 1–7.

Nor, N. A. M., Jaafar, J., Ismail, A. F., Mohamed, M. A., Rahman, M. A., Othman, M. H. D., Lau, W. J. and Yusof, N. (2016) "Preparation and performance of PVDF-based nanocomposite membrane consisting of TiO_2 nanofibers for organic pollutant decomposition in wastewater under UV irradiation", *Desalination*, 391, pp. 89–97.

Pakravan, M., Heuzey, M. and Ajji, A. (2012) "Core – shell structured PEO-chitosan nanofibers by coaxial electrospinning", *Biomacromolecules*, 13, pp. 412–421.

Partheniadis, I., Nikolakakis, I., Laidmäe, I. and Heinamaki, J. (2020) "A mini-review: Needleless electrospinning of nanofibers for pharmaceutical and biomedical applications", *Processes*, 8, p. 673.

Pepe, F., de Gennaro, B., Aprea, P. and Caputo, D. (2013) "Natural zeolites for heavy metals removal from aqueous solutions: Modeling of the fixed bed Ba2+/Na+ ion-exchange process using a mixed phillipsite/chabazite-rich tuff", *Chemical Engineering Journal*, 219, pp. 37–42.

Pereao, O., Bode-Aluko, C., Laatikainen, K., Nechaev, A. and Petrik, L. (2019) "Morphology, modification and characterisation of electrospun polymer nanofiber adsorbent material used in metal ion removal", *Journal of Polymers and the Environment*, 27(9), pp. 1843–1860.

Phan, D. N., Rebia, R. A., Saito, Y., Kharaghani, D., Khatri, M., Tanaka, T., Lee, H. and Kim, I. S. (2020) "Zinc oxide nanoparticles attached to polyacrylonitrile nanofibers with hinokitiol as gluing agent for synergistic antibacterial activities and effective dye removal", *Journal of Industrial and Engineering Chemistry*, 85, pp. 258–268.

Qiu, J., Liu, F., Yue, C., Ling, C. and Li, A. (2019) "A recyclable nanosheet of Mo/N-doped TiO_2 nanorods decorated on carbon nanofibers for organic pollutants degradation under simulated sunlight irradiation", *Chemosphere*, 215, pp. 280–293.

Qu, X., Alvarez, P. J. J. and Li, Q. (2013) "Applications of nanotechnology in water and wastewater treatment", *Water Research*, 47, pp. 3931–3946.

Ramakrishna, S., Fujihara, K., Teo, W.-E., Yong, T., Ma, Z. and Ramaseshan, R. (2006) "Electrospun nanofibers: solving global issues", *Materials Today*, 9(3), pp. 40–50.

Samitsu, S., Zhang, R., Peng, X., Krishnan, M. R., Fujii, Y. and Ichinose, I. (2013) "Flash freezing route to mesoporous polymer nanofibre networks", *Nature Communications*, 4, pp. 1–7.

Sezgin, N. and Balkaya, N. (2016) "Adsorption of heavy metals from industrial wastewater by using polyacrylic acid hydrogel", *Desalination and Water Treatment*, 57(6), pp. 2466–2480.

Sofi, H. S., Akram, T., Shabir, N., Vasita, R., Jadhav, A. H. and Sheikh, F. A. (2020) "Regenerated cellulose nanofibers from cellulose acetate: Incorporating hydroxyapatite (HAp) and silver (Ag) nanoparticles (NPs), as a scaffold for tissue engineering applications", *Materials Science and Engineering: C*, 118(November 2019), p. 111547.

Stevens, B. (2014) "Nanofibres – A maturing technology", *Membrane Technology*, pp. 8–9.

Su, C., Lu, C., Cao, H., Tang, K., Chang, J. and Duan, F. (2018) "Fabrication and post-treatment of nanofibers-covered hollow fiber membranes for membrane distillation", *Journal of Membrane Science*, 562(May), pp. 38–46.

Su, R., Wu, W., Song, C., Lui, G. and Yu, Y. (2021) "Recent progree in electrospun nanofibrous membranes for oil-water separation.pdf", *Separation and Purification Technology*, 256, p. 117790.

Su, S., Li, J. L., Zhou, L., Wan, S., Bi, H. C., Ma, Q. and Sun, L. T. (2017) "Ultra-Thin Electro-Spun PAN Nanofiber Membrane for High-Efficient Inhalable PM& lt; sub& gt; 2.5</ sub& gt; Particles Filtration", *Journal of Nano Research*, 46(March), pp. 73–81.

Tang, M., Christie, K. S. S., Hou, D., Ding, C., Jia, X. and Wang, J. (2021) "Fabrication of a novel underwater-superoleophobic/hydrophobic composite membrane for robust anti-oil-fouling membrane distillation by the facile breath figures templating method", *Journal of Membrane Science*, 617, p. 118666.

Tarus, B., Fadel, N., Al-Oufy, A. and Magdi, E.-M. (2016) "Effect of polymer concentration on the morphologyand mechanical characteristics of electrospuncellu. pdf", *Alexandria Engineering Journal*, 55, pp. 2975–2984.

Tijing, L. D., Yao, M., Ren, J., Park, C.-H., Kim, C. S. and Shon, H. K. (2019) "Nanofibers for water and wastewater treatment: Recent advances and developments", in Bui, X.-T., Chiemchaisri, C., Fujioka, T., and Varjani, S. (eds) *Water and Wastewater Treatment Technologies*. Singapore: Springer Singapore, pp. 431–468.

Varesano, A., Rombaldoni, F., Mazzuchetti, G., Tonin, C. and Comotto, R. (2010) "Multi-jet nozzle electrospinning on textile substrates: Observations on process and nanofibre mat deposition", *Polymer International*, 59(July), pp. 1606–1615.

Wang, C., Wang, J., Zeng, L., Qiao, Z., Liu, X. and Liu, H. (2019) "Fabrication of electrospun polymer nanofibers with diverse morphologies", *Molecules*, 834, p. 24050834.

Wang, J., Shen, J., Ye, D., Yan, X., Zhang, Y., Yang, W., Li, X., Wang, J., Zhang, L. and Pan, L. (2020a) "Disinfection technology of hospital wastes and wastewater: Suggestions for disinfection strategy during coronavirus Disease 2019 (COVID-19) pandemic in China", *Environmental Pollution*, 262, p. 114665.

Wang, W., Lin, J., Cheng, J., Cui, Z., Si, J., Wang, Q., Peng, X. and Turng, L. S. (2020b) "Dual super-amphiphilic modified cellulose acetate nanofiber membranes with highly efficient oil/water separation and excellent antifouling properties", *Journal of Hazardous Materials*, 385, p. 121582.

Wang, X. and Hsiao, B. S. (2016) "Electrospun nanofiber membranes", *Current Opinion in Chemical Engineering*, 12, pp. 62–81.

Wang, X., Yu, J., Sun, G. and Ding, B. (2016) "Electrospun nanofibrous materials: A versatile medium for effective oil/waterseparation", *Materials Today*, 19, p. 7.

Xiao, S., Shen, M., Ma, H., Guo, R., Zhu, M., Wang, S. and Shi, X. (2010) "Fabrication of water-stable electrospun polyacrylic acid-based nanofibrous mats for removal of copper (II) ions in aqueous solution", *Journal of Applied Polymer Science*, 116(4), pp. 2409–2417.

Xu, T., Wu, F., Gu, Y., Chen, Y., Cai, J., Lu, W., Hu, H., Zhu, Z. and Chen, W. (2015) "Visible-light responsive electrospun nanofibers based on polyacrylonitrile-dispersed graphitic carbon nitride", *RSC Advances*, 5, pp. 86505–86512.

Xue, J., Wu, T., Dai, Y., Xia, Y., States, U. and States, U. (2019) "Electrospinning and electrospun nanofibers: Methods, materials, and applications", *Chemical Reviews*, 119(8), pp. 5298–5415.

Ya, V., Martin, N., Chou, Y. H., Chen, Y. M., Choo, K. H., Chen, S. S. and Li, C. W. (2018) "Electrochemical treatment for simultaneous removal of heavy metals and organics from surface finishing wastewater using sacrificial iron anode", *Journal of the Taiwan Institute of Chemical Engineers*, 83, pp. 107–114.

Yang, B., Zhou, H., Zhang, X. and Zhao, M. (2015) "Electron spin-polarization and band gap engineering in carbon-modified graphitic carbon nitrides", *Journal of Materials Chemistry C*, 3, pp. 10886–10891.

Yang, X., Zhou, Y., Sun, Z., Yang, C. and Tang, D. (2020) "Synthesis and Cr adsorption of a super-hydrophilic polydopamine-functionalized electrospun polyacrylonitrile", *Environmental Chemistry Letters*, 19(1), pp. 743–749.

Yilmaz, E., Salem, S., Sarp, G., Aydin, S., Sahin, K., Korkmaz, I. and Yuvali, D. (2020) "TiO_2 nanoparticles and C-Nanofibers modified magnetic Fe_3O_4 nanospheres (TiO_2@ Fe_3O_4@C–NF): A multifunctional hybrid material for magnetic solid-phase extraction of ibuprofen and photocatalytic degradation of drug molecules and azo dye", *Talanta*, 213(December 2019), p. 120813.

Yu, S., Gao, Y., Khan, R., Liang, P., Zhang, X. and Huang, X. (2020) "Electrospun PAN-based graphene/SnO_2 carbon nanofibers as anodic electrocatalysis microfiltration membrane for sulfamethoxazole degradation", *Journal of Membrane Science*, 614(May), p. 118368.

Zhang, R., Ma, Y., Lan, W., Sameen, D. E., Ahmed, S. and Dai, J. (2021) "Enhanced photocatalytic degradation of organic dyes by ultrasonic-assisted electrospray TiO_2/graphene oxide on polyacrylonitrile/β-cyclodextrin nanofibrous membranes", *Ultrasonics-Sonochemistry*, 70(August 2020), p. 105343.

Zhang, S., Shi, Q., Christodoulatos, C. and Meng, X. (2019) "Lead and cadmium adsorption by electrospun PVA/PAA nanofibers: Batch, spectroscopic, and modeling study", *Chemosphere*, 233, pp. 405–413.

Zhang, Y. and Duan, X. (2020) "Chemical precipitation of heavy metals from wastewater by using the synthetical magnesium hydroxy carbonate", *Water Science and Technology: A Journal of the International Association on Water Pollution Research*, 81(6), pp. 1130–1136.

Zhang, Y., Mu, T., Huang, M., Chen, G., Cai, T., Chen, H., Meng, L. and Luo, X. (2020) "Nanofiber composite forward osmosis (NCFO) membranes for enhanced antibiotics rejection: Fabrication, performance, mechanism, and simulation", *Journal of Membrane Science*, 595(August 2019), p. 117425.

Zhuang, G. L., Wu, S. Y., Lo, Y. C., Chen, Y. C., Tung, K. L. and Tseng, H. H. (2020) "Gluconacetobacter xylinus synthesized biocellulose nanofiber membranes with super-hydrophilic and superoleophobic underwater properties for the high-efficiency separation of oil/water emulsions", *Journal of Membrane Science*, 605, p. 118091.

11 Carbon Nanomaterials for Removal of Pharmaceuticals from Wastewater

Aydin Hassani
Near East University, Nicosia, Turkey

Alireza Khataee
University of Tabriz, Tabriz, Iran
Gebze Technical University, Gebze, Turkey

CONTENTS

11.1 Introduction 333
11.2 Carbon Nanomaterials (CNMs) 339
11.2.1 Carbon Nanotubes (CNTs) 340
11.2.2 Graphene (GR) and Reduced Graphene Oxide (RGO) 341
11.2.3 Fullerene (C60) 342
11.2.4 Activated Carbon (AC) 342
11.2.5 Biochar (BC) 342
11.2.6 Carbon Quantum Dots (CQDs) 343
11.2.7 Graphitic Carbon Nitride (g-C_3N_4) 343
11.3 Photocatalysis Process 344
11.4 Sonocatalytic Process 351
11.5 Fenton and Fenton-Like Process 352
11.6 Electrochemical Process 353
11.7 Sulfate Radical-Based AOPs (SR-AOPs) 355
11.8 Hybrid AOPs 357
11.9 Summary and Future Perspectives 358
References 360

11.1 INTRODUCTION

The earth has abundant sources of water; however, what is available for human consumption is less than 1% (Adeleye et al., 2016). Moreover, the consumable portion is sold at higher costs due to energy prices rising, the populated areas spreading, and

DOI: 10.1201/9781003118749-11

climatic–environmental conditions deteriorating (Levin et al., 2002). But the disastrous episode of the story is that the potable water resources are rapidly contaminated by the newly appearing pharmaceuticals and personal care products (PPCPs) pollutants (Houtman, 2010). The explosive growth of pharmaceutical industries and other detrimental activities has resulted in tons of harmful waste being released into the environment, including organic, inorganic, biodegradable, and non-biodegradable materials. In the meantime, PPCPs have been more seriously contaminating the surface and underground water bodies, posing a direct threat to aquatic and lives (Shen and Andrews, 2011; Carmalin Sophia et al., 2016). The contamination mechanism encompasses not only inappropriate use and disposal of the aforementioned harmful toxins, but also the pharmaceutical industry complexes with their production lines. Numerous classes of PPCPs have been identified so far as contaminating agents of global consumable waters, including antibiotics, anti-acids, steroids, antidepressants, analgesics, anti-inflammatories, antipyretics, beta-blockers, lipid-lowering drugs, tranquilizers, and stimulants (Al-Khateeb et al., 2014). What is worrisome is that the pharmaceuticals increasingly polluting potable waters, cause imminent danger for human-animal health and environmental life, can lead to the spread of antibiotic resistance (Yan et al., 2014; Liu et al., 2016), and can be toxic to aquatic living organisms (Sanderson et al., 2004; Liu et al., 2016). Thus, their initiation and destiny are currently among the main issues humans are faced with that should be addressed by world scientists, thinkers, and policymakers (Jelić et al., 2012). Given the transforming effects of the pharmaceuticals on potable water and thereby the impacts they produce on the food supplies, and their adverse impact on the ecosystems functions, the governmental authorities worldwide are calling for efficient yet affordable solutions to mitigate the pharmaceutical compounds emissions. Pharmaceuticals from human consumption, excretion, pharmaceutical substances resulting from the individuals' consumption, urination-excretion, and the wastes disposed of the same origin continually enter the environment, mainly through the wastewater, and if not removed, they can enter water bodies directly. Nowadays, new methods and approaches have emerged to remove pharmaceutical pollutions from potable water (He et al., 2016; Hassani et al., 2018d). CNMs play important roles in many wastewater treatment technologies, including physicochemical (Liu et al., 2014), chemical and electrochemical (Duan et al., 2020a; Ghasemi et al., 2020), and biologically related removal (Ncibi and Sillanpää, 2017; Sbardella et al., 2018). Most recently, CNMs, due to their unmatched properties such as large surface-to-volume ratio, high porosity, hollow structure, and outstanding catalytic potentials, have attracted much attention as promising catalysts (Zhai et al., 2017; Gholami et al., 2019b).

The present chapter covers a detailed summary of the methods utilized for the elimination of PPCPs off the sewages using CNMs, particularly carbon nanotubes (CNTs), graphene (GR) and reduced graphene oxide (RGO)-based materials, fullerene (C60), activated carbon (AC), biochar (BC), carbon quantum dots (CQDs), and graphitic carbon nitride (g-C_3N_4). The latest progress for utilizing CNMs in various water treatment strategies is considered in the following sections, while a great emphasis is put on photocatalytic, sonocatalytic, Fenton and Fenton-like, EC oxidation, sulfate radical-based advanced oxidation processes (SR-AOPs), and hybrid

AOPs approaches. Moreover, the future relevant perspectives and potential advances are covered and discussed. Table 11.1 summarizes the main and most recent studies concerning eliminating PPCPs using CNMs as a catalyst, providing information about their family and chemical structure.

TABLE 11.1
Studies Reported in the Literature concerning the Elimination of PPCPs Using CNMs

Pharmaceuticals	Family	Chemical Structure
Oxytetracycline	Antibiotic	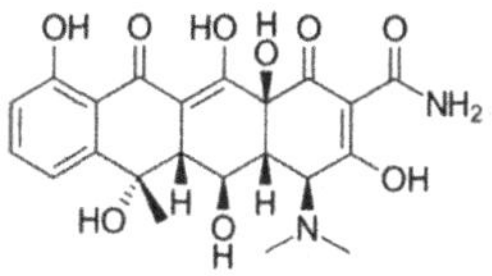
Ciprofloxacin hydrochloride	Antibiotic	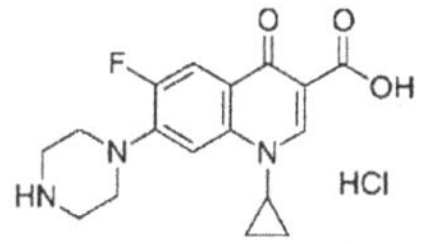
Sulfamethoxazole	Antibiotic	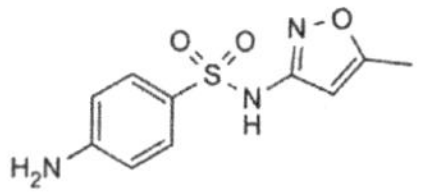
Erythromycin	Antibiotic	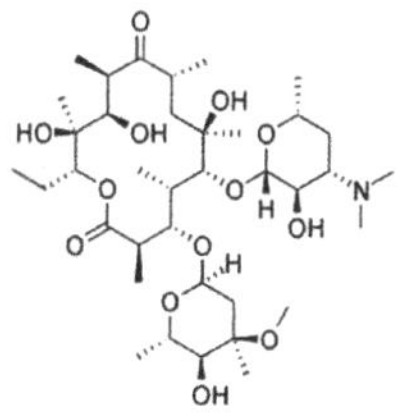
Clarithromycin	Antibiotic	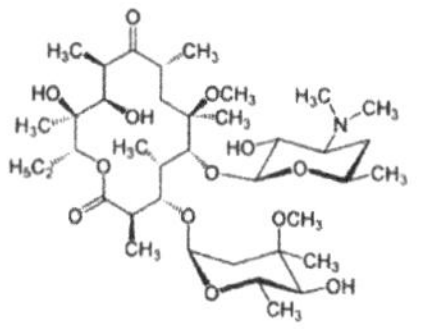
Tetracycline hydrochloride	Antibiotic	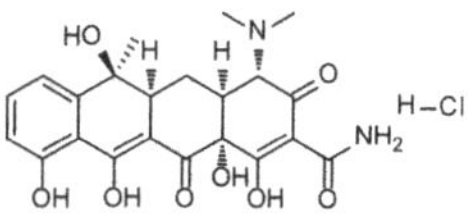
Cefixime	Antibiotic	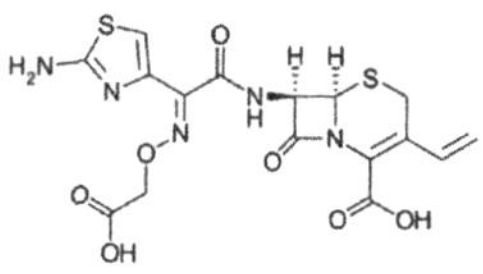
Amoxicillin	Antibiotic	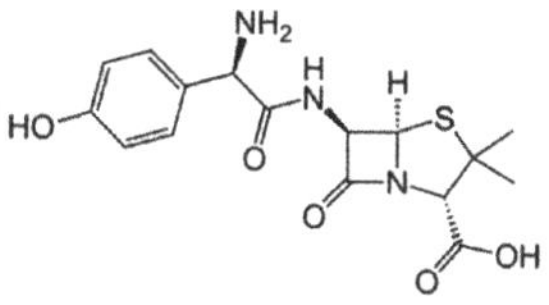

(Continued)

TABLE 11.1 (Continued)
Studies Reported in the Literature concerning the Elimination of PPCPs Using CNMs

Pharmaceuticals	Family	Chemical Structure
Ofloxacin	Antibiotic	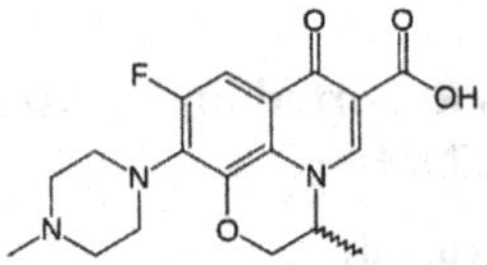
Metronidazole	Antibiotic	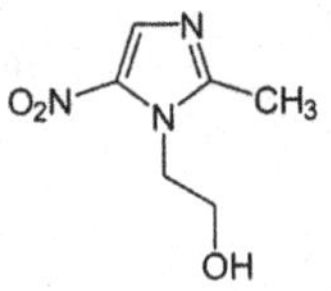
Sulfisoxazole	Antibiotic	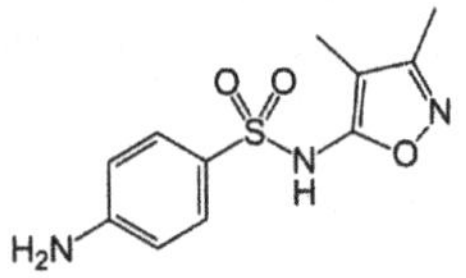
Sulfadiazine	Antibiotic	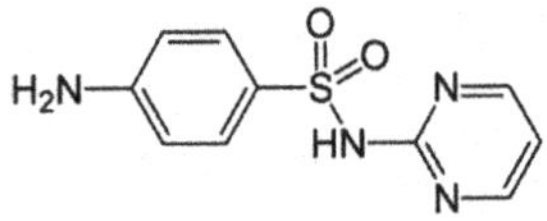
Sulfamerazine	Antibiotic	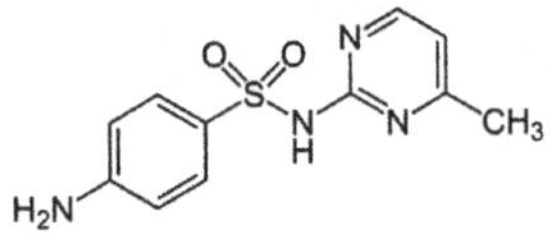
Norfloxacin	Antibiotic	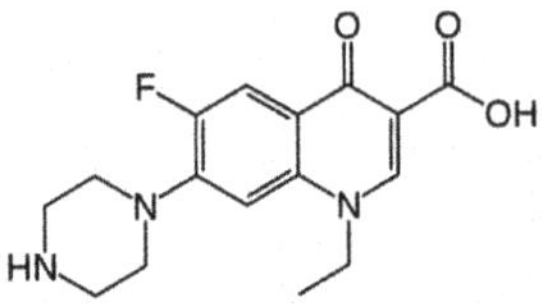
Ampicillin	Antibiotic	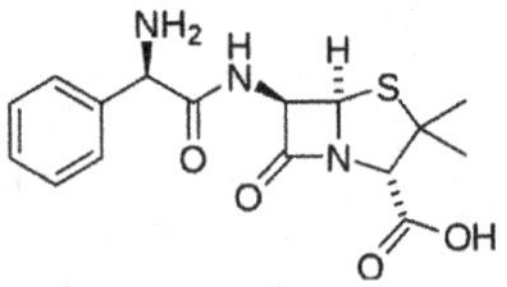
Ofloxacin	Antibiotic	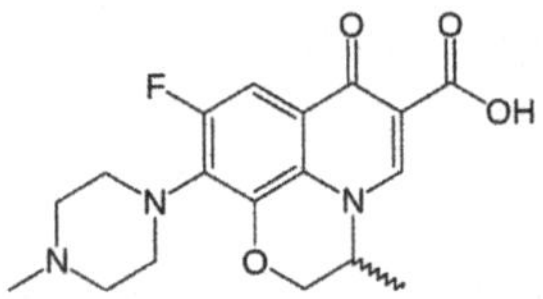
Levofloxacin	Antibiotic	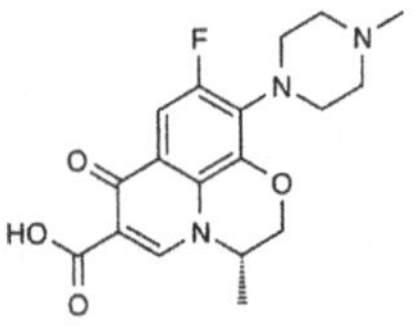

(Continued)

TABLE 11.1 (Continued)
Studies Reported in the Literature concerning the Elimination of PPCPs Using CNMs

Chloramphenicol	Antibiotic	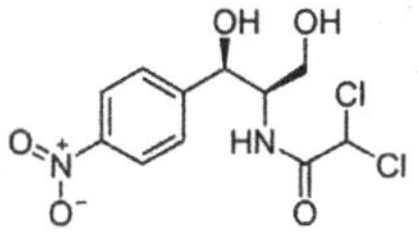
Gemifloxacin	Antibiotic	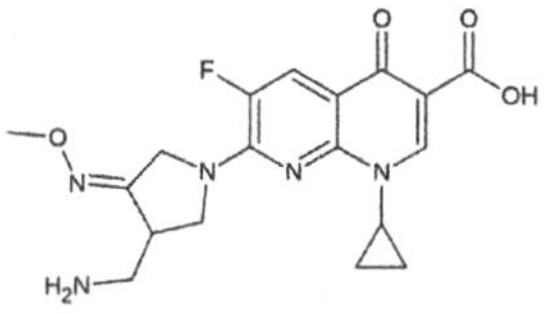
Cefazolin sodium	Antibiotic	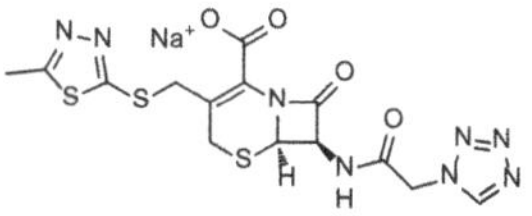
Moxifloxacin	Antibiotic	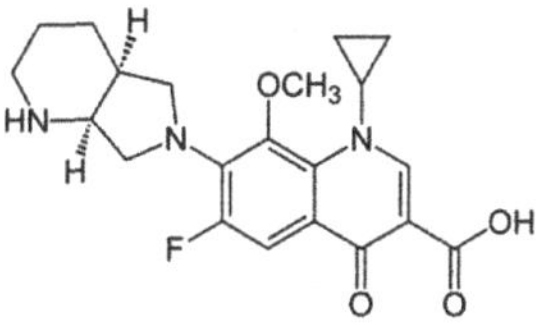
Sulfathiazole	Antibiotic	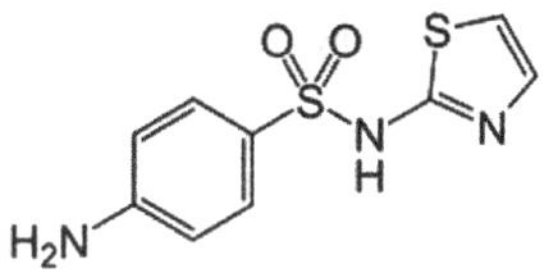
Sulfacetamide	Antibiotic	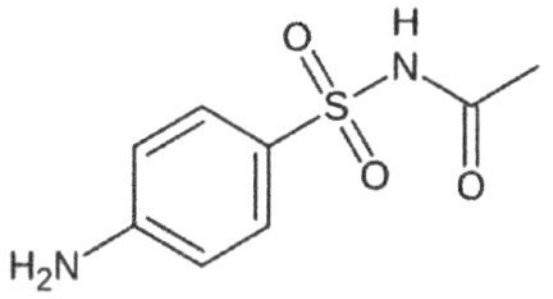
Sulfachlorpyridazine	Antibiotic	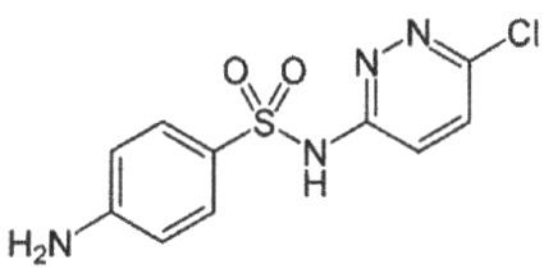
Cefixime	Antibiotic	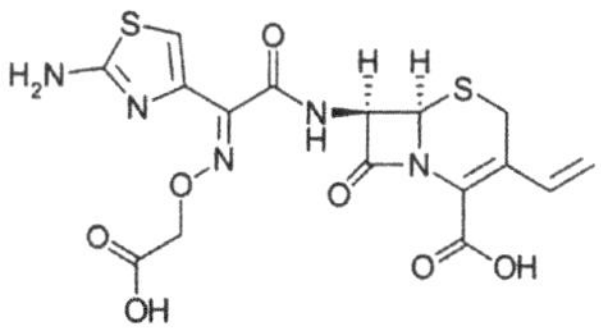
Florfenicol	Antibiotic	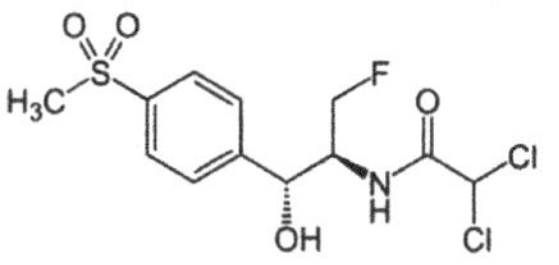

(Continued)

TABLE 11.1 (Continued)
Studies Reported in the Literature concerning the Elimination of PPCPs Using CNMs

Pharmaceuticals	Family	Chemical Structure
Ceftazidime	Antibiotic	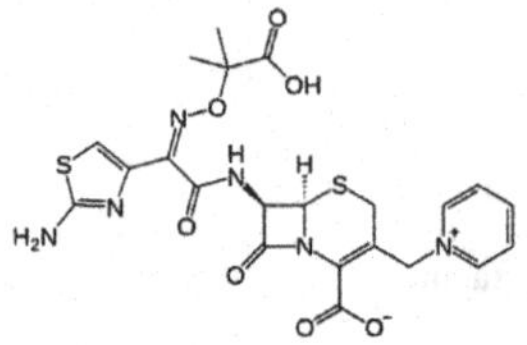
Ceftriaxone sodium	Antibiotic	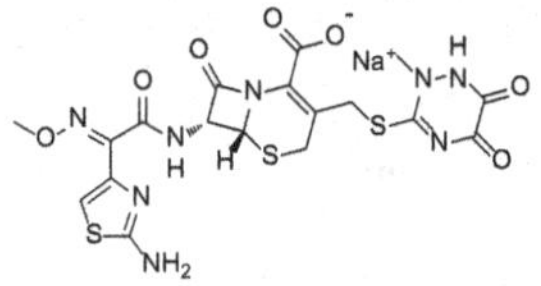
Sulfamethazine	Antimicrobial	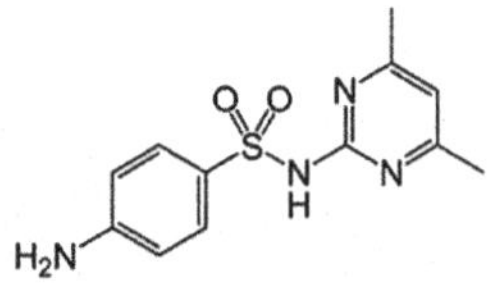
Phenazopyridine hydrochloride	Analgesic antispasmodic	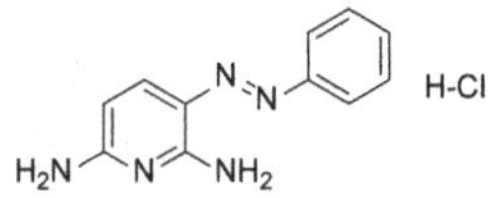
Carbamazepine	Anticonvulsant	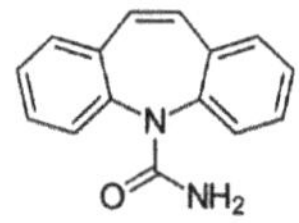
Primidone	Anticonvulsant	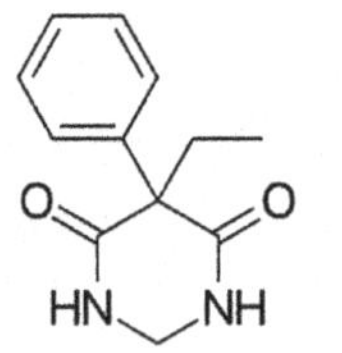
Antipyrine	Nonsteroidal anti-inflammatory	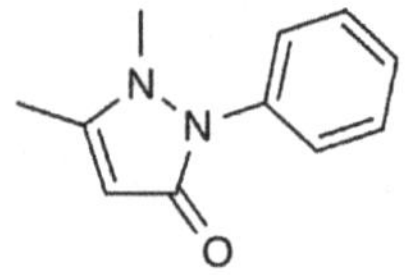
Diclofenac	Nonsteroidal anti-inflammatory	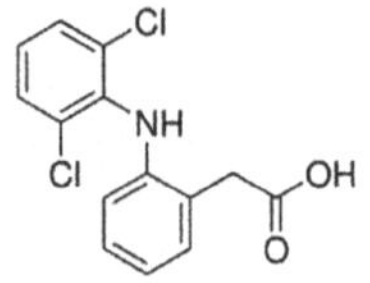

(Continued)

TABLE 11.1 (Continued)
Studies Reported in the Literature concerning the Elimination of PPCPs Using CNMs

Acetaminophen	Nonsteroidal anti-inflammatory	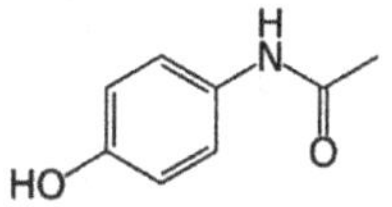
Ibuprofen	Nonsteroidal anti-inflammatory	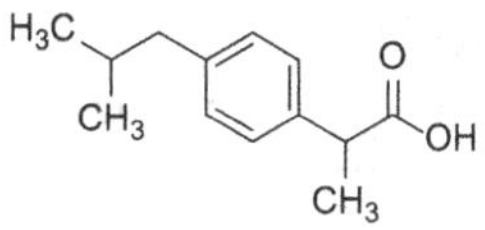
Naproxen	Nonsteroidal anti-inflammatory	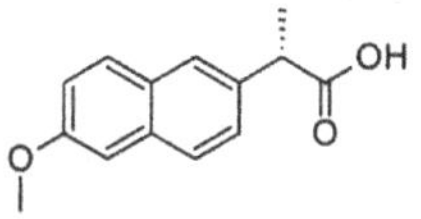
Bisphenol A	Endocrine disrupting	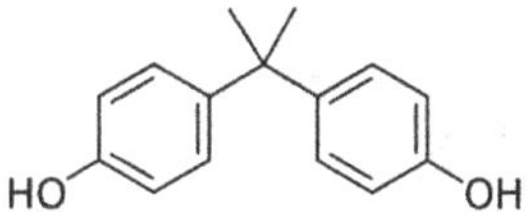

11.2 CARBON NANOMATERIALS (CNMs)

A member of the P block of the periodic table, carbon has often provoked scholars' astonishment because of its capability in forming a broad spectrum of compounds, frequently through sharing the four valence electrons it possesses. Amorphous carbon, graphite, and diamond, as the older allotropes of carbon, have been discovered during the past decades. The later ones have different crystalline structures and characteristics. Considering diamond, for example, one atom of carbon having a tetrahedral structure can establish four single bonds (sp^3 hybridization) together with four other carbon atoms, leading to the creation of a three-dimensional crystalline covalent chemical bonding. Graphite, in contrast, is formed by the accumulation of two-dimensional graphene layers resting on top of one another, as illustrated in Figure 11.1. The graphene monolayer carbon atoms undergo sp^2 hybridization, having van der Waals interaction with the remaining layers of graphite crystalline lattice. Such different structures lead to almost totally opposite specifications. Graphite is a high electricity conductor, while diamond acts as an insulator. Again, diamond is optically transparent and the hardest known natural substance, while graphite is soft and made of opaque blackish material. On the other hand, carbon has other allotropes, namely carbon nanostructures, which are classified into two major categories, depending on the type of covalent bond among their carbon atoms (Georgakilas et al., 2015). Group 1 is comprised of graphenic nanostructures – basically formed by densely packed sp^2 carbon atoms inside a hexagonal honeycomb crystal lattice – having graphene, graphenic nanosheets, CNTs, nanohorns, onion-like carbon (OLC) nanospheres, and carbon dots (CDs) as its members. Graphene is the simplest member of the group that will be dealt with briefly later. Group 2 of carbon nanostructures comprises carbon atoms with hybridization of both sp^3 and sp^2 with different ratios as well as mixes of graphitic and amorphous regions, or sometimes dominant sp^3 carbon atoms. Until

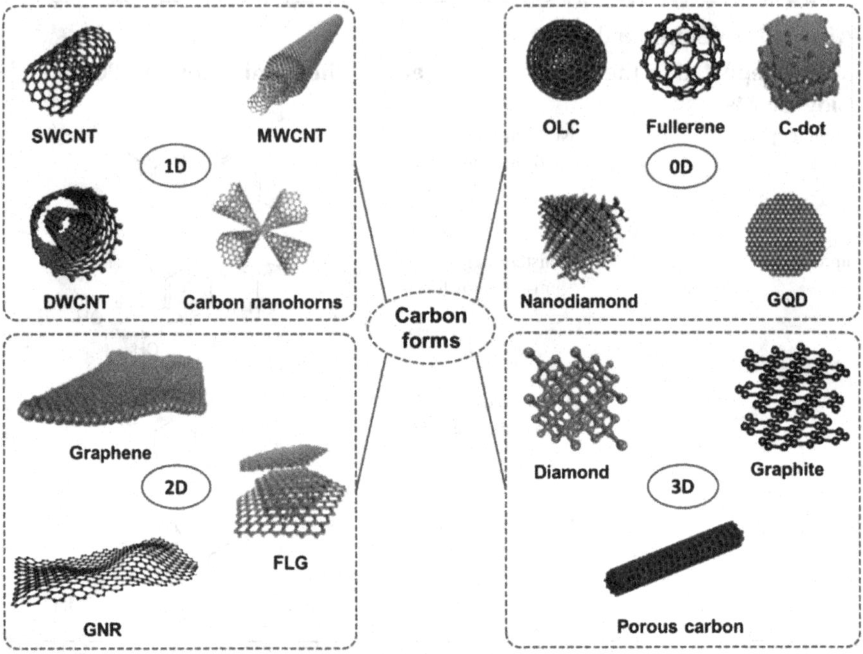

FIGURE 11.1 Classification of carbon forms with respect to their dimensions. (Adapted from reference (Georgakilas et al., 2015)) (GQD: graphene quantum dot, GNR: graphene nanoribbon, FLG: few-layer graphene.)

today, the only known members of this group are nanodiamond and some other forms of CDs. As its major property, this group's graphene parts or monolayers (e.g., CNTs) are not considered as constituting constructs. Carbon nanoallotropes may also be classified in terms of their morphological features (Georgakilas et al., 2015; Taherpour and Mousavi, 2018). Accordingly, the nanostructures with porous constructs such as fullerene, CNTs, and nanohorns constitute one category and robust nanostructures with no internal porosity, such as nanodiamonds, OLC spheres, and even graphene, as it also has no internal spaces. The carbon nanostructure dimensionality is the most recent categorization of carbon nanoallotropes (Georgakilas et al., 2015; Taherpour and Mousavi, 2018). This type of classification is represented in Figure 11.1. However, carbon nanoallotropes have a number of shard properties aside from the above categorizations (owing to the sp^2 carbon atoms present in a hexagonal lattice), and several important differences in size and shapes. Although CNMs are used nowadays in numerous applications, their application for pharmaceutical removal in wastewater treatment will be highlighted in the present chapter.

11.2.1 Carbon Nanotubes (CNTs)

Sumio Iijimam (1991) discovered the CNTs for the first time, initially in the form of graphene sheets, which were observed as tube-shaped cylinders with diameters of nano-scale measure (De Volder et al., 2013). CNT exists in three categories of single-walled carbon nanotubes (SWCNTs), which is one layer graphene sheet rolled into single-cylinder, double-walled carbon nanotubes (DWCNTs) and multi-walled carbon nanotubes (MWCNTs), which are multiconcentric cylindrical shells of graphene sheets (Yu et al., 2009). Arc discharge, laser ablation, and chemical vapor deposition (CVD) have been very popular techniques among others for the CNTs synthesis (Vajtai, 2013); they have been studied extensively for their capacity to be adopted for various applications due to their unmatched structural and physicochemical features (Xu et al., 2018). Some of the studies have demonstrated the CNTs capabilities for the destruction of PPCPs from water bodies.

11.2.2 Graphene (GR) and Reduced Graphene Oxide (RGO)

As one of the 2D and hexagonal honeycomb structured carbon allotropes, GR is synthesizable through various techniques, including mechanical cleavage, CVD, and mechanical exfoliation of graphite (Yin et al., 2017). Great surface-to-volume ratio, good chemical stability, and tunable functionality (Wang et al., 2019b; Zhang et al., 2020b) are dependable features that made the GR-based materials promising catalysts for wastewater treatment. In addition, with their different functionalities for absorbing contaminants through establishing hydrophobic, electrostatic, hydrogen bonding, and π-π stacking interactions, they are convenient for the distinctive removal of various pollutants. Graphene derivatives like graphene oxide (GO), RGO, and GO nanocomposites, as reported by some of the studies, demonstrated greater potential for removal of pharmaceuticals from sewages (Park et al., 2018; Yang et al., 2019b) due to their superior physicochemical properties thanks to the attachment of abundant functional groups on their surface compared with pristine graphene (Figure 11.2). GO is normally obtained through the Hummer's process by the oxidation of graphite utilizing powerful oxidizing agents and the GO surfaces undergoing oxidation treatment producing highly oxygenated functional chemical groups, like hydroxyl, carboxyl, etc., which are capable of altering GO van der Waals interactions. As a result, higher solubility and dispersibility properties in water and organic solvents are achieved (Yin et al., 2017). Additionally, RGO is produced by reducing the exfoliated GO.

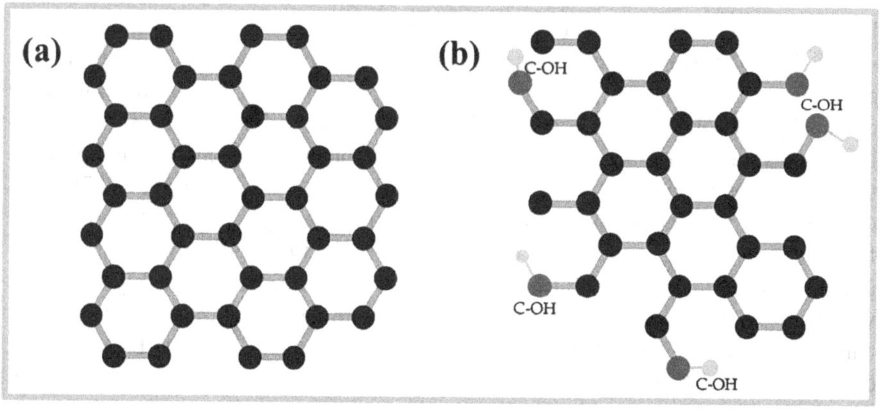

FIGURE 11.2 Structures of graphene-based materials. (a) graphene, and (b) reduced graphene oxide (RGO). (Adapted from Li et al., 2020.)

11.2.3 Fullerene (C60)

Fullerene, an allotrope of carbon also called Buckminsterfullerene (C60) or Buckyball, due to its spherical, tubular, or ellipsoid shapes were discovered by Kroto et al. (1985) (Kroto et al., 1985). Properties like unique spherical construct, the adequate surface-to-volume ratio, low aggregation tendency and great electrons (e^-) affinity of fullerenes along with their composites turn them into suitable materials to be applied as catalysts in wastewater treatment processes (Hasanzadeh et al., 2018; Elessawy et al., 2020). Fullerenes may comprise of 60, 70, 78, or even greater number of carbon atoms, the 60-atom structure though showing the most stable sphericity in shape (Kulkarni, 2014). Theoretically, carbon atoms in the fullerene family may reach up to 300 (Abo-Hamad et al., 2016). However, a fewer number of such structures have ever been synthesized, and as just mentioned, they are only theoretical models (Taherpour, 2008). Practically, the π electrons existing in the fullerene inner and outer spheres react with the pollutants in water treatment processes *via* π-π stacking; however, limited researches have been conducted on the fullerenes' adsorption potential for pharmaceuticals so far.

11.2.4 Activated Carbon (AC)

AC, a traditional adsorbent extensively used for water treatment, has also certain applicability for PPCPs adsorption from wastewater (Khataee et al., 2013; Hassani and Khataee, 2017). It appears in two forms: powdered activated carbon (PAC) and granular activated carbon (GAC). With the operation time, however, the system would encounter two problems, requiring additional studies to address: 1) decrease of adsorption capacity, and 2) the deterioration of AC operating in the complicated wastewater treatment systems (Wang and Wang, 2016).

11.2.5 Biochar (BC)

One of the substances used in agriculture for soil enrichment and carbon sequestration is biochar (BC), which is produced *via* oxygenless biomass (wood and manure) heat up at 300 to 700°C (also called as charring, and scientifically speaking, pyrolysis) (Sohi, 2012; Hassani and Khataee, 2018). BC decreases carbon release into the environment from burning or degradation through stabilizing the carbon, similar to charcoal processing (carbon negative). BC helps restore carbon of the soil after it has been buried in the ground. Furthermore, the saved bioenergy from charring prepares the possible substitution of fossil fuels, i.e., carbon-neutral (Tan et al., 2015). BC's extraordinary properties like its wide specific surface area, porous structure, rich surface functional groups, and mineral constituents make them a suitable adsorbent for purifying polluted water. With its Ac-like porous structure, BC has the highest usage worldwide as the greatest efficient adsorbent for eliminating pharmaceutical contaminants from aquatic medium (Dai et al., 2020; Huang et al., 2020). The BC, however, compared with the AC, is being gradually known as a new, low-price, and efficient adsorbent. The AC production requires a higher heat-up temperature and a more complicated activation process, whereas the BC production costs are lower because of its lower consumption of energy (Ahmad et al., 2012). On the other hand, there are abundant and cheap resources for the production of BC, primarily found among agricultural biomass and solid wastes (Qian and Chen, 2013). Furthermore, the invasive plant conversion into BC contributes to better management of invasive plants and provides better environmental protection (Zheng et al., 2010; Dong et al., 2013).

11.2.6 Carbon Quantum Dots (CQDs)

Xu et al. (2004) introduced CQDs as fluorescent carbon nanoparticles for the first time just months prior to the graphene introduction, but it was not before 2006 when Sun et al. (2006) gave it the popular name of CQD: an approx. flat or quasi-spherical carbon nanoparticles (<10 nm) formed by amorphous to nanocrystalline cores with mainly sp^2 carbon (i.e., graphitic carbon and/or GR and GO sheets) fused by diamond-like sp^3 hybridized carbon intersections (Li et al., 2012; Lim et al., 2015). It is a highly aqueous soluble owing to the dramatic amounts of carboxylic surface moieties, providing 5% and 50% oxygen content (Sun et al., 2006; Li et al., 2012; Lim et al., 2015). Studies have reported strong long-wavelength light harnessing capability. CQDs can exchange energy with solution species, which suggests great potential for their application as photocatalysts (Lim et al., 2015). Although the near IR (NIR) region is often preferable (Lim et al., 2015), it is also a highly valuable characteristic for optimizing the photoactivity of semiconductors under the fullest sunlight spectrum (Li et al., 2012). Practically, CQDs are strong additives for solar photocatalytic systems because their capability for controlling the full sunlight spectrum and up-converting visible light into shorter wavelength light would trigger the semiconductor charge separation, i.e., electrons-holes (e^-/h^+) and prevents recombination, which through reaction with H_2O/O_2 will generate the active species required to intensify pharmaceutical pollutants decomposition (Li et al., 2012; Lim et al., 2015).

11.2.7 Graphitic Carbon Nitride (g-C_3N_4)

Given the metal-free catalysts, g-C_3N_4 as a polymeric material with a graphite-like matrix is deemed a prosperous catalyst containing adequate elements such as C, N, and H (Ma et al., 2017; Naseri et al., 2017). Carbon nitride, in general, has numerous allotropes with varying characteristics, including g-C_3N_4, α-C_3N_4, β-C_3N_4, pseudo-cubic-C_3N_4, and cubic-C_3N_4, among which g-C_3N_4 is the material with the most applicability and great stability under practical conditions (Naseri et al., 2017). g-C_3N_4 is a double bonds polymeric system, made from s-triazine or tri-s-triazine units interconnected *via* tertiary amines (Figure 11.3a) (Wang et al., 2009; Zheng et al., 2015). The layers atoms are arranged in a honeycomb structure with strengthened covalent bonds, and the force existing between two-dimensional sheets is a weak van der Waals force. Varying carbon- and nitrogen-rich precursors (like melamine, cyanamide, dicyandiamide) are utilized for g-C_3N_4, synthesis *via* thermal polyaddition and polycondensation integration process (Figure 11.3b) (Wang et al., 2009; Zheng et al., 2015).

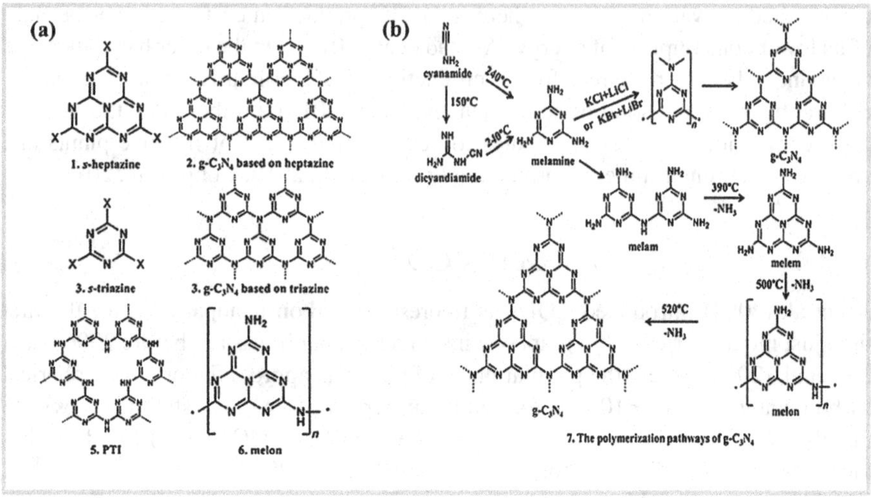

FIGURE 11.3 Structures of s-heptazine, s-triazine, hypothetical g-C_3N_4, poly(triazine imide) (PTI), and melon (a). Thermal polymerization routes of g-C_3N_4 (b). (Adapted from Reference (Naseri et al., 2017) with permission from Royal Society of Chemistry.)

11.3 PHOTOCATALYSIS PROCESS

Heterogeneous photocatalytic elimination of contaminants is a blooming and prosperous water treatment technology (Hassani et al., 2018b). Generally, in a photocatalyst, the valence band (VB) electrons could be excited to the conduction band (CB) by exerting sufficient photon energy to produce e^-/h^+ pairs. The photo-induced e^-/h^+ pairs provoke several redox reactions to generate more active radicals that strengthen the photocatalytic elimination of pollutants (Hassani et al., 2020b). A series of metal oxides like TiO_2, ZnO, SnO_2, WO_3, $SrTiO_3$, and CeO_2 are known to be the most prevailing photocatalysts because of their prevalence in nature, environmental compatibility, high stability under varying situations, and the potential to produce e^-/h^+ pairs when subjected to adequate light energy amount (Zhang et al., 2019). However, the drawback is that the broad application of such photocatalysts is unfortunately limited by their near-UV range-wide energy band and the produced e^-/h^+ pairs' fast recombination. Literature shows that numerous techniques have been tested to remove such drawbacks, concentrating on designing new photocatalysts possessing modified properties and constructs. CNMs, among others, are proper candidates to produce new composite photocatalysts due to their promising advantages like chemical stability in a wide pH range, adjustable structural and chemical specifications, and interesting electronic properties. Graphene, for example, has been investigated recently as one of the most prosperous materials for the preparation of novel photocatalysts, owing to the good mobility of its charge carriers, large surface area, higher transparency, and excellent electrical-thermal conductivity (Safarpour and Khataee, 2019). For instance, Khalid et al. (Khalid et al., 2017) compared the TiO_2/CNMs efficiency vs. bare TiO_2 and concluded that carbon doping and composites with AC, fullerene, CNTs, or GR leads to the generation of materials with optimal photocatalytic performance, adsorption potential, e^- trapping and synthesizing capability, improved visible light absorption, and simpler separation. The improved photocatalytic performance due to the construction of carbon-based semiconductors can be attributed to the use of two approaches (i) optimized separation of charge in the presence of carbon, (ii) carbon performing as a photosensitizer (Figure 11.4) (Ania et al., 2012). By adding porous carbons to semiconductors, a series of synergistic effects arising from adsorption is observed that must be taken into account. Woan et al. (2009) proposed the charge separation process *via* semiconductor/carbon composites, stressing that CNMs with metallic conductivity can trap the electrons produced by the semiconductor, thereby preventing the recombination (Figure 11.4a). The mechanism of carbon acting as the photosensitized by transferring electrons to the CB of semiconductor (Figure 11.4b) was introduced by Wang et al. (2005).

Karaolia et al. (2018) investigated the elimination of the antibiotics sulfamethoxazole (SMX), erythromycin (ERY) and clarithromycin (CLA) by TiO_2-RGO photocatalyst under solar irradiation in urban wastewaters. The results revealed that TiO_2-RGO was more effective in the photocatalytic elimination of ERY (84% ± 2%), CLA (86% ± 5%), and SMX (87% ± 4%) after 60 min of treatment.

Hu et al. (2016a) synthesized graphene/TiO_2/ZSM-5 (GTZ) nanocomposites by solid dispersion technique for maximal photocatalytic removal of oxytetracycline (OTC). A complete removal (ca. 100%) was obtained in 180 min under optimal experimental conditions (i.e., pH 7 and 25°C). The results proved that the holes had played a key role in the photocatalytic system.

Jo et al. (2017) prepared spinel-like Co_3O_4/TiO_2/GO photocatalyst through sol-gel and hydrothermal methods. This research investigated the photocatalytic efficiency of Co_3O_4/TiO_2 and Co_3O_4/TiO_2/GO photocatalyst under simulated solar radiation. According to their findings, the 2.0 wt.% Co_3O_4/TiO_2/GO was shown to have greater photocatalytic performance (~91% after 90 min) for the degradation of OTC than those without GO photocatalysts with an enhancement of 15%–20%. GO might play a role in the separation of charge carriers.

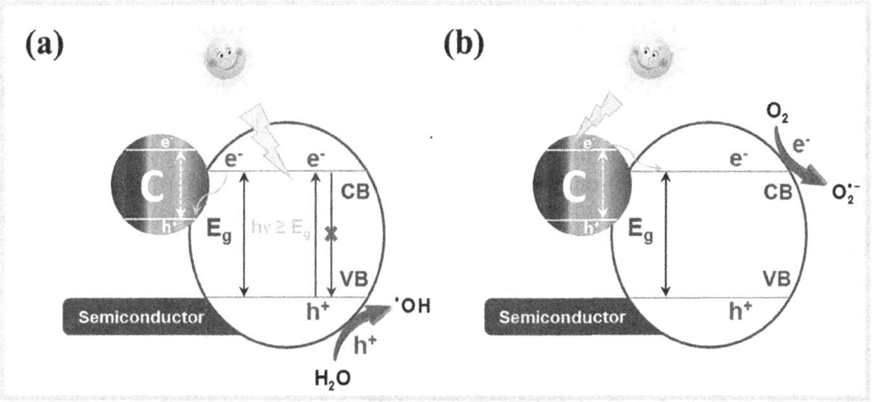

FIGURE 11.4 Schematic representation of photocatalysis system using a carbon-based semiconductor. (a) Charge separation in the presence of carbon, and (b) carbon acting as a photosensitizer. (Adapted from reference (Mestre and Carvalho, 2019.))

Tang et al. (2018) prepared N-TiO_2/RGO by a photo-reduction method and its visible light photocatalytic activity was assessed by dissociation of tetracycline hydrochloride (TC). The results revealed that N-TiO_2/RGO (98%) nanocomposites displayed effective removal performance comparing with the pure TiO_2 and N-doped TiO_2 (~79%) after 60 min.

Shaniba et al. (2020) used TiO_2/nitrogen-doped holey graphene hybrid (TiO_2/NHG) to remove cefixime under sunlight illumination. Their findings showed 92.3% of removal capacity under optimal conditions (cefixime concentration of 250 mg/L, TiO_2/NHG concentration of 0.05 g, H_2O_2 concentration of 5 mg/L, and reaction mixture of 100 mL).

Song et al. (2016) synthesized graphene and TiO_2 nanotubes (GR-TNTs) *via* a facile one-step hydrothermal technique for amoxicillin (AMX) removal. According to their findings, the photocatalytic activity of GR-TNTs was closely related to GR loading. The removal efficiency of AMX can reach up to 95% within 3 h with 0.1 g/L 10% GR-TNT at pH 9. Furthermore, kinetic-study results showed that degradation could be expressed by a first-order-reaction kinetic model. Yekan Motlagh et al. (2019) applied the chemical co-precipitation method to made ZnFe-layered double hydroxide (LDH) with different intercalated anions, including chloride and sulfate (ZnFe-SO_4-LDH and ZnFe-Cl-LDH). They were then modified with GO to obtain ZnFe-SO_4-LDH/GO and ZnFe-Cl-LDH/GO photocatalysts. The as-prepared samples were evaluated for the elimination of ofloxacin (OFX) under visible light radiation. After visible light radiation for 150 min, the decomposition rates of ZnFe-SO_4-LDH/GO and ZnFe-Cl-LDH/GO were 71.19% and 43.29%, respectively. In another work, Yekan Motlagh et al. (2020) fabricated ZnFe-LDH plates modified with GO. They immobilized them on the glass substrates (marked as ZnFe-LDH/GO/GS) and utilized them as a fixed photocatalyst for the heterogeneous photocatalytic process for easy separation of the photocatalyst. The photocatalytic implementation of the composites was assessed for the decomposition of phenazopyridine hydrochloride (PhP) from the aquatic environment under visible light illumination. The optimal elimination performance appeared under natural pH conditions (pH of 8), initial PhP concentration of 15 mg/L, and three photocatalysts plates, in which 60.01% of PhP were removed in 150 min. TiO_2/RGO was synthesized by Lin et al. (Lin et al., 2017) with 0.1% to 10% of RGO weights. They covered and examined the nanocomposites photoactivity for photocatalytic removing three pharmaceuticals such as ibuprofen (IBP), SMX, and carbamazepine (CBZ), under visible light and UV. TiO_2/RGO-2.7% was the best performing photocatalyst due to lower bandgap energy (Eg) and the creation of a heterojunction interface resulting in the enhanced charge separation. Removing SMX under UV-vis irradiation is more operative than CBZ (92% against 54%); however, similar mineralization levels (54–59%) are found by both pharmaceuticals revealing that though SMX provides faster removal, the oxidation intermediates require the same time as those of CBZ for achieving mineralization. It was proved that immobilizing the composite onto optical fibers is an auspicious strategy since the removal efficiency of IBP by TiO_2/RGO-2.7% was over 80%, followed by 15 cycles (45 hours) under UV-vis irradiation (Yang et al., 2019a). The Fe_3O_4/RGO/TiO_2 was made by Bashiri et al. (2020)through the hydrothermal technique. The as-prepared sample photocatalytic efficacy was determined by removing metronidazole (MET) under visible light irradiation. The MET highest removal rate (0.0092 min^{-1}) was attained for 0.75 g composite comprising of 5% TiO_2 and 4% RGO at pH = 5 and 20 mg/L MET concentration. Complete removal of MET (after 120 min) and total organic carbon (TOC) (after 250 min) by Fe_3O_4/RGO/TiO_2 coupled with visible light was achieved at optimal conditions.

Various catalysts of TiO_2/AC heterostructures were synthesized by Peñas-Garzón et al. (2020) through three various approaches, including sol-gel, solvothermal, and microwave-assisted (MW) for degrading three emergent contaminants (antipyrine (ANT), acetaminophen (APAP), and IBP in water under simulated solar light. For the photocatalytic activity, the best performance was yielded by the heterostructure prepared by a microwave-assisted protocol (TiO_2/AC-MW) in the solar-driven photocatalytic degrading the three examined pharmaceuticals (ANT, APAP, and IBP), both based on parent compound mineralization and disappearance (conversion). The best outcomes were found for IBP degradation completely conversed in less than 3h; however, the most recalcitrant one was ANT.

Song et al. (2017) utilized g-C_3N_4, in 2017, for photodegrading the sulfonamides, such as sulfisoxazole (SSX), SMX, sulfamerazine (SMZ), sulfadiazine (SDZ), under visible light. The findings revealed that the photocatalytic degradation efficacies of four sulfonamides were enhanced considerably in comparison to photolysis, by adding g-C_3N_4 and over 90% of photodegradation removal was achieved.

Zhang et al. (2020a) made a porous Z-scheme MnO_2/Mn-altered alkalinized g-C_3N_4 catalyst (MnO_2/CNK-OH-Mn) through the calcinating-impregnating technique. MnO_2/CNK-OH-Mn revealed greater Fenton-like photocatalytic activities in the degradation of TC compared to alkalinized g-C_3N_4 and pure g-C_3N_4. The optimum MnO_2/CNK-OH-Mn15% represented the greatest catalytic performance with 74.9% TOC elimination efficacy and 96.7% TC removal due to improved reactive oxygen species (ROS). Chakraborty and co-workers (2018) synthesized the reduced graphene oxide-zinc telluride (RGO-ZnTe) photocatalyst responsive to visible light through a single-pot one-stage solvothermal procedure. The RGO-ZnTe composite showed greater (6 times in comparison to RGO and 2.6 times in comparison to ZnTe) photocatalytic efficacy toward photodegrading TC antibiotics driven by visible light. Within the RGO-ZnTe composite, the two-dimensional wrinkled surface of RGO plays a key role in obtaining the improved behavior of ZnTe nanoparticles by minimization of the recombination likelihoods of the photo-induced e^-/h^+. ZnS-RGO nanocomposites were prepared by Bai et al. (2017) through a facile hydrothermal approach. The photocatalytic capability of the as-prepared ZnS-RGO was assessed by using norfloxacin (NOX) as a model contaminant. The ZnS-RGO nanocomposites revealed greater photocatalytic activities than the pure ZnS (75%), in which 92% of NOX can be photodegraded under similar conditions. Anirudhan and Deepa (2017) reported photocatalytic degrading of ciprofloxacin hydrochloride (CIP) from aqueous solutions through nano-zinc oxide integrated GO or nanocellulose composite (ZnO-GO/NC). Considerable degradation ability was found by ZnO-GO/NC along with 98.0% CIP degraded within 40 min. In another work, Zhu et al. (2017) discovered SMX removal feasibility *via* reduced graphene oxide-WO_3 (RGO-WO_3) nanocomposites powered under visible light. Using a facile one-step hydrothermal technique, three kinds of RGO-WO_3 nanocomposites were prepared (RW-400,

RW-200, and RW-100) as photocatalysts driven by visible light. Over 98% of SMX removal reached within 3 h by both RW-100 and RW-200 under visible light irradiation. Priya et al. (2016) considered immobilizing GSC (graphene sand composite) over nano-sized BiOCl (BiOCl/GSC) for the simultaneous photodegradation and adsorption of OTC and ampicillin (AMP) antibiotics from aqueous solutions. The considerable photocatalytic activity was exhibited by the BiOCl/GSC to mineralize AMP and OTC antibiotics under solar lights. The removal efficiencies of 84% and 90% were obtained utilizing BiOCl/GSC for OTC and AMP, respectively.

New BiOCl-$Bi_{24}O_{31}Cl_{10}$/RGO nanocomposites were fabricated by Shabani et al. (2019) using a simple sono-solvothermal technique through the photo-elimination of fluoroquinolones (FQs) antibiotics including OFX, CIP, and levofloxacin (LVX) under the simulated solar light. There are different elimination rates and photocatalytic degradation efficiency over the irradiation for 120 min for the stated FQs and the highest values were achieved at about 70.7%, 84.8%, and 57.2% for LVX, OFX, and CIP, respectively. Zhang et al. (2017) synthesized the TiO_2 supported by reed straw biochar (TiO_2/pBC) (acid pre-treated) using the sol-gel technique. Owing to the combination of photocatalysis and adsorption, the greater removal efficacy of SMX was performed by TiO_2/pBC than pure TiO_2 powder exposed to UV light irradiations. The SMX degradation rate of 91.27% was obtained with TiO_2/pBC (300), at the operational conditions (dosage of 0.2 g, pH of 4), which were much greater compared to pure TiO_2 (58.47%). Mineralizing SMX and its oxidation products are reflected by the chemical oxygen demand (COD) removal. The COD reduction of SMX at UV light and simulated sunlight was 25.64% and 10.73%, representing that degrading the intermediates fabricated from photolysis was difficult. Appavu et al. (2018) synthesized the new nitrogen-doped reduced graphene oxide/$BiVO_4$ ($BiVO_4$/N-RGO) photocatalyst through a single-step hydrothermal technique. The synthesized catalyst was examined for its catalytic activity in photodegrading chloramphenicol (CAP) and MET under visible light irradiation. The $BiVO_4$/N-RGO degradation efficiency was considerably high in degrading CAP and MET that is 93% and 95% under visible light illumination. The improved photocatalytic activity and diminished charge recombination were found because of the concerted effects between nitrogen-RGO and $BiVO_4$. Xie and co-workers (2019) obtained SMX removal of 80% within 3 hours under simulated visible light utilizing Zn-TiO_2/biochar (Zn-TiO_2/BC) nanocomposites. These nanocomposites overtook the TiO_2/BC and bare TiO_2. The reason is that Zn-doping decreased the crystal size and the agglomeration of TiO_2 effectively. Moreover, the greater charge transfer and separation are justified by the greater photocurrent of Zn-TiO_2/BC. After the first cycle, the photoactivity of Zn-TiO_2/BC is slightly decreased; however, it is almost constant in the subsequent four reuses. Li et al. (2019) studied CBZ photodegradation with magnetic Fe_3O_4/BiOBr, bare BiOBr, and Fe_3O_4/BiOBr/BC. Fe_3O_4/BiOBr/BC (0.05 g Fe_3O_4, and 10% biochar) photocatalyst has the best performance at pH 6: the greatest constant rate (0.01777 min^{-1}), CBZ elimination similar to BiOBr (95%), mineralization of 70% (only ≈ 55% for Fe_3O_4/BiOBr, ≈30% for BiOBr). The species hydroxyl radicals ($^{\bullet}OH$), superoxide radicals anion ($O_2^{\bullet -}$), and h^+ seem to take part in the CBZ photodegradation, indicating the most important role of oxygen radicals. Followed by five reuse cycles under visible irradiation, 90% of the CBZ was degraded by Fe_3O_4/BiOBr/BC, and more than 60% mineralization was obtained in the fourth cycle. TiO_2/biochar (TiO_2/BC) was investigated by Khraisheh et al. (2013) with coconut shells powder char (CNSP) for photodegrading CBZ under UV light in the presence of oxygen flow. Despite the BC quantity, higher photocatalytic activity and adsorption were presented by all nanocomposites compared to bare TiO_2, GAC, and BC. For pHs of 3–11, the CBZ degradation rate was nearly constant (0.05 min^{-1}), obtaining the CBZ removal of 89.2%–94.4%. Remarkably, followed by 1 hour of UV-C irradiation, the removal of ~90% and ~99% CBZ are obtained for composite and GAC, respectively. However, BC and bare TiO_2 only achieved CBZ removal of 42%. Whereas the 90% destruction with GAC is mainly related to adsorption, in the case of the BC, which has an incipient pore structure, the 42% CBZ removal may be caused by the combination of adsorption and photocatalytic processes. Concerning recyclability, TiO_2/BC also shows superior behavior because it eliminates 60% CBZ after 11 reuse cycles; however, the removal is <25% for the other solids. The surface area is responsible for the improved CBZ removal with the TiO_2/BC composite that provides adsorption of CBZ, which is more photodegraded by the semiconductor-created ROS.

Magnetic carbon nanotube-TiO_2 (MCNT-TiO_2) nanocomposites were synthesized by Awfa et al. (2019) and their photocatalytic activity was assessed for degrading SMX and CBZ under solar irradiation. In comparison to a TiO_2 reference catalyst, a higher photodegradation rate was exhibited by the MCNT-TiO_2 nanocomposites due to the large surface area and prolonged visible light absorption. The photodegradation rate constants for SMX and CBZ in pure water were ~1.2 and ~1.5 times higher than TiO_2, respectively. WO_3/MWCNT nanocomposites were prepared by Zhu et al. (2018) by incrementing CNT contents. It was verified that the greater CNT quantity reduces the e^-/h^+ recombination and bandgap energy. The best compromise was presented by the composite WO_3/MWCNT-4 (MWCNT of 4 g) between the bandgap energy (2.52 eV), the amount of MWCNT, and the dispersion in water. Under simulated solar light, the SMX degradation efficiency was in the order of WO_3 (25%) <WO_3/MWCNT-2 (42%)<WO_3/MWCNT-4 (65%)<WO_3/MWCNT-8 (73%). The WO_3/MWCNT-4 composite lost only the performance of 5% followed by four recycle experiments. The mechanism suggested for degrading SMX under visible light includes transferring the photo-excited electrons from the CB bands of WO_3 to the CNT acting as the e^- traps for hindering e^-/h^+ recombination. By the reaction of O_2 with electrons, $O_2^{\bullet-}$ is generated; however, the reaction of h^+ and water and/or surface hydroxyl generates $^{\bullet}OH$. Such radicals will decompose the SMX.

TiO_2-SiO_2/MWCNT nanocomposites were examined by Czech and Buda (2015) with 0.01%–17.8% of CNT vs. TiO_2-SiO_2 for photocatalytic degrading of CBZ (Czech and Buda, 2015) and diclofenac (DCF) (Czech and Buda, 2016). Adding SiO_2 enhances the TiO_2 dispersion; however, adding MWCNT steadily decreases the bandgap of the TiO_2-SiO_2/MWCNT composite from 3.2 to 2.2 eV by increasing the CNT contents from 0.15% to 17.8%, making the absorption visible (Czech and Buda, 2015). CNT plays the role of a dopant for over 3.5 wt.% CNT, however, it acts as supports for a greater concentration. TiO_2-SiO_2/MWCNT with CNT of 17.8 wt.% under UV-A irradiation leads to rapid CBZ destruction *via* a dissociation pathway preferable from that achieved by P25, and the generation of dissociation intermediates. These nanocomposites possessing TiO_2-SiO_2/MWCNT with CNT of 0.01 wt.%

were effective compared with the bare P25 for degradation and adsorption of the DCF under solar or UV-A light. MWCNTs mainly have a role in the degradation step due to the small weight percentages in nanocomposites. It is essential to emphasize that MWCNTs are useful metal-free solar-responsive photocatalysts for degrading DCF despite the light source (visible and UV-A). MWCNTs removed almost 50% of DCF through adsorption plus over 30% to 40% *via* photocatalytic degradation. Begum and Ahmaruzzaman (2018) synthesized biogenic SnO_2/AC nanocomposites through hydrothermal technique. An invincible photocatalytic property was found with the synthesized SnO_2/AC nanocomposites in degrading naproxen (NPX) under direct sunlight. The findings indicated the degradation of 94% of NPX within 2 h.

A new heterojunction photocatalyst was designed by Hu and co-workers (2020); it was also identified as carbon quantum dots-decorated BiOCOOH/ultrathin g-C_3N_4 nanosheets (CQDs/BiOCOOH/uCN) for superior photocatalytic performance for degrading sulfathiazole (STZ). Followed by a light-emitting diode (LED) lamp irradiation for 90 min, the 4-CQDs/BiOCOOH/uCN represented greater degrading efficacy (99.28%) of STZ than pure uCN, BiOCOOH, and 50% BiOCOOH/uCN. Compared to 50% BiOCOOH/uCN, introducing the CQDs improved the photocatalytic performance, and the reason is probably the fast photo-created charges transfer and separation over the 4-CQDs/BiOCOOH/uCN surface. Moreover, $^{\bullet}OH$ and $O_2^{\bullet-}$ were the main reactive types for inducing STZ degradation.

A metal-free photocatalyst of 0D GQDs was reported by Yuan et al. (2019) *via* decorated g-C_3N_4 nanorods (g-CNNR) attained through the hydrothermal treatments. Such noticeable advantages enhanced the GQDs/g-CNNR photocatalytic activity for the effective elimination of OTC antibiotics. It has a photocatalytic reaction rate of 2.03 and 3.46 times greater than the g-CNNR and the pristine g-C_3N_4, respectively. Figure 11.5 represents the proposed photocatalytic mechanism for GQDs/g-CNNR.

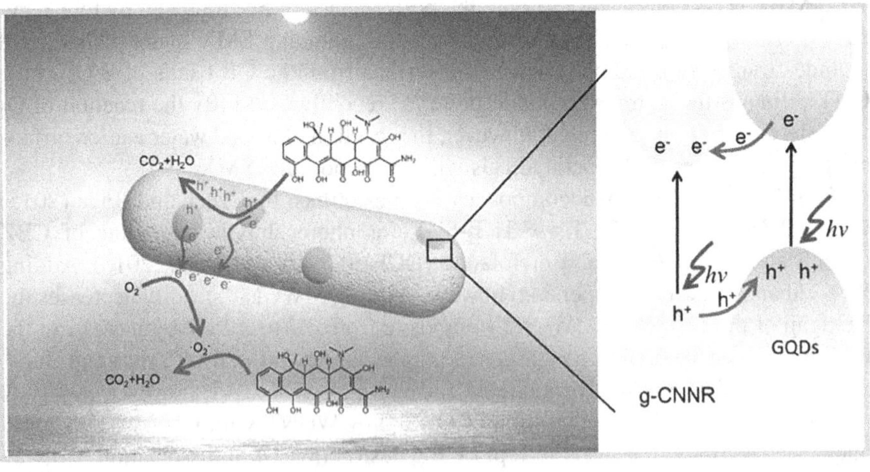

FIGURE 11.5 Proposed photocatalytic mechanism of the GQDs/g-CNNR. (Adapted from reference (Yuan et al., 2019) with permission from Elsevier.)

11.4 SONOCATALYTIC PROCESS

The ultrasound (US) technology application for the destruction of resisting organic contaminants (Khataee et al., 2018; Gholami et al., 2019b) is performed through the sonocatalytic process, which is an innovative method of water and wastewater treatment. Using ultrasound technology for treatment applications provokes the acoustic cavitation phenomenon, which in turn causes degradation of organic compounds through breakage of bonds (because of high pressure and temperature) and generation of radicals through dissociation of water and other oxidants (Hassani et al., 2018a; Anandan et al., 2020). Similar to the photocatalytic process, the US may be applied for excitation of the photocatalyst by generating e^-/h^+ pair in the sonocatalytic process (Hassani et al., 2017b; Eghbali et al., 2019); the potential catalyst, however, may encounter the recombination problem of e^- and h^+ which may be prevented through the use of CNMs, as they are applicable as an accelerator of the e^- transfer phenomena (Li et al., 2018b). The application of CNMs for sonocatalytic removal of water and wastewater is restricted to modifying the catalyst. CNMs have been utilized as supporting substances for homogeneous and heterogeneous systems. The support put forward by CNMs increased the performance of sonocatalytic systems (Nidheesh, 2017). Researchers have proven that CNTs could be utilized as catalysts and supports in ultrasonic systems since they not only increase the generation of $^{\bullet}OH$ radicals by enhancing the number of cavitation bubbles, but they also adsorb the contaminant molecules onto the surface of catalysts (Im et al., 2013). Recent studies have demonstrated satisfactory results in the sonocatalytic elimination of pharmaceuticals in aquatic solutions.

Al-Hamadani et al. (2017) have combined the sonolysis with SWCNTs and obtained positive outcomes in removing SMX and IBP. In their work, they achieved the benefits of two technologies, including adsorption and oxidation. The removal efficiency of 99% for SMX and IBP utilizing US/SWCNTs system within the contact time of 60 min was obtained. Likewise, Al-Hamadani et al. (2018) investigated the sonocatalytic removal of DCF and CBZ in the presence of GO within an aqueous solution. Removal of over 99% for DCF and CBZ was found by utilizing 45 mg/L of GO at pH 3.5 and 580 kHz within 40 min. The higher removal ration could be due to the US system's improved creation of $^{\bullet}OH$ generation and cavitation microbubbles. Lee and co-workers (Lee et al., 2018) reported the synthesis of $GO/\beta\text{-}Bi_2O_3/TiO_2/Bi_2Ti_2O_7$ heterojuncted nanocomposites, selected as GBT, through a two-phase hydrothermal procedure. The GBT sonocatalytic performance was investigated for degrading two target pharmaceutical compounds ([APAP] and [CBZ]) at various frequencies (580, 28, and 970 kHz) and a comparison was made with Ti-doped GO (GT) and Bi-doped GO (GB). GBT presented greater sonocatalytic efficiency than single-metal-doped catalysts, similar to GT and GB based on the generation of H_2O_2 and degrading the chosen pharmaceuticals. ZnO-biochar (ZnO-BC) nanocomposites were prepared by Gholami et al. (2019a) using the hydrothermal method as an effective sonocatalyst for degrading and mineralizing the gemifloxacin (GMF). Greater sonocatalytic performance of BC/ZnO (96.1%) compared to BC (23.4%) and ZnO (12.2%) was found as a result of the improved surface area of the nanocomposites sonocatalysts and the lower bandgap. Moreover, the sonoluminescence phenomena are enhanced in the presence of BC. The removal efficiency of COD of 84.0% was found over the reaction time of 90 min.

In another study, Gholami and co-workers (2020) synthesized Fe-Cu layered double hydroxide/biochar (Fe-Cu-LDH/BC) nanocomposites through hydrothermal technique and tested the potential efficiency for cefazolin sodium (CFZ) sonocatalytic destruction from the solution. The mechanisms of sonoluminescence and hot spots were utilized for explaining the CFZ sonocatalytic dissociation in the presence of Fe-Cu-LDH/BC nanocomposites. The main optimized operational parameters included Fe-Cu-LDH/BC dose of 1 g/L, pH value of 6.5 (natural), an ultrasonic power of 300 W, and an initial CFZ concentration of 0.1 mM. These optimized conditions resulted in a degradation efficacy of 97.6% within 80 minutes. The elimination of COD in the CFZ solution was 71.6%.

Li et al. (2018b) made a new Z-scheme composite sonocatalyst known as mMBiPO$_4$-MWCNT-In$_2$O$_3$, through calcination and hydrothermal approaches. The sonocatalytic activities for mMBiPO$_4$-MWCNT-In$_2$O$_3$ nanocomposites were assessed by degrading the NOX in aqueous solutions under ultrasonic irradiations. The findings indicated that the electron transfer is accelerated by adding MWCNT and recombining e^-/h^+ in Z-scheme mMBiPO$_4$-MWCNT-In$_2$O$_3$ sonocatalyst is restrained. However, the creation of the Z-scheme sonocatalytic system and the presence of MWCNT led to heightened activity of mMBiPO$_4$-In$_2$O$_3$ in the degradation of NOX. For mMBiPO$_4$-MWCNT-In$_2$O$_3$ sonocatalyst, 69.07% of degradation efficiency was found under ultrasonic irradiation for 150 minutes. The results indicated that the presence of MWCNT with superior electron conductivity and definite surface area increases the speed of the electrons transfer from CB of monoclinic monazite structure mMBiPO$_4$ to VB of In$_2$O$_3$, by further improving the sonocatalytic activities.

11.5 FENTON AND FENTON-LIKE PROCESS

As one of the conventional AOPs, Fenton and Fenton-like processes are based on the ferrous ions (Fe^{2+}, Fe^{3+}) and hydrogen peroxide (H_2O_2) to generate $^{\bullet}OH$ radicals in aqueous solution, having a high oxidation degree of pharmaceuticals removal (Hassani et al., 2018c). The use of CNMs is also applicable for the decomposition of H_2O_2 and the generation of $^{\bullet}OH$. Besides, the loading of iron or iron oxides by the CNMs may considerably enhance the catalytic performance under the Fenton system. The reactions are summarized as follows:

$$Fe^{2+} + H_2O_2 + H^+ \rightarrow Fe^{3+} + {}^{\bullet}OH + H_2O \tag{11.1}$$

$$Fe^{3+} + H_2O_2 \rightarrow Fe^{2+} + H^+ + HO_2^{\bullet} \tag{11.2}$$

$$HO_2^{\bullet} + H_2O_2 \rightarrow H_2O + {}^{\bullet}OH + O_2 \tag{11.3}$$

According to related literature, some types of CNMs with Fe catalysts were employed in Fenton and Fenton-like processes for the elimination of pharmaceuticals, such as Fe_3O_4-Mn_3O_4/RGO (Wan and Wang, 2017), Ce^0/Fe^0-RGO (Wan et al., 2016), α-FeOOH/RGO (Zhuang et al., 2019), Fe_3O_4-CeO_2/AC (Liu et al., 2019), Fe-Cu doped CNTs (Barrios-Bermúdez et al., 2020).

Graphene composite electrode Ce-Fe-graphene (Ce^0/Fe^0-RGO) Fenton-like system has 99% removal rate of antibiotic sulfamethazine (SMT) (20 mg/L) and 73% degradation rate of TOC under the condition of 8 mM H_2O_2 and 7 pH (Wan et al., 2016). OFX elimination was assessed using Fe_3O_4-CeO_2/AC as a heterogeneous Fenton catalyst (Liu et al., 2019). The best conditions were found to be pH 3.3, H_2O_2 20.0 mM, OFX 12 mg/L, and catalyst 0.5 g/L. The OFX and TOC elimination efficiencies were 95% and 54%, respectively, within 60 min.

Fe-Cu doped carbon nanotubes were synthesized using a facile way by Barrios-Bermúdez et al. (2020). They were utilized as effective catalysts for the removal of paracetamol (90%–98% within 5 h) by a hybrid process of Fenton-like oxidation and adsorption under mild reaction conditions (25°C and pH nearly neutral). The catalyst showed the best performance for both removal and mineralization of paracetamol. Complete elimination and 85.6% of TOC removal were obtained within 5 h.

11.6 ELECTROCHEMICAL PROCESS

Electrochemical (EC) methods during the past decades, among various water treatment models, have attracted growing interest owing to their advantages like high efficiency, convenience, and environmental friendliness (Brillas and Martínez-Huitle, 2015). The electrochemical advanced oxidation processes (EAOPs) like anodic oxidation, electro-Fenton (EF), and other electro-Fenton-like systems have emerged as a strong approach to successful elimination of contaminants (Moreira et al., 2017). Using EAOPs, the contaminants are deteriorated at the anode or by Fenton's reagent, partially or completely generated by electrode reactions. EF is the most prevalent method employed for EAOPs, through which H_2O_2 is produced at the cathode with O_2 or air feed, while an iron catalyst (Fe^{2+}, Fe^{3+}, or iron oxides) is applied to the liquid. Fenton process occurring between Fe^{2+} and electrogenerated H_2O_2 generates $^{\bullet}OH$ radicals as a strong oxidizing factor for the elimination of pollutants (Brillas et al., 2009). As was predicted, the electrode material introduces influential impacts on the properties and performance of EC water treatment techniques. CNMs are applicable as appropriate electrodes due to their large conductivity, great stability, high non-toxicity, and commercial accessibility. Meanwhile, modifying other electrode materials using CNMs may enhance their electroactive surface area and its oxygen mass transfer rate results in more electrogenerated species and better process effectivity. EC oxidation for organic pollutants consumes much less time than that of traditional biological wastewater purification methods (Szpyrkowicz et al., 2005). Cathode is the functional electrode of the electro-Fenton process. In-situ H_2O_2 generation and the $^{\bullet}OH$ production rate *via* reaction with catalyst is dependent mainly upon the material of cathode and its properties. CNMs for electrodes have proved their effectiveness as an EF cathode material for the production of H_2O_2 and degrading contaminants. For instance, graphene layers would be an appropriate surface for the attachment of Fe_3O_4 particles as the catalyst of the EF process. Divyapriya et al. (2017) removed bisphenol A (BPA) with a quinone-functionalized electrochemically exfoliated graphene (QEEG) and Fe_3O_4 composite electrode (QEEG/Fe_3O_4) by EF system. At pH 3, 100% removal rate could be obtained in 90 min, 98% removal rate could be obtained under neutral conditions, and the iron-leaching rate of the

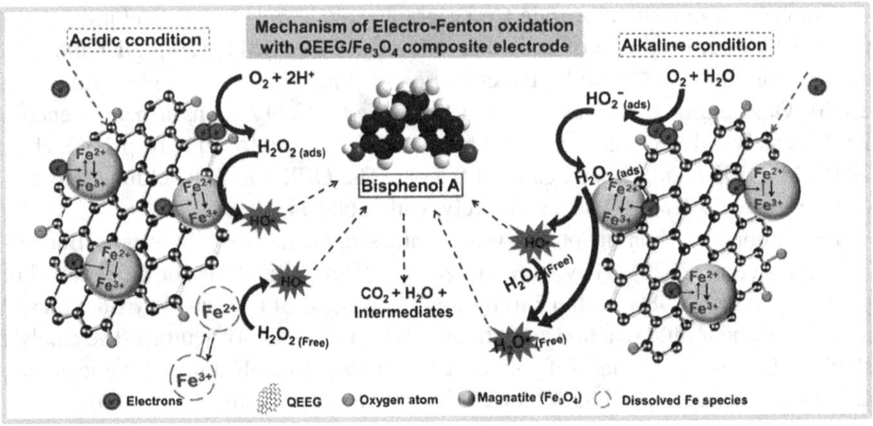

FIGURE 11.6 Simplified presentation of the mechanisms of EF oxidation in the presence of QEEG/Fe_3O_4 electrode. (Adapted from Divyapriya et al. (2017) with permission from Elsevier)

composite electrode was less than 1%. Figure 11.6 shows the simplified presentation of the mechanisms of EF oxidation in the presence of QEEG/Fe_3O_4 electrode.

Similarly, Divyapriya et al. (2018) studied the heterogeneous EF removal of CIP under neutral pH by using a graphene-based graphite felt electrode with ferrocene. They found 99.9% removal efficiency of CIP within 15 min and 120 min of reaction time at pH 3.0 and pH 7.0, respectively. The reuse test results revealed the high stability of the electrode within five consecutive tests with no significant change in its activity. Ghasemi et al. (2020) prepared CuFeNLDH-CNTs incorporated graphite cathode for elimination and mineralization of CFZ antibiotic *via* EF process. At the end of the reaction time of 100 min, the removal efficiencies of CFZ by the CuFeNLDH-CNTs-coated graphite cathode, CNTs-coated, and bare graphite cathode were obtained 89.9%, 28.7%, and 22.8%, respectively. The COD removal of the CFZ was found to be 70.1% after the reaction time of 300 min. A 3D CeO_2/RGO composite was prepared by Li et al. (2017) *via* hydrothermal self-assembly and in-situ deposition techniques. The prepared composite was utilized as cathode for elimination of CIP in EF system. 90.97% elimination efficiency for CIP (50 mg/L) was achieved after the treatment time of 6.5 h.

Dong et al. (2016) developed an effective heterogeneous metal-free cathodic electrochemical advance oxidation process (CEAOP) for decomposition of BPA from wastewater by electrochemically reduced graphene oxide (ERGO)-modified gas diffusion electrode (GDE). BPA and TOC removals were recorded as 100% and 74.6% by using modified cathode, respectively. Jiang et al. (2018) used the catalytic membrane EF system prepared by graphene to remove florfenicol, with a removal rate of 90%, much better than the single EC filtration process (50%) and the filtration process (27%).

Mi et al. (2019) modified carbon felt (CF) with an active, stable, and low-cost RGO-Ce/WO_3 nanosheets (RCW). They used the prepared RCW/CF cathode to remove CIP in EF process. About 100% removal efficiency of CIP (after 1 h) and

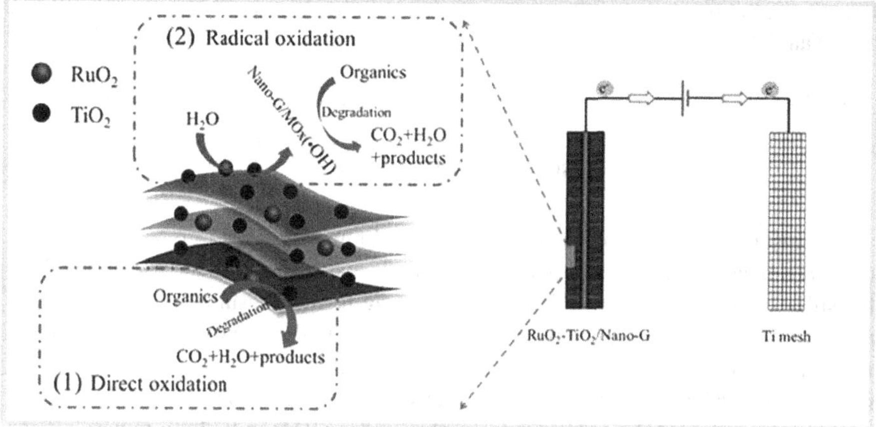

FIGURE 11.7 Schematic representation of electrochemical removal of organics using RuO_2-TiO_2/Nano-G anode. (Adapted from Li et al. (2018a) with permission from Elsevier)

98.55% mineralization (after 8 h) was achieved. Since CNMs have a high specific surface area and outstanding conductivity, the EF cathode based on graphene has great advantages in increasing the rate of electron transfer and H_2O_2 produced by a cathodic oxygen reduction reaction. They can be used as a kind of electrode material with high activity and stability to treat real wastewater. Li et al. (2018a) prepared RuO_2-TiO_2/Nano-graphite (Nano-G) composite anode for EC removal of ceftriaxone sodium. According to obtained results, RuO_2-TiO_2/Nano-G anode showed an improved EC removal efficiency toward ceftriaxone and yield of $^{\bullet}OH$, which is obtained from the synergetic impact between RuO_2, TiO_2, and Nano-G increasing the surface area, improving the EC oxidation performance, and lowering the charge transfer resistance. Based on direct and indirect oxidation, the mechanism of EC oxidation for the removal of ceftriaxone sodium by RuO_2-TiO_2/Nano-G anode is suggested and demonstrated in Figure 11.7.

11.7 SULFATE RADICAL-BASED AOPs (SR-AOPs)

Recently, SR-AOPs have attracted much attention owing to their unmatched benefits, such as broad pH range, high oxidation degree, selectivity, and efficiency to oxidize pollutants (Ghanbari and Moradi, 2017; Guo et al., 2020). SR-AOPs could generate high reactive sulfate $\left(SO_4^{\bullet-}\right)$ radicals by activating peroxymonosulfate (PMS) and peroxydisulfate (PDS) in an aquatic environment. Consequently, $SO_4^{\bullet-}$ radicals can eliminate the pharmaceuticals efficiently (Wang and Wang, 2018; Ghanbari et al., 2020). CNMs were typically used as a significant supportive medium for the production of $SO_4^{\bullet-}$ radicals (Duan et al., 2015; Ghanbari and Moradi, 2017; Nidheesh, 2017). CNMs supported metal oxides such as CoO, CuO, Fe_2O_3, MnO_2, etc., displayed efficiently boosting catalytic performance of persulfate (PS) and PMS (Duan et al., 2015; Ghanbari and Moradi, 2017). CNMs could give electrons to PS or PMS to produce the reactions of the reactive radicals (Ghanbari and Moradi, 2017; Wang and Wang, 2018).

$$S_2O_8^{2-} + e^- \rightarrow SO_4^{\bullet-} + SO_4^{2-} \tag{11.4}$$

$$HSO_5^- + e^- \rightarrow SO_4^{\bullet-} + OH^- \tag{11.5}$$

For instance, Chen et al. (2018c) synthesized nitrogen-doped graphene (NG) *via* thermal annealing of GO and urea. The prepared catalyst was applied to activate PMS for sulfacetamide (SAM) elimination. According to their results, NG600 (NG prepared at 600°C) was a high-efficient PMS activator with a superior elimination performance of SAM 100% after 50 min. Noorisepehr et al. (2020) applied SR-AOPs for removal of APAP using PDS activated by ferroferric oxide nanocatalyst (FON) coated on AC (denoted as FON@AC) and UV light. 100% APAP and 72.8% mineralization were obtained by FON@AC/PDS/UV process at the end of 60 min reaction.

Yang et al. (2020) have employed a simple hydrothermal method to prepare sludge-activated carbon-based $CoFe_2O_4$ ($CoFe_2O_4$-SAC) nanocomposites and achieved successful results for the removal of NOX *via* heterogeneously activating PMS (Li et al., 2012; Lim et al., 2015). Based on their results, NOX could almost be fully removed at pH in the range of 5–9 within 60 min reaction in the $CoFe_2O_4$-SAC/PMS system. Moreover, the radical path of NOX removal in the $CoFe_2O_4$-SAC/PMS system was mainly contributed by $SO_4^{\bullet-}$ and •OH radicals. Cherifi and co-workers (Cherifi et al., 2019) have reported removing BPA and TC by PMS activation by RGO under US irradiation. The application of RGO in the PMS/US system led to full elimination in relatively short times (40 min). TOC results demonstrated that US/PMS/RGO could mineralize (more than 85%) the investigated pollutants significantly. Kang and co-workers (2020) synthesized NG using a hydrothermal technique, which resulted in improved catalytic performance for PMS activation and sulfachlorpyridazine (SCP) removal compared with the undoped RGO and GO. The complete removal rate of SCP was achieved in a basic solution (pH = 12) after 9 min of reaction time.

A mesoporous nanocomposite ($CoFe_2O_4$/mpg-C_3N_4,CF/MCN) was prepared by a self-assembly technique and used as an activator of PMS to remove APAP (Hassani et al., 2020a). The impact of key factors of CF/MCN/PMS system on APAP removal was studied. More than 92% of APAP removal was obtained under the optimum conditions (pH of 7, PMS of 1.5 mM, CF/MCN of 40 mg/L, and time of 25 min). Wang et al. (2017) created a magnetic g-C_3N_4/$MnFe_2O_4$/graphene (C_3N_4@$MnFe_2O_4$-G) composite utilizing a facile impregnation method. The photocatalytic performances of C_3N_4@$MnFe_2O_4$-G were investigated within photo Fenton-like degrading antibiotic pollutants such as AMX, MET, TC, and CIP by utilizing PS as an oxidant under visible light radiation. The removal efficiencies of 91.5%, 94.6%, 84.3%, and 64.6% were found by TC, MET, AMX, and CIP, respectively, indicating the synergistic promoting impact between PS and C_3N_4@$MnFe_2O_4$-G. Yang et al. (2019a) examined photocatalytic removal of SMX by TiO_2/RGO nanocomposites under visible light and demonstrated that TiO_2/RGO is a PS activator for SMX and other pharmaceuticals removal *via* •OH and $SO_4^{\bullet-}$.

Duan et al. (2020b) employed a sol-gel method to prepare a novel SnO_2-Al_2O_3/CNT anode. They used the prepared anode for electrocatalytic decomposition of aqueous ceftazidime. The results showed an improvement from 40% to almost 90% with SnO_2-Al_2O_3/CNT anode. Moreover, the activation by PDS and Fenton process enhanced the mineralization degree of ceftazidime. A 40% TOC elimination was obtained by anodic oxidation alone, while 45.2% and 55.1% TOC elimination were achieved by adding 2 mM PDS and 2 mM Fe^{2+}, respectively.

11.8 HYBRID AOPs

A series of hybrid AOPs techniques were assessed to increase the removal efficiency for pharmaceuticals by CNMs: photoelectrocatalytic (Fang et al., 2013; Kadeer et al., 2019); sonophotocatalysis (Khataee et al., 2019; Reheman et al., 2019; Vinesh et al., 2019); photocatalytic ozonation (Hassani et al., 2017a; Sheydaei et al., 2018; Checa et al., 2019); and sono-electrochemical (Darvishi Cheshmeh Soltani and Mashayekhi, 2018).

Kadeer et al. (2019) prepared RGO/AgCl quantum dots (QDs) by ultrasonic-assisted technique and assessed their PEC efficiency for the elimination of TC under visible light irradiation with an applied potential of 1.5 V (vs. Ag/AgCl). The best performance (82.5%) was obtained for the PEC system, which was higher than sole photocatalysis (39.2%) and electrocatalysis (20.7%), respectively. The enhanced PEC activity of the RGO/AgCl QDs nanocomposites could be related to the intimate contact between RGO and AgCl QDs, which is efficient for the generation of more active sites and promoted e^-/h^+ separation. Vinesh et al. (2019) evaluated and compared the sonophotocatalytic removal of TC by using Au/BeTiO_2/RGO nanocomposites. The outcomes showed that the individual impact of US and photocatalysis for the removal of TC was obtained 45% and 12%, respectively. The complete removal of TC (100%) happened with 1.3-fold synergistic effect for hybrid system in 1 h. Furthermore, ~74% of TOC elimination was observed within 60 min, which further confirmed the efficient mineralization of TC by the resultant nanocomposites.

Supporting CNMs could be utilized in the photocatalytic ozonation system. Sheydaei et al. (2018) prepared N-TiO_2/GO/Ti grid sheet *via* electrophoretic deposition technique for photocatalytic ozonation of cefixime under visible light radiation. The removal efficiencies of cefixime were 17%, 29%, and 51% for adsorption, photocatalysis, and ozonation, respectively. However, they were enhanced to 80% by applying a photocatalytic ozonation process. Zeng et al. (2019) employed photocatalytic ozonation with TiO_2/CDs to treat CIP. The results revealed that the photocatalytic ozonation (99.7% in 16 min) was more effective than the sum of the single efficiency of 6 wt.% TiO_2/CDs and ozonation process. Furthermore, among the above-mentioned systems, 6 wt.% TiO_2/CDs put forward the maximum decomposition rate of CIP.

A GO/TiO_2 composite was prepared through the sol-gel technique. The prepared catalyst was used for the mineralization of primidone by using LED visible light and ozone (Checa et al., 2019). About 20 mg/L of primidone was eliminated in a relatively short time (20 min) by the ozonation process. However, the results demonstrated a synergistic impact between photocatalysis and ozonation for mineralization purposes. The integration of ozone and GO enhanced the performance of TiO_2 under visible light. Under optimization of the process (0.25 g/L catalyst load, 359 W/m light radiance, around 0.75% GO loading in the catalyst). The mineralization degree of 82% was achieved with GO/TiO_2 in comparison to bare TiO_2 (70%) within 2 h. Kakavandi et al. (2019) have reported a sonophotocatalytic system to remove TC using TiO_2 decorated on magnetic activated carbon (MAC@T). According to the suggested system, the removal efficiency of the pollutant was enhanced significantly in the coupled system (MAC@T/UV/US). More than 93% and 50% TOC were removed after the reaction time of 180 min. The results showed the involvement of h^+, $^1O_2$, and $^{\bullet}OH$ radicals in the removal of TC. Hayati et al. (2020) have employed the sonophotocatalytic process for the removal of SDZ by using hexagonal magnesium oxide (MgO) coupled with carbon nanotubes (CNTs) (marked as MCs/UV/US). The complete removal efficiency was achieved at initial SDZ concentration of 45 mg/L, MCs concentration of 0.9 g/L, pH of 11, US power of 200 W, and UV intensity of 150 W within 80 min reaction, owing to the production of $^{\bullet}OH$ radicals and h^+ under sonophotocatalytic system. MCs/UV/US system successfully removed 89% and 96% of TOC and COD in the real pharmaceutical sample, respectively.

Khataee et al. (2019) used a hydrothermal method to synthesize NiFe-layered double hydroxide/reduced graphene oxide (NiFe-LDH/RGO) nanocomposites. The sonophotocatalytic performance of NiFe-LDH/RGO nanocomposites was explored for the decomposition of moxifloxacin (MOX) from the solution. The best sonophotocatalytic performance of 90.40% was achieved after 60 min reaction time under optimal conditions ([NiFe-LDH/RGO] = 1 g/L, [MOX] = 20 mg/L, ultrasonic power = 150 W, and pH = 8 (natural)). Table 11.2 summarizes several recently published studies on the various pharmaceuticals treatment methods using CNMs.

11.9 SUMMARY AND FUTURE PERSPECTIVES

One of the emergent threats seriously endangering both humans and the environment is PPCPs pollutants. To mitigate such a threat, many scientists and policymakers at both national and international levels have shown interest in investigating the possible ways to remove PPCPs from the environment, especially from water resources. These researches revealed that CNMs have the potential to eliminate PPCPs due to their physicochemically promising properties. Based on the latest water and wastewater treatment strategies, the purification of aqueous solutions before releasing them into the environment is the top priority. Accordingly, developing a comprehensive yet cost-effective and globally applicable approach for pollutants removal from the sewages constitutes a serious challenge. Among the CNMs, several compounds including CNTs, GR, RGO, fullerene, AC, BC, CQDs, and g-C_3N_4 have shown greater potential for the PPCPs removal according to the reports found in the literature owing to their excellent physicochemical properties as well as environmental friendliness. In this chapter, initially, a summary of the most recent developments regarding the use of CNMs in the water treatment techniques is given, including photocatalysis, sonocatalysis, Fenton and Fenton-like, electrochemical, sulfate radical-based AOPs, and hybrid AOPs. The CNMs exhibit great contaminant elimination performance due to the large surface area they can provide for the adsorption process. The CNMs' better photocatalytic and

TABLE 11.2
Summary of Various Pharmaceuticals Removal Methods Using CNMs

Catalyst	Process	Pollutant	Experimental Conditions	Efficiency %	References
MWCNT/TiO_2	Photocatalytic	TC	$[TC]_0$ = 10 mg/L, $[catalyst]_0$ = 0.2 g/L, light source = 6 W UVC, pH = 5, and time = 100 min	100	Ahmadi et al. (2017)
GQD/$AgVO_3$	Photocatalytic	IBP	$[IBP]_0$ = 10 mg/L, catalyst dosage = 0.01 g/50 mL, light source = 350 W Xenon, and time = 120 min	>90	Lei et al. (2016)
BiOCOOH/CQD	Photocatalytic	DCF	$[DCF]_0$ = 4 mg/L, catalyst dosage = 0.01 g/50 mL, pH = 7, light source = 350 W Xenon, and time = 120 min	98	Chen et al. (2018b)
NCDs/TNS-001	Photocatalytic	DCF	$[DCF]_0$ = 10 mg/L, catalyst dosage = 50 mg, light source = 350 W Xenon, and time = 60 min	91.5	Wang et al. (2019a)
Fe_3O_4-Mn_3O_4/RGO	Fenton-like	SMT	$[SMT]_0$ = 0.07 mmol/L, $[catalyst]_0$ = 0.6 g/L, pH = 7, light source = 350 W Xenon, and time = 60 min	98	Wan and Wang (2017)
$CuFeO_2$/BC	Fenton-like	TC	$[TC]_0$ = 20 mg/L, $[catalyst]_0$ = 500 mg/L, pH = 5, $[H_2O_2]$ = 50 mM, and time = 300 min	88.4	Xin et al. (2020)
α-FeOOH/RGO	Fenton-like	TC	$[TC]_0$ = 100 mg/20 mL, $[catalyst]_0$ = 100 mg/L, pH = 6, $[H_2O_2]$ = 60 μL, and time = 160 min	97.3	Zhuang et al. (2019)
$CoFe_2O_4$-GO	Oxidant activation	NOX	$[NOX]_0$ = 15 μM, $CoFe_2O_4$-GO dosage = 0.3 g/L, pH = 9, [PS] = 0.5 mM, and time = 20 min	100	Chen et al. (2018a)
Granular activated carbon (GAC)	Oxidant activation	MET	$[MET]_0$ = 0.58 mM, GAC dosage = 5 g/L, pH = 3.9, PS/MET molar ratio= 100/1, and time = 240 min	80	Forouzesh et al. (2019)
Co-Fe PBAs@rGO	Oxidant activation	LVX	$[LVX]_0$ = 20 mg/L, [catalyst] = 50 mg/L, pH = 5.5, [PMS] = 1 mM, and time = 30 min	97.6	Pi et al. (2018)
MgO/ZnO/GR (MZG)	Sonophotocatalytic	SMX	$[SMX]_0$ = 55 mg/L, MZG dosage = 0.8 g/L, pH = 9.0, US power = 250 W, light source= 90 W LED, and time = 120 min	100	Moradi et al. (2020)
WO_3/CNT	Sonophotocatalytic	TC	$[TC]_0$ = 60 mg/L, WO_3/CNT dosage = 0.7 g/L, pH = 9, US power = 250 W/m^2, light intensity = 120 W/m^2, and time = 240 min,	100	Isari et al. (2020)
CeO_2/MWCNTs	Electrochemical	Ceftazidime	$[Ceftazidime]_0$ = 1 mg/L, $[Na_2SO_4]_0$ = 1 g/L, current density = 3 mA/cm^2, time = 60 min, electrode spacing =1 cm	100	Hu et al. (2016b)
Carbon black-coated carbon cloth	ECP/US	IBA	$[IBP]_0$ = 5 mg/L, $[Na_2SO_4]_0$ = 0.05 M, pH = natural, frequency = 37 kHz, applied current = 0.03 A, and time = 60 min	~85	Darvishi Cheshmeh Soltani and Mashayekhi (2018)

sonocatalytic properties are principally associated with their CNMs electron acceptance role, preventing the recombination of the photo-induced e^-/h^+ pairs. The CNMs additionally can be used in the decomposition process of hydrogen peroxide and •OH generation in the Fenton and Fenton-like processes. Moreover, the CNMs iron and iron oxide loads can dramatically improve the catalytic properties under the aforementioned conditions. Besides, the CNMs can be used as proper electrodes owing to their great conductivity, reliable stability, and non-toxicity during EC processes in addition to their market availability. Furthermore, the CNMs have already been used as an appropriate supporting medium for producing sulfate radicals. Nevertheless, there is a need for further investigations to address various CNMs challenges pertinent to their structural and physicochemical specifications, feasibility, and reproduction, in addition to improving their disposal, consistency, and constancy in working with different water treatment systems. Cost-effectiveness and long-term durability monitoring of CNMs application under practical conditions in large-scale treatment systems constitute other important issues to be considered by further research.

REFERENCES

Abo-Hamad, A., AlSaadi, M.A., Hayyan, M., Juneidi, I., Hashim, M.A. 2016. Ionic liquid-carbon nanomaterial hybrids for electrochemical sensor applications: A review. *Electrochim. Acta* 193: 321–343.

Adeleye, A.S., Conway, J.R., Garner, K., et al. 2016. Engineered nanomaterials for water treatment and remediation: Costs, benefits, and applicability. *Chem. Eng. J.* 286: 640–662.

Ahmad, M., Lee, S.S., Dou, X., et al. 2012. Effects of pyrolysis temperature on soybean stover- and peanut shell-derived biochar properties and TCE adsorption in water. *Bioresour. Technol.* 118: 536–544.

Ahmadi, M., Ramezani Motlagh, H., Jaafarzadeh, N., et al. 2017. Enhanced photocatalytic degradation of tetracycline and real pharmaceutical wastewater using MWCNT/TiO_2 nano-composite. *J. Environ. Manage.* 186: 55–63.

Al-Hamadani, Y.A.J., Jung, C., Im, J.-K., et al. 2017. Sonocatalytic degradation coupled with single-walled carbon nanotubes for removal of ibuprofen and sulfamethoxazole. *Chem. Eng. Sci.* 162: 300–308.

Al-Hamadani, Y.A.J., Lee, G., Kim, S., et al. 2018. Sonocatalytic degradation of carbamazepine and diclofenac in the presence of graphene oxides in aqueous solution. *Chemosphere* 205: 719–727.

Al-Khateeb, L.A., Almotiry, S., Salam, M.A. 2014. Adsorption of pharmaceutical pollutants onto graphene nanoplatelets. *Chem. Eng. J.* 248: 191–199.

Anandan, S., Kumar Ponnusamy, V., Ashokkumar, M. 2020. A review on hybrid techniques for the degradation of organic pollutants in aqueous environment. *Ultrason. Sonochem.* 67: 105130.

Ania, C., Velasco, L., Valdés-Solís, T. 2012. Photochemical behavior of carbon adsorbents. In *Novel Carbon Adsorbents*, edited by Tascón, J.M.D., 521–547. Oxford, UK: Elsevier.

Anirudhan, T.S., Deepa, J.R. 2017. Nano-zinc oxide incorporated graphene oxide/nanocellulose composite for the adsorption and photo catalytic degradation of ciprofloxacin hydrochloride from aqueous solutions. *J. Colloid Interface Sci.* 490: 343–356.

Appavu, B., Thiripuranthagan, S., Ranganathan, S., Erusappan, E., Kannan, K. 2018. $BiVO_4$/N-rGO nano composites as highly efficient visible active photocatalyst for the degradation of dyes and antibiotics in eco system. *Ecotoxicol. Environ. Saf.* 151: 118–126.

Awfa, D., Ateia, M., Fujii, M., Yoshimura, C. 2019. Novel magnetic carbon nanotube-TiO_2 composites for solar light photocatalytic degradation of pharmaceuticals in the presence of natural organic matter. *J. Water Process. Eng.* 31: 100836.

Bai, J., Li, Y., Jin, P., Wang, J., Liu, L. 2017. Facile preparation 3D ZnS nanospheres-reduced graphene oxide composites for enhanced photodegradation of norfloxacin. *J. Alloys Compd.* 729: 809–815.

Barrios-Bermúdez, N., González-Avendaño, M., Lado-Touriño, I., Cerpa-Naranjo, A., Rojas-Cervantes, M.L. 2020. Fe-Cu doped multiwalled carbon nanotubes for fenton-like degradation of paracetamol under mild conditions. *Nanomaterials (Basel)* 10: 749.

Bashiri, F., Khezri, S.M., Kalantary, R.R., Kakavandi, B. 2020. Enhanced photocatalytic degradation of metronidazole by TiO_2 decorated on magnetic reduced graphene oxide: Characterization, optimization and reaction mechanism studies. *J. Mol. Liq.* 314: 113608.

Begum, S., Ahmaruzzaman, M. 2018. Biogenic synthesis of SnO_2/activated carbon nanocomposite and its application as photocatalyst in the degradation of naproxen. *Appl. Surf. Sci.* 449: 780–789.

Brillas, E., Martínez-Huitle, C.A. 2015. Decontamination of wastewaters containing synthetic organic dyes by electrochemical methods: An updated review. *Appl. Catal. B Environ.* 166–167: 603–643.

Brillas, E., Sirés, I., Oturan, M.A. 2009. Electro-Fenton process and related electrochemical technologies based on Fenton's reaction chemistry. *Chem. Rev.* 109: 6570–6631.

Carmalin Sophia, A., Lima, E.C., Allaudeen, N., Rajan, S. 2016. Application of graphene based materials for adsorption of pharmaceutical traces from water and wastewater- a review. *Desalination Water Treat.* 57: 27573–27586.

Chakraborty, K., Pal, T., Ghosh, S. 2018. RGO-ZnTe: A graphene based composite for tetracycline degradation and their synergistic effect. *ACS Appl. Nano Mater.* 1: 3137–3144.

Checa, M., Figueredo, M., Aguinaco, A., Beltrán, F.J. 2019. Graphene oxide/titania photocatalytic ozonation of primidone in a visible LED photoreactor. *J. Hazard. Mater.* 369: 70–78.

Chen, L., Ding, D., Liu, C., et al. 2018a. Degradation of norfloxacin by $CoFe_2O_4$-GO composite coupled with peroxymonosulfate: A comparative study and mechanistic consideration. *Chem. Eng. J.* 334: 273–284.

Chen, P., Zhang, Q., Su, Y., et al. 2018b. Accelerated photocatalytic degradation of diclofenac by a novel CQDs/BiOCOOH hybrid material under visible-light irradiation: Dechloridation, detoxicity, and a new superoxide radical model study. *Chem. Eng. J.* 332: 737–748.

Chen, X., Oh, W.-D., Hu, Z.-T., et al. 2018c. Enhancing sulfacetamide degradation by peroxymonosulfate activation with N-doped graphene produced through delicately-controlled nitrogen functionalization via tweaking thermal annealing processes. *Appl. Catal. B Environ.* 225: 243–257.

Cherifi, Y., Addad, A., Vezin, H., et al. 2019. PMS activation using reduced graphene oxide under sonication: Efficient metal-free catalytic system for the degradation of rhodamine B, bisphenol A, and tetracycline. *Ultrason. Sonochem.* 52: 164–175.

Czech, B., Buda, W. 2015. Photocatalytic treatment of pharmaceutical wastewater using new multiwall-carbon nanotubes/TiO_2/SiO_2 nanocomposites. *Environ. Res.* 137: 176–184.

Czech, B., Buda, W. 2016. Multicomponent nanocomposites for elimination of diclofenac in water based on an amorphous TiO_2 active in various light sources. *J. Photochem. Photobiol. A.* 330: 64–70.

Dai, J., Meng, X., Zhang, Y., Huang, Y. 2020. Effects of modification and magnetization of rice straw derived biochar on adsorption of tetracycline from water. *Bioresour. Technol.* 311: 123455.

Darvishi Cheshmeh Soltani, R., Mashayekhi, M. 2018. Decomposition of ibuprofen in water via an electrochemical process with nano-sized carbon black-coated carbon cloth as oxygen-permeable cathode integrated with ultrasound. *Chemosphere* 194: 471–480.

De Volder, M.F.L., Tawfick, S.H., Baughman, R.H., Hart, A.J. 2013. Carbon nanotubes: Present and future commercial applications. *Science* 339: 535–539.

Divyapriya, G., Nambi, I., Senthilnathan, J. 2018. Ferrocene functionalized graphene based electrode for the electro–Fenton oxidation of ciprofloxacin. *Chemosphere* 209: 113–123.

Divyapriya, G., Nambi, I.M., Senthilnathan, J. 2017. An innate quinone functionalized electrochemically exfoliated graphene/Fe_3O_4 composite electrode for the continuous generation of reactive oxygen species. *Chem. Eng. J.* 316: 964–977.

Dong, H., Su, H., Chen, Z., Yu, H., Yu, H. 2016. Fabrication of electrochemically reduced graphene oxide modified gas diffusion electrode for in-situ electrochemical advanced oxidation process under mild conditions. *Electrochim. Acta* 222: 1501–1509.

Dong, X., Ma, L.Q., Zhu, Y., Li, Y., Gu, B. 2013. Mechanistic investigation of mercury sorption by brazilian pepper biochars of different pyrolytic temperatures based on X-ray photoelectron spectroscopy and flow calorimetry. *Environ. Sci. Technol.* 47: 12156–12164.

Duan, P., Gao, S., Lei, J., Li, X., Hu, X. 2020a. Electrochemical oxidation of ceftazidime with graphite/CNT-Ce/PbO_2–Ce anode: Parameter optimization, toxicity analysis and degradation pathway. *Environ. Pollut.* 263: 114436.

Duan, P., Liu, W., Lei, J., Sun, Z., Hu, X. 2020b. Electrochemical mineralization of antibiotic ceftazidime with SnO_2-Al_2O_3/CNT anode: Enhanced performance by peroxydisulfate/Fenton activation and degradation pathway. *J. Environ. Chem. Eng.* 8: 103812.

Duan, X., Sun, H., Kang, J., et al. 2015. Insights into heterogeneous catalysis of persulfate activation on dimensional-structured nanocarbons. *ACS Catal.* 5: 4629–4636.

Eghbali, P., Hassani, A., Sündü, B., Metin, Ö. 2019. Strontium titanate nanocubes assembled on mesoporous graphitic carbon nitride ($SrTiO_3$/mpg-C_3N_4): Preparation, characterization and catalytic performance. *J. Mol. Liq.* 290: 111208.

Elessawy, N.A., Elnouby, M., Gouda, M.H., et al. 2020. Ciprofloxacin removal using magnetic fullerene nanocomposite obtained from sustainable PET bottle wastes: Adsorption process optimization, kinetics, isotherm, regeneration and recycling studies. *Chemosphere* 239: 124728.

Fang, T., Liao, L., Xu, X., Peng, J., Jing, Y. 2013. Removal of COD and colour in real pharmaceutical wastewater by photoelectrocatalytic oxidation method. *Environ. Technol.* 34: 779–786.

Forouzesh, M., Ebadi, A., Aghaeinejad-Meybodi, A. 2019. Degradation of metronidazole antibiotic in aqueous medium using activated carbon as a persulfate activator. *Sep. Purif. Technol.* 210: 145–151.

Georgakilas, V., Perman, J.A., Tucek, J., Zboril, R. 2015. Broad family of carbon nanoallotropes: Classification, chemistry, and applications of fullerenes, carbon dots, nanotubes, graphene, nanodiamonds, and combined superstructures. *Chem. Rev.* 115: 4744–4822.

Ghanbari, F., Moradi, M. 2017. Application of peroxymonosulfate and its activation methods for degradation of environmental organic pollutants: Review. *Chem. Eng. J.* 310: 41–62.

Ghanbari, F., Zirrahi, F., Olfati, D., Gohari, F., Hassani, A. 2020. TiO_2 nanoparticles removal by electrocoagulation using iron electrodes: Catalytic activity of electrochemical sludge for the degradation of emerging pollutant. *J. Mol. Liq.* 310: 113217.

Ghasemi, M., Khataee, A., Gholami, P., et al. 2020. In-situ electro-generation and activation of hydrogen peroxide using a CuFeNLDH-CNTs modified graphite cathode for degradation of cefazolin. *J. Environ. Manage.* 267: 110629.

Gholami, P., Dinpazhoh, L., Khataee, A., Hassani, A., Bhatnagar, A. 2020. Facile hydrothermal synthesis of novel Fe-Cu layered double hydroxide/biochar nanocomposite with enhanced sonocatalytic activity for degradation of cefazolin sodium. *J. Hazard. Mater.* 381: 120742.

Gholami, P., Dinpazhoh, L., Khataee, A., Orooji, Y. 2019a. Sonocatalytic activity of biochar-supported ZnO nanorods in degradation of gemifloxacin: Synergy study, effect of parameters and phytotoxicity evaluation. *Ultrason. Sonochem.* 55: 44–56.

Gholami, P., Khataee, A., Soltani, R.D.C., Bhatnagar, A. 2019b. A review on carbon-based materials for heterogeneous sonocatalysis: Fundamentals, properties and applications. *Ultrason. Sonochem.* 58: 104681.

Guo, R., Wang, Y., Li, J., Cheng, X., Dionysiou, D.D. 2020. Sulfamethoxazole degradation by visible light assisted peroxymonosulfate process based on nanohybrid manganese dioxide incorporating ferric oxide. *Appl. Catal. B Environ.* 278: 119297.

Hasanzadeh, A., Khataee, A., Zarei, M., Joo, S.W. 2018. Photo-assisted electrochemical abatement of trifluralin using a cathode containing a C60-carbon nanotubes composite. *Chemosphere* 199: 510–523.

Hassani, A., Çelikdağ, G., Eghbali, P., et al. 2018a. Heterogeneous sono-Fenton-like process using magnetic cobalt ferrite-reduced graphene oxide ($CoFe_2O_4$-rGO) nanocomposite for the removal of organic dyes from aqueous solution. *Ultrason. Sonochem.* 40: 841–852.

Hassani, A., Eghbali, P., Ekicibil, A., Metin, Ö. 2018b. Monodisperse cobalt ferrite nanoparticles assembled on mesoporous graphitic carbon nitride ($CoFe_2O_4$/mpg-C_3N_4): A magnetically recoverable nanocomposite for the photocatalytic degradation of organic dyes. *J. Magn. Magn. Mater.* 456: 400–412.

Hassani, A., Eghbali, P., Kakavandi, B., Lin, K.-Y.A., Ghanbari, F. 2020a. Acetaminophen removal from aqueous solutions through peroxymonosulfate activation by $CoFe_2O_4$/mpg-C_3N_4 nanocomposite: Insight into the performance and degradation kinetics. *Environ. Technol. Innov.* 20: 101127.

Hassani, A., Faraji, M., Eghbali, P. 2020b. Facile fabrication of mpg-C_3N_4/Ag/ZnO nanowires/Zn photocatalyst plates for photodegradation of dye pollutant. *J. Photochem. Photobiol. A.* 400: 112665.

Hassani, A., Karaca, M., Karaca, S., et al. 2018c. Preparation of magnetite nanoparticles by high-energy planetary ball mill and its application for ciprofloxacin degradation through heterogeneous Fenton process. *J. Environ. Manage.* 211: 53–62.

Hassani, A., Khataee, A., Fathinia, M., Karaca, S. 2018d. Photocatalytic ozonation of ciprofloxacin from aqueous solution using TiO_2/MMT nanocomposite: Nonlinear modeling and optimization of the process via artificial neural network integrated genetic algorithm. *Process Saf. Environ. Prot.* 116: 365–376.

Hassani, A., Khataee, A., Karaca, S., Fathinia, M. 2017a. Degradation of mixture of three pharmaceuticals by photocatalytic ozonation in the presence of TiO_2/montmorillonite nanocomposite: Simultaneous determination and intermediates identification. *J. Environ. Chem. Eng.* 5: 1964–1976.

Hassani, A., Khataee, A., Karaca, S., Karaca, C., Gholami, P. 2017b. Sonocatalytic degradation of ciprofloxacin using synthesized TiO_2 nanoparticles on montmorillonite. *Ultrason. Sonochem.* 35: 251–262.

Hassani, A., Khataee, A.R. 2017. Activated carbon fiber for environmental protection. In *Activated Carbon Fiber and Textiles*, edited by Chen, J.Y., 245–280. Oxford: Woodhead Publishing.

Hassani, A., Khataee, A.R. 2018. Application of biochar in advanced oxidation processes. In *Non-Soil Biochar Applications*, edited by Dimitrios, K., Dimitrios, N., Pantelis, S. Environmental Science, Engineering and Technology, 85–126. New York: Nova Science Publishers.

Hayati, F., Isari, A.A., Anvaripour, B., Fattahi, M., Kakavandi, B. 2020. Ultrasound-assisted photocatalytic degradation of sulfadiazine using MgO@CNT heterojunction composite: Effective factors, pathway and biodegradability studies. *Chem. Eng. J.* 381: 122636.

He, Y., Sutton, N.B., Rijnaarts, H.H.H., Langenhoff, A.A.M. 2016. Degradation of pharmaceuticals in wastewater using immobilized TiO_2 photocatalysis under simulated solar irradiation. *Appl. Catal. B Environ.* 182: 132–141.

Houtman, C.J. 2010. Emerging contaminants in surface waters and their relevance for the production of drinking water in Europe. *J. Integr. Environ. Sci.* 7: 271–295.

Hu, X.-Y., Zhou, K., Chen, B.-Y., Chang, C.-T. 2016a. Graphene/TiO_2/ZSM-5 composites synthesized by mixture design were used for photocatalytic degradation of oxytetracycline under visible light: Mechanism and biotoxicity. *Appl. Surf. Sci.* 362: 329–334.

Hu, X., Yu, Y., Sun, Z. 2016b. Preparation and characterization of cerium-doped multiwalled carbon nanotubes electrode for the electrochemical degradation of low-concentration ceftazidime in aqueous solutions. *Electrochim. Acta* 199: 80–91.

Hu, Z., Xie, X., Li, S., et al. 2020. Rational construct CQDs/BiOCOOH/uCN photocatalyst with excellent photocatalytic performance for degradation of sulfathiazole. *Chem. Eng. J.* 404: 126541.

Huang, J., Zimmerman, A.R., Chen, H., Gao, B. 2020. Ball milled biochar effectively removes sulfamethoxazole and sulfapyridine antibiotics from water and wastewater. *Environ. Pollut.* 258: 113809.

Iijimam, S., 1991. Helical microtubules of graphitic carbon. *Nature*, 354: 56–58.

Im, J.-K., Heo, J., Boateng, L.K., et al. 2013. Ultrasonic degradation of acetaminophen and naproxen in the presence of single-walled carbon nanotubes. *J. Hazard. Mater.* 254–255: 284–292.

Isari, A.A., Mehregan, M., Mehregan, S., et al. 2020. Sono-photocatalytic degradation of tetracycline and pharmaceutical wastewater using WO_3/CNT heterojunction nanocomposite under US and visible light irradiations: A novel hybrid system. *J. Hazard. Mater.* 390: 122050.

Jelić, A., Petrović, M., Barceló, D. 2012. Pharmaceuticals in drinking water. In *Emerging Organic Contaminants and Human Health*, edited by Barceló, D., 47–70. Berlin, Heidelberg: Springer Berlin Heidelberg.

Jiang, W.-L., Xia, X., Han, J.-L., et al. 2018. Graphene modified electro-fenton catalytic membrane for in situ degradation of antibiotic florfenicol. *Environ. Sci. Technol.* 52: 9972–9982.

Jo, W.-K., Kumar, S., Isaacs, M.A., Lee, A.F., Karthikeyan, S. 2017. Cobalt promoted TiO_2/GO for the photocatalytic degradation of oxytetracycline and Congo Red. *Appl. Catal. B Environ.* 201: 159–168.

Kadeer, K., Reheman, A., Maimaitizi, H., et al. 2019. Preparation of rGO/AgCl QDs and its enhanced photoelectrocatalytic performance for the degradation of Tetracycline. *J. Am. Ceram. Soc.* 102: 5342–5352.

Kakavandi, B., Bahari, N., Rezaei Kalantary, R., Dehghani Fard, E. 2019. Enhanced sono-photocatalysis of tetracycline antibiotic using TiO_2 decorated on magnetic activated carbon (MAC@T) coupled with US and UV: A new hybrid system. *Ultrason. Sonochem.* 55: 75–85.

Kang, J., Zhou, L., Duan, X., Sun, H., Wang, S. 2020. Catalytic degradation of antibiotics by metal-free catalysis over nitrogen-doped graphene. *Catal. Today* 357: 341–349.

Karaolia, P., Michael-Kordatou, I., Hapeshi, E., et al. 2018. Removal of antibiotics, antibiotic-resistant bacteria and their associated genes by graphene-based TiO_2 composite photocatalysts under solar radiation in urban wastewaters. *Appl. Catal. B Environ.* 224: 810–824.

Khalid, N.R., Majid, A., Tahir, M.B., Niaz, N.A., Khalid, S. 2017. Carbonaceous-TiO_2 nanomaterials for photocatalytic degradation of pollutants: A review. *Ceram. Int.* 43: 14552–14571.

Khataee, A., Alidokht, L., Hassani, A., Karaca, S. 2013. Response surface analysis of removal of a textile dye by a Turkish coal powder. *Adv. Environ. Res.* 2: 291–308.

Khataee, A., Eghbali, P., Irani-Nezhad, M.H., Hassani, A. 2018. Sonochemical synthesis of WS_2 nanosheets and its application in sonocatalytic removal of organic dyes from water solution. *Ultrason. Sonochem.* 48: 329–339.

Khataee, A., Sadeghi Rad, T., Nikzat, S., et al. 2019. Fabrication of NiFe layered double hydroxide/reduced graphene oxide (NiFe-LDH/rGO) nanocomposite with enhanced sonophotocatalytic activity for the degradation of moxifloxacin. *Chem. Eng. J.* 375: 122102.

Khraisheh, M., Kim, J., Campos, L., et al. 2013. Removal of carbamazepine from water by a novel TiO_2-coconut shell powder/UV process: Composite preparation and photocatalytic activity. *Environ Eng Sci* 30: 515–526.

Kroto, H.W., Heath, J.R., O'Brien, S.C., Curl, R.F., Smalley, R.E. 1985. C60: Buckminsterfullerene. *Nature* 318: 162–163.

Kulkarni, S.K. *Nanotechnology: Principles and Practices.* Cham, Switzerland: Springer International Publishing, 2014.

Lee, G., Chu, K.H., Al-Hamadani, Y.A.J., et al. 2018. Fabrication of graphene-oxide/β-Bi_2O_3/TiO_2/$Bi_2Ti_2O_7$ heterojuncted nanocomposite and its sonocatalytic degradation for selected pharmaceuticals. *Chemosphere* 212: 723–733.

Lei, Z.-D., Wang, J.-J., Wang, L., et al. 2016. Efficient photocatalytic degradation of ibuprofen in aqueous solution using novel visible-light responsive graphene quantum dot/$AgVO_3$ nanoribbons. *J. Hazard. Mater.* 312: 298–306.

Levin, R.B., Epstein, P.R., Ford, T.E., et al. 2002. U.S. drinking water challenges in the twenty-first century. *Environ. Health Perspect.* 110: 43–52.

Li, D., Guo, X., Song, H., Sun, T., Wan, J. 2018a. Preparation of RuO_2-TiO_2/Nano-graphite composite anode for electrochemical degradation of ceftriaxone sodium. *J. Hazard. Mater.* 351: 250–259.

Li, D., Wang, T., Li, Z., et al. 2020. Application of graphene-based materials for detection of nitrate and nitrite in water: A review. *Sensors* 20: 54.

Li, H., Kang, Z., Liu, Y., Lee, S.-T. 2012. Carbon nanodots: Synthesis, properties and applications. *J. Mater. Chem* 22: 24230–24253.

Li, S., Wang, G., Qiao, J., et al. 2018b. Sonocatalytic degradation of norfloxacin in aqueous solution caused by a novel Z-scheme sonocatalyst, mMBIP-MWCNT-In_2O_3 composite. *J. Mol. Liq.* 254: 166–176.

Li, S., Wang, Z., Zhao, X., et al. 2019. Insight into enhanced carbamazepine photodegradation over biochar-based magnetic photocatalyst Fe_3O_4/BiOBr/BC under visible LED light irradiation. *Chem. Eng. J.* 360: 600–611.

Li, Y., Han, J., Xie, B., et al. 2017. Synergistic degradation of antimicrobial agent ciprofloxacin in water by using 3D CeO_2/RGO composite as cathode in electro-Fenton system. *J. Electroanal. Chem.* 784: 6–12.

Lim, S.Y., Shen, W., Gao, Z. 2015. Carbon quantum dots and their applications. *Chem. Soc. Rev.* 44: 362–381.

Lin, L., Wang, H., Xu, P. 2017. Immobilized TiO_2-reduced graphene oxide nanocomposites on optical fibers as high performance photocatalysts for degradation of pharmaceuticals. *Chem. Eng. J.* 310: 389–398.

Liu, F.-f., Zhao, J., Wang, S., Du, P., Xing, B. 2014. Effects of solution chemistry on adsorption of selected pharmaceuticals and personal care products (PPCPs) by graphenes and carbon nanotubes. *Environ. Sci. Technol.* 48: 13197–13206.

Liu, J., Wu, X., Liu, J., et al. 2019. Ofloxacin degradation by Fe_3O_4-CeO_2/AC Fenton-like system: Optimization, kinetics, and degradation pathways. *Mol. Catal.* 465: 61–67.

Liu, W., Sutton, N.B., Rijnaarts, H.H.M., Langenhoff, A.A.M. 2016. Pharmaceutical removal from water with iron- or manganese-based technologies: A review. *Crit. Rev. Environ. Sci. Technol.* 46: 1584–1621.

Ma, J., Yang, Q., Wen, Y., Liu, W. 2017. Fe-g-C_3N_4/graphitized mesoporous carbon composite as an effective Fenton-like catalyst in a wide pH range. *Appl. Catal. B Environ.* 201: 232–240.

Mestre, A.S., Carvalho, A.P. 2019. Photocatalytic degradation of pharmaceuticals carbamazepine, diclofenac, and sulfamethoxazole by semiconductor and carbon materials: A review. *Molecules* 24: 3702.

Mi, X., Han, J., Sun, Y., et al. 2019. Enhanced catalytic degradation by using RGO-Ce/WO_3 nanosheets modified CF as electro-Fenton cathode: Influence factors, reaction mechanism and pathways. *J. Hazard. Mater.* 367: 365–374.

Moradi, S., Sobhgol, S.A., Hayati, F., et al. 2020. Performance and reaction mechanism of MgO/ZnO/Graphene ternary nanocomposite in coupling with LED and ultrasound waves for the degradation of sulfamethoxazole and pharmaceutical wastewater. *Sep. Purif. Technol.* 251: 117373.

Moreira, F.C., Boaventura, R.A.R., Brillas, E., Vilar, V.J.P. 2017. Electrochemical advanced oxidation processes: A review on their application to synthetic and real wastewaters. *Appl. Catal. B Environ.* 202: 217–261.

Naseri, A., Samadi, M., Pourjavadi, A., Moshfegh, A.Z., Ramakrishna, S. 2017. Graphitic carbon nitride (g-C_3N_4)-based photocatalysts for solar hydrogen generation: recent advances and future development directions. *J. Mater. Chem. A.* 5: 23406–23433.

Ncibi, M.C., Sillanpää, M. 2017. Optimizing the removal of pharmaceutical drugs Carbamazepine and Dorzolamide from aqueous solutions using mesoporous activated carbons and multi-walled carbon nanotubes. *J. Mol. Liq.* 238: 379–388.

Nidheesh, P.V. 2017. Graphene-based materials supported advanced oxidation processes for water and wastewater treatment: a review. *Environ. Sci. Pollut. Res.* 24: 27047–27069.

Noorisepehr, M., Kakavandi, B., Isari, A.A., et al. 2020. Sulfate radical-based oxidative degradation of acetaminophen over an efficient hybrid system: Peroxydisulfate decomposed by ferroferric oxide nanocatalyst anchored on activated carbon and UV light. *Sep. Purif. Technol.* 250: 116950.

Park, C.M., Heo, J., Wang, D., Su, C., Yoon, Y. 2018. Heterogeneous activation of persulfate by reduced graphene oxide–elemental silver/magnetite nanohybrids for the oxidative degradation of pharmaceuticals and endocrine disrupting compounds in water. *Appl. Catal. B Environ.* 225: 91–99.

Peñas-Garzón, M., Gómez-Avilés, A., Belver, C., Rodriguez, J.J., Bedia, J. 2020. Degradation pathways of emerging contaminants using TiO_2-activated carbon heterostructures in aqueous solution under simulated solar light. *Chem. Eng. J.* 392: 124867.

Pi, Y., Ma, L., Zhao, P., et al. 2018. Facile green synthetic graphene-based Co-Fe Prussian blue analogues as an activator of peroxymonosulfate for the degradation of levofloxacin hydrochloride. *J. Colloid Interface Sci.* 526: 18–27.

Priya, B., Shandilya, P., Raizada, P., et al. 2016. Photocatalytic mineralization and degradation kinetics of ampicillin and oxytetracycline antibiotics using graphene sand composite and chitosan supported BiOCl. *J. Mol. Catal. A: Chem.* 423: 400–413.

Qian, L., Chen, B. 2013. Dual role of biochars as adsorbents for aluminum: The effects of oxygen-containing organic components and the scattering of silicate particles. *Environ. Sci. Technol.* 47: 8759–8768.

Reheman, A., Kadeer, K., Okitsu, K., et al. 2019. Facile photo-ultrasonic assisted reduction for preparation of rGO/Ag_2CO_3 nanocomposites with enhanced photocatalytic oxidation activity for tetracycline. *Ultrason. Sonochem.* 51: 166–177.

Safarpour, M., Khataee, A. 2019. Graphene-based materials for water purification. In *Nanoscale Materials in Water Purification*, edited by Thomas, S., Pasquini, D., Leu, S.-Y., Gopakumar, D.A., 383–430 New York: Elsevier.

Sanderson, H., Brain, R.A., Johnson, D.J., Wilson, C.J., Solomon, K.R. 2004. Toxicity classification and evaluation of four pharmaceuticals classes: antibiotics, antineoplastics, cardiovascular, and sex hormones. *Toxicology* 203: 27–40.

Sbardella, L., Comas, J., Fenu, A., Rodriguez-Roda, I., Weemaes, M. 2018. Advanced biological activated carbon filter for removing pharmaceutically active compounds from treated wastewater. *Sci. Total Environ.* 636: 519–529.

Shabani, M., Haghighi, M., Kahforoushan, D., Haghighi, A. 2019. Sono-solvothermal hybrid fabrication of $BiOCl-Bi_{24}O_{31}Cl_{10}$/rGO nano-heterostructure photocatalyst with efficient solar-light-driven performance in degradation of fluoroquinolone antibiotics. *Sol. Energy Mater Sol.* 193: 335-350.

Shaniba, C., Akbar, M., Ramseena, K., et al. 2020. Sunlight-assisted oxidative degradation of cefixime antibiotic from aqueous medium using TiO_2/nitrogen doped holey graphene nanocomposite as a high performance photocatalyst. *J. Environ. Chem. Eng.* 8: 102204.

Shen, R., Andrews, S.A. 2011. Demonstration of 20 pharmaceuticals and personal care products (PPCPs) as nitrosamine precursors during chloramine disinfection. *Water Res.* 45: 944–952.

Sheydaei, M., Shiadeh, H.R.K., Ayoubi-Feiz, B., Ezzati, R. 2018. Preparation of nano N-TiO_2/graphene oxide/titan grid sheets for visible light assisted photocatalytic ozonation of cefixime. *Chem. Eng. J.* 353: 138–146.

Sohi, S.P. 2012. Carbon Storage with Benefits. *Science* 338: 1034–1035.

Song, J., Zhen, X., Chang, C.-T. 2016. Hydrothermal synthesis of graphene and titanium dioxide nanotubes by a one-step method for the photocatalytic degradation of amoxicillin. *Nanosci. Nanotechnol. Lett.* 8: 113–119.

Song, Y., Tian, J., Gao, S., et al. 2017. Photodegradation of sulfonamides by g-C_3N_4 under visible light irradiation: Effectiveness, *mechanism and pathways. Appl. Catal. B Environ.* 210: 88–96.

Sun, Y.-P., Zhou, B., Lin, Y., et al. 2006. Quantum-sized carbon dots for bright and colorful photoluminescence. *J. Am. Chem. Soc.* 128: 7756–7757.

Szpyrkowicz, L., Kaul, S.N., Neti, R.N. 2005. Tannery wastewater treatment by electro-oxidation coupled with a biological process. *J. Appl. Electrochem.* 35: 381–390.

Taherpour, A. 2008. Quantitative relationship study of mechanical structure properties of empty fullerenes. *Fuller. Nanotub. Carbon Nanostructures* 16: 196–205.

Taherpour, A.A., Mousavi, F. 2018. Chapter 6-Carbon nanomaterials for electroanalysis in pharmaceutical applications. In *Fullerens, Graphenes and Nanotubes*, edited by Grumezescu, A.M., 169–225 Burlington, MA: William Andrew Publishing.

Tan, X., Liu, Y., Zeng, G., et al. 2015. Application of biochar for the removal of pollutants from aqueous solutions. *Chemosphere* 125: 70–85.

Tang, X., Wang, Z., Wang, Y. 2018. Visible active N-doped TiO_2/reduced graphene oxide for the degradation of tetracycline hydrochloride. *Chem. Phys. Lett.* 691: 408–414.

Vajtai, R. *Springer Handbook of Nanomaterials.* Berlin Heidelberg: Springer, 2013.

Vinesh, V., Shaheer, A.R.M., Neppolian, B. 2019. Reduced graphene oxide (rGO) supported electron deficient B-doped TiO_2 (Au/B-TiO_2/rGO) nanocomposite: An efficient visible light sonophotocatalyst for the degradation of Tetracycline (TC). *Ultrason. Sonochem.* 50: 302–310.

Wan, Z., Hu, J., Wang, J. 2016. Removal of sulfamethazine antibiotics using CeFe-graphene nanocomposite as catalyst by Fenton-like process. *J. Environ. Manage.* 182: 284–291.

Wan, Z., Wang, J. 2017. Degradation of sulfamethazine using Fe_3O_4-Mn_3O_4/reduced graphene oxide hybrid as Fenton-like catalyst. *J. Hazard. Mater.* 324: 653–664.

Wang, F., Wu, Y., Wang, Y., et al. 2019a. Construction of novel Z-scheme nitrogen-doped carbon dots/{001} TiO_2 nanosheet photocatalysts for broad-spectrum-driven diclofenac degradation: Mechanism insight, products and effects of natural water matrices. *Chem. Eng. J.* 356: 857–868.

Wang, J., Wang, S. 2016. Removal of pharmaceuticals and personal care products (PPCPs) from wastewater: A review. *J. Environ. Manage.* 182: 620–640.

Wang, J., Wang, S. 2018. Activation of persulfate (PS) and peroxymonosulfate (PMS) and application for the degradation of emerging contaminants. *Chem. Eng. J.* 334: 1502–1517.

Wang, W., Serp, P., Kalck, P., Faria, J.L. 2005. Visible light photodegradation of phenol on MWNT-TiO_2 composite catalysts prepared by a modified sol–gel method. *J. Mol. Catal. A: Chem.* 235: 194–199.

Wang, X., Maeda, K., Thomas, A., et al. 2009. A metal-free polymeric photocatalyst for hydrogen production from water under visible light. *Nat. Mater* 8: 76–80.

Wang, X., Wang, A., Ma, J. 2017. Visible-light-driven photocatalytic removal of antibiotics by newly designed C_3N_4@$MnFe_2O_4$-graphene nanocomposites. *J. Hazard. Mater.* 336: 81–92.

Wang, X., Yin, R., Zeng, L., Zhu, M. 2019b. A review of graphene-based nanomaterials for removal of antibiotics from aqueous environments. *Environ. Pollut.* 253: 100–110.

Woan, K., Pyrgiotakis, G., Sigmund, W. 2009. Photocatalytic carbon-nanotube–TiO_2 composites. *Adv. Mater.* 21: 2233–2239.

Xie, X., Li, S., Zhang, H., Wang, Z., Huang, H. 2019. Promoting charge separation of biochar-based Zn-TiO_2/pBC in the presence of ZnO for efficient sulfamethoxazole photodegradation under visible light irradiation. *Sci. Total Environ.* 659: 529–539.

Xin, S., Liu, G., Ma, X., et al. 2020. High efficiency heterogeneous Fenton-like catalyst biochar modified $CuFeO_2$ for the degradation of tetracycline: Economical synthesis, catalytic performance and mechanism. *Appl. Catal. B Environ.* 280: 119386.

Xu, J., Cao, Z., Zhang, Y., et al. 2018. A review of functionalized carbon nanotubes and graphene for heavy metal adsorption from water: Preparation, application, and mechanism. *Chemosphere* 195: 351–364.

Xu, X., Ray, R., Gu, Y., et al. 2004. Electrophoretic analysis and purification of fluorescent single-walled carbon nanotube fragments. *J. Am. Chem. Soc.* 126: 12736–12737.

Yan, Q., Gao, X., Chen, Y.-P., et al. 2014. Occurrence, fate and ecotoxicological assessment of pharmaceutically active compounds in wastewater and sludge from wastewater treatment plants in Chongqing, the Three Gorges Reservoir Area. *Sci. Total Environ.* 470–471: 618–630.

Yang, L., Xu, L., Bai, X., Jin, P. 2019a. Enhanced visible-light activation of persulfate by Ti^{3+} self-doped TiO_2/graphene nanocomposite for the rapid and efficient degradation of micropollutants in water. *J. Hazard. Mater.* 365: 107–117.

Yang, W., Zhou, M., Oturan, N., Li, Y., Oturan, M.A. 2019b. Electrocatalytic destruction of pharmaceutical imatinib by electro-Fenton process with graphene-based cathode. *Electrochim. Acta* 305: 285–294.

Yang, Z., Li, Y., Zhang, X., et al. 2020. Sludge activated carbon-based $CoFe_2O_4$-SAC nanocomposites used as heterogeneous catalysts for degrading antibiotic norfloxacin through activating peroxymonosulfate. *Chem. Eng. J.* 384: 123319.

Yekan Motlagh, P., Khataee, A., Hassani, A., Sadeghi Rad, T. 2020. ZnFe-LDH/GO nanocomposite coated on the glass support as a highly efficient catalyst for visible light photodegradation of an emerging pollutant. *J. Mol. Liq.* 302: 112532.

Yekan Motlagh, P., Khataee, A., Sadeghi Rad, T., Hassani, A., Joo, S.W. 2019. Fabrication of ZnFe-layered double hydroxides with graphene oxide for efficient visible light photocatalytic performance. *J. Taiwan Inst. Chem. Eng.* 101: 186–203.

Yin, F., Gu, B., Lin, Y., et al. 2017. Functionalized 2D nanomaterials for gene delivery applications. *Coord. Chem. Rev.* 347: 77–97.

Yu, O., Daoyong, L., Weiran, C., Shaohua, S., Li, C. 2009. A temperature window for the synthesis of single-walled carbon nanotubes by catalytic chemical vapor deposition of CH_4 over Mo_2-Fe_{10}/MgO catalyst. *Nanoscale Res. Lett.* 4: 574.

Yuan, A., Lei, H., Xi, F., et al. 2019. Graphene quantum dots decorated graphitic carbon nitride nanorods for photocatalytic removal of antibiotics. *J. Colloid Interface Sci.* 548: 56–65.

Zeng, Y., Chen, D., Chen, T., et al. 2019. Study on heterogeneous photocatalytic ozonation degradation of ciprofloxacin by TiO_2/carbon dots: Kinetic, mechanism and pathway investigation. *Chemosphere* 227: 198–206.

Zhai, W., Srikanth, N., Kong, L.B., Zhou, K. 2017. Carbon nanomaterials in tribology. *Carbon* 119: 150–171.

Zhang, F., Wang, X., Liu, H., et al. 2019. Recent advances and applications of semiconductor photocatalytic technology. *Appl. Sci.* 9: 2489.

Zhang, H., Wang, Z., Li, R., et al. 2017. TiO_2 supported on reed straw biochar as an adsorptive and photocatalytic composite for the efficient degradation of sulfamethoxazole in aqueous matrices. *Chemosphere* 185: 351–360.

Zhang, Q., Peng, Y., Deng, F., Wang, M., Chen, D. 2020a. Porous Z-scheme MnO_2/Mn-modified alkalinized g-C_3N_4 heterojunction with excellent Fenton-like photocatalytic activity for efficient degradation of pharmaceutical pollutants. *Sep. Purif. Technol.* 246: 116890.

Zhang, S., Li, B., Wang, X., et al. 2020b. Recent developments of two-dimensional graphene-based composites in visible-light photocatalysis for eliminating persistent organic pollutants from wastewater. *Chem. Eng. J.* 390: 124642.

Zheng, W., Guo, M., Chow, T., Bennett, D.N., Rajagopalan, N. 2010. Sorption properties of greenwaste biochar for two triazine pesticides. *J. Hazard. Mater.* 181: 121–126.

Zheng, Y., Lin, L., Wang, B., Wang, X. 2015. Graphitic carbon nitride polymers toward sustainable photoredox catalysis. *Angew. Chem. Int.* 54: 12868–12884.

Zhu, W., Li, Z., He, C., Faqian, S., Zhou, Y. 2018. Enhanced photodegradation of sulfamethoxazole by a novel WO_3-CNT composite under visible light irradiation. *J. Alloys Compd.* 754: 153–162.

Zhu, W., Sun, F., Goei, R., Zhou, Y. 2017. Facile fabrication of RGO-WO_3 composites for effective visible light photocatalytic degradation of sulfamethoxazole. *Appl. Catal. B Environ.* 207: 93–102.

Zhuang, Y., Liu, Q., Kong, Y., et al. 2019. Enhanced antibiotic removal through a dual-reaction-center Fenton-like process in 3D graphene based hydrogels. *Environ. Sci. Nano.* 6: 388–398.

12 Metal–Organic Frameworks and Their Derived Materials in Water Purification

Chizoba I. Ezugwu
University of Alcalá, Alcalá de Henares, Madrid, Spain

Srabanti Ghosh
CSIR – Central Glass and Ceramic Research Institute, Kolkata, India

Marta E.G. Mosquera
Instituto de Investigación en Química "Andrés M. del Río" (IQAR), Universidad de Alcalá, Campus Universitario, Madrid, Spain

Roberto Rosal
University of Alcalá, Alcalá de Henares, Madrid, Spain

CONTENTS

12.1 Introduction 372
12.2 Components of MOFs and Secondary Building Units (SBUs) 373
  12.2.1 Components 373
  12.2.2 Secondary Building Units 375
12.3 Synthesis of Metal–Organic Frameworks 377
12.4 MOF-Derived Materials 379
  12.4.1 Carbonization 379
  12.4.2 Deposition of Metal Nanoparticles 380
  12.4.3 Heterostructured MOFs-Derived Semiconductors Composite 380
12.5 Strategies for Using MOFs in Wastewater Treatment 381
  12.5.1 Sulfate Radical-Based Advanced Oxidation Process 381
  12.5.2 Photocatalysis 384
  12.5.3 Adsorption of Contaminants 387
    12.5.3.1 More on the Mechanisms of Adsorption 393
  12.5.4 MOF-Based Membranes 394
12.6 Conclusion and Prospects 397
References 398

DOI: 10.1201/9781003118749-12

12.1 INTRODUCTION

One of the most critical global environmental challenges threatening humanity is water pollution caused by the rapid growth in industrialization and other anthropogenic activities. The incessant release of chemicals due to agricultural production (pesticides, fungicides, herbicides, and insecticides) and industrial waste like organic (dyes) and inorganic pollutants (heavy metals) that are wastewater contaminants, constitutes serious environmental concern (Bedia et al., 2019). In addition, emerging organic contaminants comprising pharmaceuticals, detergents, personal care products, caffeine, steroids, and hormones, plasticizers, and flame retardants, among others, have been identified in wastewater in concentrations usually ranging from ppt to ppb (Rodriguez-Narvaez et al., 2017). The risk of emerging pollutants for human health and the entire biodiversity has been highlighted elsewhere (Taheran et al., 2018; Rout et al., 2020). Furthermore, the continuous increase in the world population results in an exponential rise for clean drinking water demand. Although 72% of the Earth's surface is occupied by water, only 0.5% is freshwater (Jun et al., 2020). Moreover, most water-intensive activities such as energy production and agriculture are anticipated to increase the global water demand from 60% to 85% over the next three decades (Boretti & Rosa, 2019). Global warming is another factor impacting the availability of water resources. This stresses that water resources are becoming seriously scarce and that there is an urgent need to address this global problem to avoid potential conflicts among water users (Tzanakakis et al., 2020).

Until now, researchers have prepared various materials for wastewater purification (Siegert et al., 2019). For example, zeolites, activated carbon, magnetic materials, chitosan, graphene, and metal oxides have been widely described (Mário et al., 2020). Notwithstanding that most of them have been successfully used for wastewater treatment, there is still a need to enhance their performance (Lee et al., 2012). The increased demand for efficient materials together with the need for purified water, induced the development of advanced hybrid nanomaterials, notably those based on metal–organic frameworks (MOFs). Recently, there have been a tremendous improvement in this class of porous coordination polymers (Mukherjee et al., 2018). Their pores and linkers can be easily functionalized, which is advantageous when compared to conventional inorganic and organic microporous materials (Li et al., 1999). The basic idea for the design and preparation of MOFs is derived from metal carboxylate cluster chemistry, i.e., through the interface between coordination chemistry and material sciences (Yang et al., 2020). Metal–organic frameworks were first reported by the research groups of O.M. Yaghi et al., 1995) and S. Kitagawa (Kondo et al., 1997). The main feature of these framework materials is that their crystal structures are retained after evacuation of solvent, which can be proven by X-ray single-crystal analyses (Figure 12.1). Extensive research attention has been directed toward the synthesis of new MOFs and their applications due to their promising properties (Smolders et al., 2018; Joharian & Morsali, 2019). For industrial applications, the main challenge of MOFs is their stability as the coordination bonds between the ligand and metal components are weak, hence their thermal and chemical stability are generally lower when compared to zeolites. Also, some MOFs are

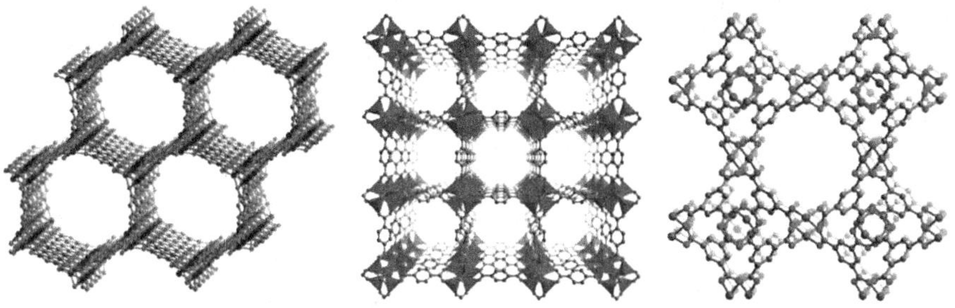

FIGURE 12.1 Crystal structures of some representative MOFs. Reproduced with Permission from *Chem. Sci.*, 2014, 5, 32–51.

moisture- and air-sensitive following evacuation of the pores and should be handled under inert atmosphere.

MOFs materials possess significantly high surface areas along with the lowest densities per gram when compared to competing materials (Furukawa et al., 2010; Farha et al., 2012). These distinctive features make them suitable for environmental applications, like the removal of pollutants by adsorption, which have been extensively investigated (Mason et al., 2014; Wang et al., 2016; Khan & Jhung, 2017; Ezugwu et al., 2018, 2019). Their shape, pore size, and functionality played an essential role in the selective adsorption of pollutants from contaminated water. The fundamental adsorption mechanisms include acid-base π-π interactions/stacking, electrostatic, hydrophobic, and hydrogen bonding (Hasan & Jhung, 2015). MOFs have proven to be good adsorbents for water contaminated with industrial and agricultural wastes such as dyes, pesticides, herbicides, pharmaceuticals and personal care products (PPCP) (Haldar et al., 2020; Hasan et al., 2020), and heavy metals (Dias & Petit, 2015; Shayegan et al., 2020). Sulfate radical-based advanced oxidation process (SR-AOPs), photocatalysis, MOFs-mixed matrix membranes (MMMs) and ion exchange are other techniques for treating wastewater using MOFs.

12.2 COMPONENTS OF MOFS AND SECONDARY BUILDING UNITS (SBUs)

12.2.1 Components

MOFs are composed of metal-based nodes coordinated with organic ligands, resulting from 1D to 3D coordination networks (Ezugwu et al., 2016; Langmi et al., 2016). They are sometimes termed hybrid organic-inorganic materials. However, this has been disallowed by the IUPAC (Batten et al., 2013). The possibility of changing these components (linking unit and metal ion) makes it possible to construct a wide range of metal–organic frameworks. The organic linkers used for MOFs synthesis are from mono- to tetravalent ligands, which can donate electron lone pairs to the

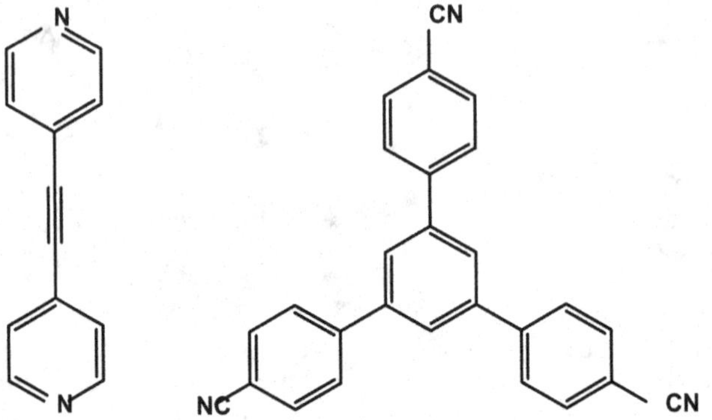

1,2-di(pyridin-4-yl)ethyne5'-(4-cyanophenyl)-[1,1':3',1"-terphenyl]-4,4"-dicarbonitrile

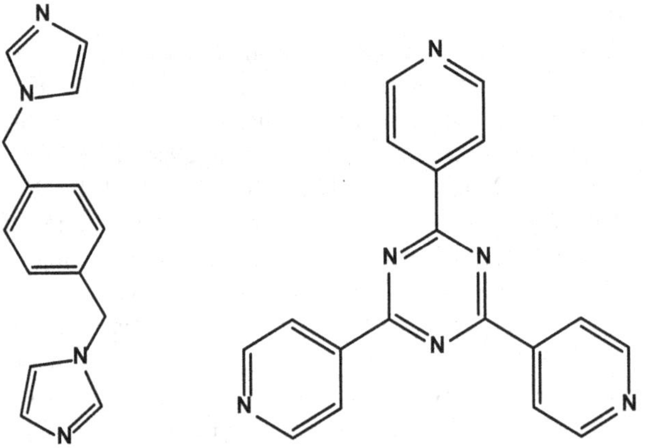

1,4-bis((1*H*-imidazol-1-yl)methyl)benzene 2,4,6-tri(pyridin-4-yl)-1,3,5-triazine

FIGURE 12.2 N-donor ligands employed for the preparation of MOFs.

metal ions that contain vacant orbitals able to accept them. The choice of the ligand and metal ion together with their rational combination determines the structure and properties of the MOFs. For example, the affinity of the metal ion to bind to a particular ligand affects the pore size and influences the number of ligands that can coordinate to the metal. Normally, the two groups of ligands that are commonly employed for the preparation of MOFs are nitrogen-and oxygen-donor ligands, dominated by pyridyl- and carboxylate-containing linkers, respectively (Cook et al., 2013).

Figure 12.2 shows some N-donor ligands. For these ligands, the nitrogen atom makes use of its lone electron pair to coordinate to the metal ions (Zhang et al., 2018). To build MOFs with a particular topology, the functionality, shape, and size of the ligand can be rationally selected. Thus, ditopic ligands (i.e., ligands able to coordinate at two different sites) with an angle of 180°, like 4,4′-bpy, may form molecular frameworks with varied dimensionality. The final structure of MOFs constructed

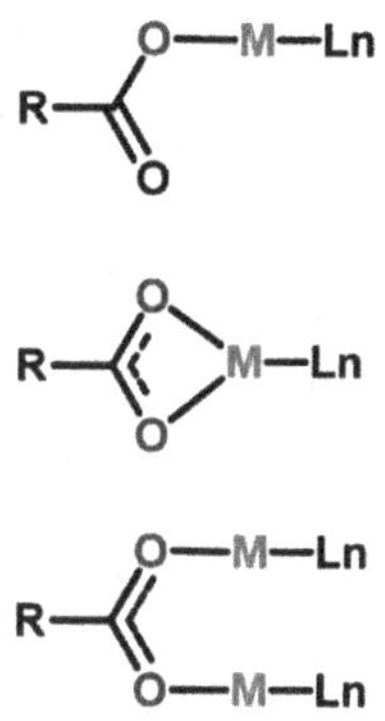

SCHEME 12.1 Binding modes of carboxylate linkers. Permission taken from *Chem. Rev.*, 2013, 113, 734–777.

from these ligands is usually pre-determined by the metal node (Jin et al., 2019). The main limitation of using N-donating ligands for MOFs synthesis is the easy decomposition of the extended structure due to the weak coordination bonds.

Carboxylate compounds are the most frequently used O-donor ligands. They have stronger chelation capacity than N-donating ligands and are the most common choice for MOF synthesis. The negatively charged carboxylate ligand neutralizes most of the positively charged metal ions in the MOFs. Importantly, this may reduce the number of counter-anions needed, which could otherwise occupy part of the available space within the internal cavities and channels (Zhao et al., 2019).

As shown in Scheme 12.1, carboxylates can display three different binding modes: monodentated, bidentated chelate, and bidentated bridging. The versatility offered by different binding modes (chelate and bridging) and their stronger coordination bonds make carboxylate-based ligands (Figure 12.3) are more suitable for MOFs synthesis.

12.2.2 Secondary Building Units

In MOF formation, the major restriction for using a single metal ion is the limitations of orientations, which results in low-quality frameworks with high defective sites. Yaghi et al. (2003) developed a versatile and fruitful design strategy, termed reticular chemistry. This approach utilizes secondary building units (SBUs) as molecular polygons for MOFs formation (Guillerm & Maspoch, 2019; Gropp et al., 2020). These SBUs are well-defined metal clusters rigid in nature, so they maintain their directionality during the self-assembly process (Tranchemontagne et al., 2009). Some common SBUs employed in MOFs synthesis are presented in Figure 12.4. These are unit of "polynuclear clusters" designed from multidentate ligands and two/more metal ions. The organic ligand extends out from these SBUs in an ordered geometry to forms the MOFs structure. Notably, when the geometry and site of SBUs are considered, the network topology can be predicted.

SBUs are not reagents in a synthetic scheme, but they are in-situ generated during MOFs synthesis and depend on the reaction conditions and precursors (Rowsell &

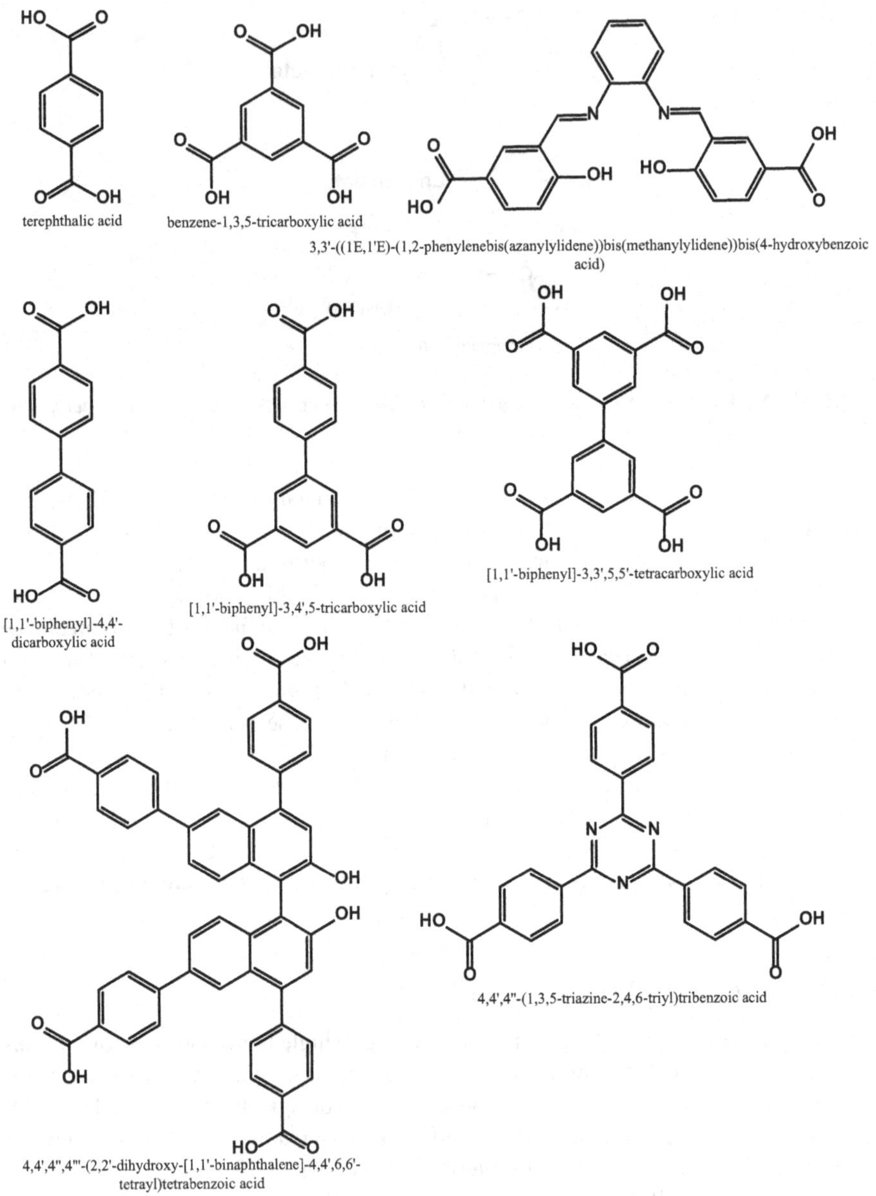

FIGURE 12.3 Some carboxylate-based ligands used for MOFs synthesis.

Yaghi, 2004). The introduction of SBUs helps to comprehend the geometry formed around the metal nodes of the frameworks. In terms of design and utility, the development of these units increases the cavity and pores of the MOFs, as well as allowing higher surface area. The SBUs can control the different coordination modes of

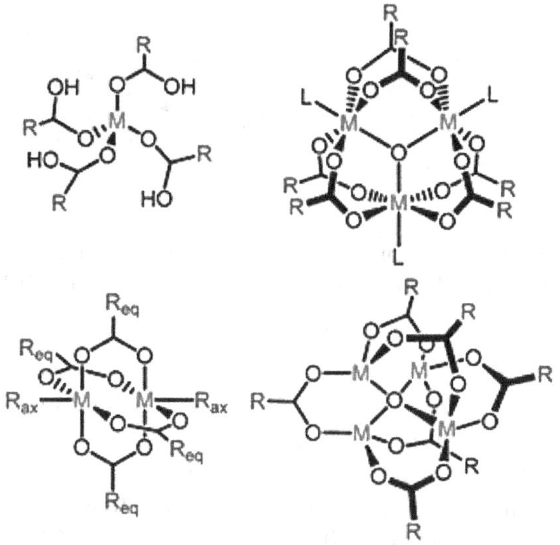

FIGURE 12.4 Structural illustration of some common SBUs. Permission taken from *Chem. Rev.*, 2013, 113, 734–777.

transition metals during the synthesis of the material. Generally, transition metal ions are mainly employed as the metal nodes of the SBUs owing to their inherent properties of adopting varied coordination numbers, which results in different geometry (Kalmutzki et al., 2018). Metals from the first row are commonly used due to their low cost, availability, and thorough understanding of their chemistry.

12.3 SYNTHESIS OF METAL–ORGANIC FRAMEWORKS

The rationale for using different synthetic methods to prepare MOFs is the need to take advantage of all the possible frameworks that can be obtained using different synthesis approaches (Stock & Biswas, 2012). Alternative synthetic methods can produce frameworks with different morphologies and particle size, thereby influencing the overall properties of the material. The particle sizes of MOFs have a direct influence on the diffusion rate/capacity of guest molecules, thus affecting the adsorption and separation efficiencies of the material.

Basically, MOFs are prepared by mixing organic ligand and metal salt in a vessel (Figure 12.5) followed by providing the required condition for their formation (Abednatanzi et al., 2019). The two common techniques adopted for the preparation of MOFs can be classified into the conventional and unconventional methods.

i. **Conventional synthesis**
   This synthetic method is usually applicable to reactions achieved by conventional heating without side reactions. One of the key parameters considered

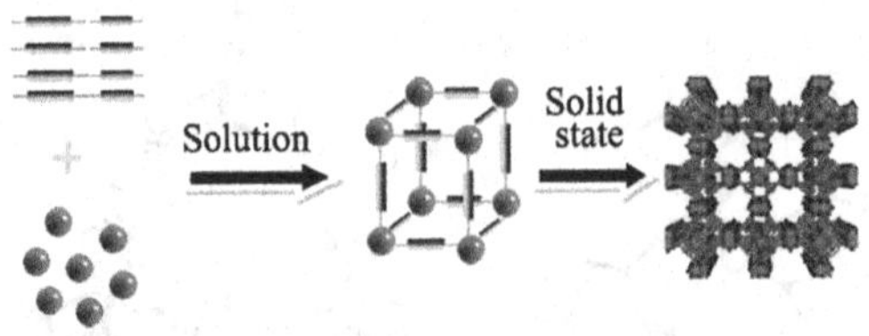

FIGURE 12.5 Metal ions combination with the ligands to synthesize MOFs materials. Reproduced with permission from Coord. *Chem. Rev.*, 2016, 307, 188–210.

in this synthesis is the reaction temperature. Conventional syntheses can follow either solvothermal or non-solvothermal route. Solvothermal synthesis simply involves heating the reactants in closed vessels (normally Teflon-lined autoclave) under pressure and above the solvent's boiling point, which usually contain formamide functionality, like dimethylformamide (DMF) and *N, N*-diethylformamide (DEF) (Safaei et al., 2019). Some prominent MOFs/MOFs composite, applied for the treatment of contaminated water, have been obtained by these methods (Joseph et al., 2019; Sule & Mishra, 2019; Olawale et al., 2020). Crystalline MOFs with high surface area prepared by this approach are often suitable for X-ray crystallography. Good crystallization is achieved by optimizing the amount of solvent and reaction temperature (Vardhan et al., 2016). Nevertheless, the limitations of solvothermal synthesis include high energy requirements, long reaction time, and nitrogen-containing organic solvents, which are not always environmentally benign. Furthermore, some frameworks prepared by this method are thermally unstable or even can react with the solvent.

ii. **Unconventional synthetic method**
The methods for preparing MOFs without using conventional heating are termed unconventional synthesis. They include electrochemical, microwave-assisted, mechanochemical, sonochemical, and solvent-free synthesis techniques. The main objective of these methods is the absence of anions like nitrate and chloride during the synthesis. In the electrochemical method, the metal ions are produced from the metal electrode (Cao et al., 2019; Wei et al., 2019), whereas in the mechanochemical method, a mechanical force is applied to the reactant mixture containing the solid reactant and little solvent (Chen et al., 2019a, b). Unlike solvothermal synthesis that takes a long reaction time (days sometimes), the reaction time is reduced to reach even seconds for microwave-assisted synthesis. The reactants are activated by microwave irradiation (Chen et al., 2019b). However, the limitation with the microwave method is the rapid nucleation and crystal growth, which results in a higher amount of crystal defects and low internal surface area (Mousavi et al., 2018). MOFs can also be prepared by ionothermal techniques that use ionic liquids, fulfilling a dual role as the solvent and templating agent, thereby minimizing the number of species involved in the synthesis (Luo et al., 2019).

12.4 MOF-DERIVED MATERIALS

MOFs have been used as a template for the preparation of different nanostructured materials, like transition metal oxides, heteroatom-doped carbons, and carbon materials by calcination and pyrolysis under appropriate conditions. For example, nanoporous metal/carbons composites can be generated by the pyrolysis of a MOF precursor under an inert atmosphere. Furthermore, direct heat treatment of the framework can form porous metal oxides (Salunkhe et al., 2017), and carbon materials can be produced by etching off the residual metal (Liu et al., 2018a). Some of the advantages of these MOF-derived materials include higher stability, enhanced performance, expanded surface area/porosity and facile synthesis (Jiang et al., 2019).

12.4.1 Carbonization

The stability of MOF materials can be improved by introducing functional groups or direct carbonization of MOF-derived materials. Porous carbons prepared from a metal azolate framework-6 (MAF-6) via different pyrolysis times were used to remove artificial sweeteners (acesulfame, saccharin, and cyclamate) from water (Song et al., 2018). Interestingly, these porous carbons have surface functionality, high porosity, stability, and hydrophobicity. The pyrolysis temperature and time influenced the adsorption process. The samples prepared for 6 h exhibited higher saccharin adsorption (93 mg/g) from the water after 12 h, compared to activated carbon (4.7 mg/g). The main interaction mechanism was assigned to H-bonding; the surface functional phenolic group in the porous carbon acts as the H-donor, while artificial sweeteners act as the H-acceptor.

Liu et al. (2018b) used a fast and facile method of short-term high-temperature approach for the preparation of a functional carbon material (TEPA-C-MOF-5), which was derived from MOF-5. During the synthesis, tetraethylenepentamine (TEPA) doped MOF-5 materials were carbonized at 500°C for several seconds under an air atmosphere, the sample changed from white to dark yellow, Scheme 12.2.

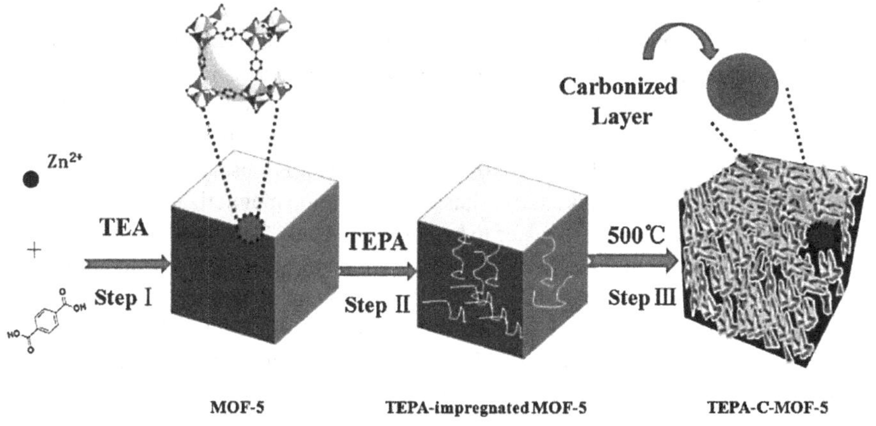

SCHEME 12.2 The formation process of the carbonized sorbent. Permission taken from *Colloids Surf. A Physicochem. Eng. Asp.*, 2018, 556, 72–80.

The prepared carbonized material demonstrated a higher adsorption of U (VI) than the parent MOF-5, 550 mg/gat a relatively low pH (3.5).

Incorporating heteroatoms, such as N, O, and S, into a porous carbon material can be used to enhance their interaction with substrates. Xu et al. (2017) employed ZIF-8 as a self-sacrificial template to synthesize nitrogen-doped porous carbons by first forming ZIF/dicyandiamide (Carbon-ZD) and ZIF/sucrose (Carbon-ZS) composites, followed by carbonization at 950°C under argon atmosphere. Using excess Carbon-ZD to treat wastewater proved an excellent candidate for methylene blue (MB) removal with almost 100% efficiency.

12.4.2 Deposition of Metal Nanoparticles

A high number of magnetic Co nanoparticles uniformly dispersed on nanoporous carbon particles (Co/NPC) were successfully prepared via one-step carbonization of ZIF-67 (Torad et al., 2014). The graphitic carbon has a strong affinity to aromatic compounds, while the zeolite-like pore structure is good for adsorbing metal ions. The obtained composite showed a well-developed graphitized wall with high surface area, thereby exhibiting fast diffusion and excellent adsorption capacity (500 $mg \cdot g^{-1}$) of MB. Interestingly, 99.9% of the MB was captured into Co/NPC sample after 24 h. Compared with the commercial activated carbon, Co/NPC particles showed impressive saturation capacity for MB dye.

Ahsan et al. (2019) prepared Ni and Cu nanoparticles-embedded carbon sheets (C@Ni and C@Cu) via annealing of Cu-BDC and Ni-BDC MOFs, at 600°C under an inert environment. SEM and TEM analyses illustrated the deposition of the metallic Cu and Ni particles on the nanoporous carbon sheet. The developed nano-catalysts were investigated for the $NaBH_4$-mediated reduction of 4-nitrophenol, methyl orange (MO), and MB. They showed efficient nature for the degradation of these water pollutants to their corresponding reduced organic molecules in about 60 s.

12.4.3 Heterostructured MOFs-Derived Semiconductors Composite

Cobalt-containing magnetic carbonaceous nanocomposite (MCN) was prepared from the carbonization of cobalt-based MOF, ZIF-67 at 600°C under N_2 atmosphere (Lin et al., 2015a). The Co_3O_4/C (MCN) composite showed catalytic activity for Oxone (a triple salt of potassium peroxymonosulfate) activation to degrade rhodamine B(RhB) in water. Efficient activation process was observed at Oxone/MCN ratio of 5/1. The heterogenous catalyst has high saturation magnetization (45 emu g^{-1}), allowing their separation from water to be easy (Lin et al., 2015b). Gong et al. (2018) fabricated heterostructured ZIF-NC/g-C_3N_4 composite, which manifested enhanced decomposition of bisphenol A under light, displaying an apparent rate constant with peroxymonosulfate (PMS) and outperforming g-C_3N_4 with a value 8.4 times higher. The improved activity was attributed to the efficient interfacial charge separation between ZIF-NC and g-C_3N_4 together with the abundant surface-active sites for the activation of PMS to $SO_4^{\bullet-}$.

Li et al. (2020) prepared a lotus leaf-like aerogel composed of MOF nanocoating and kapok fiber (a cellulosic fiber from the kapok tree) core. The aerogel was synthesized by growing hydrophobic zeolitic imidazolate frameworks-8 (ZIF-8) on the surface of kapok, which was triggered by Fe^{3+}-tannic acid networks (FTN, a kind of metal-phenolic network). The as-prepared ZIF-8FTN/kapok aerogel showed an ultrahydrophobic surface with a water contact angle of up to 162°. In addition, this composite has the capacity to adsorb oil/organic (20.0–72.0 g $(sorbent)^{-1}$), 17–40 times higher compared to ZIF-8. The superior capacity was attributed to the ZIF-8 being supported to the kapok aerogel.

12.5 STRATEGIES FOR USING MOFS IN WASTEWATER TREATMENT

12.5.1 Sulfate Radical-Based Advanced Oxidation Process

AOPs is a family of innovative wastewater treatment technologies based on the production of highly reactive radicals, like superoxide ($O_2^{\bullet-}$), hydroperoxyl radical ($OOH^{\bullet}$), hydroxyl radical ($OH^{\bullet}$), and sulfate radical ($SO_4^{\bullet-}$) (Chen et al., 2019a, b; Milenković et al., 2020; Zhao et al., 2020). Sulfate radical-based advanced oxidation process (SR-AOPs) is characterized by the in-situ production of sulfate radicals from certain salts and gained popularity in wastewater treatment due to its high selectivity even in complex matrices. $SO_4^{\bullet-}$ is a strong oxidant capable of degrading many organic pollutants. The three main reaction mechanisms of $SO_4^{\bullet-}$ with organic pollutants are: (a) abstraction of hydrogen atom from esters, ethers, alcohol and alkanes; (b) single-electron transfer from compounds with electron-donating groups like amines and hydroxyl moieties; and (c) the addition reaction of $SO_4^{\bullet-}$ to organic compounds that contain unsaturated bonds (Luo et al., 2017; Xiao et al., 2018; George et al., 2001).

It has been reported that SR-AOPs overperform the traditional $^{\bullet}OH$-based AOPs; $SO_4^{\bullet-}$ can work over a wider pH range during the degradation of organic pollutants. It displays higher efficiency to degrade pollutants containing unsaturated bonds/aromatic ring, and has higher oxidation potential (2.6–3.1 V for $SO_4^{\bullet-}$ versus 2.8 V for $^{\bullet}OH$) (Wang et al., 2019a). Moreover, the sulfate radical has a longer half-life (30–40 μs) and better selectivity when compared to its hydroxyl radical counterpart (Ahmed et al., 2012; Zhu et al., 2019).

Heterogenous catalysts are employed to form $SO_4^{\bullet-}$ by the activation (either physically or chemically) of persulfate (PS, $S_2O_8^{2-}$) or peroxymonosulfate (PMS, HSO_5^-). The activation by ultraviolet light, thermal, alkaline, metal oxides, metal-free materials, transition metal ions (Co^{2+}, Cu^{2+} and Fe^{2+}) and carbonaceous-based materials generate $SO_4^{\bullet-}$ (Ismail et al., 2017; Wang & Wang, 2018; Xiao et al., 2018). The O-O bond energy in PS is 140 kJ/mol and that of PMS is in the range of 140–213.3 kJ/mol. Therefore, the formation of sulfate radicals at least requires the minimum energy for O-O bond cleavage (Yang et al., 2010; Wang & Wang, 2018). The cleavage of O-O bond by heating can be achieved at a temperature >50°C. It can also be activated by ultrasound; in this case, the activation mechanism involves the production of cavitation bubbles, which initiate the activation process (Neppolian et al., 2010). In alkaline conditions, the activation requires a nucleophilic attack on the O-O bond, achieved by the addition of hydroxide (Furman et al., 2010).

Although iron and its oxides were the most investigated homogenous metals to activate PS and post-synthetic modification (PSM), cobalt ion proved best performance for activating PMS while PM was best activated with silver ion (Rastogi et al., 2009; Hu & Long, 2016). Compared to homogenous catalysts for PS/PMS activation, heterogeneous catalysts are a more efficient strategy owing to higher energy savings and convenient catalyst recovery. Although metal-based catalysts exhibit high efficiency for activating PMS or PS, attention must be paid to the possible leaching of potentially toxic metals out of the system. On the other hand, most carbon-based materials have insufficient activity (Hu et al., 2017). Therefore, the design of novel and more efficient catalysts for PS/PMS activation is of great technological interest.

MOFs have been successfully applied for PS/PMS activation owing to their inherent unique features. Both the metal clusters and the organic ligand of MOFs can be modulated to enhance their activation capacity. Their PS/PMS activation capacity can be enhanced by three major strategies: structural and morphological optimization; modification of the framework's metal-cluster and/or ligand via functionalization; and fabrication of MOF composites (Du & Zhou, 2020).

The degradation efficiency of MOF-based materials for organic dyes in SR-AOPs is related to particle size, structural topology, and cage size. Gao's group prepared ZIF-9 and ZIF-12 with the same composition of Co^{2+} and organic linker but having different structural topologies and particle size (Cong et al., 2017). The two MOFs were employed as PMS activation catalysts for RhB degradation. ZIF-9 had closer contact with PMS due to its nanoscale nature, allowing a higher degradation ratio (54.8%) than ZIF-12 (27.7%). Fe-based MILs such as MIL-100(Fe), MIL-101(Fe), MIL-88B(Fe), and MIL-53(Fe) were introduced as catalysts and adsorbents to generate sulfate radicals and hydroxyl radicals from PS for the degradation of acid orange 7(AO7) dye in water (Li et al., 2016). The adsorption and catalytic capacities of these materials were in the sequence MIL-101(Fe) > MIL-100(Fe) > MIL-53(Fe) > MIL-88B(Fe), which was directly related to the activity of the metal ion in catalyst sites and the different cage sizes. Furthermore, both hydroxyl radicals and sulfate radicals were the reactive species. As illustrated in Figure 12.6a, the mechanism for the catalytic degradation of AO7 by MIL-101(Fe)/PS starts by the PS diffusing into the open pore framework of MILs, which is favored by its big cage size. Then PS is activated by Fe (III) to yield $S_2O_8^{\bullet-}$ and Fe(II), after which Fe(II) is re-oxidized by PS to generate $SO_4^{\bullet-}$ (Li et al., 2016). Gao et al. (2017) further reported that the degradation of AO7 using MIL-53(Fe) under the irradiation of visible LED light was unsatisfactory due to the fast recombination of photoinduced electron-hole pairs, which was suppressed by the introduction of PS as an external electron acceptor. The addition of PS together with the formation of reactive $SO_4^{\bullet-}$ and $^{\bullet}OH$ radicals synergistically accelerated the light-induced degradation of AO7.

As aforementioned, integrating MOFs with other functional materials in order to form MOFs composites is a good strategy for preparing efficient catalysts for SR-AOPs. A yolk-shellCo_3O_4@MOFs nanoreactor was designed and used to incorporate SR-AOPs into its interior cavity (Zeng et al., 2015). The nanoreactor was applied as a catalyst for 4-chlorophenol (4-CP) degradation in the presence of PMS. The mesoporous MOFs shells promoted the fast diffusion of the reactants (4-CP and

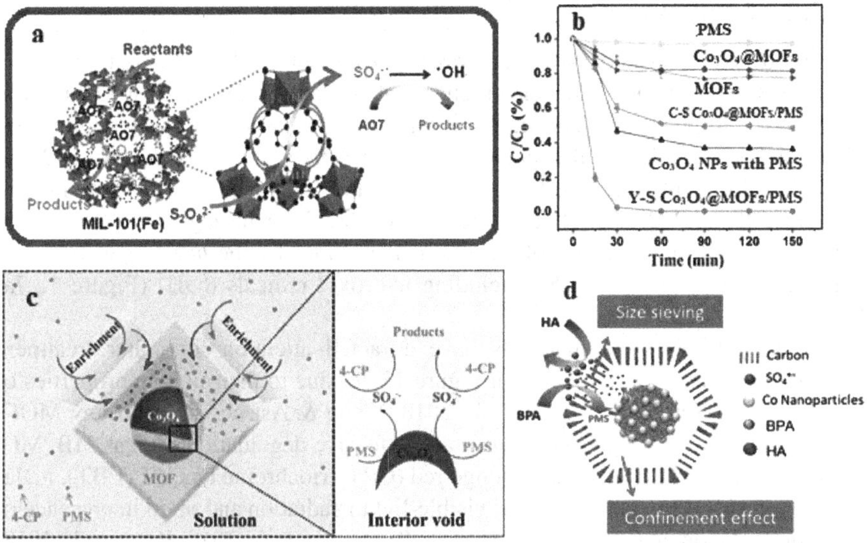

FIGURE 12.6 (a) Mechanism for the adsorptive removal of AO7 by MIL-101(Fe). Adapted with permission from *Appl. Surf. Sci.*, 2016, 369, 130–136. (b) Removal performance of 4-CP under different conditions, (c) The plausible 4-CP degradation mechanism in the yolk-shell Co_3O_4@ MOFs nanoreactor. Permission taken from *Environ. Sci. Technol.*, 2015, 49, 2350–2357. (d) The mechanisms of BPA degradation on the yolk-shell Co/C nanoreactors. Permission taken from *Environ. Sci. Technol.* 2020, 54, 10289–10300.

PMS) into the encapsulated active sites (Co_3O_4), and the confined high concentration of these reactant molecules in the local void space facilitated the SR-AOPs. The Co_3O_4@MOFs catalysts showed enhanced SR-AOPs with 4-CP degradation efficiency of almost 100% within 60 min, whereas only 59.6% was degraded for bare Co_3O_4 NPs, as shown in Figure 12.6b (Zeng et al., 2015). The underlying reaction mechanism of the nanoreactor is represented in Figure 12.6c. The high porosity and open pore network of the frameworks enable fast reactant/product molecules diffusion. Through π-π interaction, the organic linker of the MOFs takes 4-CP. The large pore cavity offers PMS access to the active Co_3O_4 catalytic sites, thus generating $SO_4^{\bullet-}$ radicals. The high concentration of this radical and 4-CP in a confined microenvironment offers a driving force that facilitates 4-CP degradation (Zeng et al., 2015). Recently, Li's research group demonstrated the structure-performance relationship between designed yolk-shell Co/C nanoreactors (YSCCNs), prepared through pyrolysis and the selective removal of bisphenol A (BPA) in the presence of 10 mg/L humic acid (HA) (Zhang et al., 2020). YSCCNs recorded an enhanced degradation rate of BPA, which is 23.1% and 45.4% higher than that of hollow and solid ZIF-67 derived Co/C nanoparticles (HCCNs and SCCNs), respectively. The improved activity can be due to the cooperative effects from size exclusion in the shell layer and confinement effect in the core-shell of the nanoreactor, as illustrated in Figure 12.6d (Zhang et al., 2020).

12.5.2 Photocatalysis

Advanced oxidation processes for water purification include Fenton processes (Fe(II)/H_2O_2), ozonation (O_3), electrochemical oxidation, UV, or solar radiation in the presence of photocatalysts, among others (Deng & Zhao, 2015). Heterogeneous photocatalysts accelerate the photochemical reaction under irradiation and the excited electrons migrate into the conduction band (CB) while the holes remain in the valence band (VB). The photoexcited carriers facilitate the degradation of organic compounds and the treatment of hazardous metal ions by the *in-situ* generation of reactive oxygen species (ROS), including hydroxyl radicals (•OH) (Figure 12.7a) (Ahmed & Haider, 2018).

Recently, MOF-based materials have attracted attention for water treatment through photocatalysis (as shown in Figure 12.7b) due to their unique properties of MOFs (Wang et al., 2012; Mon et al., 2018; Wang & Astruc, 2019). Many MOFs have been reported as efficient photocatalysts for dye degradation such as MB, MO, RhB, rhodamine 6G (R6G), AO7, congo red (CR), eriochrome black T (EBT), malachite green (MG) etc. under UV and visible-light irradiation and often in presence of an additional oxidant (for example, H_2O_2 or PS) (Jiang et al., 2018; Kaur et al., 2018; Dong et al., 2019). For example, Natarajan and co-workers reported three different MOFs, [Co_2 (4,40-bpy)](4,40-obb)$_2$, [Ni_2(4,40-bpy)$_2$], 4,40-obb)$_2$. H_2O, and [Zn_2(4,40-bpy)](4,40-obb)$_2$ having band gap values of 3.11, 3.89 and 4.02 eV, respectively, which demonstrated significant photocatalytic degradation of organic dyes, such as orange G (OG), remazol brilliant blue R (RBBR), RhB, and MB (Mahata et al., 2006) through an activated complex involving M^{2+}(Figure 12.7c–f). The fast degradation demonstrated that MOF catalysts are active in the decomposition of dyes under light irradiation. Further, a series of MOFs, such as Zn(5-aminoisophthalic acid) •H_2O, [Co_3(BPT)$_2$(DMF)(bpp)]. DMF, GR/MIL-53(Fe), Ti(IV)-based MOF structure (NTU-9), Cd(II)-imidazole MOFs, ZIF-8, and others, demonstrated high activity for dye degradation under solar light (Gao et al., 2014; Jing et al., 2014; Liu et al., 2014; Dias & Petit, 2015; Wang et al., 2015a; Zhao et al., 2015). In these MOFs, the ligand molecule donates an electron to the metal center, which facilitated the oxidation of organic molecules.

However, most of the MOFs are photocatalytically active under UV light due to large band gap and only a few, like MIL-53 or MIL-88A, absorb light in the visible region with low band gap. Even in this case, they usually need additional inorganic oxidants like H_2O_2, $KbrO_3$, and $(NH_4)_2S_2O_8$ to achieve satisfactory efficiency for organic pollutants degradation (Du et al., 2011; Xu et al., 2014).

Basically, ZIFs consist of divalent cations such as Zn^{2+} or Co^{2+} that are tetrahedrally coordinated to imidazolate ligands, with proven high thermal and chemical stability, were also tested for photocatalytic dye removal. Zhang and co-workers studied ZIF-67and Cu/ZIF-67 (band gaps 1.98 and 1.95 eV, respectively) for the photocatalytic degradation of MO under visible-light illumination (Yang et al., 2012). Additionally, a novel Cu-MOFs prepared with a mixture of O, N-donor ligands (imidazole- and pyridine- based ligands) was able to degrade about 70–80% of dye molecules, such as MO, RhB, and MB, under the irradiation of visible light in less than 4 h, as reported by Qiao et al. (2017).

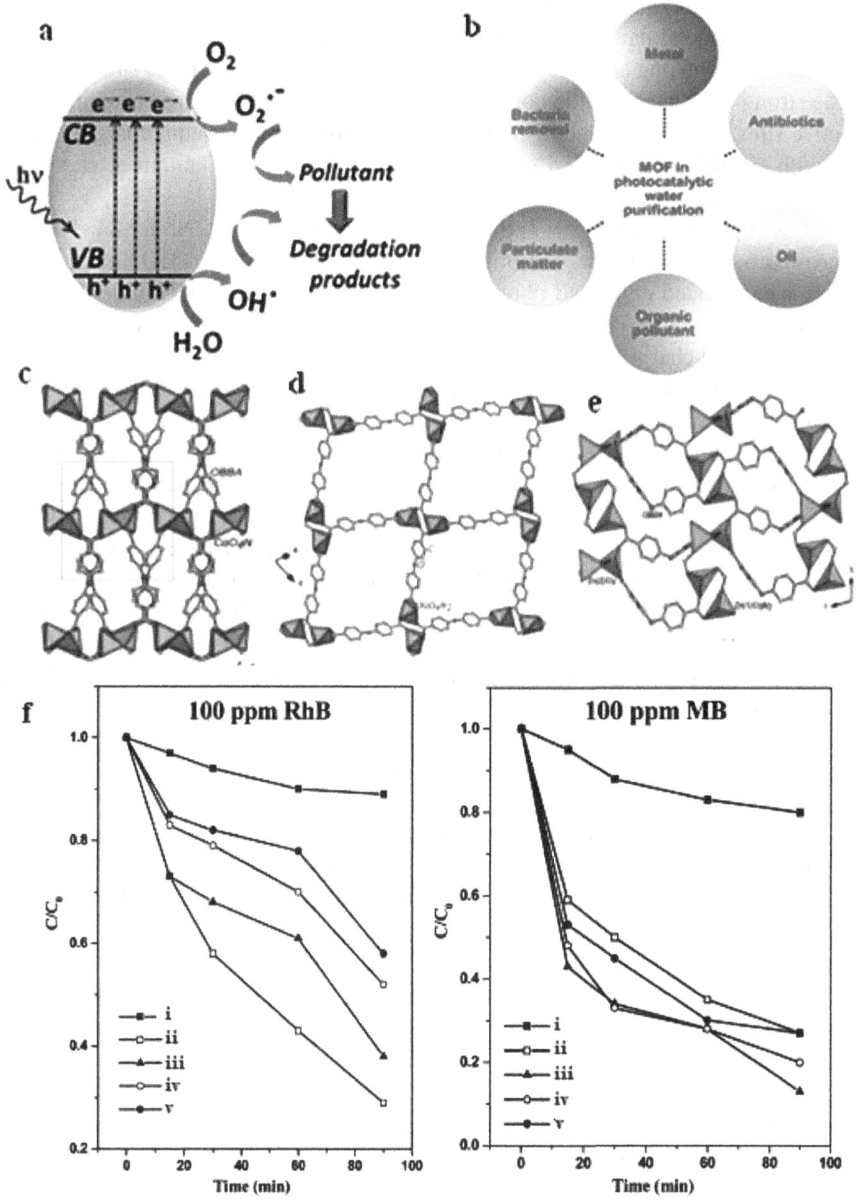

FIGURE 12.7 (a) Photocatalytic organic pollutant degradation mechanism, (b) possible application of MOF in photocatalytic water treatment. Layer structure formed by the connectivity between the (c) Co^{2+} and OBA anions, (d) Ni^{2+} ions and OBA anions, I Zn^{2+} and OBA. (f) Degradation profiles of two representative dyes with an initial concentration of 100 mg/L without catalyst (i), in presence of Degussa P-25 (ii), compound $[Co_2(4,40\text{-bpy})](4,40\text{-obb})_2$(iii), $[Ni_2(4,40\text{-bpy})_2](4,40\text{-obb})_2{\cdot}H_2O$(iv), and $[Zn_2(4,40\text{-bpy})](4,40\text{-obb})_2$,(v). Permission taken from *J. Phys. Chem. B*, 2006, 110, 13759–13768.

Furthermore, the fabrication of MOF-based heterostructures with other photoactive materials (metal nanoparticles, metal oxides decoration, or free-metal semiconductors) is a promising approach to improve the photocatalytic activity of MOFs due to the strong interaction of the components in intimate contact that create abundant reactive sites, thereby enhancing visible light absorption, improving charge separation with suitable energy band gap, and promoting stability (Gautam et al., 2020; Wang et al., 2020).

For example, Liu et al. (2017) developed a hierarchical TiO2NS@MIL-100(Fe) heterostructure through the combination of MIL-100(Fe) into TiO_2 nanosheets, which demonstrated enhanced photocatalytic decomposition of MB under the irradiation of visible light. Porous MOFs with high surface area facilitate efficient electron transfer and charge separation at the interface.

Due to the narrow band gap of bismuth-based semiconductors, MOFs are widely combined with compounds like Bi_2WO_6, Bi_2MoO_6, and BiOBr for the removal of dyes under visible light. Wu and co-workers (Sha et al., 2014) obtained Bi_2WO_6 assembled on a zirconium-based MOF, UiO-66 (Zr) and used it for the removal of RhB under visible light, which was complete after 3 h, due to a catalytic activity that strongly depended on Bi:Zr ratio. An improved version achieved the complete elimination of RhB in 15 min by using BiOBr/UiO-66(Zr) composite as photocatalysts. Its higher activity was attributed to enhanced charge separation that favored the generation of reactive radical species (Sha & Wu, 2015). A AgI/UiO-66 composite photocatalyst proved high stability when examined during repeated RhB degradation under visible light (Sha et al., 2015).

The presence of graphene can also improve the degradation of dyes using MOFs as photocatalysts. For example, MIL-53(Fe)–rGO (reduced graphene oxide) showed good efficiency for the degradation MB when using 2.5% of rGO on MIL-53(Fe), attributed to electron transfer from MOF to graphene that suppressed electron-hole recombination (Zhang et al., 2014). An interesting heterojunction was reported, using g-C_3N_4 as a metal-free semiconductor (band gap, 2.70 eV) coupled with different MOFs to increase the dye degradation rate.

Wang et al. (2015b) fabricated a g-C_3N_4/Ti-benzenedicarboxylate (MIL-125(Ti)) heterostructured photocatalyst that showed high catalytic activity for RhB degradation. They proposed that the Ti^{3+}–Ti^{4+} intervalence electron transfer and the indirect dye photosensitization may improve photodegradation performance. Similarly, other MOFs such as MIL-100(Fe) and UiO-66 were also combined with g-C_3N_4 showing remarkable photocatalytic performance due to the electron transfer in the metal-oxo-clusters, and intimate contact at the interface of heterostructures which favors charge separation (Hong et al., 2016).

Very recently, a ternary nanocomposite photocatalyst (MOF-Nd/GO/Fe_3O_4) exhibited stable, and high photocatalytic performance for the degradation of MB (~95% in 80 min) under sunlight irradiation due to fast electron migration, slow photogenerated charge recombination, and improved optical absorption (Bai et al., 2020).

Besides organic dyes, pharmaceutical products such as tetracycline (TC) can be removed through photocatalytic degradation using a BiOI/MIL-125(Ti) composite as photocatalyst (Jiang et al., 2021). It has been reported that 9 wt% BiOI/MIL-125(Ti) composite showed 80% of degradation of TC under visible-light illumination with good stability and reusability. In another study, Abazari et al. (2020) reported the photocatalytic efficiency of Zn(II)-based MOF@Ni–Ti layered double hydroxide composite for the degradation of the antibiotic sulfamethoxazole (>98% in 45 minutes) under solar light irradiation. MIL-68(In)-NH_2/graphene oxide composite displayed 93% photodegradation efficiency of amoxicillin when irradiated with visible light (Yang et al., 2017).

MOFs can also effectively remove Pb, Hg, Cd, and Cr using adsorption and photocatalytic processes (Chen et al., 2020a, b). For instance, MOFs and their composites have been applied photocatalytically to reduce Cr(VI) (Liang et al., 2015a; Shi et al., 2015). Moreover, g-C_3N_4/MIL-153(Fe) composites also showed high catalytic activity for the reduction of Cr(VI) due to the inhibition of charge recombination and the presence of highly active sites at the interface (Huang et al., 2017). Similarly, rGO modified MIL-53(Fe) also demonstrated activity for the photoreduction of Cr(VI) with up to 100% yield after 80 min under visible light attributed to the fast transfer of photogenerated electrons at the interface between both components (Liang et al., 2015b). A Cr(VI) photoreduction rate of 13.3 $mg_{Cr(VI)}/g_{catalyst}$/min was recorded with cationic Ru-UiO-dmbpy as photocatalysts under visible-light illumination (Zheng et al., 2020). Ti-benzenedicarboxylate and amino-functionalized Ti-benzenedicarboxylate showed efficient activity for Cr(VI) reduction from aqueous solution under visible light in acidic medium (Wang et al., 2015c). Recently, a core-shell structure hybrid photocatalytic material (IR-MOF3@COF-LZU1) has been tested for the removal of nitroaromatic explosives (4-nitrophenol) under visible light (Zhao et al., 2020).

Hence, MOFs can be considered a new photocatalyst with potential applications in water treatment, including organic pollutant degradation and removal of metal ions. Compared to traditional photocatalysts, MOFs possess several advantages:

i. A high degree of crystallinity and variety of morphologies
ii. Possibility of fine-tuning the structures at the molecular level, with different electronic structures
iii. Availability of active sites to harvest solar light more efficiently
iv. High porosity and surface area
v. Facile formation of composites with semiconductors

In spite of having superior features for photocatalysis, very limited semiconducting behavior occurs in MOFs. The low stability of many MOFs in the presence of water may also hamper their practical applications in solar catalysis for water treatment. However, a better understanding of the photocatalytic mechanisms in MOF materials and structure-function correlation is required to optimize photocatalytic activity.

12.5.3 Adsorption of Contaminants

The ion-exchange chemistry of metal–organic frameworks is still in an early stage and most of the reported works are still proof-of-principle studies that show the suitability of MOFs as sorbents. Unlike conventional sorbents that are mainly cationic-exchange materials (Ezugwu et al., 2013) for MOFs, most studies have been done in the exchange of anionic species rather than their cationic counterparts, due to the positive charge of most metal nodes of the frameworks. Nevertheless, cation-exchange MOFs can be prepared by choosing the suitable charged ligand or PSM strategies to provide the desired amount of negative charges to the framework. Compared to conventional ion-exchange materials, MOFs have some advantages, such as: tunable structural feature, high-porous surface area (offering more area/surface for host-guest interaction), and a high degree of crystallinity (Kumar et al., 2017). Accordingly, both cationic and anionic MOF adsorbents have been successfully applied for the removal of organic pollutants via ion-exchange interaction. Through ion-exchange mechanisms, selective capturing of toxic oxoanions of arsenic As^{V} ($HAsO_4^{2-}$) and selenium Se^{VI} (SeO_4^{2-}) with maximum adsorption capacity of 85 and 100 mg g^{-1} have been achieved using cationic MOFs material (iMOF-C) (Sharma et al., 2020).

Neutral polytopic organic ligands (1,8-diazabicyclooctan, 4,4′-bipyridine and others) together with some polycarboxylate ligands are usually employed for the synthesis of MOFs with anion-exchange characteristics. Such MOF-based materials have been used to remove toxic anionic species, like carcinogenic Cr(VI), selenium, arsenic, phosphate anions, and others.

The adsorption of arsenate from water using MIL-53(Fe) and MIL-53(Al) has been explored and adsorption in both cases was ascribed to the interaction between trivalent metal sites of the MOFs and the oxyanion (Li et al., 2014; Vu et al., 2015). By experimental and computational analysis, Jun et al. (2015) reported the effect of open metal sites of MIL-100(Fe), Figure 12.8a, in the adsorption of organo-arsenic compounds. In addition, the effect of different metal nodes was studied using the adsorption of *p*-arsanilic acid on MIL-100(Al), MIL-100(Cr) and MIL-100(Fe). Among these three analogous MIL-100 species, MIL-100(Fe) showed the highest *p*-arsanilic acid adsorption capacity, assigned to the facile desorption of water from the MIL-100(Fe) and the large replacement energy, higher than that of MIL-100-Cr and MIL-100-Al. Fourier-transform infrared (FTIR) analysis was used to confirm the Fe–O–As coordination motif. Other studies highlighted the binding of anionic species to metal nodes of MOFs facilitated by labile binding of water and hydroxyl ligands. It was the case Zr_6 nodes of NU-1000, which, together with a large 30 Å apertures, provided fast adsorption of selenite (102 mg g^{-1}) and selenate (62 mg g^{-1}), higher than to other Zr-based MOFs (Howarth et al., 2015a, 2015b). Selenium oxyanions bridged to the Zr_6 node in Zr–O–Se–O–Zr with the configuration, as shown in Figure 12.8b.

UiO-66 and UiO-66-NH_2 have been applied at different temperatures for the sorption of phosphate anions (Lin et al., 2015a, b). For an initial concentration of ~40 mg/L, the adsorption capacity of both MOFs reached~60 mg/g, whereas increased capture capacity was achieved at higher temperatures: for UiO-66-NH_2, 120 and 140 mg/g of phosphate anions were removed at 40°C and 60°C, respectively. Amine-based MOFs interact with phosphate through amine-phosphate hydrogen bonds, thus adsorbing more than UiO-66 under the same conditions.

Using MOFs for the removal of cationic species has been studied using UiO-68-P(O)$(OH)_2$and UiO-68-P(O)$(OEt)_2$ with the uranyl cation (UO_2^{2+}) from seawater showing adsorption capacities of 188 and 217 mg g^{-1}, respectively (Carboni et al., 2013). In this case, the adsorption mechanism was attributed to the coordination between one UO_2^{2+} and the phosphorylurea groups, as shown in Figure 12.9.

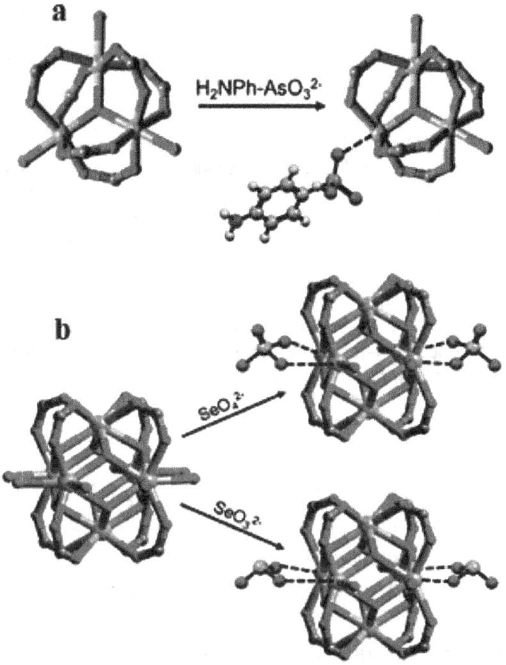

FIGURE 12.8 (a)The *p*-arsanilic acid adsorption mechanism on the metal cluster of MIL-100(Fe) (Jun et al., 2015), (b) Interaction of selenite and selenate on the Zr_6 node of NU-1000. Permission taken from *CrystEngComm*, 2015, 17, 7245–7253.

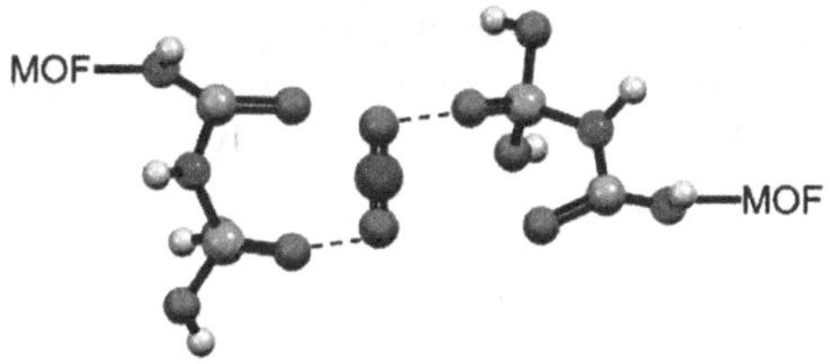

FIGURE 12.9 Interaction between the phosphorylurea functional groups and UiO-68-P(O)$(OH)_2$ for the capture of uranyl. Permission taken from *CrystEngComm*, 2015, 17, 7245–7253.

Other commonly available heavy metal/metalloid pollutants include mercury (Hg), lead (Pb), arsenic (As), cadmium (Cd) and chromium (Cr) (Chen et al., 2020a). Even at low concentrations, heavy metal ions are highly toxic (Azimi et al., 2017).

Due to the structural diversity of MOFs, different metal ion-exchange capacities ranging from 10 to 100 mg/g can be achieved. Depending on the targeted metal ion, MOFs displayed different selectivity. The incorporation of diverse functional groups via pre-, in-situ-, or post-functionalization of the organic ligand or the metal cluster in MOFs results in physical or chemical adsorptive removal of metal ions pollutants.

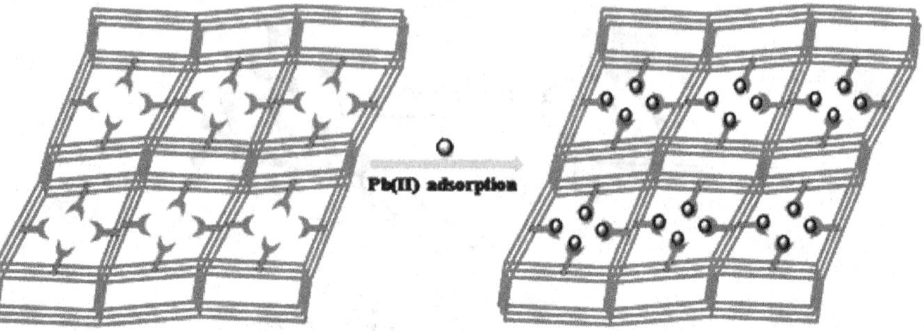

FIGURE 12.10 Carbomethoxy-functionalized MOF for Pb(II) removal. Reproduced with permission from *Cryst. Growth Des.* 2018, 18, 1474–1482.

Guo and co-workers employed MOF-derived inorganic materials for the adsorptive removal of Pb(II) from water (Chen et al., 2016a, 2016b). They fabricated nanoporous $ZnO/ZnFe_2O_4/C$ adsorbents using Fe(III)-modified MOF-5 as a precursor. The nanoporous inorganic adsorbents displayed a maximum Pb(II) adsorption capacity of 345 mg g^{-1}, and the results of the spectroscopic analysis suggested that a significant portion of Pb(II) substituted Zn(II) on the surface of the ZnO nanocrystals. Moreover, the strong magnetic behavior of Pb-containing $ZnO/ZnFe_2O_4/C$ adsorbents allowed their easy separation from water using a magnet.

Although many MOFs have proved good Pb (II) adsorption capacity, most of them showed poor removal performance at low Pb(II) concentrations (Ricco et al., 2015). However, carbomethoxy-group-based MOFs (Yu et al., 2018) had been prepared through a mixed ligand synthetic approach using (pyridin-3-yl)methyl 4-(2-(4-((pyridin-3-yl)methoxy)-phenyl)diazenyl) benzoate and 2,2′-azodibenzoic acid. The as-synthesized MOF was able to quickly and completely remove trace amounts of toxic Pb(II) from water, as shown in Figure 12.10. At Pb(II) concentration as low as 100 mg/L, the as-prepared frameworks achieved 96% removal efficiency, even reaching the drinking water standard (0.01 mg/L) recommended by World Health Organization (WHO). The high Pb(II) removal performance even at low concentration was attributed to the strong affinity the carbomethoxy groups, toward Pb(II). In addition, fast kinetic rate constant of 0.162 g mg^{-1} min^{-1} was recorded, higher than that for most Pb(II) adsorbents (Yu et al., 2018). Another example has been reported by Afshariazar et al. (2020), in this case, *N*1, N_2-di(pyridine-4-yl)oxalamide) was integrated into the pores of Zn(II)-MOF, TMU-56, for the Pb(II) ions removal. TMU-56 is densely decorated with oxamide motif (-NH-CO-CO-NH-), a dual functionality promoting host-guest interaction in the framework cavities. The large pore size, together with the high densities of strong metal chelating sites explained the uptake capacity for the adsorption of Pb(II) ions (Afshariazar et al., 2020). Interestingly, the ion-exchange reaction is favored at high Pb(II) concentration, thereby resulting in a high degree of metal exchange and high removal capacity.

Furthermore, the experimental adsorption parameters fitted well with the Langmuir isotherm, which confirmed monolayer-type adsorption (Awad et al., 2017).

In the adsorption isotherm, a maximum adsorption of 1130 mg Pb/g TMU-56 was recorded (Afshariazar et al., 2020). From the XRD and IR results, the adsorption mechanism was assigned to the chelation of Pb(II) on the oxamide moiety followed by the ion-exchange reactions. This results in the replacement of some of the Zn(II) ions in the MOFs with Pb(II) ions (Afshariazar et al., 2020). Nevertheless, the oxamide motif plays the major role for the fast and high capture of Pb(II) ions from water.

Amino-decorated MOFs, MIL-101-NH_2 have demonstrated potential not only for detecting Fe^{3+}, Pb^{2++}, and Cu^{2+} but also for their adsorptive removal from aqueous solution with a capacity of 3.5, 1.1, and 0.9 m moL/g, respectively (Lv et al., 2019). In addition, other MOFs functionalized with amino groups, such as MIL-101-NH_2(Cr), MIL-53-NH_2(Al), MOF-5-NH_2(Zn) and UiO-66-NH_2(Zr), exhibit similar sensing and capturing capacity for metal ions. Generally, MOF's fluorescence is due to the linker. Fluorescence quenching upon adsorption is based on the chemical chelation between the surface NH_2 groups and the metal ions, thereby inducing host-guest electron transfer as shown in Scheme 12.3 (Lv et al., 2019). Thus, the fluorescence amino-functionalized MIL-101(Fe) had a good sensing performance for detecting metal ions.

Clark et al. (2019) also employed Zr-based MOFs, UiO-66 containing several defects for the adsorptive removal of perfluorooctanesulfonate (PFOS) from simulated industrial wastewater. Defective UiO-66 was also effective for perfluorobutanesulfonate (PFBS), which is the shorter-chain homologue of PFOS. Defective MOFs showed higher adsorption capacity than the non-defective frameworks (UiO-66-DF), like sample UiO-66-10 (with 10% vol. HCl), exhibited a 300% increase in PFOS capture capacity over UiO-66-DF. Large-pore defects within UiO-66 reported as ~16

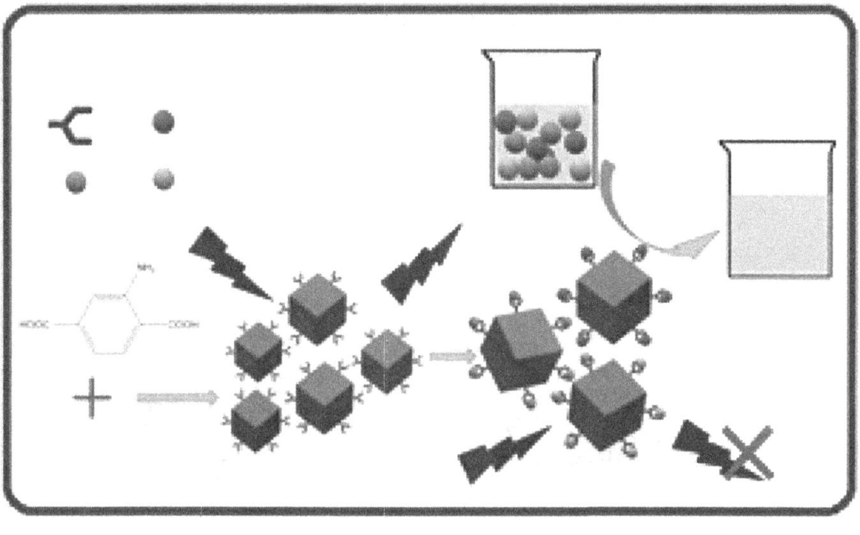

SCHEME 12.3 Illustration of the removal and detection of heavy metals by MIL-101-NH_2. Permission taken from *Chem. Eng. J.*, 2019, 375, 122111.

and ~20 Å significantly promoted PFOS adsorption from an aqueous solution. The defective UiO-66 material displayed a maximum Langmuir sorption capacity (monolayer adsorption of PFOS) of 1.24 mmol/g, which is twofold higher than activated carbon, and proceeded two-orders of magnitude faster than GAC and commercially available ion-exchange resin (Clark et al., 2019). The mechanism for PFOS adsorption on the defective UiO-66, supported by zeta potential analysis, was based on the electrostatic interaction between the sulfonate headgroup and the coordinatively unsaturated (CUS) Zr^{4+} sites together with the decreasing framework hydrophobicity (which depends on structural defectiveness) (Liang et al., 2016).

Ibuprofen (IBP) and diclofenac (DCF) are among the most useful nonsteroidal anti-inflammatory drugs used worldwide. Owing to the reason that PPCPs are used in large quantities, they are easily detected in wastewater. Although PPCPs are usually biodegradable, they are found in water bodies because of their continuous input, a phenomenon referred to as pseudo-persistency (Chen et al., 2016a). Bhadra et al. (2017) used porous carbons derived from MOF (PCDMs) to remove IBP and DCF from water. PCDMs were prepared by the pyrolysis of ZIF-8 at several temperatures (800°C, 1000°C, and 1200°C), the sample prepared at 1000°C exhibiting the best adsorption capacity for IBF and DCF: 320 and 400 mg/g, respectively. Based on the pH effect and the zeta potential of PCDM, the proposed adsorption mechanism is H-bonding interaction, as observed in other applications, which implies H-donation from the adsorbent (mainly via the phenolic group) and H-acceptance from DCF or IBP as shown in Scheme 12.4 (Bhadra et al., 2017).

MOFs together with MOF-based materials have been successfully used for Cr(VI) adsorption with many frameworks that demonstrated outstanding adsorption capacity. Yang and co-workers reported that a cationic silver-triazolate MOF, with a formular ($[Ag_8(tz)_6](NO_3)_2{\cdot}6H_2O$) n (tz = 3,5- diphenyl-1,2,4-triazolate), displayed a fast and high Cr(VI) capture via anion exchange (Li et al., 2017). The anion-exchange behavior of 1-NO_3 followed the Hofmeister bias due to the hydrophobicity of the framework pores. During the adsorption process, the nitrates in 1-NO_3 were replaced

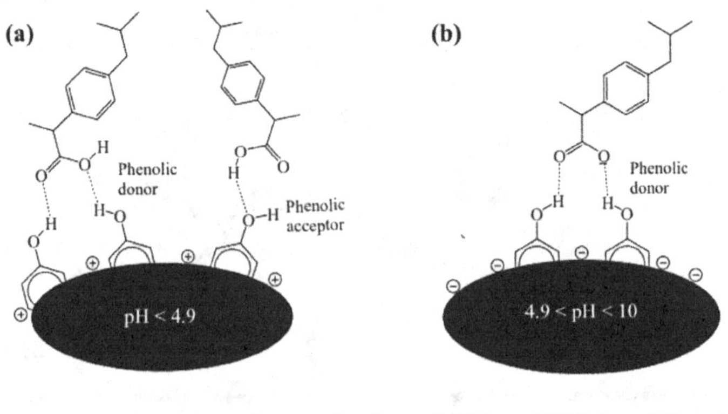

SCHEME 12.4 Proposed adsorption mechanism of IBP on PCDM at different pH range (a) pH < 4.9 and (b) 4.9 < pH < 10. Reproduced with permission from *Chem. Eng. J.*, 2017, 314, 50–58.

by the Cr(VI) oxyanion. Within 15 mins, 40% of the Cr(VI) in the aqueous solution was removed and the white powder of the silver-triazolate MOFs turned orange. At equilibrium, an adsorption capacity of 22.6 mg/g (1.35 mol mol^{-1}) was recorded. The reversibility of Cr(VI) was studied by immersing the Cr(VI)-loaded sample in an excess amount of $NaNO_3$. After 5 h, 97% of Cr(VI) was released from 1-NO_3 (Li et al., 2017). MOFs-based magnetic polydopamine@ ZIF-8 (MP@ZIF-8) afforded Cr(VI) capture capacity of 137 mg/g (Zhu et al., 2017). The interaction was explained by the synergistic adsorption and reduction of Cr(VI) by the hybrid microsphere. The Cr(VI) adsorption-reduction mechanisms were examined by X-ray photoelectron spectroscopy (XPS), showing that during the process, the N atoms on ZIF-8 and PDA were reduced, resulting in the conversion of Cr(VI) into less toxic Cr(III), further immobilized onto the MP@ZIF-8.

12.5.3.1 More on the Mechanisms of Adsorption

The underlying adsorption mechanisms for metal ion pollutants removal using MOFs absorbents are normally verified by spectroscopic measurements, such as FTIR and XPS. These measurements help to evaluate the interaction between the framework materials and the metal ion. The common mechanisms for adsorptive removal of contaminants are ion exchange, coordination with the MOFs' metal nodes, and binging with the linker's functional moiety/group.

The work of Liu et al. (2018c), explained Pb(II) adsorption based on its binding with the C=O unit from the carbomethoxy linker as shown in Figure 12.11a. The FTIR analysis of the MOFs after Pb(II) adsorption revealed a peak at 696 cm^{-1} (Figure 12.11b) ascribed to the Pb-O stretching vibration, which confirmed that Pb(II) was absorbed into the frameworks(Yu et al., 2018). In addition, new bands occurred at 798 and 1378 cm^{-1}, implying that NO_3^- is presence for charge balance (Lebrero et al., 2002). There is a red shift (1713 to 1706 cm^{-1}) attributed to the C-O vibration, confirming the coordination between the Pb(II) and the O atom of C=O unit (Zhao et al., 2015). Furthermore, the interaction of Pb(II) with the MOFs was unveiled by the XPS measurement before and after adsorption. The occurrence of Pb 4f, Pb 4d, and Pb 4p peaks undoubtedly confirmed that Pb(II) was attached to the

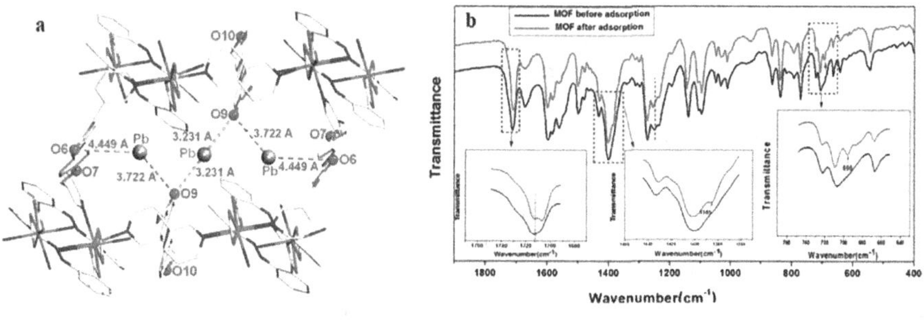

FIGURE 12.11 (a) The interaction between Pb(II) and MOF, and (b) the FT-IR spectra before and after. Reproduced with permission from *Cryst. Growth Des.* 2018, 18, 1474–1482.

framework after adsorption. Furthermore, the XPS spectrum of Pb 4fshowedPb $4f_{5/2}$ (143.1 eV) and Pb $4f_{7/2}$(138.2 eV) peaks. Comparing the Pb 4f XPS spectrum for the Pb(II)-loaded MOFs and purified $Pb(NO_3)_2$, it was unveiled that Pb $4f_{5/2}$ and Pb $4f_{7/2}$ peaks for the MOFs appeared at 143.1 and 138.2 eV, while that of $Pb(NO_3)_2$ was observed at 144.5 eV and 139.6 eV, respectively. The lower shift in the binding energy (BE) by1.4 eV for the Pb(II)-loaded MOFs was attributed to the strong interaction between Pb(II) and the MOFs material. Overall, the two spectroscopic analyses confirmed that the carbomethoxy moiety played an important role in Pb(II) adsorption coupled with the fact that methoxy and diazene (free-standing groups) did not participate in the Pb(II) adsorption process. Therefore, the underlying adsorption mechanism follows that the of C=O unit from the carbomethoxy groups preferentially complex with the Pb(II) without the participation of the N atoms of the diazene group, since O is more electronegative than N (Yu et al., 2018).

12.5.4 MOF-Based Membranes

Membrane technology in wastewater treatment has become an important solution for several processes in cosmetics, dairy, drugs, and chemical industries. Suitable approaches for the design of MOF-based membranes are important to achieve accurate and rapid wastewater purification. Numerous techniques have been established for the preparation of MOF-based membranes, such as the solvothermal method (Chen et al., 2020b) phase inversion, (Lee et al., 2015) (porous) mixed matrix, (Liu et al., 2018c; Katayama et al., 2019) thin-film nanocomposite (TFN) (Zhang et al., 2019) and electrodeposition method (Hou et al., 2019). The properties and performance of membranes can be enhanced by incorporating fillers like titanium oxide and zeolites. It is the case of MMM that can be prepared using MOFs grown on a given substrate or applied as fillers to construct MOFs-mixed matrix membranes (MOF-MMMs).

For rational selection of MOFs for MMMs, the stability, dispersibility, and hydrophilicity/hydrophobicity of the frameworks are crucial. Consequently, the features for ideal MOF-MMMs include highly efficient separation ability, satisfactory component stability, long-term operation capacity and, possibly, environmentally friendly preparation (Huang et al., 2017). Unlike the rigid nature of traditional inorganic frameworks, the organic linkers in novel MOFs facilitate the interactions with the polymer materials, thereby promoting good compatibility. Normally, the obtained MOF-MMMs exhibit better permeability and selectivity than the pristine membranes (Huang et al., 2017). The most common MOFs applied for MMMs are UiO-66 (Qian et al., 2019), green MOFs (F300, C300 and A100) (Lee et al., 2014), ZIFs (Wei et al., 2020) and MIL-type MOFs (Knebel et al., 2016). Specifically, four synthesis routes are applicable for the fabrication of MOF-MMMs: layer-by-layer (LBL), in-situ growth, blending technique, and gelatin-assisted seed growth. These techniques are aimed to improve MOF-based membranes for enhanced separation, like membrane distillation, adsorption desalination, and capacitive deionization (Jun et al., 2020).

By modifying the pore size and chemical properties, the application of MOF-based membranes for gas separation can be extended for wastewater treatment. Zhu et al. (2016) used simulation and experimental studies to demonstrate the efficiency of polycrystalline ZIF membranes for desalination favored by the small pore size of the framework, between the size of hydrate ions and water molecules. However, a recent investigation unveiled that ZIF-8 membranes in aqueous solution experience continuous Zn^{2+} leaching, thereby delimiting its application in water purification (Zhang et al., 2015). Liu et al. (2015) first used a solvothermal synthetic approach to fabricate Zr-MOF (UiO-66(Zr))polycrystalline membranes with satisfactory water stability and also shown to be efficient for water desalination. They demonstrated good multivalent ion rejection based on the size inclusion and good permeability (0.28 L m^{-2} h^{-1} bar^{-1}µm, for a membrane thickness of 2.0 µm). Based on various tests conducted up to 170 h, no degradation of the membranes was recorded.

Wan et al. (2017) prepared narrow pore size (about 0.52 ± 0.02 nm) amine-functionalized UiO-66, which stand in-between the size of water molecules and hydrated ion (0.26 nm), and used them for seawater desalination by pervaporation at a temperature of 75°C, demonstrating a high ionic sieving capacity. In another work, it was shown that the mitigation of the ligand-missing defects in polycrystallineUiO-66(Zr)-$(OH)_2$ MOF membranes helped to enhance the membranes' Na^+ rejection rate, as shown in Figure 12.12a–c (Wang et al., 2017). The mitigation was achieved by post-synthetic defect healing (PSDH), resulting in up to 74.9% Na^+ rejection rate

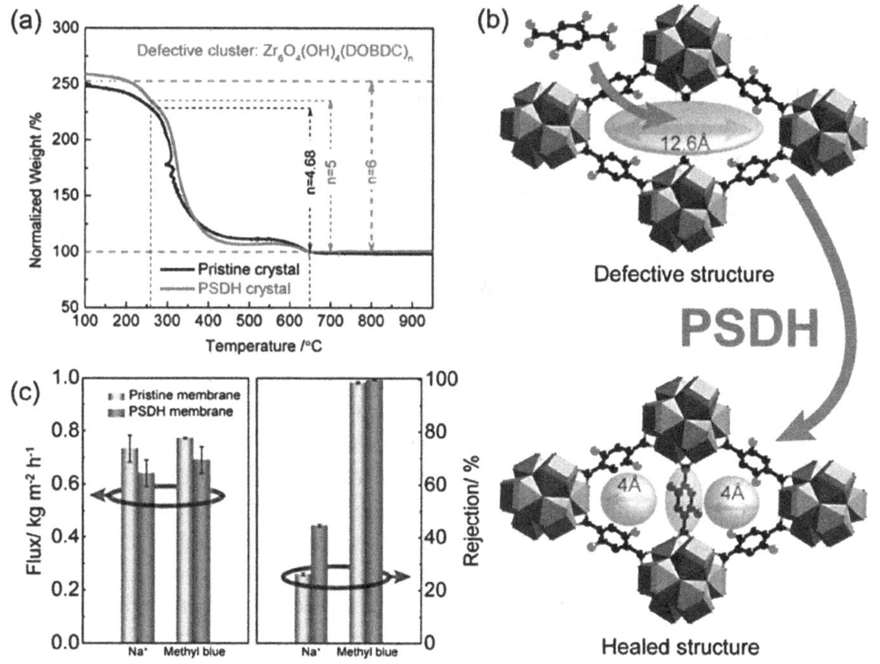

FIGURE 12.12 (a) Thermogravimetric analysis (TGA) curve of UiO-66(Zr)-$(OH)_2$ frameworks before and after post-synthetic defect healing; (b) illustrating PSDH by relinking one DOBDC ligand and two adjacent Zr_6O_4 $(OH)_4$ nodes and (c) The membrane's separation performance before and after PSDH. Permission taken from *ACS Appl. Mater. Interfaces*, 2017, 9, 37848–37855.

and excellent hydrothermal stability in water. Moreover, the healed membranes were highly selective.

ZIF-8 is the most used ZIF family framework due to its hydrophobic pores and stable tetrahedral MN4 structure, thus making it stable and resistant to organic solvents and water. Unlike conventional desalination techniques, membrane distillation (MD), with high salt rejection, low operation cost, and moderate operation temperature, can be used for wastewater treatment. A thin film composite (TFC) membrane containing a thin ZIF-8/chitosan layer plated on the surface of polyvinylidene fluoride (PVDF) membrane was prepared for water desalination (Kebria et al., 2019). The ZIF-8/chitosan layer formation resulted in a high increase in the liquid entry pressure of water for the TFC membrane. In addition, the output of the air gap membrane distillation (AGMD) tests showed that the TFC with ZIF-8)/chitosan layer increased the permeate water flux with NaCl rejection >99.5%. The TFC composite also showed good antifouling capacity and stability, attributed to the good compatibility between ZIF-8 and chitosan (Kebria et al., 2019). TFN membranes with ZIF-8 are known to display improved reverse osmosis (RO) performance due to their high specific surface area and intrinsic microporosity (Lee et al., 2019). The effect of particle size of ZIF-8 on the RO desalination performance of a polyamide TFN membrane was evaluated by Lee et al. (2019). ZIF-8 coating on the polysulfone support preceding the interfacial polymerization was strongly influenced by the particle size of the frameworks (Lee et al., 2019). It was observed that ZIF-8 of small–average size showed higher surface coverage.

MOF-5-incorporated polymeric membranes of PVDF, cellulose acetate (CA) and polyethersulfone (PES) were fabricated by phase inversion method and applied for the adsorption of Co(II) and Cu(II) ions from wastewater (Gnanasekaran et al., 2019). The incorporation of MOF-5 significantly increased the hydrophilicity and rejection of both metal ions by the polymeric membranes. A high rejection efficiency was obtained for Co(II)in CA/MOF-5 and PES/MOF-5. Both the permeability flux and the rejection of PES/MOF-5, PVDF/MOF-5 and CA/MOF-5 membranes are higher compared to those recorded for the neat polymeric materials.

MMMs with ultra-high MOF loading and good flexibility have been produced in a large scale. High-performance adsorption membrane was prepared by thermally induced phase separation-hot pressing (TIPS-HoP) with MOFs loadings up to 86 wt% (Wang et al., 2019b). The MOF-MMMs displayed 99% rejection of organic dyes due to the adsorption on porous MOFs with a high-water flux of 125.7 L m^{-2} h^{-1} bar^{-1} under crossflow filtration mode. Interestingly, the MOF-MMMs have strong chemical and thermal robustness together with interesting mechanical strength. The MOFs were dispersed and suspended during the fabrication process to melt high-density polyethylene (HDPE), ultra-high-molecular-weight polyethylene (UHMWPE), and paraffin at a 200°C. This stage was followed by roll-to-roll hot pressing to produce the MOF-MMMs.

The TIPS-HoP synthetic procedure for the fabrication of MOF PE MMMs and its application for separation are illustrated in Figure 12.13a and b, respectively. In addition, porous matrix membrane that was fabricated with MOFs as template exhibited improved porosity and enhanced separation performance (Lee et al., 2014).

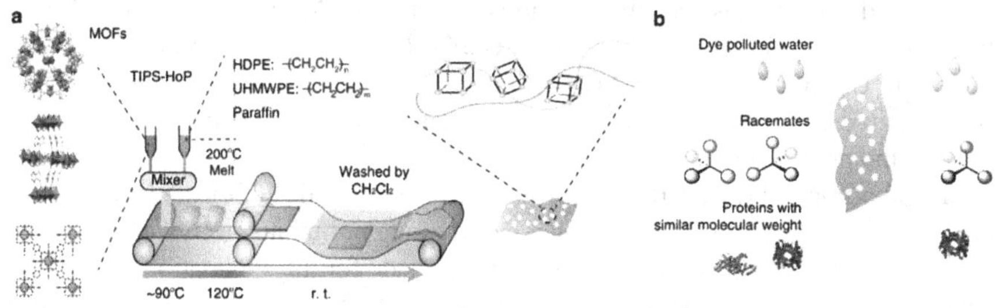

FIGURE 12.13 (a) Schematic representation for applying the TIPS-HoP technique for the fabrication of MOF PE MMMs and (b) racemates, dye, and protein separations by the as-prepared MOF PE MMMs material. Reproduced with permission from *Nat. Commun.*, 2019, 10, 1–9.

12.6 CONCLUSION AND PROSPECTS

Further developments in the treatment of wastewater may strongly depend on advances in functional materials. The various interesting features of MOFs, such as large surface area and pore volume, high crystallinity, availability of active sites, and easy structural tunability make them attractive to create materials with the potential to outperform traditional ones. Furthermore, they can be suitably and easily tailored to obtain new functionalities by PSMs. Therefore, they are promising materials that can be applied in the field of water purification.

As such, MOFs/MOF-derived materials have been applied in SR-AOPs. The organic ligands and the metal nodes can be modulated to enhance the activation capacity of the frameworks. Moreover, their performance can be further improved by structural/morphological optimization and fabrication of MOF composites.

As well, MOFs can be used as photocatalysts for water purification. However, most MOFs have a large band gap; thus, they are photocatalytically active for the degradation of organic pollutants and pharmaceutical products under the irradiation of UV light, whereas only a few with non-functionalized ligands are visible-light responsive, such as MIL-53 or MIL-88A. To achieve even a satisfactory performance, they usually need additional inorganic oxidants like $KBrO_3$, and $(NH_4)_2S_2O_8$. MOFs photocatalytic activities could also be improved by preparing MOF-based heterostructures with other photoactive materials, thereby enhancing the charge separation and light absorption.

The adsorptive removal of metals/metalloid (Hg, As, Pb, Cr, and Cd) and organic pollutants from water using MOFs and MOF-derived materials have also been extensively studied, showing good performance. The functionalization of their ligand or the metal cluster may result in an improvement of the physical or chemical adsorption. The common mechanisms for removing contaminants are coordination with the metal nodes of the MOFs, ion exchange, and binging with the linker's functional moiety/group.

The organic ligands facilitate the interactions of the framework with polymer materials, thereby promoting good compatibility, which is very important to develop

MOF-based membranes. For the judicious selection of MOFs as filler for MMMs, the frameworks' stability, dispersibility, and hydrophilicity/hydrophobicity are important. UiO-66, ZIFs, green MOFs, and MIL-type MOFs are commonly applied. The prepared MOF-MMMs usually have better selectivity and permeability compared to the pristine membranes.

Metal/carbon hybrids, carbon materials, and metal oxides fabricated from MOFs have also showcased good performance in removing pollutants from water.

Even though significant advances have been made for the application of MOFs in water treatment, for the commercial application of MOFs and MOF-derived materials, some aspects need to be improved, such as the synthetic conditions (high temperature and long duration in solvothermal synthesis). In addition, the stability of the frameworks in water should be enhanced. More developments are needed by both material scientists and chemists to overcome some technical issues and increase the basic knowledge of MOFs regarding structure, properties, and functional capabilities. The development of low-cost synthesis is another improvement required for their use in high-volume water treatment processes.

REFERENCES

Abazari, R., Morsali, A., Dubal, D. P. 2020. An advanced composite with ultrafast photocatalytic performance for the degradation of antibiotics by natural sunlight without oxidizing the source over TMU-5@ Ni–Ti LDH: Mechanistic insight and toxicity assessment. *Inorg. Chem. Front.* 7: 2287–2304.

Abednatanzi, S., Derakhshandeh, P. G., Depauw, H., Coudert, F.-X., Vrielinck, H., Van Der Voort, P., Leus, K. J. 2019. Mixed-metal metal–organic frameworks. *Chem. Soc. Rev.* 48(9): 2535--2565.

Afshariazar, F., Morsali, A., Wang, J., Junk, P. C. J. 2020. Highest and fastest removal rate of PbII ions through rational functionalized decoration of a metal–organic framework cavity. *Chem. Eur. J.* 26(6): 1355–1362.

Ahmed, M. M., Barbati, S., Doumenq, P., Chiron, S. J. 2012. Sulfate radical anion oxidation of diclofenac and sulfamethoxazole for water decontamination. *Chem. Eng. J.* 197: 440–447.

Ahmed, S. N., Haider, W. 2018. Heterogeneous photocatalysis and its potential applications in water and wastewater treatment: A review. *Nanotechnology* 29(34): 342001.

Ahsan, M. A., Jabbari, V., El-Gendy, A. A., Curry, M. L., Noveron, J. C. 2019. Ultrafast catalytic reduction of environmental pollutants in water via MOF-derived magnetic Ni and Cu nanoparticles encapsulated in porous carbon. *Appl. Surf. Sci.* 497: 143608.

Awad, F. S., Abou Zeid, K. M., El-Maaty, W. M. A., El-Wakil, A. M., El-Shall, M. S. J. 2017. Efficient removal of heavy metals from polluted water with high selectivity for mercury(II) by 2-imino-4-thiobiuret–partially reduced graphene oxide (IT-PRGO).*ACS Appl. Mater. Interfaces* 9(39): 34230–34242.

Azimi, A., Azari, A., Rezakazemi, M., Ansarpour, M. J. 2017. Removal of heavy metals from industrial wastewaters: A review. *ChemBioEng Rev.* 4(1): 37–59.

Bai, Y., Zhang, S., Feng, S., Zhu, M., Ma, S. J. 2020. The first ternary Nd-MOF/GO/Fe_3O_4 nanocomposite exhibiting excellent photocatalytic performance for dye degradation. *Dalton Trans.* 49: 10745–10754.

Batten, S. R., Champness, N. R., Chen, X.-M., Garcia-Martinez, J., Kitagawa, S., Öhrström, L., O'Keeffe, M., Suh, M. P., Reedijk, J. 2013. Terminology of metal–organic frameworks and coordination polymers (IUPAC Recommendations 2013). *Pure Appl. Chem.* 85(8): 1715–1724.

Bedia, J., Muelas-Ramos, V., Peñas-Garzón, M., Gómez-Avilés, A., Rodríguez, J. J., & Belver, C. J. 2019. A review on the synthesis and characterization of metal organic frameworks for photocatalytic water purification. *Catalysts* 9(1): 52.

Bhadra, B. N., Ahmed, I., Kim, S., Jhung, S. H. J. 2017. Adsorptive removal of ibuprofen and diclofenac from water using metal-organic framework-derived porous carbon. *Chem. Eng. J.*314: 50–58.

Boretti, A., Rosa, L. 2019. Reassessing the projections of the world water development report. *NPJ Clean Water.* 2.1: 1–6.

Cao, W., Liu, Y., Xu, F., Li, J., Li, D., Du, G., Chen, N. J. 2019. In Situ Electrochemical Synthesis of Rod-Like Ni-MOFs as Battery-Type Electrode for High Performance Hybrid Supercapacitor. *J. Electrochem. Soc.* 167(5): 050503.

Carboni, M., Abney, C. W., Liu, S., Lin, W. 2013. Highly porous and stable metal–organic frameworks for uranium extraction. *Chem. Sci.* 4.6: 2396–2402.

Chen, C., Feng, H., Deng, Y. J. 2019a. Re-evaluation of sulfate radical based–advanced oxidation processes (SR-AOPs) for treatment of raw municipal landfill leachate. *Water Res.* 153: 100–107.

Chen, C., Feng, X., Zhu, Q., Dong, R., Yang, R., Cheng, Y., He, C. J. 2019b. Microwave-assisted rapid synthesis of well-shaped MOF-74 (Ni) for CO_2 efficient capture. *Inorg. Chem.* 58(4): 2717–2728.

Chen, D., Shen, W., Wu, S., Chen, C., Luo, X., Guo, L. J. 2016a. Ion exchange induced removal of Pb (II) by MOF-derived magnetic inorganic sorbents. *Nanoscale* 8(13): 7172–7179.

Chen, D., Zhao, J., Zhang, P., Dai, S. J. 2019c. Mechanochemical synthesis of metal–organic frameworks. *Polyhedron* 162: 59–64.

Chen, X., Chen, D., Li, N., Xu, Q., Li, H., He, J., Lu, J. 2020a. Modified-MOF-808-Loaded polyacrylonitrile membrane for highly efficient, simultaneous emulsion separation and heavy metal ion removal. *ACS Appl. Mater. Interfaces* 12(35): 39227–39235.

Chen, Y., Bai, X., Ye, Z. J. 2020b. Recent progress in heavy metal ion decontamination based on metal–organic frameworks. *Nanomaterials* 10(8): 1481.

Chen, Y., Vymazal, J., Březinová, T., Koželuh, M., Kule, L., Huang, J., Chen, Z. J. 2016b. Occurrence, removal and environmental risk assessment of pharmaceuticals and personal care products in rural wastewater treatment wetlands. *Sci. Total Environ.* 566: 1660–1669.

Clark, C. A., Heck, K. N., Powell, C. D., Wong, M. S. 2019. Highly defective UiO-66 materials for the adsorptive removal of perfluorooctanesulfonate. *ACS Sustain. Chem. Eng* 7(7): 6619–6628.

Cong, J., Lei, F., Zhao, T., Liu, H., Wang, J., Lu, M., Li, Y., Xu, H., Gao, J. 2017. Two Co-zeolite imidazolate frameworks with different topologies for degradation of organic dyes via peroxymonosulfate activation. *J. Solid State Chem.* 256: 10–13.

Cook, T. R., Zheng, Y.-R., Stang, P. J. 2013. Metal–organic frameworks and self-assembled supramolecular coordination complexes: comparing and contrasting the design, synthesis, and functionality of metal–organic materials. *Chem. Rev.* 113(1): 734—777.

Deng, Y., Zhao, R. J. 2015. Advanced oxidation processes (AOPs) in wastewater treatment. *Curr. Pollut. Rep.* 1(3): 167–176.

Dias, E. M., Petit, C. J. 2015. Towards the use of metal–organic frameworks for water reuse: a review of the recent advances in the field of organic pollutants removal and degradation and the next steps in the field. *J. Mater. Chem. A* 3(45): 22484–22506.

Dong, J.-P., Shi, Z.-Z., Li, B., Wang, L.-Y. J. 2019. Synthesis of a novel 2D zinc (II) metal–organic framework for photocatalytic degradation of organic dyes in water. *Dalton Trans.* 48(47): 17626–17632.

Du, J.-J., Yuan, Y.-P., Sun, J.-X., Peng, F.-M., Jiang, X., Qiu, L.-G., Xie, A.-J., Shen, Y.-H., Zhu, J.-F. J. 2011. New photocatalysts based on MIL-53 metal–organic frameworks for the decolorization of methylene blue dye. *J. Hazard. Mater.* 190: 945–951.

Du, X., Zhou, M. J. 2020. Strategies to enhance catalytic performance of metal-organic frameworks in sulfate radical-based advanced oxidation processes for organic pollutants removal. *Chem. Eng. J.* 403: 126346.

Ezugwu, C. I., Asraf, M. A., Li, X., Liu, S., Kao, C.-M., Zhuiykov, S., Verpoort, F. 2018. Cationic nickel metal-organic frameworks for adsorption of negatively charged dye molecules. *Data Brief* 18: 1952–1961.

Ezugwu, C. I., Kabir, N. A., Yusubov, M., Verpoort, F. 2016. Metal–organic frameworks containing N-heterocyclic carbenes and their precursors. *Coord. Chem. Rev* 307: 188–210.

Ezugwu, C. I., Ujam, O. T., Ukoha, P. O., Ukwueze, N. N. 2013. Complex formation and extraction studies of *N, N'*-Bis (salicylidene)-3, 5-diaminobenzoic Acid on Hg (II) and Ag (I).*Chem. Sci. Trans.* 2: 1118–1125.

Ezugwu, C. I., Zhang, S., Li, S., Shi, S., Li, C., Verpoort, F., Yu, J., Liu, S. 2019. Efficient transformative HCHO capture by defective NH_2-UiO-66 (Zr) at room temperature. *Environ. Sci.: Nano* 6(10): 2931–2936.

Farha, O. K., Eryazici, I., Jeong, N. C., Hauser, B. G., Wilmer, C. E., Sarjeant, A. A., Snurr, R. Q., Nguyen, S. T., Yazaydin, O., Hupp, J. T. 2012. Metal–organic framework materials with ultrahigh surface areas: is the sky the limit? *J. Am. Chem. Soc.* 134(36), 15016–15021.

Furman, O. S., Teel, A. L., Watts, R. J. J. 2010. Mechanism of base activation of persulfate. *Environ. Sci. Technol.* 44(16): 6423–6428.

Furukawa, H., Ko, N., Go, Y. B., Aratani, N., Choi, S. B., Choi, E.,Yazaydin, A. O., Snurr, R. Q., O'Keeffe, M., Kim, J., Yaghi, O. M. 2010. Ultrahigh porosity in metal-organic frameworks. *Science,* 329(5990): 424–428.

Gao, J., Miao, J., Li, P.-Z., Teng, W. Y., Yang, L., Zhao, Y., Liu, B., Zhang, Q. 2014. A p-type Ti (IV)-based metal–organic framework with visible-light photo-response. *Chem. Commun.* 50(29): 3786–3788.

Gao, Y., Li, S., Li, Y., Yao, L., Zhang, H. J. 2017. Accelerated photocatalytic degradation of organic pollutant over metal-organic framework MIL-53(Fe) under visible LED light mediated by persulfate. *Appl. Catal. B* 202: 165–174.

Gautam, S., Agrawal, H., Thakur, M., Akbari, A., Sharda, H., Kaur, R., Amini, M. J. 2020. Metal oxides and metal organic frameworks for the photocatalytic degradation: A review. *J. Environ. Chem. Eng.* 8(3), 103726.

George, C., Rassy, H. E., Chovelon, J. M. J. 2001. Reactivity of selected volatile organic compounds (VOCs) toward the sulfate radical (SO_4^-). *Int. J. Chem. Kinet.* 33(9): 539-547.

Gnanasekaran, G., Balaguru, S., Arthanareeswaran, G., Das, D. B. 2019. Removal of hazardous material from wastewater by using metal organic framework (MOF) embedded polymeric membranes. *Sep. Sci. Technol.* 54(3): 434–446.

Gong, Y., Zhao, X., Zhang, H., Yang, B., Xiao, K., Guo, T., Zhang, J., Shao, H., Wang, H., Yu, G. 2018. MOF-derived nitrogen doped carbon modified g-C_3N_4 heterostructure composite with enhanced photocatalytic activity for bisphenol A degradation with peroxymonosulfate under visible light irradiation. *Appl. Catal. B*233: 35–45.

Gropp, C., Canossa, S., Wuttke, S., Gándara, F., Li, Q., Gagliardi, L., Yaghi, O. M. 2020. Standard practices of reticular chemistry. *ACS Cent. Sci.* 6, 8: 1255–1273.

Guillerm, V., Maspoch, D. 2019. Geometry mismatch and reticular chemistry: Strategies to assemble metal–organic frameworks with non-default topologies. *J. Am. Chem. Soc.* 141(42): 16517–16538.

Haldar, D., Duarah, P., Purkait, M. K. 2020. MOFs for the treatment of arsenic, fluoride and iron contaminated drinking water: A review. *Chemosphere* 251: 126388.

Hasan, Z., Jhung, S. H. 2015. Removal of hazardous organics from water using metal-organic frameworks (MOFs): Plausible mechanisms for selective adsorptions. *J. Hazard. Mater.* 283: 329–339.

Hasan, Z., Khan, N. A., Jhung, S. H. 2020. Adsorptive purification of organic contaminants of emerging concern from water with metal–organic frameworks. In *Contaminants of Emerging Concern in Water and Wastewater*. Butterworth-Heinemann 47–92.

Hong, J., Chen, C., Bedoya, F. E., Kelsall, G. H., O'Hare, D., Petit, C. 2016. Carbon nitride nanosheet/metal–organic framework nanocomposites with synergistic photocatalytic activities. *Catal. Sci. Technol.* 6(13): 5042–5051.

Hou, J., Hong, X., Zhou, S., Wei, Y., Wang, H. 2019. Solvent-free route for metal–organic framework membranes growth aiming for efficient gas separation. *AIChE J.* 65(2): 712–722.

Howarth, A. J., Katz, M. J., Wang, T. C., Platero-Prats, A. E., Chapman, K. W., Hupp, J. T., Farha, O. K. 2015a. High efficiency adsorption and removal of selenate and selenite from water using metal–organic frameworks. *J. Am. Chem. Soc.* 137(23): 7488–7494.

Howarth, A. J., Liu, Y., Hupp, J. T., Farha, O. K. 2015b. Metal–organic frameworks for applications in remediation of oxyanion/cation-contaminated water. *Cryst Eng Comm* 17(38): 7245–7253.

Hu, P., Long, M. 2016. Cobalt-catalyzed sulfate radical-based advanced oxidation: a review on heterogeneous catalysts and applications. *Appl. Catal. B* 181: 103–117.

Hu, P., Su, H., Chen, Z., Yu, C., Li, Q., Zhou, B., Alvarez, P. J. J., Long, M. 2017. Selective degradation of organic pollutants using an efficient metal-free catalyst derived from carbonized polypyrrole via peroxymonosulfate activation. *Environ. Sci. Technol.* 51(19): 11288–11296.

Huang, W., Liu, N., Zhang, X., Wu, M., Tang, L. 2017. Metal organic framework g-C_3N_4/MIL-53(Fe) heterojunctions with enhanced photocatalytic activity for Cr(VI) reduction under visible light. *Appl. Surf. Sci.* 425: 107–116.

Ismail, L., Ferronato, C., Fine, L., Jaber, F., Chovelon, J.-M. 2017. Elimination of sulfaclozine from water with SO_4^- radicals: Evaluation of different persulfate activationmethods. *Appl. Catal B.* 201: 573–581.

Jiang, D., Chen, M., Wang, H., Zeng, G., Huang, D., Cheng, M., Liu, Y., Xue, W., Wang, Z. 2019. The application of different typological and structural MOFs-based materials for the dyes adsorption. *Coord. Chem. Rev.* 380: 471–483.

Jiang, D., Xu, P., Wang, H., Zeng, G., Huang, D., Chen, M., Lai, C., Wan, J.,Xue, W. 2018. Strategies to improve metal organic frameworks photocatalyst's performance for degradation of organic pollutants. *Coord. Chem. Rev.* 376: 449–466.

Jiang, W., Li, Z., Liu, C., Wang, D., Yan, G., Liu, B., Che, G. 2021. Enhanced visible-light-induced photocatalytic degradation of tetracycline using BiOI/MIL-125 (Ti) composite photocatalyst. *J. Alloys Compd.* 854: 157166.

Jin, A.-P., Chen, Z.-W., Wang, M.-S., Guo, G.-C. 2019. [Zn (OOCH) 2 (4, 4′-bipyridine)] n: A metal-organic-framework (MOF) with x-ray-induced photochromic behaviour at room temperature. *Dyes Pigm.* 163: 656–659.

Jing, H.-P., Wang, C.-C., Zhang, Y.-W., Wang, P., Li, R. 2014. Photocatalytic degradation of methylene blue in ZIF-8. *Rsc Adv.* 4(97): 54454–54462.

Joharian, M., Morsali, A. 2019. Ultrasound-assisted synthesis of two new fluorinated metal-organic frameworks (F-MOFs) with the high surface area to improve the catalytic activity. *J. Solid State Chem.* 270: 135–146.

Joseph, L., Jun, B.-M., Jang, M., Park, C. M., Muñoz-Senmache, J. C., Hernández-Maldonado, A. J., Heyden, A., Yu, M., Yoon, Y. 2019. Removal of contaminants of emerging concern by metal-organic framework nanoadsorbents: A review. *Chem. Eng. J.* 369:928–946.

Jun, B.-M., Al-Hamadani, Y. A., Son, A., Park, C. M., Jang, M., Jang, A., Kim, N. C., Yoon, Y. 2020. Applications of metal-organic framework based membranes in water purification: A review. *Sep. Purif. Technol.* 247: 116947.

Jun, J. W., Tong, M., Jung, B. K., Hasan, Z., Zhong, C., Jhung, S. H. 2015. Effect of central metal ions of analogous metal–organic frameworks on adsorption of organoarsenic compounds from water: Plausible mechanism of adsorption and water purification. *Chem. Eur. J.*21.1: 347–354.

Kalmutzki, M. J., Hanikel, N., Yaghi, O. M. 2018. Secondary building units as the turning point in the development of the reticular chemistry of MOFs. *Sci. Adv.* 4(10): eaat 9180.

Katayama, Y., Bentz, K. C., Cohen, S. M. 2019. Defect-free MOF-based mixed-matrix membranes obtained by corona cross-linking. *ACS Appl. Mater. Interfaces* 11(13): 13029–13037.

Kaur, H., Venkateswarulu, M., Kumar, S., Krishnan, V., Koner, R. R. 2018. A metal–organic framework based multifunctional catalytic platform for organic transformation and environmental remediation. *Dalton Trans.* 47(5): 1488–1497.

Kebria, M. R. S., Rahimpour, A., Bakeri, G., Abedini, R. 2019. Experimental and theoretical investigation of thin ZIF-8/chitosan coated layer on air gap membrane distillation performance of PVDF membrane. *Desalination* 450: 21–32.

Khan, N. A., Jhung, S. H. 2017. Adsorptive removal and separation of chemicals with metal-organic frameworks: Contribution of π-complexation. *J. Hazard. Mater.* 325: 198–213.

Knebel, A., Friebe, S., Bigall, N. C., Benzaqui, M., Serre, C., Caro, J. 2016. Comparative study of MIL-96 (Al) as continuous metal–organic frameworks layer and mixed-matrix membrane. *ACS Appl. Mater. Interfaces* 8(11): 7536–7544.

Kondo, M., Yoshitomi, T., Matsuzaka, H., Kitagawa, S., Seki, K. 1997. Three-Dimensional Framework with Channeling Cavities for Small Molecules:{[M2 (4, 4′-bpy) 3 (NO 3) 4]· xH2O} n (M Co, Ni, Zn). *Angew. Chem. Int. Ed.* 36(16): 1725–1727.

Kumar, P., Pournara, A., Kim, K.-H., Bansal, V., Rapti, S., Manos, M. J. 2017. Metal-organic frameworks: challenges and opportunities for ion-exchange/sorption applications. *Prog. Mater. Sci.* 86: 25–74.

Langmi, H. W., Ren, J., Musyoka, N. M. 2016. Metal–organic frameworks for hydrogen storage. In *Compendium of Hydrogen Energy*. Woodhead Publishing 163–188.

Lebrero, M. C. G., Bikiel, D. E., Elola, M. D., Estrin, D. o. A., Roitberg, A. E. 2002. Solvent-induced symmetry breaking of nitrate ion in aqueous clusters: A quantum-classical simulation study. *J. Chem. Phys.* 117(6): 2718–2725.

Lee, J.-Y., She, Q., Huo, F., Tang, C. Y. 2015. Metal–organic framework-based porous matrix membranes for improving mass transfer in forward osmosis membranes. *J. Membr. Sci.* 492: 392–399.

Lee, J.-Y., Tang, C. Y., Huo, F. 2014. Fabrication of porous matrix membrane (PMM) using metal-organic framework as green template for water treatment. *Sci. Rep.* 4: 3740.

Lee, K. E., Morad, N., Teng, T. T., Poh, B. T. 2012. Development, characterization and the application of hybrid materials in coagulation/flocculation of wastewater: A review. *Chem. Eng. J.* 203: 370–386.

Lee, T. H., Oh, J. Y., Hong, S. P., Lee, J. M., Roh, S. M., Kim, S. H., Park, H. B. 2019. ZIF-8 particle size effects on reverse osmosis performance of polyamide thin-film nanocomposite membranes: Importance of particle deposition. *J. Membr. Sci.* 570: 23–33.

Li, H., Eddaoudi, M., O'Keeffe, M., Yaghi, O. M. 1999. Design and synthesis of an exceptionally stable and highly porous metal-organic framework. *Nature,* 402(6759): 276–279.

Li, J., Wang, H., Yuan, X., Zhang, J., Chew, J. W. 2020. Metal-organic framework membranes for wastewater treatment and water regeneration. *Coord. Chem. Rev.* 404: 213116.

Li, J., Wu, Y.-n., Li, Z., Zhu, M., Li, F. 2014. Characteristics of arsenate removal from water by metal-organic frameworks (MOFs). *Water Sci. Technol.* 70(8): 1391–1397.

Li, L.-L., Feng, X.-Q., Han, R.-P., Zang, S.-Q., Yang, G. 2017. Cr (VI) removal via anion exchange on a silver-triazolate MOF.*J. Hazard. Mater.* 321: 622–628.

Li, W., Shi, J., Zhao, Y., Huo, Q., Sun, Y., Wu, Y., Tian, Y., Jiang, Z. 2020. Superhydrophobic metal–organic framework nanocoating induced by metal-phenolic networks for oily water treatment. *ACS Sustainable Chem. Eng.* 8(4): 1831–1839.

Li, X., Guo, W., Liu, Z., Wang, R., Liu, H. 2016. Fe-based MOFs for efficient adsorption and degradation of acid orange 7 in aqueous solution via persulfate activation. *Appl. Surf. Sci.* 369: 130–136.

Liang, R., Jing, F., Shen, L., Qin, N., Wu, L. 2015a. MIL-53(Fe) as a highly efficient bifunctional photocatalyst for the simultaneous reduction of Cr(VI) and oxidation of dyes. *J. Hazard. Mater.,* 287, 364–372.

Liang, R., Shen, L., Jing, F., Qin, N., Wu, L. 2015b. Preparation of MIL-53(Fe)-reduced graphene oxide nanocomposites by a simple self-assembly strategy for increasing interfacial contact: Efficient visible-light photocatalysts. *ACS Appl. Mater. Interfaces* 7(18): 9507–9515.

Liang, W., Coghlan, C. J., Ragon, F., Rubio-Martinez, M., D'Alessandro, D. M., Babarao, R. 2016. Defect engineering of UiO-66 for CO_2 and H_2O uptake–a combined experimental and simulation study. *Dalton Trans.* 45(11): 4496–4500.

Lin, K.-Y. A., Chang, H.-A., & Chen, R.-C. 2015a. MOF-derived magnetic carbonaceous nanocomposite as a heterogeneous catalyst to activate oxone for decolorization of Rhodamine B in water. *Chemosphere* 130: 66–72.

Lin, K.-Y. A., Chen, S.-Y., Jochems, A. P. 2015b. Zirconium-based metal organic frameworks: Highly selective adsorbents for removal of phosphate from water and urine. *Mater. Chem. Phys.* 160:168–176.

Liu, C., Huang, X., Wang, J., Song, H., Yang, Y., Liu, Y., Li, J., Wang, L., Yu, C. 2018a. Hollow mesoporous carbon nanocubes: Rigid-interface-induced outward contraction of metal-organic frameworks. *Adv. Funct. Mater.* 28(6): 1705253.

Liu, F., Xiong, W., Liu, J., Cheng, Q., Cheng, G., Shi, L., Zhang, Y. 2018b. Novel amino-functionalized carbon material derived from metal organic framework: A characteristic adsorbent for U (VI) removal from aqueous environment. *Colloids Surf. A Physicochem. Eng. Asp.* 556: 72–80.

Liu, G., Chernikova, V., Liu, Y., Zhang, K., Belmabkhout, Y., Shekhah, O., Zhang, C., Yi, S., Eddaoudi, M., Koros, W. J. 2018c. Mixed matrix formulations with MOF molecular sieving for key energy-intensive separations. *Nat. Mater.* 17(3): 283–289.

Liu, L., Ding, J., Huang, C., Li, M., Hou, H., Fan, Y. 2014. Polynuclear CdII polymers: Crystal structures, topologies, and the photodegradation for organic dye contaminants. *Cryst. Growth Des.* 14(6): 3035–3043.

Liu, X., Dang, R., Dong, W., Huang, X., Tang, J., Gao, H., Wang, G. 2017. A sandwich-like heterostructure of TiO_2 nanosheets with MIL-100 (Fe): A platform for efficient visible-light-driven photocatalysis. *Appl. Catal. B* 209: 506–513.

Liu, X., Demir, N. K., Wu, Z., Li, K. 2015. Highly water-stable zirconium metal–organic framework UiO-66 membranes supported on alumina hollow fibers for desalination. *J. Am. Chem. Soc.* 137(22): 6999–7002.

Luo, S., Wei, Z., Dionysiou, D. D., Spinney, R., Hu, W.-P., Chai, L., Yang, Z., Ye, T., Xiao, R. 2017. Mechanistic insight into reactivity of sulfate radical with aromatic contaminants through single-electron transfer pathway. *Chem. Eng. J.* 327: 1056–1065.

Luo, X., Mai, Z., Lei, H. 2019. A bifunctional luminescent Zn (II)-organic framework: Ionothermal synthesis, selective Fe(III) detection and cationic dye adsorption. *Inorg. Chem. Commun.* 102: 215–220.

Lv, S.-W., Liu, J.-M., Li, C.-Y., Zhao, N., Wang, Z.-H., Wang, S. 2019. A novel and universal metal-organic frameworks sensing platform for selective detection and efficient removal of heavy metal ions. *Chem. Eng. J.* 375: 122111.

Mahata, P., Madras, G., Natarajan, S. 2006. Novel photocatalysts for the decomposition of organic dyes based on metal-organic framework compounds. *J. Phys. Chem. B.* 110(28): 13759–13768.

Mário, E. D. A., Liu, C., Ezugwu, C. I., Mao, S., Jia, F., Song, S. 2020. Molybdenum disulfide/montmorillonite composite as a highly efficient adsorbent for mercury removal from wastewater. *Appl. Clay Sci.* 184: 105370.

Mason, J. A., Veenstra, M., Long, J. R. 2014. Evaluating metal–organic frameworks for natural gas storage. *Chem. Sci.* 5(1): 32–51.

Milenković, D. A., Dimić, D. S., Avdović, E. H., Amić, A. D., Marković, J. M. D., Marković, Z. S. 2020. Advanced oxidation process of coumarins by hydroxyl radical: towards the new mechanism leading to less toxic products. *Chem. Eng. J.* 395: 124971.

Mon, M., Bruno, R., Ferrando-Soria, J., Armentano, D., Pardo, E. 2018. Metal–organic framework technologies for water remediation: Towards a sustainable ecosystem. *J. Mater. Chem.* A6(12): 4912–4947.

Mousavi, B., Chaemchuen, S., Ezugwu, C. I., Yuan, Y., Verpoort, F. 2018. The effect of synthesis procedure on the catalytic performance of isostructural ZIF-8. *Appl. Organomet. Chem.* 32(2), e 4062.

Mukherjee, S., Desai, A. V., Ghosh, S. K. 2018. Potential of metal–organic frameworks for adsorptive separation of industrially and environmentally relevant liquid mixtures. *Coord. Chem. Rev.* 367: 82–126.

Neppolian, B., Doronila, A., Ashokkumar, M. 2010. Sonochemical oxidation of arsenic (III) to arsenic (V) using potassium peroxydisulfate as an oxidizing agent. *Water Res.* 44(12): 3687–3695.

Olawale, M. D., Obaleye, J. A., & Oladele, E. O. 2020. Solvothermal synthesis and characterization of novel [Ni (II)(Tpy)(Pydc)]. $2H_2O$ metal-organic framework as an adsorbent for the uptake of caffeine drug from aqueous solution. *New J. Chem.* 44: 18780–18791.

Qian, Q., Wu, A. X., Chi, W. S., Asinger, P. A., Lin, S., Hypsher, A., Smith, Z. P. 2019. Mixed-matrix membranes formed from imide-functionalized UiO-66-NH2 for improved interfacial compatibility. *ACS Appl. Mater. Interfaces* 11(34): 31257–31269.

Qiao, Y., Zhou, Y.-F., Guan, W.-S., Liu, L.-H., Liu, B., Che, G.-B., Liu, C.-B., Lin, X., Zhu, E.-W. 2017. Syntheses, structures, and photocatalytic properties of two new one-dimensional chain transition metal complexes with mixed N, O-donor ligands. *Inorganica Chim. Acta* 466: 291–297.

Rastogi, A., Al-Abed, S. R., Dionysiou, D. D. 2009. Sulfate radical-based ferrous–peroxymonosulfate oxidative system for PCBs degradation in aqueous and sediment systems. *Appl. Catal. B*85(3–4): 171–179.

Ricco, R., Konstas, K., Styles, M. J., Richardson, J. J., Babarao, R., Suzuki, K., Scopece, P., Falcaro, P. 2015. Lead(II) uptake by aluminium based magnetic framework composites (MFCs) in water. *J. Mater. Chem. A* 3(39): 19822–19831.

Rodriguez-Narvaez, O. M., Peralta-Hernandez, J. M., Goonetilleke, A., Bandala, E. R. (2017). Treatment technologies for emerging contaminants in water: A review. *Chem. Eng. J.* 323, 361–380.

Rout, P. R., Zhang, T. C., Bhunia, P.,Surampalli, R. Y. 2020. Treatment technologies for emerging contaminants in wastewater treatment plants: A review. *Sci. Total Environ.* 753: 141990.

Rowsell, J. L., Yaghi, O. M. 2004. Metal–organic frameworks: A new class of porous materials. *Micropo. Mesopor. Mat.* 73(1): 3–14.

Safaei, M., Foroughi, M. M., Ebrahimpoor, N., Jahani, S., Omidi, A., Khatami, M. 2019. A review on metal-organic frameworks: Synthesis and applications. *Trends. Analyt. Chem.* 118: 401–425.

Salunkhe, R. R., Kaneti, Y. V., Yamauchi, Y. 2017. Metal–organic framework-derived nanoporous metal oxides toward supercapacitor applications: Progress and prospects. *ACS Nano* 11(6): 5293–5308.

Sha, Z., Sun, J., Chan, H. S. O., Jaenicke, S., Wu, J. 2014. Bismuth tungstate incorporated zirconium metal–organic framework composite with enhanced visible-light photocatalytic performance. *RSC Adv.* 4(110): 64977–64984.

Sha, Z., Sun, J., Chan, H. S. O., Jaenicke, S., Wu, J. 2015. Enhanced photocatalytic activity of the AgI/UiO-66 (Zr) composite for rhodamine B degradation under visible-light irradiation. *Chem Plus Chem* 80(8): 1321–1328.

Sha, Z., Wu, J. 2015. Enhanced visible-light photocatalytic performance of BiOBr/UiO-66 (Zr) composite for dye degradation with the assistance of UiO-66. *RSC Adv.* 5(49): 39592–39600.

Sharma, S., Desai, A. V., Joarder, B., Ghosh, S. K. 2020. A water-stable ionic MOF for the selective capture of toxic oxoanions of SeVI and AsV and crystallographic insight into the ion-exchange mechanism. *Angew. Chem.* 132(20): 7862–7866.

Shayegan, H., Ali, G. A., & Safarifard, V. 2020. Recent progress in the removal of heavy metal ions from water using metal-organic frameworks. *Chemistry Select* 5(1): 124–146.

Shi, L., Wang, T., Zhang, H., Chang, K., Meng, X., Liu, H., Ye, J. 2015. An amine-functionalized iron(III) Metal–organic framework as efficient visible-light photocatalyst for Cr(VI) reduction. *Adv. Sci.* 2(3): 1500006.

Siegert, M., Sonawane, J. M., Ezugwu, C. I., Prasad, R. 2019. Economic assessment of nanomaterials in bio-electrical water treatment. In *Advanced Research in Nanosciences for Water Technology*. Springer International Publishing AG, 1–23.

Smolders, S., Struyf, A., Reinsch, H., Bueken, B., Rhauderwiek, T., Mintrop, L., Kurz, P., Stock, N., De Vos, D. E. 2018. A precursor method for the synthesis of new Ce (IV) MOFs with reactive tetracarboxylate linkers. *Chem Comm* 54(8): 876–879.

Song, J. Y., Bhadra, B. N., Khan, N. A., Jhung, S. H. 2018. Adsorptive removal of artificial sweeteners from water using porous carbons derived from metal azolate framework-6.*Micropor. Mesopor. Mat.* 260, 1–8.

Stock, N., Biswas, S. 2012. Synthesis of metal-organic frameworks (MOFs): Routes to various MOF topologies, morphologies, and composites. *Chem. Rev.* 112(2), 933–969.

Sule, R., Mishra, A. K. 2019. Synthesis of mesoporous MWCNT/HKUST-1 composite for wastewater treatment. *Appl. Sci.* 9(20): 4407.

Taheran, M., Naghdi, M., Brar, S. K., Verma, M., Surampalli, R. Y. 2018. Emerging contaminants: Here today, there tomorrow! *Environ. Nanotechnol. Monit. Manag.* 10: 122–126.

Torad, N. L., Hu, M., Ishihara, S., Sukegawa, H., Belik, A. A., Imura, M., Ariga, K., Sakka, Y., Yamauchi, Y. 2014. Direct synthesis of MOF-derived nanoporous carbon with magnetic Co nanoparticles toward efficient water treatment. *Small* 10(10): 2096–2107.

Tranchemontagne, D. J., Mendoza-Cortés, J. L., O'Keeffe, M., Yaghi, O. M. 2009. Secondary building units, nets and bonding in the chemistry of metal–organic frameworks. *Chem. Soc. Rev.* 38(5): 1257–1283.

Tzanakakis, V. A., Paranychianakis, N. V., Angelakis, A. N. 2020. Water supply and water scarcity. *Water* 12(9): 2347.

Vardhan, H., Mehta, A., Ezugwu, C. I., Verpoort, F. 2016. Self-assembled arene ruthenium metalla-assemblies. *Polyhedron* 112: 104–108.

Vu, T. A., Le, G. H., Dao, C. D., Dang, L. Q., Nguyen, K. T., Nguyen, Q. K., Dang, P. T., Tran, H. T. K., Duong, Q. T., Nguyen, T. V., Lee, G. D. 2015. Arsenic removal from aqueous solutions by adsorption using novel MIL-53 (Fe) as a highly efficient adsorbent. *RSC Adv.* 5(7): 5261–5268.

Wan, L., Zhou, C., Xu, K., Feng, B., Huang, A. 2017. Synthesis of highly stable UiO-66-NH_2 membranes with high ions rejection for seawater desalination. *Micropor. Mesopor. Mat.* 252: 207–213.

Wang, C., Kim, J., Malgras, V., Na, J., Lin, J., You, J., Zhang, M., Li, J., Yamauchi, Y. 2019a. Metal–organic frameworks and their derived materials: Emerging catalysts for a sulfate radicals-based advanced oxidation process in water purification. *Small* 15(16): 1900744.

Wang, C., Liu, X., Demir, N. K., Chen, J. P., & Li, K. 2016. Applications of water stable metal–organic frameworks. *Chem. Soc. Rev.* 45(18): 5107–5134.

Wang, F., Dong, C., Wang, C., Yu, Z., Guo, S., Wang, Z., Zhao, Y., Li, G. 2015a. Fluorescence detection of aromatic amines and photocatalytic degradation of rhodamine B under UV light irradiation by luminescent metal–organic frameworks. *New J. Chem.* 39(6): 4437–4444.

Wang, H., Yuan, X., Wu, Y., Zeng, G., Chen, X., Leng, L., & Li, H. 2015b. Synthesis and applications of novel graphitic carbon nitride/metal-organic frameworks mesoporous photocatalyst for dyes removal. *Appl. Catal. B*174: 445–454.

Wang, H., Yuan, X., Wu, Y., Zeng, G., Chen, X., Leng, L., Wu, Z., Jiang, L., Li, H. 2015c. Facile synthesis of amino-functionalized titanium metal-organic frameworks and their superior visible-light photocatalytic activity for Cr(VI) reduction. *J. Hazard. Mater.* 286: 187–194.

Wang, H., Zhao, S., Liu, Y., Yao, R., Wang, X., Cao, Y., Ma, D., Zou, M., Cao, A., Feng, X., Wang, B. 2019b. Membrane adsorbers with ultrahigh metal-organic framework loading for high flux separations. *Nat. Commun.* 10(1): 1–9.

Wang, J., Wang, S. 2018. Activation of persulfate (PS) and peroxymonosulfate (PMS) and application for the degradation of emerging contaminants. *Chem. Eng. J.* 334: 1502–1517.

Wang, J.-L., Wang, C., Lin, W. 2012. Metal–organic frameworks for light harvesting and photocatalysis. *ACS Catal.* 2(12): 2630–2640.

Wang, Q., Astruc, D. 2019. State of the art and prospects in metal–organic framework (MOF)-based and MOF-derived nanocatalysis. *Chem. Rev.* 120(2): 1438–1511.

Wang, Q., Gao, Q., Al-Enizi, A. M., Nafady, A., Ma, S. 2020. Recent advances in MOF-based photocatalysis: environmental remediation under visible light. *Inorg. Chem. Front.* 7(2): 300–339.

Wang, X., Zhai, L., Wang, Y., Li, R., Gu, X., Yuan, Y. D., Qian, Y., Hu, Z., Zhao, D. 2017. Improving water-treatment performance of zirconium metal-organic framework membranes by postsynthetic defect healing. *ACS Appl. Mater. Interfaces* 9(43): 37848–37855.

Wei, J.-Z., Gong, F.-X., Sun, X.-J., Li, Y., Zhang, T., Zhao, X.-J., Zhang, F.-M. 2019. Rapid and low-cost electrochemical synthesis of UiO-66-NH_2 with enhanced fluorescence detection performance. *Inorg. Chem.* 58(10): 6742–6747.

Wei, W., Liu, J., Jiang, J. 2020. Atomistic simulation study of polyarylate/zeolitic-imidazolate framework mixed-matrix membranes for water desalination. *ACS Appl. Nano Mater.* 3(10): 10022–10031.

Xiao, R., Luo, Z., Wei, Z., Luo, S., Spinney, R., Yang, W., Dionysiou, D. D. 2018. Activation of peroxymonosulfate/persulfate by nanomaterials for sulfate radical-based advanced oxidation technologies. *Curr. Opin. Chem. Eng.* 19: 51–58.

Xu, S., Lv, Y., Zeng, X., Cao, D. 2017. ZIF-derived nitrogen-doped porous carbons as highly efficient adsorbents for removal of organic compounds from wastewater. *Chem. Eng. J.* 323, 502–511.

Xu, W.-T., Ma, L., Ke, F., Peng, F.-M., Xu, G.-S., Shen, Y.-H., Zhu, J.-F., Qiu, L.-G., Yuan, Y.-P. 2014. Metal–organic frameworks MIL-88A hexagonal microrods as a new photocatalyst for efficient decolorization of methylene blue dye. *Dalton Trans.* 43(9): 3792–3798.

Yaghi, O. M., Li, G., Li, H. 1995. Selective binding and removal of guests in a microporous metal–organic framework. *Nature* 378(6558): 703–706.

Yaghi, O. M., O'Keeffe, M., Ockwig, N. W., Chae, H. K., Eddaoudi, M., Kim, J. 2003. Reticular synthesis and the design of new materials. *Nature* 423(6941): 705–714.

Yang, C., You, X., Cheng, J., Zheng, H., Chen, Y. 2017. A novel visible-light-driven In-based MOF/graphene oxide composite photocatalyst with enhanced photocatalytic activity toward the degradation of amoxicillin. *Appl. Catal. B*200: 673–680.

Yang, D., Babucci, M., Casey, W. H., Gates, B. C. 2020. The surface chemistry of metal oxide clusters: From metal–organic frameworks to minerals. *ACS Cent. Sci.* 6(9): 1523–1533.

Yang, H., He, X.-W., Wang, F., Kang, Y., Zhang, J. 2012. Doping copper into ZIF-67 for enhancing gas uptake capacity and visible-light-driven photocatalytic degradation of organic dye. *J. Mater. Chem.* 22(41): 21849–21851.

Yang, S., Wang, P., Yang, X., Shan, L., Zhang, W., Shao, X., Niu, R. 2010. Degradation efficiencies of azo dye Acid Orange 7 by the interaction of heat, UV and anions with common oxidants: persulfate, peroxymonosulfate and hydrogen peroxide. *J. Hazard. Mater.* 179(1–3): 552–558.

Yu, C., Han, X., Shao, Z., Liu, L., Hou, H. 2018. High efficiency and fast removal of trace Pb (II) from aqueous solution by carbomethoxy-functionalized metal–organic framework. *Cryst. Growth Des.* 18(3): 1474–1482.

Zeng, T., Zhang, X., Wang, S., Niu, H., Cai, Y. 2015. Spatial confinement of a Co_3O_4 catalyst in hollow metal–organic frameworks as a nanoreactor for improved degradation of organic pollutants. *Environ. Sci. Technol.* 49(4): 2350–2357.

Zhang, D.-S., Zhang, Y.-Z., Gao, J., Liu, H.-L., Hu, H., Geng, L.-L., Zhang, X., Li, Y.-W. 2018. Structure modulation from unstable to stable MOFs by regulating secondary N-donor ligands. *Dalton Trans.* 47(39): 14025–14032.

Zhang, H., Liu, D., Yao, Y., Zhang, B., & Lin, Y. S. 2015. Stability of ZIF-8 membranes and crystalline powders in water at room temperature. *J. Membr. Sci.* 485: 103–111.

Zhang, M., Xiao, C., Yan, X., Chen, S., Wang, C., Luo, R., Qi,. J., Sun, X., Wang, L., Li, J. 2020. Efficient removal of organic pollutants by metal–organic framework derived Co/C Yolk–shell nanoreactors: Size-exclusion and confinement effect. *Environ. Sci. Technol.* 54(16): 10289–10300.

Zhang, X., Zhang, Y., Wang, T., Fan, Z., & Zhang, G. 2019. A thin film nanocomposite membrane with pre-immobilized UiO-66-NH_2 toward enhanced nanofiltration performance. *RSC Adv.* 9(43): 24802–24810.

Zhang, Y., Li, G., Lu, H., Lv, Q., Sun, Z. 2014. Synthesis, characterization and photocatalytic properties of MIL-53 (Fe)–graphene hybrid materials. *RSC Adv.* 4: 7594–7600.

Zhang, Y., Zhou, J., Feng, Q., Chen, X., Hu, Z. 2018. Visible light photocatalytic degradation of MB using UiO-66/g-C_3N_4 heterojunction nanocatalyst. *Chemosphere* 212: 523–532.

Zhao, J., Dong, W.-W., Wu, Y.-P., Wang, Y.-N., Wang, C., Li, D.-S., Zhang, Q.-C. 2015. Two (3, 6)-connected porous metal–organic frameworks based on linear trinuclear $[Co_3(COO)_6]$ and paddlewheel dinuclear $[Cu_2(COO)_4]$ SBUs: gas adsorption, photocatalytic behaviour, and magnetic properties. *J. Mater. Chem. A* 3(13): 6962–6969.

Zhao, J., Jin, B., Peng, R. 2020. New core–shell hybrid material IR-MOF3@COF-LZU1 for highly efficient visible-light photocatalyst degrading nitroaromatic explosives. *Langmuir* 36(20): 5665–5670.

Zhao, S.-N., Zhang, Y., Song, S.-Y., Zhang, H.-J. 2019. Design strategies and applications of charged metal organic frameworks. *Coord. Chem. Rev.* 398: 113007.

Zhao, Y., Song, M., Cao, Q., Sun, P., Chen, Y., Meng, F. 2020. The superoxide radicals' production via persulfate activated with $CuFe_2O_4$@ Biochar composites to promote the redox pairs cycling for efficient degradation of o-nitrochlorobenzene in soil. *J. Hazard. Mater.* 400: 122887.

Zheng, H.-Q., He, X.-H., Zeng, Y.-N., Qiu, W.-H., Chen, J., Cao, G.-J., Lin, R.-G., Lin, Z.-J., Chen, B. 2020. Boosting the photoreduction activity of Cr(vi) in metal–organic frameworks by photosensitiser incorporation and framework ionization. *J. Mater. Chem. A* 8(33): 17219–17228.

Zhu, C., Liu, F., Ling, C., Jiang, H., Wu, H., Li, A. 2019. Growth of graphene-supported hollow cobalt sulfide nanocrystals via MOF-templated ligand exchange as surface-bound radical sinks for highly efficient bisphenol A degradation. *Appl. Catal. B* 242: 238–248.

Zhu, K., Chen, C., Xu, H., Gao, Y., Tan, X., Alsaedi, A., Hayat, T. 2017. Cr(VI) reduction and immobilization by core-double-shell structured magnetic polydopamine@ zeolitic idazolate frameworks-8 microspheres. *ACS Sustainable Chem. Eng.* 5(8): 6795–6802.

Zhu, Y., Gupta, K. M., Liu, Q., Jiang, J., Caro, J., Huang, A. 2016. Synthesis and seawater desalination of molecular sieving zeolitic imidazolate framework membranes. *Desalination* 385: 75–82.

13 Photocatalytic Mechanism in Low-Dimensional Chalcogenide Nanomaterials

An Exciton Dynamics Insight

Yawei Yang and Wenxiu Que

International Center for Dielectric Research, Shaanxi Engineering Research Center of Advanced Energy Materials and Devices, School of Electronic Science and Engineering, Xi'an Jiaotong, China

CONTENTS

13.1 Introduction 410
13.1.1 Metal Chalcogenide Photocatalytic Nanomaterials 410
13.1.2 Quantum Confinement Effect in Low-Dimensional Chalcogenide Nanomaterials 413
13.1.3 Photocatalytic Processes in Low-Dimensional Chalcogenide Nanomaterials 414
13.1.4 Ultrafast Spectroscopy Technique for Exciton Dynamics Study 417
13.2 Exciton Dynamics in 0D Chalcogenide QDs 418
13.2.1 Exciton Dissociation Dynamics of CdX QDs 419
13.2.2 Photocatalytic Mechanisms of CdX QDs 422
13.3 Exciton Dynamics in 1D Chalcogenide NRs 425
13.3.1 Electronic Structure of CdX NRs Heterostructures 426
13.3.2 Exciton Dissociation Dynamics of CdX NRs 427
13.3.3 Photocatalytic Mechanisms of CdX NRs 432
13.4 Exciton Dynamics in 2D Chalcogenide NPLs 435
13.4.1 Electronic Structure of CdX NPLs Heterostructures 435
13.4.2 Exciton Dissociation Dynamics of CdX NPLs 436
13.4.3 Photocatalytic Mechanisms of CdS NPLs 439

DOI: 10.1201/9781003118749-13

13.5 Summary and Perspectives 443
13.5.1 Summary 443
13.5.2 Perspectives 443
Acknowledgments........ 444
References 444

13.1 INTRODUCTION

Advanced oxidation processes (AOPs), also known as deep oxidation technology, is promising for solving the issue in terms of deep degradation of organics in water. Photocatalytic oxidation by applying semiconductor nanomaterials is one kind of AOPs for low-concentration/trace sewage treatment and remediation. Metal chalcogenides are a series of representative semiconductor photocatalysts. Over the past decades, many studies have fully investigated the chemical and physical properties, and photocatalytic applications of many chalcogenide photocatalysts (Cheng et al., 2018). Moreover, recent years have consistently seen a significant enhancement in the photocatalytic performance of low-dimensional (0D, 1D, and 2D) chalcogenide nanomaterials with unique quantum confinement phenomenon compared to their bulk materials (Haque et al., 2018). Nowadays, the exciton dynamics/charge carrier properties in photocatalysts and their interfaces have become a research hotspot, thanks to the giant development of ultrafast spectroscopy techniques, providing new dynamic insight into the photocatalytic mechanism (Maiuri et al., 2020). However, the photocatalytic mechanism still poses a big challenge to the field of photocatalysis and solar energy conversion. In this chapter, the exciton dynamics/charge carrier properties of low-dimensional metal chalcogenide nanomaterials, especially cadmium chalcogenides (CdX, X = S, Se, Te), were summarized for contributing further comprehension of the photocatalytic mechanism.

13.1.1 Metal Chalcogenide Photocatalytic Nanomaterials

In photocatalytic reaction, both $O_2^{\bullet-}$ and $OH^{\bullet}$ radicals are the key active species. All efforts were made to improve the production efficiency of these active radicals. In fact, not all semiconductors can serve as a photocatalyst because they must satisfy static and dynamic conditions as follows:

1. The conduction band (CB) potential must be negative to the O_2 molecule reduction potential ($E^{\theta}(O_2/O_2^{\bullet-})$: −0.33 eV *vs*. NHE, pH = 7), so that the photogenerated electron can transfer from semiconductor CB to O_2 molecule to generate $O_2^{\bullet-}$ radical. As a result, the hole staying in the valence band (VB) directly oxides the organic matter, or transfers to a water molecule to form $OH^{\bullet}$ radical when the VB potential is positive to the water oxidation potential ($E^{\theta}(H_2O/OH^{\bullet})$: 2.3 eV *vs*. NHE, pH = 7). Therefore, for a single semiconductor, its bandgap must be in a suitable range of about 1.5–3.0 eV, because too wide or too narrow bandgap respectively leads to low utilization of visible light energy or a too positive CB potential.
2. The exciton/charge carrier in a photocatalyst must live long enough to diffuse to the photocatalyst surface to enable photocatalytic reactions. From the

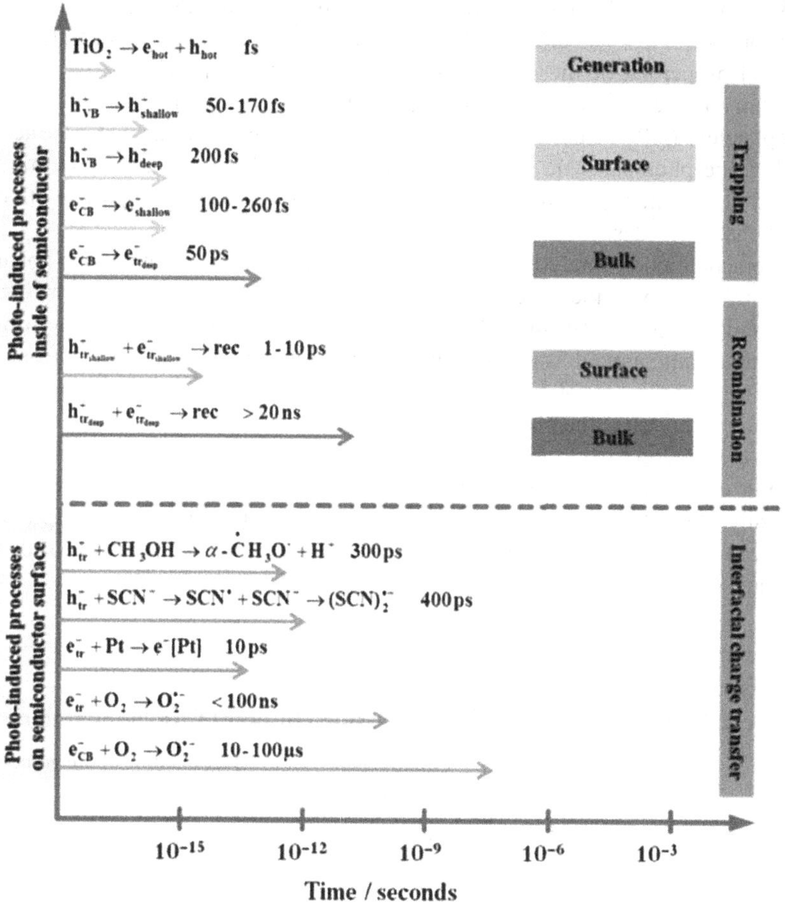

FIGURE 13.1 Time domain of some typical photocatalytic processes. (Reprinted with permission from Schneider and Bahnemann (Schneider et al., 2014). Copyright 2014 American Chemical Society.)

dynamic process perspective, it takes some time to complete all steps (Schneider et al., 2014). For example, it takes up to microseconds for electrons to form the $O_2{\bullet}^-$ radicals, as shown in Figure 13.1.

3. The photocatalyst must have sufficient adsorption and reaction active sites. Due to the limited diffusion distance of the active radicals, the O_2 molecules and organic matter must be close enough to the photocatalyst surface to enable photocatalytic reactions.
4. The chemical and physical properties of the photocatalyst must be stable.
5. Only when the above premises are met can a semiconductor have the potential for photocatalysis. Under this limitation, several metal oxides, such as TiO_2 and ZnO, metal chalcogenides, such as CdX, CuX, and ZnX, multi-metal sulfides, such as $CuInS_2$, Cu_2ZnSnS_4, and $ZnInS_2$, have been developed as representative semiconductor photocatalysts with excellent performance.

Among plenty of chalcogenides, CdX with superior optical, electronic, and transport properties has attracted broad interest in the field of photocatalysis. The physical properties of these materials can be flexibly adjusted by the degree of quantum confinement, that is 0D quantum dots (QDs), 1D nanorods (NRs), and 2D nanoplatelets (NPLs), by material composition, and by forming heterostructures. Further, the photocatalytic performance of CdX can be easily controlled through these routes, which is the most significant advantage for CdX compared with other metal chalcogenides.

In practical photocatalytic applications, various types of pollutants, i.e., pharmaceuticals, toxic dyes, metal cations, pathogenic microorganisms, and pesticides, can be degraded from water using CdX nanomaterials. In laboratory experiments, organic dyes are usually applied as the indicator for probing the photocatalytic performance of the CdX nanomaterials, as shown in Table 13.1. Although it is always hard to qualitatively compare the photocatalytic performance among the CdX nanomaterials due to the different experimental parameters in different work, the superior photocatalytic degradation behavior of the quantum confinement nanomaterials can be still generally seen. What is the reason for this phenomenon? This chapter explains the advanced photocatalytic performance of low-dimensional CdX nanomaterials from exciton dynamics insight.

TABLE 13.1
Photocatalytic Pollutants Degradation Using Different CdX Nanomaterials

	Photocatalyst	Experimental Parameters	Degradation Yield (%)	References
Bulk materials	CdS nanoparticles	MO[a], CR[b], OII[c], Rh B[d], MG[e], MB[f]: 6.0×10^{-5} M, 50 mL 300 W Xe lamp, > 420 nm 20 mg photocatalyst	16, 84, 26, 31, 95, 51@30 min	Li et al. (2012)
	CdSe microspheres	Rh B: 10 mg/L, 50 mL 300 W Xe lamp	80@300 min	Gong et al. (2019)
0D QDs	CdS	MB: 100 mL 500 W Xe lamp 2 mg photocatalyst	45@120 min	Bhuvaneswari et al. (2020)
	CdSe	Cefalexin: 15 mg/L, 100 mL 300 W mercury lamp, 365 nm 50 mg photocatalyst	65@30min	Liu et al. (2013)
1D NRs	CdS	MO: 10 mg/L, 100 mL 300 W mercury lamp 20 mg photocatalyst	95@10 min	Chen et al. (2015)
2D NPLs	CdSe	MG: 1.0×10^{-5} M, 10 mL 500 W Xe lamp, > 420 nm 10 mg photocatalyst	80@30 min	Han et al. (2014)

MO[a] = methyl orange, CR[b] = congo red, OII[c] = orange II, Rh B[d] = rhodamine B, MG[e] = malachite green, MB[f] = methylene blue.

TABLE 13.2
Photocatalytic Degradation of Various Pollutants from Water by Using Different 2D NPLs

2D NPLs	Pollutant	Experimental Parameters	Degradation Yield (*vs.* pure sample, %)	References
Ni doped CdS	MB	1 mg/L, 100 mL 8 W mercury vapor lamp/ Natural sunlight 10 mg photocatalyst	50 *vs.* 30@80 min 70 *vs.* 45@80 min	Ahmed et al. (2017)
CdS/RGO[a]	MB	1.33×10^{-5} M, 30 mL 300 W Xe lamp, > 420 nm 10 mg photocatalyst	98 *vs.* 79@60 min	Bera et al. (2015)
CdSe/CdS/Au	MB, MO	0.01 mM, 10 mL Natural sunlight 7 mg photocatalyst	90@30 min 90@3 min	Chauhan et al. (2016)
ZnSe/CdSe dots-on-plates	MB	3 mL 60 W mercury lamp 2 mg photocatalyst	80 *vs.* 40@120 min	Yadav et al. (2017)

RGO[a] reduced graphene oxide.

Despite the strong quantum confinement in 0D, QDs have been exhibited satisfactory interfacial charge transfer rates; the diameter of the QDs limits both the charge separation space and light absorption cross-section. In 1D NRs, a faster interfacial charge transfer through strong quantum confinement in a radial direction, a larger absorption cross-section, and a longer charge separation distance can be simultaneously enabled, while the charge carrier movement perpendicular to the rod is still limited. In NPLs, charge carriers can transport freely parallel to the plane, providing them with more charge separation for photocatalysis. Compared to 0D QDs and 1D NRs, a large absorption cross-section and a uniform 1D quantum confinement are displayed in 2D CdX NPLs, making them powerful photocatalysts. Photocatalytic activities of CdX NPLs can be further enhanced through doping and hybridizing with other nanomaterials to form 2D NPL heterostructures, as shown in Table 13.2.

13.1.2 Quantum Confinement Effect in Low-Dimensional Chalcogenide Nanomaterials

In case of at least one diameter of CdX nanoparticle is of the same magnitude as the de Broglie wavelength of the electron wave function, the quantum confinement would be observed. For bulk CdX materials, whose confining dimensions are large regarding to the wavelength of the particle. Their bandgap shows original energy for a continuous energy state. However, the energy spectrum turns discrete when the confining dimension decreases to the quantum size. Therefore, their electronic and optical properties actually differ from those of bulk materials. That is, the absorption and emission edges show a blue shift, or the bandgap becomes size-dependent.

Specifically, the exciton Bohr radius is used for describing the phenomenon of excitons being squeezed into a critical quantum dimension,

$$R_B = \varepsilon_r \left(\frac{m_0}{m*} \right) R_0 \tag{13.1}$$

$$m* = \frac{m_e m_h}{(m_e + m_h)} \tag{13.2}$$

where ε_r is the relative permittivity, m* is the reduced mass of the electron-hole system, m_0 is the electron mass, and m_e and m_h are the effective mass of eletron and hole, respectively, and R_0 is the Bohr radius of hydrogen atom (0.0529 nm). Therefore, the exciton Bohr radius R_B of CdX is calculated to be ~3, ~5, and ~7 nm for X = S, Se, and Te. The bandgap of nanomaterial, whose radius is comparable to its exciton Bohr radius R, can be defined by

$$\begin{aligned} E(R) = E_g + \hbar^2 \pi^2 / 2R^2 (1/m_e + 1/m_h) \\ -1.786 e^2 / \varepsilon_r R - 0.124 e^4 / \varepsilon_r^{2' \, 2} (1/m_e + 1/m_h) \end{aligned} \tag{13.3}$$

where R is the particle radius, Eg is the bulk bandgap, and ħ is the reduced Planck's constant. A QD, NR, and NPL are confined in three, two, and only one dimension, respectively. These are defined as 0D, 1D, and 2D potential wells, which refer to the number of dimensions of free carriers moving in a confined nanoparticle. Their small sizes can generate novel properties that cannot be done in bulk materials.

Compared to bulk CdX, 0D QDs, 1D NRs, and 2D NPLs consist of an emerging class of quantum effect materials which show many unique properties, such as large specific surface area, uniform quantum confinement, high exciton binding energy, long Auger lifetime, low nonradiative recombination yield, and giant oscillator strength. These properties lead to excellent performance in photocatalytic and other optoelectrical applications, for example, lasing materials with a large gain coefficient and a low threshold (Li & Lian, 2019).

Many of these properties result from the structure and band-edge exciton dynamics in these low-dimensional chalcogenide nanomaterials. The free electrons and holes usually come from the band-edge excitons. In other words, the photocatalytic performance of CdX depends on the structure and exciton dynamics to a great extent. In recent years, the properties of 0D, 1D, and 2D excitons have attracted much attention motivated by a fundamental understanding of low-dimensional chalcogenide nanomaterials and their photocatalytic applications.

This chapter provides an overview of key properties of low-dimensional excitons depending on dimensions in QDs, NRs, and NPLs, and their effects on the photocatalytic mechanism.

13.1.3 Photocatalytic Processes in Low-Dimensional Chalcogenide Nanomaterials

Photocatalytic processes in CdX photocatalysts are generally involved in light absorption, charge carrier transport, mass (O_2, radicals, and organic matter) transportation,

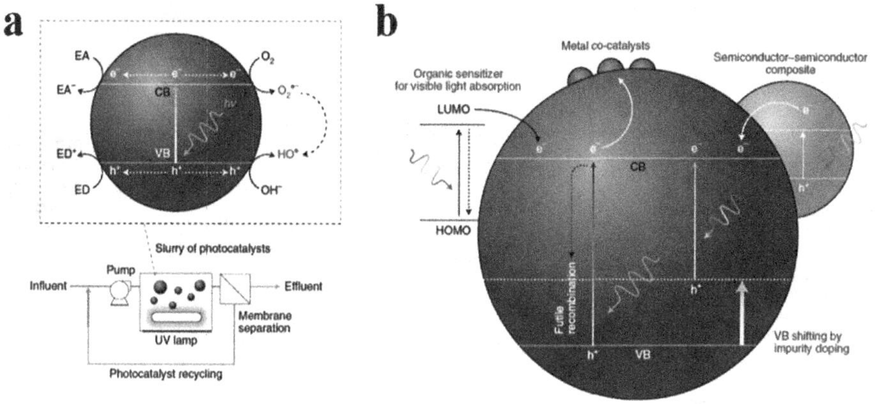

FIGURE 13.2 Scheme of photocatalytic processes and charge carrier dynamics within the photocatalyst. (Reprinted with permission from Hodges et al. (2018). Copyright 2018 Springer Nature.)

and chemical reactions. The photocatalytic processes within the CdX photocatalyst can be generally divided into the following steps, as shown in Figure 13.2a (Hodges et al., 2018).

1. Excitons/electrons-holes pairs generation in the semiconductor by photon excitation

$$CdX + hv \rightarrow h^+ + e^- \quad (13.4)$$

2. Electron transfer to O_2 for $O_2^{\bullet-}$ radical generation

$$e + O_2 \rightarrow O_2^{\bullet-} \quad (13.5)$$

3. Hole transfer to water for $OH^\bullet$ radical generation (depend on the VB potential)

$$h^+ + H_2O \rightarrow OH^\bullet + H^+ \quad (13.6)$$

$$h^+ + OH^- \rightarrow OH^\bullet \quad (13.7)$$

4. Chemical reactions of $O_2^{\bullet-}$ radical

$$O_2^{\bullet-} + H^+ \rightarrow OOH^\bullet \quad (13.8)$$

$$2\,OOH^\bullet \rightarrow O_2 + H_2O_2 \quad (13.9)$$

$$O_2^{\bullet-} + H_2O_2 \rightarrow OH^\bullet + OH^- + O_2 \quad (13.10)$$

$$H_2O_2 \rightarrow 2\,OH^\bullet \quad (13.11)$$

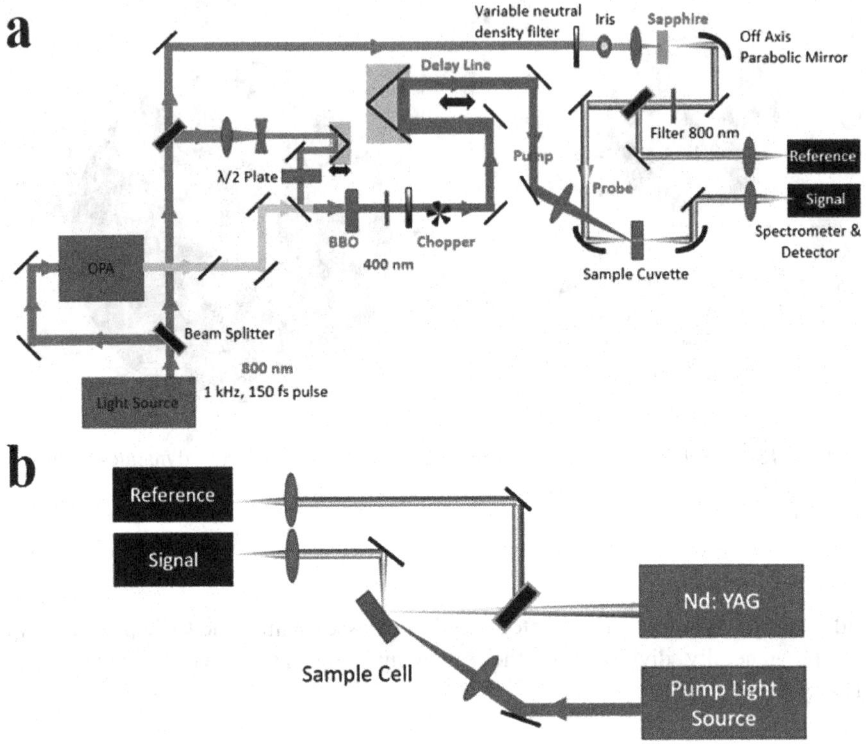

FIGURE 13.3 Scheme of the femtosecond-picosecond and nanosecond-microsecond domain TAS systems.
(Source: Yawei Yang's doctoral dissertation 2019).

5. Mineralization of organic matter by active species

$$OH^{\bullet} / h^{+} / O_2^{\bullet -} + \text{organics} \rightarrow CO_2 + H_2O + \text{micromolecule matter}$$

In the photocatalytic application, the material performance highly depends on the fundamental properties of excitons, including single and multiple exciton radiative and nonradiative decay, exciton and carrier transport, and interfacial energy and charge transfer. Hence, the traditional way of enhancing the photocatalytic performance is to improve the number of photo-generated electrons, such as doping, composite, sensitizer/co-catalyst loading, as shown in Figure 13.2b.

The timescale of the photon transmission process through the reactor and the reaction solution is at the picosecond-nanosecond range. The timescale of exciton generation, separation, and transfer processes within the photocatalysts is typically at femtosencond-microsecond range. However, the timescale of the surface catalytic process, including reactant adsorption, chemical reaction, and desorption is at microsecond-millisecond range. That means the much slower surface catalytic process will result in the photo-excited charge carriers' aggregation, thereby making charge recombination serious. Hence, one of the main restrictions of low photocatalytic efficiency is attributed to an obstacle in exciton dynamics and catalytic reaction at the time domain, namely non-coupling or mismatching. In other words, the lifetime of charge carriers/charge-separated state is too short to effectively transfer to the reactants (O_2, water, and organic matter).

The timescale of catalytic reactions cannot be easily changed, so improving the charge-separated state lifetime of CdX photocatalyst is trying to solve this issue. Specifically, making CdX to low-dimensions and forming heterostructures are possible routes that can work on in future studies. Moreover, if the exciton/charge carrier dynamics within the low-dimensional CdX nanomaterials can be fully understood, the difficulty of changing the timescale of exciton behaviors might be accordingly responded.

13.1.4 Ultrafast Spectroscopy Technique for Exciton Dynamics Study

Much research focused on advancing the fundamental understanding of interfacial structure and dynamics that governs charge transport and energy conversion within CdX photocatalysts. These fundamental advances are essential to improve the existent photocatalysts and develop novel photocatalytic technologies. The state-of-the-art ultrafast spectroscopy technique was developed and applied for directly seeing these interfacial dynamics *in situ* under photocatalytic and photoelectrocatalytic conditions. Among many ultrafast spectroscopy techniques, transient absorption spectroscopy (TAS), a commonly used ultrafast pump-probe technique with femtosecond-level resolution, is the most effective tool for exciton/charge carrier dynamics study (Miao & Tang, 2020).

The transition between energy levels in the excited state materials can be analyzed by the TAS, including the energy transfer and charge transfer in physical and chemical processes. During the TA test, a high-energy monochrome femtosecond pulse can pump the ground state material to an excited state. After a certain delay time, a low-energy broadband probe pulse is used to detect the absorption spectrum of the excited state. By adjusting the delay time, the change of optical absorption (ΔA) of the excited state material over time can be obtained, which means that the change of particle number over time in the excited energy level of the material can be studied. As a result, the relaxation processes can be clearly described, from a high-level excited state to a low-level ground state. The TA spectrum signal is defined as the differential absorption of the excited state and the ground state of the material ($\Delta A = A_{ex} - A_{ground}$) over time. Under the pump pulse, some particles absorb the photon and transit from the ground state to an excited state, followed by the physical processes of ground-state bleaching, excited-state absorption, and stimulated emission. The ΔA signal variety can be generally assigned to the following three types (Miao & Tang, 2020):

1. Ground state bleaching
   The decrease of particle number in the ground state results in a decreased absorption of the probe light, which generally results in a negative signal of $\Delta A<0$ in the relevant wavelength range.
2. Excited state absorption
   The particles in the excited state will further absorb the probe photons to transit to a higher energy level, which usually shows a positive signal of $\Delta A>0$ in the relevant wavelength range.

3. Stimulated emission
   Part of the particles in the excited state will return to the ground state by stimulated emission. The stimulated emission signal and the spontaneous emission signal of the probe light are in the same wavelength range. Since their emission intensity is proportional to the cube of the emission frequency, respectively, the intensity and wavelength range of the excited state stimulated emission will be stronger than the spontaneous emission of the probe light. As a result, a negative signal of $\Delta A<0$ appears in the relevant wavelength range. In addition to be used to study charge separation, relaxation, transfer, and recombination in semiconductors and at interfaces, for example CdSe, CdS/Pt and CdS/CdSe heterostructures, the TAS can also be applied in transient products detection in ultrafast chemical reaction, such as the generation of ligand radicals in metal complexes (Solovyev et al., 2016).

The visible TAS measurements were usually done by a regenerative amplified Ti-Sapphire laser system (typically 800 nm) and spectrometers. The 800 nm output pulse was split in two parts with a beam-splitter. The frequency of the high-energy part was doubled by a β-BaB_2O_4 (BBO) crystal to produce the 400 nm pump beam (or mixed by Optical Parametric Amplifier, OPA, to generate various wavelength), which was chopped to half frequency. The low-energy part passes a sapphire/CaF_2 window to generate a visible probe. An appropriate wavelength pump pulse was applied to selectively excite the material. The resulting ΔA signal, changing with both wavelength and time, was recorded by the white light probe. The probe pulse was continuously delayed (femtosecond-microsecond range) with respect to the pump pulse, monitoring the whole exciton separation, transfer, and recombination events. Note, the time delay is achieved differently in femtosecond-picosecond and nanosecond-microsecond domain TAS systems. For the femtosecond-picosecond system, the time delay is achieved by adjusting the optical path difference (delay line), as shown in Figure 13.3a; for the nanosecond-microsecond system, the probe light was electronically delayed, as shown in Figure 13.3b. By assigning the attribution of the ΔA signal, the whole exciton dynamics of the material can be fully understood. The single-electron acceptor, such as methyl viologen (MV^{2+}), methylene blue (MB^{+}), rhodamine B (Rh B), anthraquinone (AQ), and benzoquinone (BQ), and hole acceptor, typically phenothiazine (PTZ), are respectively added onto the photocatalysts in order to determine the contribution of electron and hole in ΔA signal.

13.2 EXCITON DYNAMICS IN 0D CHALCOGENIDE QDs

Bulk CdX materials have achieved impressive performance in highly efficient solar cells and photocatalysis. Colloidal CdX QDs offer the possibility to further tuning, improving, and optimizing the properties of these materials through the quantum confinement effect. This section aims to understand charge carrier dynamics in CdX QDs and examine its dependence on QD size and composition, and how these properties affect their photocatalytic mechanisms.

13.2.1 Exciton Dissociation Dynamics of CdX QDs

Absorption of photons in CdX semiconductors produces bound electron-hole pairs (excitons), which must be spatially dissociated into free electrons and holes in order to conduct the reduction and oxidation reactions, respectively. Early studies have shown that CB edge electron state-filling totally contributes to the exciton bleach (XB) signal in TA spectra of CdX, while VB edge hole state-filling contributes negligibly because of its small level spacing and large degeneracy (Zhu et al., 2013). Next, the charge transfer between CdX QDs and adsorbed molecular acceptors/donors or semiconductors is discussed.

1. **Conventional and Auger-assisted electron transfer from QDs to molecular acceptors**

The assignment of TA signal agrees with many TAS studies in CdX QD-acceptor complexes, in which the removal of CB electron results in XB signal recovery. The electron transfer processes from CdX QDs to three kinds of molecular acceptors, MB^+, MV^{2+}, and AQ, as shown in Figure 13.4, were investigated (Zhu et al., 2013). The electron transfer rates over an apparent driving force ($-\Delta G$) range of 0–1.3 eV were successfully obtained by combining of band edges of bulk materials, size-tunable confinement energies, and redox potentials of acceptors.

How the electron transfer rates depend on the QDs size within QD-acceptor complexes was examined (Zhu et al., 2013). The electron transfer rates are shown as a function of QD size for CdX, as shown in Figure 13.5 ABC. For all case, electron transfer rate increases with the decrease of the QD size, no matter what kind of QD compositions and molecular acceptors. Besides, the size dependence is more obvious in AQ than MB^+.

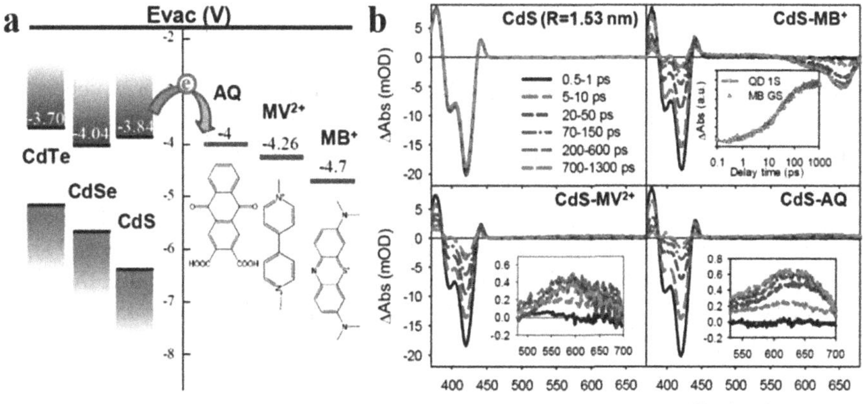

FIGURE 13.4 Schematic diagram of the electron transfer from CdX to molecular acceptors. (Reprinted with permission from Zhu et al., 2013. Copyright 2013 American Chemical Society.)

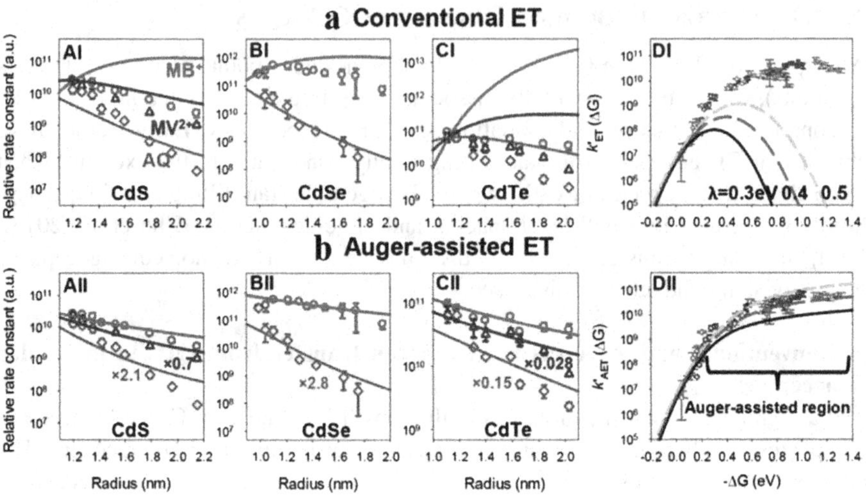

FIGURE 13.5 Size-dependent electron transfer rates in conventional and Auger-assisted electron transfer. (Reprinted with permission from Zhu et al., 2013. Copyright 2013 American Chemical Society.)

In the conventional electron transfer model, the hole remains at the $1S_h$ level of QD, while only the 1S electron transfers to the acceptor. Hence, the nonadiabatic electron transfer rate $k_{ET}(R)$ from the excited QD to adsorbate can be expressed by Marcus model

$$k_{ET}(R) = 2\pi C_H \,|\Psi_{1Se}(R)_0|^2 \left(4\pi\lambda k_B T\right)^{-1/2} \exp\left[-\left(\lambda + \Delta G(R)\right)^2 / 4\lambda k_B T\right] / \hbar \qquad (13.12)$$

where ΔG is the change of free energy, C_H is a size-independent factor which relies on the material and molecule, λ is the total reorganization energy of electron transfer between the reactant and product states, and $\Psi_{1Se}(R)_0$ is the 1S electron density at the surface of QDs. As a result, the $-\Delta G$ for electron transfer from the QDs to these acceptors meets the order of $MB^+ > MV^{2+} >$ AQ, due to the molecules' redox potentials difference. Generally, the electron transfer rates increase at decreasing QD sizes (or increasing driving force) for all QD-acceptor complexes. Especially, the −ΔG values (0.04–0.31 eV) are smaller than λ for electron transfer from CdS and CdSe QDs to AQ, falling in the Marcus normal regime. In this case, the size-dependent electron transfer rates can be accurately described by the conventional model. However, the −ΔG values (0.67–1.31 eV) obviously exceed λ in the case of electron transfer from CdS or CdSe QDs to MB^+ and from CdTe QDs to MB^+ or MV^{2+}, falling in the Marcus inverted regime, as shown in Figure 13.5 DI. So, the conventional model foretells a slower electron transfer rate at decreasing QD size.

The conventional model is in the hypothesis of hole dynamics-independent electron transfer. However, this premise is often inadequate, in fact, because the enhanced Coulomb interaction in QDs can lead to correlated electron-hole dynamics, including multi-exciton Auger recombination and Auger-assisted hot carrier thermalization. Therefore, an Auger-assisted electron transfer model was proposed. It assumed that in addition to the lattice and acceptor molecules' vibrations, the excess electron energy could also be consumed through exciting 1S holes to a deeper level $E_{h,I}$. Then, the excited holes can further relax within the VB levels. Hence, the total electron transfer rate $k_{AET}(R)$ to all product states can be expressed by

$$k_{AET}(R) = 2\pi C \,|\Psi_{1Se}(R)_0|^2 R^2 \int dE_h (4\pi\lambda k_B T)^{-1/2} \exp\left[-(\lambda + \Delta G(R) + E_h)^2 / 4\lambda k_B T\right] / \hbar \tag{13.13}$$

By using this model, the size-dependent Auger-assisted electron transfer rates can be well predicted for all QDs and acceptors. The significantly different dependencies on the driving force appearing in the Marcus inverted regime are the key difference between the above two models, as shown in Figure 13.5 DII. In a word, the unfavorable Franck–Condon overlap occurring in the Marcus inverted regime can be overcome through Auger-assisted electron transfer, enhancing the electron transfer rate.

2. Electron transfer from QDs to metal oxide semiconductors

Electron transfer from QDs to metal oxide semiconductors have been extensively studied in QD based-photovoltaics and photocatalysis. In this case, the Marcus electron transfer model is usually applied. Electron transfer from CdSe QDs with four diameters to various metal oxides (SnO_2, TiO_2, and ZnO) was examined, as shown in Figure 13.6a (Tvrdy et al., 2011). As $-\Delta G$ increased, electron transfer rate k_{MO} shows a sharp rise at first, and turns into a modest rise once $-\Delta G$ exceeds λ, as shown in Figure 13.6b. This trend is well described by

$$k_{MO} = 2\pi \int dE \rho(E) \,|H_{DA}(E)|^2 (4\pi\lambda k_B T)^{-1/2} \exp\left[-(\lambda + \Delta G + E)^2 / 4\lambda k_B T\right] / \hbar \tag{13.14}$$

where $\rho(E)$ is the accepting states density of metal oxide, and $H_{DA}(E)$ is the initial and final states electronic coupling of the donor-acceptor system. The total electron transfer rate is the sum of electron transfer rates of all available states determined by $\rho(E)$, on the basis of the multi-state electron transfer model. When $-\Delta G < \lambda$, the density of

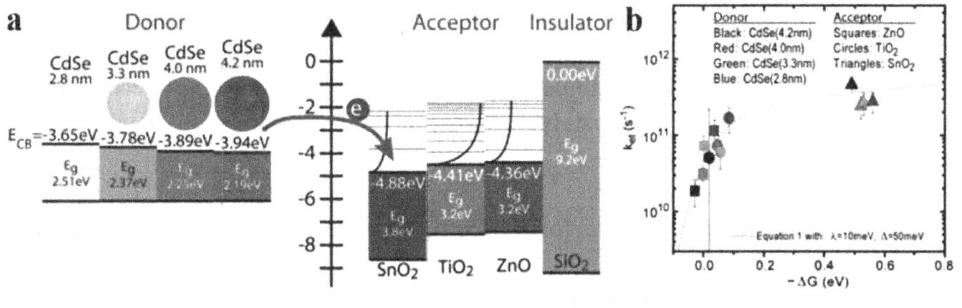

FIGURE 13.6 Size-dependent electron transfer from QDs to metal oxides. (Adapted from Tvrdy et al. (2011) with permission. Copyright 2011 National Academy of Sciences.)

states increases rapidly with driving force because the accepting states are below and near the band edge. However, as $-\Delta G > \lambda$, the electron transfer rate is mainly contributed by accepting states above the band edge, whose density changes slowly ($\sqrt{E}$) with increasing energy. Equation 13.14 describes an Auger-assisted electron transfer-like driving force-dependent electron transfer behavior. But they are fundamentally different. In QD-metal oxide systems, the continuum of product states originates from the CB, whereas it has resulted from the quasi-continuum of donor hole levels in the QD-molecule system. Although this equation is proved to fit the data well for electron transfer from QDs to metal oxide semiconductor, the Auger-assisted pathway cannot be explained by this model.

As discussed above, the electron transfer rate in the inverted regime is significantly enhanced by the Auger-assisted electron transfer, which is not present in QD-metal oxide systems. The understanding of this interfacial electron transfer may benefit the comprehension of QDs-sensitized photocatalytic process.

3. Stepwise two-photon-induced electron transfer from molecular donors to QDs

In QD-molecule systems, most of the photochemical reactions and photophysical properties were limited by one-photon-induced events. Photochemical reactions from higher excited states between QD-molecule systems were demonstrated by studying electron transfer from noncovalently bounded protoporphyrin IX (PP) to CdS QDs under visible light excitation (Uno et al., 2018). In the conventional one-photon-induced electron transfer process, the electron always transfers from the excited PP to CdS, as shown in Figure 13.7. While in CdS@ZnS core@shell QDs, the direct electron transfer from PP to CdS is suppressed by the high energy barrier of the ZnS shell. However, the successive stepwise two-photon absorption process can be expected at higher excited states, which can accumulate sufficient energy to overcome the ZnS barrier, because of the high density of states at higher energy levels in semiconductor nanocrystals and the partly resolved spin selection rule for the transitions by heavy atoms. The stepwise two-photon-induced electron transfer event can be seen from higher excited states PP to CdS in CdS@ZnS core@shell QDs by TAS. Hence, the stepwise two-photon-driven photochemical reactions become expectable since the stepwise two-photon absorption power threshold may be potentially initiated by continuous-wave light through rational intermediate transient states design. The understanding of these photon-induced electron transfers may benefit to the comprehension of dye-sensitized photocatalytic process.

13.2.2 Photocatalytic Mechanisms of CdX QDs

Based on the above systematical understanding of the 0D exciton properties, in this part, the mechanisms of photocatalytic performance from exciton dynamics insight in 0D QDs are discussed.

1. Enhanced photocatalytic performance by charge separation in S^{2-}-modified CdSe QDs

Mercaptopropionic acid (MPA) and S^{2-} ligands capped on the surface of chalcogenide QDs are always served as the hole trap for boosting hole transfer in photocatalytic

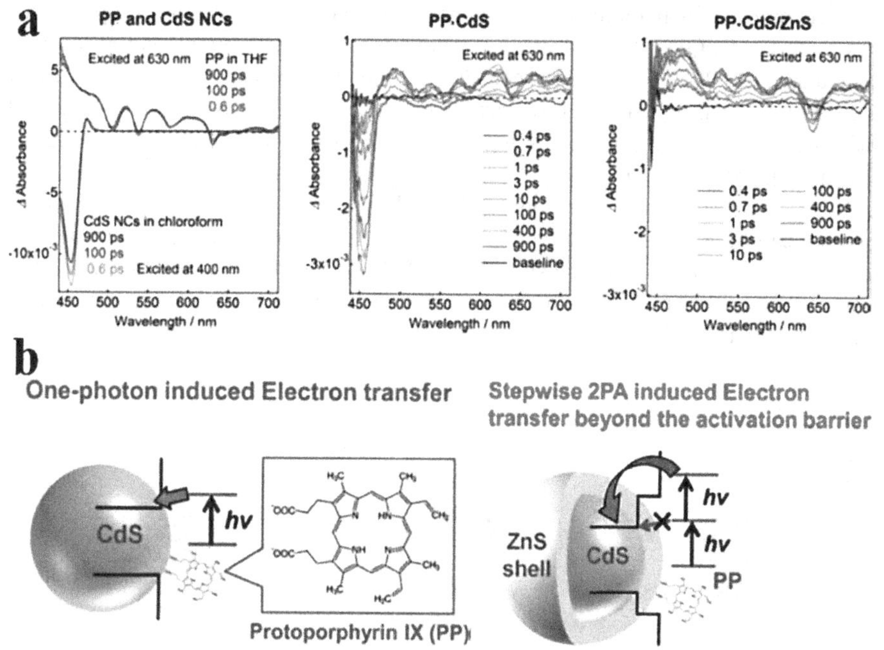

FIGURE 13.7 TA spectra and scheme of the one-photon-induced and stepwise two-photon-induced electron transfer processes. (Adapted from Uno et al. (2018) with permission. Copyright 2018 American Chemical Society.)

reaction, as shown in Figure 13.8a (Fan et al., 2018). The exciton dynamics of CdSe-MPA and CdSe-S QDs was explored by TAS, both of whose TA spectra display negative signals in 425–500 nm and positive signals > 550 nm regions, as shown in Figure 13.8b. The strong bleaching signal originates from state-filling of the 1S electrons level of CdSe QDs, and the positive photo-induced absorption (PA) signal is ascribed to photo-generated holes. The much faster PA decay of the CdSe-S QDs than that of CdSe-MPA QDs, indicates a more efficient hole removal process and a better charge separation occurred in the CdSe-S QDs. The short hole lifetime (< 3 ps) is highly attributed to the surface hole trapping process through capped MPA or S^{2-} ligands. As a result, the enhanced photocatalytic performance was achieved in CdSe-S QDs due to an improved charge-separated state.

2. Infrared-driven photocatalysis by interfacial hot electron transfer from plasmonic Cu_7S_4 to CdS QDs

Infrared (IR) light-driven photocatalysis has been achieved in plasmonic Cu_7S_4/CdS p-n junction (Lian et al., 2019). The IR-induced carrier dynamics were conducted by TAS. Upon selective excitation of a plasmonic band of Cu_7S_4 QDs at 1300 nm, a broad absorption in the visible region with a fast and a slow decay component, can be observed, as shown in Figure 13.9a. Such observation can be attributed to the trapped holes, which indicates the plamonic-induced multiple hole trapping processes in Cu_7S_4 QDs. In Cu_7S_4/CdS p-n junction, a similar broadband absorption with a dip at 450–500 nm can be seen. The dip can be assigned to the state-filling of CdS derived from interfacial hot electron transfer from Cu_7S_4. The kinetics at 460 nm only show a rising component regrading to the CdS state-filling decay, suggesting an ultrashort complete time of interfacial hot-electron transfer within the laser pulse (< 100 fs), as shown in Figure 13.9b. Hence, a ballistic hot-electron transfer process can be illustrated by this ultrafast electron transfer. The lifetime of interfacial charge separation can be estimated as 273 μs from the XB recovery (i.e., the rising component) of CdS, which presents the charge recombination process, as shown in Figure 13.9c.

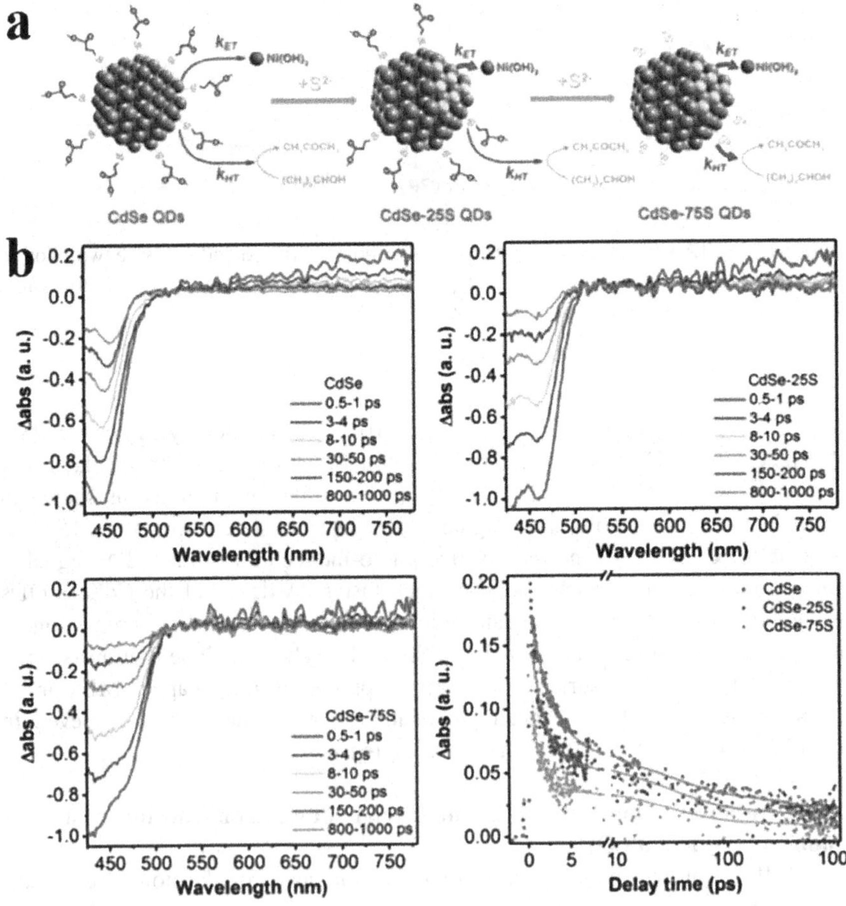

FIGURE 13.8 TA spectra and kinetics of CdSe-MPA and CdSe-S QDs. (Modified from Fan et al. (2018) with permission. Copyright 2018 WILEY-VCH Verlag GmbH & Co. KGaA, Weinheim.)

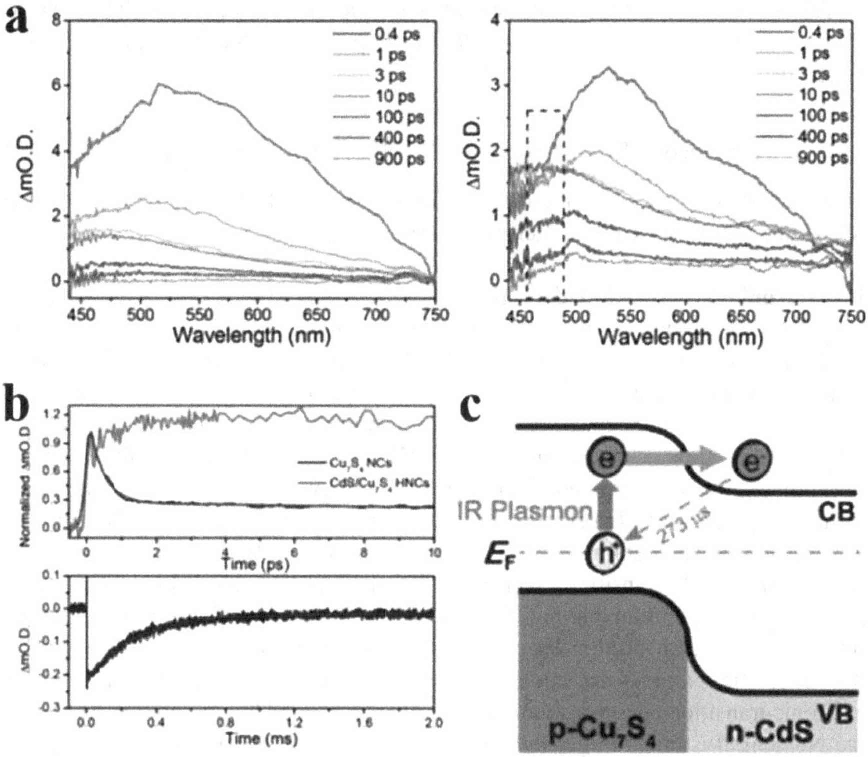

FIGURE 13.9 Hot electron transfer from Cu_7S_4 to CdS. (Reprinted with permission from Lian et al. (2019). Copyright 2019 American Chemical Society.)

13.3 EXCITON DYNAMICS IN 1D CHALCOGENIDE NRs

Unlike the exciton quantum confined in all dimensions in 0D QDs, exciton in 1D NRs is only quantum confined in the radial direction but is free along the axial direction. 1D NRs and related semiconductor-metal heterostructures are efficient photocatalysts due to their unique physical and chemical properties, namely the simultaneously exhibiting quantum confinement effects in the radial direction and bulk-like carrier movement in the axial direction. In other words, on one hand, the concepts well-established in 0D QDs, including size-tunable energetics and wave function engineering through band alignment in heterostructures, can also work in NRs. On the other hand, long-distance charge separation can be achieved in NRs with fast carrier transport along the axial direction. In general, multicomponent heterostructures with controllable charge separation distances for efficient photocatalysis can be further realized by selectively growing catalytic metallic nanoparticles, such as Pt and Au, at the tips of NRs. Therefore, an understanding of exciton and charge carrier dynamics is highly required to design and optimize such composite materials.

This section provides an overview of the basic understanding of charge carrier dynamics in CdX NRs, and how these properties affect their photocatalytic mechanisms.

13.3.1 Electronic Structure of CdX NRs Heterostructures

This part discusses the current understandings of the 1D electronic structures of CdX NRs and CdS/Pt (or Au) NR heterostructures.

1. Electronic structure of CdX NRs

CdX NRs, as shown in Figure 13.10a, have a zinc-blend structure (Zhu & Lian, 2012). The electron-hole interaction of the NRs can be expressed by an effective 1D Coulomb potential depending on their separation along the long axis, as shown in Figure 13.10b, since the much faster carrier movement in the radial direction than in the axial direction. A manifold of bound $1\Sigma(\Pi)$ exciton states with the oscillator strength largely concentrated on the lowest energy exciton state, $1\Sigma_0(1\Pi_0)$, are formed between the 1σ (π) electron and hole in this 1D potential model. By applying this model, the absorption peaks at 390 and 450 nm can be assigned to the lowest energy $1\Pi_0$ and $1\Sigma_0$ transitions, respectively. Generally, the exciton binding energy in NRs is much larger than that in QDs on account of the reduced dielectric screening. Interestingly, the absorption spectra of CdS NRs with different lengths and same diameters show almost the same, as shown in Figure 13.10c, illustrating that the excitonic transition energies should be mainly determined by the rod diameter when the NR length is much larger than its exciton Bohr radius (Wu et al., 2015a). The discrete electron and hole levels, labeled as 1σ, 1π, *etc.*, are led by quantum confinement in the radial direction.

2. Electronic structure of NR heterostructures

NR heterostructures can be formed by selectively growing catalytic metallic nanoparticles at the tips of NRs. In metal particle-tipped semiconductors, the semiconductors

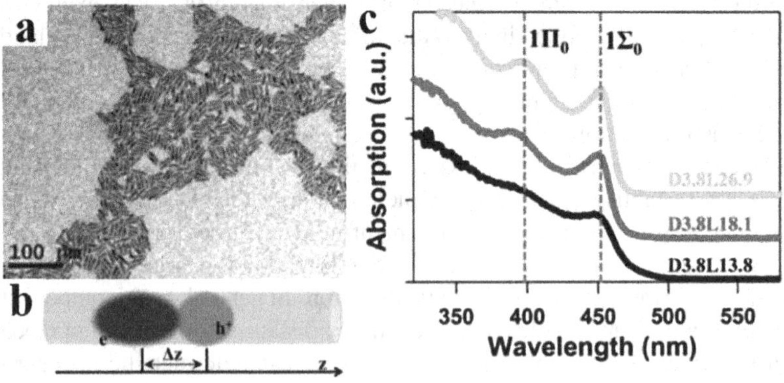

FIGURE 13.10 Electronic structure of CdS NRs. (Adapted from Zhu and Lian (2012) and Wu et al. (2015a) with permissions. Copyright 2012 and 2015 American Chemical Society.)

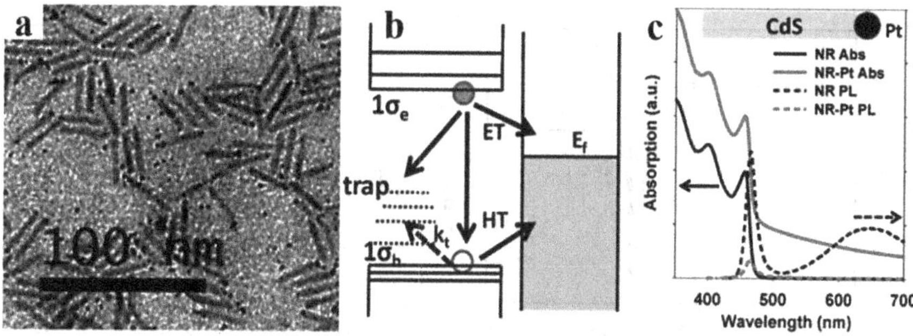

FIGURE 13.11 Electronic structure of CdS/Pt NRs. (Adapted from Wu et al. (2012) with permission. Copyright 2012 American Chemical Society.)

harvest light, while the metal tip works as an electron acceptor and the catalyst. Metal nanoparticles decorated on the tip of the NRs can facilitate the charge-separated state for photocatalysis. CdS/Pt (or Au) NRs heterostructure is formed by extending one Pt (or Au) nanoparticle at one or two tips of a CdS NR (Wu et al., 2012). The absorption of the CdS/Pt (or Au) NRs is the sum of absorption of free CdS NRs and Pt (or plasmonic Au) nanoparticles. Excitons are produced in the CdX NR after light absorption, and further quenched by electron-, hole-, and energy transfer to the metal tips. For example, the photoluminescence (PL) intensity is completely quenched in CdS/Pt NRs, indicating the effective electron extraction by Pt nanoparticles, as shown in Figure 13.11.

13.3.2 Exciton Dissociation Dynamics of CdX NRs

Excitons can move freely along the long axis of NRs because of the lack of quantum confinement in the long axis dimension. In this part, 1D exciton dissociation in CdX NRs and NR heterostructures are discussed.

1. Exciton generation and decay in NRs

Under weak 400 nm excitation, which creates excitonic states higher than 1Π, can generate the average of ~0.08 excitons per CdSe NR. The formation and decay of the positive peak at ~610 nm present a bi-exciton interaction between 1Π and 1Σ, indicating the excitons rapidly relax into and out of the 1Π band, as shown in Figure 13.12a (Zhu & Lian, 2012). Meanwhile, the excitons relaxed into the 1Σ band, leading to the 1Σ XB signal growth. The excitons reach the band edge at ~1 ps, and the TA spectrum is dominated by the lowest level ($1\Sigma_0$ ~587 nm) state-filling. The TA kinetics at 587 nm in CdSe NRs show a hot exciton relaxation time of 0.5 ps and a band edge $1\Sigma_0$ exciton lifetime time of 4.9 ns, as shown in Figure 13.12b, which are respectively comparable to that of QDs and ~3 times shorter than the typical value in CdSe QDs. Hence, this result agrees with the theoretically predicted accelerated 1D exciton decay in NRs, owing to the giant oscillator strength concentrated at the band edge.

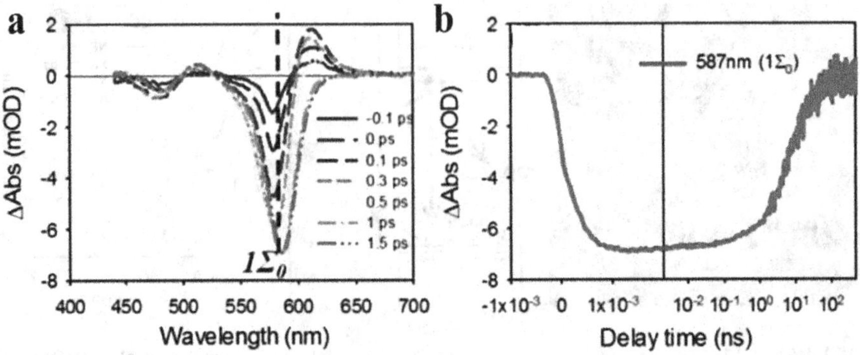

FIGURE 13.12 Exciton generation and decay kinetics in CdSe NRs. (Adapted from Zhu & Lian (2012) with permission. Copyright 2012 American Chemical Society.)

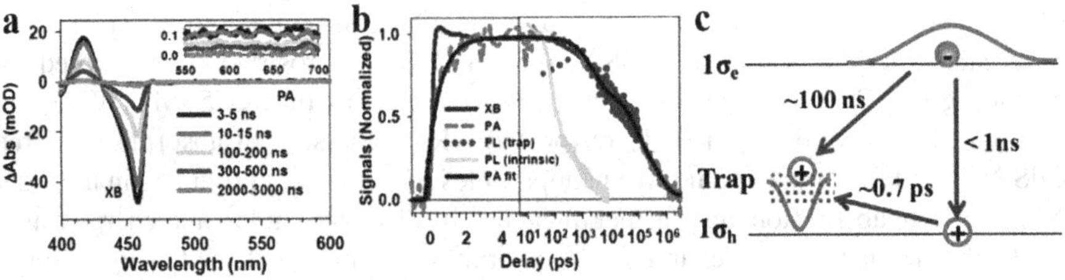

FIGURE 13.13 TA spectra and kinetics of CdS NRs. (Adapted from Wu et al. (2012) with permission. Copyright 2012 American Chemical Society.)

2. Hole trapping in NRs

Unlike reasonably well passivated CdSe NRs, CdS NRs always contain plenty of trap states. This is the main reason for the extremely low (< 0.1%) overall PL quantum yield in CdS. The PL spectrum of CdS NRs exhibits two distinct emission bands: one narrow band-edge emission at 463 nm and the other broadband trap-state emission centered at 650 nm. The impact of trap states on exciton dynamics can be checked by TAS (Wu et al., 2012). A XB of the $1\Sigma_0$ exciton band at ~456 nm and a much weaker broad PA band at > 500 nm can be observed in TA spectra of CdS NRs under 400 nm excitation. The XB and PA decay kinetics match well with the trap-state emission decay, thereby the PA signal can be indexed to trapped holes in CdS NRs. Therefore, long-lived (~100 ns) trapped excitons (i.e., trapped holes bound to CB electrons) can be achieved after ultrafast (~0.78 ps) VB holes trapping, as shown in Figure 13.13. In spite of ultrafast hole trapping, obvious band-edge PL emission can still be seen. Two reasons for this phenomenon, (i) a much larger radiative decay rate than trapped excitons, and (ii) negligible ultrafast hole trapping process in a small part of NRs.

3. Electron transfer from NRs to molecular acceptors

As discussed in Section 13.2.1, both a stronger electron-hole coupling then electron-phonon coupling, and a quasi-continuum of states are required in Auger-assisted electron transfer. These conditions can be met not only in 0D QDs, but also in 1D NRs and 2D NPLs. Hence, Auger-assisted electron transfer is a general model for interfacial charge transfer from these low-dimensional nanoparticles. Besides, similar Auger-assisted hole transfer processes can also be found. Differences between Auger-assisted electron (or hole) transfer from QDs and NRs may be notable. The density of states is low and discrete near the band edge in QDs. In NRs, continua of well-defined *k* states caused by free carrier movement along the NR axis can be observed, in addition to QD-like size-dependent quantized energy levels. Furthermore, the electron-hole Coulomb interaction is also stronger in 1D NRs. Hence, more efficient Auger-assisted charge transfer from NRs is desirable.

The electron transfer processes from CdX NRs to molecular acceptor MV^{2+} was investigated (Zhu & Lian, 2012). TA spectra of CdSe NRs with and without MV^{2+} at 1.5 ps under weak 400 nm excitation are compared, as shown in Figure 13.14a. Comparing the TA kinetics of free NRs and NR-MV^{2+} complexes at 587 nm ($1\Sigma_0$) and 670 nm ($MV^{+\bullet}$ radical) show that (i) a much smaller initial amplitude and a faster decay of the $1\Sigma_0$ XB in NR-MV^{2+} complexes in first 1 ps can be observed, compared to the long-lived $1\Sigma_0$ XB of free CdSe NRs; and (ii) a much larger initial amplitude and a slower decay at 670 nm by the formation of $MV^{+\bullet}$ radical in NR-MV^{2+} complexes can be seen, compared to that in free CdSe NRs, as shown in Figure 13.14b. These results suggest an ultrafast electron transfer from CdSe NRs to adsorbed MV^{2+}. The electron transfer time was fitted to be ~59 fs, much faster than the hot electron cooling process (0.5 ps) in NRs. Therefore, the majority of hot electrons can ultrafast transfer to MV^{2+} before cooling to band edge. This ultrafast electron transfer from

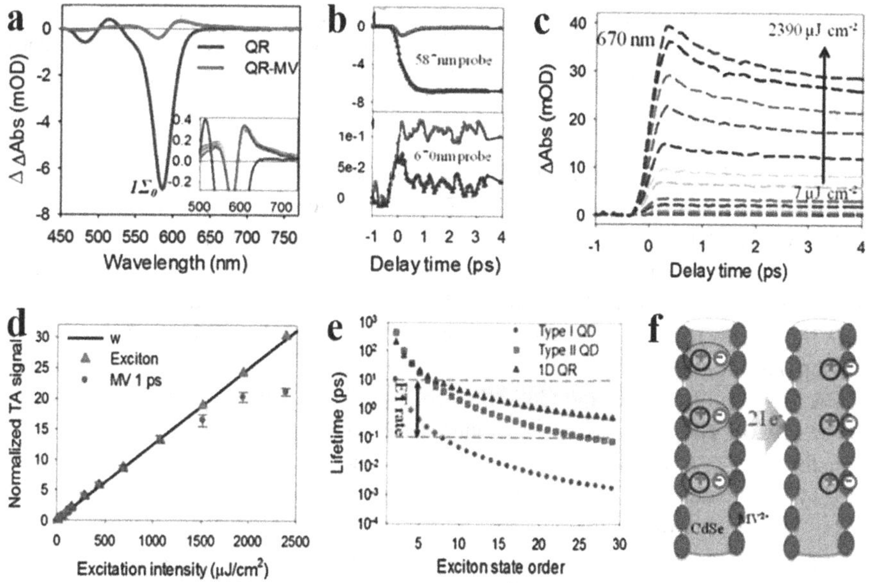

FIGURE 13.14 Single and multiple electron transfer from CdSe NRs. (Adapted from Zhu & Lian, (2012) with permission. Copyright 2012 American Chemical Society.)

NRs should be explained by two aspects: (i) there is still a large amplitude of electron wave function with strong electronic coupling with adsorbed electron acceptors at their surfaces, due to the quantum confinement in the radial direction; and (ii) a large number of electron acceptors can be adsorbed by the large surface areas. Besides, the derivative-like feature in TA spectra below 650 nm is caused by the electric field of charge-separated pair (or state) ($CdSe^{+}$-$MV^{+\bullet}$) shifting the excitonic transition in CdSe NRs through Stark effect, as shown in the inset of Figure 13.14a.

Except for the single-electron transfer process, multiple exciton generation and dissociation processes also provide a promising approach for improving the photocatalytic efficiency on the basis of a much faster multiple exciton transfer rate than that in multi-exciton Auger annihilation processes (sub-picosecond-nanosecond range). Inspired by the ultrafast interfacial electron transfer in CdSe NR-MV^{2+} complexes (~59 fs), which may effectively compete with multiple exciton annihilation, the NR-acceptor system can be applied for efficient multiple exciton dissociation from NRs. In fact, many adsorbed MV^{2+} molecules can be reduced by multiple excitons generated on CdSe NRs. Furthermore, the number of reduced MV^{2+} molecules per NR can be estimated by scaling the $MV^{+\bullet}$ radical signal in TA spectra at an early time (1 ps). Hence, more than 30 excitons per NR and 21 $MV^{+\bullet}$ radicals can be created, which can be obtained from fluence-dependent $MV^{+\bullet}$ radical kinetics, as shown in Figure 13.14c, d, and f. Besides, the longer-lived multi-excitons in NRs can be observed due to a much slower scale with exciton number of the bi-molecular Auger recombination of 1D excitons in NRs than the three-particle Auger recombination in QDs, as shown in Figure 13.14e. The simultaneous reduction of many acceptors on a single NR synergistically comes from: (i) a relatively large volume for adsorbing many acceptors on NR surfaces; and (ii) a relatively slow multi-exciton Auger recombination in NR. In photocatalytic reactions, both the electron and hole need to be extracted from nanomaterials. Regarding multiple hole transfer, better efficiency from NRs should be expected than from core@shell QDs, in which the hole is blocked by the shell.

4. Hole transfer from NRs to molecular acceptors

The hole extraction from trap-state in NRs can be investigated by TA kinetics of the CdS NR-PTZ (PTZ as a hole acceptor) complexes, as shown in Figure 13.15 (Wu et al., 2015b). The lifetime of XB feature in CdS NR-PTZ complexes is longer than that of free NRs, consistent with a prolonged CB electron lifetime caused by interfacial hole transfer to PTZ. Conceivably, above mentioned ultrafast hole trapping process (0.78 ps) is not affected by PTZ, which is confirmed by the same formation processes of PA feature in free NRs and CdS NR-PTZ complexes. The PA signal decays much faster in CdS NR-PTZ complexes than in free NRs, indicating the trapped holes transfer to PTZ. A hole transfer time of ~3.8 ns and a high hole extraction efficiency of ~94.7% are achieved, suggesting an efficient trapped holes extraction from CdS NRs. This is significant for photocatalytic reactions in CdS NRs. Besides the PA signal, the hole extraction process can also be monitored by the PTZ^{+} radical kinetics, which can be subtracted from the total TA signal (also contains the PA signal) at ~520 nm. As expected, PTZ^{+} radical formation time and hole transfer time from PA kinetics match well with each other, and PTZ^{+} radical and XB signals have the same

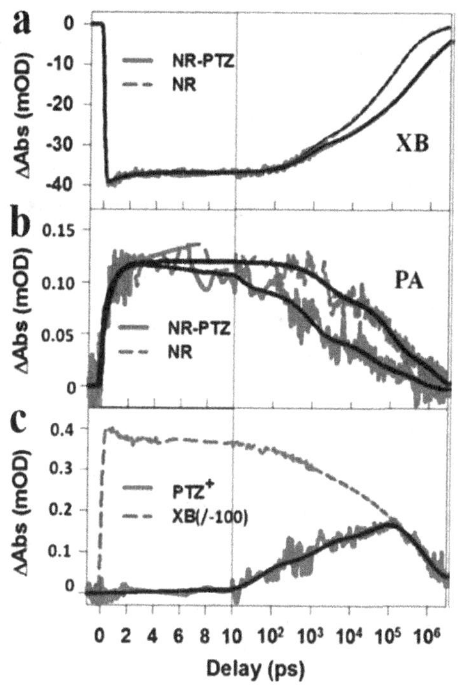

FIGURE 13.15 Comparison of TA kinetics of CdS NRs and CdS NR-PTZ complexes. (Reprinted with permission from Wu et al. (2015b). Copyright 2015 American Chemical Society.)

decay kinetics at long delay time domain. The matched XB and PTZ^+ radical decay processes represent the charge recombination between CB electrons in CdS NR and extracted holes in PTZ^+ radical. The successful study of the charge recombination process is thanks to the stability and reversibility of PTZ^+ radicals in oxidized form. In practical photocatalytic applications, oxidation is expected for suppressing the charge recombination step through irreversible decomposition or transformation of sacrificial donors.

5. Electron transfer from NRs to metallic tips

Efficient photocatalysis in NRs demands i) an efficient electron transfer from semiconductor to metal tip; and ii) a long-lived charge-separated state with holes in the semiconductor and electrons in the metal. The XB signal of CdS/Pt NRs decays much faster than that of free CdS NRs after 400 nm excitation, as shown in Figure 13.16 (Wu et al., 2012). An ultrafast electron transfer from the CdS NR to Pt tip (~3.4 ps) can be observed. On the contrary, the PA signal kinetics resulting from trapped holes remains unchanged, excluding both hole and energy transfer processes, both of which should annihilate the hole. The absence of both hole and energy transfer pathways to Pt tip should be attributed to the above-mentioned ultrafast hole trapping (~0.78 ps) on CdS NRs. As a result, the charge separation efficiency in CdS/Pt

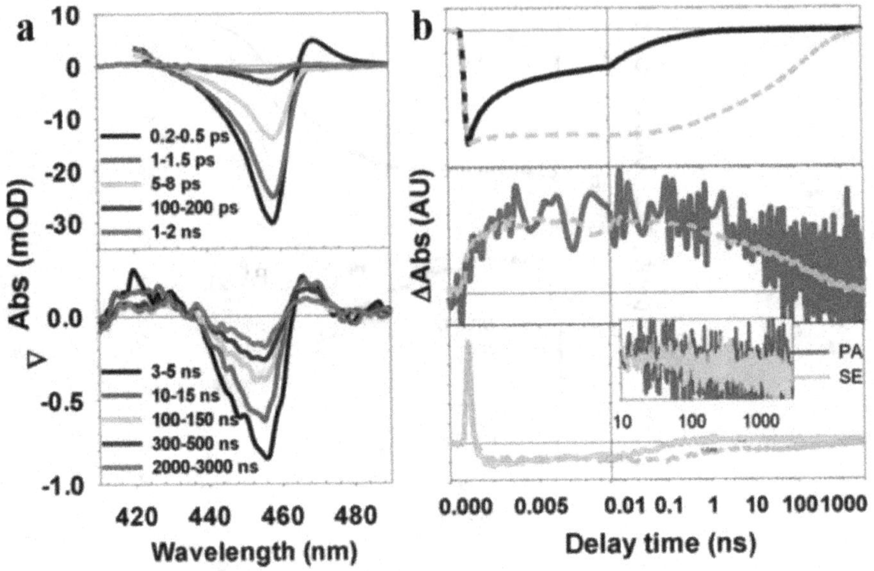

FIGURE 13.16 TA spectra and kinetics of CdS/Pt NRs. (Reprinted with permission from Wu et al. (2012). Copyright 2012 American Chemical Society.)

NRs is close to unity. Besides, the derivative-like feature of charge-separated state formed by CdS^{+}-Pt^{-} can also be seen, which is similar to the afore-mentioned $CdSe^{+}$-$MV^{+\bullet}$ feature. Hence, both the PA signal and this charge-separated state feature can be applied for tracking the charge recombination process between the trapped holes in CdS and electrons in Pt tip. A long lifetime of 1.2 μs can be obtained from both charge-separated state and PA features. The ultrafast charge separation and slow charge recombination can be attributed to the ultrafast hole trapping on CdS/Pt NRs, in which a long charge separation distance is reasonably set up.

13.3.3 Photocatalytic Mechanisms of CdX NRs

Based on the above systematical understanding of the 1D exciton properties, in this part, the mechanisms of photocatalytic performance from exciton dynamics insight in 1D NRs are discussed.

1. **Tuning interfacial electron transfer for photocatalysis by ligands length in CdS-PDQ^{2+} NR complexes**

The effect of ligand length on the electron transfer behavior in a CdS NR-molecular acceptor photocatalytic system was investigated (Yang et al., 2020). CdS NRs and propyl-bridged 2,2′-bipyridinium (PDQ^{2+}) work as the light absorber and the redox mediator, respectively. Various mercaptocarboxylate ligand lengths (n = 2–11 -CH_2-units between carboxylate moieties and thiolate) are capped on NR surfaces. Comparing TA spectra and corresponding kinetics of CdS NRs (n = 2) without and with PDQ^{2+} under 400 nm excitation shows a faster XB and a slower PA decay,

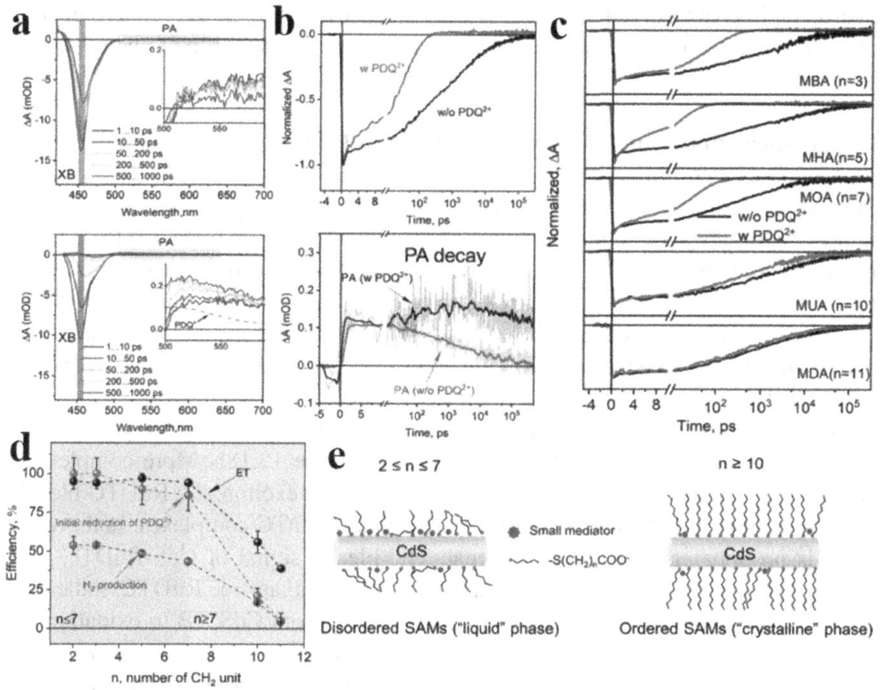

FIGURE 13.17 TA kinetics, electron transfer, and PDQ^{2+} photo-reduction efficiencies of CdS-PDQ^{2+} complexes with various length ligands. (Adapted from Yang et al. (2020) with permission. Copyright 2020 American Chemical Society.)

confirming an electron transfer pathway from CdS NR to PDQ^{2+} through ligands, as shown in Figure 13.17ab. The XB decays of other samples without and with PDQ^{2+} were compared, showing a similar accelerated XB decay with PDQ^{2+} for $n < 10$, indicating electron transfer from NR to PDQ^{2+}. However, there is negligible XB acceleration of decay at $n \geq 10$, suggesting a suppressed electron transfer by long-chain ligands, as shown in Figure 13.17c.

The trend of ligand length-dependent electron transfer and PDQ^{2+} photo-reduction efficiencies indicate a critical n between 7 to 10, above which the efficiencies all drop dramatically, as shown in Figure 13.17d. The abrupt change of interfacial electron transfer behavior induced by the ligand length indicates that the photocatalytic activity should be tuned by surface ligands. Self-assembled monolayers (SAMs) on NR surfaces can be formed by the alkanethiols and stabilized by interchain van der Waals interactions, which undergo a phase transition from the disordered "liquid" to ordered "crystalline" phase as the alkyl chain increased. For the short capping ligands ($n < 7$), the SAM layer is disordered with high PDQ^{2+} permeability into the CdS NRs surface, as shown in Figure 13.17e. In these cases, an almost unity and alkyl chain length-independent electron transfer efficiency can be observed. But for the longer ones ($n \geq 10$), crystallization of nearby mercaptocarboxylate ligands through the stronger lateral van der Waals interactions occurs. The formation of strong hydrophobic chain domains caused by the "phase transition" leads to the reduction of surface PDQ^{2+} permeability, which accounts for the sudden drop of photocatalytic performance.

2. **Enhanced photocatalytic performance by full charge separation in RuDTC-CdS/Pt NR complexes**

Simultaneous electron and hole transfer within a photocatalyst is an ideal way for improving photocatalytic performance. The photocatalysts are expected to combine reduction and oxidation co-catalysts within one NR, as shown in Figure 13.18a (Wolff et al., 2018). The Ru(tpy)(bpy)Cl_2 molecules borne by dithiocarbamate (DTC) groups (RuDTC) were loaded on the lateral side of CdS NR as the oxidation co-catalyst, and Pt nanoparticles were anchored to CdS NR tip as the reduction co-catalyst.

A bleach peak around 500 nm with a tail until 600 nm can be seen in TA spectra of free RuDTC excited at 400 nm, as shown in Figure 13.18b. More complex TA spectra with the bleaching features containing the CdS exciton, the RuDTC bleaching and a tail until 650 nm, can be observed in CdS-RuDTC complexes. In this case, there is no apparent trapped hole signal. The bleach signal of CdS-RuDTC rises much faster (< 300 fs) and reaches a stronger intensity than free RuDTC, indicating an ultrafast component stemming from hole transfer from CdS VB to oxidation co-catalyst. Functionalization of the NRs with RuDTC shows little effect on the XB decay at first 100 ps, but leads to a faster decay afterward, as shown in Figure 13.18c. This enhanced XB decay is stronger with a higher RuDTC density on NR surface. As expected, the XB decay is almost as fast as for CdS-MV^{2+} in the presence of Pt tips, indicating that the Pt nanoparticles get electrons rapidly (< 50 ps) from CdS NR. This electron transfer process is also unperturbed in CdS-RuDTC complexes. As a result,

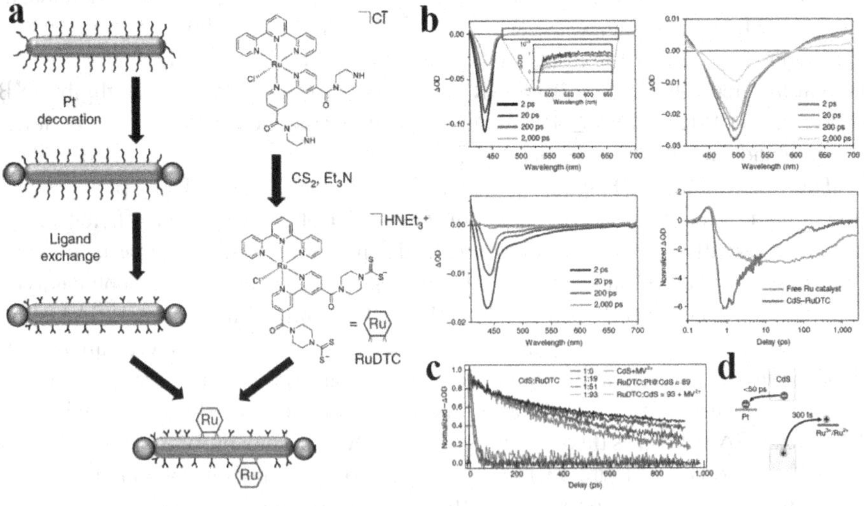

FIGURE 13.18 Structure, and TA spectra and kinetics of RuDTC-CdS/Pt NR complexes. (Adapted from Wolff et al. (2018) with permission. Copyright 2018 Springer Nature.)

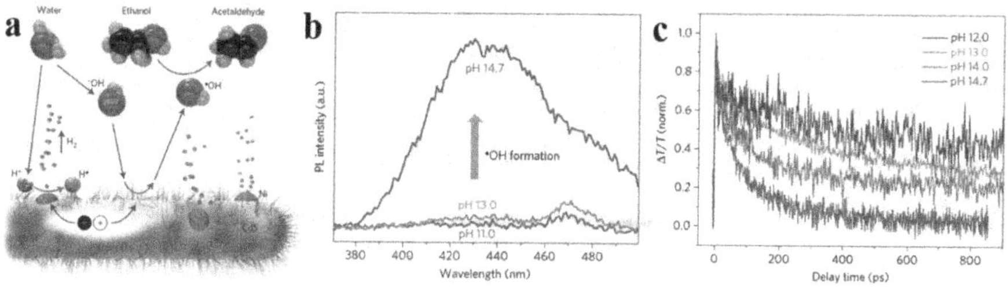

FIGURE 13.19 Structure, active radical formation, and TA kinetics of CdS@Ni NRs. (Adapted from Simon et al. (2014) with permission. Copyright 2014 Macmillan Publishers Limited.)

full charge separation in RuDTC-CdS/Pt NR complexes can be achieved by an ultrafast electron transfer to Pt (for photo-reduction) and an ultrafast hole transfer to RuDTC (for photo-oxidation), as shown in Figure 13.18d.

3. Enhanced photocatalytic performance by charge separation in CdS@Ni NRs
An efficient hole relay through $OH^-/OH\bullet$ redox couple from the semiconductor to the organic matter was achieved in Ni nanoparticles decorated CdS NRs, as shown in Figure 13.19ab (Simon et al., 2014). TA kinetics pumped at 400 nm and probed at the XB signal decay faster with increasing pH, as shown in Figure 13.19c, representing a faster electron transfer rate to Ni co-catalyst. The enhancement of photocatalytic performance depends on the much efficient electron removal and $OH^-/OH\bullet$ generation, both of which can react with organic matter.

13.4 EXCITON DYNAMICS IN 2D CHALCOGENIDE NPLs

In 0D QDs and 1D NRs, the lifetime of the charge-separated state is still not long enough for photocatalytic reactions, limiting the solar energy conversion efficiency. Compared to 0D QDs and 1D NRs, 2D NPLs have the unique and tunable advantages of large absorption cross-section and uniform quantum confinement along the *c*-crystal axis. These properties differing significantly from 0D QDs and 1D NRs, suggest many novel applications in solar energy conversion and optoelectronic devices. An enhanced band-edge transition strength can be observed by strongly bound electron-hole pairs in 2D NPLs. In a word, the photocatalytic performance of 2D NPLs can be tuned by the behavior of 2D excitons.

This section provides an overview of the basic understanding of key 2D exciton properties in CdX NPLs and how these properties affect their photocatalytic mechanisms.

13.4.1 Electronic Structure of CdX NPLs Heterostructures

This part discusses the comprehending of 2D electronic structures of CdX NPLs and CdS/Pt NPL heterostructures.

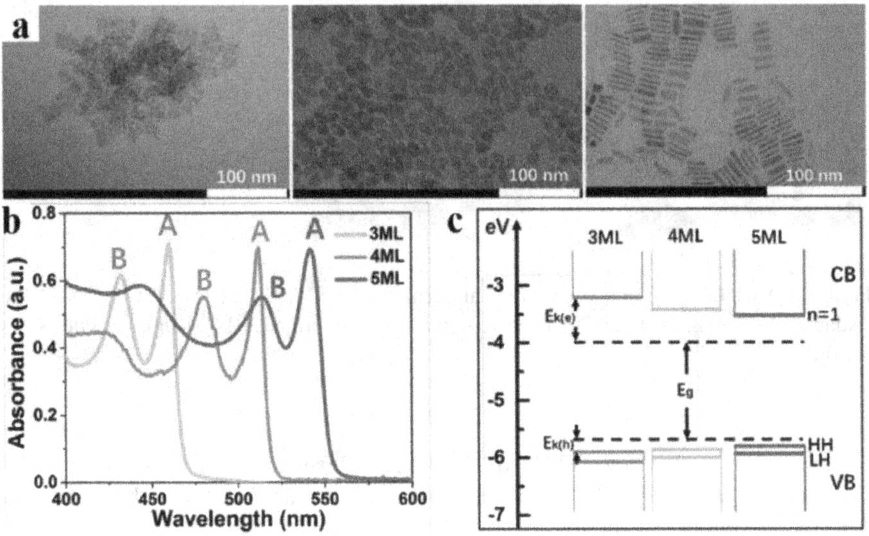

FIGURE 13.20 Electronic structure of 2D CdSe NPLs. (Reprinted with permission from Li & Lian (2012). Copyright 2017 American Chemical Society.)

1. Electronic structure of CdX NPLs

The electronic structure of 2D NPLs can be analyzed by their absorption spectra, as shown in Figure 13.20 (Li & Lian, 2012). The two lowest energy absorption peaks can be assigned to the electron-heavy hole (exciton A) and electron-light hole (exciton B) transitions. The positions of A and B are mainly determined by diameter in QDs and thickness in NRs and NPLs, and are rarely affected by the length of NRs and area of NPLs. Especially, 2D CdX NPLs with Cd atoms on both the top and bottom basal planes, containing n+1 Cd and n X layers, are often called n CdX monolayer (ML) NPLs. They usually have a nearly rectangular shape with dozens of nanometers. The CB and VB edge positions of CdSe NPLs can be found by adjusting different ML through the quantum confinement effect.

2. Electronic structure of NPL heterostructures

NPL heterostructures can be produced by anchoring co-catalysts on NPLs. CdS/Pt NPLs heterostructure is formed by extending Pt nanoparticles laterally around a CdS NPL. The total absorption of CdS/Pt NPLs consists of the individual absorption of free CdS, NPLs, and Pt nanoparticles. However, the PL intensity is completely quenched in NPL-Pt heterostructures, suggesting extra exciton quenching pathways, including electron-, hole-, and/or energy- transfer to Pt nanoparticles, as shown in Figure 13.21 (Li et al., 2018).

13.4.2 Exciton Dissociation Dynamics of CdX NPLs

Excitons or charge carriers can move laterally within NPLs due to the lack of quantum confinement in the lateral dimension. In this part, 2D exciton dissociation in CdX NPLs and NPL heterostructures are discussed.

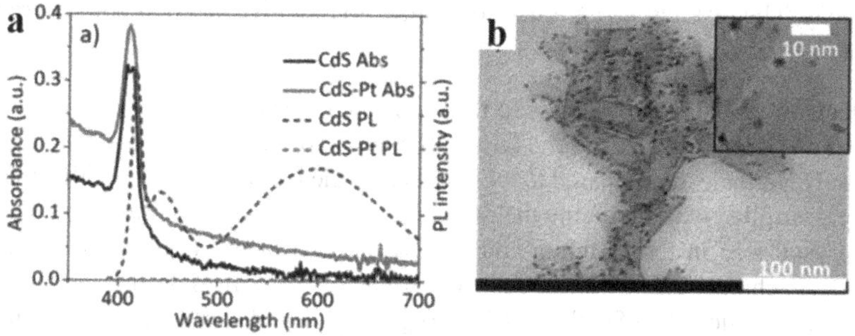

FIGURE 13.21 Electronic structure of CdS/Pt NPLs. (Adapted from Li et al. (2018) with permission. Copyright 2018 American Chemical Society.)

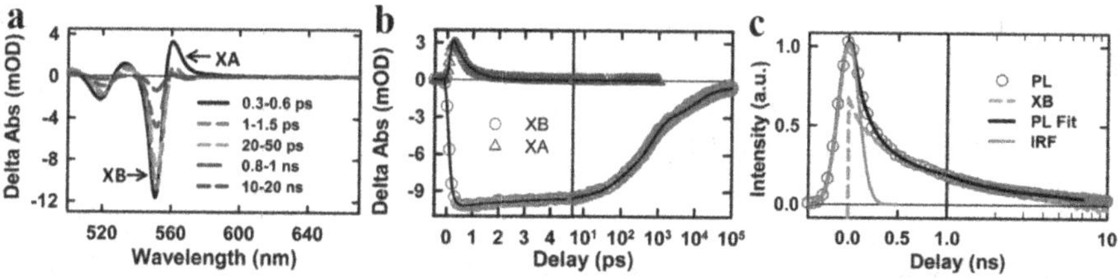

FIGURE 13.22 Exciton generation and decay kinetics in CdSe NPLs. (Adapted from Wu et al. (2015c) with permission. Copyright 2015 The Royal Society of Chemistry.)

1. Exciton generation and decay in NPLs

An XB feature at ~552 nm and an exciton absorption (XA) feature at ~560 nm can be observed in TA spectra of CdSe NPLs under weak 400 nm excitation, as shown in Figure 13.22a (Wu et al., 2015c). The formation of XB signal (~80 fs) means an ultrafast hot electron cooling from the higher excited level to the CB edge. Moreover, its decay processes can be fitted by a four-order exponential function, as shown in Figure 13.22b. The XA feature is caused by the presence of an exciton shifting the energy of all exciton transitions through exciton-exciton interaction. Hence, it contains the initial exciton relaxation dynamics due to the state-specific exciton-exciton interaction strength. The XA feature in CdSe NPLs forms instantaneously (<< 150 fs) and disappears flashily (~0.41 ps). In QDs and NRs, this XA decay contributes to the XB signal growth, which can be assigned to electron cooling from the higher excited levels to the CB edge. However, the XA decay is much slower than the XB growth in NPLs (~80 fs). In addition, a redshift of XA and XB features during the XA decay process can be seen early, suggesting a decreased bi-exciton interaction rather than the electrons population at CB edge. As a result, the XA decay in NPLs can be indexed to in-plane exciton localization (i.e., extra kinetic energy loss of in-plane movement) and trapping. The XB and band-edge PL decay match well on the timescale of 300 ps to 3 ns, as shown in Figure 13.22c. Comparison of TA and PL kinetics shows that ~32.4% of excitons decay through fast hole trapping, leading to a the long-lived (>3 ns) CB electron (XB signal) and a fast (< 240 ps)

PL decay. The residual ~67.6% excitons undergo radiative and nonradiative recombination (< 3 ns).

2. Electron transfer from NPLs to molecular acceptors

In fact, it is difficult to extend the NRs in the length direction without changing their radius. Nevertheless, compared to QDs and NRs, the lateral areas of 2D NPLs can be solely controlled without tuning their thickness, thereby bandgap. Based on this, how nanocrystal size in the quantum-confined and non-confined dimensions affects its interfacial charge transfer behavior can be clearly examined in the model of 2D NPLs. As mentioned in Section 13.3.2, Auger-assisted electron transfer can also be applied in 2D NPLs. Compared to QDs and NRs, NPLs are desirable for a faster charge transfer and a larger charge separation space based on their unique properties of more uniform quantum confinement and a larger surface area extended in two dimensions.

The electron transfer processes from CdX NPLs to MV^{2+} was investigated (Wu et al., 2015c). An XB signal centered at ~552 nm and an XA signal the peak at ~560 nm in TA spectra of CdSe NPLs under weak 400 nm excitation can be observed, as shown in Figure 13.23a. Except for XB and XA features, the absorption of $MV^{+\bullet}$ radical is also formed after 10 ps in CdSe NPL-MV^{2+} complexes, as shown in Figure 13.23b, confirming electron from CdSe to MV^{2+}. The XB completely recovers at > 200 ps, companied by the formation of a derivative-like feature of charge-separated (CS) state with holes in NPLs and electrons in MV^{2+}, similar to the case of QDs and NRs. The formation kinetics of CS and $MV^{+\bullet}$ is consistent with each other and the XB decay, proving the electron transfer process from CdSe NPL to MV^{2+} (~7.7 ps). The following matched CS and $MV^{+\bullet}$ decays represent the charge recombination process (~3.1 ns) of the charge-separated state.

3. Electron transfer from NPLs to lateral metallic terminals

The initial normalized XB amplitude of CdSe/Pt NPL takes ~36% of that of free CdSe NPLs, as shown in Figure 13.24a, implying an ultrafast electron removal within the instrument response (Wu et al., 2015c). Then, the XB amplitude further decays to ~13% swithin the first 1 ps, as shown in Figure 13.24b. However, there is no CS state

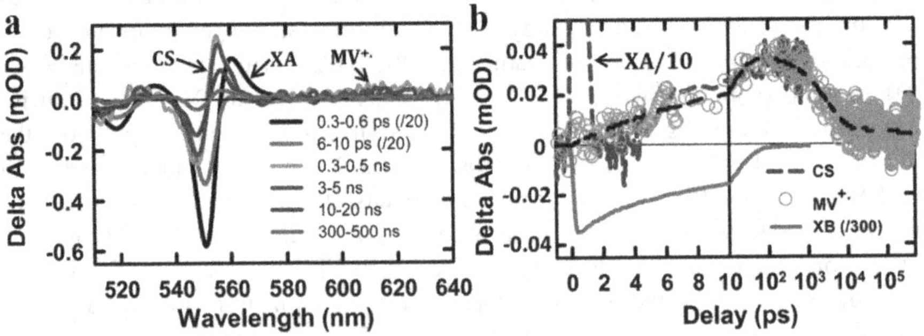

FIGURE 13.23 TA spectra and kinetics of CdSe NPL-MV^{2+} complexes. (Adapted from Wu et al. (2015c) with permission. Copyright 2015 The Royal Society of Chemistry.)

feature formation within this short timescale during these heavy electron decays. The reasonable explanation for this phenomenon is an ultrafast energy transfer (EnT, with both electrons and holes) from CdSe NPL to Pt terminal. The agreement of the XB decay and charge-separated state (CdS^+-Pt^-) formation kinetics in the range of 1–200 ps indicates electron transfer from CdSe NPL to Pt. The average time constants of electron transfer, EnT, and charge recombination can be fitted by the XB and CS kinetics to be ~9.4 ps, ~200 fs, and ~75 ns, respectively. Although electron transfer is fast and the CS state is long-lived in NPLs, most of (~87%) excitons are quenched through ultrafast diffusion to the NPL-Pt interface, followed by EnT to Pt. The remaining small part (~13%) excitons with holes trapped before the in-plane diffusion process conducts the charge separation process, as shown in Figure 13.24c. The hole trapping time in CdSe NPLs is ~280 fs, faster than that in NRs. In a word, the exciton dissociation efficiency in CdSe/Pt NPL heterostructures is so poor for photocatalysis, and the hole localization is important for suppressing EnT and realizing efficient charge separation.

13.4.3 Photocatalytic Mechanisms of CdS NPLs

Based on the above systematical understanding of the 2D exciton properties, in this part, the mechanisms of enhanced photocatalytic performance from exciton dynamics insight in 2D NPLs are discussed.

1. Enhanced photocatalytic performance by charge separation in CdS/Pt NPLs
Although with efficient CB electron transfer and VB hole transfer to the metal (Pt or Ni) tip (finally to O_2) and water (finally to organic matter), respectively, the lifetime of CS state in QDs and NRs is still not long enough, due to the slow photocatalytic reaction (the reduction of O_2) on the photocatalyst surface. Thus, hindering charge recombination loss (the recombination of transferred electrons with hole) is a key route for improving photocatalytic performance.

As discussed earlier, only 13% of excitons undergo charge separation in CdSe/Pt NPLs, leading to a relatively low photocatalytic activity. However, the exciton quenching mechanism in CdS/Pt NPLs is quite different. A much faster XB decay can be seen in CdS/Pt NPLs than in free CdS NPLs, indicating a shorter CB electron lifetime in the presence of Pt, as shown in Figure 13.25a (Li et al., 2018) The growth of CS state feature matches well with the XB decay CdS/Pt NPLs, confirming the main exciton quenching pathway of electron transfer from CdS NPL to the anchored Pt nanoparticle. The CS state decays on the timescale of 1 ns to μs, owing to the charge recombination process, namely the electron back from Pt to NPL. The electron transfer yield is estimated to ~99.4% according to the electron transfer and the intrinsic exciton decay rates. Because the initial electron transfer from CdS NPL to Pt is ultrafast and efficient, the key limitation for the overall photocatalytic performance is most probably the CS state lifetime. In fact, there is a negligible change of the CS state lifetime from pH 9 to 13, but a significant increase from pH = 13 to pH = 14. These results suggest a much more effective charge separation at pH = 14, as shown in Figure 13.25b. when pH < 13, charge recombination is the key limit step, because OH^- is not the dominant hole acceptor.

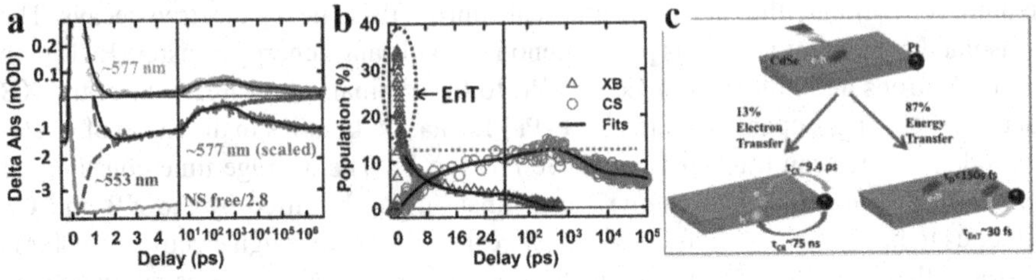

FIGURE 13.24 Electron and energy transfer from CdSe NPLs to Pt nanoparticles. (Adapted from Wu et al. (2015c) with permission. Copyright 2015 The Royal Society of Chemistry.)

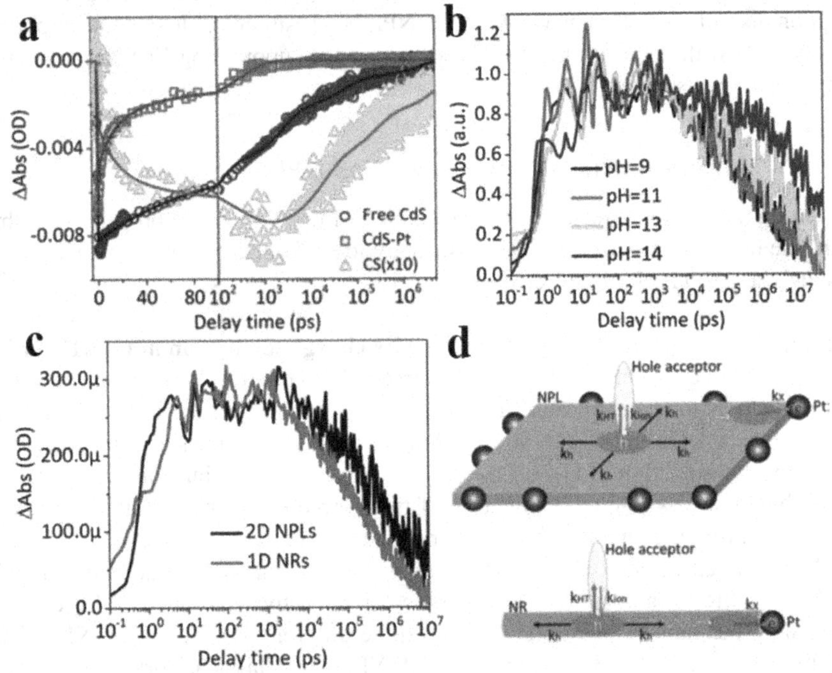

FIGURE 13.25 TA kinetics, and scheme of charge separation of CdS/Pt NPLs. (Adapted from Li et al. (2018) with permission. Copyright 2018 American Chemical Society.)

To provide insight into how morphology impacts the charge recombination process, a comparison of TA kinetics of 2D CdS NPL-MV^{2+} and 1D CdS NR-MV^{2+} complexes shows that a ~3.3-fold longer-lived CS state can be achieved in NPLs, as shown in Figure 13.25c. Therefore, the 2D morphology of NPLs can improve the CS state lifetime, which helps enhance photocatalytic performance. A model was proposed to simulate the charge recombination process, as shown in Figure 13.25d. Simulation results show that the charge recombination in 2D NPLs is over 5-fold slower than 1D NRs. Because compared to 1D NRs, many more random walk steps are required in 2D morphology before the hole finds the recombination site (Pt), resulting in much slower charge recombination. Although 2D morphology also slows down the initial electron transfer to Pt compared to 1D NRs, the charge separation efficiency cannot be affected due to a much faster electron transfer rate than intrinsic exciton quenching within CdS NPL.

2. Enhanced photocatalytic performance by charge separation in CdS/Ni NPLs
An XB signal centered at ~442 nm and an additional broad bleach feature of the defect states located at the XB red side (~457 nm) can be observed in TA spectra of CdS/Ni NPLs, as shown in Figure 13.26ab (Zhukovskyi et al., 2015). Comparison of the XB kinetics in free CdS and CdS/Ni NPLs displays a much faster XB decay in CdS/Ni NPLs, as shown in Figure 13.26c, suggesting a fast exciton quenching in CdS/Ni NPLs heterostructure. This exciton quenching can be induced by energy-, hole-, or electron- transfer from CdS to attached Ni nanoparticles. Among them, the enhanced photocatalytic performance can be ascribed to the electron transfer pathway. The PL quantum yield of CdS NPLs (a few percent) is much lower than that of CdSe NPLs (~50%), confirming a mass of trap states in CdS. As a result, exciton trapping in CdS can prevent the ultrafast and serious EnT, as usually occurred in CdSe, facilitating charge separation in CdS NPLs.

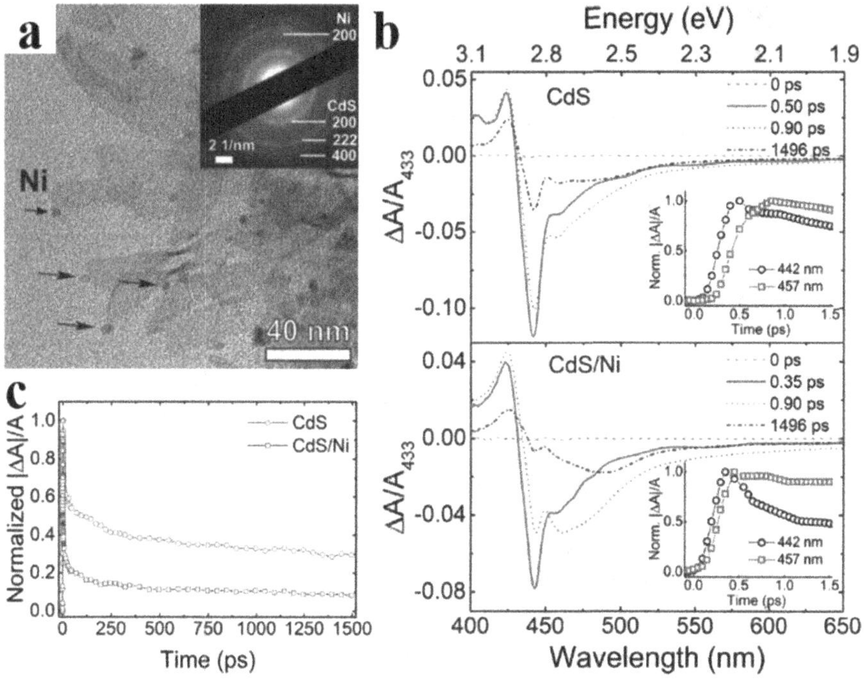

FIGURE 13.26 TA spectra and kinetics of CdS/Ni NPLs. (Adapted from Zhukovskyi et al. (2015) with permission. Copyright 2015 American Chemical Society.)

3. Enhanced photocatalytic performance by charge separation in CdSe/CdS/Au NPLs

The tiny Au nanoparticles can be observable at the lateral side of the CdSe/CdS NPLs, as shown in Figure 13.27a. The XB kinetics at 470 nm of CdSe/CdS NPLs can be fitted with a multi-exponential function. Here, the interfacial electron transfer time is estimated to 15 ps, and both of the longer components can be assigned to charge recombination. However, the XB bleach recovers much faster in CdSe/CdS/Au NPLs than in CdSe/CdS NPLs, as shown in Figure 13.27b. This faster bleach recovery can be ascribed to electron transfer from delocalized CdSe/CdS CBs to Au nanoparticles. The band offset between the CBs of CdSe and CdS is very small due to the smaller CdSe core and much thicker CdS shell, so that the excited electron can

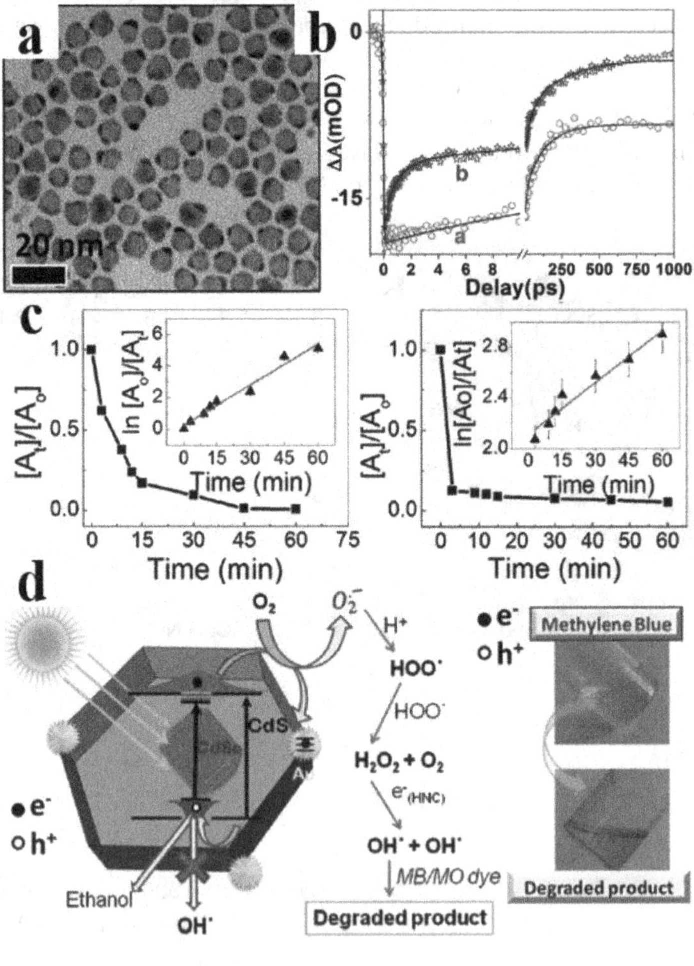

FIGURE 13.27 TA kinetics and photocatalytic dyes degradation of CdSe/CdS/Au NPLs. (Adapted from Chauhan et al. (2016) with permission. Copyright 2016 The Royal Society of Chemistry.)

be delocalized throughout the CdS and CdSe CBs. However, the hole is only localized on the CdSe VB due to the higher band offset between the VBs of CdS and CdSe. Under visible light excitation, the photo-excited electrons from the delocalized CdSe/CdS CBs transfer to the Fermi level of the Au nanoparticles, and further transfer to O_2 molecules to form $O_2{\bullet}^-$ radicals, while the holes are localized in CdSe VB, leading to a grand CS state and photocatalytic dyes degradation performance, as shown in Figure 13.27cd.

13.5 SUMMARY AND PERSPECTIVES

13.5.1 Summary

In summary, this chapter reviews recent studies on the photocatalytic mechanism of low-dimensional CdX nanomaterials, emphasizing how exciton dynamics affect their photocatalytic activity. Efficient photocatalysis requires both a long-lived CS state and an efficient exciton dissociation through ultrafast charge transfer to reactants. Heterostructures by applying charge acceptor should be a promising way for effective charge separation. Electron transfer from CdX QDs, NRs, and NPLs to electron acceptors (molecules and Pt, thus O_2) is found to be efficient. Auger-assisted charge transfer should be a general model for exciton dissociation, including both electron and hole transfer processes. Both transient absorption and static photocatalytic performance studies indicate that in CdX nanomaterials, the initial ultrafast hole localization/trapping is an essential step to get efficient charge separation and photocatalysis. Compared to 0D QDs, 1D NRs and 2D NPLs morphologies may offer richer opportunities to control charge transfer properties *via* adjusting the length for NRs and the thickness and area for NPLs, thus optimizing photocatalytic performance systematically and flexibly.

13.5.2 Perspectives

Although promising progress has been made for CdX and related heterostructures photocatalysts, many challenges remain: (i) Compared with electron transfer, systematic studies of direct hole transfer processes involving photo-oxidation half-reactions are still lacking but are much needed due to its importance in the overall photocatalytic cycles; (ii) Although the electron transfer is efficient (close to unity) and the resulting CS states are long-lived (hundreds of nanosecond) in CdX/Pt NRs and NPLs, the photocatalytic quantum efficiencies reported are still far from unity, suggesting either non-unity initial charge separation and/or undesirable recombination of separated electrons and holes before active radicals formation; (iii) All the ultrafast measurements mentioned only regard to the first electron transfer event, which is only sufficient for describing one-electron photocatalytic reactions, such as photo-reduction of O_2. Many physical and chemical processes involved in the second/multiple electron transfer and catalysis remain unclear; (iv) The stabilities of CdX photocatalytic systems are still limited by their chemical corrosion and/or photodegradation.

Developing more sophisticated heterostructures with integrated oxidation and reduction co-catalysts should be a promising direction for future studies: (i) One potential route to improve the efficiency as well as the charge separation lifetime is to construct heterostructured photocatalysts, which can effectively couple the photoreduction and photo-oxidation half-reactions to completely utilize the potential energy stored in both the photo-generated electron and hole; (ii) Another approach for suppressing charge recombination is to increase the irreversible hole removal rate by applying an efficient hole acceptor; (iii) Heterostructures of different classes of nanomaterials may give an additional material design thought for improving the photocatalytic performance; (iv) Surface coating approaches might help stabilize chalcogenide-based nanomaterials.

ACKNOWLEDGMENTS

This work was supported by the National Natural Science Foundation of China (Grant No. 61774122 to W. Q.), the China Postdoctoral Science Foundation (Grant No. BX20200266 and 2020M673400 to Y. Y.), the Natural Science Foundation of Shaanxi Province (Grant No. 2021JQ-059 to Y. Y.), and the Fundamental Research Funds for the Central Universities (Grant No. xjh012020041 to Y. Y.).

REFERENCES

Ahmed, B., Ojha, A. K., and Kumar, S. 2017. One-pot synthesis of Ni doped CdS nanosheets for near infrared emission and excellent photocatalytic materials for degradation of MB dye under UV and sunlight irradiation. *Spectrochimica Acta Part A: Molecular and Biomolecular Spectroscopy* 179:144–154.

Bera, R., Kundu, S., and Patra A. 2015. 2D hybrid nanostructure of reduced graphene oxide-CdS nanosheet for enhanced photocatalysis. *ACS Applied Materials & Interfaces* 7:13251–13259.

Bhuvaneswari, K., Vaitheeswari, V., Palanisamy, G., Maiyalagan, T., Pazhanivel, T. 2020. Glutathione capped inverted core-shell quantum dots as an efficient photocatalyst for degradation of organic dyes. *Materials Science in Semiconductor Processing* 106:104760.

Chauhan, H., Kumar, Y., Dana, J., Satpati, B., Ghosh, H., and Deka, S. 2016. Photoinduced ultrafast charge separation in colloidal 2-dimensional CdSe/CdS-Au hybrid nanoplatelets and corresponding application in photocatalysis. *Nanoscale* 8:15802–15812.

Chen, F., Jia, D., Cao, Y., Jin, X., Liu, A. 2015. Facile synthesis of CdS nanorods with enhanced photocatalytic activity. *Ceramics International* 41:14604–14609.

Cheng, L., Xiang, Q., Liao, Y., and Zhang, H. 2018. CdS-Based photocatalysts. *Energy Environmental Science* 11:1362–1391.

Fan, X., Yu, S., Wang, X., Li, Z., Zhan, F., Li, J., Gao, Y., Xia, A., Tao, Y., Li, X., Zhang, L., Tung, C., and Wu, L. 2018. Susceptible surface sulfide regulates catalytic activity of CdSe quantum dots for hydrogen photogeneration. *Advanced Materials* 31: 1804872.

Gong, J., Hao, Y., Li, L., Xue, S., Xie, P., Hou X., Feng, H., Wei, X., Liu, Z., Xu, Z., and Huang, J. 2019. The preparation and photocatalytic performance research of CdSe and wool ball-like GO/CdSe microspheres. *Journal of Alloys and Compounds* 779:962–970.

Han, G., Wang, L., Pei, C., Shi, R., Liu, B., Zhao, H., Yang, H., and Liu, S. 2014. Size-dependent optical properties and enhanced visible light photocatalytic activity of wurtzite CdSe hexagonal nanoflakes with dominant {0 0 1} facets. *Journal of Alloys and Compounds* 610: 62–68.

Haque, F., Daeneke, T., Kalantar-Zadeh, K., and Ou, J. 2018. *Nano-Micro Letters* 10:23.

Hodges, B. C., Cates E. L., and Kim J. H. 2018. Challenges and prospects of advanced oxidation water treatment processes using catalytic nanomaterials. *Nature Nanotechnology* 13:642–650.

Li, Q., and Lian, T. 2012. Area- and thickness-dependent biexciton Auger recombination in colloidal CdSe nanoplatelets: Breaking the "universal volume scaling law". *Nano Letters* 17:3152–3158.

Li, Q., and Lian, T. 2019. Exciton spatial coherence and optical gain in colloidal two-dimensional cadmium chalcogenide nanoplatelets. *Accounts of Chemical Research* 52:2684–2693.

Li, Q., Zhao, F., Qu, C., Shang, Q., Xu, Z., Yu, L., McBride, J. R., and Lian, T. 2018. Two-dimensional morphology enhances light-driven H_2 generation efficiency in CdS nanoplatelet-Pt heterostructures. *Journal of the American Chemical Society* 140:11726–11734.

Li, X., Zhu, J., and Li H. 2012. Comparative study on the mechanism in photocatalytic degradation of different-type organic dyes on SnS_2 and CdS. *Applied Catalysis B: Environmental* 123–124:174–181.

Lian, Z., Sakamoto, M., Vequizo, J. J. M., Ranasinghe, C. S. K., Yamakata, A., Nagai, T., Kimoto, K., Kobayashi, Y., Tamai, N., and Teranishi, T. 2019. Plasmonic p-n junction for infrared light to chemical energy conversion. *Journal of the American Chemical Society* 141, 6:2446–2450.

Liu, X., Ma, C., Yan, Y., Yao, G., Tang, Y., Huo, P., Shi, W., and Yan, Y. 2013. Hydrothermal synthesis of CdSe quantum dots and their photocatalytic activity on degradation of cefalexin. *Industrial & Engineering Chemistry Research* 52:15015–15023.

Maiuri, M., Garavelli, M., and Cerullo G. 2020. Ultrafast spectroscopy: state of the art and open challenges. *Journal of the American Chemical Society* 142:3–15.

Miao, T., and Tang, J. 2020. Characterization of charge carrier behavior in photocatalysis using transient absorption spectroscopy. *The Journal of Chemical Physics* 152:194201.

Schneider, J., Matsuoka, M., Takeuchi M., Zhang, J., Horiuchi, Y., Anpo, M., and Bahnemann, D. W. 2014. Understanding TiO_2 photocatalysis: mechanisms and materials. *Chemical Review* 114:9919–9986.

Simon, T., Bouchonville, N., Berr, M. J., Vaneski, A., Adrovic, A., Volbers, D., Wyrwich, R., Döblinger, M., Susha A. S., Rogach, A. L., Jäckel, F., Stolarczyk, J. K., Feldmann, J. 2014. Redox shuttle mechanism enhances photocatalytic H_2 generation on Ni-decorated CdS nanorods. *Nature Materials* 13:1013–1018.

Solovyev, A., Plyusnin, V., Shubin, A., Grivin, V., and Larionov, S. 2016. Photochemistry of dithiocarbamate $Cu(S_2CNEt_2)_2$ complex in $CHCl_3$. Transient species and TD-DFT calculations. *Journal Physical Chemistry A* 120:7873–7880.

Tvrdy, K., Frantsuzov, P. A., and Kamat, P. V. 2011. Photoinduced electron transfer from semiconductor quantum dots to metal oxide nanoparticles. *PNAS* 108:29–34.

Uno, T., Koga, M., Sotome, H., Miyasaka, H., Tama N., and Kobayashi, Y. 2018. Stepwise two-photon-induced electron transfer from higher excited states of noncovalently bound porphyrin-CdS/ZnS core/shell nanocrystals. *Journal of Physical Chemistry Letter* 9:7098–7104.

Wolff, C. M., Frischmann, P. D., Schulze, M., Bohn, B. J., Wein, R., Livadas, P., Carlson, M. T., Jäckel, F., Feldmann, J., Würthner, F. and Stolarczyk, J. K. 2018. All-in-one visible-light-driven water splitting by combining nanoparticulate and molecular co-catalysts on CdS nanorods. *Nature Energy* 3:862–869.

Wu, K., Du, Y., Tang, H., Chen, Z., and Lian, T. 2015b. Efficient extraction of trapped holes from colloidal CdS nanorods. *Journal of the American Chemical Society* 137:10224–10230.

Wu, K., Li, Q., Du, Y., Chen, Z., and Lian, T. 2015c. Ultrafast exciton quenching by energy and electron transfer in colloidal CdSe nanosheet-Pt heterostructures. *Chemical Science* 6:1049–1054.

Wu, K., Zhu, H., and Lian, T. 2015a. Ultrafast exciton dynamics and light-driven H_2 evolution in colloidal semiconductor nanorods and Pt-tipped nanorods. *Accounts of Chemical Research* 48:851–859.

Wu, K., Zhu, H., Liu, Z., Rodríguez-Córdoba, W., and Lian, T. 2012. Ultrafast charge separation and long-lived charge separated state in photocatalytic CdS-Pt nanorod heterostructures. *Journal of the American Chemical Society* 134:10337–10340.

Yadav, S., Adhikary, B., Tripathy, P., and Sapra, S. 2017. Efficient charge extraction from CdSe/ZnSe dots-on-plates nanoheterostructures. *ACS Omega* 2:2231–2237.

Yang, W., Vansuch, G. E., Liu, Y., Jin, T., Liu, Q., Ge, A., Sanchez, M. L. K., Haja, D. K., Adams, M. W. W., Dyer, R. B. and Lian, T. 2020. Surface-ligand "liquid" to "crystalline" phase transition modulates the solar H_2 production quantum efficiency of CdS nanorod/ mediator/hydrogenase assemblies. *ACS Applied Materials & Interfaces* 12:35614–35625.

Zhu, H., and Lian, T. 2012. Enhanced multiple exciton dissociation from CdSe quantum rods: The effect of nanocrystal shape. *Journal of the American Chemical Society* 134:11289–11297.

Zhu, H., Yang, Y., Kim, H. D., Califano, M., Song, N., Wang, Y., Zhang, W., Prezhdo, O. V., and Lian, T. 2013. Auger-assisted electron transfer from photoexcited semiconductor quantum dots. *Nano Letters* 14:1263–1269.

Zhukovskyi, M., Tongying, P., Yashan, H., Wang, Y., and Kuno, M. 2015. Efficient photocatalytic hydrogen generation from Ni nanoparticle decorated CdS nanosheets. *ACS Catalysis* 5:6615–6623.

14 Emerging Semiconductor Photocatalysts for Antibiotic Removal from Water/Wastewater

Mohamad Fakhrul Ridhwan Samsudin and Suriati Sufian

Universiti Teknologi PETRONAS, Perak, Malaysia

CONTENTS

14.1 Introduction 447
14.2 Fundamental Principles of Advanced Oxidation Process 448
14.3 Emerging Semiconductor Photocatalyst 450
14.3.1 Titanium Dioxide 450
14.3.2 Bismuth Vanadate 451
14.3.3 Tungsten Trioxide 452
14.3.4 Zinc Oxide 453
14.3.5 Strontium Titanate 454
14.4 Progress in Antibiotics Removal 454
14.4.1 Amoxicillin 454
14.4.2 Ciprofloxacin 456
14.4.3 Tetracycline 457
14.5 Conclusion and Perspective 460
Acknowledgment 460
References 460

14.1 INTRODUCTION

The rapid industrialization in order to improve the living standards of civilization had facilitated technological advancement, particularly in the development of antibiotics. Since Fleming's first discovery of the penicillin antibiotic in 1928, a wide range of antibiotics have been developed and have benefited thousands of people around the globe (Calvete et al., 2019). Antibiotics have been widely used in many medical applications to cure infections faced by humans or animals. Nevertheless,

DOI: 10.1201/9781003118749-14

the arbitrary discharge of antibiotics stemming from rapid industrialization and improper consumption has evoked environmental threats, particularly the contamination of clean water sources. Moreover, the presence of these antibiotic compounds cannot be easily removed and treated using the conventional water/wastewater treatment process (Suwannaruang et al., 2020). For instance, the utilization of simple flocculation and adsorption method for treating the antibiotics contaminants lead to the formation of harmful byproducts (Samsudin and Sufian, 2020). Meanwhile, membrane technology is often associated with fouling issues that abate system efficiency (Hansima et al., 2021). Thus, it is an indispensable obligation to develop sustainable frontier technology to mitigate this issue without engendering the secondary complication.

On account of this exhalation issue, semiconductor photocatalysts have been employed in many advanced oxidation processes (AOPs) to prevent this issue before it becomes irrevocable (Samsudin et al., 2020b). Recently, AOPs have attracted substantial attention among the scientific community and industrial practitioners due to their efficiency in removing the recalcitrant pollutants in water/wastewater. Notably, the AOPs via semiconductor photocatalyst can be powered using solar energy, signifying its sustainable approach compared with the conventional treatment systems that mainly depend on fossil fuel systems (Ahmad Madzlan et al., 2020; Zulfiqar et al., 2020). Additionally, it has been proven that the highly active radicals generated from the semiconductor photocatalysts can unbiasedly attack and mineralize a wide range of pollutants, including antibiotics residue, without the formation of any harmful byproducts (Nidheesh et al., 2020).

Thus, this chapter aims to recapitulate the latest advances in the AOPs system via semiconductor photocatalysts to remove antibiotics in the water/wastewater. Initially, the fundamental principles of AOPs will be introduced. Afterward, the rising star of semiconductor photocatalyst in AOPs will be highlighted. Next, several types of antibiotics residue present in the water/wastewater, such as amoxicillin, ciprofloxacin, and tetracycline, will be disclosed. The details on the degradation pathways of each of the antibiotics as mentioned will also be presented. Finally, invigorating perspectives on the future directions of this frontiers system will be provided.

14.2 FUNDAMENTAL PRINCIPLES OF ADVANCED OXIDATION PROCESS

Figure 14.1 illustrates the fundamental process of AOPs via a semiconductor photocatalyst upon light illumination. There are three basic processes involved in this system which are (i) light absorption via photocatalyst, (ii) photocharge carrier mobility, and (iii) surface reaction (Samsudin et al., 2020c). Initially, the semiconductor photocatalysts will absorb the photon energy that is less than or equivalent to its bandgap energy, resulting in the formation of the photocharge carriers composed of electrons and holes. These electrons and holes will be generated at the conduction band and valence band, respectively. Afterward, the photocharge carriers will be separated and transferred to the surface of the photocatalysts in order for the reduction and oxidation process to occur.

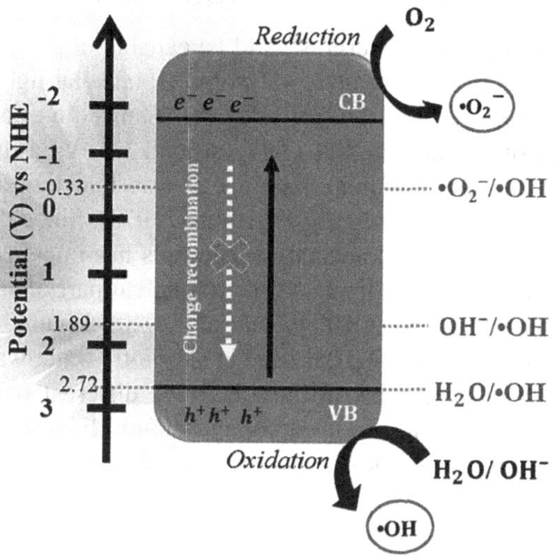

FIGURE 14.1 Schematic illustration of the fundamental concept for AOPs via semiconductor photocatalyst.

As a matter of fact, the conventional photocatalyst materials often suffer from the fast recombination of the photocharge carriers; consequently, a limited amount of photocharge carriers are available to partake in the redox reaction (Wu et al., 2020). This issue stems from the strong Coulombic force, which hindered the mobility of the photocharge carriers to the surface of the photocatalyst (Xu et al., 2018). Upon light illumination, the electron will be excited and migrate toward the conduction band, leaving the holes in the valence band. However, the photocharge carriers not only need to suppress the strong Coulombic force, but also a thermodynamic barrier upon its separation and migration process from the active sites to the targeted reactions (Younis et al., 2020). Therefore, if there is insufficient energy available to initiate the reaction and overcome these forces, the excited electrons will often recombine back with the counterpart holes, consequently dissipated.

Gratifyingly, the generated holes usually possess a strong oxidation potential that is capable of generating the superhydroxyl radical ($^{\bullet}OH$) from the water sources or OH^-. Meanwhile, the excited electrons will oxidize the available oxygen and producing the superoxide radicals ($^{\bullet}O_2^-$). In some cases, the $^{\bullet}O_2^-$ radicals will be further reacted and converted into H_2O_2. The formation of these H_2O_2 will be beneficial as an electron acceptor in which supplementary $^{\bullet}OH$ can also be produced from the generated H_2O_2 (Calvete et al., 2019; Nidheesh et al., 2020). These radicals will directly attack the targeted antibiotics, mineralize it into smaller fragments, and eventually transform the contaminated water into clean water.

14.3 EMERGING SEMICONDUCTOR PHOTOCATALYST

An efficient photocatalyst material should be capable of absorbing the visible-light spectrum, which accounted for up to 48% of the total solar light energy, corresponding to the bandgap energy of 3.0 eV or lower (Samsudin et al., 2018b). The wide bandgap energy of titanium dioxide (TiO_2) around 3.2 eV has been the intrinsic bottleneck of this conventional material since it was first introduced by Honda and Fujishima back in 1972 (Honda and Fujishima, 1972). Since then, the development of visible-light responsive photocatalyst materials has impelled heightened interest owing to its capability for utilizing the astronomical part of solar energy (Hou and Zhang, 2020; Prasad et al., 2020; Wei et al., 2020). Figure 14.2 summarized the recent semiconductor photocatalysts used in antibiotic water/wastewater degradation applications. This section aims to introduce five different types of semiconductor photocatalysts, which include conventional titanium dioxide (TiO_2), bismuth vanadate ($BiVO_4$), tungsten oxide (WO_3), zinc oxide (ZnO), and strontium titanate ($SrTiO_3$).

14.3.1 Titanium Dioxide

Titanium dioxide (TiO_2) is one of the pioneering photocatalyst materials used in photocatalytic applications since it was first discovered in the 1970s (Honda and Fujishima, 1972). TiO_2 shows promising performance in various industrial applications such as catalysis, antibacterial agents, lithium-ion batteries, dye-sensitized solar cells, and cosmetics (Chen and Mao, 2007). This is attributed to the peculiar

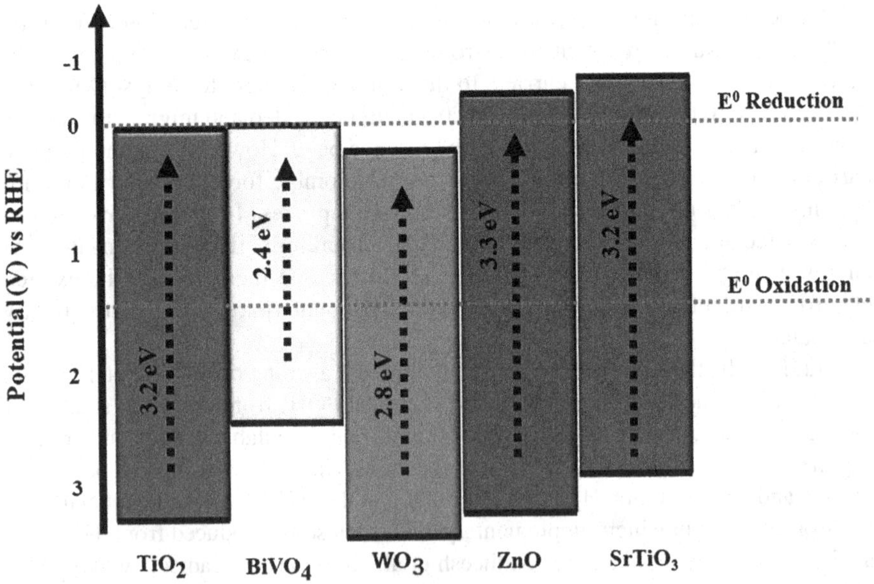

FIGURE 14.2 Emerging semiconductor photocatalyst materials along with their bandgap energy.

features that TiO_2 possesses, including non-toxicity, excellent photostability, chemical inertness, high oxidation efficiency, and environmentally friendly nature. In addition, TiO_2 materials are inexpensive owing to their abundance in the Earth's crust (approximately around 0.44%) (Westerhoff et al., 2011). Nevertheless, the photocatalytic application of TiO_2 materials is hampered by its wide bandgap energy around 3.2 eV, consequently, limit its application to the UV part of the solar light spectrum only. This limiting factor has caused a small fraction of solar energy to be absorbed while the astronomical part of the energy is wasted. Furthermore, the electron-hole pair in TiO_2 will rapidly quench despite the light energy being absorbed due to the photocharge recombination issue.

Generally, TiO_2 exists in four different crystalline polymorphs: tetragonal rutile, tetragonal anatase, orthorhombic brookite, and monoclinic TiO_2 (B) (Westerhoff et al., 2011). Among these four TiO_2 polymorphs, tetragonal rutile and tetragonal anatase are the most widely studied compared to the orthorhombic brookite and monoclinic TiO_2 (B) (Zhang et al., 2019). Both anatase and rutile have a common primary structural unit which is TiO_6 octahedron. However, their octahedron structure is slightly distorted in comparison to the real octahedron structures. Additionally, the anatase, brookite, and TiO_2 (B) are known to be metastable structures in which, under high-temperature conditions, their polymorph structure will turn into rutile phases. Moreover, all of them have different symmetry and growth behavior. For example, the rutile phase has a tetragonal structure with a = 0.459 nm and c = 0.296 nm. Since the (011) and (100) are the type of crystal facets that possess the lowest energy, hence the thermodynamical equilibrium morphology of this phase is a trunked octahedron (Wang et al., 2014). Similarly, the anatase polymorph is made up of tetragonal structure but with a longer c-axis of 0.951 nm compared to the a-axis of 0.379 nm.

In the case of anatase polymorphs, four of the eight neighbors of each octahedron share edges, and the others share corners. The corner-sharing octahedron forms (001) planes and is connected to their edges with the plane of octahedra below. For rutile polymorph, each octahedron has 10 neighbors in which two of them share edges while the remaining eight share corners. The neighbors share corners along the (110)-type direction and are stacked with their long axis alternating by 90°. The bandgap energy of the rutile and anatase polymorphs is approximately 3.0 and 3.2 eV, corresponded with a wavelength of 413 and 387 nm, respectively (Samsudin et al., 2018a). The small differences in their bandgap energy are associated with their differences in the lattice structural arrangement, particularly the bond length and the stacking of TiO_6. These differences result in slight changes in the orbital coupling between the oxygen 2p and titanium 3d orbitals.

14.3.2 Bismuth Vanadate

Bismuth vanadate ($BiVO_4$) is one of the most sought bismuth-based metal oxides owing to its excellent photocatalytic efficiency. $BiVO_4$ has low bandgap energy, which corresponds to the visible-light spectrum, excellent dispersibility, environmentally friendly materials, and good photostability and photocorrosion (Tan et al., 2017). In addition, the smooth photocharge carrier transport within the $BiVO_4$ structure and its high crystalline quality with minimal defects result in a good photocatalytic performance, as reported previously (Wang et al., 2019). Initially, the $BiVO_4$ materials have been studied as toxic-pigments-based materials for the coating and plastic industry (Tan et al., 2017). Nevertheless, the excavation of this semiconductor photocatalyst has been shifted toward the photocatalytic application field when it was first published by Kudo et al. (1998) for water splitting reaction in an aqueous $AgNO_3$ solution. Since then, $BiVO_4$ has elicited ripples of excitement among the scientific community as a frontier photocatalyst owing to its promising features, including as a non-toxic metal-free photocatalyst, an excellent chemical, photonic properties, appealing electronic band structure, and high physicochemical stability.

Naturally, the $BiVO_4$ structures exist in an orthorhombic structure that originates from a mineral pucherite (Samsudin et al., 2020a). However, the attempts to develop this orthorhombic structure led to an unsuccessful outcome. All of the synthesizing methods of $BiVO_4$ would result either in scheelite or a zircon-like structure. There are three different types of $BiVO_4$ structures: monoclinic scheelite, tetragonal scheelite, and tetragonal zircon (Wang et al., 2019). In the case of the monoclinic $BiVO_4$ structure, its bandgap energy is estimated to be around 2.4 eV in comparison to the other two crystallographic structures. The tetragonal $BiVO_4$ structure has a bandgap energy of around 2.9–3.1 eV. The small bandgap energy of a monoclinic structure is attributed to the addition of Bi 6s orbitals in the valence band, which allows a short transition of a photoexcited electron to the V 3d orbitals in the conduction band. Meanwhile, the tetragonal structures are only composed of O 2p orbitals; consequently, a longer transition of the photoexcited electron is required, resulting in a wide bandgap energy.

In 2009, Walsh et al. (2009) further supported that the monoclinic $BiVO_4$ structure is made up of direct bandgap energy, in which its conduction band consists of O 2p and Bi 6 p orbitals while the valence band consists of O 2p and Bi 6s orbitals. In addition, the monoclinic structure (space group 14, $C_{2h}{}^6$) consists of four unique lattice sites: Bi (4e), V (4e), O1 (8f) and O2 (8f). The layered structure of Bi-V-O units stacked parallel to the *c*-axis draws originates from Bi_2O_3 and V_2O_5 binary analogs. The arrangement of the Bi atom is in a distorted oxygen octahedron with nearest neighbor distances ranging from 2.35 to 2.53 Å, whereas its center is made up of V species with 2 × 1.74 Å and 2 × 1.75 Å bond lengths. Furthermore, the O1 is coordinated to one Bi and V, whereas O2 is coordinated to two Bi and a single V. The asymmetric coordination system observed within the monoclinic structure is attributed to the $6s^2$ electronic configuration of Bi, which is associated with the active lone pair commonly present with ns^2 valence cations (Payne et al., 2006).

14.3.3 Tungsten Trioxide

Tungsten trioxide (WO_3) is an important semiconductor material widely used in several different applications including sensor, dye-sensitized solar cells, photocatalyst, and heterogeneous catalysis (Kalanoor et al., 2018). The great contribution of WO_3 in vast industrial applications, particularly in the photocatalysis arena, is owing to its peculiar features such as narrow bandgap energy of 2.4 – 2.8 eV, excellent physicochemical stability, cheap, photocorrosion stability, and non-toxicity (Dong et al., 2017). One of the earliest exploratory works on WO_3 as a semiconductor photocatalyst material was done by Butler et al. (1976). They investigated the potential of this material for photocatalytic water oxidation. They observed that the sample was stable when tested with 1 M of sodium acetate with no decay of the photocurrent, and no visible deterioration of the electrode surface was monitored. Nevertheless, the flat band potential analysis measured by a photocapacitance is + 0.3 V at pH =1.75. This result indicates that the bands are essentially flat, and the depletion layer for charge separation does not exist.

In addition, there are several different types of WO_3 crystal structures such as bronze-type hexagonal, pyrochlore-type cubic, hexagonal type, monoclinic I (γ-WO_3), and others (Zheng et al., 2014; Kong et al., 2015). Among them, monoclinic I (γ-WO_3) showed a remarkable photocatalytic performance at room temperature with a stable phase condition. (Dong et al., 2017). In lieu of WO_3 good effectuation, WO_3 has suffered from its limited photocatalytic performance owing to its low conduction band and the fast recombination of electron-hole pairs. The utilization of pristine WO_3 photocatalyst without any modification would result in zero spontaneous photocatalytic reaction, although its bandgap energy corresponds well with the visible-light energy. Moreover, the negative position of the WO_3 conduction band in comparison to the potential for single-electron reduction of oxygen results in limitation efficiency ability for the utilization of this localized photoexcited electrons, and consequently promotes the recombination of the electron-hole pair (Dong et al., 2017).

14.3.4 Zinc Oxide

Zinc oxide (ZnO) is another emerging green photocatalyst material that has a broad direct bandgap energy of 3.37 eV, large free-excitation binding energy (60 meV), and strong oxidation ability (Govender et al., 2002; Xu and Wang, 2011). The excitation binding energy of ZnO is bigger than the available thermal energy at room temperature (25 meV). The large free-excitation binding energy allows the excitonic emission processes can occur above room temperature conditions. Gratifyingly, zinc oxide comprises two elements – zinc (2+) and oxygen (2-) – both of which are elements abundant in the Earth's crust. Generally, the white crystal of ZnO possesses well-defined crystal structures such as rock salt, wurtzite, or cubic (zinc blende) structure (Morkoç and Özgür, 2009). The rock salt structure is a rare type of species that can be yielded under high pressure, while the wurtzite structure has the highest thermodynamic stability in comparison to the rock salt and zinc blende (cubic) structures.

The typical hexagonal wurtzite structure is composed of two lattice parameters, which are a and c with a value of 0.3296 and 0.52065 nm, respectively (Lee et al., 2016). In addition, the Zn^{2+} and O^{2-} atoms are tetrahedrally-coordinated and occupy alternating planes along the c-axis. Furthermore, the wurtzite structure properties belong to the P63mc space group, in which it exhibits a non-centrosymmetric structure (Maggard et al., 2001). The non-centrosymmetric structure allows the ZnO to act as piezoelectric and pyroelectric materials. On the other hand, the valence band of ZnO mainly consist of the O 2p orbitals owing to a large difference in the electronegativity while the conduction band mainly consist of the empty 4 s orbitals and/or hybridization of empty s and p orbitals (Xu et al., 2003; Hisatomi et al., 2014). Moreover, ZnO is also considered an n-type semiconductor in which it exhibits a high number of electron donors.

14.3.5 Strontium Titanate

Strontium titanate ($SrTiO_3$) is an archetypical example of a perovskite structure oxide with outstanding features for various applications. The $SrTiO_3$ is a promising thermoelectric metal oxide material with a cubic perovskite structure containing a lattice parameter *a*, of around 0.3905 nm (Phoon et al., 2019). Furthermore, $SrTiO_3$ is an optically transparent insulator with a high dielectric constant. In addition, the properties of this material can be easily tuned with different degrees of elemental doping, results in a high-quality interface with other materials, including high-temperature superconducting thin films (Phoon et al., 2019). With respect to this merit, $SrTiO_3$ has garnered significant attention in surface and interface studies for various industrial applications such as nano-electronics, photonics, and photocatalysis.

In the early 1980s, Wagner and Somorjai evaluated the performance of a single crystal $SrTiO_3$ in hydrogen production application from highly alkaline conditions ([NaOH] > 5 M) under UV-light sources (Wagner and Somorjai, 1980a, 1980b). It was found that the hydroxide ions at the photocatalyst surface act as a hole acceptor in which it increases the lifetime of electrons for proton reduction. In addition, the reduced Ti^{3+} is present on the crystal surface in which it does contribute to the photocatalytic performance of the $SrTiO_3$ sample upon UV-light illumination. Similarly, Domen et al. (1980) evaluated the performance of $SrTiO_3$ with the incorporation of NiO as a cocatalyst in photocatalytic water splitting application. It was observed that the addition of a cocatalyst does significantly enhanced the overall photocatalytic performance with three times higher than the pure $SrTiO_3$ sample. All of these early discoveries of the promising potential of $SrTiO_3$ has led the researcher to further optimize the pure $SrTiO_3$ with various strategies in order to enhance its photocatalytic performance.

14.4 PROGRESS IN ANTIBIOTICS REMOVAL

As mentioned above, the presence of antibiotics residue in the water/wastewater is harmful to human consumption as well as can endanger aquatic life. Thus, substantial progress on antibiotics removal via AOPs semiconductor photocatalysts has been reported. This section aims to epitomize the prevailing progress on the antibiotic's removal, particularly on the three main common antibiotics residue reported in the literature. The three archetypes of the antibiotics include amoxicillin, ciprofloxacin, and tetracycline.

14.4.1 Amoxicillin

Amoxicillin is a common type of antibiotic which belongs to the β-lactam antibiotics family. It is estimated that around 65% of the global antibiotics is originated from this β-lactam family. Additionally, approximately around 86% ± 8% of administered antibiotics consumption is generally excreted in the urine after consumption owing to the slow rate of metabolism. Moreover, Alygizakis et al. (2016) reported that amoxicillin contamination shows the highest concentration level in the Eastern Mediterranean Sea than the other antibiotics residue. Similarly, Hernández et al. (2019) also found a significant concentration level of amoxicillin in the Antarctic.

Sparked by these growing concerns, Tang et al. (2019) synthesized the ternary CdS-MoS_2-coated ZnO nanobrush via the chemical bath deposition method. Their ternary composite photocatalyst is capable of degrading 94% of amoxicillin residue within 60 minutes. This fast amoxicillin removal is attributed to the enhanced light absorption capacity and rich active sites that enable more electron-hole pairs to partake in the degradation process. Meanwhile, Le et al. (2019) demonstrated 89.3% of amoxicillin removal via carbon dots@$BiVO_4$/Bi_3TaO_7 photocatalyst. The addition of carbon dots in the composite photocatalyst expedites the photocharge carrier separation and migration. Correspondingly, the addition of carbon dots improved the light-harvesting capacity and hindered the recombination of the photocharge carriers. Subsequently, Chahkandi and Zargazi (2020) introduce a new approach to preparing the $BiVO_4$ via a facile water-based electrophoretic deposition technique. The inexpensive and green fabrication process of $BiVO_4$ possesses a moderate size of nanorods structure in the range of 100–150 nm. The aforementioned $BiVO_4$ is capable of removing 97.45% of amoxicillin within 90 minutes. Recently, Silva et al. (2021) demonstrated that the metal-free polymeric g-C_3N_4 with controllable structural defects could achieve complete removal of amoxicillin residue within 48 hours under visible-light illumination. The controllable structural defects allow the properties of the g-C_3N_4 to be tailored accordingly, results in minimizing the charge carrier recombination.

Figure 14.3 manifests the possible photocatalytic degradation pathways of amoxicillin proposed by Li et al. (2019). A trapping experiment is conducted by employing several electron and hole scavengers to investigate the reactive species that dominantly contribute to the degradation process. They found that the •OH radicals are the

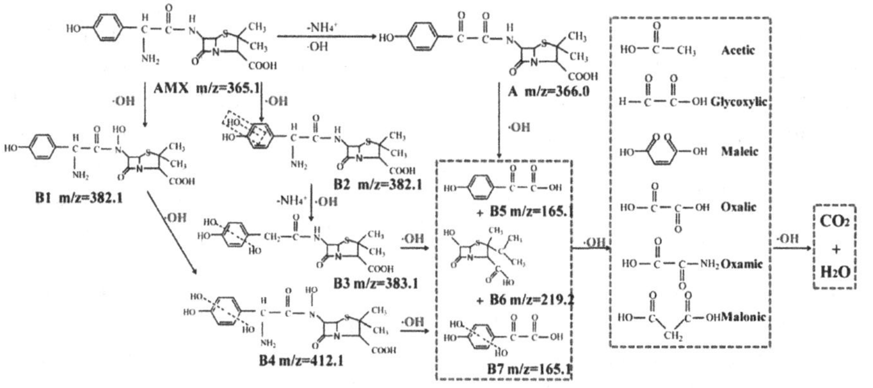

FIGURE 14.3 Proposed amoxicillin degradation pathways via semiconductor photocatalysts. (Adapted from Li et al. (2019).)

main radicals that attacked the amoxicillin residue. Based on the LC/MS analysis, there are two possible degradation pathways were found, which are (i) hydroxylation, and (ii) decomposition of the primary amino group. Initially, the active radicals will attack the amoxicillin, results in the formation of the intermediates B1 and B2 (m/z 382.1). After multiple oxidation processes, the complex amoxicillin structures are mineralized into 2-(4-hydroxyphenyl)-2-oxoacetic acid (m/z 165.1) and an intermediate-based sulfonic group (m/z 219.2). Meanwhile, in the route (ii), the radicals attacked the primary amino group, subsequently degrade the cleavage at both secondary amine and carbonyl groups. A similar oxidation process has occurred for the formation of the B5 (m/z 165.1) and B6 (m/z 219.2) intermediates. Finally, the active radicals will further attack these intermediates and mineralize them into smaller fragments of organic acids, which will eventually be converted into CO_2, H_2O, and inorganic ions.

14.4.2 Ciprofloxacin

Ciprofloxacin is one of the colorless antibiotics under the third generation of the fluoroquinolone family that has been commonly prescribed to the community and is widely used in the veterinary field. The chemical formula of ciprofloxacin can be written as $C_{17}H_{18}FN_3O_3$. The overconsumption of this type of antibiotics has led to a significant level of ciprofloxacin residue detected in the water/wastewater stream. It is known that ciprofloxacin is hardly metabolized even after consumption. Therefore, detecting a significant amount of it can promote the enhancement in the antibiotic-resistant bacteria and resistance gene; consequently, adverse impact on human health is postulated if it is consumed. Thus, it is urgently called for treating the ciprofloxacin residue in the water/wastewater stream before an unforeseen circumstance occurs.

For example, Shi et al. (2013) prepared a strawberry-like $BiVO_4$ nanocrystal with 5 nm sized mastoids incorporated with platinum nanoparticles via facile organic additive-free microwave-assisted techniques. The aforementioned Pt@$BiVO_4$ photocatalyst with a bandgap energy of 2.5 eV can degrade 91.97% of 10 mg/L ciprofloxacin residue under visible-light irradiation. The addition of Pt nanoparticles not only acts as an electron-hole pair receptors and transporters, but also formed more active sites that can partake in the photocatalytic reaction. Meanwhile, Lai et al. (2019) designed a CuS/$BiVO_4$ with a dominant (040) crystal facet to remove ciprofloxacin. They found that the binary composite photocatalysts are capable of removing 86.7% ciprofloxacin and obeyed pseudo-first-order kinetics. The significant removal of ciprofloxacin is attributed to the smooth photocharge carrier mobility stemming from the benefit of the p-n heterojunction system possess by the CuS/$BiVO_4$ sample. Additionally, the CuS/$BiVO_4$ shows excellent photostability even after being recycled four times with infinitesimal loss in degradation efficiency.

On the other hand, Karuppaiah et al. (2019) explored the performance of Gd_2WO_6/ZnO/bentonite in treating the ciprofloxacin residue. The aforementioned ternary heterostructure system can remove 97.9% of ciprofloxacin within 60 minutes, which is one of the fastest reported performances in the photocatalytic degradation of ciprofloxacin application. The fastest and profound performance observed is attributed to (i) the enhancement in the light absorption capacity within the visible-light region,

(ii) limited photocharge carrier recombination according to the PL analysis, and (iii) larger active surface area for the degradation reaction. Recently, Samsudin et al. (2020b) demonstrated the promising potential of the $Ag/AgVO_3/g\text{-}C_3N_4$ composite photocatalyst in treating the ciprofloxacin. The composite sample prepared via a wet-impregnation method can achieve 82.6% of ciprofloxacin removal under visible-light irradiation. The enhanced crystallographic properties and intimate contact between the parental photocatalyst allow a smooth photocharge carrier mobility across the heterostructure interface and minimize the photocharge carrier resistance. Hence, more available photocharge carriers can participate in the degradation process.

Correspondingly, Raja et al. (2020) improved the poor performance of the $BiVO_4$ via the formation of the ternary rGO-$BiVO_4$-ZnO heterostructure photocatalyst. The aforementioned ternary structure shows promising performance in removing the ciprofloxacin with 98.4% removal within 60 minutes under visible-light irradiation. This boosted performance compared to the pure sample indicating that the ternary structure possesses an excellent photocharge carrier transfer concerning the function of rGO as an excellent electron mediator and bridge. Figure 14.4 demonstrated the potential degradation pathways of ciprofloxacin via rGO-$BiVO_4$-ZnO heterostructure photocatalyst as proposed by Raja et al. (2020). The study on the intermediates was performed using GC-MS analysis. They detected a single intermediate peak at a retention time of 14.184, which corresponds to the 7-amino-1-cyclopropyl-2,3,5,6, 8-pentahyroxyquinolin-4(1H)-1(D1). Initially, the generated active radicals will attack the ciprofloxacin, leading to the defluoridation and decarboxylation process. These processes lead to the formation of intermediate I and II. Subsequently, the intermediates further fragmentized to become intermediate III before the extra oxidation process on the piperazine ring of the ciprofloxacin. After completely removing the piperazine ring, a quinolone derivative labeled as D1 is formed, and several oxygenated aliphatic compounds are detected as well. Finally, all of the smaller fragments will eventually be destroyed and transformed into CO_2 and clean H_2O.

14.4.3 Tetracycline

Tetracycline is another common type of antibiotic residue that is commonly studied in the arena of semiconductor photocatalysts. Tetracycline is frequently used as an antibiotic for livestock breeding and aquaculture. Similar to amoxicillin and ciprofloxacin, tetracycline is also hardly absorbed by humans or animals' bodies after being consumed. This had led to more than 80% of tetracycline residue being excreted and discharged into the environment, consequently jeopardizing the quality of the water/wastewater stream.

Motivated by this issue, Chen et al. (2016) designed a novel $AgI/BiVO_4$ photocatalyst via an in-situ precipitation method. The aforementioned composite photocatalyst can substantially remove the tetracycline with 94.91% removal within 60 minutes compared with the pure samples; $BiVO_4$ (62.68%) and AgI (75.43) % under similar experimental conditions. Additionally, the $AgI/BiVO_4$ sample can achieve 90.46% of TOC removal within 120 minutes, indicating a superior mineralization performance. This outstanding degradation performance is attributed to the accelerated photocharge carrier mobility, which hindered the potential of charge

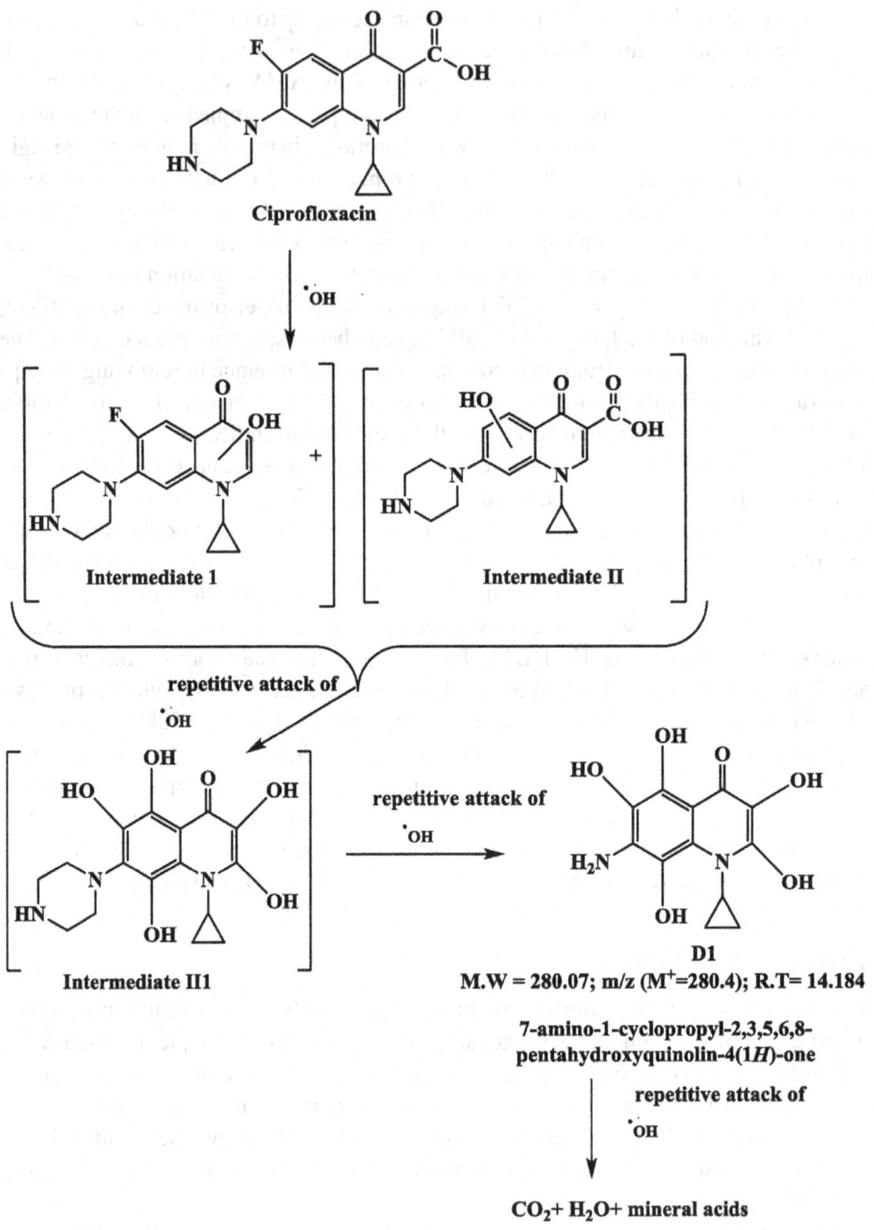

FIGURE 14.4 Proposed Ciprofloxacin degradation pathways via semiconductor photocatalysts. (Adapted from Raja et al. (2020).)

recombination, resulting in an augmented photocatalytic reaction. Subsequently, Chen et al. (2017) demonstrated the beneficial effect of combining the $BiVO_4$ photocatalyst with the Ag_3PO_4/Ag photocatalyst to form the Z-scheme heterojunction system. They deployed two-step processes in fabricating the Z-scheme system, which are (i) in-situ deposition technique and (ii) photo-reduction technique. The Z-scheme $Ag_3PO_4/Ag/BiVO_4$ photocatalyst can remove 94.96% tetracycline residue within 60 minutes. This outstanding degradation performance corresponds to the limited photocharge carrier recombination, larger active surface area, and better absorption capacity demonstrated by the Z-scheme composite photocatalyst.

Meanwhile, Deng et al. (2018) mitigated the inherent limitations of pure $BiVO_4$ by hybridizing with the phosphorus-doped ultrathin g-C_3N_4 (PCNS). The composite PCNS/$BiVO_4$ photocatalyst was synthesized via a one-pot impregnated precipitation method. It was found that the aforementioned photocatalyst was capable of removing 95.95% of 10 mg/L tetracycline concentration within 60 minutes. The hybridization of $BiVO_4$ and PCNS helps improve the photocharge carrier separation and migration and augment the the photocharge carriers' lifetime (1.65 ns). Additionally, Xiao et al. (2018) improved the performance of pure g-C_3N_4 via the formation of a ternary heterostructure system with $CdIn_2S_4$ and RGO. The ternary RGO/$CdIn_2S_4$/g-C_3N_4 heterostructure system can remove 74.02% of tetracycline in comparison to the pure g-C_3N_4 (30.2%) and $CdIn_2S_4$ (37.1%). The addition of RGO within the composite $CdIn_2S_4$/g-C_3N_4 heterostructure system helps ease the photocharge carrier mobility across the heterojunction, resulting in a profound photocatalytic reaction. On the other hand, Ahmadi et al. (2017) proposed utilizing TiO_2 decorated with MWCNT to remove tetracycline. Complete removal of tetracycline is achieved via this composite photocatalyst in which the plotted data suggest the MWCNT/TiO_2 obeyed a pseudo-first-order kinetic. Additionally, they found that 83% of mineralization occurs within

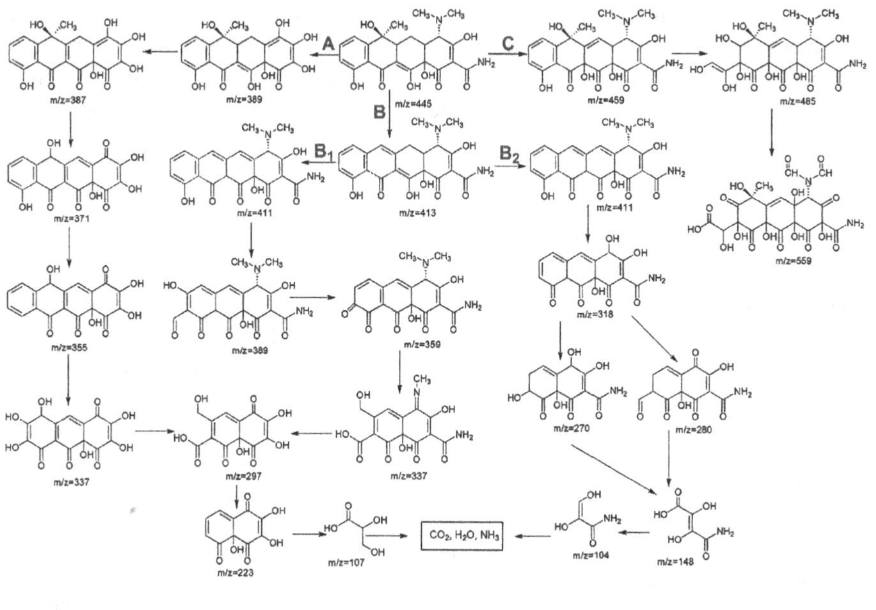

FIGURE 14.5 Proposed tetracycline degradation pathways via semiconductor photocatalysts. (Adapted from Yu et al. (2019).)

300 minutes, evidently from the TOC analysis. Recently, Yu et al. (2019) developed a Z-scheme $Ag_3PO_4/AgBr/g\text{-}C_3N_4$ photocatalyst via a facile chemical deposition technique. The aforementioned Z-scheme photocatalyst exhibited remarkable removal of tetracycline with 90% removal within 10 minutes. Additionally, a detailed degradation pathway of tetracycline is proposed, as illustrated in Figure 14.5.

14.5 CONCLUSION AND PERSPECTIVE

Substantial progress on the photocatalytic degradation of antibiotics residue from the water/wastewater stream via semiconductor photocatalyst has been reported and concisely summarized. The recent related progress on this emerging field, including the contemporary photocatalyst and common reported antibiotic residue, has been properly introduced. Although the recent progress shows a significant potential of AOPs as a compelling strategy to alleviate the antibiotic residue issues, there are still several problems that this emerging system faces. Firstly, most of the reported studies only focus on the synthetic water matrix containing a single antibiotic residue or component. However, in the view of industrial applications, a single recalcitrant pollutant is unlikely to exist. Thus, it is important to explore the performance of the proposed semiconductor photocatalyst for treating the real industrial water/wastewater. Secondly, the toxicity study of the byproducts or intermediates formed during the photocatalytic degradation process needs further exploration. It is often that some of the intermediates are more toxic than the parental antibiotics, in which it can poison the photocatalyst sample during the reaction process, consequently minimizing photocatalyst efficiency. Thus, researchers need to take into account the toxicity analysis before and after the performance evaluation. Finally, the mass-scale production of inexpensive photocatalyst along with the affordable large-scale setup of the overall photocatalytic degradation reactor are still in their infancy stage. Despite focusing on laboratory-scale production, it is important for the researcher to simultaneously aim for the real pilot-scale applications in order to further facilitate this emerging technology's readiness.

ACKNOWLEDGMENT

This work is supported by Fundamental Research Grant Scheme (FRGS) funded by the Ministry of Higher Education (MOHE) Malaysia (Ref no: FRGS/1/2020/TK0/UTP/02/22). The authors are also grateful for the support given by Chemical Engineering Department and Centre of Innovative Nanostructures and Nanodevices (COINN), Universiti Teknologi PETRONAS, Malaysia. There is no conflict of interest among authors in publishing this manuscript.

REFERENCES

Ahmad Madzlan, M. K. A., Samsudin, M. F. R., Maeght, F., Goepp, C., and Sufian, S. (2020). Enhancement of $g\text{-}C_3N_4$ via acid treatment for the degradation of ciprofloxacin antibiotic. *Malaysian J. Microsc.* 16, 105–114.

Ahmadi, M., Ramezani Motlagh, H., Jaafarzadeh, N., Mostoufi, A., Saeedi, R., Barzegar, G., et al. (2017). Enhanced photocatalytic degradation of tetracycline and real pharmaceutical wastewater using $MWCNT/TiO_2$ nano-composite. *J. Environ. Manage.* 186, 55–63.

Alygizakis, N. A., Gago-Ferrero, P., Borova, V. L., Pavlidou, A., Hatzianestis, I., and Thomaidis, N. S. (2016). Occurrence and spatial distribution of 158 pharmaceuticals, drugs of abuse and related metabolites in offshore seawater. *Sci. Total Environ.* 541, 1097–1105.

Butler, M. A., Nasby, R. D., and Quinn, R. K. (1976). Tungsten trioxide as an electrode for photoelectrolysis of water. *Solid State Commun.* 19, 1011–1014.

Calvete, M. J. F., Piccirillo, G., Vinagreiro, C. S., and Pereira, M. M. (2019). Hybrid materials for heterogeneous photocatalytic degradation of antibiotics. *Coord. Chem. Rev.* 395, 63–85.

Chahkandi, M., and Zargazi, M. (2020). New water based EPD thin $BiVO_4$ film: Effective photocatalytic degradation of amoxicillin antibiotic. *J. Hazard. Mater.* 389, 121850.

Chen, F., Yang, Q., Li, X., Zeng, G., Wang, D., Niu, C., et al. (2017). Hierarchical assembly of graphene-bridged Ag_3PO_4/Ag/$BiVO_4$(040) Z-scheme photocatalyst: An efficient, sustainable and heterogeneous catalyst with enhanced visible-light photoactivity towards tetracycline degradation under visible light irradiation. *Appl. Catal. B Environ.* 200, 330–342.

Chen, F., Yang, Q., Sun, J., Yao, F., Wang, S., Wang, Y., et al. (2016). Enhanced photocatalytic degradation of tetracycline by AgI/$BiVO_4$ Heterojunction under visible-light irradiation: Mineralization efficiency and mechanism. *ACS Appl. Mater. Interfaces* 8, 32887–32900.

Chen, X., and Mao, S. S. (2007). Titanium dioxide nanomaterials: Synthesis, properties, modifications and applications. *Chem. Rev.* 107, 2891–2959.

Deng, Y., Tang, L., Zeng, G., Wang, J., Zhou, Y., Wang, J., et al. (2018). Facile fabrication of mediator-free Z-scheme photocatalyst of phosphorous-doped ultrathin graphitic carbon nitride nanosheets and bismuth vanadate composites with enhanced tetracycline degradation under visible light. *J. Colloid Interface Sci.* 509, 219–234.

Domen, K., Naito, S., Soma, M., Onishi, T., and Tamaru, K. (1980). Photocatalytic decomposition of water vapour on an NiO–$SrTiO_3$ catalyst. *J. Chem. Soc., Chem. Commun.* 12, 543–544.

Dong, P., Hou, G., Xi, X., Shao, R., and Dong, F. (2017). WO_3-based photocatalysts: Morphology control, activity enhancement and multifunctional applications. *Environ. Sci. Nano* 4, 539–557.

Govender, K., Boyle, D. S., O'Brien, P., Binks, D., West, D., and Coleman, D. (2002). Room-temperature lasing observed from ZnO nanocolumns grown by aqueous solution deposition. *Adv. Mater.* 14, 1221–1224.

Hansima, M. A. C. K., Makehelwala, M., Jinadasa, K. B. S. N., Wei, Y., Nanayakkara, K. G. N., Herath, A. C., et al. (2021). Fouling of ion exchange membranes used in the electrodialysis reversal advanced water treatment: A review. *Chemosphere* 263, 127951.

Hernández, F., Calısto-Ulloa, N., Gómez-Fuentes, C., Gómez, M., Ferrer, J., González-Rocha, G., et al. (2019). Occurrence of antibiotics and bacterial resistance in wastewater and sea water from the Antarctic. *J. Hazard. Mater.* 363, 447–456.

Hisatomi, T., Kubota, J., and Domen, K. (2014). Recent advances in semiconductors for photocatalytic and photoelectrochemical water splitting. *Chem. Soc. Rev.* 43, 7520–7535.

Honda, A., and Fujishima, K. (1972). Electrochemical photolysis of water at a semiconductor electrode. *Nature* 238, 37–38.

Hou, H., and Zhang, X. (2020). Rational design of 1D/2D heterostructured photocatalyst for energy and environmental applications. *Chem. Eng. J.* 395, 125030.

Kalanoor, B. S., Seo, H., and Kalanur, S. S. (2018). Recent developments in photoelectrochemical water-splitting using WO_3/$BiVO_4$ heterojunction photoanode: A review. *Mater. Sci. Energy Technol.* 1, 49–62.

Karuppaiah, S., Annamalai, R., Muthuraj, A., Kesavan, S., Palani, R., Ponnusamy, S., et al. (2019). Efficient photocatalytic degradation of ciprofloxacin and bisphenol A under visible light using Gd_2WO_6 loaded ZnO/bentonite nanocomposite. *Appl. Surf. Sci.* 481, 1109–1119.

Kong, Y., Sun, H., Zhao, X., Gao, B., and Fan, W. (2015). Fabrication of hexagonal/cubic tungsten oxide homojunction with improved photocatalytic activity. *Appl. Catal. A Gen.* 505, 447–455.

Kudo, A., Ueda, K., Kato, H., and Mikami, I. (1998). Photocatalytic O_2 evolution under visible light irradiation on $BiVO_4$ in aqueous $AgNO_3$ solution. *Catal. Letters* 53, 229–230.

Lai, C., Zhang, M., Li, B., Huang, D., Zeng, G., Qin, L., et al. (2019). Fabrication of CuS/$BiVO_4$(0 4 0) binary heterojunction photocatalysts with enhanced photocatalytic activity for Ciprofloxacin degradation and mechanism insight. *Chem. Eng. J.* 358, 891–902.

Le, S., Li, W., Wang, Y., Jiang, X., Yang, X., and Wang, X. (2019). Carbon dots sensitized 2D-2D heterojunction of $BiVO_4/Bi_3TaO_7$ for visible light photocatalytic removal towards the broad-spectrum antibiotics. *J. Hazard. Mater.* 376, 1–11.

Lee, K. M., Lai, C. W., Ngai, K. S., and Juan, J. C. (2016). Recent developments of zinc oxide based photocatalyst in water treatment technology: A review. *Water Res.* 88, 428–448.

Li, Q., Jia, R., Shao, J., and He, Y. (2019). Photocatalytic degradation of amoxicillin via TiO_2 nanoparticle coupling with a novel submerged porous ceramic membrane reactor. *J. Clean. Prod.* 209, 755–761.

Maggard, P. A., Stern, C. L., and Poeppelmeier, K. R. (2001). Understanding the role of helical chains in the formation of noncentrosymmetric solids. *J. Am. Chem. Soc.* 123, 7742–7743.

Morkoç, H., and Özgür, U. (2009). *Zinc Oxide: Fundamentals, Materials And Device Technology*. First edit. Wiley-VCH Weinheim.

Nidheesh, P. V., Gopinath, A., Ranjith, N., Praveen Akre, A., Sreedharan, V., and Suresh Kumar, M. (2020). Potential role of biochar in advanced oxidation processes: A sustainable approach. *Chem. Eng. J.* 405, 126582.

Payne, D. J., Egdell, R. G., Walsh, A., Watson, G. W., Guo, J., Glans, P. A., et al. (2006). Electronic origins of structural distortions in post-transition metal oxides: Experimental and theoretical evidence for a revision of the lone pair model. *Phys. Rev. Lett.* 96, 157403.

Phoon, B. L., Lai, C. W., Juan, J. C., Show, P. L., and Pan, G. T. (2019). Recent developments of strontium titanate for photocatalytic water splitting application. *Int. J. Hydrogen Energy* 44, 14316–14340.

Prasad, C., Liu, Q., Tang, H., Yuvaraja, G., Long, J., Rammohan, A., et al. (2020). An overview of graphene oxide supported semiconductors based photocatalysts: Properties, synthesis and photocatalytic applications. *J. Mol. Liq.* 297, 111826.

Raja, A., Rajasekaran, P., Selvakumar, K., Arunpandian, M., Kaviyarasu, K., Asath Bahadur, S., et al. (2020). Visible active reduced graphene oxide-$BiVO_4$-ZnO ternary photocatalyst for efficient removal of ciprofloxacin. *Sep. Purif. Technol.* 233,.115996.

Samsudin, M. F. R., Bashiri, R., Mohamed, N. M., Ng, Y. H., and Sufian, S. (2020a). Tailoring the morphological structure of $BiVO_4$ photocatalyst for enhanced photoelectrochemical solar hydrogen production from natural lake water. *Appl. Surf. Sci.* 504, 144417.

Samsudin, M. F. R., Frebillot, C., Kaddoury, Y., Sufian, S., and Ong, W. J. (2020b). Bifunctional Z-Scheme Ag/$AgVO_3$/g-C_3N_4 photocatalysts for expired ciprofloxacin degradation and hydrogen production from natural rainwater without using scavengers. *J. Environ. Manage.* 270, 110803. doi:10.1016/j.jenvman.2020.110803.

Samsudin, M. F. R., Mahmood, A., and Sufian, S. (2018a). Enhanced photocatalytic degradation of wastewater over RGO-$TiO_2/BiVO_4$ photocatalyst under solar light irradiation. *J. Mol. Liq.* 268, 26–36.

Samsudin, M. F. R., and Sufian, S. (2020). Hybrid 2D/3D g-$C_3N_4/BiVO_4$ photocatalyst decorated with RGO for boosted photoelectrocatalytic hydrogen production from natural lake water and photocatalytic degradation of antibiotics. *J. Mol. Liq.* 314, 113530.

Samsudin, M. F. R., Sufian, S., and Hameed, B. H. (2018b). Epigrammatic progress and perspective on the photocatalytic properties of $BiVO_4$-based photocatalyst in photocatalytic water treatment technology: A review. *J. Mol. Liq.* 268, 438–459.

Samsudin, M. F. R., Ullah, H., Bashiri, R., Mohamed, N. M., Sufian, S., and Ng, Y. H. (2020c). Experimental and DFT Insights on Microflower g-C_3N_4/$BiVO_4$ Photocatalyst for Enhanced Photoelectrochemical Hydrogen Generation from Lake water. *ACS Sustain. Chem. Eng.* 8, 9393–9403.

Shi, W., Yan, Y., and Yan, X. (2013). Microwave-assisted synthesis of nano-scale $BiVO_4$ photocatalysts and their excellent visible-light-driven photocatalytic activity for the degradation of ciprofloxacin. *Chem. Eng. J.* 215–216, 740–746.

Silva, I. F., Teixeira, I. F., Rios, R. D. F., do Nascimento, G. M., Binatti, I., Victória, H. F. V., et al. (2021). Amoxicillin photodegradation under visible light catalyzed by metal-free carbon nitride: An investigation of the influence of the structural defects. *J. Hazard. Mater.* 401, 123713.

Suwannaruang, T., Hildebrand, J. P., Taffa, D. H., Wark, M., Kamonsuangkasem, K., Chirawatkul, P., et al. (2020). Visible light-induced degradation of antibiotic ciprofloxacin over Fe–N–TiO_2 mesoporous photocatalyst with anatase/rutile/brookite nanocrystal mixture. *J. Photochem. Photobiol. A Chem.* 391, 112371.

Tan, H. L., Amal, R., and Ng, Y. H. (2017). Alternative strategies in improving the photocatalytic and photoelectrochemical activities of visible light-driven $BiVO_4$: A review. *J. Mater. Chem. A* 5, 16498–16521.

Tang, Y., Zheng, Z., Sun, X., Li, X., and Li, L. (2019). Ternary CdS-MoS_2 coated ZnO nanobrush photoelectrode for one-dimensional acceleration of charge separation upon visible light illumination. *Chem. Eng. J.* 368, 448–458.

Wagner, F. T., and Somorjai, G. A. (1980a). Photocatalytic and Photoelectrochemical Hydrogen Production on Strontium Titanate Single Crystals. *J. Am. Chem. Soc.* 102, 5494–5502.

Wagner, F. T., and Somorjai, G. A. (1980b). Photocatalytic hydrogen production from water on Pt-free $SrTiO_3$ in alkali hydroxide solutions. *Nature* 285, 559–560.

Walsh, A., Yan, Y., Huda, M. N., Al-Jassim, M. M., and Wei, S. H. (2009). Band edge electronic structure of $BiVO_4$: Elucidating the role of the Bi s and V d orbitals. *Chem. Mater.* 21, 547–551.

Wang, X., Li, Z., Shi, J., and Yu, Y. (2014). One-dimensional titanium dioxide nanomaterials: Nanowires, nanorods, and nanobelts. *Chem. Rev.* 114, 9346–9384.

Wang, Z., Huang, X., and Wang, X. (2019). Recent progresses in the design of $BiVO_4$-based photocatalysts for efficient solar water splitting. *Catal. Today* 335, 31–38.

Wei, Z., Liu, J., and Shangguan, W. (2020). A review on photocatalysis in antibiotic wastewater: Pollutant degradation and hydrogen production. *Chinese J. Catal.* 41, 1440–1450.

Westerhoff, P., Song, G., Hristovski, K., and Kiser, M. A. (2011). Occurrence and removal of titanium at full scale wastewater treatment plants: Implications for TiO_2 nanomaterials. *J. Environ. Monit.* 13, 1195–1203.

Wu, H., Tan, H. L., Toe, C. Y., Scott, J., Wang, L., Amal, R., et al. (2020). Photocatalytic and photoelectrochemical systems: imilarities and differences. *Adv. Mater.* 32, 1904717.

Xiao, P., Jiang, D., Ju, L., Jing, J., and Chen, M. (2018). Construction of RGO/$CdIn_2S_4$/g-C_3N_4 ternary hybrid with enhanced photocatalytic activity for the degradation of tetracycline hydrochloride. *Appl. Surf. Sci.* 433, 388–397.

Xu, P. S., Sun, Y. M., Shi, C. S., Xu, F. Q., and Pan, H. B. (2003). The electronic structure and spectral properties of ZnO and its defects. *Nucl. Instruments Methods Phys. Res. B* 199, 286–290.

Xu, Q., Zhang, L., Yu, J., Wageh, S., Al-Ghamdi, A. A., and Jaroniec, M. (2018). Direct Z-scheme photocatalysts: Principles, synthesis, and applications. *Mater. Today* 21, 1042–1063.

Xu, S., and Wang, Z. L. (2011). One-dimensional ZnO nanostructures: Solution growth and functional properties. *Nano Res.* 4, 1013–1098.

Younis, S. A., Kwon, E. E., Qasim, M., Kim, K. H., Kim, T., Kukkar, D., et al. (2020). Metal-organic framework as a photocatalyst: Progress in modulation strategies and environmental/energy applications. *Prog. Energy Combust. Sci.* 81, 100870.

Yu, H., Wang, D., Zhao, B., Lu, Y., Wang, X., Zhu, S., et al. (2019). Enhanced photocatalytic degradation of tetracycline under visible light by using a ternary photocatalyst of Ag_3PO_4/AgBr/g-C_3N_4 with dual Z-scheme heterojunction. *Sep. Purif. Technol.* 237, 116365.

Zhang, L., Ran, J., Qiao, S. Z., and Jaroniec, M. (2019). Characterization of semiconductor photocatalysts. *Chem. Soc. Rev.* 48, 5184–5206.

Zheng, Y., Chen, G., Yu, Y., Wang, Y., Sun, J., Xu, H., et al. (2014). Solvothermal synthesis of pyrochlore-type cubic tungsten trioxide hemihydrate and high photocatalytic activity. *New J. Chem.* 38, 3071–3077.

Zulfiqar, M., Chowdhury, S., Samsudin, M. F. R., Siyal, A. A., Omar, A. A., Ahmad, T., et al. (2020). Effect of organic solvents on the growth of TiO_2 nanotubes: An insight into photocatalytic degradation and adsorption studies. *J. Water Process Eng.* 37, 101491.

Index

A

acetone, 139
acid fuchsin, 294
acridine orange (AO), 295
activated carbon, 334, 342
adsorption, 182, 184
advanced oxidation process (AOP), 164
advanced oxidation technologies, 259
aerogel, 380
AgI/$BiVO_4$ photocatalyst, 457
air gap membrane distillation (AGMD), 396
amalgam, 197
ametryn, 127
3-aminopropyl triethoxysilane (APTES), 289
amoxicillin, 74, 454
ampicillin (AMP), 348
anatase phase, 167, 172
anthraquinone, 227
antibacterial, 71
antibacterial activity, 227
antibiotics, 71, 323
antibiotics removal, 454
antimony, 292
aromatic phenanthrene, 127
arsenic, 182
arsenic removal, 188, 190
auger-assisted hole transfer process, 429
auxiliary electric fields, 312
azo dye, 223

B

Bacillus cereus, 227
barbituric acid (BA), 222, 226
BET analysis, 205
biochar, 334, 342
bioimaging, 95
bismuth, 70, 75
bismuth ferrite, 85
bismuth oxyhalide, 85, 147
bismuth vanadate ($BiVO_4$), 450–451
blending technique, 394
bottom-up approach, 135
brilliant cresyl blue (BCB), 295
bromine, 147
bromine nitride (BN) nanosheets, 148
brookite phase, 172

C

cadmium removal, 192
calcination, 99
capacitive deionization, 394
capping, 171
capsid, 263
carbamazepine, 71
carbonaceous materials, 105
carbon dots, 105, 339
carbon nanoallotropes, 340
carbon nanotubes, 334, 339
carbon nitride, 99
carbon quantum dots, 334, 343
carboxymethyl cellulose (CMC), 206
catalyst, 165
cathodic electrochemical advance oxidation process (CEAOP), 353
CdX NRs, 426
cefazolin sodium (CFZ), 352
cefixime, 346
cellulose acetate, 396
cement/mortar-based systems, 265
charge neutralization, 292
charge separation process, 345
chemical oxygen demand (COD), 348
chemical vapor deposition, 341
chemisorption, 170
chitosan, 45, 316
chloramphenicol (CAP), 348
chlorine, 147
chlorpyrifos, 127, 133
chlortetracycline, 81
chromate, 193
chromium, 182
chromium removal, 193
ciprofloxacin, 71, 106, 456
clarithromycin, 345
CNMs, 334
coaxial electrospinning, 312
cobalt crystal (CO), 324
cobalt ferrite, 206
co-doping, 169
conduction band, 73
conjugated polyelectrolytes (CPE), 232
conventional synthesis, 377
COP, 223
copper removal, 198

coprecipitation, 182, 188
core-shell electrospinning, 312
coronavirus disease, 263
covalent bonding, 132
cross-linking process, 296
cyanamide, 94
cyclodextrin, 294

D

deacetylation-cellulose acetate (d-CA) nanofiber membrane, 319
decahedral anatase particles (DAPS), 252
degradation, 74
density functional theory (DFT), 96
desalination, 127, 129
desorption-adsorption processes, 81
diclofenac, 71, 348
dicyandiamide, 94
dimethylforamide (DMF), 378
dioxygen, 173
dithiocarbamate (DTC), 434
ditopic ligands, 374
doping, 167
double-Walled carbon nanotubes (DWCNTs), 341
drawing, 310
drop casting, 130
dubinin Radushkevich (D-R) isotherm model, 186
dye effluents, 320
dye-sensitized solar cells, 450
dynamic light scattering (DLS), 205

E

electrochemical, 378
electrochemical process, 353
electrocoagulation, 2, 7
electrodeposition method, 394
electro-Fenton, 145
electrolysis, 164
electron, 71
electron affinity, 78
electron beam, 124
electronegativity, 78
electron mediators, 78
electron paramagnetic resonance, 248
electron transfer, 421
electro-oxidation, 7
electrospinning, 310–311
electrostatic force, 132
electrostatic interaction, 292
endosulfan, 133
environmental parameters, 109
environmental pollution, 103
eosin blue, 295
eriochrome, 33
erythromycin, 73, 345
ethylenediamine-reduced graphene oxide (ED-DMF-RGO) nanosheets, 195

F

faujasite, 206
femtosecond absorption spectroscopy, 248
femtosecond-picosecond system, 418
filamentous fungi, 264
filtration, 292
flash freezing, 310
flotation gravity separation, 318
fluroquinolones (FQs), 348
forward osmosis, 51
Freundlich model, 185
FTIR, 192, 201
fullerene, 334, 342
fullerenes, 4, 10, 51, 342

G

gas chromatography-mass spectrometry (GC-MS), 289
gas jet, 310
g-C_3N_4 based nanocomposites, 99
gelatin-assisted seed growth, 394
gemifloxacin, 74
genotoxicity, 74
germination test, 196
glutaraldehyde, 319
goethite nanospheres, 196
gold nanoparticles, 183
GO nanosheets, 128
grafting, 296
gram-positive bacteria, 262
granular activated carbon (GAC), 342
graphene, 122, 145, 334
graphene oxide, 146
graphene oxide framework (GOF), 128
graphene sand composite (GSC), 348
graphene sheets, 85
graphite, 339
graphitic carbon nitride (g-C_3N_4), 4, 10, 51, 94, 145, 183, 220, 334, 343
group replacement, 292

H

heavy metal removal, 184
helium, 124
hematite, 282
heterogenous catalysts, 78
heterojunction, 74, 100
hexamethylenetetramine (HMT), 194
hexavalent chromium, 182, 189
high-density polyethylene (HDPE), 396

high-performance nanofiber membranes, 318
holes, 71
humic acid, 383
hydrated manganese oxide-biochar (HMO-BC), 206
hydrochloric salt, 75
hydrogel, 207
hydrolysis, 281
hydroperoxy radicals, 166
hydrothermal, 99
hydrothermal method, 75, 201

I

ibuprofen, 323
illite, 42
infrared (IR) light-driven photocatalysis, 423
in-situ growth, 394
in-situ precipitation, 457
intercalation, 129, 296
internal diffusion, 293
iodine, 75
ion beam, 124
iron oxide, 146

K

kaolinite, 42

L

laminar membranes, 123
langmuir model, 185
layer-by-layer (LBL), 394
leaching, 281
lead, 184
lead removal, 195
lepidocrocite nanoparticles, 196
levofloxacin (LVX), 348
liquid chromatography, 81
liquid-phase exfoliation technique, 139
lithium-ion batteries, 450
localized surface plasmon resonance (LSPR), 247

M

maghemite, 288
maghemite nanoparticles, 194
magnetic carbonaceous nanocomposite (MCN), 380
magnetic graphene nanocomposites (MGNC), 195
magnetite, 286
malathion, 127, 133
manganese oxide, 147
mechanochemical, 378
melamine, 94
melt-bowling, 310
melt fibrillation, 310
membrane adsorption, 315
membrane fouling, 232
membrane technology, 182, 188
mercaptopropionic acid (MPA), 422
mercaptopropyl trimethoxysilane (MPTS), 192
mercury, 184
mercury removal, 197
metal nanoparticles, 77
metal-organic frameworks (MOFs), 372, 387
metal oxyhalides, 147
methylene blue, 168
methyl orange, 168
metronidazole (MET), 347
microfiltration, 121, 188
microspheres, 75
migration, 71
mixed matrix, 394
mixed matrix substrate (MMS), 123
MOF-based membranes, 394
molecule acceptors, 438
MoS2, 139
multiple-jet nozzle, 312
multiwalled carbon nanotubes (MWCNTs), 317
murexide, 295
MXenes, 135

N

nanocellulose, 183
nanocomposites, 95
nanocrystals, 75
nanofiber structure, 310
nanofiltration, 121, 188
nanohorns, 339
nanomaterials, 164
nanoparticles, 182
nano-plates, 130–131
nanopores, 124
nanoporous graphene (NPG), 124
nanoscience, 164
nanosecond-microsecond system, 418
nanosheets, 81, 123
N-doping method, 168
nickel acetate, 294
nickel oxide, 294
nitrogen-doped magnetic carbon particles (N-MCNPs), 206
N, N-diethylformamide (DEF), 378
noble metals (NMs), 246
nontronite, 42
norfloxacin, 71, 73
NR heterostructures, 426

O

octahedral anatase particles (OAPS), 252
oilfield produced water (OPW), 313

onion-like carbon (OLC) nanospheres, 339
optical parametric amplifier, 418
organic pollutants, 93
oxalic acid, 294
oxcarbazepine, 297
oxidation, 281
oxidative pathway, 76, 78
oxidative stress, 144
oxytetracycline, 75, 81
ozonation process, 357

P

PAMAM (polyamidoamine), 132
PAN/dopamine nanofiber, 316
particle filtration, 121
peptidoglycan, 262
perfluorooctanesulfonate (PFOS), 391
peroxymonosulfate, 380
peroxymonosulfate (PMS), 324
pharmaceuticals, 334
phase inversion, 394
phase separation, 310
phenazopyridine hydrochloride (PhP), 346
photocatalysis, 70, 78, 93, 421
photocatalyst, 70, 84
photocatalytic disinfection, 144
photocatalytic efficiency, 74–75
photocatalytic mechanisms, 432
photocataytic ozonation process, 357
photodegradation, 74–75, 84, 102
photodeposition method, 252
photo-Fenton, 145, 147
photonic band gap (PBG), 256
photonic crystals, 255
photon transmission process, 416
photo-reduction method, 346
photothermal reaction, 144
photovoltaics, 421
phyllosilicates, 41–42
physisorption, 170
plasmids, 260
plasmonic heating, 248
plasmons, 80
p-nitrosodimethylaniline, 255
poly (acrylic acid) (PAA), 316
poly(diphenylbutadiyne) (PDPB), 221
poly(thiophenes), 229
poly (vinylidene fluoride), 45
polyacrylamide (PAM), 206
polyacrylonitrile (PAN), 322
polyacrylonitrile bovine serum albumin (PAN-BSA), 321
polyamide, 45
polyamide (PI) nanofibers, 318
polyaniline, 45, 47
polyaniline (PANI), 223
polybenzoxazine (PBZ), 318
polydiphenylbutadiyne (PDPB) nanofibers, 226
polydopamine, 130
polyetherimide, 36
polyethersulfone (PES), 188, 396
polyethylene-terephthalate (PET), 146
polymer-based photocatalysis, 218
polymeric material, 122
polymeric membranes, 231
polypyrrole, 45
polytetrafluoroethylene (PTFE), 319
polyvinyl alcohol/silica particle (PVA-Si), 319
polyvinylidene difluoride (PVDF), 232
polyvinylpyrrolidone (PVP), 322
post-synthetic defect healing (PSDH), 395
post-synthetic modification (PSM), 382
powered activated carbon (PAC), 342
precipitation, 281
primary rejection layer (PRL), 123
PVA/moringa seed protein nanofiber composite, 323

Q

quanta, 70

R

radionuclide wastes, 287
reactive oxygen (ROS), 164, 225, 384
Redlich-Petersen model, 186
redox reaction, 165
reduced graphene, 293
reduced graphene oxide, 106, 334
reduction, 189, 281
reduction pathway, 71
reverse osmosis, 121, 188
rhodamine B, 169, 226
rock salt structure, 453
rotating disk, 312
rotating tube, 312
rutile phase, 172

S

salt rejection, 126
saponite, 42
sauconite, 42
sawdust, 192
S-doped TiO_2 films, 169
secondary pollution, 292
sedimentation, 292
selective degradation, 174
selective formation, 174
selenium oxyanions, 388
self-assembled monolayers (SAMs), 433
self-assembly, 310

SEM and TEM, 202
semiconductor, 93
semiconductor photocatalysis, 109
sensitization, 170
sensors, 95
silicon nanoparticles, 183
silver composite, 146
simple impregnation, 296
single-layer graphene (SLG), 126
singlet oxygen, 228, 232
single-walled carbon nanotubes (SWCNTs), 341
sips isothermal model, 196
skimming, 318
smectite, 42
SnO_2-Al_2O_3/CNT anode, 357
solar photocatalysis, 218, 246
sol-gel, 99
solvothermal method, 75
solvothermal synthesis, 378
sonocatalytic process, 351
sonochemical, 378
spin coating, 130
SPR-enhanced photocatalysts, 80
strontium titanate, 450, 454
sulfadiazine (SDZ), 347
sulfamethazine, 71
sulfamethoxazole, 71, 77, 323–324
sulfonamide, 323
supercapacitors, 95
superoxide radicals, 166
surface complexation, 292
surface located nanocomposite (SLN), 123
surface plasmon resonance (SPR), 72
surfactants, 286
suzuki polycondensation, 223

T

tartrazine dye, 147
template, 310
template synthesis, 296
tetracycline, 75, 81, 457
tetraethlyenepentamine (TEPA), 379
TFN, 123
TGA analysis, 205
thallium, 292
thermal activation, 248
thermal energy, 71
thin-film nanocomposite (TFN), 394
thiourea, 94
time-resolved microwave conductivity (TRMC), 248
TiO_2, 165
TiO_2 nanomaterials, 175
titanium carbide ($Ti_3C_2T_x$), 136
titanium dioxide, 72, 75
toluidine blue O (TDO), 321
top-down approach, 135
total organic carbon (TOC), 347
transient absorption spectroscopy (TAS), 417
transition metal dichalcogenides (TMDs), 139
trimethoprim, 77
Tris(4-carbazoyl-9-ylphenyl) amine (TCTA), 221
tungsten oxide (WO_3), 450, 452–453

U

ultrafast organic dye removal technique, 171
ultrafiltration, 121, 188
ultrasound technology, 351
unconventional synthetic method, 378
united States Environmental Protection Agency (USEPA), 298
urea, 94
urinary tract infections, 77

V

vacuum filtration, 130
valence bond, 73
vermiculite, 42, 283
visible light, 347
visible light active semiconductors, 72
vis irradiation reaction, 250

W

water filtration, 121
water pollution, 69, 75, 94
water splitting reaction, 94
water treatment, 70
wet air oxidation, 164
wrinkle layer morphology, 138

X

XPS analysis, 204
X-ray (EDS) analysis, 142
XRD analysis, 203

Z

ZIF-8, 396
zinc oxide, 183, 450
zirconium nanoparticles (ZrP), 195
Z-scheme heterojunction, 103

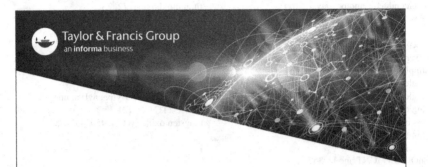

Printed in the United States
by Baker & Taylor Publisher Services